Hyperidentities

Boolean and De Morgan Structures

Hyperidentities

Boolean and De Morgan Structures

Yuri Movsisyan

Yerevan State University, Armenia
University of Bergen, Norway

World Scientific

NEW JERSEY • LONDON • SINGAPORE • BEIJING • SHANGHAI • HONG KONG • TAIPEI • CHENNAI • TOKYO

Published by

World Scientific Publishing Co. Pte. Ltd.
5 Toh Tuck Link, Singapore 596224
USA office: 27 Warren Street, Suite 401-402, Hackensack, NJ 07601
UK office: 57 Shelton Street, Covent Garden, London WC2H 9HE

Library of Congress Control Number: 2022040211

British Library Cataloguing-in-Publication Data
A catalogue record for this book is available from the British Library.

HYPERIDENTITIES
Boolean and De Morgan Structures

ISBN 978-981-125-491-8 (hardcover)
ISBN 978-981-125-492-5 (ebook for institutions)
ISBN 978-981-125-493-2 (ebook for individuals)

For any available supplementary material, please visit
https://www.worldscientific.com/worldscibooks/10.1142/12796#t=suppl

Desk Editors: Vishnu Mohan/Rok Ting Tan

Typeset by Stallion Press
Email: enquiries@stallionpress.com

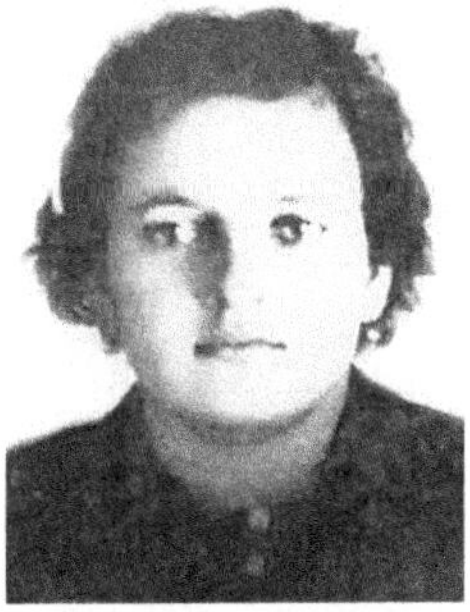

To my mother, Izabel Avagyan

Preface

The current research monograph continues my two previous books, which are the first monographs on hyperidentities that were published in 1986 and 1990.

Hyperidentities are important formulae of second-order logic, and research in hyperidentities paves way for the study of second-order logic and second-order model theory.

This book illustrates many important current trends and perspectives for the field of hyperidentities and their applications, of interest to researchers in modern algebra and discrete mathematics. It covers a number of directions, including the characterizations of the Boolean algebra of n-ary Boolean functions and the distributive lattice of n-ary monotone Boolean functions; the classification of hyperidentities of the variety of lattices, the variety of distributive (modular) lattices, the variety of Boolean algebras, and the variety of De Morgan algebras; the characterization of algebras with aforementioned hyperidentities; the functional representations of finitely-generated free algebras of various varieties of lattices and bilattices via generalized Boolean functions (De Morgan functions, quasi-De Morgan functions, super-Boolean functions, super-De Morgan functions, etc.); the structural results for De Morgan algebras, Boole–De Morgan algebras, super-Boolean algebras, bilattices, among others.

While problems of Boolean functions theory are well known, the present book offers alternative, more general problems, involving the concepts of De Morgan functions, quasi-De Morgan functions, super-Boolean functions, and super-De Morgan functions, etc. In contrast to other generalized Boolean functions discovered and investigated so far, these functions have clearly normal forms. This quality is of crucial importance for their applications in pure and applied mathematics, especially in discrete mathematics, quantum computation, quantum information theory, quantum logic, and the theory of quantum computers.

The book is self-contained. All necessary fundamental concepts and facts are introduced, but some background knowledge of abstract algebra and mathematical logic would be helpful for readers.

I am grateful to Prof. V.D. Belousov, Prof. B.I. Plotkin, Prof. L.A. Skornyakov, Prof. Yu.M. Ryabuchin, Prof. S.I. Adian, Prof. E.I. Zel'manov, Prof. H.P. Sankappanavar, Prof. A.E. Guterman, Prof. J.D.H. Smith, Prof. A. Romanowska, Prof. L. Budaghyan, Dr. A. Barchudaryan, Dr. V. Melkonian, Dr. S. Davidov, Dr. V. Aslanyan, Dr. T. Hakobyan, Dr. D. Davidova for discussions on topics and results covered in this book.

The work is partially supported by the State Committee of Science of the Republic of Armenia: grants 10-3/1-41 and 21T-1A213.

Contents

Introduction

By definition, a Boolean algebra is a complemented bounded distributive lattice. In the 19th century, George Boole's [44] attempt to formalize propositional logic led to the concept of Boolean algebra. Historically, lattice theory started with Boolean algebras which have played a fundamental role in the development of the lattice theory.

Boolean functions, i.e., $\{0,1\}$-valued functions of a finite number on $\{0,1\}$-valued variables, are among the most fundamental objects that have been investigated in pure and applied mathematics, especially in discrete mathematics, algebra, mathematical logic and computer science. The Boolean algebra of Boolean functions of n variables is a free Boolean algebra on n free generators. The bounded lattice of monotone Boolean functions of n variables is a free bounded distributive lattice on n free generators (R. Dedekind [84]). These are the first Boolean structures in both mathematics and applied mathematics, and they are used extensively in discrete mathematics.

Algebra and model theory study the connections between formal languages and their interpretations in models and algebras. The simplest and most widespread formal language is the first-order language (A. Church [66], A. I. Mal'tsev [212,214,216], G. Grätzer [142], C. Chang with H. Keisler [64], S. Burris with H. P. Sankappanavar [61], B. I. Plotkin [329]). The founders of the first-order language (logic) are Löwenheim, Skolem, Gödel, Łoś, Tarski, Mal'tsev and Birkhoff.

However, there exist very commonly encountered, classical algebraic structures that are not axiomatizable by the first-order formulas (logic). For example, rings, associative rings, commutative rings, associative–commutative rings, fields, or fields of fixed characteristics are axiomatized by first-order formulas, but their multiplicative groupoids, semigroups and groups are not, because these classes of groupoids, semigroups and groups are not closed under elementary equivalence (A. I. Mal'tsev and S. Kogalovskii [177], G. Sabbagh [350]). The situation is analogous for near-fields (M. Hall [147]), Grätzer algebras (G-fields) [26,116,140,268], topological rings and topological fields (L. S. Pontryagin [332]). However, if some class $\mathcal{K}$ of lattices is axiomatized by the first-order formulas, then the class of its multiplicative semigroups is also axiomatized by the first-order formulas [252].

Characterizations of such semigroups and groups are the most important problems in modern algebra, logic and topology. L. Fuchs [115, p. 324], called the characterization of multiplicative groups of fields a big problem.

This is why it is necessary to widen the formal language to allow to express phenomena that the first-order logic cannot capture.

An important extension of the first-order logic (language) is the second-order logic (language), described in detail in [64,66,214,216] (also see [170]). The second-order formulas consist of the same logical symbols of $\&, \vee, \neg, \rightarrow, \exists, \forall$ of individual and functional (predicate) variables, which are used in the first-order formulas. The difference is that in the second-order formulas, the quantifiers $\forall, \exists$ can be applied not only to individual variables, but also to functional (or to predicate) variables. Investigations of the second-order formulas (logic) go back to L. Henkin, A. I. Mal'tsev, A. Church, S. Kleene, A. Tarski. Many important mathematical concepts can be written in the second-order language. Consequently, investigation of the theories of the second-order languages (logic) is one of the central problems of algebra and mathematical logic (Mal'tsev [214]). For a compactness theorem for quasi-universal second-order formulas, see [170,244].

Starting with the 1960s the following second-order formulae were studied in various domains of algebra and its applications (see [25,28,31, 32,74,76–80,93,95,136,137,183,213,214,218,219,244,247,249,252,254,257,259, 268–270,281–284,288,292,299,300,302,303,309,310,314,351,352,361–363,376, 378,383,402,408,414–417,423,431]):

$$\forall X_1, \dots, X_m \forall x_1, \dots, x_n (w_1 = w_2), \tag{1}$$

$$\forall X_1, \dots, X_k \exists X_{k+1}, \dots, X_m \forall x_1, \dots, x_n (w_1 = w_2), \tag{2}$$

$$\exists x_1, \dots, x_n \forall X_1, \dots, X_m (w_1 = w_2), \tag{3}$$

$$\exists X_1, \dots, X_k \forall X_{k+1}, \dots, X_m \forall x_1, \dots, x_n (w_1 = w_2), \tag{4}$$

$$\forall X_1, \dots, X_k \exists X_{k+1}, \dots, X_t \forall X_{t+1}, \dots, X_m \forall x_1, \dots, x_n (w_1 = w_2), \tag{5}$$

where w_1, w_2 are words (respectively, hyperterms) in the functional variables $X_1, \ldots, X_m$ and the individual (respectively, object) variables $x_1, \ldots, x_n$. The first formula is called *hyperidentity* or $\forall(\forall)$-identity (see [25,31,93,179, 249,252,257,259,363,383,402], and also [377]); the second (respectively, third, fourth, fifth) formula is called an $\forall\exists(\forall)$-identity (respectively, $(\exists)\forall$-identity, $\exists\forall(\forall)$-identity, $\forall\exists\forall(\forall)$-identity). Sometimes the $\forall\exists(\forall)$-identity is called a generalized identity [25], the $(\exists)\forall$-identity is called a *coidentity* [247,249] (see also [377]) and $\exists\forall(\forall)$-identity is called a hybrid identity [28,285,363]. The satisfiability of these second-order formulas in an algebra $\mathfrak{A} = (Q; \Sigma)$ is understood by functional quantifiers $(\forall X_i)$ and $(\exists X_j)$, meaning: "for every value $X_i = A \in \Sigma$ of the corresponding arity", "there exists a value $X_j = A \in \Sigma$ of the corresponding arity". It is assumed that such a replacement is possible, that is

$$\{|X_1|, \ldots, |X_m|\} \subseteq \{|A| \mid A \in \Sigma\} = T_{\mathfrak{A}} = T_{(\Sigma)},$$

where $|S|$ is the arity of S, and $T_{\mathfrak{A}}$ is called the arithmetic type of $\mathfrak{A}$.

Hyperidentities and coidentities are usually presented without quantifiers: $\omega_1 = \omega_2$. The coidentity $\omega_1 = \omega_2$ is said to be satisfied in the algebra $(Q; \Sigma)$, if there exist values for object variables $x_1, \ldots, x_n$ from Q, such that the equality $\omega_1 = \omega_2$ holds when every functional variable in it is replaced by any operation of the corresponding arity from Σ (a possibility of such replacement is supposed, too). In addition, the object variables in the notation of the coidentity $\omega_1 = \omega_2$ are replaced by corresponding fixed values from Q.

Coidentities can be defined as second-order formulas of the following form, as well:

$$\forall X_1, \ldots, X_m (w_1 = w_2).$$

Example 1. *In any multioperator Ω-group the following coidentity is valid:*

$$X(\underbrace{0, \ldots, 0}_{n}) = 0,$$

for all $n \in T_{(\Omega)}$, where all object variables are replaced by the zero element of the Ω-group [189,328,377].

Example 2 (J. von Neumann). *Let $L(+, \cdot)$ be a modular lattice and $a, b, c \in L$. The sublattice of L, generated by the elements a, b, c, is* *distributive if and only if the following coidentity of left distributivity*

$$X\,(a,\, Y(b,\, c)) = Y\,(X(a,\, b),\, X(a,\, c))$$

holds in $L(+,\, \cdot)$.

Second-order formulas with analogous predicative quantifiers in models and algebraic systems are also often used in mathematical logic. For example, finiteness, the axiom of well-ordering, the continuum hypothesis, the property of being countable and others can be formulated within the second-order logic. Every algebra can be embedded into an algebra of functions whose identities are equivalent to hyperidentities or $\forall\exists(\forall)$-identities (A. Cayley, P. M. Cohn, unpublished).

The above semantics is different from the standard definition of the semantics of second-order formulas (logics) (see [66,214,216]). In this semantics, the compactness theorem is valid for any second-order logic (language) [257]. Studying logic via semantics is a well-established and very active branch of mathematical logic with many applications in computer science and elsewhere.

A variety (or equational class) is a class of algebras (all of the same similarity type or signature, or Ω-algebras) closed under the formation of products, subalgebras and homomorphic images. Equivalently, a variety is a class of algebras defined by a set of equations (identities). A hypervariety is a class of algebras (all of the same arithmetic type or the same signature) defined by a set of hyperidentities. Since 1954, the following-type second-order formulas were studied in algebras of term functions of various classes of varieties

$$\exists X_1, \ldots, X_m \forall x_1, \ldots, x_n (w_1 = w_2), \tag{6}$$

which are called Mal'tsev (Mal'cev) conditions (see [141,163,212,305,379, 380,400,401]), reducing to the hyperidentities of the class of term functions' algebras (termal or term algebras). (Note that the formula (6) is called functional equation in the set theory [4,5,8].)

The formulas (1)–(6) are usually written without quantifiers, if the structures of the quantifiers are understanding from the content. The formulas (2)–(6) are more general than hyperidentities. The numbers m and n in hyperidentity (1) are called the functional and object *rank*, respectively. A hyperidentity is said to be *nontrivial* if its functional *rank* is greater than 1, and it is called trivial otherwise ($m = 1$). A hyperidentity is called n-ary, if its functional variables are n-ary. For $n = 1, 2, 3$, the n-ary hyperidentity is called unary, binary, ternary. A formula (hyperidentity, coidentity, ...) is called a formula (hyperidentity, coidentity, ...) of algebra $\mathfrak{A}$, if it is satisfied in algebra $\mathfrak{A}$. Hyperidentities (coidentities, ...) are usually written without quantifiers: $w_1 = w_2$. Let V be a variety or another class of algebras. A hyperidentity (coidentity, ...) $w_1 = w_2$ is called a hyperidentity (coidentity, ...) of V if it is a hyperidentity (coidentity, ...) for any algebra $\mathfrak{A} \in V$.

The concept of hyperidentity is present in many well-known notions. For example, an algebra $\mathfrak{A} = (Q; \Sigma)$ is said to be *Abelian* (A. G. Kurosh [190]) or *entropic* (medial) if the following nontrivial hyperidentity

$$X(Y(x_{11}, \ldots, x_{1n}), \ldots, Y(x_{m1}, \ldots, x_{mn})) \\ = Y(X(x_{11}, \ldots, x_{m1}), \ldots, X(x_{1n}, \ldots, x_{mn}))$$

is valid for all $m, n \in T_{\mathfrak{A}}$. An algebra $\mathfrak{A} = (Q; \Sigma)$ is said to be *idempotent* if any of its operation is idempotent, i.e., if the following hyperidentity of idempotency

$$X(\underbrace{x, \ldots, x}_{n}) = x$$

is valid for all $n \in T_{\mathfrak{A}}$.

A *mode* is an idempotent and entropic algebra (studied in monographs [345,346]). A *distributive bisemilattice* (multisemilattice) [176] is a binary algebra with semilattice operations satisfying the following (nontrivial) hyperidentity of distributivity:

$$X(x, Y(y, z)) = Y(X(x, y), X(x, z)).$$

A *doppelsemigroup* (see [11,197,318,342,433,436–439,441]) is an algebra with two binary operations satisfying the following nontrivial hyperidentity of associativity:

$$X(x, Y(y, z)) = Y(X(x, y), z).$$

Binary algebras satisfying the hyperidentity of associativity

$$X(x, Y(y, z)) = Y(X(x, y), z)$$

are called hyperassociative algebras. These algebras under the name of Γ-semigroups, Γ-semirings (or gamma-semigroups, gamma-semirings) also were considered by various authors [20,149,172,188,200,306,358,359,364–366,369,434,435,440].

Example 3. *Let A and B be nonempty sets, Σ be the set of all mappings from B to A and Q be the set of all mappings from A to B. Then we can consider every element $\alpha \in \Sigma$ as a binary operation on Q:*

$$\alpha(a, b) = a \cdot \alpha \cdot b,$$

where $a, b \in Q$ and $a \cdot \alpha \cdot b$ is the usual superposition of mappings. So we obtain a binary algebra $(Q; \Sigma)$, which is hyperassociative.

The investigation of hyperidentities (coidentities) is a relatively new, actively developing field of pure and applied algebra. The concept of hyperidentity offers a high-level approach to algebraic questions, leading to new results, applications and problems. In particular, the investigation of hyperidentities is useful from the point of view of new technologies too, via optimization problems of block diagrams [259]. For applications of hyperidentities in discrete mathematics and topology, see [91,92,122,237,276–280,302,431]. For characterization of Sheffer functions and primal algebras via hyperidentities, see [91,92].

Any algebra $\mathfrak{A} = (Q; \Sigma)$ may be interpreted as a many-sorted algebra $(Q; \Sigma_i, \ldots, \Sigma_n, \ldots)$ (where Σ_n is a set of all n-ary operations of the given algebra) with the following operations $(f, x_1, \ldots, x_n) \to f(x_1, \ldots, x_n)$ where $f \in \Sigma_n$, $x_1, \ldots, x_n \in Q$, $n \in T_{\mathfrak{A}}$ (see [41,151]). Moreover, the hyperidentities of the given algebra become the identities of the corresponding many-sorted algebra and vice versa. In this way, the theory of hyperidentities as a second-order theory of algebras is converted into a first order theory of many-sorted algebras. Simultaneously there is a bijection between hyperidentities of termal (term) algebra $\mathcal{F}(\mathfrak{A})$ and identities of the clone $Cl(\mathfrak{A})$ of an algebra $\mathfrak{A}$ (a *clone* is also a many-sorted algebra, see [69,193,217,234,252,329,348, 385,394]). One of the specifics of a hyperidentity (coidentity) is that if a hyperidentity (coidentity) is valid in algebra $\mathfrak{A}$ then it is also valid in every reduct $\mathfrak{B}$ of $\mathfrak{A}$ with the condition $T_{\mathfrak{B}} = T_{\mathfrak{A}}$.

Hyperidentities are also "identities" of algebras in the category of bihomomorphisms $(\varphi, \tilde{\psi})$, where

$$\varphi A(x_1, \ldots, x_n) = (\tilde{\psi} A)(\varphi x_1, \ldots, \varphi x_n),$$

which were studied in the monograph [249]. Thus, we obtain the categorical definition of hyperidentities. More about the application of such morphisms in the cryptography can be found in [15]. For the categorical definition of $\forall\exists(\forall)$-identities and Birkhoff style theorem for $\forall\exists(\forall)$-identities, see [244] (Belousov's problem).

Hyperidentities for binary algebras with quasigroup operations were first considered by V. D. Belousov [25] (as a special case of $\forall\exists(\forall)$-identities which earlier is considered by R. Schauffler [361,362] in coding theory) and then by J. Aczel [6], about the classification of associative and distributive hyperidentities in binary algebras with quasigroup operations. Currently, more general results about these and other classifications of hyperidentities may be found in [249,252,257,259]. Observe that in algebras with quasigroup operations many $\forall\exists\forall(\forall)$-identities are equivalent to hyperidentities (see [252]). Coidentities for algebras were first considered in [249].

The multiplicative groups of fields have been characterized in [251] (see also [257]) via hyperidentities. On the base of these results, the concept of binary G-spaces is developed in topology [122]. The hyperidentities of varieties of lattices, modular lattices, distributive lattices, Boolean algebras, De Morgan algebras and weakly idempotent lattices have been characterized in the works [253,255–257,273–275,281,287,288,294].

Theorem A. *The variety of lattices satisfies the following hyperidentities:*

$$X(x,x) = x, \tag{7}$$

$$X(x,y) = X(y,x), \tag{8}$$

$$X(x, X(y,z)) = X(X(x,y),z), \tag{9}$$

$$X(Y(X(x,y),z), Y(y,z)) = Y(X(x,y),z). \tag{10}$$

Conversely, every hyperidentity of the variety of lattices is a consequence of the hyperidentities: (7)–(10) (see Chapter 2).

Theorem B. *The variety of modular lattices satisfies the following hyperidentities: (7)–(10) and*

$$X(Y(x, X(y,z)), Y(y,z)) = Y(X(x, Y(y,z)), X(y,z)). \tag{11}$$

Conversely, every hyperidentity of the variety of modular lattices is a consequence of the hyperidentities: (7)–(11) (see Chapter 2).

Theorem C. *The variety of distributive lattices satisfies the following hyperidentities: (7)–(9) and*

$$X(x, Y(y,z)) = Y(X(x,y), X(x,z)). \tag{12}$$

Conversely, every hyperidentity of the variety of distributive lattices is a consequence of the hyperidentities: (7)–(9), (12) (see Chapter 2).

Theorem D. *The variety of Boolean algebras satisfies the following hyperidentities: (7)–(9), (12) and*

$$F(F(x)) = x, \tag{13}$$

$$X(F(x), y) = X(F(X(x, y)), y), \tag{14}$$

$$F(X(F(X(x, y)), F(X(x, F(y))))) = x. \tag{15}$$

Conversely, every hyperidentity of the variety of Boolean algebras is a consequence of the hyperidentities: (7)–(9), (12)–(15).

All hyperidentities of the variety of Boolean algebras are consequences of one of its hyperidentities, i.e., the hyperequational theory of the variety of Boolean algebras is one-based (see Chapter 5).

Theorem E. *The variety of De Morgan algebras satisfies the following hyperidentities: (7)–(9), (12), (13) and*

$$F(x) = G(x), \tag{16}$$

$$F(Y(F(X(x, y)), z)) = X(F(Y(F(x), z)), F(Y(F(y), z))), \tag{17}$$

$$X(F(X(x, y)), F(x)) = F(x), \tag{18}$$

$$X(x, X(y, z)) = X(Y(x, Y(y, z)), F(Y(F(x), Y(F(y), F(z))))), \tag{19}$$

$$\begin{aligned} &X(F(X(Y(x, e), Y(x, Y(y, e)))), F(X(Y(z, e), Y(z, Y(t, e))))) \\ &\quad = X(F(X(Y(x, Y(z, e))), Y(x, Y(y, Y(z, Y(u, e)))))), \\ &\qquad F(X(F(Y(F(Y(x, Y(y, e))), F(Y(z, Y(t, e))))), e)))). \end{aligned} \tag{20}$$

Conversely, every hyperidentity of the variety of De Morgan algebras is a consequence of the hyperidentities: (7)–(9), (12), (13), (16)–(20).

The hyperequational theory of the variety of De Morgan algebras is not one-based (see Chapter 7).

Theorem F. *The variety of weakly idempotent lattices satisfies the following hyperidentities:*

$$X(x, x) = Y(x, x), \tag{21}$$

$$X(x, y) = X(y, x), \tag{22}$$

$$X(x, X(y, z)) = X(X(x, y), z), \tag{23}$$

$$X(x, X(y, y)) = X(x, y), \tag{24}$$

$$X(Y(X(x, y), z), Y(x, z)) = Y(X(x, y), z). \tag{25}$$

Every hyperidentity of the variety of weakly idempotent lattices is a consequence of the hyperidentities: (21)–(25) (see Chapter 13).

A hyperidentity $\omega_1 = \omega_2$ is called *termal* or t-hyperidentity of the algebra $\mathfrak{A}$ if it is valid in the term algebra $\mathcal{F}(\mathfrak{A})$. Let V be a variety. A hyperidentity $\omega_1 = \omega_2$ is called a termal hyperidentity of V if it is a termal hyperidentity for any algebra $\mathfrak{A} \in V$. Termal hyperidentities for varieties were first considered by W. Taylor [402] (as a special case of Mal'tsev conditions for varieties) for characterization of classes of varieties which are closed under formation of equivalent varieties, products of varieties, reducts of varieties and subvarieties. Since the operations of an algebra are included in the set of term operations (clone) of the algebra, the concept of termal hyperidentity of a variety is stronger than the concept of hyperidentity. In particular, the variety of rings (even commutative rings) does not have termal hyperidentities except $w = w$, but has hyperidentities.

Termal hyperidentities of varieties of groups and semigroups have been characterized by G. Bergman [31] (see also [32]). Termal hyperidentities of the variety of lattices and of the variety of semilattices were studied by R. Padmanabhan and P. Penner ([309,310,317]).

Hyperidentities in algebras as an individual direction of investigations, were first presented in the monographs [249,252]. The problem of characterization of termal hyperidentities of important classes of groups, semigroups, loops, quasigroups has been posed in the book [249, p. 129, Problem 26]. The hyperidentities of algebras and varieties, termal and essential hyperidentities, pre-hyperidentities of various varieties of groups, semigroups, quasigroups, loops and related algebras were also studied by many authors (see references of the current book).

Recall the following well-known classical results about functional representations of free Boolean algebras and free distributive lattices. On function algebras see also the classical researches of M. H. Stone [391], I. M. Gelfand and A. N. Kolmogoroff [120], I. Kaplansky [169], E. Newitt [150], L. Gillman and M. Jerison [124].

Theorem G (R. Dedekind, 1897). *The free Boolean algebra on n free generators is isomorphic to the Boolean algebra of Boolean functions of n variables (see Chapter 4).*

Theorem H (R. Dedekind, 1897). *The free bounded distributive lattice on n free generators is isomorphic to the bounded lattice of monotone Boolean functions of n variables (see Chapter 4).*

A problem posed by B. I. Plotkin in 1970s has required finding the varieties of algebras with analogous functional representations of free finitely generated algebras. From the results of the book follows a solution of this problem, too. The importance of this problem is that the generated functions are alternative to Boolean functions with their respective applications which are comparable to the applications of Boolean functions. All these functions are more general than the Boolean functions and, in contrast to the other generalized Boolean functions discovered and investigated so far (see [227,387,409] and etc.), these have clearly normal forms, which are importance for the applications of these functions in pure and applied mathematics, especially in discrete mathematics, quantum computation, quantum information theory, quantum logic and the theory of quantum computers.

As an important consequences, every finitely generated algebra is finite in these varieties. Note that even in the classical algebras (groups, semigroups, quasigroups, rings, etc.) the characterization of varieties in which every finitely generated algebra is finite is an open problem too (see [9,10,426–429]).

A variety of algebras is called Burnside variety (W. Burnside) [268] if its any finitely generated algebra is finite.[a] For instance, we have the following conditions:

(1) Any finitely generated distributive lattice is finite [143,329] (see Sec. 4.1);
(2) Any finitely generated Boolean algebra is finite [143,329] (see Sec. 4.1);
(3) Any finitely generated De Morgan algebra is finite [277] (see Sec. 4.3);
(4) Any finitely generated Boole–De Morgan algebra is finite [260,261,278] (see Sec. 3.3).
(5) Any finitely generated algebra with two binary, one unary and two nullary operations, satisfying the hyperidentities of the variety of Boolean algebras is finite [253,256,279] (see Sec. 10.3).
(6) Any finitely generated algebra with two binary and one unary operations, satisfying the hyperidentities of the variety of De Morgan algebras is finite [280] (see Sec. 11.3).
(7) Any finitely generated distributive bilattice is finite [271] (see Sec. 12.5).
(8) Any finitely generated Boole–De Morgan bilattice is finite [271].

[a]In 1902, Burnside proposed the following problem:

Is a finitely generated group satisfying an identity of the form $x^n = 1$ *necessarily finite?*

The answer in yes for $n = 1, 2, 3, 4$ and 6. The case $n \leq 2$ is trivial, the case $n = 3$ was settled by Burnside [60], the case $n = 4$ by Sanov [356] and the case $n = 6$ by M. Hall [147]. The original problem finally received a negative answer by Novikov and Adian in 1968 [307] (see also [9,201]).

(9) Any finitely generated idempotent semigroup is finite [49,50,144,192, 195,199,235,357]. A finitely generated idempotent and hyperassociative algebra with a finite set of operations is finite [265].

The material of this book is based on lectures given by the author at the Yerevan State University (Armenia) and University of Bergen (Norway).

A more detailed breakdown of the book is as follows. In Chapter 1, the basic concepts of algebras, Ω-algebras, T-algebras, free algebras, varieties of Ω-algebras and hypervarieties of T-algebras, quotient algebras, term algebras, homomorphisms and bihomomorphisms of algebras, products and superproducts, filtered products and ultraproducts of algebras, hyperidentities and termal hyperidentities of algebras, clones are considered (with a complete proof of Birkhoff's theorem on varieties).

Chapter 2 presents a sampling of techniques finding the hyperidentities of lattice varieties. These include the characterization of hyperidentities of varieties of lattices, distributive lattices, modular lattices.

Chapter 3 studies Boole–De Morgan algebras. A Stone-type and Birkhoff-type representations of these algebras are given. The subdirectly-irreducible Boole–De Morgan algebras are characterized. We define the concept of quasi-De Morgan function and characterize the free n-generated Boole–De Morgan algebras via n-ary quasi-De Morgan Functions.

Chapter 4 concerns to the De Morgan algebras. The results of this chapter include a proof of Stone-type representation of De Morgan algebras. We characterized the free n-generated Boolean algebras via n-ary Boolean functions, and the free n-generated bounded distributive lattices via n-ary monotone Boolean functions. We define the concept of De Morgan function and characterize the free n-generated De Morgan algebras via n-ary De Morgan functions.

Chapter 5 studies the hyperidentities of the variety of Boolean algebras. As a consequence, we obtain a finite base of hyperidentities of this variety. Since the variety of Boolean algebras does not has a minimal subvariety, we obtain the same result for any Boolean algebra too. The concept of super-Boolean algebra is one of fundamental concepts introduced in this chapter. A super-Boolean algebra is an algebra with the hyperidentities of the variety of Boolean algebras. For example, superproduct of Boolean algebras is a super-Boolean algebra.

In Chapter 6, results on decidability of elementary theory of super-Boolean algebras and varieties of super-Boolean algebras are discussed using the structural concept of Boolean sum introduced in [256]. In particular, the following results are proved.

Theorem K. *The elementary theory of the variety of super-Boolean algebras with two binary and one unary operations is decidable (see Chapter 6).*

Theorem L. *Any super-Boolean algebra with two binary and one unary operations has a decidability elementary theory (see Chapter 6).*

Chapter 7 discusses characterization of hyperidentities of the variety of De Morgan algebras.

In Chapter 8, the functional completeness theorem for De Morgan functions is proved.

In Chapter 9, the structural results for bilattices are proved via superproduct of lattices.

Chapter 10 provides the characterization of finitely generated free super-Boolean algebras via super-Boolean functions. Chapter 11 provides the characterization of finitely generated free super-De Morgan algebras via super-De Morgan functions and Chapter 12 provides the characterization of finitely-generated free distributive bilattices via bi-De Morgan functions. Note that another characterization of free n-generated distributive bilattices via specification of the (super) product representation of distributive bilattices is given in [63]. Advantage of our approach is that it is applicable for various varieties. For the more general result see paper [271] dedicated to the characterization of finitely generated free Boole–De Morgan bilattices.

Chapters 13 and 14 discuss a variety of topics on weakly idempotent lattices and interlaced weakly idempotent bilattices. Chapter 15 discusses the concept of bigroups and their characterizations by Grätzer algebras. In Chapter 16, the connection between the right (left) invertible medial algebras and the right (left) group of binary operations is studied. As a consequence, we prove the structural result for transitive modes (cf. [251,257]). We consider the general concept of pre-quasigroup at the end of Chapter 16.

In Chapter 17, we discuss the problem of classification of nontrivial associative and distributive hyperidentities in q-algebras. We discuss the same problem for essential associative and distributive hyperidentities for semigroups, in Chapter 18. In Chapter 19, we present a characterization of multiplicative groups of fields by classical hyperidentities of idempotency and distributivity. In Chapters 20 and 21, we discuss the problem of characterization of invertible algebras satisfying the associative and distributive formulas with functional variables.

The last chapter discusses a set of other open problems about characterization of hyperidentities and termal hyperidentities of varieties related to classical varieties of groups, rings, lattices, Boolean algebras, etc.

Solutions of these problems could serve for development of next steps of the results and concepts involved in the current book.

Readership: Undergraduates, graduates and researchers in algebra and model theory, Boolean functions and cryptology (cryptography), mathematical logic and quantum information theory. It is also of interest to researchers working on logic and category theory with applications.

Chapter 1

Some Basic Concepts

1.1. Algebras, Ω-algebras, T-algebras: Identities, hyperidentities and coidentities

Definition 1.1.1. Let A be a nonempty set and n be a nonnegative integer. We define $A^0 = \{\emptyset\}$ and, for $n > 0$, A^n is the set of n-tuples of elements from A. An *n-ary operation* (or *function*) on A is any function from A^n to A; n is called the *arity* (or *rank*) of f, and it is denoted $n = |f|$. A *finitary* operation is an n-ary operation for some n. An operation on A with arity 0 is called a *nullary* operation (or *constant*); it is completely determined by the image $f(\emptyset)$ and we identify it with the element $f(\emptyset)$ (thus a nullary operation is thought of as an element of A). An operation f on A is called *unary*, *binary*, or *ternary* if its arity is 1, 2 or 3, respectively.

In particular, if A is the two element set $\{0, 1\}$ then the n-ary operations on A are called *Boolean functions* of n variables. For example, the functions $x \& y$, $x \vee y$, $x \to y$, $\neg x$, given by table below, are classical Boolean functions:

x	y	$x\&y$	$x \vee y$	$x \to y$	$\neg x$
0	0	0	0	1	1
0	1	0	1	1	1
1	0	0	1	0	0
1	1	1	1	1	0

Boolean functions will play important role starting from the third chapter.

Definition 1.1.2. An ordered pair $(A; \Sigma)$, where A is a nonempty set and Σ is a nonempty set of finitary operations on A, is called an *algebra* (or *universal algebra*).

Let $\mathfrak{A} = (A; \Sigma)$ be an arbitrary algebra, and let $\Sigma' \subseteq \Sigma$, where $\Sigma' \neq \emptyset$. An algebra $\mathfrak{A}' = (A; \Sigma')$ is called a *reduct* (or semisubalgebra) of the algebra $\mathfrak{A}$.

An algebra $(A; \Sigma)$ is called *nontrivial* if the cordinality of A is $|A| > 1$; otherwise the algebra is called *trivial* ($|A| = 1$). An algebra $(A; \Sigma)$ is called *finite* if the set A is finite.

An algebra with unary (binary, ternary) operations is called a unary (binary, ternary) algebra.

The theory of algebras depends on the environment (a category!) in which we consider them.

Definition 1.1.3. A *type* (signature) of algebra is a set Ω of *function symbols* such that a nonnegative integer n is assigned to each member f of Ω. This integer is called the *arity* (or *rank*) of f; and f is said to be an n-ary function symbol, and it is denoted by $n = |f|$. The subset of all n-ary function symbols in Ω is denoted by Ω_n.

Definition 1.1.4. Let Ω be a type. An algebra $\mathfrak{A} = (A; \Sigma)$ is called an Ω-*algebra*, if there exists a surjective mapping $\mu : \Omega \to \Sigma$, which preserves the arity, i.e., $|\mu(f)| = |f|$ for any function symbol $f \in \Omega$. Thus, an Ω-algebra can be treated as a triple $(Q; \Sigma, \mu)$. The set Ω is called a type for the Ω-algebra $\mathfrak{A}$. The set A is called the *universe* (or *underlying set*) of $\mathfrak{A}$; and the $\mu(f)$'s are called the *fundamental operations* of $\mathfrak{A}$. For an algebra $\mathfrak{A}$ we often denote $\mu(f)$ by $f^{\mathfrak{A}}$ or simply by f (when it can cause no confusion).

If Ω is finite, say $\Omega = \{f_1, f_2, \dots, f_k\}$, we often write $(A; f_1, \dots, f_k)$ for $(A; \Sigma)$.

Thus any universal algebra can be considered as an Ω-algebra for some type of algebras Ω.

If f is a binary operation on a set A or a binary function symbol in a type Ω then sometimes we write afb for $f(a, b)$.

Basic properties of algebras are expressed by formulas of first- and second-order languages. So let us recall their definitions.

First, we define the first-order formulas (language) [61,66,214,216].

The alphabet of the first-order language consists of the following symbols:

(1) function symbols denoted by $f, g, \dots$ with or without superscripts, a nonnegative integer is assigned to each function symbol called its arity;
(2) predicate symbols denoted by $P, Q, \dots$ with or without superscripts, a nonnegative integer is assigned to each predicate symbol called its arity;
(3) object variables, denoted by $x, y, z, \dots$ with or without subscripts;

(4) logical symbols $\&, \vee, \rightarrow, \neg, =, \forall, \exists$ and also auxiliary symbols $($, $)$ — parenthesis, and comma.

A function (predicate) symbol with arity n is called an n-ary function (predicate) symbol.

Nullary function symbols are also called constants.

The first-order terms are defined inductively.

(i) All object variables are first-order terms.
(ii) All nullary function symbols are first-order terms.
(iii) If f is an n-ary function symbol ($n \geq 1$) and $t_1, \ldots, t_n$ are first-order terms then the expression $f(t_1, \ldots, t_n)$ is a first-order term too.
(iv) There are no other first-order terms.

First-order terms are called terms too.

First-order formulas are also defined inductively.

(i) All nullary predicate symbols are first-order formulas. If $t_1, \ldots, t_n$ are first-order terms and P is an n-ary predicate symbol, then $P(t_1, \ldots, t_n)$ is a first order formula. All object variables in these formulas are free.
(ii) If t_1, t_2 are first-order terms, then the expression $t_1 = t_2$ is a first-order formula. All object variables in this first-order formula are free.
(iii) If A is a first order formula and x is an object variable that is free in A, then $\forall x A$ and $\exists x A$ are first order formulas. The object variable x is bound in these formulas. All other object variables that have free (bound) occurrence in A remain free (bound) in $\forall x A$, $\exists x A$.
(iv) If A is a first order formula, then so is $\neg A$. All object variables that are free in A are also free in $\neg A$. All object variables that are bound in A are also bound in $\neg A$.
(v) Let A and B be first order formulas such that all object variables occurring both in A and B are free in both formulas. Then $A \& B$, $A \vee B$, $A \rightarrow B$ are first order formulas. All object variables that are free in at least one of A and B, are free in those formulas. All object variables that are bound in at least one of A and B, are bound in those formulas.
(vi) There are no other first-order formulas.

Satisfiability of first-order formulas is defined by the standard way.

In such languages, *identities* are defined to be first-order formulas of the form

$$\forall x_1, \ldots, \forall x_n \, (t_1 = t_2),$$

where t_1 and t_2 are first order terms with object variables from the set $\{x_1, \ldots, x_n\}$.

Note that identities are usually written without universal quantifiers [36,61,69,142,143,189,190,216,321,329,379].

A first-order formula is called a *sentence* if it contains no free object variables. If $\mathfrak{A}$ is an Ω-algebra, then the set of all sentences that are satisfied in $\mathfrak{A}$ is called the *elementary theory* of algebra $\mathfrak{A}$ and it is denoted by $ElTh\,\mathfrak{A}$. Note that here the function symbols of our language are the elements of Ω. If $\mathcal{K}$ is a class of Ω-algebras, then the set of all sentences that are satisfied in all algebras from $\mathcal{K}$ is called the *elementary theory* of class $\mathcal{K}$ and it is denoted by $ElTh\,\mathcal{K}$.

For a class $\mathcal{K}$ of Ω-algebras denote by $Eq\,\mathcal{K}$ the set of all identities that are satisfied in all algebras of $\mathcal{K}$. This set is called the *equational theory* of the class $\mathcal{K}$. When $\mathcal{K} = \{\mathfrak{M}\}$, we simply write $Eq\,\mathfrak{M}$ instead of $Eq\,\{\mathfrak{M}\}$, and call this set the *equational theory* of the algebra $\mathfrak{M}$.

Now let us define the second-order formulas (language) [64,66,213,214, 216].

The alphabet of the second-order language consists of the following symbols:

(1) functional (function) variables denoted by $X, Y, F, G, \ldots$ with or without superscripts, a nonnegative integer is assigned to each functional variable called its arity;
(2) predicate variables denoted by $P, Q, \ldots$ with or without superscripts, a nonnegative integer is assigned to each predicate variable called its arity;
(3) object variables, denoted by $x, y, z, \ldots$ with or without subscripts;
(4) logical symbols $\&, \vee, \rightarrow, \neg, =, \forall, \exists$ and also auxiliary symbols $(,)$ — parenthesis, and comma.

A functional (predicate) variable with arity n is called an n-ary functional (predicate) variable.

The second-order terms are defined inductively.

(i) All object variables are second-order terms.
(ii) All nullary functional variables are second-order terms.
(iii) If F is an n-ary functional variable and $t_1, \ldots, t_n$ are second-order terms then the expression $F(t_1, \ldots, t_n)$ is a second order term too.
(iv) There are no other second-order terms.

Second-order terms are called *hyperterms*, too.

Second-order formulas are also defined inductively.

(i) All nullary predicate variables are second-order formulas. If $t_1, \dots, t_n$ are second-order terms and P is an n-ary predicate variable, then $P(t_1, \dots, t_n)$ is a second-order formula. All variables in these formulas are free.

(ii) If t_1, t_2 are second-order terms, then the expression $t_1 = t_2$ is a second-order formula. All variables in this formula are free.

(iii) Let A be a second-order formula and $\mathfrak{x}$ be a functional or predicate or object variable that is free in A. Then the expressions $\forall\mathfrak{x}A$, $\exists\mathfrak{x}A$ are second order formulas. The variable $\mathfrak{x}$ is bound in these second order formulas. All other variables that have free (bound) occurrence in A remain free (bound) in $\forall\mathfrak{x}A$, $\exists\mathfrak{x}A$.

(iv) If A is a second-order formula, then so is $\neg A$. All variables that are free in A are also free in $\neg A$. All variables that are bound in A are also bound in $\neg A$.

(v) Let A and B be second-order formulas such that all variables occurring both in A and B are free in both formulas. Then $A \& B$, $A \vee B$, $A \to B$ are second-order formulas. All variables that are free in at least one of A and B, are free in those formulas. All variables that are bound in at least one of A and B, are bound in those formulas.

(vi) There are no other second-order formulas.

The satisfiability (truth) of these second-order formulas in an algebra $\mathfrak{A} = (Q; \Sigma)$ is understood with functional quantifiers $(\forall X_i)$ and $(\exists X_j)$ meaning: "for every value $X_i = A \in \Sigma$ of the corresponding arity" and "there exists a value $X_j = B \in \Sigma$ of the corresponding arity". Analogously we are understand the meaning of predicate quantifiers in model and algebraic system [218].

We see that the difference between first- and second-order languages is that in first-order language there are some fixed function and predicate symbols that cannot be quantified, and in second-order language there are function and predicate variables that can be quantified.

In such languages *hyperidentities* are defined to be second-order formulas with universal quantifiers of the form

$$\forall X_1, \dots, \forall X_m, \forall x_1, \dots, \forall x_n \, (t_1 = t_2),$$

where $X_1, \dots, X_m$ are the functional variables and $x_1, \dots, x_n$ are the object variables in the hyperterms (second-order terms) t_1, t_2. Generally hyperidentities are written without a quantifier prefix: $t_1 = t_2$. We say that the hyperidentity $t_1 = t_2$ holds (is valid) in an algebra $(Q; \Sigma)$ or an algebra $(Q;\Sigma)$ satisfies the hyperidentity $t_1 = t_2$ if this equality is valid when every object variable x_i is replaced by an arbitrary element of Q and every functional variable X_j is replaced by any operation of the corresponding arity from Σ. It is assumed that such a replacement is possible, that is

$$\{|X_1|, \ldots, |X_m|\} \subseteq \{|A| \mid A \in \Sigma\} = T_{(Q;\Sigma)},$$

where $|X_i|$ is the arity of X_i, and $T_{(Q;\Sigma)}$ is called the arithmetic type of $(Q;\Sigma)$ [249,252]. The hyperidentity is also called an $\forall(\forall)$-identity, where m is called the *functional rank* and n is called the *object rank* of the given hyperidentity. A hyperidentity is said to be *nontrivial* if its functional rank is >1, otherwise it is *trivial* ($m = 1$).

A hyperidentity is called a hyperidentity of algebra $\mathfrak{A}$, if it is satisfied in algebra $\mathfrak{A}$. Let V be a class of algebras. A hyperidentity $t_1 = t_2$ is called a hyperidentity of V if it is a hyperidentity for any algebra $\mathfrak{A} \in V$.

Coidentities are defined to be second-order formulas of the form

$$\exists x_1, \ldots, \exists x_n, \forall X_1, \ldots, \forall X_m \ \ (t_1 = t_2),$$

where $X_1, \ldots, X_m$ are the functional variables and $x_1, \ldots, x_n$ are the object variables in the hyperterms (second-order terms) t_1, t_2.

Coidentities are usually presented without quantifiers, too: $\omega_1 = \omega_2$. The coidentity $\omega_1 = \omega_2$ is said to be satisfied in the algebra $(Q;\Sigma)$, if there exist values for object variables $x_1, \ldots, x_n$ from Q, such that the equality $\omega_1 = \omega_2$ holds when every functional variable in it is replaced by any operation of the corresponding arity from Σ (a possibility of such replacement is supposed, too). In addition, the object variables in the notation of the coidentity $\omega_1 = \omega_2$ are replaced by corresponding fixed values from Q.

Now we give some well known examples of algebras, which are defined by the identities.

Examples of algebras

(1) A *groupoid* (magma [46,368]) is an algebra $(G;\cdot)$ (or $G(\cdot)$) with one binary operation. The *n-groupoid* is an algebra with one n-ary operation.

(2) A *group* is an algebra $\mathbf{G} = (G;\cdot,^{-1}, e)$ with a binary, a unary and a nullary operations in which the following identities hold:

$$\text{(G1)}\ (x \cdot y) \cdot z = x \cdot (y \cdot z),$$
$$\text{(G2)}\ x \cdot e = e \cdot x = x,$$
$$\text{(G3)}\ x \cdot x^{-1} = x^{-1} \cdot x = e.$$

L. E. Dickson [97] characterized groups in terms of one-sided invertibility.

A group **G** is called *Abelian* (or *commutative*) if it satisfies one more identity:

$$\text{(G4)}\ x \cdot y = y \cdot x.$$

(3) A *semigroup* is a groupoid $(G; \cdot)$, in which (G1) is true. It is *commutative* (or *Abelian*) if (G4) holds. A *monoid* is an algebra $(M; \cdot, e)$ with a binary and a nullary operation satisfying (G1) and (G2). A *semilattice* is a commutative semigroup with identity of idempotency: $x \cdot x = x$.

(4) A *quasigroup* is a groupoid $(Q; \cdot)$, in which for any $a, b \in Q$ each of the equations

$$a \cdot x = b,$$
$$y \cdot a = b,$$

has unique solution in Q. A loop is a quasigroup with the unit. A Moufang loop [243] is a loop with one of the following equivalent identities:

$$(zx \cdot y)x = z(x \cdot yx),$$
$$x(y \cdot xz) = (xy \cdot x)z,$$
$$xy \cdot zx = x(yz \cdot x).$$

On Dickson's theorem for Moufang loops, see [267].

The definition of quasigroup means that the operation $\cdot$ is invertible from left and from right. If we denote the unique solution of the equality $a \cdot x = b$ by $a \backslash b$ and the unique solution of $y \cdot a = b$ by b/a then we will get two more binary operations $\backslash$ (right division) and $/$ (left division) on the set Q that satisfy the following identities:

$$\text{(Q1)}\ x \backslash (x \cdot y) = y,$$
$$\text{(Q2)}\ (x \cdot y)/y = x,$$
$$\text{(Q3)}\ x \cdot (x \backslash y) = y,$$
$$\text{(Q4)}\ (x/y) \cdot y = x.$$

The converse is true too. If we have an algebra $(Q; \cdot, \backslash, /)$ with identities $(Q1)$–$(Q4)$, then $(Q; \cdot)$ is a quasigroup. So quasigroups are often defined as algebras $(Q; \cdot, \backslash, /)$ with identities $(Q1)$–$(Q4)$.

If we denote the operation of a quasigroup by A and the left and right divisions by ${}^{-1}A$ and A^{-1} then the identities ($Q1$)–($Q4$) will get the following form:

$$\begin{aligned}&(\mathrm{Q1'})\ A^{-1}(x, A(x, y)) = y,\\&(\mathrm{Q2'})\ {}^{-1}A(A(x, y), y) = x,\\&(\mathrm{Q3'})\ A(x, A^{-1}(x, y)) = y,\\&(\mathrm{Q4'})\ A({}^{-1}A(x, y), y) = x.\end{aligned}$$

Now let us show another approach to the definitions of quasigroups. If Q is a nonempty set, then for any two binary operations $A, B : Q^2 \to Q$ we define their *binary superpositions* by the following way:

$$(A \cdot B)(x, y) = A(x, B(x, y)),\ \ (A \circ B)(x, y) = A(B(x, y), y).$$

Let us denote by $\mathcal{F}_Q$ the set of all binary operations on the set Q. Also denote the binary operations $F(x, y) = x$, $E(x, y) = y$. Then the algebras $(\mathcal{F}_Q; \cdot, E)$ and $(\mathcal{F}_Q; \circ, F)$ will be semigroups with identity elements. The identities ($Q1'$)–($Q4'$) mean that $A \circ^{-1} A =^{-1} A \circ A = F$, $A \cdot A^{-1} = A^{-1} \cdot A = E$, i.e., A is invertible in both semigroups $(\mathcal{F}_Q; \cdot, E)$ and $(\mathcal{F}_Q; \circ, F)$. Thus, a groupoid $(Q; A)$ is a quasigroup if and only if the operation A is invertible in both semigroups $(\mathcal{F}_Q; \cdot, E)$ and $(\mathcal{F}_Q; \circ, F)$ [263].

(5) A *division groupoid* is a groupoid $(Q; \cdot)$, in which for any $a, b \in Q$ each of the equations

$$\begin{aligned}a \cdot x &= b,\\ y \cdot a &= b,\end{aligned}$$

has a solution in Q. Any division semigroup is a group (E. Huntington). Any finite division groupoid is a quasigroup.

(6) A *system of semigroups* (*groups, quasigroups, division groupoids*) is an algebra $(Q; \Sigma)$, in which $(Q; A)$ is a semigroup (group, quasigroup, division groupoid) for any $A \in \Sigma$.

(7) A *ring* is an algebra $R = (R; +, \cdot, -, 0)$ with two binary, one unary and one nullary operations satisfying the following conditions:

$$\begin{aligned}&(\mathrm{R1})\ (R; +, -, 0) \text{ is an Abelian group},\\&(\mathrm{R2})\ x \cdot (y + z) = (x \cdot y) + (x \cdot z),\ \ (x + y) \cdot z = (x \cdot z) + (y \cdot z)\end{aligned}$$

for all $x, y, z \in R$; and it is denoted by $(R; +, \cdot)$ (or $R(+, \cdot)$) as well.

A ring R is called commutative if the identity $x \cdot y = y \cdot x$ holds in $(R; \cdot)$ and associative if $(R; \cdot)$ is a semigroup.

An element $1 \in R$ is called an *identity element* (unit) of ring R if $1 \cdot x = x \cdot 1 = x$ for all $x \in R$.

An associative and commutative ring with an identity element $(F; +, \cdot, -, 0, 1)$ is called a *field* if for each $x \in F \setminus \{0\}$ there exists an element $x^{-1} \in F$ such that $x \cdot x^{-1} = 1$. So the set $F \setminus \{0\}$ is an abelian group under the operations $\cdot$, $^{-1}$ and 1, which is called the *multiplicative group* of a given field $(F; +, \cdot, -, 0, 1)$ and denoted by $F \setminus \{0\}(\cdot)$ or $F^*(\cdot)$, where $F^* = F \setminus \{0\}$.

(8) A *lattice* is an algebra $(L; +, \cdot)$ (or $L(+, \cdot)$) with two binary operations satisfying the following classical identities:

(L1) $x + x = x$, $x \cdot x = x$ (idempotency);
(L2) $(x + y) + z = x + (y + z)$ $(x \cdot y) \cdot z = x \cdot (y \cdot z)$, (associativity);
(L3) $x + y = y + x$, $x \cdot y = y \cdot x$ (commutativity);
(L4) $x \cdot (x + y) = x$, $x + (x \cdot y) = x$ (absorption).

A *distributive lattice* is a lattice that satisfies the identities:

(L5) $x \cdot (y + z) = (x \cdot y) + (x \cdot z)$, $x + (y \cdot z) = (x + y) \cdot (x + z)$.

In a lattice $Q(+, \cdot)$ one can define a partial order "$\leq$" with the $\sup\{x, y\} = x + y$ and $\inf\{x, y\} = x \cdot y$ by the following rule:

$$x \leq y \Leftrightarrow x + y = y.$$

A modular lattice is a lattice that satisfies the identity:

$$\text{(L6) } x(xy + z) = xy + xz.$$

(9) A *bounded lattice* is an algebra $(L; +, \cdot, 0, 1)$ with two binary and two nullary operations such that $(L; +, \cdot)$ is a lattice and $(L; +, \cdot, 0, 1)$ satisfies the following identities:

$$x + 0 = x, \ x \cdot 1 = x.$$

(10) An algebra $(B; +, \cdot, \ ', 0, 1)$ with two binary, one unary and two nullary operations is called a *Boolean algebra* if $(B; +, \cdot, 0, 1)$ is a bounded distributive lattice and $(B; +, \cdot, \ ', 0, 1)$ satisfies the following identities:

$$\text{(B1) } x + x' = 1,$$
$$\text{(B2) } x \cdot x' = 0.$$

If $(B; +, \cdot, ', 0, 1)$ is a Boolean algebra, then $(B; +, \cdot, ')$ is called a Boolean lattice.

(11) An algebra $(D;+,\cdot,^{-},0,1)$ with two binary, one unary and two nullary operations is called a *De Morgan algebra* if $(D;+,\cdot,0,1)$ is a bounded distributive lattice and $(D;+,\cdot,^{-},0,1)$ satisfies the following identities:

$$\text{(D1) } \overline{x+y} = \overline{x}\cdot\overline{y},$$
$$\text{(D2) } \overline{\overline{x}} = x,$$

where $\overline{\overline{x}} = \overline{(\overline{x})}$.

If $(D;+,\cdot,^{-},0,1)$ is a De Morgan algebra, then $(D;+,\cdot,^{-})$ is called a De Morgan lattice.

So, an algebra $(A;+,\cdot,^{-})$ with two binary and one unary operations is called a Boolean (De Morgan) lattice if there exist nullary operations 0 and 1 such that $(A;+,\cdot,^{-},0,1)$ is a Boolean (De Morgan) algebra.

(12) An algebra $(Q;\{+,\cdot,^{-},\ ',0,1\})$ with two binary, two unary and two nullary operations is called a *Boole–De Morgan* algebra if $(Q;\{+,\cdot,^{-},0,1\})$ is a De Morgan algebra and $(Q;\{+,\cdot,\ ',0,1\})$ is a Boolean algebra and the two unary operations commute, i.e., $(\overline{x})' = \overline{(x')}$ [260,261]. The last condition on commutativity of unary operations follows from the other conditions.

If $(Q;\{+,\cdot,^{-},\ ',0,1\})$ is a Boole–De Morgan algebra, then $(Q;\{+,\cdot,^{-},\ '\})$ is called a Boole–De Morgan lattice.

(13) The algebra $(L;\wedge)$ with one binary operation is called *weakly idempotent semilattice*, if it satisfies the following identities:

$$a\wedge b = b\wedge a, \text{ (commutativity)}$$
$$(a\wedge b)\wedge c = a\wedge(b\wedge c), \text{ (associativity)}$$
$$a\wedge(b\wedge b) = a\wedge b. \text{ (weakly idempotency)}$$

The operation $\wedge$ is called product. Adding the idempotent identity $a\wedge a = a$ to it, we obtain a semilattice. The element $a\in L$ is called an idempotent of the *weakly idempotent semilattice* $(L;\wedge)$, if $a\wedge a = a$. The set of the idempotent elements of each weakly idempotent semilattice form a semilattice, i.e., the product of any two idempotent elements in the weakly idempotent semilattice is an idempotent element.

(14) The algebra $(L;\wedge,\vee)$ with two binary operations is called *weakly idempotent lattice*, if the reducts $(L;\wedge)$ and $(L;\vee)$ are weakly idempotent

semilattices and the following identities are valid:

$$a \wedge (b \vee a) = a \wedge a, a \vee (b \wedge a) = a \vee a, \text{ (weakly absorption)}$$
$$a \wedge a = a \vee a. \text{ (equalization)}$$

For example, the following algebra $(Z\backslash\{0\}; \wedge, \vee)$, where $x \wedge y = (|x|, |y|)$ and $x \vee y = [|x|, |y|]$, for which $(|x|, |y|)$ and $[|x|, |y|]$ are the greatest common division (gcd) and the least common multiple (lcm) of $|x|$ and $|y|$, is a weakly idempotent lattice, which is not a lattice, since $x \wedge x \neq x$ and $x \vee x \neq x$ for negative x.

The operation $\vee$ is called sum. The element $a \in L$ is called an idempotent of the weakly idempotent lattice $(L; \wedge, \vee)$ if $a \wedge a = a$ and $a \vee a = a$. Note that the product (sum) of two elements of a weakly idempotent lattice $(L; \wedge, \vee)$ is an idempotent element. So, the set of all idempotent elements of a weakly idempotent lattice is a lattice.

(15) Let $\mathbf{R} = \{R; +, \cdot, -, 0\}$ be a ring. An algebra $M = (M; +, -, 0, R)$ with one binary operation $+$, one nullary operation 0, one unary operation $-$ and one unary operation for each $r \in R$ is called an **R**-*module* if $(M; +, -, 0)$ is an Abelian group and the following identities are satisfied for all $r, s \in R$ and $x, y \in M$:

$$\text{(M1) } r(x + y) = r(x) + r(y),$$
$$\text{(M2) } (r + s)(x) = r(x) + s(x),$$
$$\text{(M3) } (r \cdot s)(x) = r(s(x)).$$

If **R** also has an identity element 1, then the following additional identity must also hold in an **R**-module:

$$\text{(M4) } 1(x) = x,$$

and **R**-module M is called a unital **R**-module.

In the special case that **R** is a field, an **R**-module is usually called an **R**-*vector space*. The elements of this **R**-module are then called *vectors*, while the field elements are called *scalars*.

There is another way to describe **R**-modules and **R**-vector spaces algebraically. But it involves a modification of our definition of an algebra. We have so far considered only what are called *homogeneous* or *one-based* algebras, where we have only one base set of objects and all operations are defined on this set. One can also define what are called heterogeneous or multi-sorted or multi-based algebras, where there are two or more base sets of objects and operations are allowed on more than one kind of object. We can use this approach here, by allowing the set V of all vectors and the universe of the field $\mathbf{R}$ as two different base sets in such a heterogeneous algebra. Then we have two operations, $+ : V \times V \to V$ for the usual vector addition and $\cdot : R \times V \to V$ for the scalar multiplication of a scalar from R times a vector from V. Then $(V, R; +, \cdot)$ is an example of a heterogeneous algebra.

(16) Let $\mathbf{R}$ be a commutative ring with the identity element, then an algebra $(M; +, \cdot, -, R)$ is called a $\mathbf{R}$-linear algebra if $(M; +, \cdot)$ is a ring, $(M; +, -, 0, R)$ is a unital $\mathbf{R}$-module and

$$r(xy) = (rx)y = x(ry)$$

for all $x, y \in M$ and $r \in R$. Linear algebra is a $\mathbf{R}$-linear algebra for some $\mathbf{R}$.

A linear algebra is called associative if ring $(M; +, \cdot)$ is associative. For characteristic sequences of non-associative algebras, see [146].

(17) Let us introduce the following general concept of g-algebra (or generalized linear algebra).

Let Φ be an associative and commutative ring with identity element 1. A set R is called a g-algebra over the ring Φ if there is a defined on R structure of a unital Φ-module and there is a set Σ of binary operations defined on R connected with the module operations by the equalities:

$$\begin{aligned} X(a+b, c) &= X(a, c) + X(b, c), \\ X(a, b+c) &= X(a, b) + X(a, c), \\ \alpha(X(a, b)) &= X(\alpha a, b) = X(a, \alpha b) \end{aligned}$$

for all $a, b, c \in R$ and for all $\alpha \in \Phi, X \in \Sigma$. We denote a g-algebra R over the ring Φ by $R(+, \Sigma, \Phi)$ or, in short, R.

Let $R(+, \Sigma, \Phi)$ be a g-algebra, $A \subseteq R$. The subset A is called a subalgebra of the g-algebra $R(+, \Sigma, \Phi)$ if it is closed under Φ-module operations and binary operations of Σ.

Let us consider examples of g-algebras. Let Z be a ring of integers, X, Y be some abelian groups and $M = \mathrm{Hom}(X, Y)$ be a corresponding abelian group. For every element γ from $\mathrm{Hom}(Y, X)$ we can define a binary operation on M:

$$\overline{\gamma}(a, b) \overset{\text{def}}{=} a \circ \gamma \circ b - b \circ \gamma \circ a,$$

where $a, b \in M$ and $\circ$ is the superposition of the mappings. Denote by $\Sigma = \{\overline{\gamma} | \gamma \in \text{Hom}(Y, X)\}$. The axioms of g-algebra are valid for $M(+, \Sigma, Z)$.

Let P be a field, n be a natural number, $P^{n\times n}$ be a set of all quadratic $n\times n$-matrices with entries in P and $+$ be the ordinary matrix addition, $\cdot$ be the matrix multiplication. Define another binary operation $P^{n\times n}$ in the following way:

$$A \circ B \stackrel{\text{def}}{=} A^T \cdot B,$$

where $A, B \in P^{n\times n}$. Then $P^{n\times n}(+, \{\cdot, \circ\}, P)$ is a g-algebra.

Analogously we can define the operation $\circ$ in the following way:

$$A \circ B \stackrel{\text{def}}{=} B^T \cdot A.$$

(18) An *automaton without output*, or an *acceptor* or a *recognizer*, is a multi-based algebra $\mathcal{H} = (Z, X; \delta)$, where Z and X are nonempty sets called the sets of states and inputs, respectively, where $\delta : Z \times X \to Z$ is called the state transition function. An *automaton* (*with output*) is a quintuple $\mathcal{A} = (Z, X, B; \delta, \lambda)$, where $(Z, X; \delta)$ is an acceptor, the nonempty set B is called the set of outputs, where $\lambda : Z \times X \to B$ is called the output function. For any state $z \in Z$ and any input $x \in X$, the value $\delta(z, x)$ is the state which results when the input x is read in when the machine is in the state z. The element $\lambda(z, x)$ is the output element which is produced by input x if the automaton is in state z. If all the sets Z, X and B are finite, then the automaton is said to be finite; otherwise it is infinite. When δ and λ are functions, and so have exactly one image for each state-input pair (z, x), the automaton is called deterministic. In the non-deterministic case, we allow δ and λ to take on more than one value, or to be undefined, for a given input pair.

Automatons have wide application in theoretical computer science.

Examples of hyperidentities and coidentities

Let $(Q; \Sigma)$ be a binary algebra and $S = Q \times \Sigma$. Let us define the following binary multiplication on the set S:

$$(a, X) \cdot (b, Y) = (X(a, b), Y).$$

(1) $S(\cdot)$ is a semigroup if and only if the algebra $(Q;\Sigma)$ satisfies the following hyperidentity:

$$X(x, Y(y, z)) = Y(X(x, y), z);$$

(2) $S(\cdot)$ is idempotent (i.e., satisfies the identity $x \cdot x = x$) if and only if the algebra $(Q;\Sigma)$ is idempotent, i.e., the algebra $(Q;\Sigma)$ satisfies the hyperidentity of idempotency:

$$X(x, x) = x;$$

(3) $S(\cdot)$ is left distributive (i.e., satisfies the identity $x(yz) = (xy)(xz)$) if and only if the algebra $(Q;\Sigma)$ satisfies the following hyperidentity:

$$X(x, Y(y, z)) = Y(X(x, y), X(x, z));$$

(4) $S(\cdot)$ is right distributive (i.e., satisfies the identity $(xy)z = (xz)(yz)$) if and only if the algebra $(Q;\Sigma)$ satisfies the following hyperidentity:

$$X(Y(x, y), z) = Y(X(x, z), X(y, z));$$

(5) $S(\cdot)$ is medial (i.e., satisfies the identity $(xy)(zt) = (xz)(yt)$) if and only if the algebra $(Q;\Sigma)$ satisfies the following hyperidentity:

$$Y(X(x, y), Z(z, t)) = Z(X(x, z), Y(y, t));$$

(6) $S(\cdot)$ is transitive (i.e., satisfies the Kolmogoroff identity $xy \cdot yz = xz$) if and only if the algebra $(Q;\Sigma)$ satisfies the following hyperidentity:

$$Y(X(x, y), Y(y, z)) = X(x, z)).$$

(7) In any lattice, the following hyperidentities are satisfied:

$$X(x,\, x) = x,$$
$$X(x,\, y) = X(y,\, x),$$
$$X(x,\, X(y,\, z)) = X(X(x,\, y), z),$$
$$X(Y(X(x, y), z), Y(y, z)) = Y(X(x, y), z).$$

Hence, these hyperidentities are hyperidentities of the class of all lattices. The last nontrivial hyperidentity is called hyperidentity of interlacity (see [264]).

(8) In any commutative and associative ring, the following hyperidentities are satisfied:

$$X(x, y) = X(y, x),$$
$$X(x, X(y, z)) = X(X(x, y), z),$$
$$X\left(X\left(Y(x, x), Y(x, x)\right), Y\left(X(x, x), X(x, x)\right)\right)$$
$$= X\left(Y\left(X(x, x), X(x, x)\right), X\left(Y(x, x), Y(x, x)\right)\right).$$

(9) In the termal algebra (i.e., the algebra of term functions) of any group (semigroup, Moufang loop), the following nontrivial hyperidentity is satisfied (cf. [31])

$$X(Y(x, x), Y(x, x)) = Y(X(x, x), X(x, x)).$$

(10) Let $B = \{0, 1\}$ and P be the set of all binary Boolean functions. In algebra $(B; P)$, the following hyperidentities are satisfied:

$$X(X(X(x, y), y), y) = X(x, y),$$
$$X(x, X(x, X(x, y))) = X(x, y).$$

In particular, these hyperidentities are satisfied in two-element Boolean algebra $(\{0, 1\}; \&, \vee, ', 0, 1)$. Hence, these hyperidentities are satisfied in any Boolean algebra too (by Birkhoff's subdirect representation theorem [37]). See [253] for corresponding hyperidentities of n-ary Boolean functions. On the application of the results of [253] in modal logic see [153].

(11) In any De Morgan algebra $Q(+, \cdot, ^{-}, 0, 1)$, the following non-trivial hyperidentities are satisfied:

$$X(F(X(x, y)), F(x)) = F(x),$$
$$X(x, X(y, z)) = X(Y(x, Y(y, z)), F(Y(F(x), Y(F(y), F(z))))).$$

(12) In any multioperator Ω-group, the following coidentity is valid:

$$X(\underbrace{0, \ldots, 0}_{n}) = 0,$$

for all $n \in T_{(\Omega)}$, where all object variables are replaced by the zero element of the Ω-group [189,328].

(13) (J. von Neumann) Let $L(+, \cdot)$ be a modular lattice and $a, b, c \in L$. The sublattice of L generated by the elements a, b, c, is distributive if and only if the following coidentity of left distributivity

$$X(a, Y(b, c)) = Y(X(a, b), X(a, c))$$

holds in $L(+, \cdot)$.

Binary algebras satisfying the following hyperidentity of associativity

$$X(x, Y(y, z)) = Y(X(x, y), z)$$

are called hyperassociative algebras.

The following Artin theorem for linear algebras defined on commutative and associative ring with unit is well known: In the alternative linear algebra any two of its elements generate an associative subalgebra; moreover if $(a, b, c) = 0$ then the subalgebra generated by the elements a, b, c is associative. Let us consider a large generalization of this classical result, using the concepts of hyperidentity and coidentity.

A binary algebra $(Q; \Sigma)$ is called left (right) hyperalternative if it satisfies the following hyperidentity:

$$X(x, Y(x, y)) = Y(X(x, x), y) \qquad (alt)_l$$

$$(X(x, Y(y, y)) = Y(X(x, y), y)). \qquad (alt)_r$$

A binary algebra $(Q; \Sigma)$ is called hyperalternative if it is both left and right hyperalternative.

A g-algebra $R(+, \Sigma, \Phi)$ is called hyperalternative (hyperassociative) if the binary algebra $(R; \Sigma)$ is hyperalternative (hyperassociative).

For example, let $A(+, \cdot, \Phi)$ be an alternative (linear) algebra and c be in the nucleus of the algebra A (see [53,372]), i.e.,

$$(x \cdot c) \cdot y = x \cdot (c \cdot y),$$

for all $x, y \in A$. Defining another binary operation on A:

$$x \circ y \stackrel{\text{def}}{=} x \cdot c \cdot y,$$

we obtain that $A(+, \{\cdot, \circ\}, \Phi)$ is a hyperalternative g-algebra.

If we take an associative algebra $A(+, \cdot, \Phi)$, then the g-algebra $A(+, \{\cdot, \circ\}, \Phi)$ will be hyperassociative algebra for any selected element $c \in A$.

Theorem 1.1.5. *Let $R(+, \Sigma, \Phi)$ be a hyperalternative g-algebra and for some elements $a, b, c \in R$ the following coidentity holds in $(R; \Sigma)$:*

$$X(a, Y(b, c)) = Y(X(a, b), c).$$

Then the subalgebra of $R(+, \Sigma, \Phi)$ generated by elements a, b, c is hyperassociative.

In particular, if $R(+, \Sigma, \Phi)$ is a hyperalternative g-algebra and $a, b \in R$, then the subalgebra of $R(+, \Sigma, \Phi)$ generated by elements a, b is hyperassociative.

See [249] for the similar large generalization of Moufang theorem.

If $\mathfrak{A} = (A; \Sigma)$ is an algebra and $|f|$ is the arity of the operation $f \in \Sigma$, then the set

$$T_{\mathfrak{A}} = \{|f| \mid f \in \Sigma\}$$

is the *arithmetic type* of algebra $\mathfrak{A}$. For a set of nonnegative integers $T \subseteq \mathbb{N} = \{0, 1, 2, \ldots\}$ a *T-algebra* is an algebra with arithmetic type T.

For example, the arithmetic type of a groupoid, a semigroup or a lattice is $T = \{2\}$, and the arithmetic type of a group, a ring, a Boolean algebra, a De Morgan algebra or a Boole–De Morgan algebra is $T = \{0, 1, 2\}$.

Note that any Ω-algebra is also a T-algebra for some T and any T-algebra is an Ω-algebra for some type Ω. The difference is that Ω-algebras and T-algebras form different categories (see Sec. 1.6) or "environments". If the first-order properties of algebras are studied, i.e., the properties expressed by first-order formulas, then algebras are considered as Ω-algebras (i.e., in the category of Ω-algebras). Analogously, if the second-order properties of algebras are studied, i.e., the properties expressed by second-order formulas, then algebras are useful considered as T-algebras.[a]

1.2. Subalgebras, congruence relations and quotients

Definition 1.2.1. Let Ω be a type and $\mathfrak{A} = (A; \Sigma)$ and $\mathfrak{B} = (B; \Sigma')$ be two Ω-algebras. Then $\mathfrak{B}$ is called a *subalgebra* of $\mathfrak{A}$ if $B \subseteq A$ and every fundamental operation of $\mathfrak{B}$ is the restriction of the corresponding operation of $\mathfrak{A}$, i.e., for each function symbol $f \in \Omega$ we have $f^{\mathfrak{B}} = f^{\mathfrak{A}}|_B$. In this case, we write $\mathfrak{B} \leq \mathfrak{A}$.

[a]That is, in the category of T-algebras [249,252].

A *subuniverse* of $\mathfrak{A}$ is a subset B of A which is closed under the fundamental operations of $\mathfrak{A}$; i.e., if $f \in \Sigma$ is a fundamental n-ary operation of $\mathfrak{A}$ and $a_1, \ldots, a_n \in B$, then $f(a_1, \ldots, a_n) \in B$.

Thus if $\mathfrak{B}$ is a subalgebra of $\mathfrak{A}$, then B is a subuniverse of A. Note that the empty set is a subuniverse, but it is not the underlying set of any subalgebra. If $\mathfrak{A}$ has nullary operations, then every subuniverse contains them as well.

Subalgebra of a ring (group, semigroup, monoid, lattice) is called a subring (subgroup, subsemigroup, submonoid, sublattice).

For example, the ring $(\mathbb{Z}; +, \cdot, -, 0)$ of integers is a subring of the ring of rational numbers - $(\mathbb{Q}; +, \cdot, -, 0)$. The algebra $(\{-1, 1\}; \cdot, ^{-1}, 1)$ is a subgroup of the group $(\mathbb{Q}^*; \cdot, ^{-1}, 1)$ of nonzero rational numbers (the operations are the usual multiplication of numbers, the inverse and the number 1). If $(B; +, \cdot, \ ', 0, 1)$ is a Boolean algebra, then the two-element Boolean algebra $(\{0, 1\}; +, \cdot, \ ', 0, 1)$ is its subalgebra.

A class of Ω-algebras is called *hereditary* if it contains all the subalgebras of any algebra from itself.

If θ is an equivalence relation on a set A then for an element $a \in A$ we denote by $[a]_\theta$ the equivalence class of the element a modulo θ, i.e.,

$$[a]_\theta = \{x \in A \mid x\theta a\}.$$

We denote by A/θ the corresponding quotient set (or factor set), that is the set of all equivalence classes, i.e.,

$$A/\theta = \{[a]_\theta \mid a \in A\}.$$

Definition 1.2.2. Let $\mathfrak{A} = (A; \Sigma)$ be an algebra of type Ω. Then an equivalence relation θ of the set A is a called a *congruence relation* (or simply *congruence*) of the algebra $\mathfrak{A}$ if it satisfies the following *compatibility property*:

- for each n-ary function symbol $f \in \Omega$ and elements $a_i, b_i \in \Omega$ the condition $a_i \theta b_i,\ i = 1, \ldots, n$ implies

$$f^{\mathfrak{A}}(a_1, \ldots, a_n)\theta f^{\mathfrak{A}}(b_1, \ldots, b_n).$$

Trivial examples of congruence relations are the *zero congruence* $\mathbf{0}$ and $\mathbf{1}$, defined by

$$\mathbf{0} = \{(a, a) | a \in A\},\ \mathbf{1} = \{(a, b) | a, b \in A\} = A \times A.$$

The class of all congruences of an Ω-algebra $\mathfrak{A}$ is a lattice and denoted by $\mathrm{Con}^{(0)}(\mathfrak{A})$.

If θ is a congruence relation on an Ω-algebra $\mathfrak{A} = (A; \Sigma)$, then we can introduce an algebraic structure on the quotient set A/θ and turn this set into an Ω-algebra which is denoted by $\mathfrak{A}/\theta$. We proceed by the following way: for any n-ary function symbol $f \in \Omega$ and equivalence classes $[a_1]_\theta, \ldots, [a_n]_\theta \in A/\theta$, we define

$$f^{\mathfrak{A}/\theta}([a_1]_\theta, \ldots, [a_n]_\theta) = [f^{\mathfrak{A}}(a_1, \ldots, a_n)]_\theta.$$

This definition is correct (i.e., $f^{\mathfrak{A}/\theta}([a_1]_\theta, \ldots, [a_n]_\theta)$ does not depend on the representatives $a_1, \ldots, a_n$ of equivalence classes $[a_1]_\theta, \ldots, [a_n]_\theta$) since θ is a congruence relation.

This new Ω-algebra $\mathfrak{A}/\theta$ is called the *quotient algebra* of $\mathfrak{A}$ by the congruence relation θ.

Now we define the concepts of subalgebra, congruence relation and quotient algebra for T-algebras.

An algebra $\mathfrak{A}' = (A'; \Sigma')$ is called a *subalgebra* of the algebra $\mathfrak{A} = (A; \Sigma)$ if $A' \subseteq A$ and every operation from Σ' is the restriction of some operation from Σ (to the subset A'). For example, the additive semigroup of all natural numbers is a subalgebra of the ring of integers, every Abelian group is a subalgebra of some ring, and, a fortiori, every groupoid is a subalgebra of some ring. Every semilattice is a subalgebra of some distributive lattice and every distributive lattice is a subalgebra of some Boolean algebra, and so on. When we wish to specify the arithmetic type of subalgebras, we call them *T-subalgebras*.

An intermediate notion here is that of a subsystem. Let $\mathfrak{A} = (A; \Sigma)$ be an arbitrary T-algebra, and let $A' \subseteq A$, $\Sigma' \subseteq \Sigma$, where $A' \neq \emptyset$ and $\Sigma' \neq \emptyset$. A pair $\mathfrak{A}' = (A'; \Sigma')$ is called a *subsystem* of the T-algebra $\mathfrak{A}$ if A' is closed with respect to all the operations of Σ'. Subsystems of the form $(A'; \Sigma)$ are called principal.

A class of T-algebras is called *hereditary* if it contains all the T-subalgebras of any T-algebra from itself.

Let $\mathfrak{A} = (A; \Sigma)$ be a T-algebra and r and $\tilde{t}$ be equivalence relations defined, respectively, on the sets A and Σ. The pair $q = (r, \tilde{t})$ is called a *congruence relation* (or simply *congruence*) of the T-algebra $\mathfrak{A}$ if the following two conditions hold:

(i) $\tilde{t}$ preserves the arity of the operations, i.e., $f\tilde{t}g$ implies $|f| = |g|$, for any two operations $f, g \in \Sigma$;

(ii) the relations r and $\tilde{t}$ are compatible in the following sense:

$$x_i r y_i,\ i = 1, \ldots, n,\ f\tilde{t}g \Rightarrow f(x_1, \ldots, x_n) r g(y_1, \ldots, y_n),$$

for any elements $x_i, y_i \in A$ and any n-ary operations $f, g \in \Sigma$.

The class of all congruences $(r, \tilde{t})$ of an T-algebra $\mathfrak{A}$ is a lattice and denoted by Con$(\mathfrak{A})$.

If r is a congruence relation of $\mathfrak{A} = (A; \Sigma)$ considered as an Ω-algebra for some type Ω, then $(r, \tilde{0})$ is obviously a congruence of $\mathfrak{A}$ considered as a T-algebra and vice versa, where $\tilde{0}$ is the identity relation of Σ, i.e., $\tilde{0} = \{(f, f) | f \in \Sigma\}$.

If $q = (r, \tilde{t})$ is a congruence of a T-algebra $\mathfrak{A} = (A; \Sigma)$, then every element $[f]_{\tilde{t}}$ of the quotient set $\Sigma/\tilde{t}$ defines an operation on the quotient set Q/r in the following way:

$$[f]_{\tilde{t}}([x_1]_r, \ldots, [x_n]_r) = [f(x_1, \ldots, x_n)]_r,$$

where $f \in \Sigma,\ |f| = n,\ x_1, \ldots, x_n \in A$.

The definition of a congruence implies that the operation $[f]_{\tilde{t}}$ is well defined. As a result we shall obtain a quotient algebra of the T-algebra $\mathfrak{A}$ modulo the congruence $q = (r, \tilde{t})$, denoted by $\mathfrak{A}/q$.

A congruence $q = (r, \tilde{t}\,)$ of a T-algebra $\mathfrak{A} = \langle Q; \Sigma\rangle$ is said to be *nuclear* if different elements of $\Sigma/\tilde{t}$ define different operations on Q/r, that is

$$A\tilde{t}B \Longleftrightarrow A(x_1, \ldots, x_n) r B(x_1, \ldots, x_n), \qquad \forall x_1, \ldots, x_n \in Q,$$

where $A, B \in \Sigma,\ |A| = |B| = n \in T$.

An algebra $\mathfrak{A} = \langle Q; \Sigma\rangle$ is said to be functionally trivial if the following mapping is injective:

$$A \longrightarrow |A|,$$

i.e., the set of its n-ary operations is singleton for any $n \in T_{\mathfrak{A}}$; otherwise the algebra is called functionally nontrivial.

1.3. Homomorphisms and automorphisms, bihomomorphisms and biautomorphisms

Definition 1.3.1. Let $\mathfrak{A} = (A; \Sigma)$ and $\mathfrak{B} = (B; \Sigma')$ be two Ω-algebras. A mapping $\varphi : A \to B$ is called a *homomorphism* from $\mathfrak{A}$ to $\mathfrak{B}$ if

$$\varphi f^{\mathfrak{A}}(a_1, \ldots, a_n) = f^{\mathfrak{B}}(\varphi a_1, \ldots, \varphi a_n) \tag{1.3.1}$$

for each n-ary functional symbol $f \in \Omega$ and any elements $a_1, \ldots, a_n \in A$.

If θ is a congruence of an Ω-algebra $\mathfrak{A}$, then the *natural homomorphism* $\pi_\theta : \mathfrak{A} \to \mathfrak{A}/\theta$ is defined by $\pi_\theta(a) = [a]_\theta,\ a \in A$.

If we omit the superscripts $\mathfrak{A}$ and $\mathfrak{B}$, then the condition of homomorphism will acquire more simple form: $\varphi f(a_1, \ldots, a_n) = f(\varphi a_1, \ldots, \varphi a_n)$.

If, in addition, the mapping φ is onto then $\mathfrak{B}$ is said to be a *homomorphic image* of $\mathfrak{A}$, and φ is called an *epimorphism*.

An *isomorphism* is a homomorphism which is one-to-one and onto, and is denoted by $\mathfrak{A} \simeq \mathfrak{B}$ (or $\mathfrak{A} \cong \mathfrak{B}$) if there exists an isomorphism $\varphi : A \to B$. In case $\mathfrak{A} = \mathfrak{B}$, a homomorphism is also called an *endomorphism* and an isomorphism is referred to as an *automorphism*. Often we will write $\varphi : \mathfrak{A} \to \mathfrak{B}$ instead of $\varphi : A \to B$, to emphasize that φ is a homomorphism from $\mathfrak{A}$ to $\mathfrak{B}$.

The set of all endomorphisms (automorphisms) of the same Ω-algebra forms a semigroup with identity (group) under superposition of functions. We will denote the group of all automorphisms φ of the same Ω-algebra $\mathfrak{A}$ by $Aut^{(\circ)}\mathfrak{A}$.

Theorem 1.3.2. *For any group G, there exists a unary algebra $\mathfrak{A}$ with $G \simeq \mathrm{Aut}^{(\circ)}\mathfrak{A}$.*

Proof. We define unary operations f_a on the set G as follows:

$$f_a(x) = a \cdot x, \ x \in G.$$

We obtain the algebra $\mathfrak{A} = (G; \{f_a | a \in G\})$. Let us prove that this algebra satisfies the condition of the theorem.

First, it is easy to see that for any $b \in G$ the mappings $\varphi_b(x) = x \cdot b, \ x \in G$ are automorphisms of the algebra $\mathfrak{A}$. Further, any automorphism of the algebra $\mathfrak{A}$ is of the form φ_b for some $b \in G$. Indeed, let φ be an automorphism of $\mathfrak{A}$. Then we have $\varphi(f_a(x)) = f_a(\varphi(x))$ for all $a, x \in G$. If e is the unity of the group G, then denote $c = \varphi(e)$, we have

$$\varphi(x) = \varphi(x \cdot e) = \varphi(f_x(e)) = f_x(\varphi(e)) = f_x(c) = x \cdot c = \varphi_c(x).$$

Hence, $\varphi = \varphi_c$.

Also, $\varphi_b = \varphi_d$ if and only if $b = d$. Moreover, $\varphi_{bd} = \varphi_b \cdot \varphi_d$. Therefore, the group of all automorphisms of the algebra $\mathfrak{A}$ is isomorphic to G. □

An Ω-algebra $\mathfrak{B}$ is said to be a homomorphic image of the Ω-algebra $\mathfrak{A}$ if there exists a surjective homomorphism $\varphi : \mathfrak{A} \to \mathfrak{B}$. A class of Ω-algebras is said to be *homomorphically closed* if along with each algebra it contains every homomorphic image of that algebra under any homomorphism.

Let $\varphi : \mathfrak{A} \to \mathfrak{B}$ be a homomorphism. Then the *kernel of* φ, denoted by $\ker(\varphi)$, is defined by

$$\ker(\varphi) = \{(a, b) \in A^2 | \varphi(a) = \varphi(b)\}.$$

Obviously, $\ker(\varphi) \in \mathrm{Con}^{(0)}\mathfrak{A}$, i.e., $\ker(\varphi)$ is a congruence of the algebra $\mathfrak{A}$.

Theorem 1.3.3 (The homomorphism theorem). *Let $\mathfrak{A}$ and $\mathfrak{B}$ be two Ω-algebras and $\varphi : \mathfrak{A} \to \mathfrak{B}$ be an epimorphism. Then there exists an isomorphism $\mu : \mathfrak{A}/\ker(\varphi) \to \mathfrak{B}$ such that $\varphi = \pi \cdot \mu$, where $\pi = \pi_{\ker(\varphi)}$ is the natural homomorphism from $\mathfrak{A}$ to $\mathfrak{A}/\ker(\varphi)$.*

The homomorphism theorem can be formulated in the following more general form.

Theorem 1.3.4. *Let $\mathfrak{A}, \mathfrak{B}, \mathfrak{C}$ be Ω-algebras and $\varphi : \mathfrak{A} \to \mathfrak{B}$ and $\psi : \mathfrak{A} \to \mathfrak{C}$ be homomorphisms. Suppose also that φ is surjective and* $\mathrm{Ker}(\varphi) \subseteq \mathrm{Ker}(\psi)$. *Then there exists a unique homomorphism $\lambda : \mathfrak{B} \to \mathfrak{C}$ such that $\psi = \varphi \cdot \lambda$.*

Here is a sketch of the proof. For any $y \in B$ we take an element $x \in A$ such that $\varphi(x) = y$ and set $\lambda(y) = \psi(x)$. This mapping is well defined due to the condition $\mathrm{Ker}(\varphi) \subseteq \mathrm{Ker}(\psi)$. It is easily proved that λ is a homomorphism (in the first theorem it is also an isomorphism) and $\psi = \varphi \cdot \lambda$.

Let $\mathfrak{A} = (A; \Sigma)$ be a T-algebra and $\mathfrak{A}' = (A'; \Sigma')$ be a T'-algebra, where $T \subseteq T'$. The pair $(\varphi, \tilde{\psi})$ of maps $\varphi : A \to A'$ and $\tilde{\psi} : \Sigma \to \Sigma'$ is called a *bihomomorphism* from T-algebra $\mathfrak{A}$ into T'-algebra $\mathfrak{A}'$ and is denoted by $(\varphi, \tilde{\psi}) : \mathfrak{A} \Rightarrow \mathfrak{A}'$ or $(\varphi, \tilde{\psi}) : \mathfrak{A} \rightrightarrows \mathfrak{A}'$, if the map $\tilde{\psi}$ preserves the arity of operations, i.e., $|\tilde{\psi} f| = |f|$ for any operation $f \in \Sigma$, and the equality

$$\varphi f(x_1, \ldots, x_n) = [\tilde{\psi} f](\varphi x_1, \ldots, \varphi x_n)$$

holds for all n-ary operations $f \in \Sigma$ and for all $x_1, \ldots, x_n \in A$.

For a congruence $q = (r, \tilde{t})$ of a T-algebra $\mathfrak{A} = (A; \Sigma)$ the *natural bihomomorphism* $(\pi_q, \tilde{\pi}_q) : \mathfrak{A} \rightrightarrows \mathfrak{A}/q$ is defined by $\pi_q(a) = [a]_r$, $a \in A$, $\tilde{\pi}_q(f) = [f]_{\tilde{t}}$, $f \in \Sigma$.

A bihomomorphism $(\varphi, \tilde{\psi})$ is called a *biepimorphism* (*bimonomorphism*, *biisomorphism*) if the maps φ and $\tilde{\psi}$ are surjective (injective, bijective). A bihomomorphism $(\varphi, \tilde{\psi})$ is called a *semiepimorphism* (*semimonomorphism, semiisomorphism*) if the map φ is surjective (injective, bijective). A bihomomorphism (biisomorphism) of a T-algebra into itself is called a *biendomorphism* (*biautomorphism*). Any semimonomorphism $(\varphi, \tilde{\psi})$ of

T-algebras is a bimonomorphism:

$$\begin{aligned}\tilde{\psi} f = \tilde{\psi} g &\to \tilde{\psi} f(\varphi x_1, \ldots, \varphi x_n) = \tilde{\psi} g(\varphi x_1, \ldots, \varphi x_n)\\ &\to \varphi f(x_1, \ldots, x_n) = \varphi g(x_1, \ldots, x_n)\\ &\to f(x_1, \ldots, x_n) = g(x_1, \ldots, x_n) \to f = g,\end{aligned}$$

where $|f| = |g| = n$. We denote $\mathfrak{A} \simeq \mathfrak{A}'$ (or $\mathfrak{A} \cong \mathfrak{A}'$) for biisomorphic T-algebras $\mathfrak{A}$ and $\mathfrak{A}'$, i.e., if there exists a biisomorphism $(\varphi, \tilde{\psi}) : \mathfrak{A} \Rightarrow \mathfrak{A}'$.

A congruence $q = (r, \tilde{t})$ of a T-algebra $\mathfrak{A} = \langle Q; \Sigma \rangle$ is said to be *fully invariant* if it preserves any biendomorphism $(\varphi, \widetilde{\psi})$ of this algebra, that is,

$$xry \longrightarrow \varphi(x) r \varphi(y)$$

and

$$A\tilde{t}B \longrightarrow \widetilde{\psi}(A)\, \tilde{t}\, \widetilde{\psi}(B),$$

where $x, y \in Q$ and $A, B \in \Sigma$.

A T-algebra $\mathfrak{A}'$ is called a bihomomorphic image of the T-algebra $\mathfrak{A}$ if there exists a biepimorphism $(\varphi, \tilde{\psi}) : \mathfrak{A} \rightrightarrows \mathfrak{A}'$. A class of T-algebras is said to be *bihomomorphically closed* if along with each T-algebra it contains every bihomomorphic image of that algebra under any bihomomorphism.

The pair $(\varepsilon, \tilde{\varepsilon})$ of identical maps $\varepsilon : A \to A$ and $\tilde{\varepsilon} : \Sigma \to \Sigma$ is an biautomorphism of T-algebra $(A; \Sigma)$.

The set of all biendomorphisms (biautomorphisms) of the same T-algebra forms a semigroup with identity (group) under the componentwise multiplication of pairs:

$$(\varphi_1, \tilde{\psi}_1) \cdot (\varphi_2, \tilde{\psi}_2) = (\varphi_1 \varphi_2, \tilde{\psi}_1 \tilde{\psi}_2).$$

We will denote the group of all biautomorphisms $(\varphi, \tilde{\psi})$ of the same T-algebra $\mathfrak{A}$ by $\operatorname{Aut} \mathfrak{A}$. The biautomorphisms of the form $(\varphi, \tilde{\varepsilon})$, where $\tilde{\varepsilon}$ is the identical mapping, are ordinary automorphisms and form a group denoted by $\operatorname{Aut}^{(\circ)} \mathfrak{A}$. Obviously $\operatorname{Aut}^{(\circ)} \mathfrak{A} \trianglelefteq \operatorname{Aut} \mathfrak{A}$, i.e., $\operatorname{Aut}^{(\circ)} \mathfrak{A}$ is an invariant subgroup of the group $\operatorname{Aut} \mathfrak{A}$.

Theorem 1.3.5 ([249]). *For any group G, there exists a binary algebra $\mathfrak{A}$ with the hyperidentity of associativity*

$$X(x, Y(y, z)) = Y(X(x, y), z)$$

such that $G \simeq \operatorname{Aut} \mathfrak{A}$ and $\operatorname{Aut}^{(\circ)} \mathfrak{A}$ is singleton.

Theorem 1.3.6 ([249]). *For any group G, there exists a binary algebra $\mathfrak{A}$ with the hyperidentity of distributivity*

$$X(x, Y(y, z)) = Y(X(x, y), X(x, z))$$

such that $G \simeq \mathrm{Aut}^{(\circ)} \mathfrak{A}$ and $\mathrm{Hol}\, G(\cdot) \simeq \mathrm{Aut}\, \mathfrak{A}$ (for definition of $\mathrm{Hol}\, G(\cdot)$, see Sec. 16.2).

Below we will prove more general result.

Theorem 1.3.7 ([257]). *Let G ba an arbitrary group with an invariant subgroup H. Then there exists an algebra $\mathfrak{A}$ such that $G \simeq \mathrm{Aut}\, \mathfrak{A}$ and $H \simeq \mathrm{Aut}^{(\circ)} \mathfrak{A}$, where the last isomorphism is induced by the first isomorphism.*

Proof. Let $Q = (G\times\{0,1\})\cup\{a,b,c,d\}$, where $(G\times\{0,1\})\cap\{a,b,c,d\} = \emptyset$. Let $G\times\{0\}$ be a well-ordered set under the order "$\leq$". For any element $g \in G$, we shall define the unary operation g^* on Q by the following way:

$$\begin{aligned}
&g^*(h,0) = (g,0), \text{ if } (g,0) \leq (h,0), \ g \neq h,\\
&g^*(h,0) = b, \quad \text{if } (g,0) \geq (h,0),\\
&g^*(h,1) = (hg,1), \quad \text{for any } h \in G,\\
&g^*(a) = (g,0), \ g^*(b) = b, \ g^*(c) = c, \ g^*(d) = d.
\end{aligned}$$

Similarly, for every $S \in G/H$ we define the unary operation S^* on Q by the rule:

$$\begin{aligned}
&S^*(h,0) = (h,0), \quad \text{for any } h \in G,\\
&S^*(h,1) = c, \quad \text{if } h \in S,\\
&S^*(h,1) = d, \quad \text{if } h \notin S,\\
&S^*(a) = S^*(b) = b, \ S^*(c) = S^*(d) = d.
\end{aligned}$$

Obviously, $g_1^* = g_2^*$ if and only if $g_1 = g_2$, and $S_1^* = S_2^*$ if and only if $S_1 = S_2$. In the result, we get a unary algebra $\mathfrak{A} = (Q; \Sigma)$, where

$$\Sigma = \{g^* | g \in\} \cup \{S^* | S \in G/H\},$$

which satisfies the condition of the theorem. Let us define the mapping $\varphi_k : Q \to Q$ for every $k \in G$ by the rule:

$$\begin{aligned}
&\varphi_k(h,1) = (kh,1), \quad \text{for any } h \in G,\\
&\varphi_k(z) = z, \quad \text{for any } z \in (G \times \{0\}) \cup \{a,b,c,d\}.
\end{aligned}$$

The following equations hold for every $k, g \in G,\ S \in G/H$:

$$\varphi_k(g^*(x)) = g^*(\varphi_k(x)),$$
$$\varphi_k(S^*(x)) = (kS)^*(\varphi_k(x)).$$

This means that $(\varphi_k, \tilde{\psi}_k) \in \operatorname{Aut}\mathfrak{A}$, where $\tilde{\psi}_k(g^*) = g^*$ and $\tilde{\psi}_k(S^*) = (kS)^*$. In addition:

$$(\varphi_k, \tilde{\psi}_k) \in \operatorname{Aut}^{(\circ)}\mathfrak{A} \Leftrightarrow \tilde{\psi}_k = \tilde{\varepsilon} \Leftrightarrow (kS)^* = S^* \Leftrightarrow kS = S \Leftrightarrow k \in H \trianglelefteq G.$$

The mapping $k \to (\varphi_k, \tilde{\psi}_k)$ is an isomorphism between groups G and $\operatorname{Aut}\mathfrak{A}$. □

Let $\mathfrak{A} = (A; \Sigma)$ and $\mathfrak{A}' = (A'; \Sigma')$ be T-algebras. With each bihomomorphism $(\varphi, \tilde{\psi}) : \mathfrak{A} \Rightarrow \mathfrak{A}'$ two congruencies are associated, called the *kernel* and the *weak kernel* of that bihomomorphism and denoted by $\operatorname{Ker}(\varphi, \tilde{\psi})$ and $\operatorname{WKer}(\varphi, \tilde{\psi})$ respectively. They are defined as follows:

$$\operatorname{WKer}(\varphi, \tilde{\psi}) = (\operatorname{Ker}\varphi, \operatorname{Ker}\tilde{\psi}),$$
$$\operatorname{Ker}(\varphi, \tilde{\psi}) = (\operatorname{Ker}\varphi, \tilde{t}),$$

where

$$f\tilde{t}g \Leftrightarrow \tilde{\psi}f|_{\varphi(A)} = \tilde{\psi}g|_{\varphi(A)}, \quad f, g \in \Sigma.$$

Clearly $\operatorname{Ker}\tilde{\psi} \subseteq \tilde{t}$ and if φ is a surjection then $\operatorname{Ker}\tilde{\psi} = \tilde{t}$.

But in general $\operatorname{WKer}(\varphi, \tilde{\psi}) \neq \operatorname{Ker}(\varphi, \tilde{\psi})$. For example, let $(\varphi, \tilde{\varepsilon})$ be an endomorphism of a nonzero ring, where φ is the zero map and $\tilde{\varepsilon}$ is the identical. Then $\operatorname{WKer}(\varphi, \tilde{\psi}) \neq \operatorname{Ker}(\varphi, \tilde{\psi})$.

In general, we have $\operatorname{WKer}(\varphi, \tilde{\psi}) \leq \operatorname{Ker}(\varphi, \tilde{\psi})$, where the partial order between two congruencies is defined in the natural way:

$$(r_1, \tilde{t}_1) \leq (r_2, \tilde{t}_2) \Leftrightarrow r_1 \subseteq r_2,\ \tilde{t}_1 \subseteq \tilde{t}_2.$$

The class of all congruencies of the same T-algebra forms a complete algebraic lattice with respect to this partial order "$\leq$".

Theorem 1.3.8 (The bihomomorphism theorem). *Let $\mathfrak{A}$ and $\mathfrak{B}$ be two T-algebras and $(\varphi, \tilde{\psi}) : \mathfrak{A} \to \mathfrak{B}$ be a biepimorphism with $q = \operatorname{Ker}(\varphi, \tilde{\psi})$. Then there exists a biisomorphism $(\lambda, \tilde{\mu}) : \mathfrak{A}/q \rightrightarrows \mathfrak{B}$ such that $\varphi = \pi \cdot \lambda,\ \tilde{\psi} = \tilde{\pi} \cdot \tilde{\mu}$, where $(\pi, \tilde{\pi}) = (\pi_q, \tilde{\pi}_q)$ is the natural bihomomorphism from $\mathfrak{A}$ to $\mathfrak{A}/q$.*

1.4. Direct and subdirect products: Superproducts and filtered products

Let us start from the definitions of direct and subdirect products of algebras. First, we give the definitions of those concepts for Ω-algebras.

Definition 1.4.1. Let $\mathfrak{A}_i = (A_i; \Sigma_i)$, $i \in I$, be given Ω-algebras. Form the Cartesian product $\prod_{i\in I} A_i$ as the set of all functions of the form $f : I \to \bigcup_{i\in I} A_i$ for which $f(i) \in A_i$ for all $i \in I$. The *(direct) product* of the algebras $\mathfrak{A}_i$ is an Ω-algebra $\mathfrak{A} = (\prod_{i\in I} A_i; \Sigma)$ and it is denoted by $\mathfrak{A} = \prod_{i\in I} \mathfrak{A}_i$, where the operations are defined componentwisely, i.e., for each n-ary function symbol $F \in \Omega$ and for any $f_1, \ldots, f_n \in \prod_{i\in I} A_i$ we set

$$F^{\mathfrak{A}}(f_1, \ldots, f_n)(i) = F^{\mathfrak{A}_i}(f_1(i), \ldots, f_n(i)), \quad i \in I.$$

The direct product of two Ω-algebras $\mathfrak{A}$ and $\mathfrak{B}$ is denoted by $\mathfrak{A} \times \mathfrak{B}$. We have $\mathfrak{A} \times \mathfrak{B} \simeq \mathfrak{B} \times \mathfrak{A}$ under $\varphi(\langle a, b\rangle) = \langle b, a\rangle$.

The *projection maps*

$$\pi_j : \prod_{i\in I} A_i \to A_j$$

for each $j \in I$ are defined by

$$\pi_j(f) = f(j).$$

These maps give surjective homomorphisms

$$\pi_j : \prod_{i\in I} \mathfrak{A}_i \to \mathfrak{A}_j.$$

Definition 1.4.2. An Ω-algebra $\mathfrak{A}$ is called a *directly irreducible* (indecomposable), if $\mathfrak{A} \simeq \mathfrak{A}_1 \times \mathfrak{A}_2$ implies $|A_1| = 1$ or $|A_2| = 1$ for any Ω-algebras $\mathfrak{A}_1$ and $\mathfrak{A}_2$.

Theorem 1.4.3 ([61]). *Every Ω-algebra is isomorphic to a direct product of directly irreducible Ω-algebras.*

Definition 1.4.4. An Ω-algebra $\mathfrak{A}$ is called a *subdirect product* of the Ω-algebras $\mathfrak{A}_i$, $i \in I$, if $\mathfrak{A}$ is a subalgebra of the direct product $\prod_{i\in I} \mathfrak{A}_i$ and the projection maps π_i are surjective for all $i \in I$, i.e., $\pi_i(A) = A_i$ for each $i \in I$.

Definition 1.4.5. An Ω-algebra $\mathfrak{A} = (A; \Sigma)$ is called *subdirectly irreducible* if $|A| > 1$ and the intersection of all nonzero congruencies of $\mathfrak{A}$ is nonzero, i.e., if there exist two elements $u, v \in A$ such that $u \neq v$ and $u\alpha v$ for all nonzero congruences α of algebra $\mathfrak{A}$.

The two-element Boolean algebra (distributive lattice, semilattice) is the only subdirectly irreducible Boolean algebra (distributive lattice, semilattice). A vector space is subdirectly irreducible if and only if its dimension is equal to 1. Every subdirectly irreducible Abelian group is cyclic or quasicyclic.

A class of Ω-algebras is said to be *multiplicatively closed* if it contains the direct product of any family of Ω-algebras from this class.

Now we define the concepts of direct and subdirect products for T-algebras.

Let $\mathfrak{A}_i = (A_i; \Sigma_i)$, $i \in I$, be T-algebras of the same arithmetic type. Again, we form the Cartesian product $\hat{A} = \prod_{i\in I} A_i$. In addition we form the Cartesian product $\prod_{i\in I} \Sigma_i$ and define the subset $\widehat{\Sigma} \subseteq \prod_{i\in I} \Sigma_i$ as the set of all possible functions $F : I \to \bigcup_{i\in I} \Sigma_i$ satisfying the following two conditions:

(a) $F(i) \in \Sigma_i$ for all $i \in I$;
(b) $|F(i)| = |F(j)|$ for all $i, j \in I$,

that is, for fixed F the arity of an operation $F(i) \in \Sigma_i$ does not depend on $i \in I$.

The set $\widehat{\Sigma}$ can be identified with the set of all possible systems $F = (F_i)_{i\in I}$ of operations of the same arity, with one taken from each set Σ_i. We write $\widehat{\Sigma} = \tilde{\prod}_{i\in I}\Sigma_i$. If $F \in \widehat{\Sigma}$ and $|F(i)| = n$, then F defines an n-ary operation on the set $\hat{A} = \prod_{i\in I} A_i$ defined componentwise:

$$F(f_1, \ldots, f_n)(i) = F(i)(f_1(i), \ldots, f_n(i)),\ i \in I,$$

where $f_1, \ldots, f_n \in \hat{A}$.

As result we obtain a T-algebra $\widehat{\mathfrak{A}} = (\hat{A}; \widehat{\Sigma})$, which is called the (*direct*) *product* of the T-algebras $\mathfrak{A}_i$, $i \in I$, or their direct T-product and is written $\widehat{\mathfrak{A}} = \prod_{i\in I} \mathfrak{A}_i$.

The direct product of T-algebras $\mathfrak{A}_i$, $i \in I$ is also called their *superproduct*. The superproduct of two T-algebras $\mathfrak{A}$ and $\mathfrak{B}$ is denoted by $\mathfrak{A} \bowtie \mathfrak{B}$. For example, the superproduct $L_1 \bowtie L_2$ of the two lattices $(L_1; \wedge_1, \vee_1)$ and $(L_2; \wedge_2, \vee_2)$ is an algebra $(L_1 \times L_2; \wedge, \vee, \cdot, +)$ with four binary operations:

$$\wedge = (\wedge_1, \vee_2), \vee = (\vee_1, \wedge_2),$$
$$\cdot = (\wedge_1, \wedge_2), + = (\vee_1, \vee_2),$$

which operate component-wise.

The semisubalgebra (reduct) of direct product is called a semidirect product.

The *projection maps*

$$\pi_j : \prod_{i \in I} A_i \to A_j \quad \text{and} \quad \tilde{\pi}_j : \widehat{\Sigma} \to \Sigma_j$$

for each $j \in I$ are defined by

$$\pi_j(f) = f(j) \quad \text{and} \quad \tilde{\pi}_j(F) = F(j).$$

These maps give surjective bihomomorphisms

$$(\pi_j, \tilde{\pi}_j) : \prod_{i \in I} \mathfrak{A}_i \rightrightarrows \mathfrak{A}_j.$$

Definition 1.4.6. A T-algebra $\mathfrak{A}$ is called a *subdirect product* of the T-algebras $\mathfrak{A}_i$, $i \in I$, if $\mathfrak{A}$ is a subalgebra of the direct product $\prod_{i \in I} \mathfrak{A}_i$ and the projection maps π_i and $\tilde{\pi}_i$ are surjective for all $i \in I$.

Definition 1.4.7. A T-algebra $\mathfrak{A} = (A; \Sigma)$ is called *subdirectly irreducible* if $|A| > 1$ and the (componentwise) intersection of all nonzero congruencies of $\mathfrak{A}$ is nonzero.

If r is a congruence of an algebra $\mathfrak{A}$ considered as an Ω-algebra, then $(r, \tilde{0})$ is a congruence of $\mathfrak{A}$ considered as a T-algebra. Further, if $(r, \tilde{t})$ is a congruence of an algebra $\mathfrak{A}$ considered as a T-algebra, then $(r, \tilde{0})$ is also a congruence of $\mathfrak{A}$ considered as a T-algebra, i.e., r is a congruence of an algebra $\mathfrak{A}$ considered as an Ω-algebra. From these facts, we conclude that an algebra $\mathfrak{A}$ is subdirectly irreducible as an Ω-algebra if and only if it is subdirectly irreducible as a T-algebra.

Theorem 1.4.8 (Birkhoff's subdirect representation theorem [61]). *Every Ω-algebra $\mathfrak{A}$ can be represented as a subdirect product of subdirectly irreducible Ω-algebras, which are homomorphic images of $\mathfrak{A}$. Analogously, every T-algebra $\mathfrak{A}$ can be represented as a subdirect product of subdirectly irreducible T-algebras, which are bihomomorphic images of $\mathfrak{A}$. Every finite algebra can be subdirectly embedded into a product of finitely many finite subdirectly irreducible algebras.*

A class of T-algebras is said to be *multiplicatively closed* if it contains the direct product of any family of T-algebras from this class.

Let $\mathfrak{A}_i = \langle Q_i; \Sigma_i \rangle$, $i \in I$, be T-algebras of the same arithmetic type, and let $\mathcal{D}$ be a filter[b] on I. We introduce the relation $\equiv_{\mathcal{D}}$ defined on $\prod_{i\in I} Q_i$ and the relation $\sim_{\mathcal{D}}$ defined on $\widetilde{\prod_{i\in I}\Sigma_i}$ by setting

$$f \equiv_{\mathcal{D}} g \Longleftrightarrow \{\alpha \in I |\, f(\alpha) = g(\alpha)\} \in \mathcal{D},$$

where $f, g \in \prod_{i\in I} Q_i$, and

$$F \sim_{\mathcal{D}} G \Longleftrightarrow \{\alpha \in I |\, F(\alpha) = G(\alpha)\} \in \mathcal{D},$$

where $F, G \in \widetilde{\prod_{i\in I}\Sigma_i}$, $|F| = |G|$.

According to the definition of a filter, the relations $\equiv_{\mathcal{D}}$ and $\sim_{\mathcal{D}}$ will be equivalence relations. In addition, it is easy to prove that the pair of equivalence relations $(\equiv_{\mathcal{D}}, \sim_{\mathcal{D}})$ is a congruence of the product of the T-algebras $\mathfrak{A}_i = \langle Q_i; \Sigma_i \rangle$, $i \in I$. The corresponding quotient algebra is denoted by $\prod_{i\in I} \mathfrak{A}_i/\mathcal{D}$ and is called the *filtered product* (with respect to the filter $\mathcal{D}$) of the T-algebras $\mathfrak{A}_i = \langle Q_i; \Sigma_i \rangle$, $i \in I$. A filtered product of T-algebras with respect to an ultrafilter is called an ultraproduct of T-algebras (see [198]).

Moreover, the congruence $(\equiv_{\mathcal{D}}, \sim_{\mathcal{D}})$ is nuclear. Indeed, for any $f_1, \ldots, f_n \in \prod_{i\in I} Q_i$ we have

$$\begin{aligned} F \sim_{\mathcal{D}} G &\Longleftrightarrow \{\alpha \in I |\, F(\alpha) = G(\alpha)\} \in \mathcal{D} \\ &\Longleftrightarrow \{\alpha \in I |\, F(\alpha)(f_1(\alpha), \ldots, f_n(\alpha)) = G(\alpha)(f_1(\alpha), \ldots, f_n(\alpha))\} \in \mathcal{D} \\ &\Longleftrightarrow \{\alpha \in I |\, F(f_1, \ldots, f_n)(\alpha) = G(f_1, \ldots, f_n)(\alpha)\} \in \mathcal{D} \\ &\Longleftrightarrow F(f_1, \ldots, f_n)(\alpha) \equiv_{\mathcal{D}} G(f_1, \ldots, f_n)(\alpha). \end{aligned}$$

As a product of T-algebras the superproduct of two nonzero rings (lattices), where $T = \{2\}$, is an algebra with four binary operations, and does not a ring (lattice). However, it follows from the following proposition that an ultraproduct of a finite number of finite rings (lattices), as an ultraproduct of T-algebras, is a ring (lattice).

---

[b]A *filter* on a nonempty set I is a nonempty set $\mathcal{D}$ of subsets of I satisfying the requirements:

(a) the intersection of any two subsets from $\mathcal{D}$ belongs to $\mathcal{D}$;

(b) all the supersets of any subset belonging to $\mathcal{D}$ also belong to $\mathcal{D}$;

(c) the empty set $\emptyset$ does not belong to $\mathcal{D}$.

A maximal filter on I, that is, a filter on I that does not lie in any other filter on I, is usually called an *ultrafilter* on I [216].

Proposition 1.4.9. *Let $\mathfrak{A}_0, \ldots, \mathfrak{A}_{n-1}$ be pairwise nonisomorphic finite T-algebras with finite sets of operations, and let $\mathfrak{A}_i$, $i \in I$, be a nonempty family of T-algebras, each of which is one of the algebras $\mathfrak{A}_0, \ldots, \mathfrak{A}_{n-1}$. If $\mathcal{D}$ is an ultrafilter on I, then there exists a number j, $0 \leqslant j < n$, such that we have the isomorphism:*

$$\prod_{i\in I} \mathfrak{A}_i/\mathcal{D} \simeq \mathfrak{A}_j\,.$$

Proof. We define the sets

$$I_j = \{i \in I |\, \mathfrak{A}_i = \mathfrak{A}_j\}, \qquad 0 \leqslant j < n,$$

and notice that $I_0, \ldots, I_{n-1}$ are pairwise disjoint and $I = I_0 \cup I_1 \cup \cdots \cup I_{n-1}$. Since the filter $\mathcal{D}$ is maximal and $I \in \mathcal{D}$, there exists a unique $j \in \{0, \ldots, \text{n-1}\}$ such that $I_j \in \mathcal{D}$. Since the congruences $(\equiv_{\mathcal{D}}, \sim_{\mathcal{D}})$ are nuclear, we have

$$\prod_{i\in I} \mathfrak{A}_i/\mathcal{D} \simeq \prod_{i\in I} \mathfrak{A}_i/\mathcal{E},$$

where $\mathcal{E} = \{X \cap I_j |\, X \in \mathcal{D}\}$ is an ultrafilter on I_j.

For $a \in Q_j$ we define $f_a \in \prod_{i\in I_j} Q_i$ with the property $f_a(i) = a$ for all $i \in I_j$. For $X \in \Sigma_j$ we define $F_X \in \widetilde{\prod_{i\in I_j} \Sigma_i}$ with the property $F_X(i) = X$ for all $i \in I_j$. If $a \neq b$ and $a, b \in Q_j$, then

$$\{i \in I_j |\, f_a(i) = f_b(i)\} = \emptyset \notin \mathcal{E}.$$

Hence $f_a \not\equiv_{\mathcal{E}} f_b$ and the map $\alpha : a \to [f_a]_{\mathcal{E}}$, where $[f_a]_{\mathcal{E}}$ is an equivalence class modulo $\equiv_{\mathcal{E}}$, will be an injection. In an analogous way, we show that the map $\tilde{\alpha} : X \to [F_X]_{\mathcal{E}}$, where $[F_X]_{\mathcal{E}}$ is an equivalence class modulo $\sim_{\mathcal{E}}$, is an injection. Moreover, starting from the equality $f_{X(a_1,\ldots,a_n)} = F_X(f_{a_1}, \ldots, f_{a_n})$, we verify that, for the pair $(\alpha, \tilde{\alpha}) : \mathfrak{A}_j \Rightarrow \prod_{i\in I_j} \mathfrak{A}_i/\mathcal{E}$,

$$\begin{aligned}\alpha X(a_1, \ldots, a_n) &= \left[f_{X(a_1,\ldots,a_n)}\right]_{\mathcal{E}} = [F_X(f_{a_1}, \ldots, f_{a_n})]_{\mathcal{E}} \\ &= [F_X]_{\mathcal{E}}\left([f_{a_1}]_{\mathcal{E}}, \ldots, [f_{a_n}]_{\mathcal{E}}\right) = [\tilde{\alpha}(X)]\,(\alpha a_1, \ldots, \alpha a_n).\end{aligned}$$

It remains to verify that α and $\tilde{\alpha}$ are surjective.

Let $f \in \prod_{i \in I_j} Q_i$ and $F \in \tilde{\prod}_{i \in I_j} \Sigma_i$. Then, for each $a \in Q_j$ and $A \in \Sigma_j$ we define

$$I_{j,a} = \{i \in I_j \mid f(i) = a\},$$

and

$$I_{j,A} = \{i \in I_j \mid F(i) = A\}.$$

Since the $I_{j,a}$ are pairwise disjoint and $\bigcup(I_{j,a}|a \in Q_j) = I_j$, we conclude that there exists exactly one $a \in Q_j$ such that $I_{j,a} \in \mathcal{E}$. Now

$$\{i \in I_j \mid f(i) = f_a(i)\} = I_{j,a} \in \mathcal{E},$$

hence, $f \equiv_{\mathcal{E}} f_a$ and α is onto. The similar proof is valid for $\tilde{\alpha}$. Therefore,

$$\prod_{i \in I_j} \mathfrak{A}/\mathcal{E} \simeq \mathfrak{A}.$$

□

1.5. Ω-terms and *T*-hyperterms, Ω-identities and *T*-hyperidentities, varieties and hypervarieties

First we define the concept of Ω-term.

Let us fix a type (signature) Ω. Let X be a set of objects called object *variables. Terms of type* Ω *over* X or simply Ω-terms are defined inductively:

(i) all object variables are Ω-terms;
(ii) all nullary function symbols from Ω are terms;
(iii) if $f \in \Omega$ is an n-ary function symbol ($n \geq 1$) and $\omega_1, \ldots, \omega_n$ are Ω-terms then the expression $f(\omega_1, \ldots, \omega_n)$ is a Ω-term too;
(iv) there are no other Ω-terms.

The set of all terms of type Ω over the set of object variables X is denoted by $\mathcal{T}_\Omega(X)$.

For a term $\omega \in \mathcal{T}_\Omega(X)$ we often write $\omega(x_1, \ldots, x_n)$ to indicate that the variables occurring in ω are among $x_1, \ldots, x_n$. A term ω is n-ary if the number of object variables appearing explicitly in ω is less than or equal to n.

For example, let Ω consists of two binary function symbols $+$ and $\cdot$, one unary function symbol $'$ and two nullary function symbols 0, 1, i.e.,

$\Omega = \{+, \cdot, ', 0, 1\}$. Let $X = \{x, y, z\}$. Then the following expressions are some terms of type Ω over X:

$$x,\ y,\ z,\ 0,\ 1,\ x+y,\ x\cdot z,\ x\cdot x',\ x+(y\cdot z'),\ z',\ (x'\cdot y')\cdot(x'+z').$$

Note that these terms are first-order terms since they are terms in a first-order language.

Definition 1.5.1. Let $\mathfrak{A} = (A; \Sigma)$ be an algebra of type Ω and $\omega(x_1, \ldots, x_n)$ be an n-ary term of type Ω over some set of object variables X. We define a mapping $\widehat{\omega} : A^n \to A$ as follows:

(i) if ω is a variable x_i, then

$$\widehat{\omega}(a_1, \ldots, a_n) = a_i$$

for all $a_1, \ldots, a_n \in A$, i.e., $\widehat{\omega} = \delta_n^i$ is the ith projection map;

(ii) if $\omega = c$ is constant symbol (i.e., nullary functional symbol) from Ω, then

$$\widehat{\omega}(a_1, \ldots, a_n) = c^{\mathfrak{A}},$$

i.e., $\widehat{\omega}$ is a constant function;

(iii) if ω is of the form $f(\omega_1(x_1, \ldots, x_n), \ldots, \omega_n(x_1, \ldots, x_n))$, where $f \in \Omega$ is an n-ary function symbol, then

$$\widehat{\omega}(a_1, \ldots, a_n) = f^{\mathfrak{A}}(\widehat{\omega}_1(a_1, \ldots, a_n), \ldots, \widehat{\omega}_n(a_1, \ldots, a_n))$$

for all $a_1, \ldots, a_n \in A$.

$\widehat{\omega}$ is the *term function* (or *term operation*) on $\mathfrak{A}$ corresponding to the term ω.

Definition 1.5.2. Let Ω be a type of algebras and X be a set of object variables. An *identity* of type Ω over X or simply Ω-identity is a first-order formula of the form

$$\forall x_1, \ldots, \forall x_n\ (\omega_1 = \omega_2), \tag{$*$}$$

where $\omega_1 = \omega_1(x_1, \ldots, x_n)$ and $\omega_2 = \omega_2(x_1, \ldots, x_n)$ are terms of type Ω over X. However, identities are usually written without universal quantifiers: $\omega_1 = \omega_2$.

Let $\mathfrak{A} = (A; \Sigma)$ be an Ω-algebra. We say that $\mathfrak{A}$ *satisfies* the Ω-identity $(*)$ (or the Ω-identity *is true in* $\mathfrak{A}$ or *holds in* $\mathfrak{A}$), if $\widehat{\omega}_1(a_1, \ldots, a_n) = \widehat{\omega}_2(a_1, \ldots, a_n)$ for all $a_1, \ldots, a_n \in A$. This means that the equality $\omega_1 = \omega_2$

is true, when we replace all the object variables occurring in terms ω_1 and ω_2 by any elements from the set A. In this case we write

$$\mathfrak{A} \models (\omega_1 = \omega_2).$$

A class $\mathcal{K}$ of Ω-algebras satisfies the Ω-identity $(*)$, if each member of K satisfies $(*)$. In this case, we write

$$\mathcal{K} \models (\omega_1 = \omega_2).$$

If Λ is a set of Ω-identities, then we say that $\mathcal{K}$ satisfies Λ, written

$$\mathcal{K} \models \Lambda,$$

if each $\mathcal{K}$ satisfies each Ω-identity from Λ.

So, for a class $\mathcal{K}$ of Ω-algebras the $Eq\,\mathcal{K}$ is the set of all Ω-identities that are satisfied in all algebras of $\mathcal{K}$.

Definition 1.5.3. Let Λ be a set of Ω-identities. Define $Mod(\Lambda)$ to be the class of all Ω-algebras $\mathfrak{A}$ satisfying Λ. A class $\mathcal{V}$ of Ω-algebras is called a *variety* or an *equational class* if there is a set of Ω-identities Λ such that $\mathcal{V} = Mod(\Lambda)$. In this case, we say that the variety $\mathcal{V}$ is *defined*, or *axiomatized*, by Λ.

The classes of all groups, semigroups, quasigroups, rings, lattices, Boolean algebras, De Morgan algebras, Boole–De Morgan algebras, weakly idempotent semilattices, weakly idempotent lattices, g-algebras with the fixed set Σ of binary operations, etc. are examples of varieties. However, the class of multiplicative groups of fields and the class of semigroups embedding into groups do not form varieties (for characterization of multiplicative groups of fields see [251] and for characterization of semigroups embedding into groups see [67]). Necessary and sufficient conditions for the embeddability of a semigroup in a group were first given by Mal'tcev in his classical paper [211]. Necessary and sufficient conditions for the semigroup to be multiplicative group of a field were first given in the paper [251].

Now let us define the term algebra $\Omega(X) = (\mathcal{T}_\Omega(X); \Omega)$, where the operations are defined as follows: for any n-ary function symbol $f \in \Omega$ and for any terms $w_1, \ldots, w_n \in \mathcal{T}_\Omega(X)$ we set

$$f : (w_1, \ldots, w_n) \mapsto f(w_1, \ldots, w_n).$$

This algebra is an Ω-algebra.

If $\mathfrak{A} = (A; \Sigma)$ is an Ω-algebra, then any mapping $\lambda : X \to A$ can be extended to a homomorphism $\varphi : \Omega(X) \to \mathfrak{A}$. Indeed, for $x \in X$ we set

$\varphi(x) = \lambda(x)$, and for a constant symbol $c \in \Omega$ we set $\varphi(c) = c^{\mathfrak{A}}$. Now if $f \in \Omega$ is an n-ary function symbol and $w_1, \ldots, w_n \in \mathcal{T}_\Omega(X)$ are terms whose images under φ are already defined, then we set

$$\varphi(f(w_1, \ldots, w_n)) = f^{\mathfrak{A}}(\varphi(w_1), \ldots, \varphi(w_n)).$$

As a consequence, we conclude that an identity $w_1 = w_2$, where $w_1, w_2 \in \mathcal{T}_\Omega(X)$, holds in an Ω-algebra $\mathfrak{A}$ if and only if $\varphi(w_1) = \varphi(w_2)$ for any homomorphism $\varphi : \Omega(X) \to \mathfrak{A}$.

Definition 1.5.4. Suppose $\mathfrak{L}$ is a class of Ω-algebras and $\mathfrak{F} = (F; \Omega)$ is an Ω-algebra. Let X be a nonempty set and $\varphi_0 : X \to F$ be any mapping. Then the pair $(\mathfrak{F}, \varphi_0)$ is called *universal* for the class $\mathfrak{L}$, if the following two conditions hold:

(i) the set $\varphi_0(X)$ generates the algebra $\mathfrak{F}$,
(ii) for any algebra $\mathfrak{A} = (A; \Omega)$ from the class $\mathfrak{L}$ and for any mapping $\mu : X \to A$ there exists a homomorphism $\nu : F \to A$ such that $\mu = \varphi_0 \nu$.

Usually we take $X \subseteq F$ and we choose the mapping φ_0 to be the identical mapping. In that case the second condition above means that any mapping $\mu : X \to A$ can be extended to a homomorphism $\nu : F \to A$.

A universal pair $(\mathfrak{F}, \varphi_0)$ exists for any class $\mathfrak{L}$ and any nonempty set X. For example, we can take φ_0 to be the identical mapping and $\mathfrak{F} = \Omega(X)$.

If, in addition, $\mathfrak{F} \in \mathfrak{L}$, then $\mathfrak{F}$ is called a *free* $\mathfrak{L}$*-algebra* freely generated by the set X and mapping φ_0. The free algebra in a class of algebras $\mathfrak{L}$, if it exists, is denoted by $\mathfrak{F}_\mathfrak{L}(X)$. This algebra is unique up to isomorphism provided it exists. The free algebra $\mathfrak{F}_\mathfrak{L}(X)$ is called finitely generated if set X is finite.

Lemma 1.5.5. *Let $\mathfrak{L}$ be a class of Ω-algebras and the pair $(\mathfrak{F}, \varphi_0)$ is universal for $\mathfrak{L}$. Then $(\mathfrak{F}, \varphi_0)$ is also universal for the variety $\mathfrak{L}^*$ generated by $\mathfrak{L}$.*

Proof. Let $\mathfrak{F} = (F; \Sigma)$. We extend the mapping $\varphi_0 : X \to F$ to an epimorphism $\varphi : \Omega(X) \to \mathfrak{F}$. Denote $r = \mathrm{Ker}(\varphi)$. Consider an arbitrary algebra $\mathfrak{A} = (A; \Gamma)$ from the class $\mathfrak{L}$. Let $\lambda : X \to A$ be a mapping. We define a homomorphism $\lambda^* : \Omega(X) \to \mathfrak{A}$ such that $\lambda^* = \lambda\varphi$.

Then $\mathrm{Ker}(\lambda^*) \supseteq r$. Let $(w_1, w_2) \in r$. As λ is arbitrary and therefore, λ^* is arbitrary as well, we conclude that the identity $w_1 = w_2$ holds in $\mathfrak{A}$. Thus, the identity $w_1 = w_2$ holds in any algebra from $\mathfrak{L}$, and hence, it holds in any algebra from $\mathfrak{L}^*$. Now the homomorphism theorem implies the desirable result. □

A class of algebras is called *abstract*, if it along with any algebra contains all algebras isomorphic to that one.

Lemma 1.5.6. *If a class of Ω-algebras $\mathfrak{L}$ is abstract, hereditary and multiplicatively closed, then for any nonempty set X there exists a free algebra $\mathfrak{F}_{\mathfrak{L}}(X)$.*

Proof. Let $\mathfrak{A} = (Q; \Sigma)$ be any algebra from $\mathfrak{L}$ and $\lambda : X \to Q$. We extend λ to a homomorphism $\lambda^* : \Omega(X) \to \mathfrak{A}$. Let $\{r_j | j \in J\}$ be the set of all kernels of such homomorphisms. Denote $r = \bigcap_{j \in J} r_j$ and $\mathfrak{F} = \Omega(X)/r$. Let $\pi : \Omega(X) \to \mathfrak{F}$ be the canonical homomorphism and $\varphi_0 = \pi|_X$ be the restriction of π on the set X. Then, according to the theorem of homomorphisms, the pair $(\mathfrak{F}, \varphi_0)$ is universal for the class $\mathfrak{L}$. On the other hand, $\mathfrak{F}$ is isomorphic to a subdirect product of algebras $\Omega(X)/r_j$, $j \in J$. But each of these algebras is isomorphic to a subalgebra of an algebra from $\mathfrak{L}$. Therefore, $\mathfrak{F}$ belongs to the class $\mathfrak{L}$, and hence, it is a free algebra. □

Theorem 1.5.7 (Birkhoff [36]). *A class of Ω-algebras is a variety if and only if it is hereditary and homomorphically and multiplicatively closed.*

Proof. The necessity is obvious. Let us prove sufficiency. Suppose a class $\mathfrak{M}$ of Ω-algebras is hereditary and homomorphically and multiplicatively closed. Let $\mathfrak{M}^*$ be the variety generated by $\mathfrak{M}$. We should prove that $\mathfrak{M} = \mathfrak{M}^*$. Let $\mathfrak{A}^* = (A^*; \Sigma^*) \in \mathfrak{M}^*$. According to Lemma 1.5.6, there exists a free algebra $\mathfrak{F}^*_{\mathfrak{M}}(X)$ for any nonempty set X. Then $\mathfrak{A}^*$ is a homomorphic image of $\mathfrak{F}_{\mathfrak{M}^*}(X)$ for some X (for example, $X = A^*$). On the other hand, there exists a free algebra $\mathfrak{F}_{\mathfrak{M}}(X)$ in the class $\mathfrak{M}$, which is a free algebra for the class $\mathfrak{M}^* \supseteq \mathfrak{M}$ as well (Lemma 1.5.5). And we conclude that the algebras $\mathfrak{F}_{\mathfrak{M}}(X)$ and $\mathfrak{F}_{\mathfrak{M}^*}(X)$ are isomorphic. Hence, $\mathfrak{F}_{\mathfrak{M}^*}(X) \in \mathfrak{M}$, and so $\mathfrak{A}^* \in \mathfrak{M}$. □

Problem 1.5.8. *For which variety $\mathfrak{L}$ of Ω-algebras is any finitely generated free algebra $\mathfrak{F}_{\mathfrak{L}}(X)$ isomorphic to a function algebra on finite set?*

The trivial variety M consists exactly of all one-element Ω-algebras, i.e., we have $M = \mathrm{Mod}\{x = y\}$.

A subclass W of a variety V which is also a variety is called a *subvariety* of V. The variety V is called a *minimal*, or *equationally complete* variety if V is not trivial but the only subvariety of V not equal to V is the trivial variety.

Theorem 1.5.9 ([61]). *Every nontrivial variety contains a minimal subvariety.*

Example 1.5.10. The variety of Boolean algebras and the variety of distributive lattices are minimal varieties.

Now we define the concept of T-hyperterm.

Let X be an arbitrary countable set, $X = \{x_1, \ldots, x_n, \ldots\}$, elements of which are called object variables. Let $T \subseteq \mathbb{N} = \{0, 1, 2, \ldots\}$, $T \neq \emptyset$ and suppose U is an arbitrary T-set of symbol operations, i.e.,

$$T = \{|F| | F \in U\},$$

and $X \cap U = \emptyset$. The elements of set U are also called functional variables or hypervariables . The notion of an n-ary T-word (or T-term) or n-ary T-hyperterm is defined inductively (recursively):

(i) the object variables $x_1, \ldots, x_n \in X$ are n-ary T-hyperterms;
(ii) if $\omega_1, \ldots, \omega_n$ are n-ary T-hyperterms, and $F \in U$ is an m-ary hypervariable then the expression $F(\omega_1, \ldots, \omega_n)$ is also an n-ary T-hyperterm;
(iii) there are no other n-ary T-hyperterms.

A T-hyperterm is an n-ary T-hyperterm for some $n \in \mathbb{N}$. A hyperterm is an n-ary T-hyperterm for some set $T \subseteq \mathbb{N}$ and some $n \in \mathbb{N}$.

The set of all T-hyperterms is denoted by $X(U)$. The corresponding T-algebra $(X(U); U) = U(X)$ is the T-algebra of T-hyperterms, where if $F \in U$, $|F| = n$, then

$$F : (\omega_1, \ldots, \omega_n) \mapsto F(\omega_1, \ldots, \omega_n)$$

for any $\omega_1, \ldots, \omega_n \in (X)U$.

Recall that a hyperidentity (or $\forall(\forall)$-identity) is a second-order formula of the following form:

$$\forall X_1, \ldots, \forall X_m, \forall x_1, \ldots, \forall x_n \quad (\omega_1 = \omega_2), \tag{**}$$

where ω_1, ω_2 are hyperterms in the alphabet of functional variables $X_1, \ldots, X_m$ and object variables $x_1, \ldots, x_n$. However, hyperidentities are usually presented without universal quantifiers: $\omega_1 = \omega_2$. The hyperidentity $\omega_1 = \omega_2$ is said to be satisfied in the algebra $\mathfrak{A} = (A; \Sigma)$ if this equality holds whenever every functional variable X_i is replaced by an arbitrary operation of the corresponding arity from Σ (a possibility of such replacement is supposed, which means that $\{|X_1|, \ldots, |X_m|\} \subseteq T_{\mathfrak{A}}$) and every object variable x_j is replaced by an arbitrary element from A [244,249,252,257,259].

A hyperidentity $\omega_1 = \omega_2$ is called a T-hyperidentity (Ω-hyperidentity), if ω_1, ω_2 are T-hyperterms (T_Ω-hyperterms), where $T_\Omega = \{|f| \,|\, \text{for any}$ operation symbol $f \in \Omega\}$. If $\mathfrak{A} = (A; \Sigma)$ is a T-algebra (Ω-algebra) then the T-hyperidentity (Ω-hyperidentity) $\omega_1 = \omega_2$ will be satisfied in $\mathfrak{A}$ if and only if the T-hyperterms (T_Ω-hyperterms) ω_1, ω_2 have equal images under any bihomomorphism $(\varphi, \tilde{\psi}) : U(X) \Rightarrow \mathfrak{A}$, i.e., $\varphi(\omega_1) = \varphi(\omega_2)$.

The class $\mathcal{K}$ of T-algebras is said to be satisfy the T-hyperidentity $\omega_1 = \omega_2$ (a system of T-hyperidentities $\mathfrak{L}$), if this hyperidentity (every hyperidentity from the system $\mathfrak{L}$) is satisfied in any algebra of that class. This is denoted by $\mathcal{K} \models \omega_1 = \omega_2$ ($\mathcal{K} \models \mathfrak{L}$). In this case, the T-hyperidentity $\omega_1 = \omega_2$ is called a hyperidentity of the class $\mathcal{K}$. The set of all T-hyperidentities of the given class $\mathcal{K}$ is called the $\forall(\forall)$-theory or *hyperequational theory* of the class $\mathcal{K}$ and it is denoted by Heq $\mathcal{K}$. The hyperequational theory $\mathfrak{L}$ of the class $\mathcal{K}$ is *decidable*, if there exists an algorithm which, for every hyperidentity $\omega_1 = \omega_2$, determines whether that hyperidentity is in $\mathfrak{L}$ or not. The hyperequational theory of $\mathcal{K}$ is a part of its second-order theory. The importance of investigations of the second-order formulas (languages) for applied mathematics was first noted by L. Henkin, A. I. Maltsev, A. Church, S. Kleene and A. Tarski (see [66,214]).

Let $\mathfrak{L}$ be a system of T-hyperidentities. Denote by $\mathfrak{M}^T_\mathfrak{L}$ the class of all T-algebras each of which satisfies all T-hyperidentities from $\mathfrak{L}$. The class of T-algebras $\mathfrak{N}$ is called a *hypervariety* of T-algebras if there exists a system of T-hyperidentities $\mathfrak{L}$ such that

$$\mathfrak{N} = \mathfrak{M}^T_\mathfrak{L}.$$

In this case, the class of T-algebras $\mathfrak{N}$ is called the hypervariety of T-algebras defined by the system of T-hyperidentities $\mathfrak{L}$. Similarly, a variety of Ω-algebras is called a hypervariety of Ω-algebras if it is defined by Ω-hyperidentities [259].

A T-hyperidentity $\omega_1 = \omega_2$ is called a consequence of $\mathfrak{L}$ if for every T-algebra $\mathfrak{A}$,

$$\mathfrak{A} \models \mathfrak{L} \Longrightarrow \mathfrak{A} \models \quad (\omega_1 = \omega_2)$$

(the notation $\mathfrak{A} \models \mathfrak{L}$ means satisfiability of all hyperidentities of $\mathfrak{L}$ in the algebra $\mathfrak{A}$). The set $\mathfrak{L}'$ of T-hyperidentities is called a consequence of $\mathfrak{L}$ and denoted by $\mathfrak{L} \Longrightarrow \mathfrak{L}'$ if every hyperidentity of $\mathfrak{L}'$ is a consequence of $\mathfrak{L}$.

Every T-hyperidentity is a T'-hyperidentity for every set of positive integers $T' \supseteq T$. It is obvious that if $\omega_1 = \omega_2$ is a consequence of $\mathfrak{L}$ in the meaning of T-hyperidentities, then $\omega_1 = \omega_2$ is a consequence of $\mathfrak{L}$ in the meaning of T'-hyperidentities, for $T' \supseteq T$.

Let us define the concept of formal consequence of $\mathfrak{L}$.

(i) Any T-hyperidentity of the form $w = w$ and any hyperidentity from $\mathfrak{L}$ are formal consequences of $\mathfrak{L}$.
(ii) If T-hyperidentity $w_1 = w_2$ is a formal consequence of $\mathfrak{L}$, then the hyperidentity $w_2 = w_1$ is a formal consequence of $\mathfrak{L}$ as well.
(iii) If T-hyperidentities $w_1 = w_2$ and $w_2 = w_3$ are formal consequences of $\mathfrak{L}$, then the hyperidentity $w_1 = w_3$ is also formal consequence of $\mathfrak{L}$.
(iv) If T-hyperidentity $u_i = v_i$ is a formal consequence of $\mathfrak{L}$ for $i = 1, \ldots, n$, then for any n-ary functional variable Y (where $n \in T$) the T-hyperidentity

$$Y(u_1, \ldots, u_n) = Y(v_1, \ldots, v_n)$$

is a formal consequence of $\mathfrak{L}$ as well.
(v) If T-hyperidentity $w_1 = w_2$ is a formal consequence of $\mathfrak{L}$, then T-hyperidentity $w_1' = w_2'$ is also a formal consequence of $\mathfrak{L}$, where w_1' and w_2' are obtained from w_1 and w_2, respectively, by replacement of all occurrences of the functional variable X (the object variable x) by any functional variable Z with the same arity (respectively by any T-hyperterm ω).

A T-hyperidentity $\omega_1 = \omega_2$ is a formal consequence of a system of T-hyperidentities $\mathfrak{L}$, if it is a formal consequence of $\mathfrak{L}$ according to (i)–(v).

Theorem 1.5.11 (completeness for hyperidentities). *The T-hyperidentity $\omega_1 = \omega_2$ is a consequence of the system of T-hyperidentities $\mathfrak{L}$ if and only if it is a formal consequence of $\mathfrak{L}$* (see [249, p. 169; 252, p. 43]).

The concept of the base of a system of T-hyperidentities and the concept of equivalency of two systems of T-hyperidentities are defined in the ordinary way. The hyperequational theory $\mathfrak{L}$ has a finite base of hyperidentities, if there exists a finite subset $\mathfrak{L}_0 \subseteq \mathfrak{L}$ such that $\mathfrak{L}_0 \Longrightarrow \mathfrak{L}$. If $\mathfrak{L}_0$ consists of one hyperidentity, then hyperequational theory $\mathfrak{L}$ is called one-based.

The trivial hypervariety M consists exactly of all one-element T-algebras. A subclass W of a hypervariety V which is also a hypervariety is called a *subhypervariety* of V. The hypervariety V is a *minimal* hypervariety if V is not trivial but the only subhypervariety of V not equal to V is the trivial hypervariety.

A set of T-hyperidentities is said to be satisfiable if there exists a nontrivial T-algebra in which every T-hyperidentity of this set is true. The compactness theorem for first-order languages extends to hyperidentities using ultraproducts of T-algebras [252,257].

Theorem 1.5.12. *If every finite part of an infinite set of T-hyperidentities is satisfiable, then the whole set of T-hyperidentities is also satisfiable.*

1.6. Term (termal) algebras, clones and categories

1.6.1. *Categories*

A category is a special kind of partial semigroup.

Definition 1.6.1. A *category* $\mathfrak{C}$ consists of a class Ob$\mathfrak{C}$, whose elements are called *objects*, and a class Mor$\mathfrak{C}$, whose elements are called *morphisms* (or *arrows*). The objects and morphisms are related by the following properties:

(i) for any two objects $A, B \in \text{Ob}\mathfrak{C}$ some set $H_{\mathfrak{C}}(A, B)$ of morphisms of category $\mathfrak{C}$ is corresponded to the ordered pair (A, B);
(ii) each morphism of category $\mathfrak{C}$ belongs to exactly one of the sets $H_{\mathfrak{C}}(A, B)$;
(iii) a partial binary operation of *composition* (or *multiplication*) is defined on the class Mor $\mathfrak{C}$: the composition fg of two morphisms $f \in H_{\mathfrak{C}}(A, B)$ and $g \in H_{\mathfrak{C}}(C, D)$ is defined if and only if the objects B and C are the same, and in this case $fg \in H_{\mathfrak{C}}(A, D)$; composition is *associative*, that is, given $f \in H_{\mathfrak{C}}(A, B)$, $g \in H_{\mathfrak{C}}(B, C)$, $h \in H_{\mathfrak{C}}(C, D)$, we have $h(gf) = (hg)f$;
(iv) for every object A there is a morphism $1_A \in H_{\mathfrak{C}}(A, A)$, called the *identity morphism* of A, such that $f1_A = f$ and $1_A g = g$ for every $f \in H_{\mathfrak{C}}(X, A)$ and $g \in H_{\mathfrak{C}}(A, Y)$.

If $f \in H_{\mathfrak{C}}(A, B)$, then A is called the *domain* and B is called the *codomain* of f. One also writes $A = \text{dom}(f)$, $B = \text{cod}(f)$. In this case, it is said that f is a morphism of A into B. We shall frequently write $f : A \to B$ or $A \xrightarrow{f} B$ instead of $f \in H_{\mathfrak{C}}(A, B)$.

Examples

(1) The category **Set**. The set Ob$\mathfrak{C}$ of objects of this category consists of sets. For any two arbitrary sets $A, B \in \text{Ob}\mathfrak{C}$ we define $H_{\textbf{Set}}(A, B)$ to be the set of all functions from A into B. The composition of two morphisms is the superposition of functions.
(2) The category **Top**, whose objects are topological spaces. Morphisms of a topological space X to topological space Y are the continuous

maps. Composition of morphisms is the superpositions of maps (the superposition of two continuous maps is continuous as well).

(3) The category **Group**, whose objects are groups and morphisms are the group homomorphisms between groups. The composition of two morphisms is again the superposition of homomorphisms.

(4) The category of Ω-algebras, where Ω is a type. The objects here are Ω-algebras and the morphisms are the homomorphisms of Ω-algebras. Again, the composition of morphisms is the superposition of homomorphisms.

(5) The category of T-algebras, where $T \subseteq \mathbb{N}$. The objects here are T-algebras and the morphisms are the bihomomorphisms of T-algebras. The composition of these morphisms is the componentwise multiplication of pairs. This category we denote by $\widetilde{\text{Alg}}$. For applications of this morphisms (category) in cryptology, see [15].

Definition 1.6.2. Let A be an object of the category $\mathfrak{C}$ and I an arbitrary set of indices. We assume that there are given a family of objects $(A_i)_{i\in I}$ of the category $\mathfrak{C}$. We shall say that the object A is a *direct product* of the family of objects $(A_i)_{i\in I}$ if there exist a family of morphisms $f_i : A \to A_i$, $i \in I$, such that the following condition is satisfied: for any object B of the category $\mathfrak{C}$ and for any family $(g_i)_{i\in I}$ of morphisms $g_i : B \to A_i$, there exists a unique morphism $g : B \to A$ such that $g_i = gf_i$ for any $i \in I$.

We shall also say sometimes that the family $(A_i)_{i\in I}$ *possesses* a direct product. The morphisms f_i are called *canonical projections.*

The product in the category **Set** is the direct product of sets, in the category **Group** the product is the direct product of groups. The product in the category of Ω-algebras is the direct product of Ω-algebras. The product in the category of T-algebras is the superproduct of T-algebras.

1.6.2. *Term operations, term algebras, termal hyperidentities, clones*

Let Q be an arbitrary nonempty set, let $O_p^{(n)}Q$ be a set of all n-ary operations on Q, and

$$O_pQ = \bigcup_n O_p^{(n)}Q;$$

For every nonempty subset $\Sigma \subseteq O_pQ$, the pair $(Q;\Sigma)$ is an algebra.

Let $\mathfrak{A} = (Q;\Sigma)$ be an arbitrary algebra. n-ary term (termal) operations of algebra $\mathfrak{A}$ are defined by induction:

(1) all n-ary identical operations (or projections) of set Q

$$\delta_n^i(x_1, \ldots, x_n) = x_i, \qquad i = 1, \ldots, n,$$

are n-ary term operations of $\mathfrak{A}$;

(2) if $f_1, \ldots, f_m$ are n-ary term operations of $\mathfrak{A}$, then the superposition

$$\mu_m^n(f, f_1, \ldots, f_m)(x_1, \ldots, x_n) = f(f_1(x_1, \ldots, x_n), \ldots, f_m(x_1, \ldots, x_n))$$

is again an n-ary term operations of $\mathfrak{A}$, for every m-ary $f \in \Sigma$;

(3) there are no other n-ary term operations of $\mathfrak{A}$.

The operation $h \in O_pQ$ is called a term operation of $\mathfrak{A} = (Q; \Sigma)$, if h is an n-ary term operation of $\mathfrak{A}$ for some n.

If we denote the set of all n-ary term operations of algebra $\mathfrak{A}$ by $\mathcal{F}^n(\Sigma)$ and

$$\mathcal{F}(\Sigma) = \mathcal{F}^1(\Sigma) \cup \mathcal{F}^2(\Sigma) \cup \cdots$$

is a set of all its term operations, then the algebra $\mathcal{F}(\mathfrak{A}) = (Q, \mathcal{F}(\Sigma))$ is called an algebra of term operations for $\mathfrak{A}$ or a term (termal) algebra for $\mathfrak{A}$.

A hyperidentity $\omega_1 = \omega_2$ is called *termal* hyperidentity or t-hyperidentity of the algebra $\mathfrak{A}$ if it is valid in the term algebra $\mathcal{F}(\mathfrak{A})$. Let V be a variety. A hyperidentity $\omega_1 = \omega_2$ is called a termal hyperidentity or t-hyperidentity of V if it is a termal hyperidentity for any algebra $\mathfrak{A} \in V$ (see [85–88,90–93,136–139,249,252,253,363]). Termal hyperidentities for varieties were first considered by W. Taylor ([402]) (as a special case of Mal'tsev conditions for varieties) for characterization of classes of varieties which are closed under formation of equivalent varieties, products of varieties, reducts of varieties and subvarieties. Since the operations of an algebra are included in the set of term operations of the algebra, the concept of termal hyperidentity is stronger than the concept of hyperidentity. In particular, the variety of rings (even commutative rings) does not have termal hyperidentities except $w = w$, but has hyperidentities. The next example shows that nontrivial hyperidentities are valid in the variety of all rings.

Example 1.6.3. The following hyperidentities are satisfied in any ring:

$$\begin{aligned}
&X\left(X\left(Y(x, x), Y(x, x)\right), Y\left(X(x, x), X(x, x)\right)\right)\\
&\quad = X\left(Y\left(X(x, x), X(x, x)\right), X\left(Y(x, x), Y(x, x)\right)\right),\\
&X\left(Y\left(Y(x, x), X(x, x)\right), Y\left(X(x, x), Y(x, x)\right)\right)\\
&\quad = X\left(Y\left(X(x, x), Y(x, x)\right), Y\left(Y(x, x), X(x, x)\right)\right).
\end{aligned}$$

The hyperidentity $\omega_1 = \omega_2$ is called a termal consequence (briefly t-consequence) of a system of the hyperidentities $\mathcal{L}$ and is denoted by $\mathcal{L} \Rightarrow_t (\omega_1 = \omega_2)$ if for any algebra $\mathfrak{A}$:

$$\mathcal{F}(\mathfrak{A}) \models \mathcal{L} \Rightarrow \mathcal{F}(\mathfrak{A}) \models (\omega_1 = \omega_2).$$

The concept of termal equivalence of two systems of hyperidentities, and the concept of termal basis (briefly t-basis) arise in a natural way.

We proceed to the notion of a formal termal consequence of a system of $\mathbb{N}$-hyperidentities $\mathcal{L}$.

(i′) Any $\mathbb{N}$-hyperidentity of the form $w = w$ and any hyperidentity from $\mathfrak{L}$ are formal termal consequences of $\mathfrak{L}$.

(ii′) If $\mathbb{N}$-hyperidentity $w_1 = w_2$ is a formal termal consequence of $\mathfrak{L}$, then the hyperidentity $w_2 = w_1$ is a formal termal consequence of $\mathfrak{L}$ as well.

(iii′) If $\mathbb{N}$-hyperidentities $w_1 = w_2$ and $w_2 = w_3$ are formal termal consequences of $\mathfrak{L}$, then the hyperidentity $w_1 = w_3$ is also formal termal consequence of $\mathfrak{L}$.

(iv′) If $\mathbb{N}$-hyperidentity $u_i = v_i$ is a formal termal consequence of $\mathfrak{L}$ for $i = 1, \ldots, n$, then for any n-ary functional variable Y the $\mathbb{N}$-hyperidentity

$$Y(u_1, \ldots, u_n) = Y(v_1, \ldots, v_n)$$

is a formal termal consequence of $\mathfrak{L}$ as well.

(v′) If $\mathbb{N}$-hyperidentity $w_1 = w_2$ is a formal termal consequence of $\mathfrak{L}$, then $\mathbb{N}$-hyperidentity $w_1' = w_2'$ is also a formal termal consequence of $\mathfrak{L}$, where w_1' and w_2' are obtained from w_1 and w_2, respectively, by replacement of all occurrences of the functional variable X (the object variable x) by any functional variable Z with the same arity (respectively by any $\mathbb{N}$-hyperterm ω).

(vi′) If the $\mathbb{N}$-hyperidentity $w_1 = w_2$ is a formal termal consequence of $\mathcal{L}$, then the $\mathbb{N}$-hyperidentity $w_1'' = w_2''$ is also a formal termal consequence of $\mathcal{L}$, where w_1'' and w_2'' are obtained from w_1 and w_2 respectively by replacing all occurrences of an arbitrary functional variable X by an arbitrary hyperterm of the same arity.

An $\mathbb{N}$-hyperidentity $w_1 = w_2$ is a *formal termal consequence of a system of* $\mathbb{N}$-*hyperidentities* $\mathcal{L}$ if it is a formal termal consequence of $\mathcal{L}$ according to (i′)–(vi′).

Theorem 1.6.4 (Completeness for termal hyperidentities). *An* $\mathbb{N}$-*hyperidentity* $w_1 = w_2$ *is a termal consequence of a system of* $\mathbb{N}$-*hyperidentities* $\mathcal{L}$ *if and only if it is a formal termal consequence of* $\mathcal{L}$.

The proof of Theorem 1.6.4 is a repetition of the proof of the completeness theorem for hyperidentities using the concept of termally invariant congruence instead of the notion of a fully invariant congruence.

A congruence $r = (r, \tilde{0})$ of a T-algebra $\mathfrak{A} = \langle Q; \Sigma\rangle$ is said to be termally invariant if it preserves any bihomomorphism $(\alpha, \tilde{\beta}) : \mathfrak{A} => \mathcal{F}(\mathfrak{A})$, that is,

$$xry \longrightarrow \alpha(x)r\alpha(y),$$

where $x, y \in Q$ and $\mathcal{F}(\mathfrak{A})$ is the termal algebra of $\mathfrak{A}$.

Termal hyperidentities of varieties of groups and semigroups have been characterized by G. Bergman [31] (also see [32]). Termal hyperidentities of the variety of lattices and of the variety of semilattices were studied by R. Padmanabhan and P. Penner (see [309,310,317]).

Example 1.6.5. See (1) If $\mathfrak{A} = (Q; \Sigma)$ is an abelian group, then

$$\mathcal{F}(\Sigma) = \left\{ \sum_{i=1}^{n} c_i x_i \;\middle|\; n \geqslant 1,\ c_1, \ldots, c_n \in Z \right\}.$$

Similarly, $\mathcal{F}(\Sigma)$ is defined for the unitary module over a ring with identity.

(2) If $\mathfrak{A} = (Q; \Sigma)$ is a two-element Boolean algebra, then

$$\mathcal{F}(\Sigma) = O_p Q.$$

Similarly, $\mathcal{F}(\Sigma)$ is defined for a finite field and finite Post algebra.

(3) If $\mathfrak{A} = (Q; \Sigma)$ is a lattice or a semilattice, then

$$\mathcal{F}^2(\Sigma) = \Sigma \cup \{\delta_2^1, \delta_2^2\}.$$

(4) If $\mathfrak{A} = Q(\circ)$ is an idempotent semigroup, then

$$\mathcal{F}^2(\{\circ\}) = \{\circ, \delta_2^1, \delta_2^2, f_1, f_2, f_3\},$$

where

$$f_1(x, y) = y \circ x,$$
$$f_2(x, y) = x \circ y \circ x,$$
$$f_3(x, y) = y \circ x \circ y.$$

(5) If $\mathfrak{A} = Q(\cdot)$ is a semigroup with identity of mediality $(x \cdot y) \cdot (u \cdot v) = (x \cdot u) \cdot (y \cdot v)$, then its every binary termal operation f is defined by one of

the following types:

$$\begin{aligned} f(x,y) &= x^{k_1}, \\ f(x,y) &= y^{k_2}, \\ f(x,y) &= x^{k_1}y^{k_2}, \\ f(x,y) &= y^{k_1}x^{k_2}, \\ f(x,y) &= x^{k_1}y^{k_2}x, \\ f(x,y) &= yx^{k_1}y^{k_2}, \end{aligned}$$

where $k_1, k_2 \geqslant 1$ are natural number.

(6) The hyperidentity

$$X(Y(x,x), Y(x,x)) = Y(X(x,x), X(x,x))$$

is a termal hyperidentity of the variety of groups (semigroups, Moufang loops [27]). Moreover, here we have more general hyperidentity

$$\begin{aligned} &X(Y(Z_{11}(x), \ldots, Z_{1m}(x)), \ldots, Y(Z_{n1}(x), \ldots, Z_{nm}(x))) \\ &\quad = Y(X(Z_{11}(x), \ldots, Z_{n1}(x)), \ldots, X(Z_{1m}(x), \ldots, Z_{nm}(x))) \end{aligned}$$

for arbitrary $n, m \in \mathbb{N}$.

(7) The hyperidentity

$$\begin{aligned} &X(Y(x_{11}, \ldots, x_{1m}), \ldots, Y(x_{n1}, \ldots, x_{nm})) \\ &\quad = Y(X(x_{11}, \ldots, x_{n1}), \ldots, X(x_{1m}, \ldots, x_{nm})) \end{aligned}$$

for arbitrary $n, m \in \mathbb{N}$ is a termal hyperidentity of the variety of commutative groups (semigroups).

(8) The following hyperidentity

$$X(x_1, \ldots, x_{n-1}, X(x_1, \ldots, x_{n-1}, X(x_1, \ldots, x_n))) = X(x_1, \ldots, x_n),$$

for any positive integer n is a termal hyperidentity of the variety of Boolean algebras. For the two-element Boolean algebra, this hyperidentity means the equivalence of the corresponding two switching circuits [259].

If w is every n-ary and $w_1, \dots, w_n$ are every m-ary operations on Q, then a new m-ary operation (superposition) $\mu_n^m(w, w_1, \dots, w_n)$ on Q is made:

$$\mu_n^m(w, w_1, \dots, w_n)(x_1, \dots, x_m) = w(w_1(x_1, \dots, x_m), \dots, w_n(x_1, \dots, x_m)).$$

It is easy to prove the following identities:

$$\mu_n^m\left(\mu_p^n(w, v_1, \dots, v_p), u_1, \dots, u_n\right)$$
$$= \mu_p^m\left(w, \mu_n^m(v_1, u_1, \dots, u_n), \dots, \mu_n^m(v_p, u_1, \dots, u_n)\right), \tag{1.6.1}$$
$$\mu_n^m\left(\delta_n^i, u_1, \dots, u_n\right) = u_i, \tag{1.6.2}$$
$$\mu_n^n\left(v, \delta_n^1, \dots, \delta_n^n\right) = v. \tag{1.6.3}$$

Definition 1.6.6. A clone of operations on a set Q is any set of operations on Q which is closed under superpositions and contains all the identical operations (projections) δ_n^i for all n and $1 \leqslant i \leqslant n$.

For example, the set of all termal operations of algebra $\mathfrak{A} = (Q; \Sigma)$, the set O_pQ and the set $J(Q)$ of all identical operations (projections) are clones of operations on Q. The intersection of clones of operations on Q is also the clone of operations on Q. Hence, the set of all clones of operations on Q, ordered by set inclusion, forms a (complete) lattice, with O_pQ the greatest element and the $J(Q)$ as the least. (Moreover, it is the algebraic lattice, denoted by $Lat(Q)$.) The lattice $Lat(Q)$ is described by E. L. Post [333] for case $|Q| = 2$. Yu. I. Yanov and A. A. Muchnik [421] proved that the lattice $Lat(Q)$ is uncountably infinite if Q is finite, $|Q| > 2$, while a result of I. G. Rosenberg [349] shows that $|Lat(Q)| = 2^{2^{|Q|}}$ if Q is infinite (moreover, in the latter case even the maximal clones are $2^{2^{|Q|}}$ in number).

For arbitrary set $\Sigma \subseteq O_pQ$ there exists a least clone of operations on Q containing Σ, which is called the clone of operations generated by Σ. The clone of termal operations of algebra $\mathfrak{A} = (Q; \Sigma)$, which is the clone generated by Σ, is called the clone of algebra $\mathfrak{A}$, which will be denoted by $Cl(\mathfrak{A})$. For $Cl(\mathfrak{A})$ (see [394]).

In the paper [86] is define the partial many-sorted superposition of operations and the partial many-sorted clone is defined.

The notion of abstract clone or briefly clone generalizes the situation of clone of operations.

Definition 1.6.7. Let Q be an arbitrary set of operation symbols with a condition

$$\mathbb{N} = \{\, |\omega| \mid \omega \in Q\},$$

where $\mathbb{N}$ is the set of all natural numbers. We will denote the subset of all n-ary elements of set Q by Q_n. Such a set Q is called a clone (or an abstract clone) if

(a) for any natural $n \geqslant 1$ and $m \geqslant 1$ the following operations are defined

$$\mu_n^m : Q_n \times \underbrace{Q_m \times \cdots \times Q_m}_{n} \to Q_m$$

and nullary operations, fixing the elements $\delta_n^i \in Q_n$ for any natural $n \geqslant 1$ and for any natural i with the property $1 \leqslant i \leqslant n$;

(b) these operations satisfy the three identities mentioned above: (1.6.1), (1.6.2), (1.6.3).

Thus, all clones form varieties of graduated or heterogeneous algebra, the general theory of which is developed in [41,151,213,328]. The notion of a clone is equivalent to the concept of an algebraic theory in the sense of Lawvere [194,328,412].

Any subvariety of this variety is called a variety of clones. If Q and Q' are arbitrary clones, then the homomorphism of clones $Q \to Q'$ will be the mapping that preserves the arity of every element and is compatible with the operations of clones. Corresponding to homomorphisms of clones is understood the congruence of a clone, the fully invariant congruence of a clone, free clones etc.

The representation of clone Γ in the clone Γ' is a homomorphisms of clone Γ into a clone Γ':

$$\varphi : \Gamma \to \Gamma';$$

The representation φ is faithful if φ is injective.

Proposition 1.6.8 (Cayley's theorem for clones [252,329]). *Every clone has faithful representation in the clone of all operations of some set. Every clone Γ has a faithful representation in the clone of some Γ-algebra.*

1.6.3. *On the Birkhoff type theorems*

Let $\mathcal{L}$ be some nonempty set of T-hyperidentities, and let $\mathfrak{M}_{\mathcal{L}}^T$ be the class of all T-algebras in which every hyperidentity from $\mathcal{L}$ is valid. Recall that a class of T-algebras $\mathfrak{N}$ is called a hypervariety of T-algebras if there exists a system $\mathcal{L}$ of T-hyperidentities with the property

$$\mathfrak{N} = \mathfrak{M}\mathfrak{n}_{\mathcal{L}}^{T}.$$

In this case, $\mathcal{L}$ is called a defining system of hyperidentities for $\mathfrak{N}$.

Let Ω be a signature with an arithmetical type T_Ω, i.e.,

$$T_\Omega = \{|\omega| \mid \omega \in \Omega\},$$

and let $\mathcal{L}$ be some nonempty set of T_Ω-hyperidentities, and let $\mathfrak{N}_{\mathcal{L}}^{\Omega}$ be the class of all Ω-algebras, in which every hyperidentity from $\mathcal{L}$ is valid. It is easy to prove that $\mathfrak{N}_{\mathcal{L}}^{\Omega}$ is the variety for every Ω and $\mathcal{L} \neq \emptyset$. The variety V of Ω-algebras is called hypervariety of Ω-algebras, if there exists a system $\mathcal{L}$ of T_Ω-hyperidentities, such that

$$V = \mathfrak{N}_{\mathcal{L}}^{\Omega}.$$

Let L be some nonempty set of T-hyperidentities, where $T = \mathbb{N}$, and S_L^Ω is the class of all Ω-algebras, for which every hyperidentity from L is termal hyperidentity. It is easy to note that S_L^Ω is a variety for any Ω and $L \neq \emptyset$. The variety V of Ω-algebras is said to be solid (see [138]), if

$$V = S_L^\Omega$$

for some $L \neq \emptyset$. For characterization of all solid varieties of semigroups see [331].

The concept of solid hypervariety is defined analogously. Let $\mathcal{Z}$ be some non-empty set of hyperidentities ($\mathbb{N}$-hyperidentities), and $P_{\mathcal{Z}}^T$ be the class of all T-algebras, in which every hyperidentity from $\mathcal{Z}$ is termally valid. It is easy to note that the class $P_{\mathcal{Z}}^T$ is the hypervariety of T-algebras for every $\mathcal{Z} \neq \emptyset$ and $T \subseteq \mathbb{N}$. The hypervarieties $\mathcal{W}$ of T-algebras are called solid, if

$$\mathcal{W} = P_{\mathcal{Z}}^T$$

for some $\mathcal{Z} \neq \emptyset$ (see [259]).

Hypervarieties, solid varieties and solid hypervarieties are characterized in a natural way by the categorical notions introduced above. Let us formulate these results without proofs.

Theorem 1.6.9. *A class of T-algebras is a hypervariety of T-algebras if and only if it is hereditary, homomorphicaly and multiplicatively closed.*

Theorem 1.6.10. *For any variety V of Ω-algebras the following conditions are equivalent:*

(1) *The variety V is a hypervariety of Ω-algebras.*
(2) *Every identity of V is a hyperidentity for V.*
(3) *The variety V along with any algebra $\mathfrak{A} \in V$ contains any T_Ω-reduct of $\mathfrak{A}$.*

Theorem 1.6.11. *For any variety V of Ω-algebras the following conditions are equivalent:*

(4) *The variety V is solid.*
(5) *Every identity of V is a termal hyperidentity for V.*
(6) *The variety V along with any algebra $\mathfrak{A} \in V$ contains any Ω-reduct of termal algebra $\mathcal{F}(\mathfrak{A})$ (see [90,138]).*

Theorem 1.6.12. *For every hypervariety $\mathcal{W}$ of T-algebras, the following conditions are equivalent:*

(7) *The hypervariety $\mathcal{W}$ of T-algebras is solid.*
(8) *Every hyperidentity of $\mathcal{W}$ is termal hyperidentity for $\mathcal{W}$.*
(9) *The hypervariety $\mathcal{W}$ of T-algebras along with any algebra $\mathfrak{A} \in \mathcal{W}$ contains any T-reduct of termal algebra $\mathcal{F}(\mathfrak{A})$.*

Naturally here also arises the following notion of a quasi-solid variety of Ω-algebras.

The variety of Ω-algebras is called quasi-solid, if its any hyperidentity is its termal hyperidentity.

The Birkhoff-type theorems of completeness for $\forall\exists(\forall)$-identities, termal hyperidentities and hybrid identities were considered in [93,244,249, 252,363].

Chapter 2

Hyperidentities of Lattices

2.1. Examples

Example 2.1.1. In any lattice $Q(+,\cdot)$, the trivial hyperidentities of idempotence, commutativity and associativity are valid:

$$X(x,x) = x,$$
$$X(x,y) = X(y,x),$$
$$X(x,X(y,z)) = X(X(x,y),z).$$

In a nontrivial lattice $Q(+,\cdot)$, the hyperidentity

$$X(x,Y(x,y)) = x$$

is not valid, but in an arbitrary lattice $Q(+,\cdot)$ the following nontrivial hyperidentities are valid:

$$Y(y,x) = Y(y,X(x,Y(x,y))),$$
$$Y(X(x,z),y) = Y(X(x,z),X(y,Y(y,z))),$$
$$X(z,X(x,Y(x,y))) = X(x,X(z,Y(z,y))).$$

Example 2.1.2. A lattice is modular if and only if the following hyperidentity is valid in it

$$X(Y(x,X(y,z)),Y(y,z)) = Y(X(x,Y(y,z)),X(y,z)). \tag{2.1.1}$$

This hyperidentity is called hyperidentity of modularity.

A lattice is distributive if and only if the following nontrivial hyperidentities are valid in it:

$$X(x, Y(y, z)) = Y(X(x, y), X(x, z)), \tag{d_1}$$

$$X(Y(x, y), z) = Y(X(x, z), X(y, z)). \tag{d_2}$$

These hyperidentities are called hyperidentities of distributivity.

Example 2.1.3. In any convexor (or barycentric algebra [187,345,376]) defined as a binary algebra with one hyperidentity of idempotence and two $\forall\exists(\forall)$-identities of commutativity and associativity, the nontrivial distributivity hyperidentities (d_1), (d_2) are valid.[a]

Example 2.1.4. In any Boolean lattice (algebra) $Q(+, \cdot, ')$, the following nontrivial hyperidentity is valid:

$$F(X(F(Y(x, y)), z)) = Y(F(X(F(x), z)), F(X(F(y), z)))$$

or simply

$$X(Y(x, y)', z)' = Y(X(x', z)', X(y', z)').$$

A hyperidentity $w_1 = w_2$ is said to be homogeneous (in the sense of Mal'tsev) if the same object variables occur in the hyperterms w_1 and w_2. Obviously, any hyperidentity that is valid in a nontrivial lattice is homogenous. Therefore, any hyperidentity of the variety of lattices is also homogeneous.

2.2. Preliminary results

Following Birkhoff [38], a system Ψ of subsets of the set $\{1, \ldots, n\}$ is said to be *L-closed* over $\{1, \ldots, n\}$ if, together with any $s \in \Phi$, it also contains an arbitrary $S' \subseteq \{1, \ldots, n\}$ with the property $S \subseteq S'$. (G. Birkhoff called such systems *J-closed.*)

We consider T-hyperterms for $T = \{2\}$ with at most two binary functional variables F and G. Let

$$S = \{m_1, \ldots, m_t\} \subseteq \{1, \ldots, n\}.$$

[a] Indeed,using this fact one obtains a characterization of the lattices of ideals of convexors [252] (see a problem posed by L. A. Skornyakov in Chapter 22).

We denote by $G_{i \in S} x_i$ a hyperterm in the object variables $x_{m_1}, \ldots, x_{m_t}$ and the binary functional variable G. In an analogous way, one understands the notation:

$$\underset{S_i \in \Phi}{F} \left\{ \underset{k \in S_i}{G} x_k \right\},$$

where $\Phi = \{S_1, \ldots, S_\ell\}$, $S_1, \ldots, S_\ell \subseteq \{1, \ldots, n\}$. If $\Phi = \{S\}$ consists of one element, then

$$\underset{S_i \in \Phi}{F} \left\{ \underset{k \in S_i}{G} x_k \right\} = \underset{i \in S}{G} x_i,$$

and if $S = \{m\}$ consists of one element, then $G_{i \in S} x_i = x_m$.

Definition 2.2.1. A hyperterm of the form

$$\underset{S_i \in \Phi}{F} \left\{ \underset{k \in S_i}{G} x_k \right\}$$

of binary functional variables F, G and object variables $x_1, \ldots, x_n$ is said to be L-canonical hyperterm (form, word) and is denoted by $f_\Phi(F, G, x_1, \ldots, x_n)$ if Φ is an L-closed system over $\{1, \ldots, n\}$ (the case when Φ consists of one element is not excluded).

Two L-canonical words $f_\Phi(F, G, x_1, \ldots, x_n)$ and $f_{\mathcal{D}}(F, G, x_1, \ldots, x_n)$ are said to be *equal* if $\Phi = \mathcal{D}$.

Proposition 2.2.2. *For any hyperterm* $w = F_{S_i \in \mathcal{D}} \{G_{k \in S_i} x_k\}$ *there exists an* L*-canonical hyperterm* $\widetilde{w} = f_\Phi(F, G, x_1, \ldots, x_n)$ *with the same object variables such that the hyperidentity*

$$w = \widetilde{w}$$

holds in the variety of lattices, and with this hypothesis the L*-canonical hyperterm* $\widetilde{w}$ *is uniquely defined and is called the* L*-canonical form of* w*;* $\widetilde{w}$ *is obtained from* w *after a finite number of transitions realized by a finite number of hyperidentities of the variety of lattices that do not depend on* w*.*

Proof. Using the trivial hyperidentities of associativity, commutativity and idempotence of the variety of lattices (see Example 2.1.1), we eliminate repetitions of the object variables x_i, in the subwords $G_{k \in S_i} x_k$. In an analogous fashion, the repetitions of the subwords $G_{k \in S_i} x_k$ in the word w are eliminated.

To reduce the hyperterm

$$w = \underset{S_i \in \mathcal{D}}{F} \left\{ \underset{k \in S_i}{G} x_k \right\}$$

to L-canonical form, we use the additional hyperidentity

$$F(G(y,z), F(x, G(x,y))) = F(G(y,z), x) \tag{2.2.1}$$

and its immediate consequence (for $y = z$)

$$F(y, F(x, G(x,y))) = F(y,x), \tag{2.2.2}$$

which are valid in the variety of all lattices.

Let $S \in \mathcal{D}$, $S' \subseteq \{1, \ldots, n\}$ and $S \subseteq S'$. If $G_{k \in S'} x_k$ does not occur in the expression

$$\underset{S_i \in \mathcal{D}}{F} \left\{ \underset{k \in S_i}{G} x_k \right\},$$

then this subword can be obtained in the following way. We assume that $S' \setminus S = \{m, k, \ldots\}$; therefore the subword $G_{k \in S'} x_k$ differs from $G_{k \in S} x_k$ by the presence of the object variables $x_m, x_k, \ldots$. Since x_m occurs in the expression $F_{S_i \in \mathcal{D}} \{G_{k \in S_i} x_k\}$, by applying the above hyperidentities we will find

$$F\big(G(x_m, z), \underset{k \in S}{G} x_k\big) = F\big(G(x_m, z), F\big(\underset{k \in S}{G} x_k, G\big(\underset{k \in S}{G} x_k, x_m\big)\big)\big)$$

or

$$F\big(x_m, \underset{k \in S}{G} x_k\big) = F\big(x_m, F\big(\underset{k \in S}{G} x_k, G\big(\underset{k \in S}{G} x_k, x_m\big)\big)\big)$$

depending on the two possible positions of x_m. In both cases the new subword $G\big(G_{k \in S} x_k, x_m\big)$ arises, and it does not change the previous position of x_m. Hence, after a finite number of analogous steps we obtain the subword $G_{k \in S'} x_k$.

Thus the first part of the proposition has been proved. We shall now prove the uniqueness. Let

$$w = f_{\Phi_1}(F, G, x_1, \ldots, x_n)$$

and

$$w = f_{\Phi_2}(F, G, x_1, \ldots, x_n),$$

where $\Phi_1 \neq \Phi_2$. We assume that there exists a subset $\{i_1, \ldots, i_r\} \subseteq \{1, \ldots, n\}$ such that $\{i_1, \ldots, i_r\} \in \Phi_1$ but $\{i_1, \ldots, i_r\} \notin \Phi_2$. Since Φ_2 is

L-closed, we have the implication

$$\{\ell_1, \dots, \ell_p\} \subseteq \{i_1, \dots, i_r\} \Longrightarrow \{\ell_1, \dots, \ell_p\} \notin \Phi_2,$$

that is, for any $S \in \Phi_2$ there exists an $i \in S$ such that $i \notin \{i_1, \dots, i_r\}$. Now, considering the two-element lattice $= \{0, 1\}$, we remark that the following hyperidentity is not valid in this lattice:

$$f_{\Phi_1}(F, G, x_1, \dots, x_n) = f_{\Phi_2}(F, G, x_1, \dots, x_n).$$

For this we give the following values to all the functional and object variables: $F = (+)$, $G = (\cdot)$, $x_i = 1$ if $i \in \{i_1, \dots, i_n\}$, and $x_i = 0$ if $i \notin \{i_1, \dots, i_r\}$. Then

$$\underset{k \in \{i_1, \dots, i_r\}}{G} x_k = 1, \quad f_{\Phi_1}(F, G, x_1, \dots, x_n) = 1 + \cdots = 1.$$

On the other hand, by what was said above, for any $S \in \Phi_2$ there exists an $i \in S$ such that $x_i = 0$. Therefore

$$f_{\Phi_2}(F, G, x_1, \dots, x_n) = 0 + \cdots + 0 = 0.$$

Thus, in the lattice $= \{0, 1\}$ the following hyperidentity is not valid:

$$f_{\Phi_1}(F, G, x_1, \dots, x_n) = f_{\Phi_2}(F, G, x_1, \dots, x_n).$$

However, by hypothesis,

$$w = f_{\Phi_1}(F, G, x_1, \dots, x_n), \qquad w = f_{\Phi_2}(F, G, x_1, \dots, x_n)$$

and, hence, the following hyperidentity must be valid in the variety of lattices:

$$f_{\Phi_1}(F, G, x_1, \dots, x_n) = f_{\Phi_2}(F, G, x_1, \dots, x_n),$$

which is a Contradiction. □

Corollary 2.2.3. *In the variety of all lattices, the hyperidentity*

$$\underset{S_i \in \Phi_1}{F} \left\{ \underset{k \in S_i}{G} x_k \right\} = \underset{S_j \in \Phi_2}{F} \left\{ \underset{k \in S_j}{G} x_k \right\}$$

is valid if and only if the L-canonical forms of the hyperterms $F_{S_i \in \Phi_1}$ $\{G_{k \in S_i} x_k\}$ and $F_{S_j \in \Phi_2} \{G_{k \in S_j} x_k\}$ coincide.

Corollary 2.2.4. *In the variety of all lattices, the hyperidentity of the form*

$$\underset{S_i\in\Phi_1}{F}\left\{\underset{k\in S_i}{G} x_k\right\} = \underset{S_j\in\Phi_2}{F}\left\{\underset{k\in S_j}{G} x_k\right\}$$

is a consequence of the following four hyperidentities (used in the proof of Proposition 2.2.2):

$$X(x,x) = x,$$
$$X(x,y) = X(y,x),$$
$$X(x,X(y,z)) = X(X(x,y),z),$$
$$X(Y(x,y),X(z,Y(z,x))) = X(Y(x,y),z).$$

As an application of L-canonical forms we can prove that the hyperidentity

$$X(z,X(x,Y(x,y))) = X(x,X(z,Y(z,y)))$$

holds in any lattice. We shall reduce the left- and right-hand sides of this hyperidentity to L-canonical form and show that these L-canonical forms coincide. We start with the left-hand side:

$$\begin{aligned}
F(z,F(x,G(x,y))) &= F(G(x,y),F(z,x))\\
&\overset{(2.2.2)}{=} F(G(x,y),F(z,F(x,G(x,z))))\\
&= F(G(x,z),F(x,F(G(x,y),z)))\\
&\overset{(2.2.1)}{=} F(G(x,z),F(x,F(G(x,y),F(z,G(y,z)))))\\
&= F(G(x,z),F(G(x,y),F(z,F(x,G(y,z)))))\\
&\overset{(2.2.2)}{=} F(G(x,z),F(G(x,y),F(z,F(x,F(G(y,z),\\
&\quad \times G(x,G(y,z)))))))\\
&= F(G(G(z,y),x),F(G(x,y),F(G(x,z),F(G(y,z),\\
&\quad \times F(x,z))))).
\end{aligned}$$

Now we reduce the right-hand side of this hyperidentity to L-canonical form:

$$\begin{aligned}
F(x,F(z,G(z,y))) &= F(G(z,y),F(x,z))\\
&\overset{(2.2.2)}{=} F(G(z,y),F(x,F(z,G(x,z))))\\
&= F(G(x,z),F(z,F(G(y,z),x)))
\end{aligned}$$

$$\begin{aligned}
&\overset{(2.2.1)}{=} F(G(x,z),F(z,F(G(y,z),F(x,G(x,y)))))\\
&= F(G(x,z),F(G(x,y),F(z,F(x,G(y,z)))))\\
&\overset{(2.2.2)}{=} F(G(x,z),F(G(x,y),F(z,F(G(y,z),\\
&\quad \times F(x,G(x,G(y,z)))))))\\
&= F(G(G(z,y),x),F(G(x,y),F(G(x,z),F(G(y,z),\\
&\quad \times F(x,z))))).
\end{aligned}$$

Example 2.2.5. The following hyperidentity is not valid in the variety of lattices:

$$F(y,F(x,G(x,y))) = G(y,x).$$

Indeed, both sides of this hyperidentity, since they are L-canonical forms, do not coincide in the sense of equality of L-canonical forms. It remains to cite Corollary 2.2.3.

Theorem 2.2.6. *In the variety of distributive lattices, any hyperidentity of functional rank* $\leqslant 2$ *is a consequence of the following four hyperidentities*:

$$\begin{aligned}
X(x,x) &= x,\\
X(x,y) &= X(y,x),\\
X(x,X(y,z)) &= X(X(x,y),z),\\
X(x,Y(y,z)) &= Y(X(x,y),X(x,z)).
\end{aligned}$$

Proof. For any hyperterm w with at most two binary functional variables F, G there exists an L-canonical form $f_\Phi(F,G,x_1,\ldots,x_n)$ with the same object variables such that the following hyperidentity is valid in the variety of distributive lattices:

$$w = f_\Phi(F,G,x_1,\ldots,x_n),$$

and with this condition the L-canonical form is defined uniquely. In fact, using the hyperidentity of distributivity, the hyperterm w first reduces to the form

$$\underset{S_i\in\mathcal{D}}{F}\left\{\underset{k\in S_i}{G}\ x_k\right\}.$$

Then the proof of Proposition 2.2.2 is repeated. It remains to note that the hyperidentity (2.2.1) is a consequence of the above four hyperidentities (idempotence, commutativity, associativity and left distributivity). First, it is obvious that the hyperidentity of right distributivity

$$X(Y(y,z),x) = Y(X(y,x),X(z,x))$$

is a consequence of the hyperidentities of commutativity and left distributivity. In addition we have

$$\begin{aligned} X(x,Y(y,z)) &= Y(X(x,y),X(x,z)) \\ &= X(Y(x,X(x,z)),Y(y,X(x,z))) \\ &= X(X(Y(x,x),Y(x,z)),X(Y(y,x),Y(y,z))) \\ &= X(X(x,Y(x,z)),X(Y(y,x),Y(y,z))), \end{aligned}$$

that is,

$$X(x,Y(y,z)) = X(X(x,Y(x,z)),X(Y(y,x),Y(y,z))). \tag{2.2.3}$$

Hence,

$$\begin{aligned} X(Y(y,z),X(x,Y(x,y))) &= X(X(x,Y(y,z)),Y(x,y)) \\ &\overset{(2.2.3)}{=} X(X(X(x,Y(x,z)),X(Y(y,x), \\ &\quad \times Y(y,z))),Y(x,y)) \\ &= X(X(x,Y(x,z)),X(Y(y,x),Y(y,z))) \\ &= X(x,Y(y,z)) = X(Y(y,z),x). \end{aligned}$$

□

Remark 2.2.7. The algebra $Q(+,\cdot)$, with two binary operations and with the hyperidentities from Corollary 2.2.4, is not a lattice in general. In fact, the above four hyperidentities are valid in any nontrivial lattice $L(+,\cdot)$. We consider the product $Q = L \times L$ and the product $\Sigma \times \Sigma$, where $\Sigma = \{+,\cdot\}$. As a result we obtain the superproduct with four binary operations: $A = (+,+)$, $B = (+,\cdot)$, $C = (\cdot,+)$, $D = (\cdot,\cdot)$, and those four hyperidentities are valid in the superproduct $Q(A,B,C,D)$. However, a direct verification shows that the identities of absorption are not valid in the algebra $Q(A,B)$, and therefore $Q(A,B)$ is not a lattice, although the hyperidentities from Corollary 2.2.4 are valid in it.

Remark 2.2.8. The algebra $Q(+,\cdot)$ with two binary operations and with the hyperidentities from Theorem 2.2.6 is not a distributive lattice in general.

Let $\mathcal{L}$ be some (nonempty) set of T-hyperidentities, and let $\mathfrak{M}^T_{\mathcal{L}}$ be the class of all T-algebras in which every hyperidentity from $\mathcal{L}$ is valid.

A class of T-algebras $\mathfrak{N}$ is called a *hypervariety* (of T-algebras) if there exists a system of T-hyperidentities $\mathcal{L}$ with the property

$$\mathfrak{N} = \mathfrak{M}^T_{\mathcal{L}}.$$

In this case $\mathcal{L}$ is called a *defining system* of hyperidentities for $\mathfrak{N}$.

For example, the variety of semigroups is a hypervariety of T-algebras, where $T = \{2\}$ and $\mathcal{L}$ is the set consisting of the two hyperidentities

$$X(x, X(y, z)) = X(X(x, y), z),$$
$$X(x, y) = Y(x, y).$$

Two systems of T-hyperidentities $\mathcal{L}_1$, $\mathcal{L}_2$ are said to be equivalent if

$$\mathfrak{M}^T_{\mathcal{L}_1} = \mathfrak{M}^T_{\mathcal{L}_2},$$

that is, if every T-hyperidentity from $\mathcal{L}_2$ (respectively, from $\mathcal{L}_1$) is a consequence of $\mathcal{L}_1$ (respectively, of $\mathcal{L}_2$).

We say that a system of T-hyperidentities $\mathcal{L}$ has a finite basis (or is finitely based) if it is equivalent to a finite system of T-hyperidentities $\mathcal{L}_0$. Every class of T-algebras $\mathfrak{M}$ corresponds to the class $\Lambda^T_{\mathfrak{M}}$ of all T-hyperidentities, each of which is valid in all the algebras of $\mathfrak{M}$. The pair of maps

$$\mathcal{L} \to \mathfrak{M}^T_{\mathcal{L}} \quad \text{and} \quad \mathfrak{M} \to \Lambda^T_{\mathfrak{M}}$$

forms a Galois correspondence.

An intersection of hypervarieties of T-algebras is a hypervariety of T-algebras, namely:

$$\bigcap_{i \in I} \mathfrak{M}^T_{\mathcal{L}_i} = \mathfrak{M}^T_{\bigcup_{i \in I} \mathcal{L}_i}$$

so that for any class of T-algebras $\mathfrak{M}$ there exists a smallest hypervariety (relative to set-theoretic inclusion) of T-algebras $\mathfrak{M}^* \supseteq \mathfrak{M}$ called the *hypervariety of algebras* generated by $\mathfrak{M}$. It is clear that $\mathfrak{M}^*$ is defined by the system of T-hyperidentities $\Lambda^T_{\mathfrak{M}}$, so that the characterization of the hypervariety $\mathfrak{M}^*$ is equivalent to the characterization of all the hyperidentities of the class of T-algebras $\mathfrak{M}$.

The arithmetic type of a termal algebra is always equal to the set $\mathbb{N}$ of all natural numbers, so the termal algebras $\mathcal{F}(\mathfrak{A}_1)$ and $\mathcal{F}(\mathfrak{A}_2)$ have the same arithmetic type for any $\mathfrak{A}_1$, $\mathfrak{A}_2$.

Recall that a hyperidentity $w_1 = w_2$ is a *termal hyperidentity (or t-hyperidentity) of the algebra* $\mathfrak{A}$ if it is valid in the termal algebra $\mathcal{F}(\mathfrak{A})$.

Let V be a variety of algebras. A hyperidentity $w_1 = w_2$ is called a *termal hyperidentity of* V if it is a termal hyperidentity for any algebra $\mathfrak{A} \in V$. Termal hyperidentities of a class of varieties V_i, $i \in I$, are defined analogously.

Example 2.2.9. The hyperidentities

$$
\begin{aligned}
X(x,x) &= x, \\
X(x, X(y,z)) &= X(X(x,y),z), \\
X(X(x,y), X(u,v)) &= X(X(x,u), X(y,v)), \\
X(x,x,x) &= x, \\
X(x,x,x,x) &= x, \\
&\vdots
\end{aligned}
$$

are termal hyperidentities of the variety of lattices (semilattices).

We say that a hyperidentity $w_1 = w_2$ is termally valid in an algebra $\mathfrak{A}$ (or a variety V) if $w_1 = w_2$ is a termal hyperidentity of $\mathfrak{A}$ (respectively, of V).

Let $\mathcal{L}$ be some set of hyperidentities. We denote by $\mathcal{K}_{\mathcal{L}}$ the class of all varieties of algebras in each of which any hyperidentity from $\mathcal{L}$ is termally valid. A class $\mathcal{K}$ of varieties of algebras is called a *variety of varieties* if there exists a system of hyperidentities $\mathcal{L}$ such that

$$\mathcal{K} = \mathcal{K}_{\mathcal{L}}.$$

In this case, we say that the variety of varieties $\mathcal{K}$ is defined by the system of hyperidentities $\mathcal{L}$.

An intersection of varieties of varieties is again a variety of varieties, namely:

$$\bigcap_{i\in I} \mathcal{K}_{\mathcal{L}_i} = \mathcal{K}_{\bigcup_{i\in I} \mathcal{L}_i},$$

so that for any class $\mathcal{K}$ of varieties of algebras there exists a smallest (relative to set-theoretic inclusion) variety of varieties $\mathcal{K}^* \supseteq \mathcal{K}$ called the *variety of varieties generated by* $\mathcal{K}$. Obviously $\mathcal{K}^*$ is defined by the system of all termal hyperidentities of the class of varieties $\mathcal{K}$, so that the characterization of $\mathcal{K}^*$

reduces to the characterization of all the termal hyperidentities of the class of varieties $\mathcal{K}$.

Theorem 2.2.10 ([402]). *An abstract class $\mathcal{K}$ of varieties of algebras is a variety of varieties if and only if it is closed under the formation of subvarieties, reducts and products of varieties.*

Theorem 2.2.11 ([309]). *Any nontrivial variety of lattices and the variety of all semilattices do not have a finite basis of termal hyperidentities.*

Theorem 2.2.12 ([31]). *The variety of all groups (metabelian groups, monoids) does not have a finite basis of termal hyperidentities.*

Theorem 2.2.13 ([253]). *The variety of Boolean algebras does not have a finite basis of termal hyperidentities.*

The following assertions are proved using canonical forms (see the proofs of Proposition 2.2.2 and Theorem 2.2.6).

Proposition 2.2.14 ([255]). *Any termal hyperidentity with two binary functional variables of the variety of distributive lattices is a termal consequence of the following hyperidentities:*

$$\begin{aligned}
X(x,x) &= x,\\
X(x,X(y,z)) &= X(X(x,y),z),\\
X(x,Y(y,z)) &= Y(X(x,y),X(x,z)),\\
X(u,X(x,X(y,v))) &= X(u,X(y,X(x,v))),\\
X(Y(Y(y,x),z),Y(u,v)) &= X(Y(Y(y,x),z),X(Y(Y(u,x),v),Y(u,v))),\\
X(Y(Y(y,x),z),Y(u,x)) &= X(Y(Y(y,x),z),X(Y(Y(y,z),x),Y(u,x))).
\end{aligned}$$

Proposition 2.2.15. *Any unary termal hyperidentity (i.e., termal hyperidentity with unary functional variables) of the variety of Boolean algebras is a termal consequence of the following hyperidentities:*

$$X(Y(X(X(x)))) = X(Y(x)), \tag{2.2.4}$$

$$X(Y(X(Y(x)))) = X(Y(Y(X(x)))). \tag{2.2.5}$$

Proof. First we note that 0, 1, x, x' are all unary term operations of every Boolean algebra.

We can shortly denote any unary hyperterm (i.e., a hyperterm with unary functional variables)

$$F(\cdots(G(\cdots(H(x))\cdots))\cdots)$$

by

$$(F\bullet\cdots\bullet G\bullet\cdots\bullet H)x.$$

Every unary hyperterm contains only one object variable and if the hyperidentity $\omega_1 = \omega_2$ termally satisfies the variety of Boolean algebras, then $[\omega_1] = [\omega_2] = \{x\}$, where $[\omega]$ is the set of the object variables of ω. So we can shortly denote any unary hyperterm without an object variable:

$$F\bullet\cdots\bullet G\bullet\cdots\bullet H;$$

The unary hyperterm ω is called k-normal form on functional variables $(F_1,\dots,F_k)$ if

$$\omega = F_1\bullet\omega_1\bullet F_2\bullet\omega_2\bullet\cdots\bullet F_k\bullet\omega_k,$$

where $\omega_i = F_{i_1}\bullet\cdots\bullet F_{i_m}$, $1\leqslant i_1<\cdots<i\leqslant i_m$, $i=1,\dots,k$ or $\omega_i=\emptyset$. If functional variables $F_1,\dots,F_k$ are fixed, this k-normal form is presented as

$$\omega = (\omega_1,\dots,\omega_k).$$

The following assertion is proved by induction: Let unary hyperidentity $u = v$ be termally satisfied in the variety of Boolean algebras. If k-normal forms of u and v are $u = (u_1,\dots,u_k)$ and $v = (v_1,\dots,v_k)$, then $u_1 \equiv v_1,\dots,$ $u_k \equiv v_k$, i.e., u and v are graphically coincide.

In the next step we prove that one can reduce any unary hyperterm to the k-normal form (cf. [253]) using the following unary termal hyperidentities of the variety of Boolean algebras: (2.2.4), (2.2.5),

$$F = F\bullet F\bullet F, \tag{2.2.6}$$

$$F\bullet H\bullet G\bullet K\bullet F\bullet G = F\bullet H\bullet G\bullet K\bullet G\bullet F. \tag{2.2.7}$$

Besides, if we substitute x for Y in (2.2.4), we obtain hyperidentity (2.2.6), i.e., (2.2.4) $\Rightarrow_t$ (2.2.6), but the hyperidentity (2.2.7) follows from (2.2.4) and (2.2.5):

$$\begin{aligned}F\bullet H\bullet G\bullet K\bullet F\bullet G &\overset{(2.2.4)}{=} F\bullet H\bullet F\bullet F\bullet G\bullet K\bullet F\bullet G\\ &\overset{(2.2.4)}{=} F\bullet H\bullet F\bullet F\bullet G\bullet K\bullet F\bullet G\bullet F\bullet G\bullet F\bullet G\end{aligned}$$

$$\overset{(2.2.5)}{=} F \bullet H \bullet F \bullet F \bullet G \bullet K \bullet F \bullet G \bullet F \bullet G \bullet G \bullet F$$

$$\overset{(2.2.4)}{=} F \bullet H \bullet G \bullet K \bullet G \bullet F.$$

□

2.3. Hyperidentities of the varieties of distributive lattices, modular lattices, lattices

Theorem 2.3.1 ([256]). *Any hyperidentity of the variety of distributive lattices is a consequence of the following four hyperidentities:*

$$X(x, x) = x, \tag{2.3.1}$$

$$X(x, y) = X(y, x), \tag{2.3.2}$$

$$X(x, X(y, z)) = X(X(x, y), z), \tag{2.3.3}$$

$$X(x, Y(y, z)) = Y(X(x, y), X(x, z)). \tag{2.3.4}$$

One must prove that if the hyperidentities (2.3.1)–(2.3.4) are valid in a T-algebra $\mathfrak{A}$, where $T = \{2\}$, then any hyperidentity of the variety of distributive lattices is also valid in it.

Definition 2.3.2. A binary algebra $\mathfrak{A} = \langle Q; \Sigma \rangle$ is called

(a) a quasilattice if it satisfies hyperidentities (2.3.1)–(2.3.4);
(b) an α-quasilattice if it is a quasilattice and there exists a natural number $n \geqslant 2$ such that the following second-order formula is valid in $\mathfrak{A} = \langle Q; \Sigma \rangle$:

$$\exists X_1, \ldots, X_n \forall x, y (X_1(x, \ldots, X_{n-1}(x, X_n(x, y)) \ldots) = x)$$

(the values of the functional variables $X_1, \ldots, X_n$ are assumed to be from Σ).

If the set Σ is finite and $|\Sigma| = n$, then such quasilattices are studied in [322] under the name of *distributive n-quasilattices*. In case $n = 2$ see also [308,324].

Example 2.3.3. A superproduct of distributive lattices is an α-quasilattice.

The α-quasilattices turn out to be subalgebras of superproducts of distributive lattices. For this, we start characterizations of the subdirectly irreducible quasilattices.

Lemma 2.3.4. *If a quasilattice* $\mathfrak{A} = \langle Q; \Sigma \rangle$ *is subdirectly irreducible, then the cardinality* $|\Sigma|$ *of* Σ *is at most* 2.

Proof. In case $|\Sigma| \geqslant 3$ a quasilattice $\mathfrak{A}$ is not subdirectly irreducible. Indeed, let A_1, A_2, A_3 be distinct binary operations from Σ, and let

$$f_{i,j}(x, y) = A_i(x, A_j(x, y)).$$

We define relations $\Theta_{i,j}$ on Q in the following way:

$$x\Theta_{i,j}y \Longleftrightarrow f_{i,j}(x, y) = x, f_{i,j}(y, x) = y,$$

where $i, j = 1, 2, 3$ and $i < j$. It is obvious that the relations $\Theta_{i,j}$ are reflexive and symmetric. That they are transitive follows from the hyperidentity

$$X(x, Y(x, X(y, Y(y, z)))) = X(x, Y(x, X(z, Y(z, y)))), \tag{2.3.5}$$

which is a consequence of the hyperidentities (2.3.1)–(2.3.4) and, hence, is valid in any quasilattice. In fact, the hyperidentity (2.2.1) is a consequence of the hyperidentities (2.3.1)–(2.3.4) (see the proof of Theorem 2.2.6). Writing the hyperidentity (2.2.1) in the form (hyperidentities (2.3.1)–(2.3.3) are applied without mention)

$$X(x, X(Y(x, y), Y(y, z))) = X(x, Y(y, z))$$

and making the change of variables $y \to Y(x, y)$, we obtain the hyperidentity

$$X(x, X(Y(x, y), Y(Y(x, y), z))) = X(x, Y(Y(x, y), z)). \tag{2.3.6}$$

Hence, in any quasilattice, the following hyperidentity is also valid:

$$X(x, X(Y(x, z), Y(Y(x, y), z))) = X(x, Y(Y(x, y), z)).$$

The hyperidentity (2.3.5) follows from the last two hyperidentities if we use (2.3.4).

In what follows, we shall need the following hyperidentity:

$$X(Y(x, y), z) = X(X(Y(x, y), z), Y(Y(x, y), z)), \tag{2.3.7}$$

which is an obvious consequence of the hyperidentity (2.2.2).

We shall prove that $\Theta_{i,j} = (\Theta_{i,j}, \tilde{0})$ is a congruence of the quasilattice $\mathfrak{A} = \langle Q; \Sigma \rangle$, that is, if $x\Theta_{i,j}y$ and $u\Theta_{i,j}v$, then $A(x, u)\Theta_{i,j}A(y, v)$ for any operation $A \in \Sigma$. It suffices to show that $A(x,u)\Theta_{i,j}A(y,u)$, where $x\Theta_{i,j}y$ and $x, y, u \in Q$. We make use of the hyperidentity

$$X(Y(X(x,y),z),Y(y,z)) = Y(X(x,y),z), \tag{2.3.8}$$

which is also a consequence of the hyperidentities (2.3.1)–(2.3.4):

$$\begin{aligned} X(Y(X(x,y),z),Y(y,z)) &= X(X(Y(x,z),Y(y,z)),Y(y,z)) \\ &= X(Y(x,z),X(Y(y,z),Y(y,z))) \\ &= X(Y(x,z),Y(y,z)) = Y(X(x,y),z). \end{aligned}$$

Assume that $x\Theta_{i,j}y$, that is, that $A_i(x, A_j(x,y)) = x$ and $A_i(y, A_j(y,x)) = y$, where $A_i, A_j \in \Sigma$. Then for any operation $A \in \Sigma$ we have

$$\begin{aligned} A(x,u) &= A(A_i(x,A_j(x,y)),u) \\ &\overset{(2.3.8)}{=} A_i(A(A_i(x,A_j(x,)),u),A(A_j(x,y),u)) \\ &\overset{(2.3.4)}{=} A_i(A(x,u),A_j(A(x,u),A(y,u))). \end{aligned}$$

In an analogous way, we find the equality

$$A(y,u) = A_i(A(y,u),A_j(A(y,u),A(x,u))).$$

We notice that the proof of the congruence property of the relation $\Theta_{i,j}$ can also be realized by weaker methods, namely, instead of the hyperidentities (2.3.8), (2.3.4) (or the dual hyperidentity of right distributivity) we could use the hyperidentities (2.3.6), (2.3.7), and the hyperidentity

$$X(X(Y(X(x,y),z),Y(x,z)),Y(y,z)) = Y(X(x,y),z), \tag{2.3.9}$$

which is a consequence of (2.3.8). Indeed, if $x\Theta_{i,j}y$, then for any $u \in Q$ and $A \in \Sigma$, and setting $a = A(x,u)$, $b = A(y,u)$, $c = A(A_j(x,y),u)$, $t = A_i(x, A_j(x,y))$, $s = A(t,u)$, we find

$$\begin{aligned} A(x,u) &= A(A_i(x,A_j(x,y)),u) \\ &\overset{(2.3.9)}{=} A_i(A_i(A(A_i(x,A_j(x,y)),u),A(x,u)),A(A_j(x,y),u)) \\ &\overset{(2.3.1)}{=} A_i(A_i(s,a),A_i(A(A_j(x,y),u),c)) \end{aligned}$$

$$\begin{aligned}
&\overset{(2.3.9)}{=} A_i(A_i(s,a), A_i(A_j(A_j(A(A_j(x,),u), A(x,u)), A(y,u)), c))\\
&= A_i(A_i(s,a), A_i(A_j(A_j(,a),b),c))\\
&= A_i(s, A_i(A_i(a, A_j(A_j(a,b),)),c))\\
&\overset{(2.3.6)}{=} A_i(s, A_i(A_i(a, A_i(A_j(a,b), A_j(A_j(a,b),c))),c))\\
&= A_i(s, A_i(a, A_i(A_i(A_j(a,b),c), A_j(A_j(a,b),c))))\\
&\overset{(2.3.7)}{=} A_i(s, A_i(a, A_i(A_j(a,b),c))) = A_i(A_i(A_i(s,a),c), A_j(a,b))\\
&\overset{(2.3.9)}{=} A_i(s, A_j(a,b)) = A_i(A(x,u), A_j(A(x,u), A(y,u))).
\end{aligned}$$

The second compatibility equality is obtained in the analogous way.

We consider the quotient algebras $\mathfrak{A}/\Theta_{i,j}$ modulo the congruences $\Theta_{1,2}$, $\Theta_{1,3}$ and $\Theta_{2,3}$. The maps

$$\varphi_0 : x \longrightarrow ([x]_{1,2}, [x]_{1,3}, [x]_{2,3}), \quad \tilde{\psi}_0 : A \longrightarrow ([A]_{1,2}, [A]_{1,3}, [A]_{2,3})$$

define a monomorphism of the quasilattice $\mathfrak{A}$ into the direct product $\mathfrak{A}/\Theta_{1,2}\times \mathfrak{A}/\Theta_{1,3}\times\mathfrak{A}/\Theta_{2,3}$. It is obvious that this pair is a homomorphism, and the fact that φ_0, and hence $\tilde{\psi}_0$, is injective follows from the following implication: if

$$f_{1,2}(x,y) = f_{1,3}(x,y) = f_{2,3}(x,y) = x,$$

then

$$x = A_1(A_2(x,y), A_3(x,y)).$$

The last implication follows from the following hyperidentity (for $X = A_1$, $Y = A_2$, $Z = A_3$):

$$X(Y(x,y), Z(x,y)) = X(X(Y(x, Z(x,y)), Y(x,y)), Z(x,y)), \tag{2.3.10}$$

which is valid in any quasilattice $\mathfrak{A} = \langle Q; \Sigma\rangle$ since it is a consequence of the hyperidentities (2.3.1)–(2.3.4). Indeed, it suffices to write the hyperidentity (2.2.1) in the form

$$X(Y(x,y),z) = X(X(Y(x,z), Y(x,y)), z),$$

and to set $Z(x,y)$ in place of z.

As a result, the quasilattice $\mathfrak{A} = \langle Q; \Sigma\rangle$ turns out to be isomorphic to the subdirect product $(\varphi_0 Q; \tilde{\psi}_0\Sigma)$ of the quasilattices $\mathfrak{A}/\Theta_{1,2}$, $\mathfrak{A}/\Theta_{1,3}$, $\mathfrak{A}/\Theta_{2,3}$, whose projections are not isomorphisms. We shall show, for example, that the first projection of $(\varphi_0 Q; \tilde{\psi}_0\Sigma)$ is not an isomorphism. Since $A_1 \neq A_2$, there exist $x, y \in Q$ such that $A_1(x,y) \neq A_2(x,y)$, where $x \neq y$. However, $[A_1(x,y)]_{1,2} = [A_2(x,y)]_{1,2}$, since

$$f_{1,2}(A_1(x,y), A_2(x,y)) = A_1(x,y),$$
$$f_{1,2}(A_2(x,y), A_1(x,y)) = A_2(x,y).$$

In fact, we use the hyperidentity

$$X(X(x,y), Y(X(x,y), Y(x,y))) = X(x,y), \tag{2.3.11}$$

which is a consequence of the hyperidentity (2.2.2):

$$\begin{aligned} X(x,y) &\overset{(2.2.2)}{=} X(X(x,y), Y(x,y)) \\ &\overset{(2.2.2)}{=} X(X(X(x,y), Y(x,y)), Y(X(x,y), Y(x,y))) \\ &\overset{(2.2.2)}{=} X(X(x,y), Y(X(x,y), Y(x,y))). \end{aligned}$$

We have

$$\begin{aligned} f_{1,2}(A_1(x,y), A_2(x,y)) &= A_1(A_1(x,y), A_2(A_1(x,y), A_2(x,y))) \\ &\overset{(2.3.11)}{=} A_1(x,y). \end{aligned}$$

By analogy, using the hyperidentity

$$Y(x,y) \overset{(2.3.1)}{=} X(Y(x,y), Y(x,y)) \overset{(2.2.2)}{=} X(Y(x,y), Y(Y(x,y), X(x,y))) \tag{2.3.12}$$

we prove the second equality. This proves Lemma 2.3.4. □

Corollary 2.3.5. *If a binary algebra* $\mathfrak{A} = \langle Q; \Sigma \rangle$ *with hyperidentities (2.3.1)–(2.3.3), (3.5)–(3.7), (3.9)–(3.12) is subdirectly irreducible in the category of* T*-algebras, where* $T = \{2\}$*, then the cardinality* $|\Sigma|$ *of* Σ *is at most 2.*

Corollary 2.3.6. *If a binary algebra* $\mathfrak{A} = \langle Q; \Sigma \rangle$ *with hyperidentities (2.2.1), (2.3.1)–(2.3.3), (2.3.5), (2.3.9) is subdirectly irreducible in the category of* T*-algebras, where* $T = \{2\}$*, then the cardinality* $|\Sigma|$ *of* Σ *is at most* 2.

Corollary 2.3.7. *If a binary algebra* $\mathfrak{A} = \langle Q; \Sigma \rangle$ *with hyperidentities (2.2.1), (2.3.1)–(2.3.3), (2.3.5), (2.3.8) is subdirectly irreducible in the category of* T*-algebras, where* $T = \{2\}$*, then the cardinality* $|\Sigma|$ *of* Σ *is at most* 2.

Remark 2.3.8. The system of hyperidentities (2.2.1), (2.3.1)–(2.3.3), (2.3.5), (2.3.8) has some redundancy. The hyperidentities (2.2.1) and (2.3.5) are consequences of (2.3.1)–(2.3.3), (2.3.8). We will show, for example, that (2.2.1) is a consequence of (2.3.1)–(2.3.3), (2.3.8).

According to (2.3.8), we have

$$\begin{aligned} &X(Y(X(Y(z,y),x),X(y,x)),Y(x,X(y,x))) \\ &\quad = Y(X(Y(z,y),x),X(y,x)), \qquad (2.3.13) \\ &X(Y(X(x,y),x),Y(y,x)) = Y(X(x,y),x). \qquad (2.3.14) \end{aligned}$$

Hence,

$$\begin{aligned} &Y(X(Y(z,y),x),X(y,x)) \\ &\quad \overset{(2.3.13)}{=} X(Y(X(Y(z,y),x),X(y,x)),Y(x,X(y,x))) \\ &\quad \overset{(2.3.14)}{=} X(Y(X(Y(z,y),x),X(y,x)),X(Y(X(x,y),x),Y(y,x))) \\ &\quad \overset{(2.3.13)}{=} X(Y(X(Y(z,y),x),X(y,x)),Y(y,x)) \end{aligned}$$

and

$$\begin{aligned} X(x,X(Y(x,y),Y(y,z))) &= X(X(Y(z,y),x),Y(y,x)) \\ &\overset{(2.3.8)}{=} X(Y(X(Y(z,y),x),X(y,x)),Y(y,x)) \\ &= Y(X(Y(z,y),x),X(y,x)) \\ &\overset{(2.3.8)}{=} X(Y(z,y),x). \end{aligned}$$

Lemma 2.3.9. *The subdirectly irreducible α-quasilattices are distributive lattices.*

Proof. Let $\mathfrak{A} = \langle Q; \Sigma \rangle$ is an α-quasilattice, that is, a quasilattice with the additional condition

$$A_1(x, \ldots A_{n-1}(x, A_n(x,y)) \ldots) = x,$$

where $A_1, \ldots, A_{n-1}, A_n \in \Sigma$. If $\mathfrak{A}$ is subdirectly irreducible, then by Lemma 2.3.4 the set of operations Σ contains at most two elements. Let $A_1 = \cdots = A_{n-1} = A_n$; then $A_1(x,y) = x$ and $A_1(x,y) \neq A_1(y,x)$, which contradicts condition (2.3.2) of the definition of a quasilattice ($|Q| \neq 1$).

Therefore $|\Sigma| = 2$, and (because the equality of operations $A_1 = \cdots = A_{n-1} = A_n$ is excluded) we assume that $A_{i_1} = \cdots = A_{i_k} = (\cdot)$ and $A_{i_{k+1}} = \cdots = A_{i_n} = (+)$, where $i_1, \ldots, i_n$ is a permutation of the numbers $1, \ldots, n$. Then according to (2.3.4) and (2.3.1) we have

$$\begin{aligned} &A_1(x, \ldots A_{n-1}(x, A_n(x, y)) \ldots) \\ &\quad = A_{i_1}(x, \ldots A_{i_k}(x, A_{i_{k+1}}(x, \ldots A_{i_n}(x, y) \ldots)) \ldots) = x, \end{aligned}$$

that is,

$$x(x + y) = x,$$

and a subdirectly irreducible α-quasilattice $\mathfrak{A} = \langle Q, \Sigma \rangle$ will be a lattice. □

Corollary 2.3.10. *In the category of T-algebras, where $T = \{2\}$, any α-quasilattice $\mathfrak{A} = \langle Q; \Sigma \rangle$ is isomorphic to a subdirect product of distributive lattices.*

Corollary 2.3.11. *Any hyperidentity of the variety of distributive lattices is valid in any α-quasilattice $\mathfrak{A} = \langle Q; \Sigma \rangle$.*

Let $\mathfrak{A}$ be a T-algebra. Principal subsystems $\langle Q_i; \Sigma \rangle$, $i \in I$ of $\mathfrak{A}$ are said to be disjoint if $Q_i \cap Q_j = \emptyset$ for any $i, j \in I$, $i \neq j$.

Definition 2.3.12. A T-algebra $\mathfrak{A} = \langle Q; \Sigma \rangle$ is called the sum of its disjoint subsystems $\langle Q_i; \Sigma \rangle$, $i \in I$, if $Q = \bigcup_{i \in I} Q_i$ and one can define a partial order "$\leqslant$" on the index set I such that the following conditions hold (see [323]):

(1) $I(\leqslant)$ is an (upper) semilattice;
(2) if $i \leqslant j$, then there exists a homomorphism $(\varphi_{i,j}, \widetilde{\varepsilon}) : \langle Q_i; \Sigma \rangle \Longrightarrow \langle Q_j; \Sigma \rangle$, where $\varphi_{i,i} = \varepsilon$ is the identity map and $\varphi_{i,j} \cdot \varphi_{j,k} = \varphi_{i,k}$, where $i \leqslant j \leqslant k$;
(3) for any operation $A \in \Sigma$, $|A| = n$, and, for any $x_1, \ldots, x_n \in Q$,

$$A(x_1, \ldots, x_n) = A(\varphi_{i_1,i_0}(x_1), \ldots, \varphi_{i_n,i_0}(x_1)),$$

where $x_1 \in Q_{i_1}, \ldots, x_n \in Q_{i_n}$, $i_1, \ldots .i_n \in I$, $i_0 = i_1 + \cdots + i_n$.

The hyperidentities of nontrivial lattices are homogeneous hyperidentities, and if the same homogeneous hyperidentity is valid in the disjoint subsystems $\langle Q_i; \Sigma \rangle$, $i \in I$ (as subalgebras), then this hyperidentity will obviously hold in their sum $\mathfrak{A} = \langle Q; \Sigma \rangle$. We also recall the notion of a P-function, introduced in [323].

A function $f : Q \times Q \to Q$ is called a *P-function* for a T-algebra $\mathfrak{A} = \langle Q; \Sigma\rangle$ if the following identities are valid:

$$\begin{aligned} f(x,x) &= x, \\ f(x, f(y,z)) &= f(f(x,y),z), \\ f(x, f(y,z)) &= f(x, f(z,y)), \\ f(F(x_1,\ldots,x_n),y) &= F(f(x_1,y),\ldots,f(x_n,y)), \\ f(y, F(x_1,\ldots,x_n)) &= f(y, F(f(y,x_1),\ldots,f(y,x_n))), \\ f(F(x_1,\ldots,x_n),x_k) &= F(x_1,\ldots,x_n), \quad 1 \leqslant k \leqslant n, \\ f(y, F(y,\ldots,y)) &= y \end{aligned}$$

for any n-ary $F \in \Sigma$, $n \in T$. If there exists a P-function f for a T-algebra $\mathfrak{A} = \langle Q; \Sigma\rangle$, then either $\mathfrak{A}$ satisfies the condition $f(x,y) = x$ for arbitrary $x, y \in Q$ or there exists a representation of $\mathfrak{A}$ as a sum of disjoint subsystems $\langle Q_i; \Sigma\rangle$, $i \in I$, with the condition $f(x,y) = x$ for arbitrary $x, y \in Q_i$, $i \in I$ (see [323]).

Lemma 2.3.13 ([322]). *Any quasilattice with two binary operations is either a distributive lattice or the sum of its disjoint subsystems which are distributive lattices.*

Proof. The function $f(x,y) = x + xy$ is a P-function for any quasilattice $Q(+,\cdot)$. □

Lemma 2.3.14 ([308]). *Any binary algebra with two binary operations and the hyperidentities* (2.3.1)–(2.3.3), (2.3.8) *is either a lattice or the sum of its disjoint subsystems, which are lattices.*

Proof. The function $f(x,y) = x+xy$ is a P-function for any binary algebra $Q(+,\cdot)$ with hyperidentities (2.3.1)–(2.3.3), (2.3.8). □

Lemma 2.3.15. *Any binary algebra with two binary operations and the hyperidentities* (2.1.1), (2.3.1)–(2.3.3), (2.3.8) *is either a modular lattice or the sum of its disjoint subsystems, which are modular lattices.*

The proof follows from Lemma 2.3.14 and Example 2.1.2.

Proof of Theorem 2.3.1. According to Lemma 2.3.4 any quasilattice $\mathfrak{A} = \langle Q; \Sigma\rangle$ in the category of T-algebras is isomorphic to a subdirect product of quasilattices with one or two binary operations. In any quasilattice with one binary operation (that is, in a semilattice), any homogeneous hyperidentity (identity) is obviously valid. According to Lemma 2.3.13 any hyperidentity of the variety of distributive lattices is valid in any quasilattice with two binary operations.

Theorem 2.3.16. *Any hyperidentity of the variety of modular lattices is a consequence of the hyperidentities (2.1.1), (2.3.1)–(2.3.3), (2.3.8) (see also Remark 2.3.8).*

Proof. This repeats the proof of Theorem 2.3.1 with an application of Corollary 2.3.7 and Lemma 2.3.15. □

Theorem 2.3.17. *Any hyperidentity of the variety of lattices is a consequence of the hyperidentities (2.3.1)–(2.3.3), (2.3.8) (see also Remark 2.3.8).*

Proof. This repeats the proof of Theorem 2.3.1 with the use of Corollary 2.3.7 and Lemma 2.3.14. □

Since the variety of distributive lattices does not have proper subvarieties (Example 1.5.10), Theorem 2.3.1 yields a characterization of the hyperidentities of an arbitrary individually considered distributive lattice.

Corollary 2.3.18. *Any hyperidentity of a nontrivial distributive lattice is a consequence of the hyperidentities (2.3.1)–(2.3.4).*

Proof. Assume that a hyperidentity (h) holds in a distributive lattice L, and that this hyperidentity is equivalent to a finite system of identities $(h_1), \ldots, (h_m)$ that are valid in L. Therefore the identities $(h_1), \ldots, (h_m)$ are valid in any distributive lattice. Hence, the hyperidentity (h) is valid in any distributive lattice, that is, (h) is a hyperidentity of the variety of distributive lattices, and thus according to Theorem 2.3.1 it is a consequence of the hyperidentities (2.3.1)–(2.3.4). □

Chapter 3

Boole–De Morgan Algebras: Quasi-De Morgan Functions

3.1. Stone-type representation for Boole–De Morgan algebras

Definition 3.1.1. An algebra $(Q; \{+, \cdot, ^{-}, \ ', 0, 1\})$ with two binary, two unary and two nullary operations is called a *Boole–De Morgan* algebra if $(Q; \{+, \cdot, ^{-}, 0, 1\})$ is a De Morgan algebra and $(Q; \{+, \cdot, \ ', 0, 1\})$ is a Boolean algebra and the two unary operations commute, i.e., $(\overline{x})' = \overline{(x')}$. So, Boole–De Morgan algebras form a variety.

This concept is introduced in [260,261] under the name of Boolean bisemigroup. It is easy to note that the last condition $(\overline{x})' = \overline{(x')}$ of the definition follows from the other conditions:

$$\overline{(x')} + \overline{x} = \overline{x' \cdot x} = 1, \quad \overline{(x')} \cdot \overline{x} = \overline{x' + x} = 0,$$

and

$$\overline{(x')} = (\overline{x})'$$

in the Boolean algebra $(Q; \{+, \cdot, \ ', 0, 1\})$.

Let us consider some natural examples of Boole–De Morgan algebras. First note that every Boolean algebra can be considered as a Boole–De Morgan algebra with two equal unary operations. In particular if $\mathbf{B} = (B; \{+, \cdot, \ ', 0, 1\})$ is the two-element Boolean algebra and $\overline{x} = x'$ (i.e., the unary operations $^{-}$ and $'$ are equal) then the algebra $(B; \{+, \cdot, ^{-}, \ ', 0, 1\})$ is a Boole–De Morgan algebra and we will denote it by BM_2.

Now we define a Boole–De Morgan algebra on the four-element set $D = \{0, a, b, 1\}$ which will be used in the proof of the main theorem of the next

Sec. 3.2. Defining $0+x=x+0=x$, $0\cdot x=x\cdot 0=0$ and $1\cdot x=x\cdot 1=x$, $1+x=x+1=1$ and $x+x=x$, $x\cdot x=x$ for all $x\in D$, and $a+b=b+a=1$, $a\cdot b=b\cdot a=0$, $\overline{0}=1$, $\overline{1}=0$, $\overline{a}=a$, $\overline{b}=b$, $1'=0$, $0'=1$, $a'=b$, $b'=a$ we get the Boole–De Morgan algebra $BM_4=(D;\{+,\cdot,^-,\ ',0,1\})$.

For a Boolean algebra $\mathfrak{B}=(Q;\{+,\cdot,\ ',0,1\})$ consider the direct product $\mathfrak{B}\times\mathfrak{B}$. Defining one more unary operation - on the set $Q\times Q$ by $\overline{(x,y)}=(y',x')$ we get the Boole–De Morgan algebra $(Q\times Q;\{+,\cdot,\ ',-,(0,0),(1,1)\})$.

The set $\mathcal{T}(\mathfrak{A})$ of binary term operations of the nontrivial lattice $\mathfrak{A}$ is a Boole–De Morgan algebra (of order 4) where the operations are defined below. For any two binary terms $f(x,y)$ and $g(x,y)$ the binary operations are defined as the following binary superpositions:

$$(f+g)(x,y)=f(x,g(x,y)),\ (f\cdot g)(x,y)=f(g(x,y),y).$$

The nullary operations are the terms y and x. The unary operations are the commutation and dualization. The commutation is defined by $\overline{f}(x,y)=f(y,x)$ and for a binary term $f(x,y)$ to get its dual term $f'(x,y)$ we shall change all variables x by y and vice versa, and also change all operations $+$ by $\cdot$ and vice versa. So we obtain the Boole–De Morgan algebra $\mathcal{T}(\mathfrak{A})$: $\overline{x+y}=x+y$, $\overline{x\cdot y}=x\cdot y$, $\overline{x}=y$, $\overline{y}=x$, $(x+y)'=x\cdot y$, $(x\cdot y)'=x+y$, $x'=y$, $y'=x$. Note that every lattice identity of the Boole–De Morgan algebra $\mathcal{T}(\mathfrak{A})$ is equivalent to certain hyperidentity. For applications of mentioned binary superpositions, see [25,29,30,43,122,221,251,252,257,263,268,295,296, 298,311,315,336,388,413].

Let $\mathfrak{B}=(Q;\{+,\cdot,\ ',0,1\})$ be a Boolean algebra. Denote its dual Boolean algebra by $\mathfrak{B}^{\mathrm{op}}$, i.e., $\mathfrak{B}^{\mathrm{op}}=(Q;\{\cdot,+,\ ',1,0\})$. Consider the direct product $\mathfrak{B}\times\mathfrak{B}^{\mathrm{op}}=(Q\times Q;\{\vee,\wedge,\ ',(0,1),(1,0)\})$ where $(x_1,y_1)\vee(x_2,y_2)=(x_1+x_2,y_1\cdot y_2)$, $(x_1,y_1)\wedge(x_2,y_2)=(x_1\cdot x_2,y_1+y_2)$, $(x_1,y_1)'=(x_1',y_1')$ for any $x_1,x_2,y_1,y_2\in Q$. Defining one more unary operation $^-$ on the set $Q\times Q$ by $\overline{(x,y)}=(y,x)$ we get the Boole–De Morgan algebra $\mathfrak{B}\times\mathfrak{B}^{\mathrm{op}}$. Every Boole–De Morgan algebra $\mathfrak{B}$ is an embedding into $\mathfrak{B}\times\mathfrak{B}^{\mathrm{op}}$. We define an embedding of the Boole–De Morgan algebra $\mathfrak{B}$ into Boole–De Morgan algebra $\mathfrak{B}\times\mathfrak{B}^{\mathrm{op}}$ by the following rule:

$$x\rightarrow(x,\bar{x}).$$

Now we prove a Stone-type representation theorem for Boole–De Morgan algebras.

Let $\mathbf{2}^{\boldsymbol{I}}$ be the Boolean algebra of all subsets of the set I.

Theorem 3.1.2 ([260,261]). *Every Boole–De Morgan algebra* $\mathfrak{A}$ *is isomorphic to a subalgebra of the Boole–De Morgan algebra* $\mathfrak{B} \times \mathfrak{B}^{\mathrm{op}}$ *for some Boolean algebra* $\mathfrak{B} = \mathbf{2}^{\boldsymbol{I}}$.

Proof. Suppose $\mathfrak{A} = (Q; \{+, \cdot, ^-, ', 0, 1\})$ is a Boole–De Morgan algebra. From Stone's representation theorem for Boolean algebras [143], it follows that there exists a set I such that the Boolean algebra $(Q; \{+, \cdot, ', 0, 1\})$ is isomorphic to a subalgebra of the Boolean algebra $\mathbf{2}^{\boldsymbol{I}} = (2^I; \{\cup, \cap, ', \emptyset, I\}) = \mathfrak{B}$ where for a set $X \subseteq I$ we define $X' = I \setminus X$. Let $\sigma : Q \to 2^I$ be an embedding of the mentioned Boolean algebra in $\mathfrak{B}$. We define an embedding of the Boole–De Morgan algebra $\mathfrak{A}$ in the Boole–De Morgan algebra $\mathfrak{B} \times \mathfrak{B}^{\mathrm{op}}$ by the following rule:

$$\varphi(s) = (\sigma(s), \sigma(\overline{s})), \ s \in Q.$$

Indeed, for all $s, t \in Q$ we have

$$\begin{aligned}
\varphi(s+t) &= (\sigma(s+t), \sigma(\overline{s+t})) = (\sigma(s+t), \sigma(\overline{s} \cdot \overline{t})) \\
&= (\sigma(s) \cup \sigma(t), \sigma(\overline{s}) \cap \sigma(\overline{t})) \\
&= (\sigma(s), \sigma(\overline{s})) \vee (\sigma(t), \sigma(\overline{t})) = \varphi(s) \vee \varphi(t), \\
\varphi(s \cdot t) &= (\sigma(s \cdot t), \sigma(\overline{s \cdot t})) \\
&= (\sigma(s \cdot t), \sigma(\overline{s} + \overline{t})) = (\sigma(s) \cap \sigma(t), \sigma(\overline{s}) \cup \sigma(\overline{t})) \\
&= (\sigma(s), \sigma(\overline{s})) \wedge (\sigma(t), \sigma(\overline{t})) = \varphi(s) \wedge \varphi(t), \\
\varphi(\overline{s}) &= (\sigma(\overline{s}), \sigma(\overline{\overline{s}})) = (\sigma(\overline{s}), \sigma(s)) = \overline{\varphi(s)}, \\
\varphi(s') &= (\sigma(s'), \sigma(\overline{s'})) = (\sigma(s'), \sigma((\overline{s})')) = ((\sigma(s))', (\sigma(\overline{s}))') \\
&= (\sigma(s), \sigma(\overline{s}))' = (\varphi(s))'.
\end{aligned}$$

These equalities show that φ is a homomorphism. Obviously, φ is injective, hence it is an embedding. □

In the following section, we will establish a Birkhoff-type representation for Boole–De Morgan algebras.

For a Boole–De Morgan algebra $(Q; \{+, \cdot, ^-, ', 0, 1\})$ we define one more unary operation * by the following way: $x^* = (\overline{x})' = \overline{(x')}$. It is easy to see that $(x+y)^* = x^* + y^*$, $(x \cdot y)^* = x^* \cdot y^*$, $\overline{x^*} = (\overline{x})^*$, $(x^*)' = (x')^*$. Thus the mapping $x \to x^*$ is an automorphism of the Boole–De Morgan algebra $(Q; \{+, \cdot, ^-, ', 0, 1\})$. Also it is easy to see that $x^* = x$ if and only if $x' = \overline{x}$.

3.2. Subdirectly irreducible Boole–De Morgan algebras

Recall that the zero congruence (denoted by O) is defined by the following rule:

$$xOy \Leftrightarrow x = y.$$

Definition 3.2.1. An algebra $\mathfrak{A}$ is called *subdirectly irreducible* if any system of its nonzero congruence relations has a nonzero intersection, i.e., if there exist two elements u, v such that $u \neq v$ and $u\alpha v$ for all nonzero congruences α of that algebra (see [38,61,143]).

Definition 3.2.2. Let the algebra $\mathfrak{A} = (A; \Sigma)$ be a direct product of the algebras $\mathfrak{A}_i = (A_i; \Sigma)$ for $i \in I$ with the projection maps $\pi_i : A \to A_i$ for $i \in I$. A subalgebra $\mathfrak{B} = (B; \Sigma)$ of $\mathfrak{A}$ is called a *subdirect product* of the algebras $\mathfrak{A}_i$, for $i \in I$, if the projection map π_i maps B onto A_i for all $i \in I$.

Theorem 3.2.3 (Birkhoff's subdirect representation theorem for algebras of varieties). *In a variety $\mathcal{V}$, every algebra can be represented as a subdirect product of subdirectly irreducible algebras in $\mathcal{V}$ (see* [38,61,143]).

Following [167,302,408], in this section we find all subdirectly irreducible Boole–De Morgan algebras up to isomorphism. But first we establish some preliminary results.

In what follows $\mathfrak{A} = (Q; \{+, \cdot, ^-, ', 0, 1\})$ is a Boole–De Morgan algebra.

Lemma 3.2.4. *For any element $p \in Q$, we can define a congruence relation θ_p on algebra $\mathfrak{A}$ by the following way:*

$$x\theta_p y \Leftrightarrow x \cdot p = y \cdot p, \ \overline{x} \cdot p = \overline{y} \cdot p, \ x' \cdot p = y' \cdot p, \ x^* \cdot p = y^* \cdot p; \quad x, y \in Q.$$

Proof. Obviously θ_p is an equivalence relation. Let us prove that it is a congruence relation. Let $x\theta_p y$ and $z\theta_p t$. We should prove that

$$(x + z)\theta_p(y + t), \ (x \cdot z)\theta_p(y \cdot t), \ \overline{x}\theta_p\overline{y}, \ x'\theta_p y'.$$

The last two statements immediately follow from the definition of θ_p. Let us verify the first two statements. We have

$$(x + z) \cdot p = x \cdot p + z \cdot p = y \cdot p + t \cdot p = (y + t) \cdot p,$$

$$\overline{x + z} \cdot p = \overline{x} \cdot \overline{z} \cdot p = (\overline{x} \cdot p) \cdot (\overline{z} \cdot p) = (\overline{y} \cdot p) \cdot (\overline{t} \cdot p) = \overline{y + t} \cdot p,$$

$$(x + z)' \cdot p = x' \cdot z' \cdot p = (x' \cdot p) \cdot (z' \cdot p) = (y' \cdot p) \cdot (t' \cdot p) = (y + t)' \cdot p,$$

$$\begin{aligned}(x + z)^* \cdot p &= (x^* + z^*) \cdot p = (x^* \cdot p) + (z^* \cdot p) \\ &= (y^* \cdot p) + (t^* \cdot p) = (y + t)^* \cdot p.\end{aligned}$$

These equalities prove that $(x+z)\theta_p(y+t)$.
Now we have

$$(x \cdot z) \cdot p = (x \cdot p) \cdot (z \cdot p) = (y \cdot p) \cdot (t \cdot p) = (y \cdot t) \cdot p,$$

$$\overline{x \cdot z} \cdot p = (\overline{x} + \overline{z}) \cdot p = (\overline{x} \cdot p) + (\overline{z} \cdot p) = (\overline{y} \cdot p) + (\overline{t} \cdot p) = \overline{y \cdot t} \cdot p,$$

$$(x \cdot z)' \cdot p = (x' + z') \cdot p = (x' \cdot p) + (z' \cdot p) = (y' \cdot p) + (t' \cdot p) = (y \cdot t)' \cdot p,$$

$$(x \cdot z)^* \cdot p = (x^* \cdot z^*) \cdot p = (x^* \cdot p) \cdot (z^* \cdot p) = (y^* \cdot p) \cdot (t^* \cdot p) = (y \cdot t)^* \cdot p.$$

These equalities prove that $(x \cdot z)\theta_p(y \cdot t)$. □

Lemma 3.2.5. *For any elements $p, q \in Q$, the equality $\theta_p \cap \theta_q = \theta_{p+q}$ is valid.*

Proof. Let $x\theta_p y$ and $x\theta_q y$. Then we have

$$x \cdot (p+q) = x \cdot p + x \cdot q = y \cdot p + y \cdot q = y \cdot (p+q),$$

$$\overline{x} \cdot (p+q) = \overline{x} \cdot p + \overline{x} \cdot q = \overline{y} \cdot p + \overline{y} \cdot q = \overline{y} \cdot (p+q),$$

$$x' \cdot (p+q) = x' \cdot p + x' \cdot q = y' \cdot p + y' \cdot q = y' \cdot (p+q),$$

$$x^* \cdot (p+q) = x^* \cdot p + x^* \cdot q = y^* \cdot p + y^* \cdot q = y^* \cdot (p+q).$$

Hence $x\theta_{p+q}y$.

Let now $x\theta_{p+q}y$. Then

$$x \cdot p = x \cdot (p+q) \cdot p = y \cdot (p+q) \cdot p = y \cdot p,$$

$$\overline{x} \cdot p = \overline{x} \cdot (p+q) \cdot p = \overline{y} \cdot (p+q) \cdot p = \overline{y} \cdot p,$$

$$x' \cdot p = x' \cdot (p+q) \cdot p = y' \cdot (p+q) \cdot p = y' \cdot p,$$

$$x^* \cdot p = x^* \cdot (p+q) \cdot p = y^* \cdot (p+q) \cdot p = y^* \cdot p.$$

This shows that $x\theta_p y$. Analogously, we prove that $x\theta_q y$. □

Lemma 3.2.6. *For any element $p \in Q$ we have $\theta_p \cap \theta_{\overline{p}} = O$, where O is the zero congruence relation on algebra $\mathfrak{A}$.*

Proof. If $x\theta_p y$ and $x\theta_{\overline{p}} y$ then we have

$$x = x \cdot (x+p) = x \cdot \overline{\overline{x} \cdot \overline{p}} = x \cdot \overline{\overline{y} \cdot \overline{p}} = x \cdot (y+p) = x \cdot y + x \cdot p =$$

$$x \cdot y + y \cdot p = y \cdot (x+p) = y \cdot \overline{\overline{x} \cdot \overline{p}} = y \cdot \overline{\overline{y} \cdot \overline{p}} = y \cdot (y+p) = y.$$

This proves the statement of the lemma. □

Lemma 3.2.7. *For any element $p \in Q$, the equality $\theta_p = O$ is true if and only if $p \geq \overline{p}$.*

Proof. If $p \geq \overline{p}$ then $O = \theta_p \cap \theta_{\overline{p}} = \theta_{p+\overline{p}} = \theta_p$. Now let $\theta_p = O$. We claim that $p\theta_p(p+\overline{p})$. This follows from the following equalities:

$$(p+\overline{p}) \cdot p = p \cdot p,$$
$$\overline{p+\overline{p}} \cdot p = \overline{p} \cdot p,$$
$$(p+\overline{p})' \cdot p = p' \cdot p^* \cdot p = 0 = p' \cdot p,$$
$$(p+\overline{p})^* \cdot p = p^* \cdot p + p' \cdot p = p^* \cdot p.$$

Now $\theta_p = O$ implies $p = p + \overline{p}$, hence $p \geq \overline{p}$. □

Lemma 3.2.8. *Let $\mathfrak{A} = (Q; \{+, \cdot, ^-, \ ', 0, 1\})$ be a subdirectly irreducible Boole–De Morgan algebra. Then for any elements $x, y, z \in Q$ the following statements are true:*

(1) *x and $\overline{x}$ are comparable, i.e., $x \leq \overline{x}$ or $\overline{x} \leq x$,*
(2) *if $x > \overline{x}$ and $y > \overline{y}$, then $x \cdot y > \overline{x \cdot y}$,*
(3) *if $y > \overline{y}$ and $z \leq \overline{z}$, then $y > z$.*

Proof.

(1) As $\mathfrak{A}$ is subdirectly irreducible and $\theta_x \cap \theta_{\overline{x}} = O$, we have $\theta_x = O$ or $\theta_{\overline{x}} = O$. Now (1) follows immediately from Lemma 3.2.7.
(2) Now let $x > \overline{x}$ and $y > \overline{y}$, but $x \cdot y \not> \overline{x \cdot y}$. Then $x \cdot y \leq \overline{x \cdot y}$ by (1). Lemma 3.2.7 implies $\theta_{\overline{x}} \cap \theta_{\overline{y}} = \theta_{\overline{x}+\overline{y}} = \theta_{\overline{x \cdot y}} = O$. Hence $\theta_{\overline{x}} = O$ or $\theta_{\overline{y}} = O$. Again using Lemma 3.2.7 we deduce that $\overline{x} \geq x$ or $\overline{y} \geq y$, which contradicts the hypothesis.
(3) Now let $y > \overline{y}$ and $z \leq \overline{z}$. If $\overline{y \cdot \overline{z}} > y \cdot \overline{z}$, then from (2) we conclude that $y \cdot (\overline{y} + z) = y \cdot \overline{y \cdot \overline{z}} > \overline{y \cdot \overline{y \cdot \overline{z}}} = \overline{y} + y \cdot \overline{z} = y \cdot (\overline{y} + \overline{z}) \geq y \cdot (\overline{y} + z)$, which is a contradiction. Hence $\overline{y \cdot \overline{z}} \leq y \cdot \overline{z}$ (here we used (1)). So $y \geq y \cdot \overline{z} \geq \overline{y \cdot \overline{z}} = \overline{y} + z \geq z$. But $y \neq z$, therefore $y > z$. □

Lemma 3.2.9. *Let $\mathfrak{A} = (Q; \{+, \cdot, ^-, \ ', 0, 1\})$ be a subdirectly irreducible Boole–De Morgan algebra. Then we can define congruence relation $\sim$ on $\mathfrak{A}$ by setting $x \sim y$ if and only if one of the following statements is true:*

(a) $x > \overline{x}$ *and* $y > \overline{y}$, (b) $x < \overline{x}$ *and* $y < \overline{y}$, (c) $x = \overline{x} = y = \overline{y}$.

Proof. Obviously $\sim$ is an equivalence relation. We prove that $\sim$ is a congruence relation. Let $x \sim y$ and $z \sim t$. Obviously $\overline{x} \sim \overline{y}$. Using the fact that $u > \overline{u}$ implies $u' < (\overline{u})' = \overline{(u')}$ we conclude that $x' \sim y'$. Also note that $u \geq v$ implies $\overline{u} \leq \overline{v}$. Now we shall prove that $x+z \sim y+t$ and $x \cdot z \sim y \cdot t$. If $x > \overline{x}, y > \overline{y}$ and $z > \overline{z}, t > \overline{t}$ then using Lemma 3.2.8 we get

$$x + z \geq x \cdot z > \overline{x \cdot z} \geq \overline{x + z},$$

$$y + t \geq y \cdot t > \overline{y \cdot t} \geq \overline{y + t}.$$

Therefore $(x + z) \sim (y + t)$ and $(x \cdot z) \sim (y \cdot t)$.

If $x > \overline{x}, y > \overline{y}$ and $z < \overline{z}, t < \overline{t}$ then using Lemma 3.2.8 we conclude that $x > z$ and $y > t$. Therefore

$$x + z = x > \overline{x} = \overline{x + z},$$

$$y + t = y > \overline{y} = \overline{y + t},$$

$$x \cdot z = z < \overline{z} = \overline{x \cdot z},$$

$$y \cdot t = t < \overline{t} = \overline{y \cdot t}.$$

Hence $(x + z) \sim (y + t)$ and $(x \cdot z) \sim (y \cdot t)$. The other cases are considered in an analogous way. □

Lemma 3.2.10. *Let* $\mathfrak{A} = (Q; \{+, \cdot, ^{-}, ', 0, 1\})$ *be a subdirectly irreducible Boole–De Morgan algebra. Then for any element* $p \in Q$ *we can define a congruence relation* σ_p *on* $\mathfrak{A}$ *setting*

$$x \sigma_p y \Leftrightarrow x \sim y \text{ and } (x + \overline{x}) \cdot p = (y + \overline{y}) \cdot p \text{ and } (x \cdot \overline{x})' \cdot p = (y \cdot \overline{y})' \cdot p.$$

Proof. It is clear that σ_p is an equivalence relation. Let $x \sigma_p y$ and $z \sigma_p t$. Obviously $\overline{x} \sigma_p \overline{y}$ and $x' \sigma_p y'$. Let us prove that $(x+z)\sigma_p(y+t)$ and $(x \cdot z)\sigma_p(y \cdot t)$. Suppose $x > \overline{x}, y > \overline{y}$ and $z > \overline{z}, t > \overline{t}$. Then $x + z \geq x \cdot z > \overline{x \cdot z} \geq \overline{x + z}$ and $y + t \geq y \cdot t > \overline{y \cdot t} \geq \overline{y + t}$. Also from $x \sigma_p y$ and $z \sigma_p t$ it follows that $x \cdot p = y \cdot p$, $x^* \cdot p = y^* \cdot p$, $z \cdot p = t \cdot p$, $z^* \cdot p = t^* \cdot p$. So

$$(x + z + \overline{x + z}) \cdot p = (x + z) \cdot p = (y + t) \cdot p = (y + t + \overline{y + t}) \cdot p,$$

$$((x + z) \cdot \overline{x + z}))' \cdot p = (x + z)^* \cdot p = (y + t)^* \cdot p = ((y + t) \cdot \overline{y + t}))' \cdot p,$$

$$(x \cdot z + \overline{x \cdot z}) \cdot p = (x \cdot z) \cdot p = (y \cdot t) \cdot p = (y \cdot t + \overline{y \cdot t}) \cdot p,$$

$$((x \cdot z) \cdot \overline{x \cdot z}))' \cdot p = (x \cdot z)^* \cdot p = (y \cdot t)^* \cdot p = ((y \cdot t) \cdot \overline{y \cdot t}))' \cdot p.$$

These equalities show that $(x + z)\sigma_p(y + t)$ and $(x \cdot z)\sigma_p(y \cdot t)$.

Now let $x > \overline{x}, y > \overline{y}$ and $z \leq \overline{z}, t \leq \overline{t}$. Then $x > z$ and $y > t$. Also $x \cdot p = y \cdot p$, $x^* \cdot p = y^* \cdot p$, $\overline{z} \cdot p = \overline{t} \cdot p$, $z' \cdot p = t' \cdot p$. Therefore

$$\begin{aligned}
&(x + z + \overline{x + z}) \cdot p = (x + \overline{x}) \cdot p = x \cdot p = y \cdot p \\
&\quad = (y + \overline{y}) \cdot p = (y + t + \overline{y + t}) \cdot p, \\
&((x + z) \cdot \overline{x + z}))' \cdot p = (x \cdot \overline{x})' \cdot p = x^* \cdot p = y^* \cdot p = (y \cdot \overline{y})' \cdot p \\
&\quad = ((y + t) \cdot \overline{y + t}))' \cdot p, \\
&(x \cdot z + \overline{x \cdot z}) \cdot p = (z + \overline{z}) \cdot p = \overline{z} \cdot p = \overline{t} \cdot p = (t + \overline{t}) \cdot p = (y \cdot t + \overline{y \cdot t}) \cdot p, \\
&((x \cdot z) \cdot \overline{x \cdot z}))' \cdot p = (z \cdot \overline{z})' \cdot p = z' \cdot p = t' \cdot p = (t \cdot \overline{t})' \cdot p \\
&\quad = ((y \cdot t) \cdot \overline{y \cdot t}))' \cdot p.
\end{aligned}$$

Again we get $(x+z)\sigma_p(y+t)$ and $(x \cdot z)\sigma_p(y \cdot t)$. The other cases are considered analogously. □

Lemma 3.2.11. *Let $\mathfrak{A} = (Q; \{+, \cdot, ^-, \ ', 0, 1\})$ be a subdirectly irreducible Boole–De Morgan algebra. Then for any element $p \in Q$ we have $\sigma_p = O$ or $\sigma_{\overline{p}} = O$.*

Proof. Let $x\sigma_p y$ and $x\sigma_{\overline{p}} y$. Without loss of generality we can suppose that $x \geq \overline{x}$, $y \geq \overline{y}$. Then

$$x \cdot p = y \cdot p, \ x^* \cdot p = y^* \cdot p, \ x \cdot \overline{p} = y \cdot \overline{p}, \ x^* \cdot \overline{p} = y^* \cdot \overline{p}.$$

Hence $x \cdot p' = (x^* \cdot \overline{p})^* = (y^* \cdot \overline{p})^* = y \cdot p'$ and $x = x \cdot (p + p') = x \cdot p + x \cdot p' = y \cdot p + y \cdot p' = y \cdot (p + p') = y$. We conclude that $\sigma_p \cap \sigma_{\overline{p}} = O$. Now as $\mathfrak{A}$ is subdirectly irreducible, $\sigma_p = O$ or $\sigma_{\overline{p}} = O$. □

Theorem 3.2.12. *The Boole–De Morgan algebras BM_2 and BM_4 are subdirectly irreducible, and those algebras are the only nontrivial subdirectly irreducible Boole–De Morgan algebras up to isomorphism.*

Proof. The algebras BM_2 and BM_4 are clearly subdirectly irreducible. Now let $\mathfrak{A} = (Q; \{+, \cdot, ^-, \ ', 0, 1\})$ be a subdirectly irreducible Boole–De Morgan algebra. Let $u \in Q$ be an element with $u > \overline{u}$ and $u \neq 1$. Denote $v = \overline{u} + u'$. We claim that $\overline{v} = u \cdot u^* < v$. Indeed, if $v \leq \overline{v}$ then $u' \leq u$. Hence $1 = u + u' = u$, which is a contradiction. Thus $v > \overline{v}$, so $u \sim v$. Also $v^* = v$.

We claim that $1\sigma_{\overline{v}}u$ and $1\sigma_v v$. It follows from the following statements:

$$(u+\overline{u})\cdot\overline{v}=\overline{v}=(1+\overline{1})\cdot\overline{v},$$
$$(u\cdot\overline{u})'\cdot\overline{v}=u^*\cdot\overline{v}=\overline{v}=(1\cdot\overline{1})'\cdot\overline{v},$$
$$(v+\overline{v})\cdot v=v=(1+\overline{1})\cdot v,$$
$$(v\cdot\overline{v})'\cdot v=v^*\cdot v=v=(1\cdot\overline{1})'\cdot v.$$

Now from Lemma 3.2.11 it follows that $1=u$ or $1=v$. As the first equality is not true, the second must be valid, i.e., $\overline{u}+u'=1$. Hence $u=u\cdot(\overline{u}+u')=u\cdot\overline{u}=\overline{u}$. But this is a contradiction.

Thus we have $\overline{x}=x$ for all $x\in Q\setminus\{0,1\}$. Now let $x,y\in Q\setminus\{0,1\}$ and $x\neq y$. If $x+y\neq 1$ then $x\cdot y=\overline{\overline{x}+\overline{y}}=x+y$ and therefore $x=y$. This contradicts our assumption, so $x+y=1$. Similarly for any $x,y\in Q\setminus\{0,1\}$ the equality $x\cdot y=0$ is true. Now if there were three distinct elements $u,v,w\in Q\setminus\{0,1\}$ then $u=u\cdot(v+w)=u\cdot v+u\cdot w=0+0=0$. We arrived at a contradiction, therefore there are at most two elements different from 0 and 1, i.e., $|Q|\leq 4$. But $|Q|\neq 3$ as $(Q;\{+,\cdot,{}',0,1\})$ is a Boolean algebra. If $|Q|=2$ then $\mathfrak{A}$ is isomorphic to BM_2; if $|Q|=4$ then $\mathfrak{A}$ is isomorphic to BM_4. This completes the proof. □

Now from the Birkhoff's subdirect representation theorem we conclude the following result.

Corollary 3.2.13. *Every Boole–De Morgan algebra is isomorphic to a subdirect product of algebras BM_2 and BM_4.*

As BM_2 is a subalgebra of BM_4, we conclude that every Boole–De Morgan algebra is isomorphic to a subalgebra of a direct power of the algebra BM_4.

For a variety $\mathcal{V}$ of algebras denote by $Eq(\mathcal{V})$ the set of all identities that are satisfied in all algebras of $\mathcal{V}$ (the equational theory of the variety $\mathcal{V}$). Also for $\mathfrak{M}\in\mathcal{V}$, $Eq(\mathfrak{M})$ denotes the set of all identities of that algebra (the equational theory of the algebra $\mathfrak{M}$). Also $\mathcal{BM}$ denotes the variety of all Boole–De Morgan algebras.

Corollary 3.2.14. $Eq(\mathcal{BM})=Eq(BM_4)$.

Proof. Obviously $Eq(\mathcal{BM})\subseteq Eq(BM_4)$. Now if $\mathfrak{M}\in\mathcal{V}$ then $\mathfrak{M}$ is isomorphic to a subalgebra of direct power of algebra BM_4. But any identity of the algebra BM_4 is valid also in a direct power of that algebra, and hence, in $\mathfrak{M}$. Therefore $Eq(\mathcal{BM})\supseteq Eq(BM_4)$. □

Corollary 3.2.15. *The free n-generated Boole–De Morgan algebra is isomorphic to the subalgebra of $BM_4^{BM_4^n}$ generated by the projections.*

Proof. For any element (term) ω of the free n-generated Boole–De Morgan algebra we correspond the term operation (function) $\hat{\omega} : D^n \to D$. Corollary 3.2.14 implies that this mapping $\omega \to \hat{\omega}$ is injective. The homomorphism of the correspondence follows immediately from the definitions of the operations. □

Corollary 3.2.16. *Every finitely generated free Boole–De Morgan algebra is finite.*

Corollary 3.2.17. *Every finitely generated Boole–De Morgan algebra is finite, i.e. the variety of Bool–De Morgan algebras is a Burnside variety.*

3.3. Quasi-De Morgan functions

Recall that $D = \{0, a, b, 1\}$ and $B = \{0, 1\}$. Let us construct a one-to-one correspondence between the sets D and $B \times B$ as follows:

$$0 \leftrightarrow (0,0), \ a \leftrightarrow (1,0), \ b \leftrightarrow (0,1), \ 1 \leftrightarrow (1,1).$$

We define the operations $+, \cdot, ^-, '$ on the set $B \times B$ as follows:

$$\overline{(u,v)} = (v', u'), \ (u,v)' = (u', v'),$$

$$(u_1, v_1) + (u_2, v_2) = (u_1 + u_2, v_1 + v_2), \ (u_1, v_1) \cdot (u_2, v_2) = (u_1 \cdot u_2, v_1 \cdot v_2)$$

(here the operations on the right-hand side are the operations of the Boole–De Morgan algebra BM_2). We get the Boole–De Morgan algebra $(B\times B; \{+, \cdot, ^-, ', 0, 1\}) = \mathbf{B}\times\mathbf{B}$, which is isomorphic to the algebra BM_4 (the one-to-one correspondence described above is an isomorphism). However, if the ordered pair $(y, z) \in B \times B$ corresponds to $x \in D$ then we will write $x = (y, z)$ (this causes no confusion).

For $x \in D$ let

$$x^* = \begin{cases} x, & \text{if } x = 0, 1, \\ a, & \text{if } x = b, \\ b, & \text{if } x = a. \end{cases}$$

The unary operation * can also be defined on $B \times B$ taking into account the isomorphism described above. In result we get $(u, v)^* = (v, u) = (v^*, u^*)$, $u, v \in B$. It is clear that $x^* = (\overline{x})' = \overline{x'}$ (which agrees with the notation from the previous section).

Now we can introduce the following two concepts of quasi-De Morgan and De Morgan functions.

Definition 3.3.1. A function $f : D^n \to D$ is called a *quasi-De Morgan function* if the following conditions hold:

(1) if $x_i \in \{0,1\}$, $i = 1, \dots, n$, then $f(x_1, \dots, x_n) \in \{0,1\}$,
(2) if $x_i \in D$, $i = 1, \dots, n$, then $f(x_1^*, \dots, x_n^*) = (f(x_1, \dots, x_n))^*$.

In terms of clone theory, condition (1) means that the function f preserves the unary relation $\{0,1\} \subseteq D$, and condition (2) means that f preserves the binary relation $\{(0,0),(a,b),(b,a),(1,1)\} \subseteq D^2$, which is the graph of the automorphism $x \mapsto x^*$.

For $c = (c_1, \dots, c_n), d = (d_1, \dots, d_n) \in D^n$ we say that d is a *permitted modification* of c if for some k $(1 \le k \le n)$ we have $d_i = c_i$ for all $1 \le i \le n$, $i \ne k$, and

$$d_k = \begin{cases} a & \text{if } c_k = 0, \\ 1 & \text{if } c_k = b. \end{cases}$$

Definition 3.3.2. A quasi-De Morgan function $f : D^n \to D$ is called a *De Morgan function* if it satisfies the following condition:

(3) if $x, y \in D^n$ with $f(x) \ne b$ and y is a permitted modification of x then $f(y) \in \{f(x), a\}$.

In terms of clone theory, condition (3) means that f preserves the order relation $\rho = \{(b,b),(b,0),(b,1),(b,a),(0,0),(0,a),(1,1),(1,a),(a,a)\} \subseteq D^2$.

Notice that condition (1) is a consequence of condition (2), however it is more convenient to write it as a separate condition.

Note that it follows from condition (1) that every quasi-De Morgan function is an extension of some Boolean function. Notice that the constant functions $f = 1$ and $f = 0$ are quasi-De Morgan functions, but the constant functions $f = a$ and $f = b$ are not. This means that 0 and 1 are the only constant quasi-De Morgan functions. Further examples of quasi-De Morgan functions are $f(x) = x$, $g(x) = \overline{x}$, $h(x,y) = x \cdot y$, $q(x,y) = x + y$, $p(x) = x'$, where the operations on the right-hand side are the operations of the Boole–De Morgan algebra BM_4. Also note that the function p is an example of quasi-De Morgan function which is not a De Morgan function.

Below, for $x_i \in D$ we denote by (y_i, z_i) the pair from $B \times B$ which corresponds to x_i, i.e., $x_i = (y_i, z_i)$.

Theorem 3.3.3. *A function $f : D^n \to D$ is a quasi-De Morgan function if and only if there exists a Boolean function $\varphi : B^{2n} \to B$ such that*

$$f(x_1, \ldots, x_n) = (\varphi(y_1, \ldots, y_n, z_1', \ldots, z_n'), \varphi(z_1, \ldots, z_n, y_1', \ldots, y_n')), \quad (3.3.1)$$

for all $x_1, \ldots, x_n \in D$.

Proof. Let f satisfy the condition (3.3.1) for some Boolean function φ. If $x_i \in \{0, 1\}$, then $y_i = z_i$ and $\varphi(y_1, \ldots, y_n, z_1', \ldots, z_n') = \varphi(z_1, \ldots, z_n, y_1', \ldots, y_n')$. Hence, $f(x_1, \ldots, x_n) \in B$. Thus, condition (1) holds for f. Now let us check condition (2). To do this recall that $(u, v)^* = (v, u)$. Hence,

$$\begin{aligned} f(x_1^*, \ldots, x_n^*) &= (\varphi(z_1, \ldots, z_n, y_1', \ldots, y_n'), \varphi(y_1, \ldots, y_n, z_1', \ldots, z_n')) \\ &= (\varphi(y_1, \ldots, y_n, z_1', \ldots, z_n'), \varphi(z_1, \ldots, z_n, y_1', \ldots, y_n'))^* = (f(x_1, \ldots, x_n))^*. \end{aligned}$$

Now suppose f is a quasi-De Morgan function, and let us prove that there exists a Boolean function φ satisfying (3.3.1). First we prove that there are at most 2^{4^n} functions for which conditions (1) and (2) hold. To see this notice that there are 2^n n-tuples $(u_1, \ldots, u_n) \in B^n$. For such n-tuples f can take only two values (by condition (1)). Further, if the n-tuple $(v_1, \ldots, v_n) \in D^n$ contains a or b, then $(v_1^*, \ldots, v_n^*) \neq (v_1, \ldots, v_n)$ and $f(v_1^*, \ldots, v_n^*)$ is uniquely determined by $f(v_1, \ldots, v_n)$ (by condition (2)). There are $4^n - 2^n$ such n-tuples. Thus, the number of quasi-De Morgan functions does not exceed $2^{2^n} \cdot 4^{\frac{4^n - 2^n}{2}} = 2^{2^n} \cdot 2^{4^n - 2^n} = 2^{4^n}$. It is clear that for a function f with (3.3.1) there exists exactly one Boolean function φ that satisfies (3.3.1). Therefore, there are $2^{2^{2n}} = 2^{4^n}$ such functions. And all those functions satisfy conditions (1) and (2). Hence, all functions $f : D^n \to D$ satisfying (1) and (2) also satisfy (3.3.1) for some Boolean function φ. □

As we mentioned in the proof, for a quasi-De Morgan function $f : D^n \to D$ there exists a unique Boolean function $\varphi : B^{2n} \to B$ which satisfies (3.3.1). To emphasize that φ is the unique Boolean function corresponding to f, we denote it by φ_f.

Corollary 3.3.4. *There are exactly 2^{4^n} quasi-De Morgan functions of n variables.*

Denote the set of all quasi-De Morgan functions of n variables by $\mathcal{BM}_n$. For the functions $f, g : D^n \to D$ define $f + g$, $f \cdot g$, $\overline{f}$ and f' by the standard way, i.e., $(f + g)(x) = f(x) + g(x)$, $(f \cdot g)(x) = f(x) \cdot g(x)$, $\overline{f}(x) = \overline{f(x)}$, $f'(x) = (f(x))'$, $x \in D^n$, where the operations on the right-hand side are the operations of the Boole–De Morgan algebra BM_4.

Theorem 3.3.5. *The set $\mathcal{BM}_n$ is closed under operations $+, \cdot, ^-, '$, i.e., if $f, g \in \mathcal{BM}_n$, then $f+g, f\cdot g, \overline{f}, f' \in \mathcal{BM}_n$.*

Proof. We will use the facts that $(x+y)^* = x^*+y^*$, $(x\cdot y)^* = x^*\cdot y^*$, $\overline{x^*} = (\overline{x})^*$, $(x^*)' = (x')^*$. We have:

$$\begin{aligned}
(f+g)(x_1^*,\dots,x_n^*) &= f(x_1^*,\dots,x_n^*) + g(x_1^*,\dots,x_n^*)\\
&= (f(x_1,\dots,x_n))^* + (g(x_1,\dots,x_n))^* = ((f+g)(x_1,\dots,x_n))^*,\\
(f\cdot g)(x_1^*,\dots,x_n^*) &= f(x_1^*,\dots,x_n^*)\cdot g(x_1^*,\dots,x_n^*)\\
&= (f(x_1,\dots,x_n))^*\cdot (g(x_1,\dots,x_n))^* = ((f\cdot g)(x_1,\dots,x_n))^*,\\
\overline{f}(x_1^*,\dots,x_n^*) &= \overline{f(x_1^*,\dots,x_n^*)} = \overline{(f(x_1,\dots,x_n))^*}\\
&= \left(\overline{f(x_1,\dots,x_n)}\right)^* = \left(\overline{f}(x_1,\dots,x_n)\right)^*,\\
f'(x_1^*,\dots,x_n^*) &= (f(x_1^*,\dots,x_n^*))' = ((f(x_1,\dots,x_n))^*)'\\
&= ((f(x_1,\dots,x_n))')^* = (f'(x_1,\dots,x_n))^*.
\end{aligned}$$

These equalities prove the statement of the theorem. □

Thus, we get an algebra: $\mathfrak{BM}_n = (\mathcal{BM}_n, \{+, \cdot, ^-, ', 0, 1\})$ (here 0 and 1 are the constant quasi-De Morgan functions), which obviously is a Boole–De Morgan algebra.

For a set $S \subseteq 2^{\{1,\dots,n\}} \times 2^{\{1,\dots,n\}}$ define the function $f_S : D^n \to D$ by the following way:

$$f_S(x_1,\dots,x_n) = \sum_{(s_1,s_2)\in S}\left(\prod_{i\in s_1} x_i \cdot \prod_{i\in \overline{s}_1} x_i' \cdot \prod_{i\in s_2} \overline{x}_i \cdot \prod_{i\in \overline{s}_2} x_i^*\right), \tag{3.3.2}$$

where the operations on the right-hand side are the operations of BM_4 (cf. [253,256]).

Notice that f_S does not depend on the order of the elements in the set S.

Also we set $f_\emptyset = 0$ and $f_{\{(\emptyset,\emptyset)\}} = 1$.

Let us consider the projection functions

$$\delta_n^i(x_1,\dots,x_n) = x_i, \quad i = 1,\dots,n,$$

as functions $D^n \to D$. Obviously, δ_n^i is a quasi-De Morgan function for each i. According to (3.3.2), for any set $S \subseteq 2^{\{1,\dots,n\}} \times 2^{\{1,\dots,n\}}$ we have:

$$f_S = \sum_{(s_1,s_2)\in S}\left(\prod_{i\in s_1} \delta_n^i \cdot \prod_{i\in \overline{s}_1} (\delta_n^i)' \cdot \prod_{i\in s_2} \overline{\delta_n^i} \cdot \prod_{i\in \overline{s}_2} (\delta_n^i)^*\right).$$

Hence, $f_S \in \mathcal{BM}_n$, i.e., f_S is a quasi-De Morgan function for any set $S \subseteq 2^{\{1,\dots,n\}} \times 2^{\{1,\dots,n\}}$.

For $s = (s_1, s_2) \in 2^{\{1,\dots,n\}} \times 2^{\{1,\dots,n\}}$ let $s' = s_1 \cup \{n+i : i \in s_2\} \in 2^{\{1,\dots,2n\}}$, and for $S \subseteq 2^{\{1,\dots,n\}} \times 2^{\{1,\dots,n\}}$ let $S' = \{s' : s \in S\} \subseteq 2^{\{1,\dots,2n\}}$. In this way we give a one-to-one correspondence between the sets $\mathcal{P}\left(2^{\{1,\dots,n\}} \times 2^{\{1,\dots,n\}}\right)$ and $\mathcal{P}\left(2^{\{1,\dots,2n\}}\right)$.

Now, for any quasi-De Morgan function $f \in \mathcal{BM}_n$ from Theorem 3.3.3 we conclude that there exists a set $S' \subseteq 2^{\{1,\dots,2n\}}$ such that

$$
\begin{aligned}
f(x_1,\dots,x_n) &= (\varphi_f(y_1,\dots,y_n,z_1',\dots,z_n'), \varphi_f(z_1,\dots,z_n,y_1',\dots,y_n')) \\
&= \left(\sum_{s'\in S'}\left(\prod_{\substack{i\in s'\\ 1\le i\le n}} y_i \cdot \prod_{\substack{i\in \overline{s'}\\ 1\le i\le n}} y_i' \cdot \prod_{\substack{i\in s'\\ n+1\le i\le 2n}} z_{i-n}' \cdot \prod_{\substack{i\in \overline{s'}\\ n+1\le i\le 2n}} z_{i-n}\right),\right. \\
&\qquad \left.\sum_{s'\in S'}\left(\prod_{\substack{i\in s'\\ 1\le i\le n}} z_i \cdot \prod_{\substack{i\in \overline{s'}\\ 1\le i\le n}} z_i' \cdot \prod_{\substack{i\in s'\\ n+1\le i\le 2n}} y_{i-n}' \cdot \prod_{\substack{i\in \overline{s'}\\ n+1\le i\le 2n}} y_{i-n}\right)\right) \\
&= \sum_{s'\in S'}\left(\prod_{\substack{i\in s'\\ 1\le i\le n}} (y_i,z_i) \cdot \prod_{\substack{i\in \overline{s'}\\ 1\le i\le n}} (y_i,z_i)' \cdot \prod_{\substack{i\in s'\\ n+1\le i\le 2n}} \overline{(y_{i-n},z_{i-n})}\right. \\
&\qquad \left. \cdot \prod_{\substack{i\in \overline{s'}\\ n+1\le i\le 2n}} (y_{i-n},z_{i-n})^*\right) \\
&= \sum_{(s_1,s_2)\in S}\left(\prod_{i\in s_1} x_i \cdot \prod_{i\in \overline{s}_1} x_i' \cdot \prod_{i\in s_2} \overline{x}_i \cdot \prod_{i\in \overline{s}_2} x_i^*\right) = f_S(x_1,\dots,x_n),
\end{aligned}
$$

where S is the subset of $2^{\{1,\dots,n\}} \times 2^{\{1,\dots,n\}}$ corresponding to S'.

Moreover, the number of all quasi-De Morgan functions of n variables is the same as the number of all subsets of $2^{\{1,\dots,n\}} \times 2^{\{1,\dots,n\}}$. Hence, we get the following result.

Theorem 3.3.6. *For any quasi-De Morgan function f of n variables there exists a unique set $S \subseteq 2^{\{1,\dots,n\}} \times 2^{\{1,\dots,n\}}$ such that $f = f_S$.*

In particular, $f_{S_1} \neq f_{S_2}$ if $S_1 \neq S_2$.

Thus every quasi-De Morgan function can be uniquely presented in the form (3.3.2). This form is called the disjunctive normal form (or briefly — DNF) of quasi-De Morgan function f. Notice that from Theorems 3.3.3 and 3.3.6 we get an algorithm which, given a quasi-De Morgan function, gives its disjunctive normal form. We can also prove that every quasi-De Morgan function can be uniquely presented in conjunctive normal form (CNF), i.e., in the following form:

$$\prod_{(s_1,s_2)\in S}\left(\sum_{i\in s_1} x_i + \sum_{i\in \overline{s}_1} x_i' + \sum_{i\in s_2} \overline{x}_i + \sum_{i\in \overline{s}_2} x_i^*\right).$$

Below we will use the concept of essential variable (and essential dependence) of quasi-De Morgan functions. The definitions are the same as in case of Boolean functions and so we do not give them here.

Using arguments similar to those given above we can prove the following theorem.

Theorem 3.3.7. *For a quasi-De Morgan function f the corresponding Boolean function φ_f does not essentially depend on the last n variables if and only if f can be represented as a term function with functional symbols $+, \cdot, '$, i.e., f is a term function of the Boolean algebra $(D; \{+, \cdot, ', 0, 1\})$.*

Now let us prove the following result, which is related to Plotkin's problem (see Introduction).

Theorem 3.3.8 (Functional representation theorem for free Boole–De Morgan algebras ([278])). *The algebra $\mathfrak{BM}_n$ is a free Boole–De Morgan algebra with the system of free generators $\Delta = \{\delta_n^1, \dots, \delta_n^n\}$. Hence every free n-generated Boole–De Morgan algebra is isomorphic to the Boole–De Morgan algebra $\mathfrak{BM}_n$.*

Proof. Let $\mathfrak{F} = (Q; \{+, \cdot, ^-, ', 0, 1\})$ be a Boole–De Morgan algebra and $\mu : \Delta \to Q$ be a mapping. We prove that there exists a unique homomorphism $\nu : \mathcal{BM}_n \to \mathfrak{F}$ with $\nu|_\Delta = \mu$. According to the Theorem 3.3.6, any element $f \in \mathcal{BM}_n$ can be presented in the form

$$f = \sum_{(s_1,s_2)\in S}\left(\prod_{i\in s_1} \delta_n^i \cdot \prod_{i\in \overline{s}_1} (\delta_n^i)' \cdot \prod_{i\in s_2} \overline{\delta_n^i} \cdot \prod_{i\in \overline{s}_2} (\delta_n^i)^*\right)$$

for the uniquely determined set $S \subseteq 2^{\{1,\dots,n\}} \times 2^{\{1,\dots,n\}}$. Set

$$\nu(f) = \sum_{(s_1,s_2)\in S}\left(\prod_{i\in s_1} \mu(\delta_n^i) \cdot \prod_{i\in \overline{s}_1} (\mu(\delta_n^i))' \cdot \prod_{i\in s_2} \overline{\mu(\delta_n^i)} \cdot \prod_{i\in \overline{s}_2} (\mu(\delta_n^i))^*\right).$$

Obviously, $\nu(\delta_n^i) = \mu(\delta_n^i)$, $i = 1, \ldots, n$. Let us prove that ν is a homomorphism. Let $f_1, f_2 \in \mathcal{BM}_n$. Then $f_1 = f_{S_1}$, $f_2 = f_{S_2}$ for some sets $S_1, S_2 \subseteq 2^{\{1,\ldots,n\}} \times 2^{\{1,\ldots,n\}}$. Let the terms w_1 and w_2 be the disjunctive normal forms of quasi-De Morgan functions f_1 and f_2 respectively; and let w be the disjunctive normal form of quasi-De Morgan function f_1+f_2. Then the identity $w = w_1 + w_2$ holds in the Boole–De Morgan algebra BM_4. From Corollary 3.2.14 it follows that the identity $w = w_1 + w_2$ holds in the Boole–De Morgan algebra $\mathfrak{F} = (Q; \{+, \cdot, ^-, ', 0, 1\})$. Hence $\nu(f_1+f_2) = \nu(f_1)+\nu(f_2)$. Analogously we prove that $\nu(f_1 \cdot f_2) = \nu(f_1) \cdot \nu(f_2)$, $\nu(\overline{f}) = \overline{\nu(f)}$, $\nu(f') = (\nu(f))'$. Obviously $\nu(0) = 0$, $\nu(1) = 1$. Therefore ν is a homomorphism. Uniqueness of ν is evident. □

Problem 3.3.9. *Prove the Post style functional completeness theorem for quasi-De Morgan functions* (see [193,276,333,334]).

Now let us prove the following characterization of quasi-De Morgan functions.

Theorem 3.3.10. *Quasi-De Morgan functions are precisely the term functions of the Boole–De Morgan algebra BM_4, i.e., the set of all quasi-De Morgan functions is the clone of term functions of the Boole–De Morgan algebra BM_4.*

Proof. As every quasi-De Morgan function can be represented in DNF, we conclude that quasi-De Morgan functions are term functions of the algebra BM_4. We shall prove the converse, i.e., we shall prove that any term function of the algebra BM_4 is a quasi-De Morgan function. We induct on the structure of terms. (Again for a term ω we will denote by $\widehat{\omega}$ the term function of the algebra BM_4 corresponding to that term.) If $\omega = x_i$ is a variable, then the corresponding term function $\widehat{\omega}$ is a projection function, which is a quasi-De Morgan function. If $\omega = 0$ or $\omega = 1$ then $\widehat{\omega}$ is either the constant quasi-De Morgan function 0 or 1. Now suppose $\omega = \omega_1 + \omega_2$ or $\omega = \omega_1 \cdot \omega_2$, where $\widehat{\omega}_1$ and $\widehat{\omega}_2$ are quasi-De Morgan functions. In the first case $\widehat{\omega} = \widehat{\omega_1 + \omega_2} = \widehat{\omega}_1 + \widehat{\omega}_2$, in the second case $\widehat{\omega} = \widehat{\omega_1 \cdot \omega_2} = \widehat{\omega}_1 \cdot \widehat{\omega}_2$. And as we know the sum and product of two quasi-De Morgan functions are quasi-De Morgan functions too (see Theorem 3.3.5). Hence $\widehat{\omega}$ is a quasi-De Morgan function. Finally, suppose $\omega = \overline{\lambda}$ or $\omega = \lambda'$, and $\widehat{\lambda}$ is a quasi-De Morgan function. Then in the first case $\widehat{\omega} = \widehat{\overline{\lambda}} = \overline{\widehat{\lambda}}$ and in the second case $\widehat{\omega} = \widehat{(\lambda')} = (\widehat{\lambda})'$. Both of these functions are quasi-De Morgan functions due to Theorem 3.3.5. This finishes the proof. □

3.4. Representation of quasi-De Morgan functions by polynomials

In this section, we will give some representations of quasi-De Morgan functions by polynomials analogous to Zhegalkin polynomials (or polynomials modulo 2) of Boolean functions.

First recall that in the theory of Boolean functions the *exclusive-or* function is defined as follows:

$$x \oplus y = xy' + x'y.$$

This is a binary operation on the set B and the algebra $(B; \{\oplus, \cdot\})$ is a field (it is the two-element Galois field $\mathbf{GF(2)} = \mathbb{Z}_2$).

Recall the following well known result.

Theorem 3.4.1 ([72]). *For every Boolean function* $f : B^n \to B$*, there exists a unique mapping* $\alpha : \mathcal{P}(\{1, \ldots, n\}) \to \{0, 1\}$ *such that*

$$f(x_1, \ldots, x_n) = \bigoplus_{A \in \mathcal{P}(\{1,\ldots,n\})} \alpha(A) \cdot \prod_{i \in A} x_i, \tag{3.4.1}$$

where the operations on the right-hand side are the operations of the two-element field $\mathbf{GF(2)} = \mathbb{Z}_2$.

The right-hand side of (3.4.1) is a polynomial modulo 2. Those polynomials are called also Zhegalkin polynomials for Boolean functions [432].

We will try to generalize the operation $\oplus$ and define an analogous operation on the set D which satisfies some properties of the operation $\oplus$ on the set B. We will designate the operation on D by the same symbol $\oplus$. As we want to consider polynomials with respect to operations $\oplus$ and $\cdot$, it will be natural to require the operation $\oplus$ to be associative and commutative and also distributive with respect to $\cdot$. As we consider quasi-De Morgan functions we also require that $x \oplus y$ be a quasi-De Morgan function. Also we want this operation to be an extension of the analogous operation on B, i.e. the restriction of the function $x \oplus y$ on B be the exclusive-or function defined above.

Let us recall the one-to-one correspondence between the sets D and $B \times B$ described in Sec. 3.3:

$$0 \leftrightarrow (0,0), \;\; a \leftrightarrow (1,0), \;\; b \leftrightarrow (0,1), \;\; 1 \leftrightarrow (1,1).$$

Now we can define an operation $\oplus$ on $B \times B$ componentwise (i.e., $(u_1, v_1) \oplus (u_2, v_2) = (u_1 \oplus u_2, v_1 \oplus v_2)$) and taking into account the above one-to-one correspondence we can define the operation $\oplus$ on D. We get the following table of that operation:

$\oplus$	0	1	a	b
0	0	1	a	b
1	1	0	b	a
a	a	b	0	1
b	b	a	1	0

The algebras $(D; \{\oplus, \cdot\})$ and $(B \times B; \{\oplus, \cdot\})$ are isomorphic. Hence the operation $\oplus$ on D satisfies the required conditions. Also notice that $x \oplus y = x'y + xy'$, where the operations on the right-hand side are the operations of the Boole–De Morgan algebra BM_4.

Now we can define *Zhegalkin polynomials of first kind* on the set D. Those are the polynomials of the form (3.4.1), only the domain of variables now is the four-element set D, and $\oplus$ is the operation on the set D defined above and $\cdot$ is the multiplication of the Boole–De Morgan algebra BM_4. The functions $x \oplus y$, $x \cdot y$ and constant functions 0, 1 are quasi-De Morgan functions, hence all Zhegalkin polynomials of first kind considered as functions from D^n into D are quasi-De Morgan functions. Here we restrict the coefficients to be only 0 and 1 (and not a or b) because the constant functions a and b are not quasi-De Morgan functions and so if we let the coefficients to be a or b then the polynomial could be a non-quasi-De Morgan function. Two Zhegalkin polynomials of first kind are called equal if the corresponding coefficients are equal.

Theorem 3.4.2. *A quasi-De Morgan function $f : D^n \to D$ can be represented as a Zhegalkin polynomial of first kind if and only if the corresponding Boolean function $\varphi_f(t_1, \ldots, t_{2n})$ does not essentially depend on the variables $t_{n+1}, \ldots, t_{2n}$. If this is the case then the representation is unique.*

Proof. Let $\varphi_f(t_1, \ldots, t_{2n})$ do not essentially depend on the last n variables. Thus we can assume that $\varphi_f = \varphi_f(t_1, \ldots, t_n)$ is a Boolean function of n variables. From Theorem 3.4.1, it follows that there exists a unique mapping $\alpha : \mathcal{P}(\{1, \ldots, n\}) \to \{0, 1\}$ such that

$$\varphi_f(x_1, \ldots, x_n) = \bigoplus_{A \in \mathcal{P}(\{1,\ldots,n\})} \alpha(A) \cdot \prod_{i \in A} x_i. \tag{3.4.2}$$

Therefore

$$\begin{aligned} f(x_1,\dots,x_n) &= (\varphi_f(y_1,\dots,y_n),\varphi_f(z_1,\dots,z_n)) \\ &= \left(\bigoplus_{A\in\mathcal{P}(\{1,\dots,n\})} \alpha(A)\cdot\prod_{i\in A} y_i\,,\ \bigoplus_{A\in\mathcal{P}(\{1,\dots,n\})} \alpha(A)\cdot\prod_{i\in A} z_i \right) \\ &= \bigoplus_{A\in\mathcal{P}(\{1,\dots,n\})} \alpha(A)\cdot\prod_{i\in A}(y_i,z_i) = \bigoplus_{A\in\mathcal{P}(\{1,\dots,n\})} \alpha(A)\cdot\prod_{i\in A} x_i. \end{aligned}$$

Thus the "if"-part of the theorem is proved. Let us prove the "only if"-part.

There are exactly 2^{2^n} functions for which $\varphi_f(t_1,\dots,t_{2n})$ does not essentially depend on the variables $t_{n+1},\dots,t_{2n}$, and also there are exactly 2^{2^n} Zhegalkin polynomials of first kind, therefore only such functions can be represented as Zhegalkin polynomials of first kind and the representation is unique. □

This theorem shows that two Zhegalkin polynomials of first kind are equal (as polynomials) if and only if they are equal as functions, i.e., their values are equal for any values of variables.

It is easy to verify that for a quasi-De Morgan function f the corresponding Boolean function φ_f does not essentially depend on the last n variables if and only if f preserves the binary relation $\sigma = \{(0,0),(0,b),(b,0),(b,b),(a,a),(a,1),(1,1),(1,a)\} \subseteq D^2$. Thus a quasi-De Morgan function is a Zhegalkin polynomial of first kind if and only if it preserves the binary relation σ.

From Theorem 3.3.10 we deduce that the term functions of the four-element Boolean algebra $(D;\{+,\cdot,{}',0,1\})$ and only they can be represented as Zhegalkin polynomials of first kind. This result also follows from the following equalities: $u' = 1\oplus u,\ u\in D$ and $u+v = (u'\cdot v')' = 1\oplus(1\oplus u)\cdot(1\oplus v) = u\oplus v\oplus u\cdot v,\ u,v\in D$. These equalities give us an algorithm which, given a quasi-De Morgan function, gives its representation as a Zhegalkin polynomial of first kind.

Let us represent the quasi-De Morgan function $f(x,y,z) = xy' + yz'$ as a Zhegalkin polynomial of first kind. Taking into account the equality $u\oplus u = 0,\ u\in D$, we get:

$$\begin{aligned} f(x,y,z) &= ((xy')'\cdot(yz')')' = 1\oplus(1\oplus x(1\oplus y))(1\oplus y(1\oplus z)) \\ &= 1\oplus 1\oplus y\oplus yz\oplus x\oplus xy\oplus xyz\oplus xy\oplus xy\oplus xyz \\ &= x\oplus y\oplus xy\oplus yz. \end{aligned}$$

A quasi-De Morgan function f is called *monotone* if it preserves the order of the lattice $(D; \{+, \cdot\})$.

Theorem 3.4.3 ([276]). *A De Morgan function $f(x_1, \ldots, x_n)$ is monotone if and only if the corresponding Boolean function $\varphi_f(t_1, \ldots, t_{2n})$ does not essentially depend on the variables $t_{n+1}, \ldots, t_{2n}$.*

Now we deduce the following result.

Corollary 3.4.4. *A De Morgan function can be represented as a Zhegalkin polynomial of first kind if and only if it is monotone.*

Now we define Zhegalkin polynomials of second kind.

It is easy to see that the algebra $(D; \{\oplus, \cdot\})$ is a Boolean ring isomorphic to the ring $\mathbb{Z}_2 \times \mathbb{Z}_2$ (here $\cdot$ is the multiplication of the algebra BM_4). Now we define binary operation $\otimes$ on D such that the algebra $(D; \{\oplus, \otimes\})$ is the four-element Galois field **GF(4)** (as it is known, there is only one finite field with the given number of elements up to isomorphism). The operation $\oplus$ is defined above. We give the table of $\otimes$ below:

$\otimes$	0	1	a	b
0	0	0	0	0
1	0	1	a	b
a	0	a	b	1
b	0	b	1	a

It is clear that the function $x \otimes y$ is a quasi-De Morgan function. Denote $I = \{0, 1, 2, 3\}$. Now we can consider polynomials of the following form:

$$P_\alpha(x_1, \ldots, x_n) = \bigoplus_{(k_1,\ldots,k_n)\in I^n} \alpha(k_1, \ldots, k_n) \otimes \bigotimes_{i=1}^{n} x_i^{k_i}, \tag{3.4.3}$$

where $\alpha : I^n \to \{0, 1\}$ and

$$x^k = \begin{cases} \underbrace{x \otimes \ldots \otimes x}_{k} & \text{if } 1 \leq k \leq 3, \\ 1 & \text{if } k = 0. \end{cases}$$

Those polynomials are also generalizations of Zhegalkin polynomials for Boolean functions, i.e., if we consider them as functions from D^n into D then their restrictions on B are precisely Zhegalkin polynomials for Boolean functions. We call those polynomials *Zhegalkin polynomials of second kind.* Two such polynomials P_{α_1} and P_{α_2} are called equal if $\alpha_1 = \alpha_2$.

Clearly the number of such polynomials is 2^{4^n}. So if we consider them as functions then their number will not exceed 2^{4^n}. But we claim that their number (as functions) is exactly 2^{4^n}. This means that for different mappings $\alpha : I^n \to \{0,1\}$ the corresponding Zhegalkin polynomials of second kind are not equal (as functions).

It is well known (from the Lagrange's interpolation theorem for functions of several variables) that if $\mathbb{F} = (F; \{\oplus, \otimes\})$ is a finite field then all functions from F^n into F are polynomials. But here the coefficients of a polynomial can be any elements of F. Thus if we consider functions $\alpha : I^n \to D$ instead of functions $\alpha : I^n \to \{0,1\}$, then all functions from D^n into D will be polynomials of the form (3.4.3). As there are 4^{4^n} such polynomials and 4^{4^n} functions from D^n into D, we conclude that different polynomials are different as functions.

Further, $x \oplus y$ and $x \otimes y$ are quasi-De Morgan functions, hence all Zhegalkin polynomials of second kind are quasi-De Morgan functions as well. There are 2^{4^n} quasi-De Morgan functions of n variables and 2^{4^n} Zhegalkin polynomials of second kind, therefore all quasi-De Morgan functions are Zhegalkin polynomials of second kind. Thus, we proved the following result.

Theorem 3.4.5. *For every quasi-De Morgan function $f : D^n \to D$ there exists a unique mapping $\alpha : I^n \to \{0,1\}$ such that $f(x_1, \ldots, x_n) = P_\alpha(x_1, \ldots, x_n)$, i.e., every quasi-De Morgan function can be uniquely represented as a Zhegalkin polynomial of second kind.*

Now notice that

$$\begin{aligned}
&x' = 1 \oplus x, \ \overline{x} = 1 \oplus (x \otimes x) = 1 \oplus x^2, \ x^* = x \otimes x = x^2, \\
&x \cdot y = (x^2 \otimes y^2) \oplus (x^2 \otimes y) \oplus (x \otimes y^2), \\
&x + y = (x' \cdot y')' = 1 \oplus (1 \oplus x) \cdot (1 \oplus y) = x \oplus y \oplus x \cdot y \\
&\qquad = x \oplus y \oplus (x^2 \otimes y^2) \oplus (x^2 \otimes y) \oplus (x \otimes y^2).
\end{aligned}$$

These equalities give another proof of Theorem 3.4.5. Indeed, replacing the terms of the form $u', \overline{u}, u^*, u + v, u \cdot v$ in the DNF of f by corresponding polynomials according to the above equalities, we get the desired polynomial. The uniqueness of representation follows from the fact that the number of quasi-De Morgan functions of n variables is equal to the number of Zhegalkin polynomials of second kind of n variables.

Let us find the representation of the quasi-De Morgan function $g(x, y) = x' \cdot y^*$ as a Zhegalkin polynomial of second kind. We have:

$$\begin{aligned} x' \cdot y^* &= (1 \oplus x) \cdot y^2 = ((1 \oplus x)^2 \otimes y^4) \oplus ((1 \oplus x)^2 \otimes y^2) \oplus ((1 \oplus x) \otimes y^4) \\ &= ((1 \oplus x^2) \otimes y) \oplus ((1 \oplus x^2) \otimes y^2) \oplus ((1 \oplus x) \otimes y) \\ &= (y \oplus (x^2 \otimes y)) \oplus (y^2 \oplus (x^2 \otimes y^2)) \oplus (y \oplus (x \otimes y)) \\ &= (x^2 \otimes y^2) \oplus (x^2 \otimes y) \oplus (x \otimes y) \oplus y^2. \end{aligned}$$

Here we used the identity $y^4 = y$ of the four-element field $(D; \{\oplus, \otimes\})$.

Problem 3.4.6. *Develop the quasi-De Morgan analogue of the theory of Boolean functions.*

The KKL theorem, named after Kahn, Kalai, and Linial, says that for any Boolean function f on n variables, there is some variable that has non-trivial influence on the value of f, and this amount of influence depends on the variance of the function. The paper of Kahn, Kalai, and Linial [168] has been one of the most influential works applying Fourier analysis to theoretical computer science.

Problem 3.4.7. *Prove the KKL style theorem for quasi-De Morgan functions.*

Problem 3.4.8. *Prove the KKL style theorem for De Morgan functions.*

The KKL theorem and its strengthenings by Friedgut [114] and Talagrand [396] are foundational results in analysis of Boolean functions. These have found several applications, e.g., to the threshold phenomena, computational learning theory, extremal combinatorics, communication complexity, hardness of approximation, non-embeddability results in metric geometry, and coding theory.

Chapter 4

De Morgan Algebras: De Morgan Functions

4.1. Preliminaries

Let us start from the definitions of Boolean algebras.

Definition 4.1.1. An algebra $(Q; \{+, \cdot, ', 0, 1\})$ with two binary, one unary and two nullary operations is called a *Boolean algebra* if $(Q; \{+, \cdot, 0, 1\})$ is a bounded distributive lattice with least element 0 and greatest element 1 and the algebra $(Q; \{+, \cdot, ', 0, 1\})$ satisfies the following identities:

$$x + x' = 1,$$
$$x \cdot x' = 0.$$

For a lattice $(L; \{+, \cdot\})$ its partial order $\leq$ is defined in the following way:

$$x \leq y \Leftrightarrow x + y = y, \ \ x, y \in L.$$

For the definition and existence of free algebras $F_V(X)$ of the given variety V with the set of free generators X also see [32,36,61,69,142, 143,189,234,321,329,385]. If V is the variety of Boolean algebras then the free algebra $F_V(X)$ is called the free Boolean algebra with the set of free generators X. For the variety V of distributive lattices (bounded distributive lattices) the free algebra $F_V(X)$ is called the free distributive lattice (free bounded distributive lattice) with the set of free generators X. The problem of determining the cardinality of free distributive lattices goes back to R. Dedekind [84] (for a modern survey of the field, see [180]).

Definition 4.1.2. A Boolean algebra (distributive lattice) $\mathcal{F}$ is called a free Boolean algebra (free distributive lattice) with the system of free generators

$X \subseteq \mathcal{F}$ if the algebra $\mathcal{F}$ is generated by the subset $X \subseteq \mathcal{F}$ and for every Boolean algebra (distributive lattice) $\mathfrak{S}$ and for every mapping $\mu : X \to \mathfrak{S}$ there exists a unique homomorphism: $\nu : \mathcal{F} \to \mathfrak{S}$ with $\nu|_X = \mu$.

Let $B = \{0, 1\}$. Define the operations $+, \cdot, {}'$, on B in the following way: $0+0 = 0$, $0+1 = 1+0 = 1+1 = 1$, $0 \cdot 1 = 0 \cdot 0 = 1 \cdot 0 = 0$, $1 \cdot 1 = 1$, $0' = 1$, $1' = 0$. We get the two-element Boolean algebra $\mathbf{B} = (B; \{+, \cdot, {}', 0, 1\})$.

The two-element Boolean algebra is the only subdirectly irreducible Boolean algebra up to isomorphism, the two-element lattice is the only subdirectly irreducible distributive lattice up to isomorphism, and the two-element semilattice is the only subdirectly irreducible semilattice up to isomorphism [143,189,234].

For a set X denote the set of all its subsets by 2^X or $\mathcal{P}(X)$. If we consider subsets of a given set X then for a subset $s \subseteq X$ we denote $s' = X \setminus s$.

A function $f : B^n \to B$ is called a *Boolean function* of n variables, where B^n is the set of all n-element sequences of B. The following result is well known.

Theorem 4.1.3 ([72]). *For every Boolean function $f : B^n \to B$, there exists a unique set $S \subseteq 2^{\{1,\dots,n\}}$ such that*

$$f(x_1, \dots, x_n) = \sum_{s \in S} \left(\prod_{i \in s} x_i \cdot \prod_{i \in \overline{s}} x_i' \right),$$

where the operations on the right-hand side are the operations of Boolean algebra $\boldsymbol{B}$.

Note that those terms are called disjunctive normal forms for Boolean functions.

For $u = (u_1, \dots, u_n)$, $v = (v_1, \dots, v_n) \in B^n$ we define: $u \leq v$ if and only if $u_i \leq v_i$ for all $i = \overline{1, n}$. Here and afterwards $n \geq 1$ is a positive integer.

Definition 4.1.4. A Boolean function $f : B^n \to B$ is called monotone if

$$x \leq y \Rightarrow f(x) \leq f(y),$$

where $x, y \in B^n$.

If $u = (u_1, \dots, u_n)$, $v = (v_1, \dots, v_n) \in B^n$ then we will say $u \preceq v$ if there exists k $(1 \leq k \leq n)$ such that $u_i = v_i$ for all $i \neq k$ and $u_k = 0$, $v_k = 1$. A Boolean function $f : B^n \to B$ is monotone if and only if

$$x \preceq y \Rightarrow f(x) \leq f(y),$$

where $x, y \in B^n$.

Denote the set of all monotone Boolean functions of n variables by $\mathcal{M}_n$.

We can define $f+g$ and $f \cdot g$ for any two Boolean functions of n variables by the standard way. It is obvious that if f and g are monotone Boolean functions, then $f+g$ and $f \cdot g$ are monotone, too. Thus, we get the Boolean algebra of Boolean functions of n variables, and the algebra $\mathfrak{L}_n = \mathcal{M}_n(+,\cdot)$ which obviously is a bounded distributive lattice. Also let $m_n = |\mathcal{M}_n|$ be the number of monotone Boolean functions of n variables. (Note that the numbers m_n are called Dedekind's numbers.) For instance, $m_1=3$, $m_2 = 6, m_3 = 20, m_4 = 168, m_5 = 7581, m_6 = 7828354$ (see [180]).

It is commonly known that the free Boolean algebra on n free generators is isomorphic to the Boolean algebra of Boolean functions of n variables (see [38,84,143,329]). The free bounded distributive lattice on n free generators is isomorphic to the bounded lattice of monotone Boolean functions of n variables (see [38,84,143,329]). Let us present the last result in detailed.

Now let $S \subseteq 2^{\{1,\dots,n\}}$ be an antichain (or Sperner set [84,386]) with respect to the order $\subseteq$. It means that S consists of subsets of $\{1,\dots,n\}$, none of which is contained in any other subset from S. Note that the empty set is also considered an antichain. For an antichain $S \subseteq 2^{\{1,\dots,n\}}$ define the following monotone Boolean function:

$$f_S(x_1,\dots,x_n) = \sum_{s\in S}\prod_{i\in s} x_i. \tag{4.1.1}$$

For $S = \emptyset$ we set $f_\emptyset = 0$, and for $S = \{\emptyset\}$ we set $f_{\{\emptyset\}} = 1$. Notice that f_S does not depend on the order of the elements in the set S. It is easy to see that if $S_1 \neq S_2$ are two antichains, then $f_{S_1} \neq f_{S_2}$. To see this without loss of generality suppose that there exists $s \in S_1$ such that $s \notin S_2$. We can also suppose that there does not exist $s' \in S_2$ with $s' \subseteq s$. Otherwise, we would take s' instead of s (in that case $s' \notin S_1$, because S_1 is an antichain). Take the following values of the variables:

$$x_i = \begin{cases} 1 & \text{if } i \in s, \\ 0 & \text{if } i \notin s. \end{cases}$$

For that values of variables we have: $f_{S_1} = 1$ and $f_{S_2} = 0$.

The form (4.1.1) is uniquely determined by the antichain $S \subseteq 2^{\{1,\dots,n\}}$. And conversely, every monotone Boolean function is obtained in that way.

Proposition 4.1.5. *For every monotone Boolean function of n variables, there exists a unique antichain $S \subseteq 2^{\{1,\dots,n\}}$ such that $f = f_S$.*

Proof. For $a = (a_1, \ldots, a_n) \in B^n$ let $s_a = \{i : a_i = 1\}$. Consider the set $A = \{s_a : a \in B^n,\ f(a) = 1\}$. Let S be the subset of A, consisting exactly of all minimal sets in A. Then $S \subseteq 2^{\{1,\ldots,n\}}$ is an antichain. Notice that $f(a_1, \ldots, a_n) = 1$ if and only if for some $s \in S$ we have $a_i = 1$ for all $i \in s$. The same is valid for f_S. Therefore, $f(a_1, \ldots, a_n) = f_S(a_1, \ldots, a_n)$ for all $(a_1, \ldots, a_n) \in B^n$, and so $f = f_S$. The uniqueness follows from the argument stated above. □

Define the Boolean functions:

$$\delta_n^i = x_i, \ i = \overline{1, n}.$$

Theorem 4.1.6 (Functional representation theorem for free bounded distributive lattices (R. Dedekind [84])). *The algebra $\mathfrak{L}_n$ is a free bounded distributive lattice with the system of free generators: $\Delta = \{\delta_n^1, \ldots, \delta_n^n\}$. Hence, every n-generated free bounded distributive lattice is isomorphic to the distributive lattice $\mathfrak{L}_n$.*

Proof. Let $\mathfrak{L}$ be any bounded distributive lattice and $\varphi : \Delta \to \mathfrak{L}$ be a mapping. We show that there exists a unique homomorphism $\psi : \mathfrak{L}_n \to \mathfrak{L}$ such that $\psi|_\Delta = \varphi$. For any $f \in \mathfrak{L}_n$, there exists a unique antichain S such that

$$f(x_1, \ldots, x_n) = f_S(x_1, \ldots, x_n) = \sum_{s \in S} \prod_{i \in s} x_i.$$

Take

$$\psi(f) = \sum_{s \in S} \prod_{i \in s} \varphi(\delta_i).$$

Obviously $\varphi(f) = \psi(f)$ for $f \in \Delta$ and ψ is a homomorphism. The uniqueness of ψ is also obvious. □

Theorem 4.1.7 (Functional representation theorem for free Boolean algebras (R. Dedekind [84])). *The algebra $\mathfrak{B}_n$ of Boolean function of n variables is a free Boolean algebra with the system of free generators: $\Delta = \{\delta_n^1, \ldots, \delta_n^n\}$. Hence, every n-generated free Boolean algebra is isomorphic to the Boolean algebra $\mathfrak{B}_n$.*

Corollary 4.1.8. *Every finitely generated free Boolean algebra is finite.*

Corollary 4.1.9. *Every finitely generated Boolean algebra is finite.*

4.2. Stone-type representation for De Morgan algebras

Definition 4.2.1. An algebra $(Q; \{+, \cdot, ^-, 0, 1\})$ with two binary, one unary and two nullary operations is called a *De Morgan algebra* if $(Q; \{+, \cdot, 0, 1\})$ is a bounded distributive lattice with least element 0 and greatest element 1 and the algebra $(Q; \{+, \cdot, ^-\})$ satisfies the following identities:

$$\overline{x+y} = \overline{x} \cdot \overline{y},$$
$$\overline{\overline{x}} = x,$$

where $\overline{\overline{x}} = \overline{(\overline{x})}$ ([19,35,38,54,55,57,103,167,226,241,339,353,355]).

Apart from mathematical logic (see [16,21,45,118,119,178,226,239,240]) and algebra, De Morgan algebras (and De Morgan bisemilattices) have applications in multi-valued simulations of digital circuits too (see [54,55]).

For example, the standard fuzzy algebra $F = ([0,1]; max(x,y), min(x,y), 1-x, 0, 1)$ is a De Morgan algebra.

Note that in any De Morgan algebra $(Q; \{+, \cdot, ^-, 0, 1\})$: $\overline{0} = 1,\ \overline{1} = 0$ and:

$$x \le y \Leftrightarrow \overline{y} \le \overline{x},\ x, y \in Q.$$

A characterization of De Morgan algebras can be found in [35,55].

Let $\mathfrak{B} = (Q; \{+, \cdot, 0, 1\})$ be a bounded distributive lattice. Denote its dual bounded distributive lattice by $\mathfrak{B}^{op}$, i.e., $\mathfrak{B}^{op} = (Q; \{\cdot, +, 1, 0\})$. Consider the direct product $\mathfrak{B} \times \mathfrak{B}^{op} = (Q \times Q; \{\vee, \wedge, (0,1), (1,0)\})$ where $(x_1, y_1) \vee (x_2, y_2) = (x_1 + x_2, y_1 \cdot y_2)$, $(x_1, y_1) \wedge (x_2, y_2) = (x_1 \cdot x_2, y_1 + y_2)$, for any $x_1, x_2, y_1, y_2 \in Q$. Defining one more operation $^-$ on the set $Q \times Q$ by $\overline{(x,y)} = (y, x)$, we convert the bounded distributive lattice $\mathfrak{B} \times \mathfrak{B}^{op}$ into the De Morgan algebra $\mathfrak{B} \times \mathfrak{B}^{op}$. Every De Morgan algebra $\mathfrak{B}$ is an embedding into $\mathfrak{B} \times \mathfrak{B}^{op}$. We define an embedding of the De Morgan algebra $\mathfrak{B}$ into De Morgan algebra $\mathfrak{B} \times \mathfrak{B}^{op}$ by the following rule:

$$x \to (x, \overline{x}).$$

Now we prove a Stone-type representation theorem for De Morgan algebras.

Let $\mathbf{2}^{\boldsymbol{I}}$ be the bounded distributive lattice of all subsets of set I.

Theorem 4.2.2 ([260,261]). *Every De Morgan algebra $\mathfrak{A}$ is isomorphic to a subalgebra of the De Morgan algebra $\mathfrak{B} \times \mathfrak{B}^{op}$ for some bounded distributive lattice $\mathfrak{B} = \mathbf{2}^{\boldsymbol{I}}$.*

Proof. Suppose $\mathfrak{A} = (Q; \{+, \cdot, ^-, 0, 1\})$ is a De Morgan algebra. From Birkhoff's representation theorem for distributive lattices (see [143]), it follows that there exists a set I such that the distributive lattice $(Q; \{+, \cdot, 0, 1\})$ is isomorphic to a subalgebra of the distributive lattice $\mathbf{2^I} = (2^I; \{\cup, \cap, \emptyset, I\}) = \mathfrak{B}$. Let $\sigma : Q \to 2^I$ be an embedding of the mentioned distributive lattice in $\mathfrak{B}$. We define an embedding of the De Morgan algebra $\mathfrak{A}$ into De Morgan algebra $\mathfrak{B} \times \mathfrak{B}^{\mathrm{op}}$ by the following rule:

$$\varphi(s) = (\sigma(s), \sigma(\overline{s})), \ \ s \in Q.$$

Indeed, for all $s, t \in Q$ we have

$$\begin{aligned}
\varphi(s+t) &= (\sigma(s+t), \sigma(\overline{s+t})) = (\sigma(s+t), \sigma(\overline{s} \cdot \overline{t})) \\
&= (\sigma(s) \cup \sigma(t), \sigma(\overline{s}) \cap \sigma(\overline{t})) \\
&= (\sigma(s), \sigma(\overline{s})) \vee (\sigma(t), \sigma(\overline{t})) = \varphi(s) \vee \varphi(t), \\
\varphi(s \cdot t) &= (\sigma(s \cdot t), \sigma(\overline{s \cdot t})) = (\sigma(s \cdot t), \sigma(\overline{s} + \overline{t})) = (\sigma(s) \cap \sigma(t), \sigma(\overline{s}) \cup \sigma(\overline{t})) \\
&= (\sigma(s), \sigma(\overline{s})) \wedge (\sigma(t), \sigma(\overline{t})) = \varphi(s) \wedge \varphi(t), \\
\varphi(\overline{s}) &= (\sigma(\overline{s}), \sigma(\overline{\overline{s}})) = (\sigma(\overline{s}), \sigma(s)) = \overline{\varphi(s)}.
\end{aligned}$$

These equalities show that φ is a homomorphism. Obviously, φ is injective, hence it is an embedding. □

Let us consider the following De Morgan algebras: $\mathbf{2} = (\{0, 1\}; \{+, \cdot, ^-, 0, 1\})$, $\mathbf{3} = (\{0, a, 1\}; \{+, \cdot, ^-, 0, 1\})$, where $\overline{a} = a$, and $\mathbf{4} = (\{0, a, b, 1\}; \{+, \cdot, ^-, 0, 1\})$, where $\overline{a} = a$, $\overline{b} = b$, $a + b = 1$, $a \cdot b = 0$.

Let us remind the following result.

Theorem 4.2.3 ([167]). *Every non-trivial subdirectly irreducible De Morgan algebra is isomorphic to one of the following algebras:* **2**, **3**, **4**, *where* **2** *is the unique non-trivial subdirectly irreducible Boolean algebra.*

4.3. Free De Morgan algebras and De Morgan functions

If V is the variety of De Morgan algebras, then the free algebra $F_V(X)$ is called free De Morgan algebra with the set of free generators X. Thus, De Morgan algebra $\mathcal{F} = F(+, \cdot, ^-, 0, 1\})$ is called a free De Morgan algebra with the system of free generators $X \subseteq F$ if the algebra $\mathcal{F}$ is generated by the subset $X \subseteq F$ and for every De Morgan algebra $\mathfrak{S} = S(+, \cdot, ^-, 0, 1\})$ and for every mapping $\mu : X \to S$ there exists a unique homomorphism: $\nu : \mathcal{F} \to \mathfrak{S}$ with $\nu|_X = \mu$.

Denote $D = B \times B = \{(0,0),(1,0),(0,1),(1,1)\} = \{0,a,b,1\}$, where $0 = (0,0), a = (1,0), b = (0,1), 1 = (1,1)$. Defining $0 + x = x,\ 1 \cdot x = x,\ x \in D,\ a + b = 1,\ ab = 0,\ \overline{0} = 1,\ \overline{1} = 0,\ \overline{a} = a,\ \overline{b} = b$ we get the De Morgan algebra $\mathbf{4} = D(+,\cdot,^{-},0,1)$. Notice that $\overline{(u,v)} = (\overline{v},\overline{u}),\ (u_1,v_1) + (u_2,v_2) = (u_1+u_2, v_1+v_2)$ and $(u_1,v_1)\cdot(u_2,v_2) = (u_1 \cdot u_2, v_1 \cdot v_2)$ (here the operations on the right-hand side are the operations of the De Morgan algebra **2**). For $x \in D$ let

$$x^* = \begin{cases} x & \text{if } x = 0,1, \\ a & \text{if } x = b, \\ b & \text{if } x = a. \end{cases}$$

Also for $c = (c_1,\ldots,c_n), d = (d_1,\ldots,d_n) \in D^n$ we say that d is a permitted modification of c if for some k $(1 \le k \le n)$ we have $d_i = c_i$ for all $1 \le i \le n,\ i \ne k$ and

$$d_k = \begin{cases} a & \text{if } c_k = 0, \\ 1 & \text{if } c_k = b. \end{cases}$$

Definition 4.3.1. A function $f : D^n \to D$ is called a De Morgan function of n variables if the following conditions hold:

(1) if $x_i \in \{0,1\},\ i = \overline{1,n}$, then $f(x_1,\ldots,x_n) \in \{0,1\}$,
(2) if $x_i \in D,\ i = \overline{1,n}$ then $f(x_1^*,\ldots,x_n^*) = (f(x_1,\ldots,x_n))^*$,
(3) if $x \in D^n$ with $f(x) \ne b$ and y is a permitted modification of x then $f(y) \in \{f(x), a\}$.

Note that it follows from the first condition that every De Morgan function is an extension of some Boolean function. And notice that the constant functions $f = 1$ and $f = 0$ are De Morgan functions, but the constant functions $f = a$ and $f = b$ are not. This means that 0 and 1 are only constant De Morgan functions. Other examples of De Morgan functions are $f(x) = x,\ g(x) = \overline{x},\ h(x,y) = x \cdot y,\ q(x,y) = x + y$, where the operations on the right-hand side are the operations of the De Morgan algebra **4**.

As Boolean functions, De Morgan functions can be given by tables. Also note that there exists an algorithm which for a given table of a function $f : D^n \to D$ determines whether f is a De Morgan function.

Below, for $x_i \in D$ by (y_i, z_i) we mean the couple from $B \times B$ which is equal to x_i. But often we will consider $B = \{0,1\}$ as a subset of D (when it can cause no confusion).

The equivalency of the following two definitions follows from Theorem 3.3.3.

Definition 4.3.2. The function $f : D^n \to D$ is called a quasi-De Morgan function of n variables if there exists a Boolean function $\varphi : B^{2n} \to B$ such that

$$f(x_1, \dots, x_n) = (\varphi(y_1, \dots, y_n, \overline{z}_1, \dots, \overline{z}_n), \varphi(z_1, \dots, z_n, \overline{y}_1, \dots, \overline{y}_n)), \quad (4.3.1)$$

for all $x_1, \dots, x_n \in D$, $x_i = (y_i, z_i)$, $i = 1, \dots, n$.

Definition 4.3.3. The function $f : D^n \to D$ is called a quasi-De Morgan function if it satisfies conditions (1) and (2).

For any quasi-De Morgan function $f : D^n \to D$, there exists a unique Boolean function $\varphi : B^{2n} \to B$ which satisfies (4.3.1) and denoted by φ_f (see Chapter 3).

Theorem 4.3.4. *The function $f : D^n \to D$ is a De Morgan function if and only if it is a quasi-De Morgan function and φ_f is a monotone Boolean function.*

Proof. If f is a De Morgan function, then by Definition 4.3.3 it is a quasi-De Morgan function. Let us prove that φ_f is monotone. Let $u = (u_1, \dots, u_{2n}), v = (v_1, \dots, v_{2n}) \in B^{2n}$ and for some k $(1 \le k \le 2n)$ $u_i = v_i$, if $i \ne k$, $u_k = 0$, $v_k = 1$. We show that $\varphi_f(u) \le \varphi_f(v)$. Suppose it is not true, i.e., $\varphi_f(u) = 1$, $\varphi_f(v) = 0$. For $1 \le i \le n$ denote:

$$c_i = \begin{cases} (u_i, \overline{u}_{n+i}) & \text{if } 1 \le k \le n, \\ (\overline{u}_{n+i}, u_i) & \text{if } n+1 \le k \le 2n, \end{cases}$$

and

$$d_i = \begin{cases} (v_i, \overline{v}_{n+i}) & \text{if } 1 \le k \le n, \\ (\overline{v}_{n+i}, v_i) & \text{if } n+1 \le k \le 2n. \end{cases}$$

Suppose $1 \le k \le n$. Then $d = (d_1, \dots, d_n)$ is a permitted modification of $c = (c_1, \dots, c_n)$.

$$\begin{aligned} f(c) &= (\varphi_f(u_1, \dots, u_n, u_{n+1}, \dots, u_{2n}), \varphi_f(\overline{u}_{n+1}, \dots, \overline{u}_{2n}, \overline{u}_1, \dots, \overline{u}_n)) \\ &= (1, \varphi_f(\overline{u}_{n+1}, \dots, \overline{u}_{2n}, \overline{u}_1, \dots, \overline{u}_n)) \ne b. \end{aligned}$$

Analogously:

$$f(d) = (0, \varphi_f(\overline{v}_{n+1}, \dots, \overline{v}_{2n}, \overline{v}_1, \dots, \overline{v}_n)).$$

By condition (3) we have $f(d) = f(c)$ or $f(d) = a$. Contradiction!

Now suppose $n+1 \leq k \leq 2n$. Then c is a permitted modification of d. We have

$$\begin{aligned} f(d) &= (\varphi_f(\overline{v}_{n+1}, \dots, \overline{v}_{2n}, \overline{v}_1, \dots, \overline{v}_n), \varphi_f(v_1, \dots, v_n, v_{n+1}, \dots, v_{2n})) \\ &= (\varphi_f(\overline{v}_{n+1}, \dots, \overline{v}_{2n}, \overline{v}_1, \dots, \overline{v}_n), 0) \neq b. \end{aligned}$$

And also

$$f(c) = (\varphi_f(\overline{u}_{n+1}, \dots, \overline{u}_{2n}, \overline{u}_1, \dots, \overline{u}_n), 1).$$

Again, by condition (3), we have $f(c) = f(d)$ or $f(c) = a$. Contradiction!

In both cases, we arrived at a contradiction. Consequently, φ_f is a monotone Boolean function.

The first part of the theorem is proved. Now let us prove the second part. Suppose that f is an quasi-De Morgan function and φ_f is a monotone Boolean function. We verify that condition (3) holds for f. To see this let $d = (d_1, \dots, d_n) \in D^n$ be a permitted modification of $c = (c_1, \dots, c_n) \in D^n$. It means that for some k $(1 \leq k \leq n)$ $c_i = d_i$ if $i \neq k$ and

$$d_k = \begin{cases} a & \text{if } c_k = 0, \\ 1 & \text{if } c_k = b. \end{cases}$$

Let $c_i = (u_i, v_i)$, $d_i = (p_i, q_i)$. Then $u_i \leq p_i$ and $v_i = q_i$ for all $i = \overline{1, n}$. Therefore, $(u_1, \dots, u_n, \overline{v}_1, \dots, \overline{v}_n) \leq (p_1, \dots, p_n, \overline{q}_1, \dots, \overline{q}_n)$ and $(v_1, \dots, v_n, \overline{u}_1, \dots, \overline{u}_n) \geq (q_1, \dots, q_n, \overline{p}_1, \dots, \overline{p}_n)$. Hence,

$$\varphi_f(u_1, \dots, u_n, \overline{v}_1, \dots, \overline{v}_n) \leq \varphi_f(p_1, \dots, p_n, \overline{q}_1, \dots, \overline{q}_n)$$

and

$$\varphi_f(v_1, \dots, v_n, \overline{u}_1, \dots, \overline{u}_n) \geq \varphi_f(q_1, \dots, q_n, \overline{p}_1, \dots, \overline{p}_n).$$

Thus the first coordinate of $f(c)$ is less than (or equal to) the first coordinate of $f(d)$ and the second coordinate of $f(c)$ is greater than (or equal to) the second coordinate of $f(d)$. Thus, if $f(c) = 0$, then $f(d) \in \{0, a\}$; if $f(c) = a$, then $f(d) = a$; and if $f(c) = 1$, then $f(d) \in \{1, a\}$. □

Corollary 4.3.5. *There are m_{2n} De Morgan functions of n variables, where m_k is the number of monotone Boolean functions of k variables.*

Denote the set of all of n-variable De Morgan functions by $\mathcal{D}_n$. For the functions $f, g : D^n \to D$ define $f+g$, $f \cdot g$ and $\overline{f}$ by the standard way, i.e., $(f+g)(x) = f(x)+g(x)$, $(f \cdot g)(x) = f(x) \cdot g(x)$, $\overline{f}(x) = \overline{f(x)}$, $x \in D^n$, where the operations on the right-hand side are the operations of De Morgan algebra

4. Notice that $\mathcal{D}_n$ is closed under those operations, i.e., if $f, g \in \mathcal{D}_n$, then $f+g$, $f \cdot g$, $\overline{f} \in \mathcal{D}_n$. We can verify it straightforwardly, using the definition of De Morgan function. But it is easier to prove that, using Theorem 4.3.4. If $f, g \in \mathcal{D}_n$, then

$$f(x_1,\dots,x_n) = (\varphi_f(y_1,\dots,y_n,\overline{z}_1,\dots,\overline{z}_n), \varphi_f(z_1,\dots,z_n,\overline{y}_1,\dots,\overline{y}_n)),$$

and

$$g(x_1,\dots,x_n) = (\varphi_g(y_1,\dots,y_n,\overline{z}_1,\dots,\overline{z}_n), \varphi_g(z_1,\dots,z_n,\overline{y}_1,\dots,\overline{y}_n)).$$

Hence:

$$(f+g)(x_1,\dots,x_n) = ((\varphi_f+\varphi_g)(y_1,\dots,y_n,\overline{z}_1,\dots,\overline{z}_n), (\varphi_f+\varphi_g)\\ (z_1,\dots,z_n,\overline{y}_1,\dots,\overline{y}_n))$$

and

$$(f\cdot g)(x_1,\dots,x_n) = ((\varphi_f\cdot\varphi_g)(y_1,\dots,y_n,\overline{z}_1,\dots,\overline{z}_n), (\varphi_f\cdot\varphi_g)\\ (z_1,\dots,z_n,\overline{y}_1,\dots,\overline{y}_n)).$$

As φ_f, φ_g are monotone Boolean functions, $\varphi_f \cdot \varphi_g$ and $\varphi_f + \varphi_g$ are monotone, as well. So by Theorem 4.3.4 $f+g$ and $f \cdot g$ are De Morgan functions.

Further, we get:

$$\begin{aligned}
&\overline{f}(x_1,\dots,x_n)\\
&= \overline{(\varphi_f(y_1,\dots,y_n,\overline{z}_1,\dots,\overline{z}_n), \varphi_f(z_1,\dots,z_n,\overline{y}_1,\dots,\overline{y}_n))}\\
&= (\overline{\varphi_f(z_1,\dots,z_n,\overline{y}_1,\dots,\overline{y}_n)}, \overline{\varphi_f(y_1,\dots,y_n,\overline{z}_1,\dots,\overline{z}_n)})\\
&= (\varphi'_f(y_1,\dots,y_n,\overline{z}_1,\dots,\overline{z}_n), \varphi'_f(z_1,\dots,z_n,\overline{y}_1,\dots,\overline{y}_n)).
\end{aligned}$$

And φ_f is monotone; therefore, φ'_f is also monotone. Hence, $\overline{f}$ is a De Morgan function.

Thus, we get the algebra: $\mathfrak{D}_n = \mathcal{D}_n(+,\cdot,^{-},0,1)$, which obviously is a De Morgan algebra. Also, for $f, g \in \mathcal{D}_n$ we have: $\varphi_{f+g} = \varphi_f + \varphi_g$, $\varphi_{f\cdot g} = \varphi_f \cdot \varphi_g$, $\varphi_{\overline{f}} = \varphi'_f$.

Let $a = (a_1, a_2)$, $b = (b_1, b_2) \in 2^{\{1,\dots,n\}} \times 2^{\{1,\dots,n\}}$. We say that $a \subseteq b$, if $a_1 \subseteq b_1$ and $a_2 \subseteq b_2$. In this way, we get a partially ordered set $2^{\{1,\dots,n\}} \times$

$2^{\{1,\ldots,n\}}(\subseteq)$. For the antichain, $S \subseteq 2^{\{1,\ldots,n\}} \times 2^{\{1,\ldots,n\}}$, define the function, $f_S : D^n \to D$, by the following way:

$$f_S(x_1, \ldots, x_n) = \sum_{s=(s_1,s_2)\in S} \left(\prod_{i\in s_1} x_i \cdot \prod_{i\in s_2} \overline{x}_i \right). \tag{4.3.2}$$

Notice that f_S does not depend on the order of the elements in the set S (see [253]).

Note that we set $f_\varnothing = 0$ and $f_{\{(\varnothing,\varnothing)\}} = 1$.

Let us consider the functions

$$\delta_n^i = x_i, \; i = \overline{1, n},$$

as functions $D^n \to D$. Obviously, δ_n^i is a De Morgan function. And according to (4.3.2), for any antichain $S \subseteq 2^{\{1,\ldots,n\}} \times 2^{\{1,\ldots,n\}}$ we have:

$$f_S = \sum_{s=(s_1,s_2)\in S} \left(\prod_{i\in s_1} \delta_n^i \cdot \prod_{i\in s_2} \overline{\delta_n^i} \right).$$

Hence, $f_S \in \mathcal{D}_n$, i.e., f_S is a De Morgan function for any antichain $S \subseteq 2^{\{1,\ldots,n\}} \times 2^{\{1,\ldots,n\}}$.

For $s = (s_1, s_2) \in 2^{\{1,\ldots,n\}} \times 2^{\{1,\ldots,n\}}$ let $s' = s_1 \cup \{n + i : i \in s_2\} \in 2^{\{1,\ldots,2n\}}$, and for $S \subseteq 2^{\{1,\ldots,n\}} \times 2^{\{1,\ldots,n\}}$ let $S' = \{s' : s \in S\} \subseteq 2^{\{1,\ldots,2n\}}$. In this way, we give a bijective mapping from the set of all antichains of $2^{\{1,\ldots,n\}} \times 2^{\{1,\ldots,n\}}(\subseteq)$ to the set of all antichains of $2^{\{1,\ldots,2n\}}(\subseteq)$. And so the number of all antichains of the partially ordered set $2^{\{1,\ldots,n\}} \times 2^{\{1,\ldots,n\}}(\subseteq)$ is m_{2n}.

Now, for any De Morgan function $f \in \mathcal{D}_n$ from Proposition 4.1.5 and Definition 4.3.2 we conclude that there exists an antichain $S' \subseteq 2^{\{1,\ldots,2n\}}$ such that

$$f(x_1, \ldots, x_n) = (\varphi_f(y_1, \ldots, y_n, \overline{z}_1, \ldots, \overline{z}_n), \varphi_f(z_1, \ldots, z_n, \overline{y}_1, \ldots, \overline{y}_n))$$
$$= \left(\sum_{s'\in S'} \left(\prod_{\substack{i\in s' \\ 1\leq i\leq n}} y_i \cdot \prod_{\substack{i\in s' \\ n+1\leq i\leq 2n}} \overline{z}_{i-n} \right), \sum_{s'\in S'} \left(\prod_{\substack{i\in s' \\ 1\leq i\leq n}} z_i \cdot \prod_{\substack{i\in s' \\ n+1\leq i\leq 2n}} \overline{y}_{i-n} \right) \right)$$

$$= \sum_{s' \in S'} \left(\prod_{i \in s'\ 1 \le i \le n} (y_i, z_i) \cdot \prod_{i \in s'\ n+1 \le i \le 2n} (\overline{z}_{i-n}, \overline{y}_{i-n}) \right)$$

$$= \sum_{s=(s_1,s_2) \in S} \left(\prod_{i \in s_1} x_i \cdot \prod_{i \in s_2} \overline{x}_i \right) = f_S(x_1, \dots, x_n),$$

where S is the antichain of $2^{\{1,\dots,n\}} \times 2^{\{1,\dots,n\}}(\subseteq)$, corresponding to S'.

Moreover, the number of all De Morgan functions of n variables is the same as the number of all antichains of $2^{\{1,\dots,n\}} \times 2^{\{1,\dots,n\}}(\subseteq)$. Hence, we get the following result.

Theorem 4.3.6. *For any De Morgan function f of n variables, there exists a unique antichain $S \subseteq 2^{\{1,\dots,n\}} \times 2^{\{1,\dots,n\}}$ such that $f = f_S$.*

In particular, $f_{S_1} \neq f_{S_2}$ if $S_1 \neq S_2$.

Thus every nonconstant De Morgan function can be uniquely presented in the form (4.3.2). This form is called the canonical form (or disjunctive normal form (or briefly — DNF)) of De Morgan function f. Notice that from Definition 4.3.2 and from the proofs of Theorem 4.3.6 and Proposition 4.1.5 we get an algorithm which, for a De Morgan function, gives its disjunctive normal form. From this we conclude that every nonconstant De Morgan function can be uniquely presented in conjunctive normal form (CNF), i.e., in the following form:

$$\prod_{(s_1,s_2) \in S} \left(\sum_{i \in s_1} x_i + \sum_{i \in s_2} \overline{x}_i \right).$$

For a De Morgan function CNF is unique (we can prove this analogously to DNF).

The next result is also related to Plotkin's problem (see Introduction).

Theorem 4.3.7 (Functional representation theorem for free De Morgan algebras ([277])). *The algebra $\mathfrak{D}_n$ is a free De Morgan algebra with the system of free generators: $\Delta = \{\delta_n^1, \dots, \delta_n^n\}$. Hence, every free n-generated De Morgan algebra is isomorphic to the De Morgan algebra $\mathfrak{D}_n$.*

Proof. Let $\mathfrak{F} = Q(+, \cdot, ^{-})$ be a De Morgan algebra and $\mu : X \to Q$ be a mapping. We prove that there exists a unique homomorphism: $\nu : \mathfrak{D}_n \to \mathfrak{F}$

with $\nu|_\Delta = \mu$. Any element $f \in \mathfrak{D}_n$ can be presented in the form

$$f = \sum_{s=(s_1,s_2)\in S} \left(\prod_{i\in s_1} \delta_n^i \cdot \prod_{i\in s_2} \overline{\delta_n^i} \right)$$

for the uniquely determined antichain $S \subseteq 2^{\{1,\ldots,n\}} \times 2^{\{1,\ldots,n\}}$. Set

$$\nu(f) = \sum_{s=(s_1,s_2)\in S} \left(\prod_{i\in s_1} \mu(\delta_n^i) \cdot \prod_{i\in s_2} \overline{\mu(\delta_n^i)} \right).$$

Obviously, ν is a homomorphism and $\nu(\delta_n^i) = \mu(\delta_n^i)$, $i = \overline{1,n}$. Uniqueness of ν is evident too. □

Corollary 4.3.8. *Every finitely generated free De Morgan algebra is finite.*

Corollary 4.3.9. *Every finitely generated De Morgan algebra is finite.*

A similar result is valid for the finitely generated free algebras with two binary operations of the hypervariety defined by the system of hyperidentities of the variety of Boolean algebras (distributive lattices, De Morgan algebras) (see [253,255–257,273]).

Problem 4.3.10. *Develop the De Morgan analogue of the theory of Boolean functions.*

The idea of De Morgan functions is also applicable in quantum computation and the theory of quantum computers.

Chapter 5

Hyperidentities of Boolean Algebras: Super-Boolean Algebras

5.1. Preliminaries

In this chapter, we characterize the hyperidentities of the variety of Boolean algebras [253]. A consequence of this characterization is the determination of a finite base (basis) of hyperidentities for the variety of Boolean algebras.[a] However, the class of termal algebras of Boolean algebras does not have a finite basis of hyperidentities (Proposition 5.1.1).

By $\mathcal{F}(\mathfrak{A})$ we denote the termal algebra of an algebra $\mathfrak{A}$. Let V be a variety of algebras, and let $\mathcal{F}(V) = \{\mathcal{F}(\mathfrak{A}) \mid \mathfrak{A} \in V\}$.

Proposition 5.1.1 ([253]). *If V is the variety of Boolean algebras, then $\mathcal{F}(V)$ does not have a finite basis of hyperidentities.*

Proof. If V is the variety of Boolean algebras, then the hyperidentity

$$F(x_1, \dots, x_{n-1}, F(x_1, \dots, x_{n-1}, F(x_1, \dots, x_n))) = F(x_1, \dots, x_n) \qquad (0.n)$$

holds in the class $\mathcal{F}(V)$ of termal algebras for any natural number n.

Denote by $H^{(m)}$ the system of hyperidentities that hold in the class $\mathcal{F}(V)$ of algebras and have function variables of arity $\leqslant m$. Then for any m there exists a hyperidentity $w_1 = w_2$ that holds in the class $\mathcal{F}(V)$ and is not a consequence of $H^{(m)}$. As $w_1 = w_2$ we can take the hyperidentity $(0.k)$, where $k = 2^m$. Indeed, if $\mathfrak{A} = \langle Q; \Sigma \rangle$, where Q is the free semilattice of rank

[a] The equivalence of two systems of T-hyperidentities and the concept of a basis (of a system) of T-hyperidentities are defined in the first chapter. For termal algebras of Boolean algebras, the set T coincides with the set of all natural numbers.

k generated by $\{a_1, \ldots, a_k\}$, and the k-ary operation A is defined according to the rule

$$A(x_1, \ldots, x_k) = \begin{cases} x_1 & \text{if } \{x_1, \ldots, x_k\} = \{a_1, \ldots, a_k\}, \\ x_1 + \cdot + x_k & \text{otherwise,} \end{cases}$$

then the hyperidentity (0.k) does not hold in the algebra $\mathcal{F}(\mathfrak{A})$, although any hyperidentity in $H^{(m)}$ is true in $\mathcal{F}(\mathfrak{A})$. □

5.2. Hyperidentities of the variety of Boolean algebras: The case of one binary functional variable

Denote by $[w]$ the set of object variables of a hyperterm (word) w.

A hyperterm (word) in the object variables $x_1, \ldots, x_n$ and in a binary functional variable F is denoted by

$$F_{\{1,\ldots,n\}} x_i.$$

The notation $F_{s \in S}\{y_s\}$ is understood similarly. Below we set $\bar{x} = x'$, and if $w = X(\omega_1, \omega_2)$, then $w' = \overline{w} = \overline{X}(\omega_1, \omega_2)$. Here it will be useful to represent hyperterms (words) as trees. For example,

$X(x, \bar{Y}(y, \bar{z})) =$ 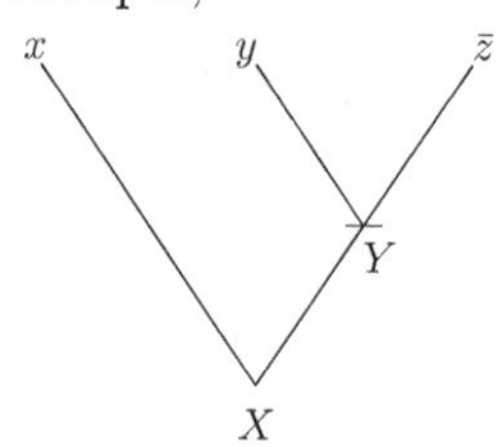

A Boolean hyperterm (word) is a hyperterm with binary and unary functional variables.

Definition 5.2.1. A hyperterm of the form $F_{s \in S}\{y_s\}$, where $S \neq \emptyset$, $S \subseteq 2^{\{1,\ldots,n\}}$, $y_s = \overline{F_{\{1,\ldots,n\}} x_i^*}$ and

$$x_i^* = \begin{cases} x & \text{if } i \in s, \\ x_i' = \bar{x}_i & \text{if } i \notin s, \end{cases}$$

is said to be b_1*-canonical word (form) over* $\{x_1, \ldots, x_n, F\}$ and is denoted by $f_S(F, -, x_1, \ldots, x_n)$, or briefly by $f_S(F)$ or f_S.

For $S = \emptyset$ let

$$f_\emptyset(F, -, x_1, \ldots, x_n) = f_\emptyset(F) = \bar{F}(x_1, \bar{x}_1).$$

Two b_1-canonical words $f_{S_1}(F)$ and $f_{S_2}(F)$ are said to be *equal* if $S_1 = S_2$.

Theorem 5.2.2. *For any Boolean hyperterm* w, $[w] \subseteq \{x_1, \ldots, x_n\}$, *with a single binary functional variable* F *there exists a* b_1*-canonical word* $\widetilde{w}$ *over* $\{x_1, \ldots, x_n, F\}$ *such that the hyperidentity* $w = \widetilde{w}$ *holds in the variety of Boolean algebras. Further, under this condition the* b_1*-canonical word* $\widetilde{w}$ *is uniquely determined and is called the* b_1*-canonical form of the word* w *over* $\{x_1, \ldots, x_n, F\}$; $\widetilde{w}$ *is obtained from* w *after finitely many transitions realized by finitely many hyperidentities of the variety of Boolean algebras, which are independent of* w.

Proof. First, all the unary functional variables in w (if there are any) are replaced by a single unary function variable (negation), and all the double negations are eliminated with the help of the (hyper)identity $\bar{\bar{x}} = x$.

If $w = X(u_1, u_2)$, then X is called the *root* of the word (tree) w. The branch joining the root of w with an object variable $x_i \in [w]$ or $\bar{x}_i$ is called the *branch* (corresponding to) x_i. With the help of hyperidentities satisfied in the variety of Boolean algebras the word w can be reduced to a form in which each branch x_i has at most one negation (not counting the negation $\bar{x}_i$). Indeed, if this condition is not satisfied for w, then w contains a subword (subtree) w^0 in which there exists a branch x_i having at least two negations. Let $w^0 = \bar{F}(1, 2)$, and suppose that there are p negations in all in the subwords 1 and 2 (not counting the negations over object variables). With the help of the hyperidentities of associativity and commutativity

$$F(x, F(y, z)) = F(F(x, y), z), \qquad F(x, y) = F(y, x)$$

we represent w^0 in the form

$$w^0 = \bar{F}\left(\overline{\underset{k=1,\ldots,\ell}{F} x^*_{i_k}}, 3\right),$$

where the subword 3 now contains $p-1$ negations, and $x^*_{i_k} = x_{i_k}$ or $x^*_{i_k} = \bar{x}_{i_k}$.

We use the following hyperidentity of the variety of Boolean algebras:

$$\bar{F}(\bar{F}(x_1, x_2), x_3) = F(\bar{F}(\bar{x}_1, x_3), \bar{F}(\bar{x}_2, x_3)). \tag{5.2.1}$$

Taken into account this hyperidentity we get that

$$w^0 = \bar{F}\left(\overline{\underset{k=1,\dots,\ell}{F} x^*_{i_k}}, 3\right) = \bar{F}\left(\bar{F}\left(\underset{k=1,\dots,\ell-1}{F} x^*_{i_k}, x^*_{i_k}\right), 3\right)$$
$$= F\left(\bar{F}\left(\overline{\underset{k=1,\dots,\ell-1}{F} x^*_{i_k}}, 3\right), \bar{F}(\bar{x}^*_{i_k}, 3)\right).$$

The subtree (subword)

$$\bar{F}\left(\overline{\underset{k=1,\dots,\ell-1}{F} x^*_{i_k}}, 3\right)$$

also has the form of w^0, and hence an analogous process can be applied to it. Consequently, after finitely many steps we get a word of the form in Fig. 5.1, where $\bar{x}^*_i = x_i$ or $\bar{x}^*_i = \bar{x}_i$.

Here there are now as many negations above any negation as there are negations in 3, i.e., exactly $p-1$ of them. Moreover, each subtree (subword) $\bar{F}(x^*_{i_k}, 3)$ has the form of w^0. Therefore, applying the indicated process to them, we ensure that above any negation there are $p-2$ negations in all, and so on. Consequently, after finitely many of the indicated steps w^0 is reduced to a form in which each branch x_i has at most one negation. As a result the original word w is also reduced to such a form.

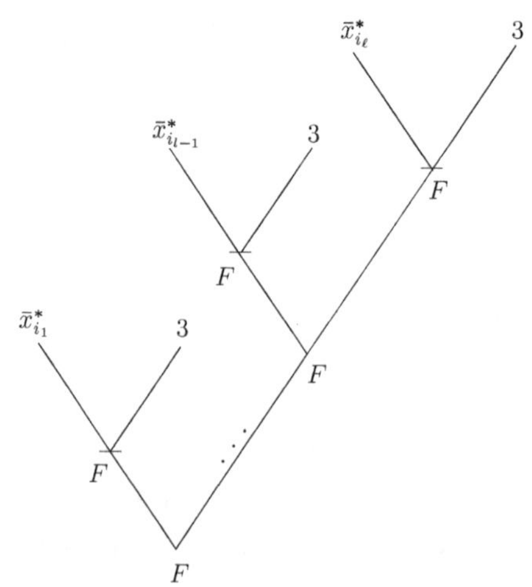

Fig. 5.1.

After this, using the associativity, commutativity, and idempotency hyperidentities we obtain the following form of the word w:

$$w = F\left(\overline{\underset{i\in S_1}{F} x_i^*}, F\left(\overline{\underset{i\in S_1}{F} x_i^*}, \dots F\left(\overline{\underset{i\in S_m}{F} x_i^*}, \underset{i\in S}{F} x_i^*\right) \cdots\right)\right), \tag{5.2.2}$$

or

$$w = \overline{\underset{i\in S_1}{F} x_i^*}. \tag{5.2.2'}$$

Then we use the following hyperidentities of the variety of Boolean algebras:

$$F(x, \bar{x}) = F(y, \bar{y}), \tag{5.2.3}$$

$$F(x, \bar{F}(y, \bar{y})) = x. \tag{5.2.4}$$

Note that (5.2.3) and the idempotency hyperidentity imply the hyperidentity

$$F(x, \bar{x}) = F(z, F(y, \bar{y})); \tag{5.2.3'}$$

and (5.2.3′) and (5.2.4) imply the hyperidentity

$$F(x, \bar{F}(z, F(y, \bar{y}))) = x. \tag{5.2.4'}$$

Now if in w some $\overline{F_{i\in S_t} x_i^*}$ simultaneously contains x_i, and $\bar{x}_i$ for some $i \in S_t$, then, writing

$$\overline{\underset{i\in S_t}{F} x_i^*} = \bar{F}(z, F(x_i, \bar{x}_i))$$

if w has the form (5.2.2), we eliminate $\overline{F_{i\in S_t} x_i^*}$ from (5.2.2) with the help of the hyperidentity (5.2.4′). But if w has the form (5.2.2′), then by (5.2.3′) we have

$$\bar{F}(z, F(x_i, \bar{x}_i)) = \bar{F}(x_1, \bar{x}_1),$$

the right-hand side of which is $f_\emptyset(F)$.

In the next step, the remaining subwords $\overline{F_{i\in S_t} x_i^*}$, along with each variable x_i^* in $F_{i\in S} x_i^*$, are replaced by subwords of the form

$$\bar{F}(x_1^*, F(x_2^*, \ldots, F(x_{n-1}^*, x_n^*) \cdots)), \tag{5.2.5}$$

where $x_i^* = x_i$ or $x_i^* = \bar{x}_i$, $(i = 1, \ldots, n)$. For this we use the hyperidentity

$$F(\bar{F}(x, y), \bar{F}(x, \bar{y})) = \bar{x}, \tag{5.2.6}$$

which is a consequence of (5.2.1) and (5.2.4) (with the commutativity hyperidentity taken into account). Indeed,

$$F(\bar{F}(x, y), \bar{F}(x, \bar{y})) \overset{(5.2.1)}{=} \bar{F}(\bar{F}(y, \bar{y}), x) \overset{(5.2.4)}{=} \bar{x}.$$

Take the subword $\overline{F_{i\in S_t} x_i^*}$, and let $r \in \{1, \ldots, n\} \backslash S_t$. We replace this subword by two similar subwords in which the variable x_r now appears. According to (5.2.6), we get

$$\overline{\underset{i\in S_t}{F} x_i^*} = F\left(\bar{F}\left(\underset{i\in S_t}{F} x_i^*, x_r\right), \quad \bar{F}\left(\underset{i\in S_t}{F} x_i^*, \bar{x}_r\right)\right)$$

Continuing this process, we reconstruct similarly the other missing variables in $\overline{F_{i\in S_t} x_i^*}$. At the same time, each variable x_i^* in $F_{i\in S_t} x_i^*$ is also replaced by subwords of the form (5.2.5) with the help of the hyperidentities (5.2.6). As a result, w takes a form that is a b_1-canonical word over $\{x_1, \ldots, x_n, F\}$. The first part of the theorem is proved.

We prove its second part-uniqueness. Let

$$w = f_{S_1}(F, \,-\,, x_1, \ldots, x_n), \qquad w = f_{S_2}(F, \,-\,, x_1, \ldots, x_n)$$

and let $S_1 \neq S_2$. Then, on the one hand, the hyperidentity

$$f_{S_1}(F, \,-\,, x_1, \ldots, x_n) = f_{S_2}(F, \,-\,, x_1, \ldots, x_n)$$

must hold in any Boolean algebra, and, on the other hand, this hyperidentity does not hold in the two-element Boolean algebra $= \{0, 1\}$. Indeed, since $S_1 \neq S_2$, there exists an $\overline{F_{\{1,\ldots,n\}} x_i^*}$ that is contained in, for example, the word $f_{S_1}(F, -, x_1, \ldots, x_n)$ but not in $f_{S_2}(F, -, x_1, \ldots, x_n)$. Let

$$\overline{\underset{\{1,\ldots,n\}}{F} x_i^*} = \bar{F}\left(\underset{\{i_1,\ldots,i_k\}}{F} x_i, \underset{\{j_1,\ldots,j_\ell\}}{F} \bar{x}_i\right)$$

where

$$\{i_1, \ldots, i_k\} \cap \{j_1, \ldots, j_\ell\} = \{1, \ldots, n\}, \qquad \{i_1, \ldots, i_k\} \cap \{j_1, \ldots, j_\ell\} = \emptyset$$

Making the substitutions $F \to +$ and

$$x_i \to \begin{cases} 0 & \text{if } i \in \{i_1, \ldots, i_k\}, \\ 1 & \text{if } i \in \{j_1, \ldots, j_\ell\}, \end{cases}$$

we get the following value for the chosen $\overline{\underset{\{1,\ldots,n\}}{F} x_i^*}$:

$$\overline{0 + \cdots + 0 + \bar{1} + \cdots + \bar{1}} = 1$$

Consequently, $f_{S_1}(F, -, x_1, \ldots, x_n)$ takes the same value. At the same time, for the same values the word $f_{S_1}(F)$ obviously takes the value 0. Theorem 5.2.2 is proved. □

Corollary 5.2.3. *A hyperidentity $w_1 = w_2$ with one binary function variable F holds in the variety of Boolean algebras if and only if the b_1-canonical forms of the hyperterms w_1 and w_2 over $\{x_1, \ldots, x_n, F\}$ coincide, where $\{x_1, \ldots, x_n\} = [w_1] \cup [w_2]$.*

Corollary 5.2.4. *In the variety of Boolean algebras, any hyperidentity with one binary function variable is a consequence of the following hyperidentities (which are used in the proof of Theorem 5.2.2):*

(1) $X(x) = Y(x)(= \bar{x})$,
(2) $F(x, x) = x$,
(3) $F(x, y) = F(y, x)$,
(4) $F(x, F(y, z)) = F(F(x, y), z)$,
(5) $\bar{\bar{x}} = x$,
(6) $F(x, \bar{F}(y, \bar{y})) = x$,
(7) $F(x, \bar{x}) = F(y, \bar{y})$,
(8) $\bar{F}(\bar{F}(x, y), z) = F(\bar{F}(x, z), \bar{F}(\bar{y}, z))$,

where $\bar{F}(x, y) = \overline{F(x, y)}$.

5.3. Hyperidentities of the variety of Boolean algebras: The case of two binary functional variables

In this section, we characterize hyperidentities with two binary function variables F and G in the variety of Boolean algebras.

Definition 5.3.1. A hyperterm of the form

$$f_{(S_1,S_2)}(G,F,\,-\,,x_1,\ldots,x_n) = F(\bar{F}(f_{S_l}(G),\bar{G}(x_1,\bar{x}_1)),\bar{F}(G(x_1,\bar{x}_1),f_{S_2}(G))),$$

where

$$f_{S_1}(G) = f_{S_1}(G,\,-\,,x_1,\ldots,x_n), \qquad f_{S_2}(G) = f_{S_2}(G,\,-\,,x_1,\ldots,x_n)$$

are b_1-canonical words in the sense of Definition 5.2.1, is called a b_2-*canonical word (form) over* $\{x_1,\ldots,x_n,G,F\}$.

We sometimes abbreviate $f_{(S_1,S_2)}(G,F,\,-,\,x_1,\ldots,x_n)$ as $f_{(S_1,S_2)}(G,F)$ or $f_{(S_1,S_2)}$.

If $S_1 = S_3$ and $S_2 = S_4$, the two b_2-canonical words $f_{(S_1,S_2)}(G,F,\,-\,,x_1,\ldots,x_n)$ and $f_{(S_3,S_4)}(G,F,\,-\,,x_1,\ldots,x_n)$ are said to be *equal.*

Theorem 5.3.2. *For any Boolean hyperterm w, $[w] \subseteq \{x_1,\ldots,x_n\}$, with at most two binary function variables G and F, a b_2-canonical word $\tilde{w}$ over $\{x_1,\ldots,x_n,G,F\}$ exists such that the hyperidentity $w = \tilde{w}$ holds in the variety of Boolean algebras.*

Moreover, under this condition the b_2-canonical word $\tilde{w}$ is uniquely determined and is called the b_2-canonical form of the hyperterm w over $\{x_1,\ldots,x_n,G,F\}$; $\tilde{w}$ is obtained from w after finitely many transitions realized with the help of finitely many hyperidentities of the variety of Boolean algebras, which are independent of w.

Proof. First, all the unary function variables in w (if there are any) are replaced by a single unary function variable (negation). We eliminate all repeated negations. At the first step we pass from w to a new word in whose tree all the binary function symbols G are located above all the binary function symbols F. To do this, we use here, in addition to the distributivity hyperidentity

$$G(F(x,y),z) = F(G(x,z),G(y,z)),$$

also the following hyperidentity of the variety of Boolean algebras:

$$\bar{G}(\bar{F}(x_1,x_2),x_3) = F(\bar{G}(\bar{x}_1,x_3),\bar{G}(\bar{x}_2,x_3)). \tag{5.3.1}$$

Further, using Theorem 5.2.2, we reduce all the G-branches in the word obtained to b_1-canonical form. Denote these b_1-canonical words by

$$f_{S_1}(G,\,-\,,x_1,\ldots,x_n),\ldots,f_{S_\ell}(G,\,-,x_1,\ldots,x_n),$$

where $\ell \leqslant 2^{2n}$. Regarding $f_{S_1}(G,\,-\,,x_1,\ldots,x_n),\ldots,f_{S_\ell}(G,\,-,x_1,\ldots,x_n)$ as the object variables $y_1,\ldots,y_\ell$, we again reduce the word obtained to b_1-canonical form. After this, expressions of the form

$$\overline{f_{S_\ell}(G,\,-,x_1,\ldots,x_n)}$$

can arise, and each of them is again reduced to b_1-canonical form according to the following hyperidentity of the variety of Boolean algebras:

$$\overline{f_{S_\ell}(G,\,-,x_1,\ldots,x_n)} = f_{\bar{S}_\ell}(G,\,-,x_1,\ldots,x_n)$$

where $\bar{S}_\ell = 2^{(1,\ldots,n)} \setminus S_i$. As a result we arrive at a word of the form in Fig. 5.2.

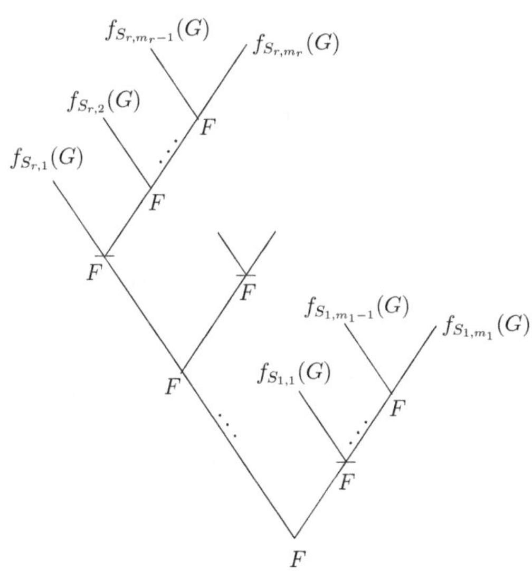

Fig. 5.2.

All branches of the form $\bar{F}(f_{S_m}(G), \ldots, F(f_{S_2}(G), f_{S_1}(G)) \cdots)$ are simplified in the next step. We begin with $F(f_{S_2}(G), f_{S_\ell}(G))$, using the following hyperidentity of the variety of Boolean algebras:

$$F(G(x,), G(x, z)) = F(G(\bar{G}(\bar{z}, \bar{y}), x), G(x, G(y, z))), \tag{5.3.2}$$

$$\bar{f}_S(G) = f_{\bar{S}}(G),$$

$$f_{S_1 \cup S_2}(G) = G(f_{S_l}(G), f_{S_2}(G)), \quad f_{S_1 \cap S_2}(G) = \bar{G}(\overline{f_{S_l}(G)}, \overline{f_{S_2}(G)}).$$

The last hyperidentity follows from the preceding two hyperidentities, whose validity follows from the definition of $f_S(G)$:

$$\begin{aligned}\bar{G}(\overline{f_{S_l}(G)}, \overline{f_{S_2}(G)}) &= \bar{G}(f_{\bar{S}_l}(G), f_{\bar{S}_2}(G)) = \overline{f_{\bar{S}_l \cup \bar{S}_2}(G)} \\ &= f_{\overline{\bar{S}_l \cup \bar{S}_2}}(G) = f_{S_l \cap S_2}(G).\end{aligned}$$

We have

$$\begin{aligned}F(f_{S_2}(G), f_{S_1}(G)) &= F(G(f_{S_1 \cap S_2}(G), f_{S_2 \setminus S_1}(G)), G(f_{S_1 \cap S_2}(G), f_{S_1 \setminus S_2}(G))) \\ &\overset{(5.3.2)}{=} F(G(\bar{G}(\overline{f_{S_2 \setminus S_1}(G)}, \overline{f_{S_1 \setminus S_2}(G)}), f_{S_1 \cap S_2}(G)), G(f_{S_1 \cap S_2}(G), \\ &\qquad G(f_{S_1 \setminus S_2}(G), f_{S_2 \setminus S_1}(G)))) \\ &= F(G(\overline{f_{\overline{S_2 \setminus S_1} \cup \overline{S_1 \setminus S_2}}(G)}, f_{S_1 \cap S_2}(G)), f_{(S_1 \cap S_2) \cup (S_2 \setminus S_1) \cup (S_1 \setminus S_2)}(G)) \\ &= F(G(f_\emptyset(G), f_{S_1 \cap S_2}(G)), f_{S_1 \cup S_2}(G)) = F(f_{S_1 \cap S_2}(G), f_{S_1 \cup S_2}(G)),\end{aligned}$$

i.e.,

$$F(f_{S_1}(G), f_{S_1}(G)) = F(f_{S_1 \cap S_2}(G), f_{S_1 \cup S_2}(G)).$$

In the next step we use the consequence

$$F(x, F(G(x, y), G(x, G(y, z)))) = F(x, G(x, G(y, z))) \tag{5.3.3$'$}$$

of the hyperidentity

$$F(G(y, z), F(x, G(x, y))) = F(G(y, z), x) \tag{5.3.3}$$

of the variety of Boolean algebras to simplify the term of the form $F(f_{S_2}(G), F(f_{S_2}(G), f_{S_1}(G)))$. Namely,

$$\begin{aligned}
&F(f_{S_3}(G), F(f_{S_2}(G), f_{S_1}(G)))\\
&\quad = F(f_{S_3}(G), F(f_{S_1\cap S_2}(G), f_{S_1\cup S_2}(G)))\\
&\quad = F(f_{S_1\cap S_2\cap S_3}(G), F(f_{S_3\cup(S_1\cap S_2)}(G), F(f_{S_3\cap(S_1\cup S_2)}(G), f_{S_1\cup S_2\cup S_3}(G))))\\
&\quad = F(f_{S_3\cup(S_1\cap S_2)}, F(f_{S_1\cap S_2\cap S_3}, F(f_{S_3\cap(S_1\cap S_2)}, f_{S_1\cup S_2\cup S_3})))\\
&\quad = F(f_{S_3\cup(S_1\cap S_2)}, F(f_{S_1\cap S_2\cap S_3}, F(G(f_{S_1\cap S_2\cap S_3}, f_{S_3\cap(S_1\cup S_2)}),\\
&G(f_{S_1\cap S_2\cap S_3}, G(f_{S_3\cap(S_1\cup S_2)}, f_{S_1\cup S_2\cup S_3})))))\\
&\quad \overset{(5.3.3')}{=} F(f_{S_3\cup(S_1\cap S_2)}, F(f_{S_1\cap S_2\cap S_3}, f_{S_1\cup S_2\cup S_3}))\\
&\quad = F(f_{S_1\cap S_2\cap S_3}, F(f_{S_3\cup(S_1\cap S_2)}, f_{S_3\cup S_2\cup S_1}))\\
&\quad = F(f_{S_1\cap S_2\cap S_3}, F(G(f_{S_1\cap S_2\cap S_3}, f_{S_3\cap(S_1\cup S_2)}),\\
&G(f_{S_1\cap S_2\cap S_3}, G(f_{S_3\cap(S_1\cup S_2)}), f_{S_1\cup S_2\cup S_3}))))\\
&\quad \overset{(5.3.3')}{=} F(f_{S_1\cap S_2\cap S_3}, G(f_{S_1\cap S_2\cap S_3}, G(f_{S_3\cap(S_1\cup S_2)}, f_{S_1\cup S_2\cup S_3})))\\
&\quad = F(f_{S_1\cap S_2\cap S_3}, f_{S_1\cup S_2\cup S_3}).
\end{aligned}$$

Analogously, we get the equality (hyperidentity)

$$\bar{F}(f_{S_m}(G), F(f_{S_2}(G), f_{S_1}(G))\cdots) = \bar{F}(f_{S_1\cap\cdots\cap S_m}, f_{S_1\cup\cdots\cup S_m}).$$

As a result, the original word is reduced to one of the following forms: $f_S(G)$, $\bar{F}(f_{S_1}, f_{S_1^*})$, or to a word of the form in Fig. 5.3, where $S_i \subseteq S_i^*$, $i = 1, \ldots, r$.

After this, we use the hyperidentity

$$\begin{aligned}
&F(\bar{F}(x, G(x, y)), \bar{F}(z, G(z, u)))\\
&\quad = F(\bar{F}(G(x, z), G(x_1, \bar{x}_1)), \bar{F}(\bar{G}(x_1, \bar{x}_1), \bar{G}(\bar{G}(x, y), \bar{G}(z, u)))), \qquad (5.3.4)
\end{aligned}$$

which is valid in the variety of Boolean algebras. Since $S_i \subseteq S^*$, it follows that

$$F(f_{S_i}(G), f_{S_1^*}(G)) = F(fS_i(G), G(f_{S_i}(G), f_{S_i^*\setminus S_i}(G))),$$

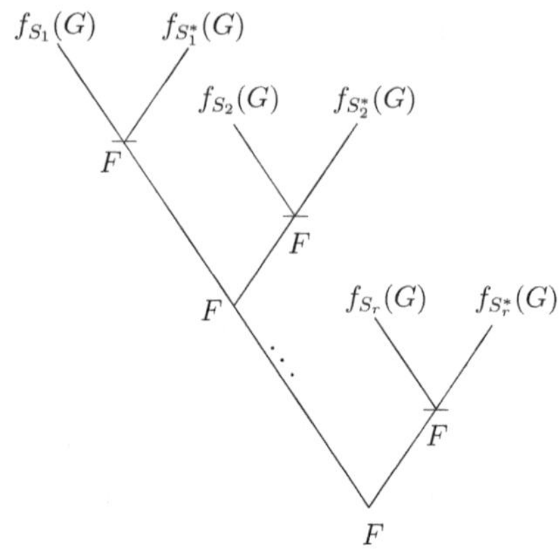

Fig. 5.3.

and so, by (5.3.4),

$$F(\bar{F}(f_{S_1}(G), f_{S_1^*}(G)), \bar{F}(f_{S_2}(G), f_{S_2^*}(G))))$$
$$= F(\bar{F}(G(f_{S_1}(G), f_{S_2}(G)), G(x_1, \bar{x}_1)), \bar{F}(\bar{G}(x_1, \bar{x}_1), \bar{G}(\overline{f_{S_1^*}(G)}, \overline{f_{S_2^*}(G)})))$$
$$= F(\bar{F}(f_{S_1 \cup S_1}(G), G(x_l, \bar{x}_1)), \bar{F}(f_{S_1^* \cap S_2^*}(G), \bar{G}(x_1, \bar{x}_1))).$$

Using the hyperidentity just obtained and the hyperidentity

$$G(x_1, \bar{x}_1) = f_{S_I}(G),$$

where $S_I = 2^{\{1,\ldots,n\}}$, we compute further:

$$F(F(\bar{F}(f_{S_1}(G), f_{S_1^*}(G)), \bar{F}(f_{S_2}(G), f_{S_2^*}(G))), \bar{F}(f_{S_3}(G), f_{S_3^*}(G)))$$
$$= F(\bar{F}(\bar{G}(x_1, \bar{x}_1), f_{S_1^* \cap S_2^*}), F(\bar{F}(f_{S_1 \cup S_2}, G(x_1, \bar{x}_1)), \bar{F}(f_{S_3}, f_{S_3^*})))$$
$$= F(\bar{F}(\bar{G}(x_1, \bar{x}_1), f_{S_1^* \cap S_2^*}), F(\bar{F}(f_{S_1 \cup S_2 \cup S_3}, G(x_1, \bar{x}_1)), \bar{F}(f_{S_3^*}, G(x_1, \bar{x}_1))))$$
$$= F(\bar{F}(f_{S_1 \cup S_2 \cup S_3}, G(x_1, \bar{x}_1)), F(\bar{F}(G(x_1, \bar{x}_1), \bar{G}(x_1, \bar{x}_1)),$$
$$\bar{F}(G(x_1, \bar{x}_1), f_{S_1^* \cap S_2^* \cap S_3^*})))$$
$$\overset{(5.2.4)}{=} F(\bar{F}(f_{S_1^* \cap S_2^* \cap S_3^*}, \bar{G}(x_1, \bar{x}_1)), \bar{F}(G(x_1, \bar{x}_1), f_{S_1 \cup S_2 \cup S_3})).$$

Continuing this process, we arrive after finitely many steps at the b_2-canonical word

$$F(\bar{F}(f_{S_1^* \cap \cdots \cap S_r^*}, \bar{G}(x_1, \bar{x}_1)), \bar{F}(G(x_1, \bar{x}_1), f_{S_1 \cup \cdots \cup S_r}))$$

with the required property.

It remains to reduce the words $\overline{f_S(G)}$ and $\bar{F}(f_{S_1}(G), f_{S_1^*}(G))$ to b_2-canonical form. Since

$$\overline{f_S(G)} = \bar{F}(f_{S_1}(G), f_{S_1}(G)),$$

it suffices to consider only the second case:

$$\begin{aligned}
&\bar{F}(f_{S_1}(G), f_{S_1^*}(G))\\
&\quad\overset{(5.2.4)}{=} F(\bar{F}(f_{S_1}, f_{S_1^*}), \bar{F}(\bar{G}(x_1, \bar{x}_1), G(x_1, \bar{x}_1)))\\
&\quad= F(\bar{F}(f_{S_1}, f_{S_1^*}), \bar{F}(f_\emptyset, f_{S_\ell})) = F(\bar{F}(f_{S_1^*}, \bar{G}(x_1, \bar{x}_1)), \bar{F}(G(x_1, \bar{x}_1), f_{S_1})).
\end{aligned}$$

We prove uniqueness. Let

$$w = f_{(S_1,S_2)}(G, F, -, x_1, \ldots, x_n), \quad w = f_{(S_3,S_4)}(G, F, -, x_1, \ldots, x_n),$$

and suppose that $S_1 \neq S_3$. Let us consider the two-element Boolean algebra $= \{0, 1\}$. We make the substitutions $F \to (+)$ and $G \to (+)$. We know from Theorem 5.2.2 that there exist values $x_1, \ldots, x_n \in \{0, 1\}$ such that the values $f_{S_1}(G, -, x_1, \ldots, x_n)$ and $f_{S_3}(G, -, x_1, \ldots, x_n)$ are distinct. Further, let

$$f_{S_1}(G, -, x_1, \ldots, x_n) = 0, \qquad f_{S_2}(G, -, x_1, \ldots, x_n) = 1.$$

Then for these values of the function and object variables we get

$$\begin{aligned}
f_{(S_1,S_2)}(G, F, -, x_1, \ldots, x_n) &= \overline{0+0} + \overline{1 + f_{S_2}} = 1,\\
f_{(S_3,S_4)}(G, F, -, x_1, \ldots, x_n) &= \overline{1+0} + \overline{1 + f_{S_4}} = 0,
\end{aligned}$$

which is a contradiction. □

Corollary 5.3.3. *A hyperidentity $w_1 = w_2$ with two binary function variables F and G holds in the variety of Boolean algebras if and only if the b_2-canonical forms of the hyperterms w_1 and w_2 over $\{x_1, \ldots, x_n, G, F\}$ coincide, where $\{x_1, \ldots, x_n\} = [w_1] \cup [w_2]$.*

Corollary 5.3.4. *In the variety of Boolean algebras any hyperidentity with two binary function variables is a consequence of the following hyperidentities (which are used in the proof of Theorem 5.3.2): the hyperidentities (1)–(7) listed in Corollary 5.2.4, along with*

(8) $F(x, G(y, z)) = G(F(x, y), F(x, z))$,
(9) $F(G(y, z), F(x, G(x, y))) = F(G(y, z), x)$,
(10) $\bar{G}(\bar{F}(x, y), z) = F(\bar{G}(\bar{x}, z), \bar{G}(\bar{y}, z))$,
(11) $F(G(x, y), G(x, z)) = F(G(\bar{G}(\bar{z}, \bar{y}), x), G(x, G(y, z)))$,

$$
\begin{aligned}
(12)\ \ & F(\bar{F}(x, G(x, y)), \bar{F}(z, G(z, u))) \\
& = F(\bar{F}(G(x, z), G(x_1, \bar{x}_1)), \bar{F}(\bar{G}(x_1, \bar{x}_1), \bar{G}(\bar{G}(x, y), \bar{G}(z, u)))).
\end{aligned}
$$

Example 5.3.1. To find the b_2-canonical forms of the words 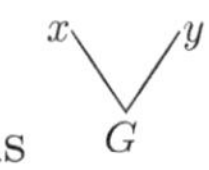

and 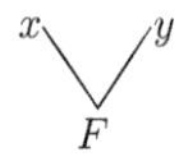 on $\{x, y, G, F\}$:

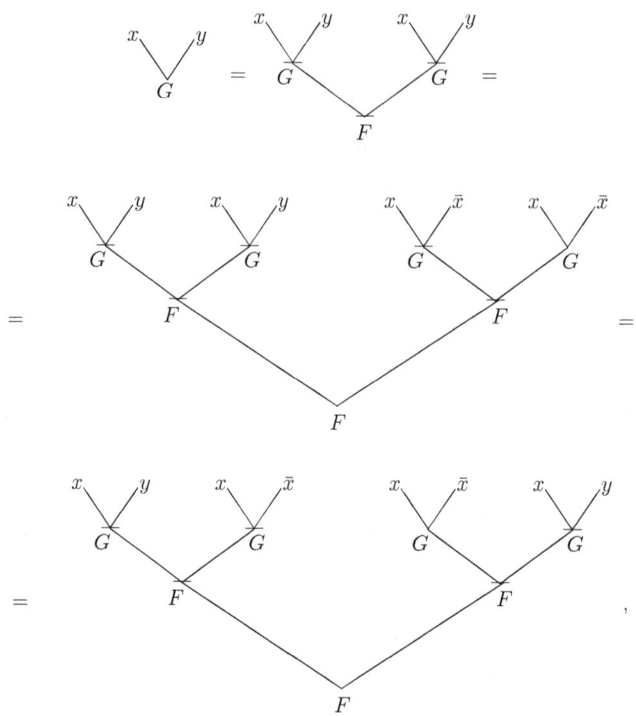

where

$$f_{S_1}(G, {}', x, y) = f_{S_2}(G, {}', x, y) =$$ 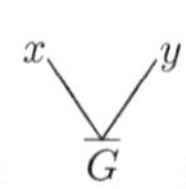 ;

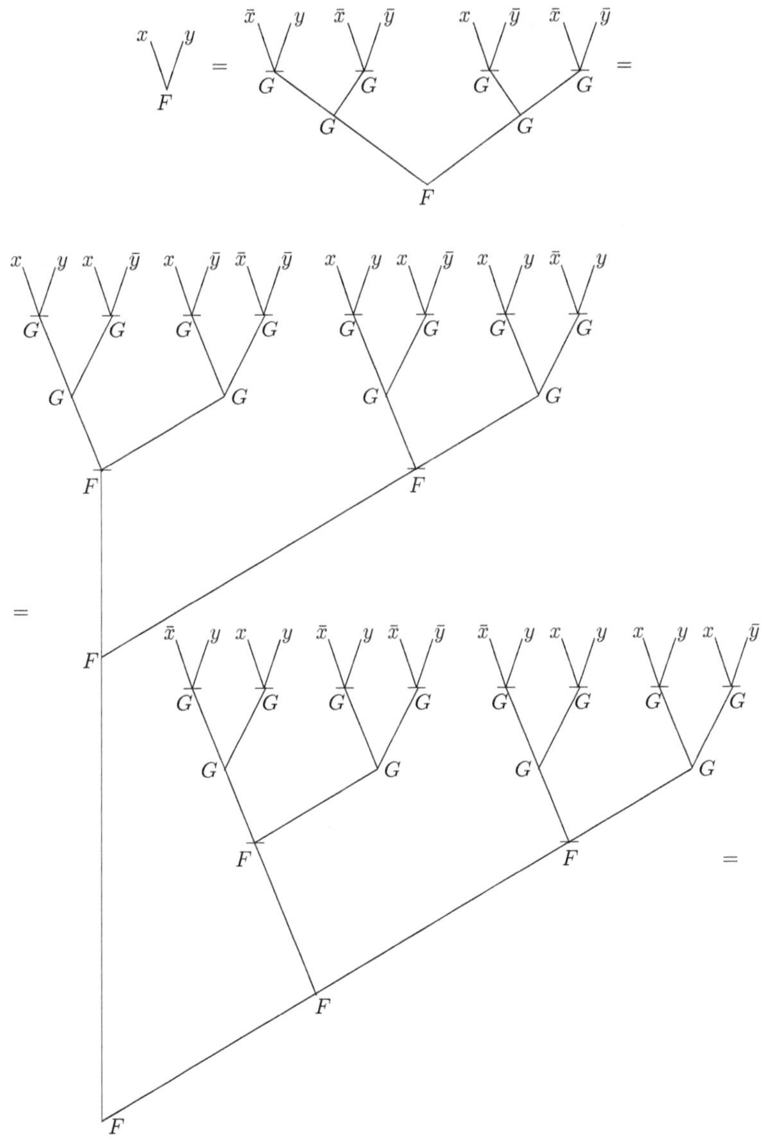

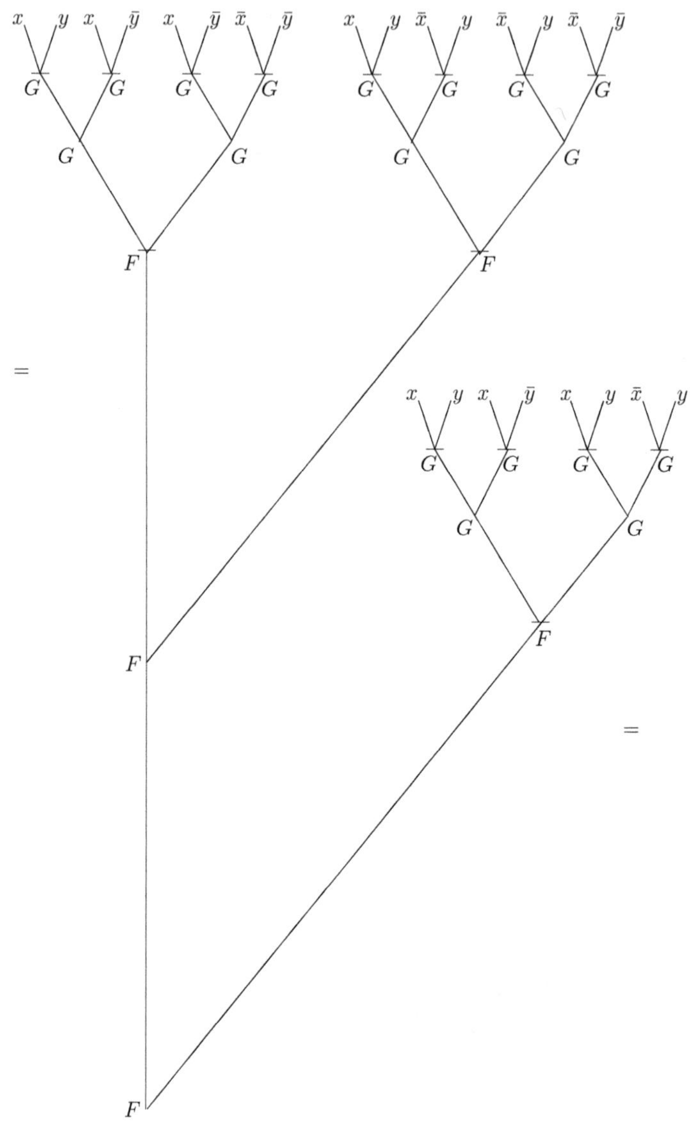
x y x ȳ x ȳ x̄ ȳ
x y x̄ y x̄ y x̄ ȳ
G G G G G G G G
G G G G
F F
=
x y x ȳ x y x̄ y
G G G G
G G
F
F
=
F

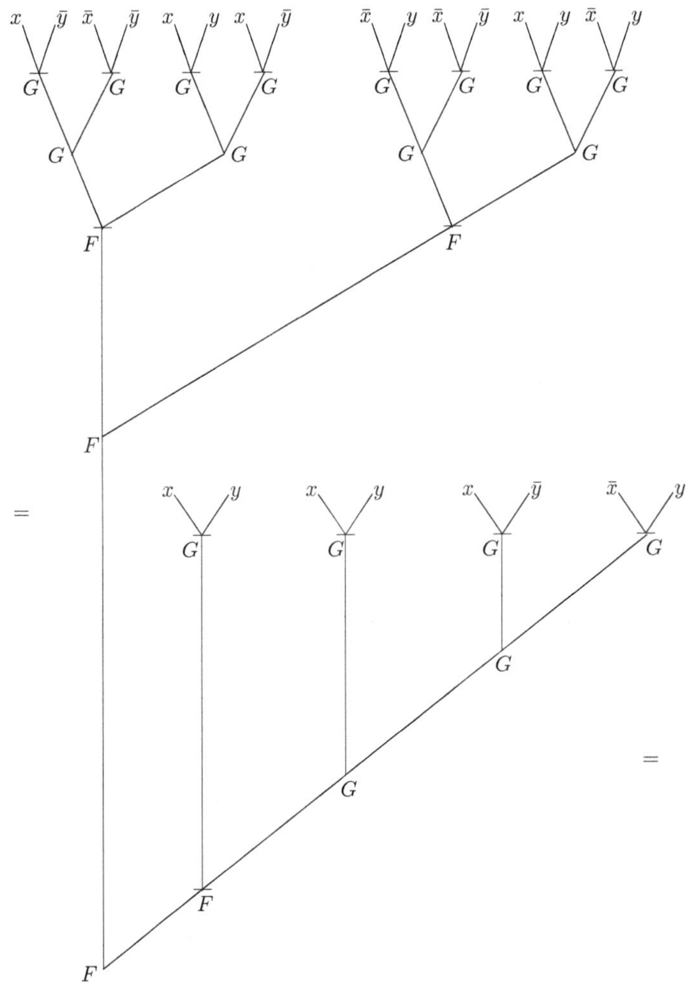
x ȳ x̄ ȳ x y x ȳ x̄ y x̄ ȳ x y x̄ y
G G G G G G G G
G G G G
F F
F
=
x y x y x ȳ x̄ y
G G G G
G
G
=
F
F

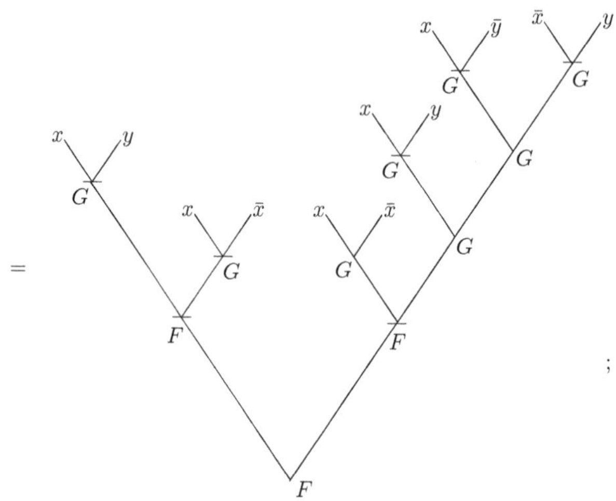

Example 5.3.2. The word 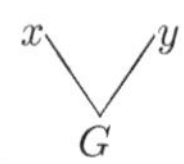 has the following b_2-canonical form on $\{x, y, z, G, F\}$:

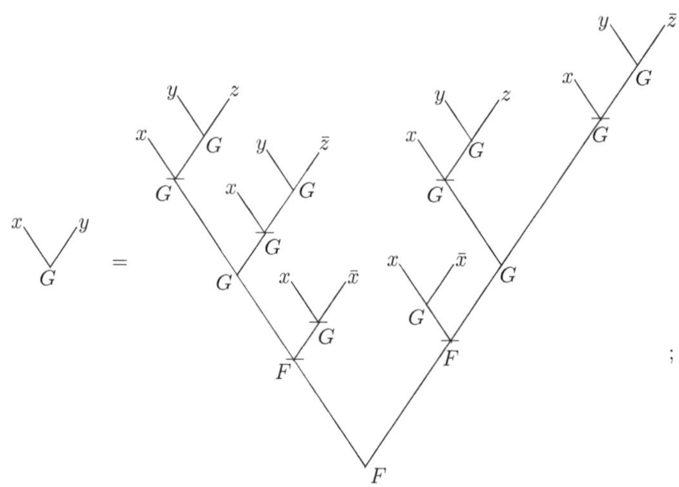

5.4. Hyperidentities of the variety of Boolean algebras: The case of m binary functional variables

Corollary 5.3.4 is valid also for any hyperidentity of the variety of Boolean algebras. For a proof the concept of a b_m-canonical word is defined inductively, and hyperidentities with m binary function variables $F_1, \ldots, F_m$ in the variety of Boolean algebras are defined.

Definition 5.4.1. A hyperterm of the form

$$F_m(\bar{F}_m(f_{(S_1,\ldots,S_{2^{m-2}})},\bar{F}_1(x_1,\bar{x}_1)),\bar{F}_m(F_1(x_1,\bar{x}_1),f_{(S_{2^{m-2}+1},\ldots,S_{2^{m-1}})}))$$

where $F_1,\ldots,F_m$ are binary function variables and

$$f_{(M_1,\ldots,M_{2^{m-2}})}=f_{(M_1,\ldots,M_{2^{m-2}})}(F_1,\ldots,F_{m-1},-,x_1,\ldots,x_n)$$

is a b_{m-1}-canonical word over $\{x_1,\ldots,x_n,F_1,\ldots,F_{m-1}\}$, is called b_m-canonical word (form) over $\{x_1,\ldots,x_n,F_1,\ldots,F_m\}$ and denoted by

$$f_{(S_1,\ldots,S_{2^{m-1}})}(F_1,\ldots,F_m,-,x_1,\ldots,x_n),$$

or, briefly, by $f_{(S_1,\ldots,S_{2^{m-1}})}(F_1,\ldots,F_m)$, by $f_{S[1,k]}\,(F_{[1,m]})$ or by $f_{S[1,k]}$, where $k=2^{m-1}$.

The equality of two b_m-canonical words has a meaning analogous to the equality of two b_2-canonical words.

A theorem of the type of Theorem 5.3.2 holds.

Theorem 5.4.2. *For any Boolean hyperterm w, $[w]\subseteq\{x_1,\ldots,x_n\}$, with at most m binary function variables $F_1,\ldots,F_m$ there exists a b_m-canonical word $\tilde{w}$ over $\{x_1,\ldots,x_n,F_1,\ldots,F_m\}$ such that the hyperidentity $w=\tilde{w}$ holds in the variety of Boolean algebras.*

Furthermore, under this condition the b_m-canonical word $\tilde{w}$ is uniquely determined and is called the b_m-canonical form of the hyperterm w over $\{x_1,\ldots,x_n,F_1,\ldots,F_m\}$; w is obtained from w after finitely many transitions realized with the help of finitely many hyperidentities, which are independent of w, of the variety of Boolean algebras.

Proof. For $m=1$ the theorem is proved. Assume that the theorem has been proved in the case of $m-1$ function variables.

In the word w we first replace all unary function variables (if there are any) by a single unary function variable (negation) and eliminate all double negations. In the second step, as in the proof of Theorem 5.3.2, we pass from the word w to a new word in whose tree all the binary function variables F_m are located below all the remaining binary function variables F_i, $i=1,\ldots,m-1$. This step is realized with the help of the distributivity hyperidentity and the hyperidentity (5.3.1).

Further, each subtree not containing F_m can be reduced by the induction hypothesis to b_m-canonical form over $\{x_1,\ldots,x_n,F_1,\ldots,F_{m-1}\}$. We denote these b_{m-1}-canonical words by $y_1,\ldots,y_\ell$ and, regarding the word obtained as a word in the single function variable F_m, we reduce it further to b_1-canonical form over $\{y_1, \ldots, y_\ell, F_m\}$. If after this we get $\bar{y}_j$ for some j, then we again reduce it to b_{m-1}-canonical form according to the equality

$$\overline{f_{(S_1,\ldots,S_{2m-1})}(F_1,\ldots,F_m)} = f_{(\bar{S}_1,\ldots,\bar{S}_{2m-1})}(F_1,\ldots,F_m). \tag{5.4.1}$$

where $\bar{S}_i = 2^{\{1,\ldots,n\}} \setminus S_i$. As a result we arrive at a word of the form in Fig. 5.4.

By induction we prove that (5.4.1) holds in the variety of Boolean algebras. For $m = 1$ we have

$$\overline{f_{S_1}(F_1)} = f_{\bar{S}_1}(F_1) \tag{5.4.2}$$

under the assumption that (5.4.1) is valid for the positive integer $m-1$. We use the following two hyperidentities:

$$\begin{aligned}
&\bar{F}(\bar{G}(x_1,x_2),\bar{G}(x_3,x_4)) \\
&\quad = G(G(\bar{F}(\bar{x}_i,\bar{x}_3),\bar{F}(\bar{x}_1,\bar{x}_4)),G(\bar{F}(\bar{x}_2,\bar{x}_3),\bar{F}(\bar{x}_2,\bar{x}_4))), && (5.4.3)\\
&F(F(\bar{F}(x,\bar{G}(x,\bar{x})),\bar{F}(G(x,\bar{x}),y)),\bar{F}(x,y)) \\
&\quad = F(\bar{F}(x,\bar{G}(x,\bar{x})),\bar{F}(G(x,\bar{x}),y)), && (5.4.4)
\end{aligned}$$

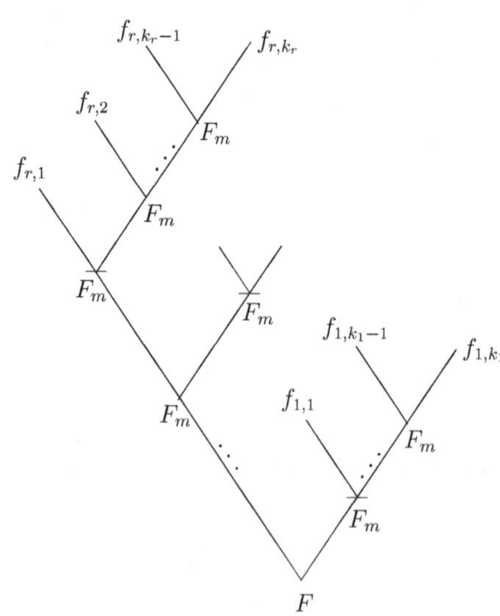

Fig. 5.4.

which are valid in the variety of Boolean algebras. For $k = 2^{m-1}$ we get ($\frac{IH}{(5.2.4)}$ — means "=(by the induction hypothesis and (5.2.4)")

$$
\begin{aligned}
&\overline{f_{S[1,k]}(F_1,\ldots,F_m)}\\
&\quad = \bar{F}_m(\bar{F}_m(f_{S[1,k/2]}(F_{[1,m-1]}),\bar{F}_1(x_1,\bar{x}_1))\bar{F}_m(F_1(x_1,\bar{x}_1),\\
&\qquad f_{S[k/2+1,k]}(F_{[1,m-1]})))\\
&\quad \overset{(5.4.3)}{=} F_m(F_m(\bar{F}_m(\overline{f_{S[1,k/2]}},\bar{F}_1(x_1,\bar{x}_1))),\bar{F}_m(F_1(x_1,\bar{x}_1),\bar{F}_1(x_1,\bar{x}_1))),
\end{aligned}
$$

$$
\begin{aligned}
&F_m(\bar{F}_m(F_1(x_1,\bar{x}_1),\overline{f_{S[k/2+1.k]}}),\bar{F}_m(\overline{f_{S[1,k/2]}},\overline{f_{S[k/2+1,k]}}))\\
&\quad \underset{(5.2.4)}{\overset{IH}{=}} F_m(F_m(\bar{F}_m(f_{\bar{S}[1,k/2]},\bar{F}_1(x_1,\bar{x}_1)),\bar{F}_m(F_1(x_1,\bar{x}_1),f_{\bar{S}[k/2+1,k]}))),
\end{aligned}
$$

$$
\begin{aligned}
&\bar{F}_m(f_{\bar{S}[1,k/2]},f_{\bar{S}[k/2+1,k]})\\
&\quad \overset{(5.4.4)}{=} F_m(\bar{F}_m(f_{\bar{S}[1,k/2]},\bar{F}_1(x_1,\bar{x}_1)),\bar{F}_m(F_1(x_1,\bar{x}_1),f_{\bar{S}[k/2+1,k]}))\\
&\quad = f_{\bar{S}[1,k]}(F_{[1,m]}),
\end{aligned}
$$

where $f_{\bar{S}[1,k]} = f_{(\bar{S}_1,\ldots,\bar{S}_k)}$

We also need the following property. For $k = 2^{m-1}$ the hyperidentities

$$
F_1(f_{S[1,k]}(F_{[1,m]}),f_{S'[1,k]}(F_{[1,m]})) = \begin{cases} f_{S\cup S'[1,k]}(F_{[1,m]}) & \text{if } m \text{ is odd,}\\ f_{S\cap S'[1,k]}(F_{[1,m]}) & \text{if } m \text{ is even}\end{cases} \tag{5.4.5}
$$

are valid in the variety of Boolean algebras, where

$$
\begin{aligned}
f_{S'[1,k]}(F_{[1,m]}) &= f_{S'_1,\ldots,S'_k}(F_{[1,m]}),\\
f_{S\cup S'[1,k]}(F_{[1,m]}) &= f_{S_1\cup S'_1,\ldots,S_k\cup S'_k}(F_{[1,m]}),\\
f_{S\cap S'[1,k]}(F_{[1,m]}) &= f_{S_1\cap S'_1,\ldots,S_k\cap S'_k}(F_{[1,m]}).
\end{aligned}
$$

The validity of (5.4.5) is proved by induction. We have

$$
F_1(f_{S_1}(F_1),f_{S'_1}(F_1)) = f_{S_1\cup S'_1}(F_1),
$$

therefore, (5.4.5) is valid for $m = 1$. Assume that it is valid for the positive integer $m-1$. We prove it for odd m (the case of even m is handled similarly).

The following hyperidentity in the variety of Boolean algebras is used:

$$F(G(\bar{G}(x,\bar{F}(x,\bar{x})),\bar{G}(F(x,\bar{x}),y)),G(\bar{G}(z,\bar{F}(x,\bar{x})),\bar{G}(F(x,\bar{x}),u)))$$
$$= G(\bar{G}(\bar{F}(\bar{x},\bar{z}),\bar{F}(x,\bar{x})),\bar{G}(F(x,\bar{x}),\bar{F}(\bar{y},\bar{u}))). \quad (5.4.6)$$

Then

$$F_1(f_{S[1,k]}(F_{[1,m]}),f_{S'[1,k]}(F_{[1,m]}))$$
$$= F_1(F_m(\bar{F}_m(f_{S[1,k/2]},\bar{F}_1(x_1,\bar{x}_1)),\bar{F}_m(F_1(x_1,\bar{x}_1),f_{S'[k/2+1,k]})))$$

$$F_m(\bar{F}_m(f_{S'[1,k/2]},\bar{F}(x_1,\bar{x}_1)),\bar{F}_m(F_1(x_1,\bar{x}_1),f_{S'[k/2+1,k]}))$$
$$\overset{(5.4.6)}{=} F_m(\bar{F}_m(\bar{F}_1(\overline{f_{S[1,k/2]},\overline{f_{S'[1,k/2]}}}),\bar{F}_1(x_1,\bar{x}_1)),\bar{F}_m(F_1(x_1,\bar{x}_1),$$
$$\bar{F}_1(\overline{f_{S[k/2+1,k]}},\overline{f_{S'[k/2+1,k]}})))$$
$$\underset{(5.4.1)}{\overset{IH}{=}} F_m(\bar{F}_m(f_{S\cup S'[1,k/2]},\bar{F}_1(x_1,\bar{x}_1)),\bar{F}_m(F_1(x_1,\bar{x}_1),f_{S\cup S'[k/2+1,k]}))$$
$$= f_{S\cup S'[1,k]}(F_{[1,m]}).$$

In the word in Fig. 5.4 each subtree beginning from $\bar{F}_m$ has the form

$$\bar{F}_m(f_{\tilde{S}[1,k/2]}(F_{[1,m-1]}),F_m(\cdots F_m(f_{S[1,k/2]}(F_{[1,m-1]}),f_{S'[1,k/2]}(F_{[1,m-1]}))\cdots)).$$

We begin the transformations with a subword of the form

$$F_m(f_{S[1,k/2]}(F_{[1,m-1]}),f_{S'[1,k/2]}(F_{[1,m-1]})).$$

If m is odd, then by (5.4.5)

$$f_{S[1,k/2]}(F_{1,m-1}) = F_1(f_{S\cup S'[1,k/2]}(F_{1,m-1}),f_{\overline{S'\setminus S}[1,k/2]}(F_{1,m-1}));$$

if m is even, then

$$f_{S[1,k/2]}(F_{1,m-1}) = F_1(f_{S\cap S'[1,k/2]}(F_{1,m-1}),f_{S\setminus S'[1,k/2]}(F_{1,m-1}));$$

If m is odd (the case of even m is handled similarly), then

$$F_m(f_S,f_{S'}) = F_m(F_1(f_{S\cup S'},f_{\overline{S'\setminus S}}),F_1(f_{S\cup S'},f_{\overline{S\setminus S'}}))$$
$$\overset{(5.3.2)}{=} F_m(F_1(\bar{F}_1(\overline{f_{\overline{S'\setminus S}}},\overline{f_{\overline{S\setminus S'}}}),f_{S\cup S'}),F_1(f_{S\cup S'},F_1(f_{\overline{S'\setminus S}},f_{\overline{S\setminus S'}})))$$
$$\underset{(5.4.5)}{\overset{(5.4.1)}{=}} F_m(f_{\overline{(S'\setminus S)\cap(S\setminus S')\cap(S\cup S')}},f_{(S\cup S')\cap\overline{(S'\setminus S)}\cap\overline{(S\setminus S')}})$$
$$= F_m(f_{S\cup S'},f_{S\cap S'}),$$

where $f_S = f_{S[1,k/2]}$ and $f_{S'} = f_{S'[1,k/2]}$.

To simplify expressions of the form $F_m(f_{\tilde{S}}, F_m(F_S, F_{S'}))$ we use the hyperidentity (5.3.3′) (in a way analogous to the way it was used in the preceding section):

$$\begin{aligned} F_m(f_{\tilde{S}}, F_m(f_S, f_{S'})) &= F_m(f_{\tilde{S}}, F_m(f_{S\cup S'}, f_{S\cap S'})) \\ &= F_m(f_{\tilde{S}\cup S\cup S'}, F_m(f_{\tilde{S}\cap(S\cup S')}, F_m(f_{\tilde{S}\cup(S\cap S')}, f_{\tilde{S}\cap S\cap S'}))) \\ &\overset{(5.3.3')}{=} F_m(f_{\tilde{S}\cup S\cup S'}, f_{\tilde{S}\cap S\cap S'}) \end{aligned}$$

(when using (5.3.3′) we take $x = f_{\tilde{S}\cup S\cup S'}$, $z = f_{\tilde{S}\cap S\cap S'}$, $= f_{\tilde{S}\cap(S\cup S')}$, and for repeated use $y = f_{\tilde{S}\cup(S\cap S')}$). After finitely many similar steps we get that

$$\bar{F}_m(f_{\tilde{S}}, F_m(\cdots F_m(f_S, f_{S'})\cdots)) = \bar{F}_m(f_{\tilde{S}\cup\cdots\cup S\cup S'}, f_{\tilde{S}\cap\cdots\cap S\cap S'}),$$

and the word in Fig. 5.4 take the form in Fig. 5.5, where the condition $S_i \subseteq S_i^*$, $i = 1, \ldots, r$, holds componentwise, i.e., $S_i^j \subseteq (S_i^j)^*$, $j = 1, \ldots, k/2$.

With the conditions $S_i^j \subseteq (S_i^j)^*$ taken into account, the subword $F_m(x, F_1(x, y))$ can be replaced in (5.3.4) by the subword $F_m(f_{S_i}, f_{S_i^*})$. Then, as in the preceding section, we get

$$\begin{aligned} &F_m(\bar{F}_m(f_{S_1}, f_{S_1^*}), \bar{F}_m(f_{S_2}, f_{S_2^*})) \\ &\quad = F_m(\bar{F}_m(f_{S_1\cup S_2}, \bar{F}_1(x_1, \bar{x}_1)), \bar{F}_m(F_1(x_1, \bar{x}_1), f_{S_1^*\cap S_2^*})) \end{aligned} \tag{5.4.7}$$

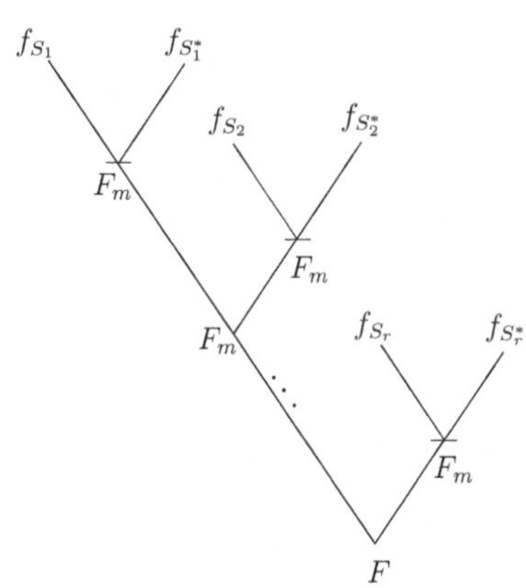

Fig. 5.5.

if m is odd, and

$$\begin{aligned}&F_m(\bar{F}_m(f_{S_1}, f_{S_1^*}), \bar{F}_m(f_{S_2}, f_{S_2^*}))\\&\quad = F_m(\bar{F}_m(f_{S_1 \cup S_2}, \bar{F}_1(x_1, \bar{x}_1)), \bar{F}_m(\bar{F}_1(x_1, \bar{x}_1), f_{S_1^* \cap S_2^*}))\end{aligned} \tag{5.4.8}$$

if m is even.

Further, we prove by induction that

$$f_{S_\emptyset[1,k]}(F_{[1,m]}) = \begin{cases} F_1(x_1, \bar{x}_1) & \text{if } m \text{ is even,} \\ \bar{F}_1(x_1, \bar{x}_1) & \text{if } m \text{ is odd,} \end{cases} \tag{5.4.9}$$

$$f_{S_\ell[1,k]}(F_{[1,m]}) = \begin{cases} \bar{F}_1(x_1, \bar{x}_1) \text{ if } m \text{ is even,} \\ F_1(x_1, \bar{x}_1) \text{ if } m \text{ is odd,} \end{cases} \tag{5.4.10}$$

where

$$S_\emptyset[1,k] = (\underbrace{\emptyset, \ldots, \emptyset}_{k}), \quad S_I[1,k] = (\underbrace{I, \ldots, I}_{k}), \quad I = 2^{\{1,\ldots,n\}}.$$

Let m be odd (the case of even m is handled similarly). Using (5.4.7)–(5.4.10) and the evenness of $m-1$, we get

$$\begin{aligned}
&F_m(\bar{F}_m(f_{S_1}, f_{S_1^*}), F_m(\bar{F}_m(f_{S_2}, f_{S_2^*}), \bar{(F)}_m(f_{S_2}, f_{S_3^*})))\\
&\overset{(5.4.8)}{=} F_m(\bar{F}_m(\bar{F}_1(x_1, \bar{x}_1), f_{S_1 \cup S_2}), F_m(\bar{F}_m(F_1(x_1, \bar{x}_1), f_{S_1^* \cap S_2^*}),\\
&\qquad \bar{F}_m(f_{S_3}, f_{S_3^*})))\\
&\overset{(5.4.9)}{=} F_m(\bar{F}_m(\bar{F}_1(x_1, \bar{x}_1), f_{S_1 \cup S_2}), F_m(\bar{F}_m(f_{S_\emptyset}, f_{S_1^* \cap S_2^*}), \bar{F}_m(f_{S_3}, f_{S_3^*})))\\
&\overset{(5.4.8)}{=} F_m(\bar{F}_m(\bar{F}_1(x_1, \bar{x}_1), f_{S_1 \cup S_2}), F_m(\bar{F}_m(f_{S_3}, \bar{F}_1(x_1, \bar{x}_1)), \bar{F}_m(F_1(x_1, \bar{x}_1),\\
&\qquad f_{S_1^* \cap S_2^* \cap S_3^*})))\\
&\overset{(5.4.10)}{=} F_m(\bar{F}_m(f_{S_1^* \cap S_2^* \cap S_3^*}, F_1(x_1, \bar{x}_1)), F_m(\bar{F}_m(f_{S_1}, f_{S_1 \cup S_2}), \bar{F}_m(f_{S_I}, f_{S_3})))\\
&\overset{(5.4.8)}{=} F_m(\bar{F}_m(f_{S_1^* \cap S_2^* \cap S_3^*}, F_1(x_1, \bar{x}_1)), F_m(\bar{F}_m(\bar{F}_1(x_1, \bar{x}_1), f_{S_1 \cup S_2 \cup S_3}),\\
&\qquad \bar{F}_m(F_1(x_1, \bar{x}_1), f_{S_I})))\\
&\underset{(5.4.4)}{\overset{(5.4.10)}{=}} F_m(\bar{F}_m(f_{S_1^* \cap S_2^* \cap S_3^*}, F_1(x_1, \bar{x}_1)), \bar{F}_m(F_1(x_1, \bar{x}_1), f_{S_1 \cup S_2 \cup S_3})).
\end{aligned}$$

Repeating the indicated process the necessary number of times, we reduce the word in Fig. 5.5 to the required form:

$$F_m(\bar{F}_m(f_{S_1\cup\cdots\cup S_r}(F_{[1,m-1]})), \bar{F}_1(x_1,\bar{x}_1), \bar{F}_m(F_1(x_1,\bar{x}_1), f_{S_1^*\cap\cdots\cap S_r^*}(F_{[1,m-1]}))),$$

if $r > 1$. But if $r = 1$, then, for example, for odd m, we have that

$$\begin{aligned}\bar{F}_m(f_{S_1}, f_{S_1^*}) &\overset{(5.2.4)}{=} F_m(\bar{F}_m(f_{S_1}, f_{S_1^*}), \bar{F}_m(\bar{F}_1(x_1,\bar{x}_1), F_1(x_1,\bar{x}_1)))\\ &\underset{(5.4.10)}{\overset{(5.4.9)}{=}} F_m(\bar{F}_m(f_{S_1}, f_{S_1^*}), \bar{F}_m(f_{S_\emptyset}, f_{S_I}))\\ &\overset{(5.4.8)}{=} F_m(\bar{F}_m(f_{S_1}, \bar{F}_1(x_1,\bar{x}_1)), \bar{F}_m(F_1(x_1,\bar{x}_1), f_{S_1^*})).\end{aligned}$$

Let us prove uniqueness. In the case $m = 1$ the uniqueness was proved in Theorem 5.2.2. We assume that uniqueness has been proved for a positive integer $m-1$, and prove it for m.

Suppose that in the variety of Boolean algebras we have the hyperidentity

$$\begin{aligned}&f_{(S_1,\ldots,S_{2^{m-1}})}(F_1,\ldots,F_m, -, x_1,\ldots,x_n)\\ &\quad= f_{(S'_1,\ldots,S'_{2^{m-1}})}(F_1,\ldots,F_m, -, x_1,\ldots,x_n),\end{aligned}$$

but $S_i \neq S'_i$ for some $i = 1,\ldots,2^{m-1}$. We consider two possible cases.

(1) $i \leqslant 2^{m-1}/2 = 2^{m-2}$. In this case there exists by the induction hypothesis a Boolean algebra B such that the hyperidentity

$$\begin{aligned}&f_{(S_1,\ldots,S_{2^{m-2}})}(F_1,\ldots,F_{m-1}, -, x_1,\ldots,x_n)\\ &\quad= f_{(S'_1,\ldots,S'_{2^{m-2}})}(F_1,\ldots,F_{m-1}, -, x_1,\ldots,x_n),\end{aligned}$$

holds in it, i.e., in B there exist values of the function and object variables $F_1,\ldots,F_{m-1}$ and $x_1,\ldots,x_n$ such that the indicated equality fails to hold. Denoting these values by $\varphi_1,\ldots,\varphi_{m-1}$ and $a_1,\ldots,a_n$, we get

$$\begin{aligned}&f_{(S_1,\ldots,S_{2^{m-1}})}(\varphi_1,\ldots,\varphi_{m-1},\varphi_1, -, a_1,\ldots,a_n)\\ &\quad= \overline{f_{(S_1,\ldots,S_{2^{m-2}})}(\varphi_1,\ldots,\varphi_{m-1}, -, a_1,\ldots,a_n)}\\ &\quad\neq \overline{f_{(S'_1,\ldots,S'_{2^{m-2}})}(\varphi_1,\ldots,\varphi_{m-1}, -, a_1,\ldots,a_n)}\\ &\quad= f_{(S'_1,\ldots,S'_{2^{m-1}})}(\varphi_1,\ldots,\varphi_{m-1},\varphi_1, -, a_1,\ldots,a_n).\end{aligned}$$

Contradiction.

(2) The case $i > 2^{m-1}/2 = 2^{m-2}$ is handled similarly, except that here in choosing the value of F_m we set $F_m = (+)$ if $\varphi_1 = (\cdot)$ and $F_m = (\cdot)$ if $\varphi = (+)$.

The theorem is proved. □

Corollary 5.4.3. *A hyperidentity $w_1 = w_2$ with binary function variables $F_1, \ldots, F_m$ holds in the variety of Boolean algebras if and only if the b_m-canonical forms of the hyperterms w_1 and w_2 over $\{x_1, \ldots, x_n, F_1, \ldots, F_m\}$ coincide, where $\{x_1, \ldots, x_n\} = [w_1] \cap [w_2]$.*

Corollary 5.4.4. *The variety of Boolean algebras has a decidable hyperequational theory.*

Corollary 5.4.5. *The variety of Boolean algebras has a finite basis of hyperidentities. In the variety of Boolean algebras any hyperidentity is a consequence of the hyperidentities (1)–(12) in Corollary 5.3.4.*

Proof. Denote by $H_{(m)}$ the set of hyperidentities with at most m binary function variables that hold in the variety of Boolean algebras. Any hyperidentity in $H_{(m)}$ is a consequence of all the hyperidentities that are used in the proof of Theorem 5.4.2, i.e., of the hyperidentities in $H_{(m-1)} \cup H_{(2)} \cup H_{(1)}$. As a result any hyperidentity in $H_{(m)}$ is a consequence of the hyperidentities in $H_{(2)}$. It remains to invoke Corollary 5.3.4. □

Remark 5.4.6. An algebra $Q(+, \cdot, ')$ with two binary operations and one unary operation that satisfies the hyperidentities of the variety of Boolean algebras is not necessarily a Boolean lattice. Indeed, in any Boolean lattice $Q(+, \cdot, ')$ the hyperidentities (1)–(12) hold. We consider the superproduct $Q = Q_1 \times Q_1$, as an algebra with the following five operations (acting componentwise):

$$A = (+, +), \quad B = (+, \cdot), \quad C = (\cdot, +), \quad D = (\cdot, \cdot), \quad E = (', ').$$

In this superproduct $Q(A, B, C, D, E)$ the indicated hyperidentities also hold. However, a direct verification shows that the absorption identity does not hold in the algebra $Q(A, B)$; therefore $Q(A, B, E)$ is not a Boolean lattice, although the hyperidentities (1)–(12) hold in it. Thus an algebra with two binary operations and one unary operation and one nullary operation that satisfies the hyperidentities of the variety of Boolean algebras is not necessarily a Boolean algebra.

5.5. Super-Boolean algebras

We have proved the following result.

Theorem 5.5.1. *Any hyperidentity of the variety of Boolean algebras is a consequence of the following eleven*[b] *hyperidentities:* (2.3.1)–(2.3.4) *and also*

$$F(x) = G(x) \qquad (= \overline{x}), \tag{5.5.1}$$

$$\overline{\overline{x}} = x, \tag{5.5.2}$$

$$X(x, \overline{x}) = X(y, \overline{y}), \tag{5.5.3}$$

$$X\left(x, \overline{X(y, \overline{y})}\right) = x, \tag{5.5.4}$$

$$\overline{X\left(\overline{Y(x, y)}, z\right)} = Y\left(\overline{X(\overline{x}, z)}, \overline{X(\overline{y}, z)}\right), \tag{5.5.5}$$

$$X(Y(x, y), Y(x, z)) = X\left(Y(\overline{Y(\overline{z}, \overline{y})}, x), Y(x, Y(y, z))\right), \tag{5.5.6}$$

$$\begin{aligned} &X\left(\overline{X(x, Y(x, y))}, \overline{X(z, Y(z, u))}\right) \\ &\quad = X\left(\overline{X(Y(x, z), Y(u, \overline{u}))}, \overline{X(Y(u, \overline{u}), \overline{Y(\overline{Y(x, y)}, \overline{Y(z, u)})})}\right). \end{aligned} \tag{5.5.7}$$

Since the variety of Boolean algebras has no proper subvarieties (Example 1.5.10), this yields a characterization of the hyperidentities of an arbitrary Boolean algebra.

Corollary 5.5.2. *Any hyperidentity of a Boolean algebra is a consequence of the hyperidentities (2.3.1)–(2.3.4), (5.5.1)–(5.5.7).*

Proof. See the proof of Corollary 2.3.18. □

As a consequence of Theorem 5.5.1 we also obtain a characterization of the regular identities of Boolean algebras (or the variety of Boolean algebras).

---

[b]Hyperidentity (9) from [253], which is a consequence of our hyperidentities (2.3.1)–(2.3.4) (according to Theorem 2.3.1) is omitted here. Moreover, we have written $Y(u, \overline{u})$ in the last hyperidentity (5.5.7) in place of the term $Y(v, \overline{v})$ (according to the hyperidentity (5.5.3)).

An identity of a Boolean algebra $\mathfrak{A} = Q(+, \cdot, ^{-})$ is said to be regular, if it remains true in $\mathfrak{A}$ when $(\cdot)$ is replaced by $(+)$. For example, the identity

$$\overline{x+y} = \overline{x} \cdot \overline{y}$$

is not regular, while the identity

$$\overline{x + \overline{y \cdot \overline{y}}} = \overline{x} \cdot \overline{y + \overline{y}}$$

is regular. The identities in which the operation $(\cdot)$ does not appear are also assumed to be regular. Using a duality principle we pass to the following result.

Corollary 5.5.3. *Any regular identity of a Boolean algebra (or the variety of Boolean algebras) is a consequence of the finite system of identities obtained from the hyperidentities (2.3.1)–(2.3.4), (5.5.2)–(5.5.7) under the changes of variables* $X \to (+), (\cdot)$ *and* $Y \to (+), (\cdot)$.

An algebra with two binary operations and one unary operation (and one nullary operation) satisfying the hyperidentities of the variety of Boolean algebras is not a Boolean lattice (Boolean algebra) in general.

A complete description of the algebras with two binary operations and one unary operation that satisfy the hyperidentities of the variety of Boolean algebras is given in Theorem 5.5.8 (see below).

Definition 5.5.4. An algebra $\mathfrak{A} = \langle Q; \Sigma \rangle$ is called a super-Boolean algebra if it satisfies the hyperidentities of the variety of Boolean algebras, that is, the hyperidentities (2.3.1)–(2.3.4) and (5.5.1)–(5.5.7) (see [269]).

Example 5.5.5. A super-product of Boolean algebras is a super-Boolean algebra.

Lemma 5.5.6. *If a super-Boolean algebra* $\mathfrak{A} = \langle Q; \Sigma \rangle$ *with binary and unary operations is subdirectly irreducible, then the cardinality of its set of binary operations is at most* 2.

Proof. Here it suffices to add to the proof of Lemma 2.3.4 that the relations $\Theta_{i,j}$, where $i, j = 1, 2, 3$ and $i < j$, are also compatible with the unary operation "$-$" of the super-Boolean algebra $\mathfrak{A} = \langle Q; \Sigma \rangle$:

$$x \Theta_{i,j} y \Longrightarrow \overline{x} \Theta_{i,j} \overline{y}, \qquad x, y \in Q.$$

Indeed, in any Boolean algebra (that is, in the variety of Boolean algebras) we have the hyperidentity

$$X\left(\overline{X(x,Y(x,y))}, Y(\overline{X(x,Y(x,y))}, \overline{X(y,Y(y,x))})\right) = \overline{X(x,Y(x,y))}, \tag{5.5.8}$$

and therefore this hyperidentity is also valid in any super-Boolean algebra $\mathfrak{A} = \langle Q; \Sigma \rangle$. Hence,

$$\begin{aligned} f_{i,j}(\overline{x},\overline{y}) &= A_i(\overline{x}, A_j(\overline{x},\overline{y})) \\ &= A_i\left(\overline{A_i(x,A_j(x,y))}, A_j(\overline{A_i(x,A_j(x,y))}, \overline{A_i(y,A_j(y,x))})\right) \\ &\overset{(5.5.8)}{=} \overline{A_i(x,A_j(x,y))} = \overline{x}. \end{aligned}$$

By an analogous computation we obtain the equality

$$f_{i,j}(\overline{y},\overline{x}) = \overline{y}.$$

Thus, $\overline{x}\Theta_{i,j}\overline{y}$, and the relation $\Theta_{i,j}$ is a congruence for the super-Boolean algebra $\mathfrak{A}$. □

Corollary 5.5.7. *In the category of T-algebras, where $T = \{1, 2, \ldots\}$, any super-Boolean algebra $\mathfrak{A} = \langle Q; \Sigma \rangle$ is isomorphic to a subdirect product of super-Boolean algebras with one or two binary operations.*

Thus, the description of the super-Boolean algebras reduces to those of the super-Boolean algebras with one or two binary operations.

In the case of one binary operation the hyperidentities (2.3.1)–(2.3.4), (5.5.1)–(5.5.7) of a super-Boolean algebra become identities, and this system of identities is redundant. For example, the identities corresponding to the hyperidentities (5.5.6) and (5.5.7) follow from the remaining ones (see also Corollary 5.2.4). Thus, a super-Boolean algebra with one binary operation is an algebra $Q(+,^-)$ with the following identities:

$$x + x = x, \tag{5.B_1}$$

$$x + y = y + x, \tag{5.B_2}$$

$$x + (y + z) = (x + y) + z, \tag{5.B_3}$$

$$\overline{\overline{x}} = x, \tag{5.B_4}$$

$$x + \overline{x} = y + \overline{y}, \tag{5.B_5}$$

$$x + (\overline{y + \overline{y}}) = x, \tag{5.B_6}$$

$$\overline{\overline{x + y} + z} = \overline{\overline{x} + z} + \overline{\overline{y} + z}. \tag{5.B_7}$$

It is obvious that if $Q(+, \cdot, ^{-})$ is a Boolean algebra, then the subalgebra (reduct) $Q(+, ^{-})$ is a super-Boolean algebra. The converse is also true: if $Q(+, ^{-})$ is a super-Boolean algebra, then, defining the following multiplication on Q:

$$x \cdot y = \overline{\overline{x} + \overline{y}},$$

we obtain a Boolean algebra $Q(+, \cdot, ^{-})$.

It remains to characterize the super-Boolean algebras with two binary operations. The representation of the super-Boolean algebra as a sum of disjoint subsystems is not possible, since in a super-Boolean algebras there exist the hyperidentities (and hence, identities) which are not homogeneous.

In order to characterize the super-Boolean algebra with two binary operations we introduce the following concept of a Boolean sum and we shall prove that any super-Boolean algebra with two binary and one unary operations is a Boolean sum of Boolean lattices [256].

Let $\mathfrak{A} = \langle Q; \Omega \cup \{F\}\rangle$ be a T-algebra with one unary operation F. Let $\langle Q_i; \Omega\rangle$, $i \in I$, be subsystems (subalgebras) for the algebra $\mathfrak{A}$, and let $\mathfrak{A}_i = \langle Q_i; \Omega \cup \{F_i\}\rangle$ be an algebra with one unary operation F_i for any $i \in I$. The algebra $\mathfrak{A} = \langle Q; \Omega \cup \{F\}\rangle$ is called the *Boolean sum* of the algebras $\mathfrak{A}_i = \langle Q_i; \Omega \cup \{F_i\}\rangle$, $i \in I$, if:

(a) $Q_i \cap Q_j = \emptyset$, where $i, j \in I$, $i \neq j$;
(b) $Q = \bigcup_{i\in I} Q_i$;
(c) one can define a partial order "$\leqslant$" on the index set I such that $I(\leqslant)$ is a Boolean algebra;
(d) if $i \leqslant j$, then there exists a (bi)isomorphism

$$(\varphi_{i,j}, \widetilde{\varepsilon}) : \mathfrak{A}_i \Longrightarrow \mathfrak{A}_j, \qquad i, j \in I,$$

where $\widetilde{\varepsilon}(F_i) = F_j$, $\widetilde{\varepsilon}(A) = A$ for any $A \in \Omega$, and $\varphi_{i,i} = \varepsilon$ and $\varphi_{i,j} \cdot \varphi_{j,k} = \varphi_{i,k}$, where $i \leqslant j \leqslant k$;
(e) for any $i \in I$ there exists a (bi)isomorphism

$$(h_{i,\overline{i}}, \widetilde{\varepsilon}) : \mathfrak{A}_i \Longrightarrow \mathfrak{A}_{\overline{i}},$$

where $h_{i,\overline{i}}^{-1} = h_{\overline{i},i}$ and $\varphi_{i,1} = h_{i,\overline{i}} \cdot \varphi_{\overline{i},1}$ and 1 is the identity element of the Boolean algebra $I(\leqslant)$ and $\overline{i}$ is the complement of i in this lattice;

(f) for any $A \in \Omega$, $|A| = n \geqslant 2$, and for any $x_1, \ldots, x_n \in Q$

$$A(x_1, \ldots, x_n) = A(\varphi_{i_1,i_0}(x_1), \ldots, \varphi_{i_n,i_0}(x_n)),$$

where $x_1 \in Q_{i_1}, \ldots, x_n \in Q_{i_n}$, $i_1, \ldots, i_n \in I$, $i_0 = i_1 + \cdots + i_n$;

(g) for the unary operation F and any $x \in Q$

$$F(x) = h_{i,\bar{i}}[F_i(x)],$$

where $x \in Q_i$.

Theorem 5.5.8 ([256]). *An algebra $\mathfrak{A} = Q(+, \cdot, ^-)$ with two binary and one unary operations satisfies the hyperidentities of the variety of Boolean algebras (that is, is a super-Boolean algebra) if and only if $\mathfrak{A}$ is a Boolean lattice or a Boolean sum of Boolean lattices.*

First we shall prove an assertion about the values of Boolean terms. Here a Boolean term is a term of (at most) two binary operation symbols "+", "·" and one unary operation symbol. If ω is a Boolean term of the object variables $x_1, \ldots, x_n$, then we write $\omega = \omega(x_1, \ldots, x_n)$.

Lemma 5.5.9. *Let $\omega = \omega(x_1, \ldots, x_n)$ be a Boolean term. If the algebra $\mathfrak{A} = Q(+, \cdot, ^-)$ is a Boolean sum of Boolean lattices Q_i, $i \in I$, then for the values $x_1 = a_1 \in Q_{i_1}, \ldots, x_n = a_n \in Q_{i_n}$ we have $\omega(a_1, \ldots, a_n) \in Q_k$, where $k = \omega(i_1, \ldots, i_n)$ is the value of $\omega = \omega(x_1, \ldots, x_n)$ in the Boolean lattice $I = I(+, \cdot, ^-)$ for $x_1 = i_1 \in I, \ldots, x_n = i_n \in I$. Moreover,*

$$\omega(\varphi_{i_1,1}(a_1), \ldots, \varphi_{i_n,1}(a_n)) = \varphi_{k,1}(\omega(a_1, \ldots, a_n)).$$

Proof (By induction). If $\omega = x$, then the assertion is obvious. Assume that $\omega = \omega_1 + \omega_2$ and that the assertion has been proved for ω_1 and ω_2. Without loss of generality we may assume that $\omega = \omega(x_1, \ldots, x_n)$, $\omega_1 = \omega_1(x_1, \ldots, x_n)$, $\omega_2 = \omega_2(x_1, \ldots, x_n)$. Then by the induction hypothesis

$$\omega_1(\varphi_{i_1,1}(a_1), \ldots, \varphi_{i_n,1}(a_n)) = \varphi_{k_1,1}(\omega_1(a_1, \ldots, a_n)),$$
$$\omega_2(\varphi_{i_1,1}(a_1), \ldots, \varphi_{i_n,1}(a_n)) = \varphi_{k_2,1}(\omega_2(a_1, \ldots, a_n)),$$

where $k_1 = \omega_1(i_1, \ldots, i_n)$, $k_2 = \omega_2(i_1, \ldots, i_n)$, $\omega_1(a_1, \ldots, a_n) \in Q_{k_1}$, $\omega_2(a_1, \ldots, a_n) \in Q_{k_2}$. Since

$$k = \omega(i_1, \ldots, i_n) = \omega_1(i_1, \ldots, i_n) + \omega_2(i_1, \ldots, i_n) = k_1 + k_2,$$

we have

$$\begin{aligned}
&\omega(\varphi_{i_1,1}(a_1),\dots,\varphi_{i_n,1}(a_n))\\
&\quad=\omega_1(\varphi_{i_1,1}(a_1),\dots,\varphi_{i_n,1}(a_n))+\omega_2(\varphi_{i_1,1}(a_1),\dots,\varphi_{i_n,1}(a_n))\\
&\quad=\varphi_{k_1,1}(\omega_1(a_1,\dots,a_n))+\varphi_{k_2,1}(\omega_2(a_1,\dots,a_n))\\
&\quad=\varphi_{k,1}(\varphi_{k_1,k}(\omega_1(a_1,\dots,a_n)))+\varphi_{k,1}(\varphi_{k_2,k}(\omega_2(a_1,\dots,a_n)))\\
&\quad=\varphi_{k,1}(\varphi_{k_1,k}(\omega_1(a_1,\dots,a_n))+\varphi_{k_2,k}(\omega_2(a_1,\dots,a_n)))\\
&\quad=\varphi_{k,1}(\omega_1(a_1,\dots,a_n)+\omega_2(a_1,\dots,a_n))=\varphi_{k,1}(\omega(a_1,\dots,a_n))
\end{aligned}$$

where $\omega(a_1,\dots,a_n)\in Q_k$.

The case $\omega=\omega_1\cdot\omega_2$ is considered analogously.

Let $\omega(x_1,\dots,x_n)=\overline{\pi(x_1,\dots,x_n)}$, and let

$$\pi(\varphi_{i_1,1}(a_1),\dots,\varphi_{i_n,1}(a_n))=\varphi_{\bar{k},1}(\pi(a_1,\dots,a_n)),$$

where $\bar{k}=\pi(i_1,\dots,i_n)$ and $\pi(a_1,\dots,a_n)\in Q_{\bar{k}}$. Then

$$\begin{aligned}
&\omega(\varphi_{i_1,1}(a_1),\dots,\varphi_{i_n,1}(a_n))=(\pi(\varphi_{i_1,1}(a_1),\dots,\varphi_{i_n,1}(a_n)))'\\
&\quad=(\varphi_{\bar{k},1}(\pi(a_1,\dots,a_n)))'=\varphi_{\bar{k},1}((\pi(a_1,\dots,a_n))')\\
&\quad=\varphi_{k,1}(h_{\bar{k},k}((\pi(a_1,\dots,a_n))'))=\varphi_{k,1}(\overline{\pi(a_1,\dots,a_n)})\\
&\quad=\varphi_{k,1}(\omega(a_1,\dots,a_n)),
\end{aligned}$$

where $\omega(a_1,\dots,a_n)\in Q_k$, $\bar{k}=\overline{\pi(a_1,\dots,a_n)}=\omega(i_1,\dots,i_n)$, completing the proof of Lemma 5.5.9.

Proof of Theorem 5.5.8 (*Sufficiency*). Assume that the algebra $\mathfrak{A}=Q(+,\cdot,^-)$ with the two binary operation and one unary operation is a Boolean sum of Boolean lattices Q_i, $i\in I$, and let $w_1=w_2$ be an arbitrary hyperidentity of the variety of Boolean algebras. We shall show that this hyperidentity is also valid in $\mathfrak{A}$. Under a change of the binary functional variables by the binary operations of $\mathfrak{A}$ the equality $w_1=w_2$ becomes an identity, which is valid in any Boolean algebra. We must show that this identity is also valid in $\mathfrak{A}$.

Without loss of generality we may assume that $w_1 = w_1(x_1, \ldots, x_n)$ and $w_2 = w_2(x_1, \ldots, x_n)$. According to Lemma 5.5.9 we have

$$\omega_1(\varphi_{i_1,1}(a_1), \ldots, \varphi_{i_n,1}(a_n)) = \varphi_{k_1,1}(w_1(a_1, \ldots, a_n)),$$
$$\omega_2(\varphi_{i_1,1}(a_1), \ldots, \varphi_{i_n,1}(a_n)) = \varphi_{k_2,1}(w_2(a_1, \ldots, a_n)),$$

where $k_1 = w_1(i_1, \ldots, i_n)$, $k_2 = w_2(i_1, \ldots, i_n)$. Since the identity $w_1(x_1, \ldots, x_n) = w_2(x_1, \ldots, x_n)$ is also valid in the Boolean algebra I, we have $k_1 = k_2 = k$. In addition, the identity $w_1(x_1, \ldots, x_n) = w_2(x_1, \ldots, x_n)$ is also valid in the Boolean algebra Q_1, where 1 is the identity of the Boolean algebra I. Therefore

$$\omega_1(\varphi_{i_1,1}(a_1), \ldots, \varphi_{i_n,1}(a_n)) = \omega_2(\varphi_{i_1,1}(a_1), \ldots, \varphi_{i_n,1}(a_n)).$$

Hence,

$$\varphi_{k_1,1}(w_1(a_1, \ldots, a_n)) = \varphi_{k_1,1}(w_2(a_1, \ldots, a_n))$$

and, because $\varphi_{k_1,1}$ is injective,

$$w_1(a_1, \ldots, a_n) = w_2(a_1, \ldots, a_n)$$

for any $a_1, \ldots, a_n \in Q = \bigcup_{i \in I} Q_i$.

(*Necessity*). We must prove that any super-Boolean algebra $\mathfrak{A} = Q(+, \cdot, ^-)$ with two binary operations is a Boolean sum of Boolean lattices. In the proof, in addition to hyperidentity (5.5.8), we also use the following hyperidentities of the variety of Boolean algebras:

$$X(x, Y(x, X(Y(x, \overline{X(x, \bar{x})}), \overline{Y(x, X(x, \bar{x}))}))) = x, \tag{5.5.9}$$
$$X(X(Y(x, \overline{X(x, \bar{x})}), \overline{Y(x, X(x, \bar{x}))}), Y(x, X(Y(x, \overline{X(x, \bar{x})}),$$
$$\overline{Y(x, X(x, \bar{x}))}))) = X(Y(x, \overline{X(x, \bar{x})}), \overline{Y(x, X(x, \bar{x}))}), \tag{5.5.10}$$
$$X(Y(y, \overline{X(y, \bar{y})}), \overline{Y(y, X(y, \bar{y}))}) = x, \tag{5.5.11}$$

where y denotes the hyperterm

$$X(Y(x, \overline{X(x, \bar{x})}), \overline{Y(x, X(x, \bar{x}))}).$$

Furthermore,

$$\begin{aligned}&X(Y(X(x,y),\overline{X(X(x,y),\overline{X(x,y)})}),\overline{Y(X(x,y),X(X(x,y),\overline{X(x,y)}))})\\&\quad=Y(X(Y(x,\overline{X(x,\bar{x})}),\overline{Y(x,X(x,\bar{x}))}),\\&\qquad X(Y(y,\overline{X(y,\bar{y})}),\overline{Y(y,X(y,\bar{y}))})),\end{aligned}\tag{5.5.12}$$

$$X(x,Y(y,X(Y(y,\overline{X(y,\bar{y})}),\overline{Y(y,X(y,\bar{y}))})))=X(x,Y(x,y)),\tag{5.5.13}$$

$$X(z,Y(y,\overline{X(x,\bar{x})}))=X(z,Y(z,y)),\tag{5.5.14}$$

$$Y(Y(z,y),\overline{X(x,\bar{x})})=Y(X(y,Y(y,z)),\overline{X(x,\bar{x})}),\tag{5.5.15}$$

$$X(z,X(y,\overline{Y(x,\bar{x})}))=X(X(y,Y(y,z)),\overline{Y(x,\bar{x})}),\tag{5.5.16}$$

$$Y(z,X(y,\overline{Y(x,\bar{x})}))=X(z,Y(z,y)),\tag{5.5.17}$$

$$\overline{X(y,\overline{Y(x,\bar{x})})}=Y(\bar{y},\overline{X(x,\bar{x})}),\tag{5.5.18}$$

$$\begin{aligned}&X(X(x,y),Y(X(x,y),X(u,v)))\\&\quad=X(X(x,Y(x,u)),X(y,Y(y,v))),\end{aligned}\tag{5.5.19}$$

$$X(Y(x,y),Y(Y(x,y),X(x,y)))=Y(x,y),\tag{5.5.20}$$

$$\overline{X(\bar{x},\overline{X(y,z)})}=X(\overline{X(\bar{x},\bar{y})},\overline{X(\bar{x},\bar{z})}),\tag{5.5.21}$$

$$\begin{aligned}&Y(X(Y(x,\overline{X(x,\bar{x})}),\overline{Y(x,X(x,\bar{x}))}),\\&X(Y(x,y),X(Y(Y(x,y),\overline{X(Y(x,y),\overline{Y(x,y)})}),\\&\overline{Y(Y(x,y),X(Y(x,y),\overline{Y(x,y)}))})))\\&\quad=X(Y(Y(x,y),\overline{X(Y(x,y),\overline{Y(x,y)})}),\\&\qquad\overline{Y(Y(x,X(x,y)),X(Y(x,y),\overline{Y(x,y)}))}),\end{aligned}\tag{5.5.22}$$

$$Y(X(y,\overline{Y(x,\bar{x})}),X(X(y,z),\overline{Y(x,\bar{x})}))=X(X(y,z),\overline{Y(x,\bar{x})}),\tag{5.5.23}$$

$$\begin{aligned}&X(Y(\bar{x},\overline{X(\bar{x},x)}),\overline{Y(\bar{x},X(\bar{x},x))})\\&\quad=\overline{X(Y(x,\overline{X(x,\bar{x})}),\overline{Y(x,X(x,\bar{x}))})},\end{aligned}\tag{5.5.24}$$

$$\begin{aligned}&\overline{X(Y(z,\overline{X(z,\bar{z})}),\overline{Y(z,X(z,\bar{z}))})}\\&\quad=X(\overline{X(Y(u,\overline{X(u,\bar{u})}),\overline{Y(u,X(u,\bar{u}))})},\\&\qquad\overline{X(Y(v,\overline{X(v,\bar{v})}),\overline{Y(v,X(v,\bar{v}))})}),\end{aligned}\tag{5.5.25}$$

where z, u and v denote the hyperterms $X(x,y)$, $X(x,Y(x,y))$ and $X(y,Y(y,x))$,

$$Y(\overline{X(Y(x,\overline{X(x,\bar{x})}),\overline{Y(x,X(x,\bar{x}))})},X(x,\bar{x}))=Y(x,X(x,\bar{x})), \tag{5.5.26}$$

$$Y(\overline{Y(y,X(x,\bar{x}))},X(x,\overline{Y(x,\bar{x})}))=x, \tag{5.5.27}$$

where y denotes the hyperterm $X(Y(x,\overline{X(x,\bar{x})}),\overline{Y(x,X(x,\bar{x}))})$.

A direct check shows that each of the hyperidentities (5.5.9)–(5.5.27) is valid in any Boolean algebra. Hence, according to the definition of a super-Boolean algebra each of them is valid in any super-Boolean algebra.

On the underlying set Q of a super-Boolean algebra $\mathfrak{A}=Q(+,\cdot,-)$ we define a relation Θ in the following way:

$$x\Theta y \Longleftrightarrow x+xy=x, \qquad y+yx=y.$$

We see from the proof of Lemma 2.3.4 that Θ is an equivalence relation. Let $Q=\bigcup_{i\in I}Q_i$ be the corresponding decomposition. Each subset Q_i is closed under addition and multiplication:

$$\begin{aligned}a,b\in Q_i &\Longrightarrow a\Theta b\Longrightarrow a+ab=a, \quad b+ba=b\\ &\Longrightarrow \begin{cases}a+a(a+b)=a+a+ab=a+ab+a,\\ a+b(a+b)a=a+b+a+ab=a+b\end{cases}\\ &\Longrightarrow a\Theta a+b\Longrightarrow a+b\in Q_i.\end{aligned}$$

Similarly,

$$a,b\in Q_i\Longrightarrow a\cdot b\in Q_i.$$

We define a new unary operation x' on Q as follows:

$$x'=x\cdot\overline{x+\bar{x}}+\overline{x\cdot(x+\bar{x})},$$

and we will show that each subset Q_i is also closed under this unary operation, that is, if $a\in Q_i$, then $a'\in Q_i$. This uses the hyperidentities (5.5.9) and (5.5.10). Making the substitutions $X\to(+)$ and $Y\to(\cdot)$ we obtain from (5.5.9) and (5.5.10) the identities $x+x\cdot x'=x$, $x'+x'\cdot x=x'$. Hence $x\Theta x'$ for any $x\in Q$.

Now we check the axioms for Boolean algebras for each $Q_i(+,\cdot,')$, $i\in I$. Idempotence, commutativity, associativity, and distributivity of the binary operations obviously hold. The condition $(a')'=a$ follows from (5.5.11) under the substitution $X\to(+)$, $Y\to(\cdot)$. The condition $(a+b)'=a'\cdot b'$ follows from (5.5.12), and the condition $a+b\cdot b'=a$ for arbitrary $a,b\in Q_i$

follows from (5.5.13) under the substitution $X \to (+)$, $Y \to (\cdot)$. We notice that the zero and unity elements for $Q_i(+, \cdot, ')$ will respectively be:

$$0_i = a \cdot \overline{x + \bar{x}}, \qquad 1_i = a + \overline{x \cdot \bar{x}},$$

where $a \in Q_i$, $x \in Q$. Hyperidentities (5.5.14)–(5.5.17) show that these are well-defined.

We define operations on the quotient set $Q/\Theta = \{Q_i |\ i \in I\}$, relative to which it also becomes a Boolean algebra:

(a) $Q_i + Q_j = Q_k \Longleftrightarrow 1_i + 1_j \in Q_k$, where $i, j, k \in I$ and $1_i + 1_j = 1_k$ indeed, if $1_i + 1_j \in Q_k$, then

$$1_i + 1_j = 1_i + \overline{x \cdot \bar{x}} + 1_j + \overline{x \cdot \bar{x}} = 1_i + 1_j + \overline{x \cdot \bar{x}} = 1_k;$$

(b) $\bar{Q}_i = Q_j \Longleftrightarrow \bar{1}_i \in Q_j$, where $i, j \in I$ and $\bar{1}_i = 0_j$, indeed, if $\bar{1}_i \in Q_j$, then according to (5.5.18) we have

$$\bar{1}_i = \overline{1_i + \overline{x \cdot \bar{x}}} = \bar{1}_i + \overline{x \cdot \bar{x}} = 0_j;$$

(c) $Q_i + Q_j = \overline{\overline{Q}_i + \overline{Q}_j}$.

Hyperidentities (5.5.19) and (5.5.20) imply the inclusions $a + b \in Q_i + Q_j$ and $a \cdot b \in Q_i + Q_j$, where $a \in Q_i$, $b \in Q_j$. Hyperidentity (5.5.8) implies the inclusion $\bar{a} \in \overline{Q}_i$, where $a \in Q_i$.

We check the Boolean algebra axioms of $Q/\Theta(+, \cdot, -)$:

B$_1$. $1_i + 1_i = 1_i \Longrightarrow Q_i + Q_i = Q_i$;
B$_2$. $1_i + 1_j = 1_j + 1_i \Longrightarrow Q_i + Q_j = Q_j + Q_i$;
B$_3$. $(1_i + 1_j) + 1_k = 1_i + (1_j + 1_k) \Longrightarrow (Q_i + Q_j) + Q_k = Q_i + (Q_j + Q_k)$;
B$_4$. $Q_i \cdot (Q_j + Q_k) = \overline{\overline{Q}_i + \overline{Q_j + Q_k}}$,

$$(Q_i \cdot Q_j) + (Q_i \cdot Q_k) = \overline{\overline{Q}_i + \overline{Q}_j} + \overline{\overline{Q}_i + \overline{Q}_k}.$$

From hyperidentity (5.5.21) we have

$$\overline{\bar{1}_i + \overline{\overline{1_j + 1_k}}} = \overline{\bar{1}_i + \bar{1}_j} + \overline{\bar{1}_i + \bar{1}_k},$$

and therefore

$$Q_i(Q_j + Q_k) = Q_iQ_j + Q_iQ_k.$$

B$_5$. $\overline{\overline{1}}_i = 1_i \Longrightarrow \overline{\overline{Q}}_i = Q_i$;
B$_6$. $\overline{Q_i \cdot Q_j} = \overline{\overline{\overline{Q}_i + \overline{Q}_j}} = \overline{Q}_i + \overline{Q}_j$;
B$_7$. $a + \overline{b + \overline{b}} = a \Longrightarrow Q_i = Q_i + \overline{Q_j + \overline{Q}_j} = Q_i + Q_j\overline{Q}_j$.

Identifying each Q_i with its index, we obtain the Boolean algebra $I(\leqslant)$, where $i \leqslant j \Longleftrightarrow Q_i \leqslant Q_j$. Assume that $i \leqslant j$. We define $\varphi_{i,j} : Q_i \to Q_j$ according to the rule

$$\varphi_{i,j}(x) = x \cdot 1_j \in Q_i + Q_j = Q_j,$$

where $x \in Q_i$, $1_j \in Q_j$. The map $\varphi_{i,j}$ is a homomorphism of Boolean algebras:

$$\varphi_{i,j}(x + y) = (x + y) \cdot 1_j = (x \cdot 1_j) + (y \cdot 1_j) = \varphi_{i,j}(x) + \varphi_{i,j}(y),$$
$$\varphi_{i,j}(x \cdot y) = (x \cdot y) \cdot 1_j = (x \cdot 1_j) \cdot (y \cdot 1_j) = \varphi_{i,j}(x) \cdot \varphi_{i,j}(y).$$

In order to verify the identity

$$\varphi_{i,j}(x') = (\varphi_{i,j}(x))'$$

we use the hyperidentity (5.5.22), from which via the substitution $X \to (+)$, $Y \to (\cdot)$, we obtain

$$x'(xy + (xy)') = xy \cdot \overline{xy + \overline{xy}} + \overline{x(x + y)(xy + \overline{xy})}.$$

Hence,

$$\begin{aligned}
\varphi_{i,j}(x') &= x' \cdot 1_j = x'((x \cdot 1_j) + (x \cdot 1_j)') \\
&= x \cdot 1_j \cdot \overline{((x \cdot 1_j) + \overline{x \cdot 1_j})} + \overline{x(x + 1_j)((x \cdot 1_j) + \overline{x \cdot 1_j})} \\
&= x \cdot 1_j \cdot \overline{((x \cdot 1_j) + \overline{x \cdot 1_j})} + \overline{x \cdot 1_j \cdot ((x \cdot 1_j) + \overline{x \cdot 1_j})} \\
&= (x \cdot 1_j)' = (\varphi_{i,j}(x))'.
\end{aligned}$$

If $i \leqslant j \leqslant k$, then, in particular, $Q_j + Q_k = Q_k$, that is, $1_j + 1_k = 1_k$. We verify the equality $\varphi_{i,j} \cdot \varphi_{j,k} = \varphi_{i,k}$:

$$\begin{aligned}
(\varphi_{i,j} \cdot \varphi_{j,k})x &= \varphi_{j,k}(\varphi_{i,j}(x)) = \varphi_{j,k}(x \cdot 1_j) \\
&= x \cdot 1_j \cdot 1_k = x \cdot 1_k = \varphi_{i,k}(x),
\end{aligned}$$

where $x \in Q_i$. Here we have used hyperidentity (5.5.23), from which for $X = (+)$, $Y = (\cdot)$ and $y = 1_j$, $z = 1_k$ we obtain

$$\begin{aligned}
1_j \cdot 1_k &= (1_j + \overline{x\overline{x}})(1_k + \overline{x\overline{x}}) = (1_j + \overline{x\overline{x}})(1_j + 1_k + \overline{x\overline{x}}) \\
&= 1_j + 1_k + \overline{x\overline{x}} = 1_k + \overline{x\overline{x}} = 1_k.
\end{aligned}$$

For any $i \in I$ we define a map $h_{i,\bar{i}} : Q_i \to Q_{\bar{i}}$ according to the rule

$$h_{i,\bar{i}}(x) = (\bar{x})' = \overline{(x')}$$

Hyperidentity (5.5.24) of the algebra $Q(+,\cdot,^-)$ implies that this map is well-defined.

We verify that $h_{i,\bar{i}}$ is a homomorphism, starting from hyperidentity (5.5.25) for the algebra $Q(+,\cdot,^-)$. Then, making the substitution $X \to (+)$, $Y \to (\cdot)$, we obtain the identity

$$\overline{(x+y)'} = \overline{(x+xy)'} + \overline{(y+yx)'}.$$

Hence,

$$\begin{aligned} h_{i,\bar{i}}(x+y) &= \overline{(x+y)'} = \overline{(x+xy)'} + \overline{(y+yx)'} \\ &= \overline{x'} + \overline{y'} = h_{i,\bar{i}}(x) + h_{i,\bar{i}}(y), \end{aligned}$$

where $x, y \in Q_i$. Similarly, using the identity

$$\overline{(x\cdot y)'} = \overline{(x(x+y))'} \cdot \overline{(y(y+x))'}$$

we shall find

$$h_{i,\bar{i}}(x\cdot y) = h_{i,\bar{i}}(x)\cdot h_{i,\bar{i}}(y).$$

Moreover,

$$h_{i,\bar{i}}(x') = (\overline{x'})' = (h_{i,\bar{i}}(x))'.$$

The obvious equality

$$h_{i,\bar{i}}\cdot h_{\bar{i},i} = h_{\bar{i},i}\cdot h_{i,\bar{i}} = \varepsilon$$

implies that the map $h_{i,\bar{i}}$ is bijective.

Thus, $h_{i,\bar{i}}$ is an isomorphism of Boolean algebras for any $i \in I$, and

$$\begin{aligned} (h_{i,\bar{i}}\cdot \varphi_{\bar{i},1})x &= \varphi_{\bar{i},1}(h_{i,\bar{i}}(x)) = \varphi_{\bar{i},1}(\overline{x'}) = \overline{x'}\cdot 1_1 \\ &= \overline{x'}(x+\bar{x}) = x(x+\bar{x}) = x\cdot 1_1 = \varphi_{i,1}(x). \end{aligned}$$

Here we have used the identity

$$\overline{x'}(x+\bar{x}) = x(x+\bar{x}),$$

which arises from hyperidentity (5.5.26) for the algebra $Q(+,\cdot,^-)$.

We shall prove that $\varphi_{i,j}$ is bijective. Let $i \leqslant j$ and let $a_1, a_2 \in Q_i$. Then

$$\begin{aligned}\varphi_{i,j}(a_1) = \varphi_{i,j}(a_2) &\Longrightarrow \varphi_{i,1}(a_1) = \varphi_{j,1}(\varphi_{i,j}(a_1))\\ &= \varphi_{j,1}(\varphi_{i,j}(a_2)) = \varphi_{i,1}(a_2) \Longrightarrow a_1 \cdot 1_1 = a_2 \cdot 1_1\\ &\Longrightarrow a_1(a_1 + \bar{a}_1) = a_2(a_2 + \bar{a}_2) \Longrightarrow a_1'\\ &= a_1 \cdot \overline{a_1 + \bar{a}_1} + \overline{a_1(a_1 + \bar{a}_1)} = 0_i + \overline{a_1(a_1 + \bar{a}_1)}\\ &= 0_i + \overline{a_2(a_2 + \bar{a}_2)} = a_2 \cdot \overline{a_2 + \bar{a}_2} + \overline{a_2(a_2 + \bar{a}_2)}\\ &= a_2' \Longrightarrow a_1 = a_2.\end{aligned}$$

For any $b \in Q_j$ we write $a = \overline{b' \cdot 1_1} \cdot 1_i \in Q_i$. Then

$$\varphi_{i,j}(a) = \overline{b' \cdot 1_1} \cdot 1_i \cdot 1_j = \overline{b' \cdot 1_1} \cdot 1_j = \overline{b'(b + \bar{b})} \cdot (b + b\bar{\bar{b}}) = b$$

according to hyperidentity (5.5.27).

Finally, let $x, y \in Q = \bigcup_{i \in I} Q_i$. If $x \in Q_i$, $y \in Q_j$ and $Q_i + Q_j = Q_k$, that is, $i + j = k$, then

$$\begin{aligned}x + y &= (x + y) \cdot 1_k = (x \cdot 1_k) + (y \cdot 1_k) = \varphi_{i,k}(x) + \varphi_{j,k}(y),\\ x \cdot y &= x \cdot y \cdot 1_k = x \cdot 1_k \cdot y \cdot 1_k = \varphi_{i,k}(x) \cdot \varphi_{j,k}(y),\\ \bar{x} &= \overline{(x')'} = h_{i,\bar{i}}(x').\end{aligned}$$

Thus, the super-Boolean algebra $Q(+, \cdot, ^-)$ is a Boolean sum of the Boolean lattices Q_i, $i \in I$. This completes the proof of Theorem 5.5.8. □

Corollary 5.5.10. *An algebra $\mathfrak{A} = Q(+, \cdot, ^-)$ with two binary and one unary operations satisfies the regular identities of Boolean algebras if and only if $\mathfrak{A}$ is a Boolean lattice or a Boolean sum of Boolean lattices.*

The functional rank of each of the hyperidentities (2.3.1)–(2.3.4), (5.5.1)–(5.5.7) is at most 3 (assuming also the unique unary operation symbol) in view of Theorem 5.5.1, and the object rank of each of them is at most 4. As a consequence here we shall mention the existence of a finite basis of hyperidentities of the variety of Boolean algebras such that the hyperidentities have functional and object ranks at most 3.

Theorem 5.5.11. *Any hyperidentity of the variety of Boolean lattices is a consequence of the hyperidentities (2.3.1)–(2.3.4), (5.5.1)–(5.5.4), (5.5.8), (5.5.9)–(5.5.18), (5.5.20)–(5.5.27).*

Theorem 5.5.12. *Any hyperidentity of the variety of Boolean algebras is a consequence of the hyperidentities (2.3.1)–(2.3.4), (5.5.1)–(5.5.4), (5.5.8), (5.5.9)–(5.5.18), (5.5.20)–(5.5.27).*

Proof. We must prove that if the hyperidentities mentioned in the theorem hold in a T-algebra $\mathfrak{A} = \langle Q; \Sigma \rangle$, where $T = \{1, 2\}$, then any hyperidentity of the variety of Boolean algebras is also valid in $\mathfrak{A}$, that is, $\mathfrak{A}$ is a super-Boolean algebra.

First we notice that if such an algebra $\mathfrak{A} = \langle Q; \Sigma \rangle$ is subdirectly irreducible in the category of T-algebras, then the cardinality of its set of binary operations is $\leqslant 2$. In fact, only hyperidentities (2.3.1)–(2.3.4) and (5.5.8) were used in the proof of Lemma 5.5.6. Therefore, a result of the type of Corollary 5.5.3 is also valid for our algebra $\mathfrak{A}$, that is, in the category of T-algebras $\mathfrak{A}$ is isomorphic to a subdirect product of T-algebras $\mathfrak{B}$ with the hyperidentities (2.3.1)–(2.3.4), (5.5.1)–(5.5.4), (5.5.8), (5.5.9)–(5.5.18), (5.5.20)–(5.5.27), whose sets of binary operations have either one or two elements. If $\mathfrak{B}$ has one binary operation $(+)$, then hyperidentities (2.3.1)–(2.3.3), (5.5.2)–(5.5.4), (5.5.21) become identities ($5.\mathrm{B}_1$)–($5.\mathrm{B}_7$), which ensure that this algebra can be contained in a Boolean algebra. Therefore in such an algebra $\mathfrak{B}$ every hyperidentity of the variety of Boolean algebras will be valid. And if $\mathfrak{B}$ has two binary operations, then $\mathfrak{B}$ is a Boolean lattice or it can be represented as a Boolean sum of Boolean lattices, since in proving the "necessity" part of Theorem 5.5.8 we in fact only used hyperidentities (2.3.1)–(2.3.4), (5.5.1)–(5.5.4), (5.5.8), (5.5.9)–(5.5.18), (5.5.20)–(5.5.27) or consequences of them (for example, hyperidentity (5.5.19), which has object rank equal to 4, is according to Theorem 2.3.1 a consequence of the hyperidentities (2.3.1)–(2.3.4)). Furthermore, according to the "sufficiency" part of Theorem 5.5.8 any hyperidentity of the variety of Boolean algebras will be valid in $\mathfrak{B}$. As a result any hyperidentity of the variety of Boolean algebras will also be valid in the algebra $\mathfrak{A}$ under consideration, completing the proof of Theorem 5.5.12. □

Corollary 5.5.13. *Any hyperidentity of a Boolean algebra is a consequence of the hyperidentities (2.3.1)–(2.3.4), (5.5.1)–(5.5.4), (5.5.8), (5.5.9)–(5.5.18), (5.5.20)–(5.5.27), whose functional and object ranks are not greater than 3.*

Corollary 5.5.14. *Any regular identity of a Boolean algebra (or the variety of Boolean algebras) is a consequence of the finite system of identities obtained from the hyperidentities (2.3.1)–(2.3.4), (5.5.2)–(5.5.4), (5.5.8),*

(5.5.9)–(5.5.18), (5.5.20)–(5.5.27) under the substitutions $X \to (+), (\cdot)$ *and* $Y \to (+), (\cdot)$.

5.6. Subdirectly irreducible super-Boolean algebras

Theorem 5.6.1. *Let* $\mathcal{A} = \langle Q; \Sigma \rangle$ *be a subdirectly irreducible binary algebra and suppose that* $\mathcal{A}$ *satisfies the hyperidentities*

$$X(x, x) = x, \tag{5.6.1}$$

$$X(x, y) = X(y, x), \tag{5.6.2}$$

$$X(x, X(y, z)) = X(X(x, y), z), \tag{5.6.3}$$

$$X(x, Y(y, z)) = Y(X(x, y), X(x, z)). \tag{5.6.4}$$

Then $|\Sigma| \leqslant 2$, *and* $|Q| \leqslant 3$.

Proof. When $|\Sigma| \geqslant 3$, the algebra $\mathcal{A}$ turns out not to be subdirectly irreducible. For, let A_1, A_2 and A_3 be pairwise different binary operations from Σ, and let

$$F_{i,j}(x, y) = A_i(x, A_j(x, y)).$$

We define maps $\Theta_{i,j}$ on Q as follows:

$$x\,\Theta_{i,j}\,y \iff f_{i,j}(x, y) = x, \quad f_{i,j}(y, x) = y,$$

where $i, j = 1, 2, 3$ and $i < j$. The maps $\Theta_{i,j}$ are obviously reflexive and symmetric; transitivity follows from the hyperidentity

$$X(x, Y(x, X(y, Y(y, z)))) = X(x, Y(x, X(z, Y(z, y)))), \tag{5.6.5}$$

which is a corollary of (5.6.1)–(5.6.4). Let us prove this. Firstly, it is clear that the right distributive hyperidentity

$$X(Y(y, z), x) = Y(X(y, x), X(z, x)) \tag{5.6.6}$$

is a consequence of (5.6.1)–(5.6.4). Further, we have

$$\begin{aligned} X(x, Y(y, z)) &\overset{(5.6.4)}{=} Y(X(x, y), X(x, z)) \\ &\overset{(5.6.6)}{=} X(Y(x, X(x, z)), Y(y, X(x, z))) \\ &\overset{(5.6.4)}{=} X(X(Y(x, x), Y(x, z)), X(Y(y, x), Y(y, z))) \\ &= X(X(x, Y(x, z)), X(Y(y, x), Y(y, z))), \end{aligned}$$

that is,

$$X(x,Y(y,z)) = X(X(x,Y(x,z)),X(Y(y,x),Y(y,z))). \tag{5.6.7}$$

Therefore,

$$\begin{aligned}
&X(x,X(Y(x,y),Y(y,z))) \\
&\overset{(5.6.2)}{=} X(x,X(Y(y,z),Y(x,y))) \\
&\overset{(5.6.3)}{=} X(X(x,Y(y,z)),Y(x,y)) \\
&\overset{(5.6.7)}{=} X(X(X(x,Y(x,z)),X(Y(y,x),Y(y,z))),Y(x,y)) \\
&\overset{(5.6.1)-(5.6.3)}{=} X(X(x,Y(x,z)),X(Y(y,x),Y(y,z))) \\
&\overset{(5.6.7)}{=} X(x,Y(y,z)),
\end{aligned}$$

that is,

$$X(x,Y(y,z)) = X(x,X(Y(x,y),Y(y,z))), \tag{5.6.8}$$

whence, on replacing y by $Y(x,y)$, we get the hyperidentity

$$X(x,X(Y(x,y),Y(Y(x,y),z))) = X(x,Y(Y(x,y),z))). \tag{5.6.9}$$

Further, interchanging y and z at this point and applying (5.6.2) and (5.6.3), we get the following hyperidentity:

$$X(x,X(Y(x,z),Y(Y(x,y),z))) = X(x,Y(Y(x,y),z)). \tag{5.6.10}$$

On comparing (5.6.9) and (5.6.10), we arrive at the hyperidentity

$$X(x,X(Y(x,y),Y(Y(x,y),z))) = X(x,X(Y(x,z),Y(Y(x,y),z))),$$

from which we get (5.6.5) by using (5.6.4).

We claim that $\Theta_{i,j} = (\Theta_{i,j},\tilde{O})$ is a congruence on the algebra $\mathcal{A} = \langle Q;\Sigma\rangle$, that is, if $x\,\Theta_{i,j}\,y$ and $u\,\Theta_{i,j}\,v$, then $A(x,u)\,\Theta_{i,j}\,A(y,v)$ for every operation A in Σ. It suffices to check that $A(x,u)\,\Theta_{i,j}\,A(y,u)$, where $x\,\Theta_{i,j}\,y$ and $x,y,u \in Q$. We use the hyperidentity

$$X(Y(X(x,y),z),Y(y,z)) = Y(X(x,y),z), \tag{5.6.11}$$

which is also a consequence of (5.6.1)–(5.6.4), to do that:

$$\begin{aligned} X(Y(X(x,y),z),Y(y,z)) &\overset{(5.6.6)}{=} X(X(Y(x,z),Y(y,z)),Y(y,z)) \\ &\overset{(5.6.3)}{=} X(Y(x,z),X(Y(y,z),Y(y,z))) \\ &\overset{(5.6.1)}{=} X(Y(x,z),Y(y,z)) \\ &\overset{(5.6.6)}{=} Y(X(x,y),z). \end{aligned}$$

Suppose that $x\,\Theta_{i,j}\,y$, that is, $A_i(x, A_j(x,y)) = x$ and $A_i(y, A_j(y,x)) = y$, where $A_i, A_j \in \Sigma$. For every operation A in Σ we get

$$\begin{aligned} A(x,u) &= A(A_i(x, A_j(x,y)), u) \\ &\overset{(5.6.11)}{=} A^i(A(A_i(x, A_j(x,y)), u), A(A_j(x,y), u)) \\ &\overset{(5.6.4)}{=} A_i(A(x,u), A_j(A(x,u), A(y,u))). \end{aligned}$$

Similarly, we get

$$A(y,u) = A_i(A(y,u), A_j(A(y,u), A(x,u))).$$

Consider the factor-algebra $\mathcal{A}/\Theta_{i,j}$ of $\mathcal{A}$ by the congruence $\Theta_{i,j}$. The maps

$$\begin{aligned} &\varphi_0 : x \to ([x]_{1,2}, [x]_{1,3}, [x]_{2,3}), \\ &\tilde{\psi}_0 : A \to ([A]_{1,2}, [A]_{1,3}, [A]_{2,3}), \end{aligned}$$

in which $[\alpha]_{i,j}$ is the $\Theta_{i,j}$-class of the element x, define a bimonomorphism of $\mathcal{A}$ into the direct product $A/\Theta_{1,2} \times \mathcal{A}/\Theta_{1,3} \times \mathcal{A}/\Theta_{2,3}$. It is clear that this pair of maps is a bihomomorphism. The fact that φ_0, and therefore $\tilde{\psi}_0$, is injective follows from this implication: if

$$f_{1,2}(x,y) = f_{1,3}(x,y) = f_{2,3}(x,y) = x,$$

then

$$x = A_1(A_2(x,y), A_3(x,y)).$$

This last implication follows from the following hyperidentity, in which $X = A_1$, $Y = A_2$ and $Z = A_3$:

$$X(Y(x,y), Z(x,y)) = X(X(Y(x, Z(x,y)), Y(x,y)), Z(x,y));$$

it is satisfied in $\mathcal{A}$ since it is also a consequence of (5.6.1)–(5.6.4). To see this, it is enough to rewrite (5.6.8) in the form

$$X(Y(x,y),z) = X(X(Y(x,z),Y(x,y)),z)$$

and replace z by $Z(x,y)$.

As a result, $\mathcal{A}$ is isomorphic to the subdirect product $(\varphi_0 Q; \tilde{\psi}_0 \Sigma)$ of the algebras $\mathcal{A}/\Theta_{1,2}$, $\mathcal{A}/\Theta_{1,3}$ and $\mathcal{A}/\Theta_{2,3}$, whose projections are not isomorphisms. For example, we show that the first projection is not an isomorphism. Since $A_1 \neq A_2$, there exist $x, y \in Q$ such that $A_1(x,y) \neq A_2(x,y)$ and $x \neq y$. However, $[A_1(x,y)]_{1,2} = [A_2(x,y)]_{1,2}$, because

$$f_{1,2}(A_1(x,y), A_2(x,y)) = A_1(x,y),$$
$$f_{1,2}(A_2(x,y), A_1(x,y)) = A_2(x,y).$$

To prove these last equalities, we use the hyperidentity

$$X(X(x,y), Y(X(x,y), Y(x,y))) = X(x,y), \tag{5.6.12}$$

which is a consequence of (5.6.8). To see this, when $y = z$ note that (5.6.8) becomes the hyperidentity

$$X(x,y) = X(X(x,y), Y(x,y)). \tag{5.6.13}$$

We get, therefore:

$$\begin{aligned} X(x,y) &= X(X(x,y), Y(x,y)) \\ &\overset{(5.6.13)}{=} X(X(X(x,y), Y(x,y)), Y(X(x,y), Y(x,y))) \\ &\overset{(5.6.13)}{=} X(X(x,y), Y(X(x,y), Y(x,y))), \end{aligned}$$

which is the hyperidentity (5.6.12). Further,

$$\begin{aligned} f_{1,2}(A_1(x,y), A_2(x,y)) &= A_1(A_1(x,y), A_2(A_1(x,y), A_2(x,y))) \\ &\overset{(5.6.12)}{=} A_1(x,y). \end{aligned}$$

The second equality is proved in a similar way, this time using the hyperidentity

$$Y(x,y) \overset{(5.6.1)}{=} X(Y(x,y), Y(x,y)) \overset{(5.6.13)}{=} X(Y(x,y), Y(Y(x,y), X(x,y))).$$

This completes the proof of the first part of Theorem 5.6.1. We now establish the second part. Let f be an element of Σ and I a non-empty f-closed subset of Q, that is,

$$x, y \in I \longrightarrow f(x, y) \in I;$$

then the relation $\tilde{f}_I$ defined by the formula

$$a\, \tilde{f}_I\, b \iff f(a, I) \cap f(b, I) \neq \emptyset,$$

in which $f(a, I) = \{f(a, x) \,|\, x \in I\}$, is a congruence on any algebra $\mathcal{A}$ satisfying the hyperidentities (5.6.1)–(5.6.4). Reflexivity and symmetry of $\tilde{f}_I$ is clear, and we prove now that it is transitive. Suppose that $a\, \tilde{f}_I b$ and $b\, \tilde{f}_I c$, that is, there exist elements x, y, u, v in I such that $f(a, x) = f(b, y)$ and $f(b, u) = f(c, v)$. We have

$$\begin{aligned} f(a, f(x, u)) &= f(f(a, x), u) = f(f(b, y), u) \\ &= f(b, f(y, u)) = f(b, f(u, y)) = f(f(b, u), y) \\ &= f(f(c, v), y) = f(c, f(v, y)). \end{aligned}$$

where $f(x, u), f(v, y) \in I$, so that $a\tilde{f}_I c$.

Suppose that $g \in \Sigma$ and that $a_1 \tilde{f}_I a_2$ and $b_1 \tilde{f}_I b_2$; we claim that $g(a_1, b_1)\, \tilde{f}_I\, g(a_2, b_2)$. There exist $x, y, u, v \in I$ such that $f(a_1, x) = f(a_2, y)$ and $f(b_1, u) = f(b_2, v)$. Without loss of generality we may assume that x, y, u and v are all different. If $\omega = f(x, f(y, f(u, v)))$, then $\omega \in I$ and from (5.6.1)–(5.6.3) we get:

$$\omega = f(x, \omega) = f(y, \omega) = f(u, \omega) = f(v, \omega).$$

so that

$$\begin{aligned} f(a_1, \omega) &= f(a_1, f(x, \omega)) = f(f(a_1, x), \omega) \\ &= f(f(a_2, y), \omega) = f(a_2, f(y, \omega)) = f(a_2, \omega). \end{aligned}$$

Similarly, $f(b_1, \omega) = f(b_2, \omega)$. Further, by (5.6.4),

$$\begin{aligned} f(\omega, g(a_1, b_1)) &= g(f(\omega, a_1), f(\omega, b_1)) \\ &= g(f(\omega, a_2), f(\omega, b_2)) = f(\omega, g(a_2, b_2)). \end{aligned}$$

that is, $g(a_1, b_1)\, \tilde{f}_I\, g(a_2, b_2)$.

If $I = \{a\}$ for some $a \in Q$, it is f-closed for every $f \in \Sigma$, and we shall denote the resulting congruence by $\tilde{f}_a$. Consider the intersection $\tilde{f} = \bigcap_{a \in Q} \tilde{f}_a$. If $x\tilde{f}y$, then $x\tilde{f}_x y$ and $x\tilde{f}_y y$, so that $f(x, x) = f(y, x)$ and $f(x, y) = f(y, y)$, that is, $x = y$. Thus $\tilde{f}$ is the null congruence; hence, since $\mathcal{A}$ is subdirectly irreducible, $\tilde{f}_a$ is the null congruence for some a in Q.

Thus $f(x,a) = x$ for all x in Q (since $f(x,a)\tilde{f}_a x$), that is, a is an identity element for $f : a = O_f$. Therefore, every operation f of a subdirectly irreducible algebra $\mathcal{A}$ satisfying (5.6.1)–(5.6.4) has an identity element $O_f \in Q$.

Further, for any operations f and g in Σ, we define a relation $f_g \subseteq Q \times Q$ as follows:

$$x f_g y \Longleftrightarrow x = y \quad \text{or} \quad f(x, O_g) \neq O_g, \quad f(y, O_g) \neq O_g.$$

It is easily shown that f_g is a congruence on the algebra $(Q; \{f, g\})$. For example, f_g is clearly an equivalence relation. Further, if $x f_g y$ and $z \in Q$, then $f(x,z)\, f_g\, f(y,z)$ and $g(x,z)\, f_g\, g(y,z)$, since if $f(x, O_g) \neq O_g$ and $f(y, O_g) \neq O_g$, then $f(f(x,z), O_g) \neq O_g$, $f(f(y,z), O_g) \neq O_g$, and $f(g(x,z), O_g) \neq O_g$, $f(g(y,z), O_g) \neq O_g$. To see this, suppose that $f(f(x,z), O_g) = O_g$; then

$$\begin{aligned}
O_f &= g(O_f, O_g) = g(O_f, f(f(x,z), O_g)) \\
&= f(g(O_f, f(x,z)), g(O_f, O_g)) = f(g(O_f, f(x,z)), O_f) \\
&= g(O_f, f(x,z)) = f(g(O_f, x), g(O_f, z)), \\
g(O_f, x) &= f(g(O_f, x), O_f) = f(g(O_f, x), f(g(O_f, x), g(O_f, z))) \\
&= f(f(g(O_f, x), g(O_f, x)), g(O_f, z)) = f(g(O_f, x), g(O_f, z)) = O_f; \\
O_g &= f(O_g, Of) = f(O_g, g(O_f, x)) = g(f(O_g, O_f), f(O_g, x)) \\
&= g(O_g, f(O_g, x)) = f(O_g, x),
\end{aligned}$$

which contradicts the fact that $f(x, O_g) = O_g$.

Thus, $f(f(y,z), O_g) = O_g$ provided that $f(O_g, y) = O_g$. If $f(g(x,z), O_g) = O_g$, we have

$$\begin{aligned}
O_f &= g(O_f, O_g) = g(O_f, f(g(x,z), O_g)) \\
&= f(g(O_f, g(x,z)), g(O_f, O_g)) \\
&= f(g(O_f, g(x,z)), O_f) = g(O_f, g(x,z)),
\end{aligned}$$

that is, $g(g(x,z), O_f) = O_f$, so that $g(O_f, x) = O_f$ as above. And of course $f(O_g, x) = O_g$, another contradiction. Similarly, the assertion that $f(g(y,z), O_g) = O_g$ leads to the contradiction that $f(O_g, y) = O_g$.

If $\mathcal{A} = (Q; \Sigma)$ is a subdirectly indecomposable T-algebra with $T = \{2\}$, $\Sigma = \{f, g\}$ and $Q = \{O_f, O_g\}$, then $\mathcal{A}$ is the two-element lattice when $f \neq g$, while it is a singleton when $f = g$ since then $(O_f = O_g)$. On the other hand, if $Q = \{O_f, O_g\}$ in this situation, so that there is an element a in Q other than O_f and O_g, we get

$$\begin{aligned} y = g(y, O_g) &= (y, f(a, O_g)) = f(g(y, a), g(y, O_g)) \\ &= f(g(y, a), y) = f(y, g(y, a)). \end{aligned}$$

Thus, if $y(\tilde{f}_a \cap \tilde{g}_a)x$, we have

$$\begin{aligned} x &= f(x, g(x, a)) = f(x, g(y, a)) = g(f(x, y), f(x, a)) = g(f(x, y), f(y, a)) \\ &= gf(y, x), f(y, a)) = f(y, g(x, a)) = f(y, g(y, a)) = y, \end{aligned}$$

so that, as $\mathcal{A}$ is subdirectly irreducible, it follows that $\tilde{f}_a$ or $\tilde{g}_a$ is the null congruence and thus $a = O_f$ or $a = O_g$. This is a contradiction! Thus $f(a, O_g) \neq O_g$ if $a \in Q$ and $a \neq O_f, O_g$.

We consider the congruence

$$q_f^g = f_g \cap \Big(\bigcap_{a \neq O_f, O_g} \big(\tilde{f}_a \cap \tilde{g}_a\big)\Big)$$

of the algebra $\mathcal{A} = (Q; \{f, g\})$. Since $\tilde{f}_a$ is null if and only if $a = O_f$, the congruence $\bigcap_{a \neq O_f, O_g} \big(\tilde{f}_a \cap \tilde{g}_a\big)$ is not null; this is because $\mathcal{A}$ is subdirectly irreducible. The following implication also holds:

$$x \, q_f^g \, y \longrightarrow x = y.$$

This is because $f(x, O_g) = f(O_f, O_g) = O_g$ if $x = O_f$, so that $x = y$ since $x f_g y$. Similarly, if $x = O_g$ we get $f(x, O_g) = f(O_g, O_g) = O_g$, and this again means that $x = y$. In exactly the same way, $x = y$ if $y = O_f$ or O_g. Thus, to prove the displayed implication we have to assume that x and y are both different from O_f and O_g. Thus $x\big(\tilde{f}_a \cap \tilde{g}_a\big)y$, and so

$$\begin{aligned} f(x, g(x, x)) &= f(y, g(y, x)), \\ x &= f(y, g(y, x)), \\ f(y, x) = f(y, f(y, g(y, x))) &= f(y, g(y, x)) = x. \end{aligned}$$

Similarly, we get $x\big(\tilde{f}_a \cap \tilde{g}_a\big)y$ and $f(x, y) = y$, so that $x = y$. Therefore, f_g is the null congruence and so Q contains at least one element a different from O_f and O_g. Thus, either $O_f \neq O_g$, in which case $|Q| = 3$, or else $O_f = O_g$ (and hence $f = g$), in which case $|Q| = 2$.

We note also that when $|Q| = 3$ the algebra $\mathcal{A} = (Q; \{f, g\})$ satisfies the coidentity

$$X(x, a) = X(a, x) = a$$

for all $x \in \{O_f, O_g\}$. For example, $f(a, O_g) \neq O_f$, and when $f(a, O_g) = O_g$ we get the contradiction

$$a = f(a, O_f) = f)a, f(O_g, a)) = f(O_g, a) = O_f.$$

Suppose that $g(a, O_f) = O_f$. Then

$$\begin{aligned} O_g &= f(O_g, O_f) = f(O_g, g(a, O_f)) = g(f(O_g, a), f(O_g, O_f)) \\ &= g(f(O_g, a), O_g) = f(O_g, a). \end{aligned}$$

Another contradiction! On the other hand, if $g(a, O_f) = O_g$, we have

$$a = g(a, O_g) = g(a, g(a, O_f)) = g(a, O_f) = O_g$$

which is yet another contradiction. □

Corollary 5.6.2. *Up to isomorphism, the only subdirectly irreducible T-algebras with $T = \{2\}$ contained in the hypervariety of T-algebras generated by an arbitrary distributive lattice (the variety of all distributive lattices) are as follows: the two-element semilattice, the two-element lattice, and the three-element algebra $(\{0, 1, a\}, \{+, \cdot\})$, which is an extension of the two-element lattice $(\{0, 1\}, \{+, \cdot\})$ by an element a satisfying the coidentity*

$$X(x, a) = X(a, x) = a$$

for all $x \in \{0, 1, a\}$. Therefore, every hyperidentity satisfied in the variety of distributive lattices (in any distributive lattice) is a consequence of the hyperidentities (5.6.1), (5.6.2), (5.6.3) *and* (5.6.4) *valid in this variety.*

Theorem 5.6.3. *Let $\mathcal{A} = (Q; \Sigma)$ be a T-algebra with $T = \{1, 2\}$ satisfying the hyperidentities* (5.6.1)–(5.6.4) *as well as the hyperidentities*

$$F(F(x)) = x, \tag{5.6.14}$$

$$X(F(x), y) = X(F(X(x, y)), y), \tag{5.6.15}$$

$$F(X(F(X(x, y)), F(X(x, F(y))))) = x. \tag{5.6.16}$$

If $\mathcal{A}$ is subdirectly irreducible, then the set of unary operations of $\mathcal{A}$ is a singleton, the cardinality of the set of binary operations is at most 2 and $|Q| \leqslant 2$.

Proof. For every binary operation f in Σ and every non-empty f-closed subset I of Q, the relation $\tilde{f}_I$ defined in the proof of Theorem 5.6.1 is a congruence for the T-algebra under consideration. All that remains to be checked is that, for every unary operation F in Σ,

$$a\,\tilde{f}_I\,b \longrightarrow F(a)\,\tilde{f}_I\,F(b).$$

This goes as follows. If $f(a,x) = f(b,x)$ for $x \in I$, then from the hyperidentity (5.6.15) we get

$$f(F(a),x) = f(F(f(a,x)),x) = f(F(f(b,x)),x) = f(F(b),x),$$

that is, $f(a)\tilde{f}_I F(b)$. Thus, as in the proof of Theorem 5.6.1, we conclude that every binary operation f of a subdirectly indecomposable T-algebra $\mathcal{A} = (Q;\Sigma)$, where $T = \{1,2\}$ and $\mathcal{A}$ satisfies (5.6.1)–(5.6.4) and (5.6.15), possesses an identity element $O_f \in Q$. Clearly,

$$I \subseteq J \longrightarrow \tilde{f}_I \subseteq \tilde{f}_J.$$

An f-closed subset I of Q is said to be simple if

$$f(x,y) \in I \longrightarrow x, y \in I.$$

It is clear that Q is f-closed and simple for every binary operation f in Σ, and that the intersection of f-closed simple subsets is f-closed and simple. For every a in Q and every binary operation f in Σ, the non-empty subset

$$(a)_f = \{x \in Q \mid f(a,x) = a\}$$

is f-closed and simple. If $x, y \in (a)_f$, that is, $f(a,x) = a$ and $f(a,y) = a$, then

$$f(a, f(x,y)) = f(f(a,x),y) = f(a,y) = a,$$

that is, $f(x,y) \in (a)_f$. If $f(x,y) \in (a)_f$, then $x, y \in (a)_f$. For example,

$$\begin{aligned} f(a,x) &= f(f(a,f(x,y)),x) = f(a,f(f(y,x),x)) \\ &= f(a,f(y,f(x,x))) = f(a,f(y,x)) = f(a,f(x,y)) = a. \end{aligned}$$

Moreover, it is clear that

$$(a)_f = \{O_f\} \Longleftrightarrow a = O_f,$$

and that the congruence $\tilde{f}_{(a)_f}$ is null if and only if $a = O_f$.

If I and J are f-closed simple subsets, then

$$\tilde{f}_I \subseteq \tilde{f}_J \longrightarrow I \subseteq J,$$

since for $\tilde{f}_I \subseteq \tilde{f}_J$ we have

$$\begin{aligned} I &= \{x \in Q \mid \exists a, b \in I,\ f(x,a) = f(O_f, b)\} \\ &= \{x \in Q \mid x\tilde{f}_I O_f\} \subseteq \{x \in Q \mid x\tilde{f}_J O_f\} \\ &= \{x \in Q \mid \exists c, d \in J, f(x,c) = f(O_f, d)\} = J. \end{aligned}$$

Thus

$$\tilde{f}_{(a)_f} \cap \tilde{f}_{(b)_f} = \tilde{f}_{(a)|_f \cap (b)_f}. \qquad (5.6.17)$$

Suppose that $|Q| \geqslant 3$, and let F and f be fixed unary and binary operations in Σ. There exists $a \in Q$ with $a \neq O_f$ and $a \neq F(O_f)$. Because of the hyperidentity (5.6.14), it follows that $F(a) \neq O_f$. Thus $(a)_f \neq \{O_f\}$ and $(F(a))_f \neq \{O_f\}$. If $x \in (a)_f \cap (F(a))_f$, then $f(a,x) = a$, $f(F(a),x) = F(a)$ and

$$\begin{aligned} x &\overset{(5.6.16)}{=} F(f(F(f(x,a)), F(f(x, F(a))))) \\ &= F(f(F(a), F(F(a)))) = F(f(F(a), a)). \end{aligned}$$

Thus, the intersection $(a)_f \cap (F(a))_f$ is a singleton, so that

$$(a)_f \cap (F(a))_f = \{O_f\}$$

and by (5.6.17) the intersection $\tilde{f}_{(a)_f} \cap \tilde{f}_{(F(a))_f}$ of two non-null congruences is null. This is a contradiction, so that $|Q| \leqslant 2$.

If $|Q| = 2$, there exist in all two binary operations satisfying (5.6.1) and (5.6.2), and as many unary operations satisfying (5.6.14). However, the identity map ε_Q does not satisfy the hyperidentity (5.6.16):

$$\begin{aligned} x &= \varepsilon_Q X(\varepsilon_Q X(x,y), \varepsilon_Q X(x, \varepsilon_Q(y))) \\ &= \varepsilon_Q X(\varepsilon_Q X(y,x), \varepsilon_Q X(y, \varepsilon_Q(x))) = y. \end{aligned}$$

□

Corollary 5.6.4. *Suppose that $T = \{1,2\}$. Up to isomorphism, the only subdirectly-irreducible T-algebras in the hypervariety of T-algebras generated by any Boolean algebra (the variety of Boolean algebras) are the two-element Boolean algebra and its $\{+,'\}$-reduced algebra. Therefore, every hyperidentity satisfies by the variety of Boolean algebras (by any Boolean algebra) is a consequence of the hyperidentities (5.6.1), (5.6.2), (5.6.3), (5.6.4), (5.6.14), (5.6.15) and (5.6.16) satisfied in that variety.*

Corollary 5.6.5. *Suppose that $T = \{2\}$. If $\mathcal{A} = (Q;\Sigma)$ is a subdirectly irreducible T-algebra satisfying the hyperidentities (5.6.1)–(5.6.3) and also the hyperidentity*

$$X(Y(X(x,y),z),Y(y,z)) = Y(X(x,y),z) \tag{5.6.18}$$

then $|\Sigma| \leqslant 2$, and therefore

(1) *every hyperidentity satisfied in the variety of all lattices is a consequence of the hyperidentities (5.6.1)–(5.6.3) and (5.6.18) satisfied in that variety;*
(2) *every hyperidentity satisfied in the variety of all modular lattices is a consequence of the hyperidentities (5.6.1)–(5.6.3), (5.6.18) and (2.1.1) satisfied in that varicty.*

5.7. Ideals and congruences of super-Boolean algebras: Some combinatorial properties

In this section, we establish a connection between congruence lattices and ideals of super-Boolean algebras. We need the following hyperidentities:

$$F(x, XF(y, Xy)) = x, \tag{5.7.1}$$

$$F(x,y) = F(y,x), \tag{5.7.2}$$

$$F(x,F(y,z)) = F(F(x,y),z), \tag{5.7.3}$$

$$F(x,x) = x, \tag{5.7.4}$$

$$F(G(x,y),z) = G(F(x,z),F(y,z)), \tag{5.7.5}$$

$$F(Xy,x) = F(XF(y,x),x), \tag{5.7.6}$$

$$X(X(x)) = x, \tag{5.7.7}$$

$$x = XF(XF(x,y),XF(x,Xy)), \tag{5.7.8}$$

$$F(x,y) = G(x,y). \tag{5.7.9}$$

The first eight hyperidentities are hyperidentities of the variety of Boolean algebras, but the last hyperidentity satisfied in any functionally trivial Boolean algebra.

Let Φ be a congruence of algebra A, $f \in \Sigma$, and $\Lambda_f^{\Phi} = \{x \in Q | \, x\Phi\Lambda_f\}$, where $\Lambda_f = \overline{f(a,\bar{a})} = \overline{f(b,\bar{b})}$ for any $a, b \in Q$ (according to the hyperidentity (5.7.1)).

Let SB be a class (hypervariety) of T-algebras with hyperidentities of the variety of Boolean algebras, where arithmetic type $T = \{0, 1, 2\}$, and SB-algebra is a super-Boolean algebra from SB.

Definition 5.7.1. Let $A = (Q; \Sigma) \in SB$ and $f \in \Sigma$, with the arity $n(f) = 2$. A non-empty subset $I \subseteq Q$ is called an f-ideal if

$$f(x, y) \in I \Leftrightarrow x, y \in I.$$

The lattice of all f-ideals of A is denoted by $I_f(A)$.

Note that if I is an f-ideal then $\Lambda_f \in I$.

Lemma 5.7.2. *Let $A = (Q; \Sigma) \in SB$ and let Φ be a congruence on A. Then $\mathfrak{I}_f(\Phi) = \Lambda_f^{\Phi}$ is an f-ideal for any binary operation $f \in \Sigma$.*

Proof. First observe that $\Lambda_f \in \Lambda_f^{\Phi}$, hence $\Lambda_f^{\Phi} \neq \emptyset$. Now let $x, y \in \Lambda_f^{\Phi}$. Then $x\Phi\Lambda_f$ and $y\Phi\Lambda_f$, therefore $f(x, y)\Phi f(\Lambda_f, \Lambda_f) = \Lambda_f$ (using (5.7.1)) and $f(x, y) \in \Lambda_f^{\Phi}$.

Assume now that $f(x, y) \in \Lambda_f^{\Phi}$, i.e., $f(x, y)\Phi\Lambda_f$. By (5.7.1) and the hyperidentities (5.7.3) and (5.7.4), we have

$$x = f(x, \Lambda_f)\Phi f(x, f(x, y)) = f(f(x, x), y) = f(x, y)\Phi\Lambda_f.$$

Therefore $x \in \Lambda_f^{\Phi}$. Similarly, using the hyperidentity of commutativity (5.7.2) we can show that $y \in \Lambda_f^{\Phi}$. □

Lemma 5.7.3. *Let $A = (Q; \Sigma) \in SB$ and $f \in \Sigma$, with the arity $n(f) = 2$. Let also $I \subseteq Q$ be an f-ideal. Then, the relation $\Phi_f(I)$ on Q defined as*

$$a\Phi_f(I)b \Leftrightarrow f(a, I) \cap f(b, I) \neq \emptyset$$

is a congruence on A, where $f(a, I) = \{f(a, x) : x \in I\}$.

Proof. For ease of notation, in this proof we will denote $\Phi_f(I)$ by $\sim$. We show first that it is an equivalence relation. Reflexivity and symmetry are trivial. Let $\sim b, b \sim c$, that is, there are $x, y, u, v \in I$ such that $f(a, x) = f(b, y)$ and $f(b, u) = f(c, v)$. By (5.7.2) and (5.7.3) we have

$$\begin{aligned} f(a, f(x, u)) &= f(f(a, x), u) = f(f(b, y), u) = f(b, f(y, u)) \\ &= f(b, f(u, y)) = f(f(b, u), y) = f(f(c, v), y) = f(c, f(v, y)). \end{aligned}$$

Since $f(x, u), f(v, y) \in I$, we get $a \sim c$.

Now let $g \in \Sigma$ with $n(g) = 2$. We show that $a_1 \sim a_2, b_1 \sim b_2$ implies $g(a_1, b_1) \sim g(a_2, b_2)$. By definition there are $x, y, u, v \in I$ with $f(a_1, x) = f(a_2, y)$ and $f(b_1, u) = f(b_2, v)$. Observe that x, y, u, v can be chosen to be equal; for if $w = f(x, f(y, f(u, v))) \in I$ then by associativity, commutativity and idempotence we have

$$w = f(x, w) = f(y, w) = f(u, w) = f(v, w),$$

hence

$$\begin{aligned} f(a_1, w) &= f(a_1, f(x, w)) = f(f(a_1, x), w) = f(f(a_2, y), w) \\ &= f(a_2, f(y, w)) = f(a_2, w). \end{aligned}$$

Similarly $f(b_1, w) = f(b_2, w)$. By distributivity (5.7.5) we deduce that $f(g(a_1, b_1), w) = g(f(a_1, w), f(b_1, w)) = g(f(a_2, w), f(b_2, w)) = f(g(a_2, b_2), w)$, so $g(a_1, b_1) \sim g(a_2, b_2)$.

Now pick a unary operation $^{-} \in \Sigma$ and assume $a_1 \sim a_2$. As above we can find $x \in I$ with $f(a_1, x) = f(a_2, x)$. Using the hyperidentity (5.7.6) we obtain

$$f(\bar{a}_1, x) = f(\overline{f(a_1, x)}, x) = f(\overline{f(a_2, x)}, x) = f(\bar{a}_2, x).$$

Hence $\bar{a}_1 \sim \bar{a}_2$. □

Thus, Lemmas 5.7.2 and 5.7.3 define mappings $\mathfrak{I}_f$ and Φ_f between algebraic lattices $I_f(A)$ and $\mathrm{Con}^0(A)$.

Theorem 5.7.4. *Let $A = (Q; \Sigma) \in SB$ and $f \in \Sigma, n(f) = 2$. The maps $\mathfrak{I}_f : \mathrm{Con}^{(0)}(A) \to I_f(A)$ and $\Phi_f : I_f(A) \to \mathrm{Con}^{(0)}(A)$ are mutually inverse isomorphisms between those lattices.*

Proof. We will prove that $\mathfrak{I}_f \Phi_f = 1_{I_f(A)}$ and $\Phi_f \mathfrak{I}_f = 1_{\mathrm{Con}^{(0)}(A)}$. Let $I \in I_f(A)$ and $\sim = \Phi_f(I)$. By (5.7.1) and (5.7.2) $f(\Lambda_f, I) = I$, hence

$$\mathfrak{I}_f(\sim) = \Lambda_f^{\sim} = \{a \in Q : \exists x \in I(f(a, x) \in I)\} = I.$$

This shows that $\mathfrak{I}_f \Phi_f = 1_{I_f(A)}$.

Let now $\sim \in \mathrm{Con}^{(0)}(A), I = \mathfrak{I}_f(\sim)$ and $\approx = \Phi_f(I)$. We need to show that $\sim = \approx$. Assume $a \approx b$. This means that there are $x \sim \Lambda_f$ and $y \sim \Lambda_f$ with $f(a, x) = f(b, y)$. Then

$$a = f(a, \Lambda_f) \sim f(a, x) = f(b, y) \sim f(b, \Lambda_f) = b.$$

Now by transitivity we must have $a \sim b$.

Conversely, if $a \sim b$ then for $^- \in \Sigma, n(^-) = 1$ we have

$$\overline{f(a,\bar{b})} \sim \overline{f(a,\bar{a})} = \Lambda_f,$$
$$\overline{f(\bar{a},b)} \sim \overline{f(\bar{a},a)} = \overline{f(a,\bar{a})} = \Lambda_f.$$

Therefore $\overline{f(a,\bar{b})}, \overline{f(\bar{a},b)} \in I$. By the hyperidentity (5.7.7), and by (5.7.2) and (5.7.6) we get

$$f(b,\overline{f(\bar{a},b)}) = f(\overline{f(\bar{a},b)},b) = f(\bar{\bar{a}},b) = f(\bar{\bar{b}},a) = f(\overline{f(\bar{b},a)},a) = f(a,\overline{f(a,\bar{b})}),$$

hence $a \approx b$.

To finish the proof of the theorem we notice that for $I, J \in I_f(A)$ the condition $I \subseteq J$ is equivalent to $\Phi_f(I) \subseteq \Phi_f(J)$. □

First we consider principal f-ideals of a SB-algebra $A = (Q; \Sigma)$ (as always $f \in \Sigma, n(f) = 2$), that is, f-ideals generated by a single element. It is easy to see that

$$(a)_f = \{x \in Q : f(a,x) = a\}$$

is an f-ideal for all $a \in Q$. Indeed, if $f(a,x) = a = f(a,y)$ then by (5.7.3)

$$f(a, f(x,y)) = f(f(a,x),y) = f(a,y) = a.$$

Further, if $f(a, f(x,y)) = a$ then using (5.7.2), (5.7.3) and (5.7.4) we get

$$f(a,x) = f(f(a,f(x,y)),x) = f(a,f(f(y,x)),x) = f(a,f(y,f(x,x)))$$
$$= f(a,f(y,x)) = f(a,f(x,y)) = a.$$

Similarly we get $f(a,y) = a$.

According to (5.7.4) $a \in (a)_f$. On the other hand, if $a \in I \in I_f(A)$, then $(a)_f \subseteq I$. Thus, $(a)_f$ is the principal f-ideal generated by a. In order to avoid cumbersome notations we will write $\sim_{a,f}$ for $\Phi_f((a)_f)$.

Remark 5.7.5. Observe that $x \sim_{a,f} y$ if and only if $f(a,x) = f(a,y)$, i.e., the equivalence classes of $\sim_{a,f}$ are determined by their largest elements with respect to the order

$$u \leq_f v \Leftrightarrow f(u,v) = v.$$

Note further that if $b \in Q$ determines an equivalence class of $\sim_{a,f}$ (this simply means that there is $x \in Q$ with $b = f(a,x)$), then $a \leq_f b$ and vice versa. Thus, the quotient algebra $A/\sim_{a,f}$ is isomorphic to the principal filter generated by a (with a unary operation $b' = f(\bar{b},a)$).

Denote the two-element Boolean lattice by $B_2 := (\{0,1\}; +, \cdot, ^-)$, and let B_2^+ be its reduct $(\{0,1\}; +, ^-)$. Let SB^+ be the class of all functionally trivial SB-algebras.

Now we can present the following new proof of the Theorem 5.6.3

Lemma 5.7.6. *The only subdirectly irreducible SB-algebras are B_2 and B_2^+. Moreover, B_2^+ is the only subdirectly irreducible SB^+-algebra.*

Proof. It is clear that B_2 and B_2^+ are subdirectly irreducible.

Let $A = (Q; \Sigma) \in SB$ with $|Q| \geq 3$. Pick $f, ^- \in \Sigma$ with $n(f) = 2, n(\bar{}) = 1$. There is an element $a \in Q \setminus \{\Lambda_f, \bar{\Lambda}_f\}$. By (5.7.7) $\bar{a} \neq \Lambda_f$, hence $(a)_f$ and $(\bar{a})_f$ are non-zero f-ideals. Assume $x \in (a)_f \cap (\bar{a})_f$. According to the hyperidentity (5.7.8) we have

$$x = \overline{f(\overline{f(x,a)}, \overline{f(x,\bar{a})})} = \overline{f(a,\bar{a})} = \Lambda_f.$$

Therefore $(a)_f \cap (\bar{a})_f = \{\Lambda_f\}$ and A is not subdirectly irreducible.

Assume now that $|Q| = 2$ and write $Q = \{a, b\}$. There are two commutative and idempotent binary operations given by the following tables:

$\vee$	a	b
a	a	b
b	b	b

$\wedge$	a	b
a	a	a
b	a	b

There are also two unary operations satisfying (5.7.7); the identity function id_Q and $^-: a \leftrightarrow b$. If $id_Q \in \Sigma$ then for every $f \in \Sigma$ with $n(f) = 2$ the equality

$$x = f(x,x) = f(x, id_Q(x)) = id_Q(f(x, id_Q(x))) = \Lambda_f$$

holds for all $x \in Q$ which contradicts the fact that Q has two elements. Hence, $^-$ the only unary operation of A. Notice also that if $\vee, \wedge \in \Sigma$ then $A \cong B_2$ and if $\vee \in \Sigma, \wedge \notin \Sigma$ or $\vee \notin \Sigma, \wedge \in \Sigma$ then $A \cong B_2^+$. This finishes the proof. □

According to Birkhoff's theorem on representation of algebras as a subdirect product of subdirectly irreducible algebras, every SB-algebra is a subdirect product of isomorphic copies of B_2 and B_2^+. Hence, SB-algebras are precisely subalgebras of direct powers of B_2.

Note that every hypervariety is uniquely determined by its subdirectly irreducible algebras as the class of all subdirect products of those.

Corollary 5.7.7. *$Heq(SB) = Heq(A)$ for any functionally non-trivial $A \in SB$. In particular, a hyperidentity is a hyperidentity of the variety of Boolean algebras if and only if it is true in B_2.*

It is clear that SB^+ is characterized in SB by the hyperidentity

$$F(x,y) = G(x,y).$$

Theorem 5.7.8. *The only non-trivial proper subhypervariety of SB is SB^+, and $Heq(SB^+) = Heq(A)$ for every non-trivial $A \in SB^+$. In particular, $Heq(SB^+) = Heq(B_2^+)$.*

Proof. Let $\mathcal{M}$ be a hypervariety satisfying hyperidentities (5.7.1)–(5.7.8). Note that so far we have used only those hyperidentities, hence Lemmas 5.7.2, 5.7.3, 5.7.6 and Theorem 5.7.4 are true in $\mathcal{M}$. Therefore, the only non-trivial subdirectly irreducible algebras of $\mathcal{M}$ are B_2 and B_2^+. If $B_2^+ \notin \mathcal{M}$, and hence $B_2 \notin \mathcal{M}$, then $\mathcal{M}$ is trivial. If $B_2^+ \in \mathcal{M}$ but $B_2 \notin \mathcal{M}$ then $\mathcal{M} = SB^+$. Finally, if $B_2^+, B_2 \in \mathcal{M}$ then $\mathcal{M} = SB$.

Now pick a non-trivial algebra $A \in SB$. Since $Heq(SB) \subseteq Heq(A)$, the hypervariety $\mathcal{M}$ defined by $Heq(A)$ is a non-trivial subhypervariety of SB. So, either $\mathcal{M} = SB$ or $\mathcal{M} = SB^+$. In the former case $Heq(A) = Heq(SB)$ and A is functionally non-trivial, while in the latter case $Heq(A) = Heq(SB^+)$ and A is functionally trivial. □

The first part of this theorem can be restated as follows: SB is a minimal strongly non-trivial hypervariety and SB^+ is a minimal non-trivial hypervariety.

Recall that in any SB-algebra there is a unique unary operation which will be denoted by $\bar{\ }$.

If $A = (Q;\Sigma) \in SB, |Q| \geq 3$ and $a \in Q \setminus \{\Lambda_f, \bar{\Lambda}_f\}$ for a fixed binary operation $f \in \Sigma$ then $A_1 = A/(a)_f$ and $A_2 = A/(\bar{a})_f$ are non-trivial SB-algebras. Consider the product of natural homomorphisms

$$\varepsilon = \varepsilon_{(a)_f} \times \varepsilon_{(\bar{a})_f} : A \to A_1 \times A_2 : x \mapsto (x_{a,f}, x_{\bar{a},f}).$$

Let $(B, C) \in A_1 \times A_2$ and let b and c be the largest elements of B and C respectively, that is, $x \in B \Leftrightarrow f(x,a) = b$ and $y \in C \Leftrightarrow f(y,\bar{a}) = c$. By the hyperidentities

$$F(x, X(\Lambda_f)) = X(\Lambda_f),$$

$$F(XF(Xx, Xy), z) = XF(XF(x,z), XF(y,z))$$

we get

$$\begin{aligned} f(\overline{f(\bar{b},\bar{c})}, a) &= \overline{f(\overline{f(b,a)}, \overline{f(c,a)})} = \overline{f(\bar{b}, \overline{f(f(c,\bar{a}),a)})} \\ &= \overline{f(\bar{b}, \overline{f(c,\bar{\Lambda}_f)})} = \overline{f(\bar{b},\bar{\Lambda}_f)} = b, \end{aligned}$$

and similarly $f(\overline{f(\bar{b},\bar{c})},\bar{a}) = c$. Thus, $\overline{f(\bar{b},\bar{c})} \in B \cap C$, hence $\varepsilon(\overline{f(\bar{b},\bar{c})}) = (B, C)$ and ε is a semiepimorphism. If $x \in B \cap C$ then (8) implies

$$x = \overline{f(\overline{f(x,a)}, \overline{f(x,\bar{a})})} = \overline{f(\bar{b},\bar{c})}$$

so ε is a monomorphism and, therefore, a semiisomorphism.

We showed that any SB-algebra with more than two elements is a semidirect product of two non-trivial SB-algebras. In particular, every finite SB-algebra is a semidirect power of B_2.

The preceding remark implies that for every finite SB-algebra $(Q;\Sigma)$ we have $|\Sigma| \leq |Q| + 1$.

Assume a SB-algebra $A = (Q;\Sigma)$ is a subdirect product of SB-algebras $\{A_i\}_{i\in I} = \{(Q_i;\Sigma_i)\}_{i\in I}$. Then for any $f_i \in \Sigma_i, n(f_i) = 2$ we have $\Lambda_{\{f_i\}} = \{\Lambda_{f_i}\}$ (provided that $\{f_i\} \in \Sigma$). To this end observe that for each $i_0 \in I$ there is $\{a_i\} \in Q$ such that $a_{i_0} = \Lambda_{f_{i_0}}$. Therefore

$$\Lambda_{f_{i_0}} = (\{a_i\})_{i_0} = (\{f_i\}(\{a_i\}, \Lambda_{\{f_i\}}))_{i_0} = f_{i_0}(a_{i_0}, (\Lambda_{f_i})_{i_0}) = (\Lambda_{f_i})_{i_0}.$$

This proves that $\Lambda_{\{f_i\}} = \{\Lambda_{f_i}\}$.

Further, notice that in B_2 and B_2^+ every binary operation is uniquely determined by its zero. So if all A_i's are either B_2 or B_2^+ then every binary operation of the algebra A is uniquely determined by its zero (i.e., distinct operations have distinct zeroes) due to the aforementioned equality. This shows that for infinite Q we have $|\Sigma| \leq |Q|$.

Thus every finite SB-algebra has 2^k elements for some $k \in \mathbb{N}$. This can also be deduced from the analogous fact for finite Boolean algebras using the following simpler tools.

Let $A = (Q;\Sigma) \in SB$. For every binary operation $f \in \Sigma$ we define a conjugate operation f^* as follows:

$$f^*(x, y) = \overline{f(\bar{x}, \bar{y})}, \;\; x, y \in Q.$$

The algebra $A' = (Q; \Sigma \cup \{f^* : f \in \Sigma, n(f) = 2\})$ is a SB-algebra too (we need to check this property only for hyperidentities (5.7.1)–(5.7.8)). For any $f \in \Sigma, n(f) = 2$ the algebra $A'' = (Q; f, f^*, ^-)$ is a Boolean lattice.

In particular, every functionally trivial SB-algebra is a semisubalgebra of a Boolean algebra.

A hyperequational theory T of the class $\mathcal{M}$ is said to be one-based if T has a base consisting of one hyperidentity. It is interesting that both hypervarieties SB and SB^+ are one-based [281]. We need the following fact to prove it.

Theorem 5.7.9 (cf. [233, Theorem 1.2], [281, Theorem 3.1]). *Let Θ be a hyperequational theory of the class $\mathcal{M}$. Assume for some hyperterm $\tau(x_1, \ldots, x_k)(k \geq 2)$ Θ contains all of the following hyperidentities:*

$$x = \tau(y, x, \ldots, x), \ldots, x = \tau(x, \ldots, x, y).$$

If Θ has a finite base consisting of hyperidentities of the form $x = \sigma$ then Θ is one-based.

Let us check that the conditions of this theorem hold for hyperequational theories $T = Heq(SB)$ and $T^+ = Heq(SB^+)$ (i.e., for the hyperequational theories of super-Boolean algebras and functionally trivial super-Boolean algebras, respectively). Observe that using (5.7.1), (5.7.7), (5.7.8),

$$x = F(XF(y, Xy), x), \quad \text{and} \quad x = XF(XF(y, x), XF(Xy, x))$$

we can represent every hyperidentity $w_1 = w_2$ in terms of the hyperidentities

$$x = F(XF(w_1, Xw_2), x) \text{ and } x = F(x, XF(Xw_1, w_2)),$$

provided that x does not occur in w_1 and w_2. Hence $T = Heq(SB)$ and $T^+ = Heq(SB^+)$ have bases consisting of 14 and 16 hyperidentities respectively of the form $x = \sigma$. For τ, we can take the hyperterm $F(XF(Xx, Xy), F(XF(Xy, Xz), XF(Xz, Xx)))$. Then the above theorem yields the desired result.

Chapter 6

Elementary Theories of Super-Boolean Algebras

6.1. Preliminaries

In this chapter, we present the results of [282,283] on elementary theories for super-Boolean algebras, i.e., algebras with the hyperidentities of the variety of Boolean algebras. For elementary theories of Boolean algebras and variety of Boolean algebras see [101] (also see [398,399]).

Denote by ω the set of all natural numbers $\{0, 1, 2, \ldots\}$ and $\omega^+ = \{1, 2, \ldots\}$ (note that we will use ω as the cardinal number, too) in this chapter.

A **signature** σ (of algebraic system) is a pair consists of three in pairs disjoint sets $\sigma_p, \sigma_F, \sigma_c$ and a map $\mu : \sigma_p \cup \sigma_F \mapsto \omega^+$. If $p \in \sigma_p$ and $\mu(P) = n$, then P is called an n**-ary predicate symbol**; if $F \in \sigma_F$ and $\mu(F) = m$, then F is called an m**-ary functional symbol**; elements of σ_c are called **constant symbols.** Often signature is defined without specifying σ, i.e., signature can be defined as $\sigma = (P_1^{n_1}, \ldots, P_k^{n_k}; F_1^{m_1}, \ldots, F_s^{m_s}; c_1, \ldots, c_t)$, where the upper indexes are values of map μ on the corresponding symbols.

An **algebraic system** U of signature σ is a pair of non-empty set $|U|$, called a **carrier set** of system U, and collections of predicates $P^U \subseteq |U|^{\mu(P)}$ $(P \in \sigma_P)$ and functions $F^U : |U|^{\mu(F)} \mapsto |U|$ $(F \in \sigma_F)$ and selected elements $c^U \in |U|$ $(c \in \sigma_c)$, called **predicates, operations** and **constants** respectively.

Concepts of subsystem, quotient-system with respect to a given congruence, homomorphism, isomorphism and etc are defined in standard (usual) way for algebraic systems of fixed signature.

Denote by dom φ and range φ the domain and the range of map φ. The identity map of set A on itself we denote by id_A.

Let U_i, $i \in I$ be a family of algebraic systems with signature σ, $\prod_{i\in I} |U_i|$ be a Cartesian product of sets $|U_i|, i \in J$, i.e., a set of all functions $f : I \mapsto \bigcup_{i\in I} |U_i|$ such that $f(i) \in |U_i|$ for every $i \in I$. A **direct product** of algebraic systems $U_i, i \in I$ is a system U with a carrier set $\prod_{i\in I} |U_i|$ and predicates, operations and constants defined in the following way:

- if $P^n \in \sigma_P$, then $P^U = \{(f_1, \ldots, f_n)|(f_1(i), \ldots, f_n(i)) \in P^{U_i}$ for all $i \in I\}$;
- if $F^m \in \sigma_F$, $f_1, \ldots, f_m \in \prod_{i\in I} |U_i|$, then $F^U(f_1, \ldots, f_m)(i) = F^{U_i}(f_1(i), \ldots, f_m(i)), i \in I$;
- if $c \in \sigma_c$, then $c^U(i) = c^{U_i}, i \in I$.

The direct product of systems U_i, $i \in I$ we denote by $\prod_{i\in I} U_i$.

A map $\pi_i : \prod_{i\in I} |U_i| \mapsto |U_i|$ such that $\pi_i(f) = f(i)$, $f \in \prod_{i\in I} |U_i|$ is a homomorphism and is called a **projection on** U_i, for any $i \in I$.

We say that algebraic system U is a **subdirect product** of algebraic systems U_i, $i \in I$, if U is isomorphic to a subsystem V of the direct product $\prod_{i\in I} |U_i|$ such that $\pi_i(|V|) = |U_i|$ for any $i \in I$.

Recall that a **preorder** θ on set A is a subset of A^2 such that

- $(a, a) \in \theta$ for any $a \in A$ (reflexivity),
- if $(a, b) \in \theta$ and $(b, c) \in \theta$, then $(a, c) \in \theta$ for any $a, b, c \in A$ (transitivity).

A preorder θ on a non-empty set A is called **oriented**, if for any $a, b \in A$ there exists $c \in A$ such that $(a, c) \in \theta$ and $(b, c) \in \theta$.

Let $(I, \leq)$ be an oriented preorder. A **straight spectrum** R **on** $(I, \leq)$ of algebraic systems is a family of algebraic systems U_i, $i \in I$ and a family of homomorphisms $\psi_{ij} : U_i \mapsto U_j, i, j \in I, i \leq j$ such that $\psi_{ii} = id_{|U_i|}$ and $\psi_{ik} = \psi_{ij}\psi_{jk}$ for all $i \leq j \leq k$. Denote it by $R = \{U_i, \psi_{jk}|i, j, k \in I, j \leq k\}$.

It is possible to construct an algebraic system U and a system of homomorphisms $\mu_i : U_i \mapsto U$ such that $\mu_i = \psi_{ij}\mu_j$ for any $i \leq j$ in a canonical way by any straight spectrum R.

Let $A^* = \{(a, i)|i \in J, a \in |U_i|\}$. Define on A^* an equivalence relation $\eta = \{((a, i), (b, j))|$ there exists $k \in J$ such that $i \leq k, j \leq k, \psi_{ik}(a) = \psi_{jk}(b)\}$. Set $A = A^*/\eta$. The equivalent class $[(a, i)]_\eta = \{(b, j)|((a, i), (b, j)) \in \eta\}$ denote by $[a, i]$. Note that if $i \leq j$, then $[a, i] = [\psi_j(a), j]$. Therefore since $(I, \leq)$ is oriented, then for any finite set of elements $[a_0, i_0], \ldots, [a_n, i_n]$ of A, without lost of generality, we can say that $i_0 = \cdots = i_n$. Define on A an algebraic system with signature σ in the following way:

- if $P^n \in \sigma_P$, then $P^U = \{([a_1, i], \ldots, [a_n, i]) | [a_1, i], \ldots, [a_n, i] \in A$, there exists $j \geq i$ such that $(\psi_{ij}(a_1), \ldots, \psi_{ij}(a_n)) \in P^{U_j}\}$;
- if $F^m \in \sigma_F$, $[a_1, i], \ldots, [a_m, i] \in A$, then $F^U([a_1, i], \ldots, [a_m, i]) = [F^{U_i}(a_1, \ldots, a_m), i]$;
- if $c \in \sigma_c$ and $j \in J$, then $c^U = [c^{U_i}, j]$.

Obviously, the definition is correct. Homomorphisms are defined as $\mu_i(a) = [a, i], a \in |U_i|$. An algebraic system U together with system of homomorphisms $\mu_i : U_i \to U$ we call a **straight limit of spectrum** R and denote it by $\lim_{\to} U_i$.

Let I be a non-empty set, F be a filter over I, i.e., a family of subsets of I such that the following conditions are fulfilled:

(1) $I \in F, \emptyset \notin F$;
(2) $J, K \in F \to J \cap K \in F$;
(3) $J \subseteq K \subseteq J, J \in F \to K \in F$.

Let $U_i, i \in I$ be a family of algebraic systems of signature σ; let $U_J = \prod_{j \in J} U_j$ for $J \in F$ and let $\pi_{K,J} : U_K \mapsto U_J$ be a homomorphism, which is a projection on U_J. Define an oriented order $\preceq$ on F dually to the relation of inclusion (i.e., $Y \preceq K$ and only if $K \subseteq J$), then, it is easy to see that $R = \{U_J, \pi_{K,L} | \, J, K, L \in F, \, K \preceq L\}$ is a straight spectrum over $(F, \preceq)$. The straight limit of this straight spectrum we denote by $\prod_{i \in I} U_i/F$ and call a **filtered product of family** $U_i, i \in I$ **by filter** F **over** I. If F is an ultrafilter over I, i.e., a filter over I such that for any $J \subseteq I$ either $J \in J$ or $I \setminus J \in F$, then filtered product $\prod_{i \in I} U_i/F$ is called **an ultraproduct**. A filter is called principle, if it contains the intersection of all its subsets. There exist non-principle ultrafilter over each infinite set [216].

The definition of formulas of language of signature σ of predicate calculus, the concept of a free and connected entries of variables in a formula is well known [66,216]. By $\Phi(x_1, \ldots, x_n)$ we mean that all free object variables of the formula Φ belong to the set $\{x_1, \ldots, x_n\}$.

A concept of truth of formula Φ on U for substitution $x_i \mapsto a_i \in |U|, i = 1, \ldots, n$ is defined by standard way for any formula $\Phi(x_1, \ldots, x_n)$ of signature σ and for any algebraic system U of signature σ. $U \models \Phi(a_1, \ldots, a_n)$ means that Φ is true on U on elements $a_1, \ldots, a_n$. If T is a system of propositions (i.e., a system of formulas without free variables), then $U \models T$ means that $U \models \Phi$ for all $\Phi \in T$.

Theorem 6.1.1 ([216]). *Let Φ be an arbitrary proposition of signature σ, F be an arbitrary ultrafilter over I and U_i, $i \in I$ be a family of algebraic systems of signature σ. Φ is true on filtered product $U = \prod_{i \in I} U_i/F$ if and only if the collection of numbers of factors on which Φ is true belongs to ultrafilter F, i.e., $U \models \Phi \iff \{i \in I | U_i \models \Phi\} \in F$.*

Universal closure of atomic formulas are called **identities**. A class of algebraic systems defined by a set of identities is called a **variety**.

A family T of propositions of signature σ is called a **theory** (in this chapter) if T is closed under implications, i.e., if for any proposition Φ of signature σ from $U \models T \to U \models \Phi$ follows $\Phi \in T$. It is clear that for any class K of algebraic systems of signature σ the family $Th(K)$ of all propositions Φ such that $U \in K \to U \models \Phi$ is a theory. $Th(K)$ is called an **elementary theory of class** K. A subset A of theory T is called a **system of axioms** of T, if for any system U from $U \models A$ follows that $U \models T$.

A theory T is called **consistent**, if it is different from the set of all propositions. A theory T is called **complete**, if T is consistent and for any proposition Φ either $\Phi \in T$ or $\neg\Phi \in T$.

Two algebraic systems U_1 and U_2 of signature σ are called an **elementary equivalent** if $\mathrm{Th}(U_1) = \mathrm{Th}(U_2)$ or in other words, if $U_1 \models \Phi \iff U_2 \models \Phi$ for any proposition Φ of signature σ. Writing $U_1 \equiv U_2$ means elementary equivalent of U_1 and U_2.

The following theorem is a criterium for algebraic systems to be elementary equivalent.

Theorem 6.1.2 ([101]). *Two algebraic systems U_1 and U_2 of signature σ are elementary equivalent if and only if there exists sequence $S_0 \supseteq S_1 \supseteq \cdots \supseteq S_n \supseteq S_{n+1} \supseteq \ldots$ of non-empty sets of finite partially isomorphisms from U_1 to U_2 such that for any $n \in \omega$ and $\varphi \in S_{n+1}$, any $a \in |U_1|, b \in |U_2|$ there exist $\psi_1, \psi_2 \in S_n$ such that $\varphi \subseteq \psi_1$, ψ_2 and $a \in \psi_1$, $b \in \mathrm{range}\,\psi_2$.*

Corollary 6.1.3. *Let U_i, $i \in I$ and $V_i, i \in I$ be families of algebraic systems of signature σ such that $U_i \equiv V_i$ for all $i \in I$. Then*

(1) $\prod_{i \in I} U_i \equiv \prod_{i \in I} V_i$;
(2) *if F is a filter over I, then $\prod_{i \in I} U_i/F \equiv \prod_{i \in I} V_i/F$.*

A theory T is called **recursively axiomatizable** if there exist a system of axioms A and an algorithm which determines by any proposition whether it is an axiom or not.

A theory T is called **recursively enumerable axiomatized** if there exist a system of axioms A and an algorithm which lists all axioms from A and only them.

A theory T is called **decidable** if there exists an algorithm which allows in a finite number of steps to detect by any proposition Φ whether Φ belongs to the theory T or not.

Theorem 6.1.4 ([101]). *Consistent and complete theory T is decidable if and only if it is recursively axiomatizable.*

A set A is called recursively enumerable if $A = \emptyset$ or if there exists a general recursive function [215,344] f such that A is a set of values of function f.

n-ary predicate P is called recursively enumerable if the set of $(x_1, \ldots, x_n)$ for which the predicate P is true, is recursively enumerable.

Let $T_0, T_1, \ldots$ be a sequence of theories of signature σ and g be a Gödel numbering of all formulas of signature σ. A sequence $T_0, T_1, \ldots$ is called **effective** if (2-ary) predicate

$$R(x, y) \iff g(x) \in T_y$$

is recursively enumerated.

Theorem 6.1.5 ([101]). *A theory T is decidable if and only if the following two conditions are valid:*

(1) *T is recursively enumerable axiomatized;*
(2) *there exists an effective sequence of complete theories $T_0, T_1, \ldots$ such that $T = \bigcap_{i\in\omega} T_i$.*

Non-zero element a of a Boolean algebra (lattice with 0) U is called **an atom**, if there is only 0 below a, i.e., $\forall b\ (b < a \to b = 0)$. A Boolean algebra (lattice with 0) U is called an **atomic**, if there is an atom below for every non-zero element of U. An element a is called an **atomic**, if there is an atom below for any non-zero element $x \leq a$ and **non-atomic**, if there is no atom in the lattice $\{x | x \leq a\}$. A Boolean algebra (lattice with 0) U is called non-atomic if its every element is non-atomic. Note that 0 is atomic and non-atomic element.

For a Boolean algebra U define an ideal $I(U)$, which we call an **Ershov–Tarski ideal**:

$$I(U) = \{x \in U | \text{there exist an atomic element } a \text{ and an non-atomic element } b \text{ of } U \text{ such that } x = a + b\}.$$

Let's make the following designations $I_0(U) = \{0\}$, ${}^0U = U$, ${}^0p = id_U, I_1(U) = I(U)$, ${}^1U = U/I_1(U)$, ${}^1p$ is the natural projection of U on ${}^1U$; assume that for $n \in \omega$ an ideal $I_n(U)$ of U is already defined, then ${}^nU = U/I_n(U)$, ${}^np : U \mapsto^n U$ is a natural projection; suppose $I_{n+1} = {}^np^{-1}(I({}^nU))$. In such a way a sequence $I_n(U)$, $n \in \omega$ of Ershov–Tarski ideals of Boolean algebra U is defined.

Following to Ershov [101] let us define an **elementary characteristic** of Boolean algebra U as a triple $\mathrm{ch}(U) = (\mathrm{ch}_1(U), \mathrm{ch}_2(U), \mathrm{ch}_3(U))$. The smallest n such that $I_{n+1}(U) = U$ is called the first characteristic of Boolean algebra and denote it by $\mathrm{ch}_1(U)$. If such n does not exist, assume that $\mathrm{ch}_1(U) = \omega$. When $\mathrm{ch}_1(U) = \omega$, we assume that the second characteristic $\mathrm{ch}_2(U)$ and the third characteristic $\mathrm{ch}_3(U)$ of Boolean algebra U are equal to 0. Otherwise (i.e. when $\mathrm{ch}_1(U) = n$), the second characteristic is the number of atoms of Boolean algebra ${}^nU$, if this number is finite and $\mathrm{ch}_2(U) = \omega$, if the number of atoms in ${}^nU$ infinite. The third characteristic equals to 1, if there exists a non-zero nonatomic element in ${}^nU$, otherwise it is equal to 0.

The elementary characteristic of Boolean algebra $(B; +, \cdot, ', 0, 1)$ is called the elementary characteristic of Boolean lattice $(B; +, \cdot, ')$, too.

Theorem 6.1.6 ([101]). *Boolean algebras U_1 and U_2 are elementary equivalent if and only if they have the same elementary characteristics (i.e., $\mathrm{ch}U_1 = \mathrm{ch}U_2$).*

Proof. Construct by induction on $n \in \omega$ formulas $A_n(x)$, $B_n(x)$, $C_n(x)$ and $D_n(x)$ such that for any Boolean algebra U and for any elements $a, b \in |U|$ the following is true (a^* denote the image of a in ${}^nU$):
$U \models C_n(a) \iff a \in \mathcal{J}_n(U)$,
$U \models A_n(a) \iff a^*$ is an atom of Boolean algebra ${}^n(U)$,
$U \models B_n(a) \iff a^*$ is a nonatomic element of Boolean algebra ${}^n(U)$,
$U \models D_n(a, b) \iff a^* = b^*$.
The formulas we construct in the following way. For $n = 0$ set $A_0(x)$ is $x \neq 0 \& \forall y(xy = 0 \vee xy = x)$; $B_0(x)$ is $\forall y(\neg A_0(xy))$; $C_n(x)$ is $x = 0$; $D_0(x, y)$ is $x = y$. Suppose $A_n(x)$, $B_n(x)$, $C_n(x)$, $D_n(x, y)$ formulas are constructed for n such that the above Conditions are satisfied. Set:
$A_{n+1}(x)$ is $\neg C_n(x) \& \forall y(C_n(x, y)) \vee D_n(x, y))$,
$B_{n+1}(x)$ is $\forall y \neg A_{n+1}(x, y)$,
$C_{n+1}(x)$ is $\exists y \exists z(x = yz \& B_{n+1}(y) \& \forall u \exists v(C_n(zu) \vee A_{n+1}(zuv)))$,
$D_{n+1}(x, y)$ is $C_n(xy' \& C_n(x'y)$.

The constructed formulas satisfy the above conditions.

Let U be a Boolean algebra and $\mathrm{ch}(U) = (\alpha, \beta, \gamma)$; then the following equivalences are true:

(1) $n \leq \alpha \iff U \models \exists x C_n(x)$;
(2) if $\alpha = n$, then $0 < m \leq \beta \iff U \models \exists x_1 \dots x_m(\&_{i \neq j} \neg D_n(x_i, x_j) \& \&_{i=1}^n A_n(x_i))$;
(3) if $\alpha = n$, then $\gamma = 1 \iff U \models \exists x(\neg C_n(x) \& B_n(x))$.

From assertions (1)–(3) immediately follow that if U_1 and U_2 are elementary equivalent, then $\mathrm{ch}(U_1) = \mathrm{ch}(U_2)$. □

By the definition of elementary characteristics, if $\alpha \leq \omega$, $\beta \leq \omega$, $\gamma \leq 1$ and (α, β, γ) is an elementary characteristics of an Boolean algebra, then when $\alpha = \omega$ the values of β and γ are equal to 0. Triples satisfying this condition we call **permissible**.

Theorem 6.1.7 ([101]). *For any permissible triple (α, β, γ) there exists a Boolean algebra U such that $\mathrm{ch}(U) = (\alpha, \beta, \gamma)$.*

Theorem 6.1.8 ([101]). *Every Boolean algebra has decidable elementary theory.*

Proof. Let U be a Boolean algebra and $\mathrm{ch}(U) = (\alpha, \beta, \gamma)$, then from the proof of Theorem 6.1.6 we know that it is possible to write down a recursive system of axioms (which is finite if $\alpha, \beta < \omega$) by a triple (α, β, γ) for theory $\mathrm{Th}(U)$. Since the theory $\mathrm{Th}(U)$ is complete and has a recursive system of axioms, then $\mathrm{Th}(U)$ is decidable by Theorem 6.1.4. □

Theorem 6.1.9 ([101]). *The elementary theory T_B of the variety of all Boolean algebras is decidable.*

Proof. Let $(\alpha_0, \beta_0, \gamma_0), (\alpha_1, \beta_1, \gamma_1), \dots$ is an effective enumeration of triples (α, β, γ), $\alpha, \beta < \omega, \gamma \leq 1$ such that $\beta + \gamma > 0$. Let T_i be a complete elementary theory of Boolean algebra of characteristic $(\alpha_i, \beta_i, \gamma_i)$, where $T_0, T_1, \dots$ is an effective sequence. Let's show that $T_B = \bigcap_{i \in \omega} T_i$. Let U_i be a Boolean algebra of characteristic $(\alpha_i, \beta_i, \gamma_i)$, then $\bigcap_{i \in \omega} T_i = \mathrm{Th}(\{U_0, U_1, \dots\})$. So we have to prove that $T_B = \mathrm{Th}(\{U_0, U_1, \dots\})$. It is enough to show that for any Boolean algebra U there exists an ultraproduct of Boolean algebras from $\{U_i, i \in \omega\}$ elementary equivalent to U. if the elementary characteristic (α, β, γ) of U is finite, i.e., $\alpha, \beta < \omega$, then $U \equiv U_i$ for some $i < \omega$. If $(\alpha, \beta, \gamma) = (n, \omega, \gamma)$, $n < \omega$, then suppose $m_j < \omega$ are such that $(\alpha_{m_j}, \beta_{m_j}, \gamma_{m_j}) = (n, j+1, \gamma)$, $\gamma < \omega$; then the ultraproduct $\prod_{j \in \omega} U_{m_j}/F$ by any non-principle ultrafilter F has the elementary characteristic (n, ω, γ). If $(\alpha, \beta, \gamma) = (\omega, 0, 0)$ and $n_j < \omega$ are such that $(\alpha_{n_j}, \beta_{n_j}, \gamma_{n_j}) = (j+1, 1, 0)$, $j \in \omega$, the ultraproduct $\prod_{j \in \omega} U_{m_j}/F$ has the elementary characteristic $(\omega, 0, 0)$. So, $T_B = \mathrm{Th}(\{U_0, U_1, \ldots\}) = \bigcap_{i\in\omega} T_i$ and T_B is decidable by Theorem 6.1.5. □

In the proof of the previous theorem the following result was used.

Theorem 6.1.10 ([101]). *Let U_1 and U_2 be Boolean algebras and let $\mathrm{ch}(U_1) = (\alpha_1, \beta_1, \gamma_1)$, $\mathrm{ch}(U_2) = (\alpha_2, \beta_2, \gamma_2)$. Then the elementary characteristic (α, β, γ) of Boolean algebra $U_1 \times U_2$ can be described in the following way:*

$$\alpha = \max\{\alpha_1, \alpha_2\},$$

$$\beta = \begin{cases} \beta_1 & \text{if } \alpha_2 < \alpha_1, \\ \beta_2 & \text{if } \alpha_1 < \alpha_2, \\ \beta_1 \oplus \beta_2 & \text{if } \alpha_1 = \alpha_2, \end{cases}$$

$$\gamma = \begin{cases} \gamma_1 & \text{if } \alpha_2 < \alpha_1, \\ \gamma_2 & \text{if } \alpha_1 < \alpha_2, \\ \max\{\gamma_1, \gamma_2\} & \text{if } \alpha_1 = \alpha_2, \end{cases}$$

where $\oplus$ is the sum of cardinal numbers.

Let us recall some results and concepts from the previous chapter.

Theorem 6.1.11. *Every hyperidentity of the variety of Boolean algebras is a consequence of the following seven hyperidentities (see Sec. 5.6):*

$$X(x, x) = x, \tag{6.1.1}$$

$$X(x, y) = X(y, x), \tag{6.1.2}$$

$$X(x, X(y, z)) = X(X(x, y), z), \tag{6.1.3}$$

$$X(x(Y(y, z)) = Y(X(x, y), X(x, z)), \tag{6.1.4}$$

$$F(F(x)) = x, \tag{6.1.5}$$

$$X(F(x), y) = X(F(X(x, y)), y), \tag{6.1.6}$$

$$F(X(F(X(x, y)), F(X(x, F(y))))) = x. \tag{6.1.7}$$

Theorem 6.1.12. *Every hyperidentity of a Boolean algebra is a consequence of the hyperidentities (6.1.1)–(6.1.7) (see Sec. 5.6).*

Definition 6.1.13. A T-algebra U, where $T = \{1, 2\}$, is called a super-Boolean algebra, if it satisfies the hyperidentities of the variety of Boolean algebras, i.e., hyperidentities (6.1.1)–(6.1.7) (see Chapter 5).

Let $U = (A; \Sigma)$ be a super-Boolean algebra, then for any unary operations $F, G \in \Sigma$ hold $F(a) = G(a)$ for any $a \in A$, i.e., a super-Boolean algebra does not have two different unary operations.

Example 5.4.6 shows that super-Boolean algebra with two binary and one unary operations is not necessarily a Boolean lattice.

The following result follows from the Birkhoff's subdirect representation theorem.

Theorem 6.1.14. *Any super-Boolean algebra $U = (A; \Sigma)$ is isomorphic to a subdirect product of super-Boolean algebras with one or two binary operations in the category of T-algebras, where $T = \{1, 2\}$ (see Secs. 5.5–5.7).*

So, the description of super-Boolean algebras reduced to the description of super-Boolean algebras with one or two binary operations.

In the case of one binary operation, the hyperidentities (6.1.1)–(6.1.7) of super-Boolean algebras turns to identities, moreover this system of identities is redundant. So, super-Boolean algebra with one binary operation is an algebra $U = (A; +, ^{-})$ with the following identities:

$$
\begin{aligned}
x + x &= x, \\
x + y &= y + x, \\
x + (y + z) &= (x + y) + z, \\
\bar{\bar{x}} &= x, \\
\bar{x} + y &= \overline{x + y} + y, \\
\overline{x + y} + \overline{x + \bar{y}} &= \bar{x}.
\end{aligned}
$$

Obviously, if $U = (A; +, \cdot, ^{-})$ is a Boolean lattice, then the reduct $U' = (A; +, ^{-})$ is a super-Boolean algebra. The converse is also true. Let $U = (A; +, ^{-})$ be a super-Boolean algebra. Then defining the operation $\cdot$ on the set A in the following way:

$$x \cdot y = \overline{\bar{x} + \bar{y}}.$$

The obtaining algebra $(A; +, \cdot, ^{-})$ is a Boolean lattice, we denote it by U^* and call a Boolean extension of super-Boolean algebra U.

For a characterization of super-Boolean algebras with two binary operations we introduced the following concept of Boolean sum in previous chapter.

Let $U = (A; \Sigma \cup \{F\})$ be a T-algebra with one unary operation F. Suppose $(A_i; \Sigma), i \in I$ are subalgebras of U and let $U_i = (A_i; \Sigma \cup \{F_i\})$ be algebras with one unary operation F_i for any $i \in I$. An algebra

$U = (A; \Sigma \cup \{F\})$ is called a Boolean sum of algebras $U_i = (A_i; \Sigma \cup \{F_i\})$, $i \in I$, if:

(a) $A_i \cap A_j = \emptyset$, where $i, j \in I$, $i \neq j$;
(b) $A = \bigcup_{i \in I} A_i$;
(c) on the set of indices I is defined a partial order $\leq$ such that $(I; \leq)$ is a Boolean algebra;
(d) if $i \leq j$, then there exists a (bi)isomorphism

$$(\varphi_{i,j}, \tilde{\varepsilon}) : U_i \Longrightarrow U_j, \ i, j \in I,$$

where $\tilde{\varepsilon}(F_i) = F_j, \tilde{\varepsilon}(G) = G$ for any $G \in \Sigma$ and $\varphi_{i,i} = id_{A_i}$ and $\varphi_{i,k} \cdot \varphi_{k,j} = \varphi_{i,j}$, where $i \leq k \leq j$;
(e) for any $G \in \Sigma$, $|G| = n \geq 2$ and for any $x_1, \ldots, x_n \in A$

$$G(x_1, \ldots, x_n) = G(\varphi_{i_1,i_0}(x_1), \ldots, \varphi_{i_n,i_0}(x_n)),$$

where $x_1 \in A_{i_1}, \ldots, x_n \in A_{i_n}$, $i_1, \ldots, i_n \in I$, $i_0 = i_1 + \cdots + i_n$;
(f) for any $i \in I$ there exists a (bi)isomorphism

$$(h_{i,\bar{i}}, \tilde{\varepsilon}) : U_i \Longrightarrow U_{\bar{i}},$$

where $h_{i,\bar{i}}^{-1} = h_{\bar{i},i}$, $\varphi_{i,1} = h_{i,\bar{i}} \cdot \varphi_{\bar{i},1}$ and 1 is the identity element of the Boolean algebra $\langle I; \leq \rangle$;
(g) for the unary operation F and for any $x \in A$

$$F(x) = h_{i,\bar{i}}[F_i(x)],$$

where $x \in A_i$.

The following result is proved in previous chapter.

Theorem 6.1.15. *The algebra $U = (A; +, \cdot, ^-)$ with two binary and one unary operations is a super-Boolean algebra if and only if U is a Boolean lattice or it is a Boolean sum of Boolean lattices.*

A Boolean structure for each super-Boolean algebra $U = (A; +, \cdot, ^-)$ is defined in the following manner.

A relation θ on the set A is defined as follows:

$$x\theta y \iff x + xy = x,\ y + yx = y,$$

which is an equivalence relation. If $A = \bigcup_{i\in I} A_i$ is a partition corresponding to θ, then algebra $U_i = (A_i; +, \cdot, ')$, $i \in I$, where the operation $'$ is defined as

$$x' = x \cdot \overline{x + \overline{\overline{x}}} + \overline{x \cdot (x + \bar{x})},$$

is a Boolean lattice. The quotient-set $A/\theta = \{A_i, i \in I\}$ is a Boolean lattice with respect to the following operations:

(a) $A_i + A_j = A_k \iff 1_i + 1_j \in A_k$, where $i, j \in I$ and $1_i + 1_j = 1_k$;
(b) $\overline{A_i} = A_j \iff \overline{1_i} \in A_j$, where $i, j \in I$ and $\overline{1_i} = 0_j$;
(c) $A_i \cdot A_j = \overline{\overline{A_i} + \overline{A_j}}$.

Moreover, the following is true: $a + b \in A_i + A_j$, $ab \in A_i + A_j, \bar{a} \in \bar{A}_i$ where $a \in A_i, b \in A_j$.

Identifying every A_i with their indexes we will get a Boolean algebra $(I; \leq)$, where $i \leq j \iff A_i \leq A_j$.

Isomorphisms $\varphi_{i,j}, h_{i,\bar{i}}$ are defined as follows:

$$h_{i,\bar{i}}(x) = (\bar{x})', i \in I;$$

$$\varphi_{i,j}(x) = x \cdot 1_j \in A_i + A_j = A_j, i \leq j, i, j \in I.$$

Thus, a super-Boolean algebra U is a Boolean sum of Boolean lattices $U_i, i \in I$. This representation we call **a canonical representation** of a super-Boolean algebra U as a Boolean sum of Boolean lattices in this chapter.

6.2. Decidability and a classification of elementary theories of super-Boolean algebras with one binary and one unary operations

Let $U = (A; +, ^{-})$ be a super-Boolean algebra with one unary and one binary operations. We defined a Boolean lattice $U^* = (A; +, \cdot, ^{-})$ which is called a boolean extension of super-Boolean algebra U, where the operation $\cdot$ is defined as follows:

$$x \cdot y = \overline{\bar{x} + \bar{y}}. \tag{6.2.1}$$

Theorem 6.2.1. *Two super-Boolean algebras $U = (A; +, ^{-})$ and $V = (B; +, ^{-})$ with one binary and one unary operations are elementary equivalent if and only if the elementary characteristics of their boolean extensions are the same, i.e.,* $\mathrm{ch}(U^*) = \mathrm{ch}(V^*)$.

Proof. Let $\mathrm{ch}(U^*) = \mathrm{ch}(V^*)$, then by Theorem 6.1.6 we have $U^* \equiv V^*$. Obviously, if $\Phi \in \mathrm{Th}\, U$, then $\Phi \in \mathrm{Th}\, U^*$ for every proposition Φ of signature $(+,^-)$. Since $U^* \equiv V^*$, then $\Phi \in \mathrm{Th}\ V^*$. Because Φ is a proposition of signature $(+,^-)$, then from $\Phi \in \mathrm{Th}\, V^*$ follows that $\Phi \in \mathrm{Th}\, V$. In the same way, if $\Phi \in \mathrm{Th}\, V$, then $\Phi \in \mathrm{Th}\, U$ for any proposition Φ of signature $(+,-)$. So, from $\mathrm{ch}(U^*) = \mathrm{ch}(V^*)$ follows that $U \equiv V$.

Let $U \equiv V$, then obviously $U^* \equiv V^*$ and by Theorem 6.1.6, we have $\mathrm{ch}(U^*) = \mathrm{ch}(V^*)$. □

Definition 6.2.2. The elementary characteristic of super-Boolean algebra with one unary and one binary operations we call the elementary characteristic of its boolean extension.

Theorem 6.2.3. *A super-Boolean algebras $U = (A;+,^-)$ and $V = (B;+,^-)$ with one binary and one unary operations are elementary equivalent if and only if their elementary characteristics coincide.*

Theorem 6.2.4. *Every super-Boolean algebra with one unary and one binary operations has a decidable elementary theory.*

Proof. Let $U = (A;+,^-)$ be a super-Boolean algebra and U^* be its boolean extension. The Boolean lattice U^* has a recursive system of axioms. Substituting every entrance of operation $\cdot$ in this system of axioms by (6.2.1) we get a recursive system of axioms for algebra U. Thus, $\mathrm{Th}\, U$ is a complete theory with recursive system of axioms and the $\mathrm{Th}\, U$ is a decidable by Theorem 6.1.4. □

Theorem 6.2.5. *The elementary theory of the class of all super-Boolean algebras with one binary and one unary operations is decidable.*

Proof. The proof of this theorem repeats the proof of Theorem 6.1.9. □

6.3. A criteria of isomorphism for super-Boolean algebras with two binary and one unary operations

Every super-Boolean algebra $U = (A;+,\cdot,')$ with two binary and one unary operations is canonically represented as a Boolean sum of Boolean lattices $U_i = (A_i;+,\cdot,')$, $i \in I_U$ (Theorem 6.1.15). Using this result, let us prove the following necessary and sufficient conditions for isomorphism of super-Boolean algebras with two binary operations.

Theorem 6.3.1. *Let $U = (A;+,\cdot,')$ and $V = (B;+,\cdot,')$ be super-Boolean algebras with two binary and one unary operations and let $U_i = (A_i;+,\cdot,')$,*

$i \in I_U$, $V_k = (B_k; +, \cdot, ')$, $k \in I_V$ be a canonical representations as a Boolean sum of Boolean lattices corresponding to given super-Boolean algebras. Then

$$U \cong V \iff I_U \cong I_V \text{ and } U_i \cong V_k \text{ for some } i \in I_U, k \in V_i,$$

i.e., algebras U and V are isomorphic if and only if the algebras I_U and I_V are isomorphic and for some $i \in I_U$ and $k \in I_V$ the algebras U_i and V_k are also isomorphic.

Proof (*Necessity*). Suppose that $U \cong V$ and φ is an isomorphism between algebras U and V. By the definition of the canonical representation of super-Boolean algebras into the Boolean sum of Boolean lattices, we have: $a, b \in A_i, i \in I_U \iff a\theta_U b \iff a + ab = a, b + ba = b \iff \varphi(a) + \varphi(a)\varphi(b) = \varphi(a), \varphi(b) + \varphi(b)\varphi(a) = \varphi(b)(\varphi$ is an isomorphism $) \iff \varphi(a)\theta_V\varphi(b) \iff \varphi(a), \varphi(b) \in B_k, k \in I_V$, for arbitrary $a, b \in A$.

Thus, two elements are in the same set A_i, $i \in I_U$ if and only if their images with respect to the isomorphism φ belong to the same set B_k, $k \in I_v$.

For arbitrary $i \in I_U$ denote the restriction of φ on A_i by φ^i, i.e., $\varphi^i = A_i$ and $\varphi^i(a) = \varphi(a)$ for any $a \in A_i$.

Let a be an arbitrary fixed element from A_i and $\varphi^i(a) = b, b \in B$. By the definition of Boolean sum, $B = \bigcup_{k \in I_V} B_k$ and $B_k \cap B_s = \emptyset$, $k \neq s$. Hence $b \in B_k$ for some $k \in I_V$. Let's show that the range of values of φ^i coincides with B_k, i.e. range $\varphi^i = B_k$.

If $c \in A_i$, then from $a, c \in A_i$ follows that $\varphi(a), \varphi(c) \in B_k$ and thus $\varphi^i(a), \varphi^i(c) \in B_k$. So, range $\varphi^i \subseteq B_k$. On the other hand, since φ is an isomorphism, then for any $d \in B_k$ there exists $c \in A$ such that $\varphi(c) = d$. We have $\varphi(a), \varphi(c) \in B_k$ and hence $a, c \in A_i$. So, range $\varphi^i \supseteq B_k$ and therefore range $\varphi^i = B_k$.

Since φ^i is a restriction of isomorphism $\varphi : (A; +, \cdot,') \mapsto (B; +, \cdot,')$, then φ^i is an isomorphism from $U_i = (A_i; +, \cdot,')$ and $V_k = (B_k; +, \cdot,')$. Recall that the operation $'$ on U_i and V_k is defined as follows: $x' = x \cdot \overline{x + \overline{x}} + \overline{x \cdot (x + \bar{x})}$. So, $\varphi(x') = \varphi(x \cdot \overline{x + \overline{x}} + \overline{x \cdot (x + \bar{x})}) = \varphi(x \cdot \overline{x + \overline{x}}) + \varphi(\overline{x \cdot (x + \bar{x})}) = \varphi(x) \cdot \varphi(\overline{x + \overline{x}}) + \overline{\varphi(x \cdot (x + \bar{x}))} = \varphi(x) \cdot \overline{\varphi(x + \overline{x})} + \overline{\varphi(x \cdot (x + \bar{x}))} = \varphi(x) \cdot \overline{\varphi(x) + \varphi(\overline{x})} + \overline{\varphi(x) \cdot (\varphi(x) + \varphi(\bar{x}))} = \varphi(x) \cdot \overline{\varphi(x) + \overline{\varphi(x)}} + \overline{\varphi(x) \cdot (\varphi(x) + \overline{\varphi(x)})} = (\varphi(x))'$.

If $c \in A_i$ then $c' \in A_i$, therefore $c, c' \in$ range φ^i and $\varphi^i(c') = \varphi(c') = (\varphi(c))' = (\varphi^i(c))'$.

Thus, φ^i is an isomorphism of algebras U_i into V_k.

Let's construct an isomorphism ψ from I_U and I_V as follows. For every $i \in I_U$ suppose $\psi(i) = k$ if $\varphi(a) \in B_k$ for some $a \in A_i$. Since if $a, b \in A_i$, then as it was shown $\varphi(a), \varphi(b) \in B_k$ and it means that every $i \in I_U$ corresponds a unique $k \in I_V$.

For any $i, j \in I_U$ and for any $a \in A_i$, $b \in A_j$ we have $a + b \in A_i + A_j = A_{i+j}$ and by the definition of ψ, $\varphi(a+b) \in B_{\psi(i+j)}, \varphi(a) + \varphi(b) \in B_{\psi(i)} + B_{\psi(j)}$. By the definition of the canonical representation into Boolean sum, $B_{\psi(i)} + B_{\psi(j)} = B_s$ for some $s \in I_V$. Since $\varphi(a+b) \in B_{\psi(i+j)}$ and $\varphi(a+b) = \varphi(a)+\varphi(b) \in B_s$, then $B_{\psi(i+j)} = B_s = B_{\psi(i)+\psi(j)}$. Thus, $\psi(i+j) = \psi(i) + \psi(j)$.

For any $i \in I_U$ and for any $a \in A_i$, $\overline{a} \in A_{\overline{i}}, \varphi(\overline{a}) \in B_{\psi(\overline{i})}, \varphi(a) \in B_{\psi(i)}, \overline{\varphi(a)} \in \overline{B}_{\psi(i)} = B_{\overline{\psi(i)}}$. From $\overline{\varphi(a)} = \varphi(\overline{a})$ follows $B_{\overline{\psi(i)}}$ and $B_{\psi(\overline{i})}$ have intersection, therefore they coincide, i.e., $B_{\overline{\psi(i)}} = B_{\psi(\overline{i})}$ and $\overline{\psi(i)} = \psi(\overline{i})$.

Since I_U and I_v are Boolean algebras, then: $\psi(i \cdot j) = \psi(\overline{\overline{i} + \overline{j}}) = \overline{\psi(\overline{i}) + \psi(\overline{j})} = \overline{\overline{\psi(i)} + \overline{\psi(j)}} = \psi(i) \cdot \psi(j)$. So, ψ is an homomorphism between I_U and I_V.

Assume $\psi(i) = \psi(j), i, j \in I_U, i \neq j$. Then $U_i \neq U_j$ and $V_{\psi(i)} = V_{\psi(j)}$. For any $a \in A_i$ and $b \in A_j$ we have $\varphi(a) \in B_{\psi(i)}$ and $\varphi(b) \in B_{\psi(j)}$. Since $B_{\psi(i)} = B_{\psi(j)}$, then $\varphi(a), \varphi(b) \in B_{\psi(i)}$ and this means that a and b cannot belong to different sets A_i and A_j, i.e., i should be equal to j. So we get a contradiction, and we conclude that ψ is a monomorphism.

For every $k \in I_V$ and every $c \in B_k$ there exists $a \in A$ such that $\varphi(a) = c$. The element a belongs to A_i for some $i \in I_U$ and since $\varphi(a) \in B_k$, then $\psi(i) = k$. Thus ψ is an epimorphism.

So, $\psi : I_U \mapsto I_V$ is an epimorphism and monomorphism, which yields ψ is an isomorphism.

Thus, by given isomorphism $\varphi : U \mapsto V$ we constructed isomorphisms $\psi : I_U \mapsto I_V$ and $\varphi' : U_i \mapsto V_k$ for some $i \in I_U$, $k \in I_V$. Hence, $U \cong V \Rightarrow I_U \cong I_V$, $U_i \cong V_k$ for some $i \in I_U, k \in I_V$.

(*Sufficiency*). Let $I_U \cong I_V$ and $U_i \cong V_k$ for some $i \in I_U, k \in I_V$. By the definition of Boolean sum for every $i \in I_U, k \in I_V$ we have $U_i \cong U_1$ and $V_k \cong V_1$, then $U_1 \cong V_1$. Suppose $\varphi : U_1 \mapsto V_1$ and $\psi : I_U \mapsto I_V$ are the corresponding isomorphisms.

Consider a map $\alpha : A \mapsto B$ defined as follows:

$$\alpha(a) = \varphi^U_{i,1}\varphi\varphi^V_{1,\psi(i)}(a) = \varphi^V_{1,\psi(i)}(\varphi(\varphi^U_{i,1}(a))),$$

where $a \in A_i$, and $\varphi^U_{i,1} : U_i \mapsto U_1$, $\varphi^V_{1,\psi(i)} : V_1 \mapsto V_{\psi(i)}$ are isomorphisms defined for Boolean sums. The map α is defined correct, since the product of maps is a map.

Let's show that α is a homomorphism, i.e., for any $a, b \in A$, $\alpha(\overline{a}) = \overline{\alpha(a)}, \alpha(a+b) = \alpha(a) + \alpha(b)$, $\alpha(a \cdot b) = \alpha(a) \cdot \alpha(b)$.

From the definition of Boolean sum, we have

$$\begin{aligned}
\varphi^V_{1,\overline{\psi(s)}} &= \left(\varphi^V_{\overline{\psi(s)},1}\right)^{-1} = \left(h^V_{\overline{\psi(s)},\psi(s)}\varphi^V_{\psi(s),1}\right)^{-1} = \left(\varphi^V_{\overline{\psi(s)},1}\right)^{-1}\left(h^V_{\overline{\psi(s)},\psi(s)}\right)^{-1} \\
&= \varphi^V_{1,\overline{\psi(s)}}\ h^V_{\overline{\psi(s)},\psi(s)}\ \varphi^U_{\overline{s},1} = h^U_{\overline{s},s}\varphi^U_{s,1}, \\
\alpha(\overline{a}) &= \varphi^U_{\overline{s},1}\varphi\varphi^V_{1,\overline{\psi}(s)}(\overline{a}) \\
&= h^U_{\overline{s},s}\varphi^U_{s,1}\varphi\varphi^V_{1,\psi(s)}h^V_{\psi(s),\overline{\psi}(s)}(\overline{a}) \\
&= h^V_{\psi(s),\overline{\psi}(s)}(\varphi^U_{s,1}\varphi\varphi^V_{1,\psi(s)}(h^U_{\overline{s},s}(\overline{a})) = h^V_{\psi(s),\overline{\psi}(s)}(\varphi^U_{s,1}\varphi\varphi^V_{1,\psi(s)}(a')) \\
&= h^V_{\psi(s),\overline{\psi}(s)}(\varphi^U_{s,1}\varphi\varphi^V_{1,\psi(s)}(a))' = h^V_{\psi(s),\overline{\psi}(s)}(\alpha(a')), \\
\alpha(a+b) &= \varphi^U_{s+j,1}\varphi\varphi^V_{1,\psi(s+j)}(a+b) = \varphi^V_{1,\psi(s+j)}(\varphi(\varphi^U_{s+j,1}(a+b))) \\
&= \varphi^V_{1,\psi(s+j)}(\varphi(1_{U_1}\cdot(a+b))) = \varphi^V_{1,\psi(s+j)}(\varphi(1_{U_1}\cdot a + 1_{U_1}\cdot b)) \\
&= \varphi^V_{1,\psi(s+j)}(\varphi(\varphi^U_{s,1}(a) + \varphi^U_{j,1}(b))) = \varphi^V_{1,\psi(s+j)}(\varphi^U_{s,1}\varphi(a) + \varphi^U_{j,1}\varphi(b)), \\
\alpha(a\cdot b) &= \varphi^U_{s+j,1}\varphi\varphi^V_{1,\psi(s+j)}(a\cdot b) = \varphi^V_{1,\psi(s+j)}(\varphi(\varphi^U_{s+j,1}(a\cdot b))) \\
&= \varphi^V_{1,\psi(s+j)}(\varphi(1_{U_1}\cdot(a\cdot b))) = \varphi^V_{1,\psi(s+j)}(\varphi((1_{U_1}\cdot a)\cdot(1_{U_1}\cdot b))) \\
&= \varphi^V_{1,\psi(s+j)}(\varphi(\varphi^U_{s,1}(a)\cdot\varphi^U_{j,1}(b))) = \varphi^V_{1,\psi(s+j)}(\varphi^U_{s,1}\varphi(a)\cdot\varphi^U_{j,1}\varphi(b)),
\end{aligned}$$

where $a \in A_s, b \in A_j$, $1_{U_1}, 1_{V_1}$ are units of Boolean algebras U_1 and V_1 correspondingly. In the previous transformations we used that $a+b, a\cdot b \in A_{s+j}$.

Let's make the following designations $c = \varphi^U_{s,1}\varphi(a)$ and $d = \varphi^U_{j,1}\varphi(b)$ and prove the following equalities:

$$\begin{aligned}
\varphi^V_{1,\psi(s+j)}(c+d) &= \varphi^V_{1,\psi(s)}(c) + \varphi^V_{1,\psi(j)}(d), \\
\varphi^V_{1,\psi(s+j)}(c\cdot d) &= \varphi^V_{1,\psi(s)}(c)\cdot\varphi^V_{1,\psi(j)}(d).
\end{aligned}$$

Really,

$$\begin{aligned}
&\varphi^V_{\psi(s+j),1}(\varphi^V_{1,\psi(s+j)}(c+d)) = c+d,\\
&\varphi^V_{1,\psi(s)}(\varphi^V_{1,\psi(s)}(c) + \varphi^V_{1,\psi(j)}(d)) = (\varphi^V_{1,\psi(s)}(c))1_{V_1} + (\varphi^V_{1,\psi(j)}(d))1_{V_1}\\
&\quad = \varphi^V_{\psi(s),1}(\varphi^V_{1,\psi(s+j)}(c)) + \varphi^V_{\psi(j),1}(\varphi^V_{1,\psi(j)}(d)) = c+d,\\
&\varphi^V_{\psi(s+j),1}(\varphi^V_{1,\psi(s+j)}(c+d)) = c+d,\\
&\varphi^V_{\psi(s+j),1}(\varphi^V_{1,\psi(s+j)}(c\cdot d)) = c\cdot d,\\
&\varphi^V_{1,\psi(s)}(\varphi^V_{1,\psi(s)}(c)\cdot \varphi^V_{1,\psi(j)}(d)) = 1_{V_1}\cdot(\varphi^V_{1,\psi(s)}(c)\cdot \varphi^V_{1,\psi(j)}(d))\\
&\quad = 1_{V_1}\cdot \varphi^V_{1,\psi(s)}(c)\cdot 1_{V_1}\cdot \varphi^V_{1,\psi(j)}(d)\\
&\quad = \varphi^V_{\psi(s),1}(\varphi^V_{1,\psi(s)}(c))\cdot \varphi^V_{\psi(s),1}(\varphi^V_{1,\psi(j)}(d)) = c\cdot d.
\end{aligned}$$

Hence,

$$\begin{aligned}
\alpha(a+b) &= \varphi^V_{1,\psi(s+j)}(\varphi^U_{s,1}\varphi(a) + \varphi^U_{j,1}\varphi(b))\\
&= \varphi^V_{1,\psi(s)}(\varphi^U_{s,1}\varphi(a)) + \varphi^V_{1,\psi(j)}(\varphi^U_{j,1}\varphi(b)\\
&= \varphi^U_{s,1}\varphi\varphi^V_{1,\psi(s)}(a) + \varphi^U_{j,1}\varphi\varphi^V_{1,\psi(j)}(b) = \alpha(a)+\alpha(b),\\
\alpha(a\cdot b) &= \varphi^V_{1,\psi(s+j)}(\varphi^U_{s,1}\varphi(a)\cdot \varphi^U_{j,1}\varphi(b))\\
&= \varphi^V_{1,\psi(s)}(\varphi^U_{s,1}\varphi(a))\cdot \varphi^V_{1,\psi(j)}(\varphi^U_{j,1}\varphi(b)\\
&= \varphi^U_{s,1}\varphi\varphi^V_{1,\psi(s)}(a)\cdot \varphi^U_{j,1}\varphi\varphi^V_{1,\psi(j)}(b) = \alpha(a)\cdot\alpha(b),
\end{aligned}$$

which shows that α is a homomorphism.

The next step is to prove that α is an epimorphism. For any $b\in B$ there exists $k\in I_k$ such that $b\in B_k$. Since ψ is an isomorphism, then for given k there exists $i\in I_U$ such that $\psi(i)=k$. The map $\varphi^U_{i,1}\varphi\varphi^V_{1,\psi(i)} : A_i \mapsto B_k$ is a product of bijective maps, hence it is a bijection itself and therefore for given $b\in B_k$ there exists $a\in A_i$ such that $h(a)=b$.

Let's show that α is a monomorphism. First, let's assume that there exists $i\in I_U$ such that $a,b\in A_i$. Since $\varphi^U_{i,1}\varphi\varphi^V_{1,\psi(i)} : A_i\mapsto B_{\psi(i)}$ is a bijection, then from $a\neq b$ follows that $\alpha(a) = \varphi^U_{i,1}\varphi\varphi^V_{1,\psi(i)}(a) \neq \varphi^U_{i,1}\varphi\varphi^V_{1,\psi(i)}(b) = \alpha(b)$.

Suppose that $a\in A_i$, $b\in A_j$, $i,j\in I_U$, $i\neq j$. Then $\alpha(a) = \varphi^U_{i,1}\varphi\varphi^V_{1,\psi(i)}(a)\in B_{\psi(i)}$ and $\alpha(b) = \varphi^U_{i,1}\varphi\varphi^V_{1,\psi(i)}(b)\in B_{\psi(j)}$. Since ψ is an isomorphism, then from $i\neq j$ follows that $\psi(i)\neq\psi(j)$ and therefore $B_{\psi(i)}\cap B_{\psi(j)} = \emptyset$ and $\alpha(a)\neq\alpha(b)$.

Thus, we prove that $\alpha : U\mapsto V$ is an isomorphism. □

Remark. If for some $i \in I_U$ and $k \in I_V$ we have $U_i \cong V_k$, then for any $s \in I_U$ and $t \in I_V$, $U_s \cong V_t$. By construction of a Boolean sum, $U_s \cong U_1$ and $V_t \cong V_1$ for any $s \in I_U$ and $t \in I_V$, where U_1 and V_1 are algebras corresponding to the identity elements of Boolean algebras I_U and I_V.

6.4. A criteria of elementary equivalency of super-Boolean algebras with two binary and one unary operations

Theorem 6.4.1. *Let* $U = (A; +, \cdot, ^{-})$ *and* $V = (A; +, \cdot, ^{-})$ *be super-Boolean algebras with two binary and one unary operations and* $U_i = (A_i; +, \cdot, ')$, $i \in I_U$, $V_k = (B_k; +, \cdot, ')$, $k \in I_V$ *are canonical representations of this algebra as a Boolean sum of Boolean lattices. Then*

$$U \equiv V \iff I_U \equiv I_V \text{ and } U_i \equiv V_k \text{ for some } i \in I_U, k \in I_V,$$

i.e., algebras U *and* V *are elementary equivalent if and only if algebras* I_U *and* I_V *are elementary equivalent and for some* $i \in I_U, k \in I_V$ *so are algebras* U_i *and* V_k.

Proof (*Necessity*). Let $U \equiv V$. There exists a sequence $S_0 \supseteq S_1 \supseteq \cdots \supseteq S_n \supseteq \cdots$ of non-empty sets of partially isomorphisms from U into V such that for any n, any $\varphi \in S_{n+1}$ and any $a \in A$, $b \in B$ there exist $\varphi_1, \varphi_2 \in S_n$ such that $\varphi \subseteq \varphi_1, \varphi_2$, $a \in \operatorname{dom} \varphi_1, b \in \operatorname{range} \varphi_2$ (Theorem 6.1.2).

Recall that a partial isomorphism from algebraic system U_1 of signature σ on algebraic system U_2 of signature σ is called any isomorphism φ of subsystem U_1' of U_1 into U_2. For φ_1 and φ_2, the notation $\varphi_1 \subseteq \varphi_2$ means that $\operatorname{dom} \varphi_1 \subseteq \operatorname{dom} \varphi_2$ and $\varphi_1(a) = \varphi_2(a)$ for any $a \in \operatorname{dom} \varphi_1$.

The restriction of mapping φ on A_1 is denoted by φ°, for each n and each $\varphi \in S_n$, i.e., $\operatorname{dom} \varphi^\circ = A_1 \cap \operatorname{dom} \varphi$ and $\varphi^\circ(a) = \varphi(a)$ for all $a \in \operatorname{dom} \varphi^\circ$. Using the statement of Theorem, we will show that $\operatorname{range} \varphi^\circ = B_k \cap \operatorname{range} \varphi$ for some $k \in I_V$.

Let $a \in \operatorname{dom} \varphi^\circ$ and $\varphi^\circ(a) \in B_k$, $k \in I_V$ and let $b \in \operatorname{range} \varphi^\circ$. Then from $a, (\varphi^\circ)^{-1}(b) \in \operatorname{dom} \varphi^\circ \cap \operatorname{range} \varphi$ follows that $a, (\varphi^\circ)^{-1}(b) \in A_1$ and $\varphi^\circ(a)$, $b \in B_k$. It means that $\operatorname{range} \varphi^\circ \subseteq B_k \cap \operatorname{range} \varphi$.

If $b \in B_k \cap \operatorname{range} \varphi$, then $\varphi^\circ(a), b \in B_k$ and $a, \varphi^{-1}(b) \in A_1$. Thus, $\varphi^{-1}(b) \in A_1 \cap \operatorname{dom} \varphi = \operatorname{dom} \varphi^\circ$ and $\varphi^\circ(\varphi^{-1}(b)) = \varphi(\varphi^{-1}(b)) = b$. It means that $b \in \operatorname{range} \varphi^\circ$ and therefore $\operatorname{range} \varphi^\circ \supseteq B_k \cap \operatorname{range} \varphi$.

Since $\operatorname{range} \varphi^\circ \subseteq B_k \cap \operatorname{range} \varphi$ and $\operatorname{range} \varphi^\circ \supseteq B_k \cap \operatorname{range} \varphi$, then $\operatorname{range} \varphi^\circ = B_k \cap \operatorname{range} \varphi$.

Since φ is a partial isomorphism from U into V, then $(\operatorname{dom}\varphi; +, \cdot, -)$ is a subalgebra of U and $\operatorname{dom}\varphi$ is closed under the operations $+, \cdot$ and $'$. Therefore $\operatorname{dom}\varphi^\circ = A_1 \cap \operatorname{dom}\varphi$ is also closed under the same operations and $(\varphi^\circ; +, \cdot, ')$ is a subalgebra of U_1.

In order to show that φ° is a partial isomorphism from U_1 into V_k, it is left to prove that φ° satisfies to the conditions of homomorphism and monomorphism (see the proof of the previous theorem).

Let $\varphi^V_{k,1} : V_k \mapsto V_1$ be an isomorphism from the definition of Boolean sum. The mapping $\varphi' = \varphi^\circ \varphi^V_{k,1}$ is a partial isomorphism from U_1 into V_1 (as a product of isomorphisms).

Thus, the partial isomorphism φ' from U_1 into V_1 is defined for each n and each $\varphi \in S_n$. In this way, a sequence $S'_0 \supseteq S'_1 \supseteq \cdots \supseteq S'_n \supseteq \cdots$ of partial isomorphisms from U_1 into V_1 is defined.

Sets $S'_n, n \in \omega$ are not empty. Really, for any $n \in \omega$, S_{n+1} is not empty. For any $\varphi \in S_{n+1}$ and $a \in A_1$ there exists $\varphi_1 \in S_n$ such that $\varphi \subseteq \varphi_1$ and $a \in \operatorname{dom}\varphi_1$. So, $\operatorname{dom}\varphi'_1 \neq \emptyset$ and $\varphi'_1 \in S'_n$. For any n, any $\varphi' \in S'_{n+1}$ and $a \in A_1$, $b \in B_1$, (since φ' corresponds $\varphi \in S_{n+1}$) there exist $\varphi_1, \varphi_2 \in S_n$ such that $\varphi \subseteq \varphi_1, \varphi_2$ and $a \in \operatorname{dom}\varphi_1$, $\varphi^V_{1,k}(b) \in \operatorname{range}\varphi_2$ (where k is defined from the condition $\operatorname{range}\varphi^\circ_2 = B_k \cap \operatorname{range}\varphi_2$. It is evident that $\varphi'_1, \varphi'_2 \in S'_n$, $\varphi' \subseteq \varphi'_1, \varphi'_2$ and $a \in \operatorname{dom}\varphi'_1 = A_1 \cap \operatorname{dom}\varphi_1$.

It is rest to check that $b \in \operatorname{range}\varphi'_2$.

From $\varphi^V_{1,k}(b) \in \operatorname{range}\varphi_2$ and $\varphi^V_{1,k}(b) \in B_k$ follows that $\varphi^V_{1,k}(b) \in \operatorname{range}\varphi^\circ_2 = B_k \cap \operatorname{range}\varphi_2$. Hence, there exists $c \in \operatorname{dom}\varphi^\circ_2$ such that $\varphi^\circ_2(c) = \varphi^V_{1,k}(b)$. Thus, $\varphi'_2(c) = \varphi^\circ_2\varphi^V_{k,1}(c) = \varphi^V_{k,1}(\varphi^\circ_2(c)) = \varphi^V_{k,1}(\varphi^V_{1,k}(b)) = b$, and yields $b \in \operatorname{range}\varphi'_2$.

So, for given U_1 and V_1 we constructed a sequence of non-empty sets of partial isomorphisms $S'_0 \supseteq S'_1 \supseteq \cdots \supseteq S'_n \supseteq \cdots$ from U_1 into V_1 such that the conditions of Theorem 6.1.2 are fulfilled. Thus, $U_1 \equiv V_1$.

Analogously, for every n and every $\varphi \in S_n$ a partial map ψ is defined in the following manner: for $i \in I_U$ we define $\psi(i) = k$, $k \in I_v$, if for some $a \in A_i \cap \operatorname{dom}\varphi$, $\varphi(a) \in B_k$, and ψ is a partial isomorphism from I_U into I_V.

In such a way is defined a sequence $z_0 \supseteq z_1 \supseteq \cdots \supseteq z_n \supseteq \cdots$ of sets of partial isomorphisms from I_U into I_V is defined.

The set $z_n, n \in \omega$ is non-empty. Indeed, for any $n \in \omega$ the set S_n is non-empty and every $\varphi \in S_n$ corresponds $\psi \in z_n$.

Let's show that for any n, any $\psi \in z_{n+1}$ and $i \in I_u, j \in I_V$ there exist $\psi_1, \psi_2 \in z_n$ such that $\psi \subseteq \psi_1, \psi_2$ and $i \in \operatorname{dom}\psi_1$, $j \in \operatorname{range}\psi_2$. The partial isomorphism ψ corresponds to some defined $\varphi \in S_{n+1}$, moreover for any $a \in A_i$, $b \in B_j$ there exist $\varphi_1, \varphi_2 \in S_n$ such that $\varphi \subseteq \varphi_1$,

φ_2, $a \in \operatorname{dom}\varphi_1, b \in \operatorname{range}\varphi_2$. Given φ_1, φ_2 are define partial isomorphisms $\psi_1, \psi_2 \in z_n$. Obviously, $\psi \subseteq \psi_1, \psi_1$. Since $a \in A_i \cap \operatorname{dom}\varphi_1$, then $i \in \operatorname{dom}\psi_1$ and since $n \in B_j \cap \operatorname{range}\varphi_2$, then $j \in \operatorname{range}\psi_2$.

Constructed sequence $z_0 \supseteq z_1 \supseteq \cdots \supseteq z_n \supseteq \cdots$ of non-empty sets of partial isomorphism from I_U into I_V satisfies the conditions of Theorem 6.1.2 and thus $I_U \equiv I_V$.

(*Sufficiency*). Suppose $I_U \equiv I_V$ and $U_i \equiv V_k$ for some $i \in I_U$ and $k \in I_V$. From $U_i \equiv V_k$ particularly follows that $U_1 \equiv V_1$.

From $I_U \cong I_V$ follows that there exists a sequence $z_0 \supseteq z_1 \supseteq \cdots \supseteq z_n \supseteq \cdots$ of non-empty sets of partial isomorphism from I_U into I_V and from $U_1 \cong V_1$, follows that there exists a sequence $S_0 \supseteq S_1 \supseteq \cdots \supseteq S_n \supseteq \cdots$ of non-empty sets of partial isomorphism from U_1 into V_1 (by Theorem 6.1.2).

Let's for every n and every $\varphi \in S_n$, $\psi \subset z_n$ construct a partial map α from A into B in the following way:

for any $a \in A$ we define $\alpha(a) = \varphi^U_{i,1}\varphi\varphi^V_{1,\psi(i)}(a) = \varphi^V_{1,\psi(1)}(\varphi(\varphi^U_{i,1}(a)))$,

where $a \in A_i$, $i \in \operatorname{dom}\psi$, $\varphi^U_{i,1}(a) \in \operatorname{dom}\varphi$.

We prove that α is a partial isomorphism from U into V in the same way as in the proof of previous theorem.

So each n corresponds a non-empty set of partial isomorphism from U on V, which we will denote by Φ. Obviously $\Phi_0 \supseteq \Phi_1 \supseteq \cdots \supseteq \Phi_n \supseteq \cdots$. Let us show that for any n, $\alpha \in \Phi_{n+1}$, $a \in A$, $b \in B$ there exist $\alpha_1, \alpha_2 \in \Phi_n$ such that $\alpha \subseteq \alpha_1, \alpha_2$ and $a \in \operatorname{dom}\alpha_1, b \in \operatorname{range}\alpha_2$.

Let $a \in A_i$, $b \in B_j$ and given α correspond $\varphi \in S_{n+1}$ and $\psi \in z_{n+1}$. For $\psi \in z_{n+1}$, $i \in I_U$ and $j \in I_V$ there exist $\psi_1, \psi_2 \in z_n$ such that $\psi \subseteq \psi_1, \psi_2$ and $i \in \operatorname{range}\psi_1$, $j \in \operatorname{range}\psi_2$. In the same way, for $\varphi \in S_{n+1}$, $\varphi^U_{i,1}(a) \in A_1, \varphi^V_{j,1}(b) \in B_1$ there exist $\varphi_1, \varphi_2 \in S_n$ such that $\varphi \subseteq \varphi_1, \varphi_2$ and $\varphi^U_{i,1}(a) \in \operatorname{dom}\varphi_1$, $\varphi^V_{j,1}(b) \in \operatorname{range}\varphi_2$. Maps φ_1 and ψ_1 correspond to some $\alpha_1 \in \Phi_n$ and maps φ_2, ψ_2 correspond to $\alpha_2 \in \Phi_2$. Obviously $\alpha \subseteq \alpha_1, \alpha_2$ and $a \in \operatorname{dom}\alpha_1$. From $j \in \operatorname{range}\psi_2$ and $\varphi^V_{j,1}(b) \in \operatorname{range}\varphi_2$ we can find $s \in I_U$ and $c \in A_1$ such that $\psi_2(s) = j$ and $\varphi_2(c) = \varphi^V_{j,1}(b)$. Therefore $\varphi_2(c) = \varphi^V_{\psi_2(s),1}(b)$ and since $\varphi^U_{s,1} : U_s \mapsto U_1$ is an isomorphism, then for some $d \in A_s$ we have $\varphi^U_{s,1}(d) = c$. So,

$$\begin{aligned}\alpha_2(d) &= \alpha_2(\varphi^U_{1,s}(c)) = \varphi^V_{1,\psi_2(s)}(\varphi_2(\varphi^U_{s,1}(\varphi^U_{1,s}(c))) = \varphi^V_{1,\psi_2(s)}(\varphi_2(c)))\\ &= \varphi^V_{1,\psi_2(s)}(\varphi^V_{\psi_2(s),1}(b)) = b,\end{aligned}$$

i.e. $b \in \operatorname{range}\alpha_2$. From Theorem 6.1.2, we have $U \cong V$. □

6.5. The elementary characteristics of super-Boolean algebras with two binary and one unary operations

Theorem 6.5.1. *Let $U = (A; +, \cdot, ^-)$ and $V = (B; +, \cdot, ^-)$ be super-Boolean algebras with two binary and one unary operations, $U_i = (A_i; +, \cdot, ')$, $i \in I_U$ and $V_i = (B_i; +, \cdot, ')$, $k \in I_V$ be the canonical representation of this algebras into Boolean sum of Boolean lattices. Then*

$$U \equiv V \iff \mathrm{ch}(I_U) = \mathrm{ch}(I_V), \mathrm{ch}(U_i) = \mathrm{ch}(V_k), \text{ for some } i \in I_V,\ \ k \in I_V,$$

i.e., U and V are elementary equivalent if and only if the elementary characteristics of Boolean algebras I_U and I_V coincide and for some $i \in I_V, k \in I_V$ the elementary characteristics of U_i and V_k are equal.

Proof. By previous theorem, $U \equiv V \iff I_U \equiv I_V$ and $U_i \equiv V_k$, for some $i \in I_V, k \in I_V$. Since $I_U, I_V, U_i, i \in I_V$ and $V_k, k \in I_V$ are Boolean algebras, then by Theorem 6.1.6 we have

$$I_U \equiv I_V \iff \mathrm{ch}(I_U) = \mathrm{ch}(I_V),$$
$$U_i \equiv V_k \iff \mathrm{ch}(U_i) = \mathrm{ch}(V_k).$$

□

Definition 6.5.2. Let $U = (A; +, \cdot, ^-)$ be super-Boolean algebra with two binary and one unary operations and $U_i = (A_i; +, \cdot, '), i \in I_U$ is the canonical representation of U as Boolean sum of Boolean lattices. The pair $(\mathrm{ch}(U_1), \mathrm{ch}(I_U))$ is called the elementary characteristic of super-Boolean algebra U and we will denote it by $\mathrm{ch}(U)$.

Hence, we have the following theorem:

Theorem 6.5.3. *Let $U = (A; +, \cdot, ^-)$ and $V = (B; +, \cdot, ^-)$ be super-Boolean algebras with two binary and one unary operations. Algebras U and V are elementary equivalent if and only if their elementary characteristics are coincide, i.e.,*

$$U \equiv V \iff \mathrm{ch}(U) = \mathrm{ch}(V).$$

By the definition, the elementary characteristics of super-Boolean algebra U with two binary and one operations is the six

$$\mathrm{ch}(U) = ((\alpha_1, \beta_1, \gamma_1), (\alpha_2, \beta_2, \gamma_2)),$$

where $\alpha_1, \alpha_2, \beta_1, \beta_2 \leq \omega, \gamma_1, \gamma_2 \leq 1$.

Definition 6.5.4. The six $((\alpha_1, \beta_1, \gamma_1), (\alpha_2, \beta_2, \gamma_2))$ is called permissible, if each of the triples $(\alpha_1, \beta_1, \gamma_1)$ and $(\alpha_2, \beta_2, \gamma_2)$ are permissible.

Each permissible triple is the elementary characteristic of some Boolean algebra. By the following result, any permissible six is the elementary characteristic of some super-Boolean algebra.

Theorem 6.5.5. *Let $U = (A; +, \cdot, ')$ and $I = (I; +, \cdot, \bar{})$ be Boolean algebras (lattices). Then there exists a super-Boolean algebra $V = (B; +, \cdot, \bar{})$ such that* $\mathrm{ch}(V) = (\mathrm{ch}(U), \mathrm{ch}(I))$.

Proof. Assume $U_i = (A_i; +, \cdot, '), i \in I$ are Boolean lattices isomorphic to U where $A_i \cap A_j = \emptyset$, $i \neq j$, $i, j \in I$. Denote by 0 and 1 the neutral and identity elements of Boolean algebra I.

An isomorphism from U_0 into U_1 denote by $\varphi_{0,i}$, $i \in I$. Let's define an isomorphism $\varphi_{i,k} : U_i \mapsto U_k$ as

$$\varphi_{i,k} = \varphi_{0,i}^{-1} \cdot \varphi_{0,k},$$

for $i \leq k, i, k \in I$.

For every $i \in I$ an isomorphism $h_{i,\bar{i}} : U_i \mapsto U_{\bar{i}}$ defined as

$$h_{i,\bar{i}} = \varphi_{i,1} \cdot \varphi_{\bar{i},1}^{-1}. \tag{6.5.1}$$

For any $i \in I$ we have

$$\varphi_{i,i} = \varphi_{0,i}^{-1} \circ \varphi_{0,i} = id_{A_i}.$$

If $i \leq j \leq k$, $i, j, k \in I$, then

$$\varphi_{i,j} \cdot \varphi_{j,k} = (\varphi_{0,i}^{-1} \cdot \varphi_{0,j}) \cdot (\varphi_{0,j}^{-1} \cdot \varphi_{0,k}) = \varphi_{0,i}^{-1} \cdot \varphi_{0,j} \cdot \varphi_{0,j}^{-1} \cdot \varphi_{0,k} = \varphi_{0,i}^{-1} \cdot \varphi_{0,k} = \varphi_{i,k}.$$

Thus, $\varphi_{i,j} \cdot \varphi_{j,k} = \varphi_{i,k}$ for $i \leq j \leq k$, $i, j, k \in I$.

From the definition of isomorphism $h_{i,\bar{i}}$, we have $\varphi_{i,1} = h_{i,\bar{i}} \cdot \varphi_{\bar{i},1}$.

Let's show that $h_{i,\bar{i}}^{-1} = h_{\bar{i},i}$.

$$h_{\bar{i},i} = \varphi_{\bar{i},1} \cdot \varphi_{i,1}^{-1} = ((\varphi_{\bar{i},1}^{-1})^{-1} \cdot \varphi_{i,1}^{-1})^{-1} = (\varphi_{i,1} \cdot \varphi_{\bar{i},1}^{-1})^{-1} = h_{i,\bar{i}}^{-1}.$$

Our defined algebras $U_i, i \in I$ and isomorphisms $\varphi_{ij}, h_{i,\bar{i}}, i, j \in I$ satisfy the conditions (a), (c), (d), (f) from the definition of Boolean sum.

Define an algebra V, whose basic set is $B = \bigcup_{i \in I} A_i$ and for any $a \in B$ define the unary operation $-$ as

$$\bar{a} = h_{i,\bar{i}}(a'), \tag{6.5.2}$$

where $a \in A_i, i \in I$.

The binary operations $+$ and $\cdot$ is defined as follows: for any $a, b \in B$

$$a + b = \varphi_{i,i+j}(a) + \varphi_{j,i+j}(b), \tag{6.5.3}$$

$$a \cdot b = \varphi_{i,i+j}(a) \cdot \varphi_{j,i+j}(b), \tag{6.5.4}$$

where $a \in A_i, b \in A_j, i, j \in I$.

Now obviously the algebra $V = (B; +, \cdot, ^-)$ is a Boolean sum of Boolean lattices $U_i = (A_i; +, \cdot, ')$, $i \in I$, therefore by Theorem 6.1.15 V is a super-Boolean algebra and has the canonical representation as a Boolean sum of Boolean lattices. Let's show that exactly $U_i, i \in I$ is the canonical representation of V as a Boolean sum.

Consider the following relation θ on B

$$a\theta b \iff a + (a \cdot b) = a, b + (b \cdot a) = b.$$

Since $U_i, i \in I$ are Boolean algebras, then for any $i \in I$, $a, b \in A_i$ follows that $a\theta b$. If $a\theta b$ is true for some $i, j \in I$, $i \neq j$ and some $a \in A_i, b \in A_j$, then

$$\begin{aligned} a + (a \cdot b) &= a \in A_i, b + (b \cdot a) = b \in A_j, \\ a + (a \cdot b) &= \varphi_{i,i+j}(a) + \varphi_{i,i+j}(a) \cdot \varphi_{j,i+j}(b) \in A_{i+j}, \\ b + (b \cdot a) &= \varphi_{j,i+j}(b) + \varphi_{j,i+j}(b) \cdot \varphi_{i,i+j}(a) \in A_{i+j}. \end{aligned}$$

So, A_i and A_{i+j} as well as A_j and A_{i+j} have a common element. Therefore $A_{i+j} = A_i = A_j$ which is a contradiction, because $i \neq j$. Thus, for any $a, b \in A$, $a\theta b$ is true if and only if a and b belong to the same set A_i for some $i \in I$.

It is rest to check that the operation $'$ satisfies the equality

$$a' = a \cdot \overline{a + \overline{a}} + \overline{a \cdot (a + \overline{a})},$$

for any $a \in B$.

Using the properties of Boolean sum as well as taking into account that $\varphi_{i,j}$ is an isomorphism between Boolean lattices $U_i = (A_i; +, \cdot, ')$ and $U_j = (A_j; +, \cdot, ')$, for any $a \in B$ we get

$$\begin{aligned} a \cdot \overline{a + \overline{a}} + \overline{a \cdot (a + \overline{a})} &= a \cdot (h_{1,0}(a + h_{i,\overline{i}}(a'))') + h_{1,0}(a(a + h_{i,\overline{i}}(a')))' \\ &= a \cdot (\varphi_{1,1}\varphi_{1,0}(a + \varphi_{i,1}\varphi_{1,\overline{i}}(a'))') \\ &\quad + \varphi_{1,1}\varphi_{1,0}(a(a + \varphi_{i,\overline{i}}(a')))' \\ &= a \cdot (\varphi_{i,0}(a + \varphi - i, 1 \cdot \varphi_{1,\overline{i}}(a\prime))') \\ &\quad + \varphi_{1,0}(a \cdot (a + \varphi_{i,1}\varphi_{i,\overline{i}}(a\prime)))' \end{aligned}$$

$$
\begin{aligned}
&\varphi_{i,i}(a)\cdot\varphi_{0,i}\cdot(\varphi_{i,0}(\varphi_{1,0}(\varphi_{i,1}(a)+\varphi_{i,1}(\varphi_{i,1}\varphi_{1,\bar{i}}(a'))))')\\
&\quad+\varphi_{i,0}(\varphi_{i,1}(a)\cdot(\varphi_{i,1}(a)+\varphi_{\bar{i},1}(\varphi_{i,1}\varphi_{1,\bar{i}}(a'))))'\\
&=\varphi_{i,i}(a)\cdot\varphi_{1,i}(\varphi_{i,1}(a)+\varphi_{i,1}(a+a'))'\\
&\quad+\varphi_{i,0}\cdot(\varphi_{i,1}(a\cdot(a+a')))'\\
&=\varphi_{i,i}(a)\cdot\varphi_{i,i}(a+a')'+\varphi_{i,0}(a(a+a'))'\\
&=\varphi_{i,i}(a(a+a')')+\varphi_{i,0}(a(a+a'))'=\varphi_{i,i}(a(a+a')')\\
&\quad+\varphi_{0,i}(\varphi_{i,0}(a(a+a'))')\\
&=a(a+a')'+(a(a+a'))'=a\cdot(a'\cdot a'')+(a'+(a'\cdot a''))\\
&=a\cdot a'\cdot a+a'+a'\cdot a=a\cdot a'+a'=0_i+a'=a',
\end{aligned}
$$

where $a \in A_i$, $i \in I$ and 0_i is the neutral element of U_i.

Thus, $U_i, i \in I$ is the canonical representation of super-Boolean algebra V as a Boolean sum of Boolean lattices. By the definition of elementary characteristic of super-Boolean algebra we have

$$\mathrm{ch}(V) = (\mathrm{ch}(U_1), \mathrm{ch}(I)) = (\mathrm{ch}(U), \mathrm{ch}(I)).$$

□

Theorem 6.5.6. *For any permissible six* $((\alpha_i, \beta_1, \gamma_1), (\alpha_2, \beta_2, \gamma_2))$, *there exists a super-Boolean algebra* U *such that*

$$\mathrm{ch}(U) = ((\alpha_i, \beta_1, \gamma_1), (\alpha_2, \beta_2, \gamma_2)).$$

Proof. Let $((\alpha_i, \beta_1, \gamma_1), (\alpha_2, \beta_2, \gamma_2))$ be an arbitrary permissible six. By the definition of permissible six $(\alpha_i, \beta_1, \gamma_1)$ and $(\alpha_2, \beta_2, \gamma_2)$ are permissible triples. By the definition of permissible triples $(\alpha_i, \beta_1, \gamma_1)$ and $(\alpha_2, \beta_2, \gamma_2)$, there exist Boolean algebras U and I such that $\mathrm{ch}(U) = (\alpha_i, \beta_1, \gamma_1)$ and $\mathrm{ch}(I) = (\alpha_2, \beta_2, \gamma_2)$. For Boolean algebras U and I, there exists a super-Boolean algebra V such that $\mathrm{ch}(V) = (\mathrm{ch}(U), \mathrm{ch}(I))$ (by the previous theorem). Thus, there exists a super-Boolean algebra V with elementary characteristic $\mathrm{ch}(V) = ((\alpha_i, \beta_1, \gamma_1), (\alpha_2, \beta_2, \gamma_2))$. □

6.6. Decidability of elementary theory of super-Boolean algebras with two binary and one unary operations

Theorem 6.6.1. *Any super-Boolean algebra* $U = (Q; +, \cdot, ^-)$ *with two binary and one unary operations has a decidability elementary theory.*

Proof. The formulas $A_n(x), B_n(x), C_n(x), D_n(x,y)$, $n \in \omega$ were constructed in Theorem 6.1.6 such that for any Boolean algebra $U = (A; +, \cdot, ')$ with elementary characteristic $\mathrm{ch}(U) = (\alpha, \beta, \gamma)$ is defined a recursive system of axioms using the equivalences (1)–(3) from the proof of Theorem 6.1.6.

Let's construct formulas $A'_n(x), B'_n(x), C'_n(x), D'_n(x,y)$ by induction on $n \in \omega$ such that for any super-Boolean algebra $U = (Q; +, \cdot, ^{-})$ and for any $a, b \in Q$ the following is true:

$$\begin{cases} U \models A'_n(a) \iff U_i \models A_n(a), & \text{if } a \in Q_i; \\ U \models B'_n(a) \iff U_i \models B_n(a), & \text{if } a \in Q_i; \\ U \models C'_n(a) \iff U_i \models C_n(a), & \text{if } a \in Q_i; \\ U \models D'_n(a,b) \iff U_i \models D_n(a,b), & \text{if } a, b \in Q_i; \\ U \models \bar{A}_n(a) \iff I \models A_n(i), & \text{if } a \in Q_i; \\ U \models \bar{B}_n(a) \iff I \models B_n(i), & \text{if } a \in Q_i; \\ U \models \bar{C}_n(a) \iff I \models C_n(i), & \text{if } a \in Q_i; \\ U \models \bar{D}_n(a,b) \iff I \models D_n(i,j), & \text{if } a \in Q_i, b \in Q_j, \end{cases} \tag{6.6.1}$$

where $U_i = (Q_i; +, \cdot, '), i \in I$ is a canonical representation of U as a Boolean sum of Boolean lattices.

For any $a, b \in Q$ we have

$a \neq a \cdot a' \iff a \in Q_i,\ a \neq 0_i = a \cdot a'$;
$a \cdot a' \neq b \cdot b' \iff a \in Q_i, b \in Q_j, i \neq j$;
$a + a' + b + b' = b + b' \iff a \in Q_i, b \in Q_j, 1_i + 1_j = 1_j$ (i.e., $i \leq j$);
if $a \in Q_i, b \in Q_j$, then $\overline{\overline{a} + \overline{b}} \in Q_{\overline{\overline{i}+\overline{j}}} = Q_{i \cdot j}$;
if $a \in Q_i, b \in Q_j, i \cdot j = i$, then $\overline{\overline{a} + \overline{b}} \in Q_{i \cdot j} = Q_i$ and $\overline{\overline{a} + \overline{b}} = \overline{\overline{a} + \overline{b}} + 0_i = \overline{\overline{a} + \overline{b}} + a \cdot a'$.

We denote:

$$\begin{aligned} A'_0(x) &\rightleftharpoons x \neq x \cdot x' \& \forall y (x \cdot x' \neq y \cdot y' \vee x \cdot y = x \cdot x' \vee x \cdot y = x), \\ B'_0(x) &\rightleftharpoons \forall y (x \cdot x' \neq y \cdot y' \vee \neg A'_0(x \cdot y)), \\ C'_0(x) &\rightleftharpoons x \cdot x', \\ D'_0(x,y) &\rightleftharpoons x = y, \\ \bar{D}_0(x,y) &\rightleftharpoons x \cdot x' = y \cdot y', \\ \bar{C}_0(x) &\rightleftharpoons \forall y (x + x' + y + y' = y + y'), \\ \bar{A}_0(x) &\rightleftharpoons \neg \bar{C}_0(x) \& \forall y (C_0(\overline{\overline{x} + \overline{y}}) \vee \overline{\overline{x} + \overline{y}} + x \cdot x' = \overline{\overline{x} + \overline{y}}), \\ \bar{B}_0(x) &\rightleftharpoons \forall y (\neg \bar{A}_0(\overline{\overline{x} + \overline{y}}). \end{aligned}$$

For $n = 0$ the condition (6.6.1) is satisfied. Let formulas $A'_n(x)$, $B'_n(x)$, $C'_n(x)$, $D'_n(x,y)$, $\bar{A}_n(x)$, $\bar{B}_n(x), \bar{C}_n(x)$, $\bar{D}_n(x,y)$ be constructed for n such that the condition (6.6.1) are satisfied, then set

$$
\begin{aligned}
A'_{n+1}(x) &\rightleftharpoons \neg C'_n(x) \& \forall y(x \cdot x' \neq y \cdot y' \vee C'_n(x \cdot y) \vee D'_n(x,y)),\\
B'_{n+1}(x) &\rightleftharpoons \forall y(x \cdot x' \neq y \cdot y' \vee \neg A'_{n+1}(x \cdot y)),\\
C'_{n+1}(x) &\rightleftharpoons \exists y \exists z(x \cdot x' = y \cdot y' = z \cdot z' \& x = y + z \& B'_{n+1}(y) \& \forall u(x \cdot x' \neq u \cdot u' \vee\\
&\qquad \exists v(x \cdot x' = v \cdot v' \& (C'_n(z \cdot u) \vee A'_{n+1}(z \cdot u \cdot v))))),\\
D'_{n+1}(x,y) &\rightleftharpoons x \cdot x' = y \cdot y' \& C'_n(x \cdot y') \& C'_n(x' \cdot y),\\
\bar{A}_{n+1}(x) &\rightleftharpoons \neg \bar{C}_n(x) \& \forall y(\bar{C}_n(\overline{\overline{x} + \overline{y}}) \vee \bar{D}_n(x,y)),\\
\bar{B}_{n+1}(x) &\rightleftharpoons \forall y(\neg \bar{A}_{n+1}(\overline{\overline{x} + \overline{y}})),\\
\bar{C}_{n+1}(x) &\rightleftharpoons \exists y \exists z(y + z = y + z + x \cdot x' \& \bar{B}_{n+1}(y) \& \forall u \exists v(\bar{C}_n(\overline{\overline{z} + \overline{u}})\\
&\qquad \wedge \bar{A}_{n+1}(\overline{\overline{z} + \overline{u} + \overline{v}}),\\
\bar{D}_{n+1}(x,y) &\rightleftharpoons \bar{C}_n(\overline{x \cdot \overline{y}}) \& C_n(\overline{\overline{x} \cdot y}).
\end{aligned}
$$

It is easy to show that constructed formulas satisfy the condition (6.6.1). We are going to show that for formulas $A'_{n+1}(x)$ and $\bar{A}_{n+1}(x)$.

Let $U \models A'_{n+1}(a)$ for $a \in Q$. Then by the definition of formula $A'_{n+1}(x)$, we have

$$
\begin{aligned}
&U \models \neg C'_n(a),\\
&U \models a \cdot a' \neq b \cdot b' \vee C'_n(a \cdot b) \vee D'_n(a,b), \text{ for } b \in Q.
\end{aligned}
$$

By the inductive assumption we get

$$U \models \neg C'_n(a) \iff U_i \models \neg C_n(a), a \in Q_i, \tag{6.6.2}$$

and for $b \in Q_i$ we have

$$U \models a \cdot a' = b \cdot b' \& C'_n(a \cdot b') \& C'_n(a' \cdot b).$$

Therefore, since

$$
\begin{aligned}
&U \models C'_n(a \cdot b') \iff U_i \models C_n(a \cdot b'),\\
&U \models C'_n(a' \cdot b) \iff U_i \models C_n(a' \cdot b),
\end{aligned}
$$

then for some $i \in I, a, b \in Q_i, U \models D'_n(a,b) \iff U_i \models D_n(a,b)$.

For $a \in Q$, $U \models \bar{A}_{n+1}(a)$ means that for any $b \in Q$, $U \models \neg \bar{C}_n(a) \& (\neg \bar{C}_n(\overline{\overline{a} + \overline{b}}) \vee \bar{D}_n(a,b))$. Taking into account that if $a \in Q_i, b \in Q_j$, then $\overline{\overline{a} + \overline{b}} \in Q_{i \cdot j}$ and by inductive assumption, we get

$$
\begin{aligned}
&U \models \neg \bar{C}_n(a) \iff I \models \neg C_n(i),\\
&U \models \bar{C}_n(\overline{\overline{a} + \overline{b}}) \iff I \models C_n(i \cdot j),\\
&U \models \neg D_n(a,b) \iff I \models D_n(i,j).
\end{aligned}
$$

Thus, it is easy to see that

$$U \models \bar{A}_{n+1}(a) \iff I \models A_{n+1}(i), a \in Q_i. \tag{6.6.3}$$

Let $U \models \bar{B}_{n+1}(a)$ for $a \in Q$, i.e., $U \models \neg \bar{A}_{n+1}(\overline{\overline{a} + \overline{b}})$ for any $b \in Q$. By (6.6.3) we have

$$U \models \bar{A}_{n+1}(\overline{\overline{a} + \overline{b}}) \iff I \models \neg A_{n+1}(i \cdot j), a \in Q_i, b \in Q_j,$$

therefore

$$U \models \bar{B}_{n+1}(a) \iff I \models B_{n+1}(i), a \in Q_i.$$

Let $U \models \bar{C}_{n+1}(a)$ for $a \in Q$.

Thus, there exist $b, c \in Q$ such that for any $d \in Q$ one can find $e \in Q$ satisfying the following condition

$$U \models b + c = b + c + a \cdot a' \& \bar{B}_{n+1}(b) \& (\bar{C}_n(\overline{\overline{c} + \overline{d}}) \vee \bar{A}_{n+1}(\overline{\overline{c} + \overline{d} + \overline{e}})).$$

For $a \in Q_i, b \in Q_j, c \in Q_s, d \in Q_k, e \in Q_m$ we have

$$b + c + 0_i = b + c + a \cdot a' = b + c \in Q_{j+s}$$

and $j + s = i$. Moreover

$U \models C'_n(a \cdot b) \iff U_i \models C_n(a \cdot b)$,
$U \models D'_n(a, b) \iff U_i \models D_n(a, b)$. Thus,

$$U \models C'_n(a \cdot b) \vee D'_n(a, b) \iff U_i \models C_n(a \cdot b) \vee D_n(a, b). \tag{6.6.4}$$

The condition $b \in Q_i$ means that $a \cdot a' = b \cdot b'$, therefore from (6.6.2) and (6.6.4) it follows that

$$U \models A'_{n+1}(a) \iff U_i \models A_{n+1}(a). \tag{6.6.5}$$

If $U \models B'_{n+1}(a)$, for $a \in Q$ then for any $b \in Q$

$$U \models a \cdot a' \neq b \cdot b' \vee \neg A'_{n+1}(a \cdot b). \tag{6.6.6}$$

If a and b belong to the same set Q_i, then $a \cdot a' = b \cdot b'$ and the condition (6.6.6) takes the form

$$U \models \neg A'_{n+1}(a \cdot b),$$

therefore

$$U \models B'_{n+1}(a) \iff U_i \models B_{n+1}(a).$$

Let $U \models C'_{n+1}(a)$ for $a \in Q$. Thus there exist $b, c \in Q$ such that for any $d \in Q$ one can find $e \in Q$ satisfying the following condition:

$U \models a \cdot a' = b \cdot b' = c \cdot c' \& a = b + c \& B'_{n+1}(b) \& (a \cdot a' \neq d \cdot d' \vee (a \cdot a' = e \cdot e' \& (C'_n(c \cdot d) \vee A'_{n+1}(c \cdot d \cdot e))))$.

If $a \in Q_i$, then the condition $a \cdot a' = b \cdot b' = c \cdot c' = e \cdot e'$ means that $b, c, e \in Q_i$ and if $d \in Q_i$, then by inductive assumption we get

$$U \models B'_{n+1}(b) \iff U_i \models B_{n+1}(b),$$
$$U \models C'_n(c \cdot d) \iff U_i \models C_n(c \cdot d),$$
$$U \models A'_{n+1}(c \cdot d \cdot e) \iff U_i \models A_{n+1}(c \cdot d \cdot e).$$

It is easy to note that $U \models C'_{n+1}(a) \iff U_i \models C_{n+1}(a)$.
Let $U \models D'_{n+1}(a, b)$ for $a, b \in Q$, i.e.,

$$U \models \bar{B}_{n+1}(\underline{b}) \iff I \models B_{n+1}(\jmath),$$
$$U \models \bar{C}_n(\overline{\bar{c} + \bar{d}}) \iff I \models C_n(s \cdot k),$$
$$U \models \bar{A}_{n+1}(\overline{\bar{c} + \bar{d} + \bar{e}}) \iff I \models A_{n+1}(s \cdot k \cdot m).$$

Therefore,
$U \models \bar{C}_{n+1}(a) \iff I \models C_{n+1}(\imath), a \in Q_i$.
Let $U \models \bar{D}_{n+1}(a, b)$ for $a, b \in Q$. It means that

$$U \models \bar{C}_n(\overline{a \cdot \bar{b}}),$$
$$U \models \bar{C}_n(\overline{\bar{a} \cdot b}).$$

Since $\overline{a \cdot \bar{b}} \in Q_{\overline{i + \bar{j}}} = Q_{\bar{\imath} \cdot j}$ and $\overline{\bar{a} \cdot b} \in Q_{i \cdot \bar{j}}$, then for $a \in Q_i$ and $b \in Q_j$ we get

$$U \models \bar{C}_n(\overline{a \cdot \bar{b}}) \iff I \models C_n(\bar{\imath} \cdot \jmath),$$
$$U \models \bar{C}_n(\overline{\bar{a} \cdot b}) \iff I \models C_n(\imath \cdot \bar{\jmath}).$$

Therefore,

$$U \models \bar{D}_{n+1}(a, b) \iff I \models D_{n+1}(\imath, \jmath).$$

Thus, the formulas $A'_n(x), B'_n(x), C'_n(x), D'_n(x, y), \overline{A}_n(x), \overline{B}_n(x), \overline{C}_n(x), \overline{D}_n(x, y)$ satisfy the condition (6.6.1) for any $n \in \omega$.

Suppose $\mathrm{ch}(U) = ((\alpha_1, \beta_1, \gamma_1), (\alpha_2, \beta_2, \gamma_2))$. From (6.6.1) and the conditions (1)–(3) from the proof of Theorem 6.1.6, it is easy to see that

(1) $n \leq \alpha_1 \iff U \models \exists x(\neg C'_n(x))$;
(2) if $\alpha_1 = n$, then
$0 < m \leq \beta_1 \iff U \models \exists x_1 \ldots \exists x_m(\&_{i \neq j} \neg D'_n(x_i, x_j) \& \&_{i=1}^m A'_n(x_i) \& \&_{i,j=1}^m x_i \cdot x'_i = x_j \cdot x'_j)$;

(3) if $\alpha_1 = n$, then
$\gamma_1 = 1 \iff U \models \exists x(\neg C'_n(x)\&B'_n(x))$;
(4) $n \leq \alpha_2 \iff U \models \exists x\, \bar{C}_n(x)$;
(5) if $\alpha_2 = n$ then
$0 < m \leq \beta_2 \iff U \models \exists x_1 \ldots \exists x_m(\&_{i\neq j}\neg\bar{D}_n(x_i, x_j)\&\&_{i=1}^m \overline{A}_n(x_i))$;
(6) if $\alpha_2 = n$ then
$\gamma_2 = 1 \iff U \models \exists x(\neg\overline{C}_n(x)\&\overline{B}_n(x))$.

Thus, it is possible to effectively rewrite a system of recursive axioms by $((\alpha_1, \beta_1, \gamma_1), (\alpha_2, \beta_2, \gamma_2))$ (which is finite, if $\alpha_1, \alpha_2, \beta_1.\beta_2 < \omega$) for theory $\mathrm{Th}(U)$. Since theory $\mathrm{Th}(U)$ is complete and has a recursive system of axioms, then $\mathrm{Th}(U)$ is decidable by Theorem 6.1.4. □

6.7. Decidability of elementary theory of the variety of super-Boolean algebras with two binary and one unary operations

Proposition 6.7.1. *A super-Boolean algebra U with two binary and one unary operations is a Boolean lattice if and only if it has the following elementary characteristic:*

$$\mathrm{ch}(U) = ((\alpha, \beta, \gamma), (0, 0, 0)).$$

Proof. Let $U_i, i \in I$ be the canonical representation of super-Boolean algebra U as a Boolean sum of Boolean lattice. Then the identity of absorption does not hold for any two elements $a, b \in |U|$ such that $a \in |U_i|, b \in |U_j|, i \neq j, i, j \in I$. Therefore U is a Boolean lattice if and only if $I = \{0\}$, i.e., $\mathrm{ch}(I) = (0, 0, 0)$. Thus, U is a Boolean lattice if and only if $\mathrm{ch}(U) = ((\alpha, \beta, \gamma), (0, 0, 0))$. □

So, if U is a super-Boolean algebra with the elementary characteristic $\mathrm{ch}(U) = ((\alpha, \beta, \gamma), (0, 0, 0))$ then U is a Boolean lattice, moreover, it is obvious that

$$\mathrm{ch}(U) = (\mathrm{ch}_B(U), (0, 0, 0)),$$

where $\mathrm{ch}_B(U)$ is the elementary characteristic of U as a Boolean lattice.

Proposition 6.7.2. *Let $U = (A; +, \cdot, ^-)$ and $V = (B; +, \cdot, ^-)$ be super-Boolean algebras and $\mathrm{ch}(U) = ((\alpha_1, \beta_1, \gamma_1), (0, 0, 0)), \mathrm{ch}(V) = ((\alpha_2, \beta_2, \gamma_2), (0, 0, 0))$. Then a super-Boolean algebra $U \times V$, which is a direct* *product of super-Boolean algebras U and V, has the elementary characteristic* $\mathrm{ch}(U \times V) = ((\alpha, \beta, \gamma), (0,0,0))$, *where* $\alpha = \max\{\alpha_1, \alpha_2\}$;

$$\beta = \begin{cases} \beta_1 & \text{if } \alpha_2 < \alpha_1, \\ \beta_2 & \text{if } \alpha_1 < \alpha_2, \\ \beta_1 \oplus \beta_2 & \text{if } \alpha_1 = \alpha_2; \end{cases}$$

$$\gamma = \begin{cases} \gamma_1 & \text{if } \alpha_2 < \alpha_1, \\ \gamma_2 & \text{if } \alpha_1 < \alpha_2, \\ \max\{\gamma_1, \gamma_2\} & \text{if } \alpha_1 = \alpha_2. \end{cases}$$

Proof. Since $\mathrm{ch}(U) = ((\alpha_1, \beta_1, \gamma_1), (0,0,0))$ and $\mathrm{ch}(V) = ((\alpha_2, \beta_2, \gamma_2), (0,0,0))$, then by Proposition 6.7.1, U and V are Boolean lattices and $\mathrm{ch}(U) = (\mathrm{ch}_B(U), (0,0,0)), \mathrm{ch}(V) = (\mathrm{ch}_B(V), (0,0,0))$. By Theorem 6.1.10, we have $\mathrm{ch}_B(U \times V) = (\alpha, \beta.\gamma)$, therefore $\mathrm{ch}(U \times V) = ((\alpha, \beta, \gamma), (0,0,0))$. □

Proposition 6.7.3. *Let* $U = (A; +, \cdot, ^-)$ *and* $V = (B; +, \cdot, ^-)$ *be super-Boolean algebras with two binary and one unary operations and* $\mathrm{ch}(U) = ((\alpha, \beta, \gamma), (0,0,0))$, $\mathrm{ch}(V) = ((0,0,0), (\xi, \eta, \varepsilon))$, *then the elementary characteristic of the direct product* $U \times V$ *is* $\mathrm{ch}(U \times V) = ((\alpha, \beta, \gamma), (\xi, \eta, \varepsilon))$.

Proof. From $\mathrm{ch}(U) = ((\alpha, \beta, \gamma), (0,0,0))$ by Proposition 6.7.1 follows that U is a Boolean lattice. From $\mathrm{ch}(V) = ((0,0,0), (\xi, \eta, \varepsilon))$ follows that $\mathrm{ch}_B(V_i) = (0,0,0), i \in I_V$ and $\mathrm{ch}_B(I_V) = (\xi, \eta, \varepsilon)$ for the canonical representation of V as a Boolean sum of Boolean lattices, $V_i, i \in I_V$, i.e., $V_i = \{0\}, i \in I_V$ and $\mathrm{ch}(V) = ((0,0,0), \mathrm{ch}_B(I_V))$. The unique element of the Boolean algebra V_i denote by b_i and the direct product of the sets A and $\{b_i\}$ denote by C_i, i.e., $C_i = A \times \{b_i\}$. Obviously, $A \times B = \bigcup_{i \in I_V} C_i$ and $C_i \cap C_j = \emptyset, i \neq j, i, j \in I_V$.

The operation $'$ is defined as (by the definition of the canonical representation of super-Boolean algebras):

$$x' = x \cdot \overline{(x + \overline{x})} + \overline{x \cdot (x + \overline{x})}.$$

Since U is a Boolean lattice then $a' = \overline{a}$ for every $a \in A$ and since $V_i = (\{b_i\}; +, \cdot, ^-)$ then $b_i' = b_i$. Therefore $(a, b_i)' = (\overline{a}, b_i)$ on $A \times B$, moreover $(a, b_i), (a, b_i)' \in C_i$. Let's show that C_i, $i \in I_V$ is a decomposition of the set $A \times B$ by congruence θ:

$$a\theta b \iff a + (a \cdot b) = a, \quad b + (b \cdot a) = b.$$

Since $W_i = (C_i; +, \cdot, ')$ (as the direct product of Boolean algebras U and V_i) is a Boolean algebra (lattice), then $a\theta b$ for any $a, b \in C_i$. If $a \in C_i, b \in C_j$

and $a\theta b$, then $a+ab=a\in C_i, b+ba=b\in C_j$. On the other hand, because $V_i=(\{b_i\};+,\cdot,')$, $i\in I_V$, is the canonical representation of super-Boolean algebra V into Boolean sum of Boolean lattices, we have that $b_i+b_j, b_i\cdot b_j\in |V_{i+j}|$, i.e., $b_i+b_j=b_i\cdot b_j=b_{i+j}$. Therefore, $a+ab\in C_{i+j}$ and $b+ba\in C_{i+j}$. Thus, $a+ab\in C_i$ and $a+ab\in C_{i+j}$, $b+ba\in C_j$ and $b+ba\in C_{i+j}$ which means that $C_i=C_j$.

So, $W_i, i\in I_V$ is the canonical representation of super-Boolean algebra $U\times V$ as a Boolean sum of Boolean lattices. By Theorem 6.1.10, $\mathrm{ch}_B(W_i)=\mathrm{ch}_B(U), i\in I_V$, therefore $\mathrm{ch}(U\times V)=(\mathrm{ch}_B(W_i),\mathrm{ch}_B(I_V))=(\mathrm{ch}_B(U),\mathrm{ch}_B(I_V))=((\alpha,\beta,\gamma),(\xi,\eta,\varepsilon))$. □

Proposition 6.7.4. *Let $U=(A;+,\cdot,^-)$ and $V=(B;+,\cdot,^-)$ be super-Boolean algebras with two binary and one unary operations and let $\mathrm{ch}(U)=((0,0,0),(\alpha_1,\beta_1,\gamma_1))$ and $\mathrm{ch}(V)=((0,0,0),(\alpha_2,\beta_2,\gamma_2))$. Then the super-Boolean algebra $U\times V$ has the elementary characteristic $\mathrm{ch}(U\times V)=((0,0,0),(\alpha,\beta,\gamma))$, where α,β,γ are defined in Proposition 6.7.2.*

Proof. From $\mathrm{ch}(U)=((0,0,0),(\alpha_1,\beta_1,\gamma_1))$ and $\mathrm{ch}(V)=((0,0,0),(\alpha_2,\beta_2,\gamma_2))$ follows that for the canonical representations U_i, $i\in I_U$ and V_i, $i\in I_V$ of super-Boolean algebras U and V as a Boolean sum of Boolean lattices, $\mathrm{ch}_B(U_i)=\mathrm{ch}_B(V_k)=(0,0,0), \mathrm{ch}_B(I_U)=(\alpha_1,\beta_1,\gamma_1), \mathrm{ch}_B(I_V)=(\alpha_2,\beta_2,\gamma_2)$, i.e., $U_i=\{0\}, i\in I_U$, $V_k=\{0\}, k\in I_V$, $\mathrm{ch}(U)=((0,0,0,),\mathrm{ch}_B(I_U))$ and $\mathrm{ch}_V=((0,0,0,),\mathrm{ch}_B(I_V))$. Let $U_i=(\{a_i\};+,\cdot,')$, $i\in I_U$ and $V_k=(\{b_k\};+,\cdot,')$, $k\in I_V$. The set $\{(a_i,b_j)\}$ is denoted by C_{ij}, $i\in I_U$, $j\in I_V$. Obviously, $A\times B=\bigcup_{i\in I_U, j\in I_V}C_{ij}$ and $C_{ij}\cap C_{kt}=\emptyset$, $(i,j)\neq(k,t),(i,j),(k,t)\in I_U\times I_V$. $W_{ij}=(C_{ij};+,\cdot,')$, $(i,j)\in I_U\times I_V$, is the canonical representation of super-Boolean algebra $U\times V$ as a Boolean sum of Boolean lattices. Indeed, $(a_i,b_j)\theta_{U\times V}(a_k,b_t)$ means that $(a_i,b_j)+(a_i,b_j)(a_k,b_t)=(a_i,b_j)$ and $(a_k,b_t)+(a_k,b_t)(a_i,b_j)=(a_k,b_t)$, i.e., $(a_i+a_ia_k,b_j+b_jb_t)=(a_i,b_j)$ and $(a_k+a_ka_i,b_t+b_tb_j)=(a_k,b_t)$ for any $(a_i,b_j),(a_k,b_t)\in A\times B$. So, $a_i+a_ia_k=a_i, a_k+a_ka_i=a_k, b_j+b_jb_t=b_j$, $b_t+b_tb_j=b_t$ and $a_i\theta_Ua_k$, $b_j\theta_Vb_t$. Since $U_i=\{0\}, i\in I_U$ and $V_k=\{0\}, k\in I_V$ are the canonical representations of U and V, then $a_i=a_k$, $b_j=b_t$, and $\theta_{U\times V}$ is a zero congruence of $A\times B$.

Thus $\mathrm{ch}_B(W_{ij})=(0,0,0),(i,j)\in I_U\times I_V$ and $\mathrm{ch}(U\times V)=((0,0,0),(\mathrm{ch}_B(I_U\times I_V))$. From $\mathrm{ch}_B(I_U)=(\alpha_1,\beta_1,\gamma_1)$ and $\mathrm{ch}_B(I_V)=(\alpha_2,\beta_2,\gamma_2)$ follows that $\mathrm{ch}_B(I_U\times I_V)=(\alpha,\beta,\gamma)$, where α,β,γ are defined in Proposition 6.7.2, i.e., $\mathrm{ch}(U\times V)=((0,0,0),(\alpha,\beta,\gamma))$. □

Let $R=\{U_s,\psi_{jk}|s,j,k\in S, j\leq k\}$ be a straight spectrum of super-Boolean algebras $U_s=(A_s;+,\cdot,\bar{}), s\in S$ with two binary and one unary operations, let $U_s = (A_s^i; +, \cdot, ')$, $i \in I_s$, $s \in S$ be a canonical representation of super-Boolean algebras $U_s, s \in S$ as a Boolean sum of Boolean lattices and let $\mathrm{ch}(U_s) = ((0,0,0,), \mathrm{ch}(I_s))$, $s \in S$. Then U_s^i, $i \in I_S, s \in S$ are one-element algebras and therefore we can consider them as sets $\{a_s^i\}$, $i \in I_s$, $s \in S$. Using mappings ψ_{jk} from the definition of straight spectrum let us consider mappings $\psi'_{jk} : I_j \mapsto I_k$, $j, k \in S$, $j \leq k$ as follows $\psi'_{jk}(i) = l$ if $\psi_{jk}(a_j^i) = a_i^l$ for any $i \in I_j$. Let's show that $\psi'_{jk}, j, k \in S, j \leq k$ are homomorphisms. Since ψ_{jk} are homomorphisms, then

$$\psi_{jk}(a_{i+t}^j) = \psi_{jk}(a_i^j + a_t^j) = \psi_{jk}(a_i^j) + \psi_{jk}(a_t^j) = a_l^j + a_p^j = a_{l+p}^j,$$
$$\overline{\psi_{jk}(a_i^j)} = \overline{a}_l^j = a_{\overline{l}}^j,$$
$$\overline{\psi_{jk}(a_i^j)} = \psi_{jk}(\overline{a}_i^j) = \psi_{jk}(a_{\overline{i}}^j),$$
$$\psi_{jk}(a_{i\cdot t}^j) = \psi_{jk}(a_{\overline{\overline{i}+\overline{t}}}^j) = \psi_{jk}(\overline{\overline{a}_i^j + \overline{a}_t^j}) = \overline{\overline{\psi_{jk}(a_i^j)} + \overline{\psi_{jk}(a_t^j)}} = \overline{\overline{a}_l^j + \overline{a}_p^j} = a_{l\cdot p}^j.$$

So,

$$\psi'_{jk}(i+t) = l + p = \psi'_{jk}(i) + \psi'_{jk}(t),$$
$$\overline{\psi'_{jk}(i)} = \overline{l} = \psi'_{jk}(\overline{s}),$$
$$\psi'_{jk}(i \cdot t) = l \cdot p = \psi'_{jk}(i) \cdot \psi'_{jk}(t).$$

Obviously that $\psi'_{ii} = id_{|I_i|}$, $\psi'_{ik} = \psi'_{ij} \cdot \psi'_{jk}$ for all $i \leq j \leq k, i, j, k \in S$.

Hence, $R' = \{I_s, \psi'_{jk} |\, s, j, k \in S,\ j \leq k\}$ is a straight spectrum.

Proposition 6.7.5. *Let a super-Boolean algebra* $U = (A; +, \cdot, ^{-})$ *with two binary and one unary operations be a straight of the limit spectrum* $R = \{U_s, \psi_{jk} | s, j, k \in S, j \leq k\}$ *of super-Boolean algebras* $U_s = (A_s; +, \cdot, ^{-})$, $s \in S$ *with elementary characteristics* $\mathrm{ch}(U_s) = ((0,0,0), \mathrm{ch}(I_s))$, $s \in S$. *Then* U *has the elementary characteristic* $\mathrm{ch}(U) = ((0,0,0), \mathrm{ch}(I))$, *where* I *is a straight limit of the straight spectrum* $R' = \{I_s, \psi'_{jk} |\, s, j, k \in S,\ j \leq k\}$.

Proof. Recall the definition of straight spectrum. Let $A^* = \{(a_i^s, i) | i \in S, a_i^s \in |U_i|\}$ and $P^* = \{(s, i) | i \in S, s \in I_i\}$. On A^* and P^* we define the following equivalence relations:

$$\eta = \{((a_i^s, i), (a_j^l, j)) | \text{ there exists } k \in S \text{ such that } i, j \leq k, \psi_{ik}(a_i^s) = \psi_{jk}(a_j^l)\},$$
$$\eta' = \{((s, i), (l, j)) | \text{ there exists } k \in S \text{ such that } i, j \leq k, \psi'_{ik}(s) = \psi'_{jk}(l)\}.$$

Obviously,
$((a_i^s, i), (a_j^l, j)) \in \eta \iff ((s, i), (l, j)) \in \eta'$.
Assume $A = A^*/\eta$ and $P = P^*/\eta'$ and

$$[a_i^s, i] = \{(a_j^l, j) | ((a_i^s, i), (a_j^l, j)) \in \eta\},$$
$$[s, i] = \{(l, j) | ((s, i), (l, j)) \in \eta'\}.$$

On A and P are defined following operations $+, \cdot, -$:

$$[a_i^s, i] + [a_i^l, i] = [a_i^s + a_i^l, i],$$
$$[a_i^s, i] \cdot [a_i^l, i] = [a_i^s \cdot a_i^l, i],$$
$$\overline{[a_i^s, i]} = [\overline{a}_i^s, i],$$
$$[s, i] + [l, i] = [s + l, i],$$
$$[s, i] \cdot [l, i] = [s \cdot l, i],$$
$$\overline{[s, i]} = [\overline{s}, i].$$

Algebras $U = (A; +, \cdot, ^-)$ and $I = (P; +, \cdot, ^-)$ are straight limits of straight spectrums R and R', respectively.

Show that U is a Boolean sum of Boolean lattices $\{[a_i^s, i]\}, [s, i] \in I$. Obviously, $|U| = \bigcup_{[s,i] \in I} \{[a_i^s, i]\}$. If $[s, i] \neq [l, i]$, $[s, i], [l, i] \in I$, then $\{[a_i^s, i]\} \cap \{[a_i^l, i]\} = \emptyset$. Indeed, assume that $[s, i] \neq [l, i]$ and $[a_i^s, i] = [a_i^l, i]$, then there exists $k \in S, i \leq k$ such that $\psi_{ik}(a_i^s) = \psi_{ik}(a_i^l)$, hence $\psi'_{ik}(s) = \psi'_{ik}(l), i \leq k$ and therefore $[s, i] = [l, i]$. So we get a contradiction.

Consider mappings $\varphi_{[k,i],[j,i]} : \{[a_i^k, i]\} \mapsto \{[a_i^j, i]\}$ and $h_{[k,i],\overline{[k,i]}} : \{[a_i^k, i]\} \mapsto \{[\bar{a}_i^k, i]\}$, $[k, i], [j, i] \in I$ and $[k, i] \leq [j, i]$. It is obvious that these mappings from one-element set to one-element set satisfy conditions (d) and (f) from the definition of Boolean sum. Let us check conditions (e) and (g):

$$[a_i^k, i] + [a_i^s, i] = [a_i^k + a_i^s, i] = [a_i^{k+s}, i],$$
$$\varphi_{[k,i],[k+s,i]}([a_k^i, i]) + \varphi_{[s,i],[k+s,i]}([a_i^s, i]) = [a_i^{k+s}, i] + [a_i^{k+s}, i] = [a_i^{k+s}, i],$$
$$[a_i^k, i] \cdot [a_i^s, i] = [a_i^k \cdot a_i^s, i] = [a_i^{k+s}, i],$$
$$\varphi_{[k,i],[k+s,i]}([a_k^i, i]) \cdot \varphi_{[s,i],[k+s,i]}([a_i^s, i]) = [a_i^{k+s}, i] \cdot [a_i^{k+s}, i] = [a_i^{k+s}, i],$$
$$h_{[k,i],\overline{[k,i]}}([a_i^k, i]') = h_{[k,i],\overline{[k,i]}}([a_i^k, i]) = [\bar{a}_i^k, i] = \overline{[a_i^k, i]}.$$

Thus, U is a Boolean sum of Boolean lattices $\{[a_i^s, i]\}, [s, i] \in I$, and therefore $\mathrm{ch}(U) = ((0, 0, 0), \mathrm{ch}(I))$. □

Proposition 6.7.6. *Let F be a filter on S, $U_s = (A_s; +, \cdot, ')$, $s \in S$ be super-Boolean algebras with two binary and one unary operations and let $U_s^i = (A_s^i; +, \cdot, '), i \in I_s, s \in S$ be the canonical representation of U_s as a Boolean sum of Boolean lattices. Let $\mathrm{ch}(U_s) = ((0, 0, 0), ch(I_s)), s \in S$. Then a filtered product $U = \prod_{s \in S} U_s/F$ is a super-Boolean algebra with elementary characteristic $\mathrm{ch}(U) = ((0, 0, 0), ch(I))$, where I is a filtered product $\prod_{s \in S} I_s/F$ of Boolean algebras $I_s, s \in S$.*

Proof. A filtered product of algebras $U_s, s \in S$ is a straight limit of straight spectrum $R = \{U_k, \pi_{T,L} | K, T, L \in F, T \preceq L\}$, where $\preceq$ is a dual order of inclusion, $U_K = \prod_{k \in K} U_k$ and $\pi_{T,L}$ is a projection on U_k.

If $K \in F$, then from $\text{ch}(U_s) = ((0,0,0), ch(I_s)), s \in S$ follows $\text{ch}(U_K) = ((0,0,0), \text{ch}(I_K))$, where $I_K = \prod_{k \in K} I_k$. A right limit U of spectrum R has an elementary characteristic $\text{ch}(U) = ((0,0,0), \text{ch}(I))$, where I is a straight limit of spectrum $R' = \{I_k; \pi'_{T,L} | K, T, L \in F, T \preceq L\}$, i.e., $I = \prod_{s \in S} I_s / F$. □

Theorem 6.7.7. *The elementary theory T_{SB} of the variety of all super-Boolean algebras with two binary and one unary operations is decidable.*

Proof. Let $((\alpha_0, \beta_0, \gamma_0), (0,0,0))$, $((0,0,0), (\alpha_0, \beta_0, \gamma_0))$, $((\alpha_1, \beta_1, \gamma_1), (0,0,0))$ and $((0,0,0), (\alpha_1, \beta_1, \gamma_1)), \ldots$ be effective conversion of all six $((\alpha, \beta, \gamma), (0,0,0))$ and $((0,0,0), (\alpha, \beta, \gamma))$, $\alpha, \beta < \omega, \gamma \leq 1, \beta + \gamma > 0$. Let T_{2i} be an elementary theory of super-Boolean algebra D_{2i} of elementary characteristic $((\alpha_i, \beta_i, \gamma_i), (0,0,0))$ and T_{2i+1} be an elementary theory of super-Boolean algebra D_{2i+1} of elementary characteristic $((0,0,0), (\alpha_i, \beta_i, \gamma_i))$. From the prove of Theorem 6.6.1 it follows that every T_i has a finite recursive system of axioms, therefore $T_0, T_1, \ldots$ is an effective sequence. Let's show that $T_{SB} = \bigcap_{i \in \omega} T_i$ and due to $\bigcap_{i \in \omega} T_i = \text{Th}(\{D_0, D_1, \ldots\})$ we will prove that $T_{SB} = \text{Th}(\{D_0, D_1, \ldots\})$.

For any super-Boolean algebra D with an elementary characteristic $((\alpha, \beta, \gamma), (0,0,0))$ or $((0,0,0), (\alpha, \beta, \gamma))$ there exists an ultraproduct of super-Boolean algebras from $\{D_i | i \in \omega\}$ elementary equivalent to D. If the elementary characteristic of algebra D is finite, then $D \equiv D_k$ for some $k < \omega$. If $(\alpha, \beta, \gamma) = (n, \omega, \gamma)$, $n < \omega$, then let $m_j < \omega$ such that $(\alpha_{m_j}, \beta_{m_j}, \gamma_{m_j}) = (n, j+i, \gamma), j \in \omega$, then the ultraproducts $\prod_{j \in \omega} D_{2m_j}/F$ and $\prod_{j \in \omega} D_{2m_j+1}/F$ will have elementary characteristics $((n, \omega, \gamma), (0,0,0))$ and $((0,0,0), (n, \omega, \gamma))$ respectively by any non-principal ultrafilter F on ω. If $(\alpha, \beta, \gamma) = (\omega, 0, 0)$ and for $n_j < \omega$ we have $(\alpha_{n_j}, \beta_{n_j}, \gamma_{n_j}) = (j+1, 1, 0)$, $j \in \omega$. Then ultraproducts $\prod_{j \in \omega} D_{2n_j}/F$ and $\prod_{j \in \omega} D_{2n_j+1}/F$ have elementary characteristics $((\omega, 0, 0), (0,0,0))$ and $((0,0,0), (\omega, 0, 0))$, respectively. So, for a super-Boolean algebra D with elementary characteristic $((\alpha, \beta, \gamma), (0,0,0))$ or $((0,0,0), (\alpha, \beta, \gamma))$ we have that $\text{Th}(D) \supseteq \text{Th}(\{D_0, D_1, \ldots\})$.

Every super-Boolean algebra U with elementary characteristic $((\alpha, \beta\gamma), (\xi, \eta, \zeta))$ is elementary equivalent to a direct product of super-Boolean algebras U_1 and U_2 with elementary characteristics $((\alpha, \beta, \gamma), (0,0,0))$ and $((0,0,0),(\xi, \eta, \zeta))$. We have $\mathrm{Th}(U_i) \supseteq \mathrm{Th}(\{D_0, D_1, \ldots\}), i = 1,2$. On the other hand, from the prove of Theorem 6.6.1 it is obviously that axioms of theories $\mathrm{Th}(U_1)$ and $\mathrm{Th}(U_2)$ are axiom of theory $\mathrm{Th}(U)$. So, $\mathrm{Th}(U) \supseteq \mathrm{Th}(U_i)$ and $\mathrm{Th}(U) \supseteq \mathrm{Th}(U_1) \cap \mathrm{Th}(U_2) \supseteq \mathrm{Th}(\{D_0, D_1, \ldots\})$.

Thus $T_{SB} = \mathrm{Th}(\{D_0, D_1, \ldots\}) = \bigcap_{i \in \omega} T_i$ and T_{SB} is decidable by Theorem 6.6.1. □

Chapter 7

Hyperidentities of De Morgan Algebras

7.1. Preliminaries

A hyperterm with binary and unary functional variables is called a De Morgan hyperterm (word) in this chapter.

In this chapter, we characterize hyperidentities of the variety of De Morgan algebras using the methodology of [253] (see Chapter 5). As a consequence of this characterization we give a finite base (basis) of hyperidentities of the hyperequational theory of this variety [273]. For the proof the concept of the canonical forms for the De Morgan words is introduced (cf. [253]) and it is proved that after a finite number of steps implemented by hyperidentities every De Morgan word can be reduced to the unique canonical form. As a result, we also obtain a characterization of free objects of the hypervariety of the T-algebras satisfying the hyperidentities of the variety of De Morgan algebras. In particular, we obtain that the variety of De Morgan algebras has a decidable hyperequational theory. And finally we prove that the hyperequational theory of this variety is not one-based.

7.2. Hyperidentities of the variety of De Morgan algebras: The case of one binary functional variable

Let $a, b \in 2^{\{1,\dots,n\}} \times 2^{\{1,\dots,n\}}$, i.e., $a = (a_1, a_2)$, $b = (b_1, b_2)$, where $a_i, b_i \subseteq \{1, \dots, n\}$, $i = 1, 2$. We say that $a \subseteq b$ if $a_1 \subseteq b_1$ and $a_2 \subseteq b_2$. Define $I = (\{1, \dots, n\}, \{1, \dots, n\})$, $E = \{I\}$ and $U = 2^{\{1,\dots,n\}} \times 2^{\{1,\dots,n\}} \setminus \{(\emptyset, \emptyset)\}$. We call a set $S \subseteq U$ ($S \neq \emptyset$) complete with respect to $\{1, \dots, n\}$ if

$$s \in S,\ s \subseteq s' \subseteq I \Rightarrow s' \in S,$$

i.e., the set S with any couple $(s_1, s_2) \in S$ contains all couples that contain (s_1, s_2) and are contained in I. Consider the object variables $x_1, \ldots, x_n$. For the set $s_1 = \{i_1, \ldots, i_k\} \subseteq \{1, \ldots, n\}$ $(s_1 \neq \emptyset)$ and binary functional variable F define

$$F_{i\in s_1}\{x_i\} = F_{s_1}\{x_i\} = \begin{cases} F(x_{i_1}, F(x_{i_2}, \ldots, F(x_{i_{k-1}}, x_{i_k}) \ldots)) & \text{if } k > 1, \\ x_{i_1} & \text{if } k = 1; \end{cases}$$

and for $s = (s_1, s_2) \in U$ define

$$F_s\{x_i\} = \begin{cases} F(F_{s_1}\{x_i\}, F_{s_2}\{\overline{x_i}\}) & \text{if } s_1 \neq \emptyset, s_2 \neq \emptyset, \\ F_{s_1}\{x_i\} & \text{if } s_1 \neq \emptyset, s_2 = \emptyset, \\ F_{s_2}\{\overline{x_i}\} & \text{if } s_1 = \emptyset, s_2 \neq \emptyset. \end{cases}$$

It is clear that these notations do not depend on the order of the elements in the sets s_1 and s_2 due to the axioms of De Morgan algebras (namely, commutativity and associativity). For example, for $s_1 = \{i_1, \ldots, i_k\}$, $s_2 = \{j_1, \ldots, j_m\}$, and $F = +$ we have: $F_s\{x_i\} = (x_{i_1} + (x_{i_2} + \cdots + (x_{i_{k-1}} + x_{i_k}) \ldots)) + (\overline{x}_{j_1} + (\overline{x}_{j_2} + \cdots + (\overline{x}_{j_{m-1}} + \overline{x}_{j_m}) \ldots))$. When we say that S is complete, we mean that S is complete with respect to $\{1, \ldots, n\}$.

The set of all object variables of a hyperterm (word) ω is denoted by $[\omega]$.

Also define $\overline{x} = x'$, and if $\omega = X(\omega_1, \omega_2)$ then $\omega' = \overline{\omega} = \overline{X}(\omega_1, \omega_2)$.

Definition 7.2.1. Let $S = \{s_1, \ldots, s_t\} \subseteq U$ be a complete set. A hyperterm of the form

$$F_{s\in S}\{\overline{F_s\{x_i\}}\} = F(\overline{F_{s_1}\{x_i\}}, F(\overline{F_{s_2}\{x_i\}}, \ldots, F(\overline{F_{s_{t-1}}\{x_i\}}, \overline{F_{s_t}\{x_i\}}) \ldots))$$

is called a d_1-canonical word (form) with respect to $\{F, x_1, \ldots, x_n\}$ and it is denoted by $f_S(F, -, x_1, \ldots, x_n)$ or simply by $f_S(F)$ or f_S.

Note that from the axioms of De Morgan algebras it follows that this definition does not depend on the order of elements in the set S.

For example, the hyperterm: $F(\overline{F}(x_1, F(x_2, F(x_3, F(\bar{x}_1, \bar{x}_2)))), F(\overline{F}(x_2, F(x_3, F(\bar{x}_1, F(\bar{x}_2, \bar{x}_3)))), \overline{F}(x_1, F(x_2, F(x_3, F(\bar{x}_1, F(\bar{x}_2, \bar{x}_3)))))))$ is a d_1-canonical word with respect to $\{F, x_1, x_2, x_3\}$.

Two d_1-canonical words $f_{S_1}(F)$ and $f_{S_2}(F)$ are called equal if $S_1 = S_2$.

First note that if S is complete then $I \in S$ and therefore the intersection of any two complete sets is not empty. Also the intersection and union of complete sets are complete sets too.

It is useful to present hyperterms as trees (see Chapter 5).

Note that we will use various hyperidentities of the variety of De Morgan algebras in this and successive sections, which are easy to check to hold in this variety. Also notice that from the principle of duality it follows that we need to check the truth of those hyperidentities with two functional variables F, G only for $F = G = +$ and $F = +,\ G = \cdot$.

Theorem 7.2.2. *For any De Morgan word ω, $[\omega] \subseteq \{x_1, \ldots, x_n\}$, with one binary functional variable F there exists a unique d_1-canonical word (form) $\widetilde{\omega}$ with respect to $\{F, x_1, \ldots, x_n\}$ such that the hyperidentity $\omega = \widetilde{\omega}$ holds in the variety of De Morgan algebras. The word $\widetilde{\omega}$ is called the d_1-canonical form of the De Morgan word ω with respect to $\{F, x_1, \ldots, x_n\}$.*

Moreover, $\widetilde{\omega}$ is obtained from the word ω after a finite number of transitions realized by the following finite number of hyperidentities of the variety of De Morgan algebras, that do not depend on ω:

(1) $X(x) = Y(x)$,
(2) $F(x, x) = x$,
(3) $F(x, y) = F(y, x)$,
(4) $F(x, F(y, z)) = F(F(x, y), z)$,
(5) $X(X(x)) = x$,
(6) $\overline{F}(\overline{F}(x, y), z) = F(\overline{F}(\overline{x}, z), \overline{F}(\overline{y}, z))$,
(7) $F(\overline{F}(x, y), \overline{x}) = \overline{x}$,

where $\bar{F}(x, y) = \overline{F(x, y)}$.

Proof. First we replace all unary functional variables in the De Morgan word ω by negation and exclude all double negations by taking into account the (hyper)identity $\overline{\overline{x}} = x$.

If $\omega = X(u_1, u_2)$ then X is called the root of the word (tree) ω. The branch which connects the root of the word ω with an objective variable $x_i \in [\omega]$ or $\overline{x}_i$ is called the branch for that occurrence of x_i.

We prove that using hyperidentities of the variety of De Morgan algebras the De Morgan word ω can be modified to a De Morgan word in which every branch of x_i includes not more than one negation (not counting negations on objective variables, i.e., $\overline{x}_i$). To show this, suppose that there exists a branch of x_i that contains two negations, at least. Then there exists a subword ω_0 on that branch such that

$$\omega_0 = \overline{F}(\omega_1, \omega_2),$$

where the subwords ω_1 and ω_2 contain at least one negation together (not counting the negations of objective variables). We also suppose that ω_0 is a minimal subword of ω with that property, i.e., ω_0 does not contain a proper subword of the form $\overline{F}(\omega_1', \omega_2')$, where ω_1' and ω_2' contain at least one negation together (not counting the negations of objective variables). Let ω_1 and ω_2 contain p negations together, where $p \geq 1$. Taking into account the hyperidentities (2.3.2) and (2.3.3) and the minimality of ω_0, we can present ω_0 in the form $\omega_0 = \overline{F}(\overline{F_s\{x_i\}}, \omega_3)$, where $s = (s_1, s_2) \in U$ and the subword ω_3 contains $p-1$ negations.

In the next step, we use the following hyperidentity of the variety of De Morgan algebras:

$$\overline{F}(\overline{F}(x_1, x_2), x_3) = F(\overline{F}(\overline{x}_1, x_3), \overline{F}(\overline{x}_2, x_3)). \tag{7.2.1}$$

This hyperidentity allows us to rewrite a subtree with two adjacent occurrences of $\overline{F}$, one of them at the root, as a tree with F at its root and one $\overline{F}$ per branch.

Suppose that $i_1 \in s_1 \cup s_2$ and denote $s' = (s_1 \backslash \{i_1\}, s_2)$, if $i_1 \in s_1$, and $s' = (s_1, s_2 \backslash \{i_1\})$, if $i_1 \in s_2 \backslash s_1$ (we suppose that $F_s\{x_i\}$ contains two objective variables at least and therefore $s' \neq (\varnothing, \varnothing)$). We have:

$$\omega_0 = \overline{F}(\overline{F}(F_{s'}\{x_i\}, x^*_{i_1}), \omega_3) = F(\overline{F}(\overline{F_{s'}\{x_i\}}, \omega_3), \overline{F}(\overline{x^*_{i_1}}, \omega_3)),$$

where

$$x^*_{i_1} = \begin{cases} x_{i_1} & \text{if } i_1 \in s_1, \\ \overline{x}_{i_1} & \text{if } i_1 \in s_2 \backslash s_1. \end{cases}$$

Subword $\overline{F}(\overline{F_{s'}\{x_i\}}, \omega_3)$ has the form of ω_0 and so an analogous process can be applied to it. After a finite number of such steps, the word ω_0 is modified to the form of Fig. 7.1.

Here $l = |s_1| + |s_2|$ and $x^*_{i_j} = x_{i_j}$ or $x^*_{i_j} = \overline{x}_{i_j}$. And now there are exactly $p-1$ negations above every negation (as in ω_3). Every subword $\overline{F}(\overline{x^*_{it}}, \omega_3)$ has the form of ω_0 and therefore by applying the same process to those subwords,

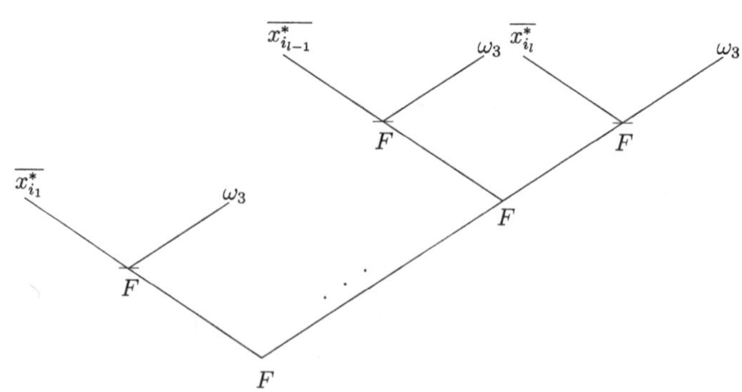

Fig. 7.1.

after a finite number of steps the word ω_0 is modified to the desired form (in each step we reduce the number of negations above every negation). Thus, ω is modified to a similar form.

Then using the hyperidentities (2.3.3), (2.3.2) and (2.3.1) we modify ω to one of the following forms:

$$\omega = F(\overline{F_{s_1}\{x_i\}}, F(\overline{F_{s_2}\{x_i\}}, \ldots, F(\overline{F_{s_m}\{x_i\}}, F_s\{x_i\}) \ldots))$$

or

$$\omega = F(\overline{F_{s_1}\{x_i\}}, F(\overline{F_{s_2}\{x_i\}}, \ldots, F(\overline{F_{s_{m-1}}\{x_i\}}, \overline{F_{s_m}\{x_i\}}) \ldots))$$

or

$$\omega = F_s\{x_i\},$$

where $s_1, \ldots, s_m, s \in U$.

In the first and third cases x_{i_j} (without negations) in $F_s\{x_i\}$ can be replaced by $\overline{\overline{x}}_{i_j}$ and in the result the word ω is modified to the second form.

Now if ω contains some subword $\overline{F_s\{x_i\}}$ then using hyperidentity

$$F(\overline{F}(x, y), \overline{x}) = \overline{x}, \tag{7.2.2}$$

we can add all subwords of the form $\overline{F_{s'}\{x_i\}}$ to ω, where $s \subseteq s'$ (since $F(\overline{F_s\{x_i\}}, \overline{F_{s'}\{x_i\}}) = F(\overline{F}(F_s\{x_i\}, F_{s'}\{x_i\}), \overline{F_s\{x_i\}}) \overset{(7.2.2)}{=} \overline{F_s\{x_i\}})$.

Applying hyperidentities (2.3.3) and (2.3.2), the word ω is modified to the form

$$\omega = F(\overline{F_{s_1}\{x_i\}}, F(\overline{F_{s_2}\{x_i\}}, \ldots, F(\overline{F_{s_{r-1}}\{x_i\}}, \overline{F_{s_r}\{x_i\}}) \ldots)),$$

where $s_i \in U$, $i = \overline{1, r}$. The set $S = \{s_1, \ldots, s_r\}$ is obviously complete. So in this way ω is modified to d_1-canonical form.

Now let us prove uniqueness. Suppose that $\omega = f_S(F, \bar{}, x_1, \ldots, x_n)$ and $\omega = f_{S'}(F, \bar{}, x_1, \ldots, x_n)$, where $S \neq S'$. Then the hyperidentity $f_S(F) = f_{S'}(F)$ must hold in the variety of De Morgan algebras.

Consider the De Morgan algebra $\mathbf{4} = (\{0, 1, a, b\}, \{+, \cdot, \bar{}\})$, where $a + b = 1, ab = 0,\ \bar{a} = a,\ \bar{b} = b$.

We claim that the hyperidentity $f_S(F) = f_{S'}(F)$ does not hold in the algebra $\mathbf{4}$. To see this take $F = \cdot$. Let $S = \{s_1, \ldots, s_m\}$, $S' = \{s'_1, \ldots, s'_t\}$. Since $S \neq S'$, we can suppose that there exist $s_k \in S$, $s_k \notin S'$.

Then there does not exist $s'_j \in S'$ such that $s'_j \subseteq s_k$. Otherwise, according to the fact that S' is complete, we have $s_k \in S'$, which is a contradiction.

Now let $s_k = (s_{k,1}, s_{k,2})$, where $s_{k,1} = \{i_1, \dots, i_r\} \subseteq \{1, \dots, n\}$, $s_{k,2} = \{j_1, \dots, j_q\} \subseteq \{1, \dots, n\}$. We take:

$$x_i = \begin{cases} a & \text{if } i \in s_{k,1} \cap s_{k,2}, \\ 1 & \text{if } i \in s_{k,1} \setminus s_{k,2}, \\ 0 & \text{if } i \in s_{k,2} \setminus s_{k,1}, \\ b \text{ if } i \in \{1, \dots n\} \setminus (s_{k,1} \cup s_{k,2}). \end{cases}$$

Then $\overline{F_{s_k}\{x_i\}} = \overline{x_{i_1} \dots x_{i_r} \bar{x}_{j_1} \dots \bar{x}_{j_q}} = a$ or 0 and therefore $f_S(F) = a$ or 0 since $ax = a$ or 0 for all $x \in \mathbf{4}$.

Now let for each l $(1 \leq l \leq t)$ we have $s'_l = (s'_{l,1}, s'_{l,2})$. We saw that $s'_l \nsubseteq s_k$. Without loss of generality suppose that $s'_{l,1} \nsubseteq s_{k,1}$. Hence, there exist $p \in s'_{l,1}$, $p \notin s_{k,1}$.

If $p \in s_{k,2}$ then $p \in s_{k,2} \setminus s_{k,1}$ and so $x_p = 0$. Hence, $F_{s'_l}\{x_i\} = 0$.

If $p \notin s_{k,2}$ then $p \in \{1, \dots, n\} \setminus (s_{k,1} \cup s_{k,2})$. Therefore $x_p = b$ and $F_{s'_l}\{x_i\} = 0$ or b as $F = \cdot$ and $bx = 0$ or b for all $x \in \mathbf{4}$.

Thus $F_{s'_l}\{x_i\} = 0$ or b for all l. Hence, $\overline{F_{s'_l}\{x_i\}} = 1$ or b, therefore $f_{S'}(F) = 1$ or b (because the product of 1's and b's is 1 or b, there can be only 1's or only b's in the product). So in all cases $f_{S'}(F) \neq f_S(F)$. □

As an example of Theorem 7.2.2 let us modify the following word to d_1-canonical form: $\omega = \overline{F}(x_2, F(x_3, F(\bar{x}_1, F(\bar{x}_2, \overline{F}(x_2, F(x_3, \bar{x}_1))))))$.

We have: $\omega = F(\overline{F}(x_2, F(x_3, F(\bar{x}_1, F(\bar{x}_2, \bar{x}_2)))), F(\overline{F}(x_2, F(x_3, F(\bar{x}_1, F(\bar{x}_2, \bar{x}_3)))), \overline{F}(x_2, F(x_3, F(\bar{x}_1, F(\bar{x}_2, x_1)))))) = F(\overline{F}(x_2, F(x_3, F(\bar{x}_1, \bar{x}_2))), F(\overline{F}(x_2, F(x_3, F(\bar{x}_1, F(\bar{x}_2, \bar{x}_3)))), F(\overline{F}(x_1, F(x_2, F(x_3, F(\bar{x}_1, \bar{x}_2)))), \overline{F}(x_1, F(x_2, F(x_3, F(\bar{x}_1, F(\bar{x}_2, \bar{x}_3)))))))) = f_S(F, \bar{}, x_1, x_2, x_3)$, where $S = \{(\{2,3\}, \{1,2\}), (\{2,3\}, \{1,2,3\}), (\{1,2,3\}, \{1,2\}), (\{1,2,3\}, \{1,2,3\})\}$ is complete set with respect to $\{1, 2, 3\}$.

Corollary 7.2.3. *A hyperidentity $\omega_1 = \omega_2$ with one binary functional variable F holds in the variety of De Morgan algebras if and only if the d_1-canonical forms of the hyperterms ω_1 and ω_2 with respect to $\{F, x_1, \dots, x_n\}$ are the same, where $\{x_1, \dots, x_n\} = [\omega_1] \cup [\omega_2]$.*

Corollary 7.2.4. *In the variety of De Morgan algebras every hyperidentity with one binary functional variable is a consequence of the hyperidentities (1)–(7) listed in Theorem 7.2.2.*

7.3. Hyperidentities of the variety of De Morgan algebras: The case of two binary functional variables

First recall that $U = 2^{\{1,\dots,n\}} \times 2^{\{1,\dots,n\}} \setminus \{(\emptyset,\emptyset)\}$, $E = \{I\} = \{(\{1,\dots,n\},\{1,\dots,n\})\}$.

Definition 7.3.1. Let S_1 and S_2 be complete sets, F and G be binary functional variables. A hyperterm of the form

$$f_{(S_1,S_2)}(G,F,\bar{~};\ x_1,\dots,x_n) = F(\overline{F}(f_{S_1}(G), f_E(G)), \overline{F}(f_{S_2}(G), f_U(G))),$$

where $f_{S_1}(G)$, $f_{S_2}(G)$, $f_E(G)$ and $f_U(G)$ are d_1-canonical words with respect to $\{G, x_1,\dots,x_n\}$, is called a d_2-canonical word (form) with respect to $\{G,F,x_1,\dots,x_n\}$ and it is denoted by $f_{(S_1,S_2)}(G,F)$ or $f_{(S_1,S_2)}$.

Two d_2-canonical words $f_{(S_1,S_2)}$ and $f_{(S_3,S_4)}$ are said to be equal if $S_1 = S_3$, $S_2 = S_4$. Notice that E and U are complete sets and so $f_E(G)$ and $f_U(G)$ are meaningful.

The analogue of Theorem 7.2.2 holds, too.

Theorem 7.3.2. *For any De Morgan hyperterm ω, $[\omega] \subseteq \{x_1,\dots,x_n\}$, with at most two binary functional variables F, G there exists a unique d_2-canonical word (form) $\widetilde{\omega}$ with respect to $\{G,F,x_1,\dots,x_n\}$ such that the hyperidentity $\omega = \widetilde{\omega}$ holds in the variety of De Morgan algebras. The word $\widetilde{\omega}$ is called the d_2-canonical form of the hyperterm ω with respect to $\{G,F,x_1,\dots,x_n\}$.*

Moreover, $\widetilde{\omega}$ is obtained from the hyperterm ω with a finite number of transitions realized by the following finite number of hyperidentities of the variety of De Morgan algebras, which do not depend on ω: (1)–(7) *(listed in Theorem 7.2.2) and*

(8) $G(F(x,y),z) = F(G(x,z),G(y,z))$,
(9) $\overline{G}(\overline{F}(x_1,x_2)x_3) = F(\overline{G}(\overline{x_1},x_3),\overline{G}(\overline{x_2},x_3))$,
(10) $F(x,F(y,z)) = F(G(x,G(y,z)),\overline{G}(\overline{x},G(\overline{y},\overline{z})))$,
(11) $F(\overline{F}(G(x,e),G(x,G(y,e))),\overline{F}(G(z,e),G(z,G(t,e)))) =$
$F(\overline{F}(G(x,G(z,e)),G(x,G(y,G(z,G(u,e))))),$
$\overline{F}(\overline{G}(\overline{G}(x,G(y,e)),\overline{G}(z,G(t,e))),e))$.

Proof. First we replace all unary functional variables in the hyperterm ω by negation and exclude all double negations by taking into account the (hyper)identity $\overline{\overline{x}} = x$.

Then, using the hyperidentities

$$G(F(x,y),z) = F(G(x,z),G(y,z)) \tag{7.3.1}$$

and

$$\overline{G}(\overline{F}(x,y),z) = F(\overline{G}(\overline{x},z),\overline{G}(\overline{y},z)), \tag{7.3.2}$$

we can modify the hyperterm ω to a form whose tree contains all binary functional symbols G above all binary functional symbols F. In this way we get a tree with all branches containing only G. Call such branches G-branches. Using Theorem 7.2.2 we modify all G-branches to the d_1-canonical forms with respect to $\{G, x_1, \ldots, x_n\}$. Denote those d_1-canonical words by $f_{S_1}(G, \bar{}, x_1, \ldots, x_n), \ldots, f_{S_l}(G, \bar{}, x_1, \ldots, x_n)$. Then consider them as objective variables $y_1, \ldots, y_l$ and modify that word to d_1-canonical form with respect to $\{F, y_1, \ldots, y_l\}$. Subwords of the form $\overline{f_{S_i}(G)}$ can appear, which are words with one binary functional variable, and so they can be modified to d_1-canonical forms with respect to $\{G, x_1, \ldots, x_n\}$. Now we can simplify the subwords of the form:

$$\overline{F}(f_{S_m}(G), F(\ldots F(f_{S_2}(G), f_{S_1}(G))\ldots)).$$

We claim that $F(f_{S_1}(G), f_{S_2}(G))$ can be modified to the word $F(f_{S_1 \cap S_2}(G), f_{S_1 \cup S_2}(G))$. First notice that $f_{S_1 \cup S_2}(G) = G(f_{S_1}(G), f_{S_2}(G))$.

If $s_1 = (a_1, b_1), s_2 = (a_2, b_2) \in 2^{\{1,\ldots,n\}} \times 2^{\{1,\ldots,n\}}$, then denote $s_1 \cup s_2 = (a_1 \cup a_2, b_1 \cup b_2)$. Notice that $G(G_{s_1}\{x_i\}, G_{s_2}\{x_i\}) = G_{s_1 \cup s_2}\{x_i\}$ (for definitions of those words see the second section).

Now use the following hyperidentity of the variety of De Morgan algebras:

$$\begin{aligned}&\overline{G}(\overline{G}(\overline{y_1}, G(\ldots G(\overline{y_{k-1}}, \overline{y_k})\ldots)), \overline{G}(\overline{z_1}, G(\ldots G(\overline{z_{t-1}}, \overline{z_t})\ldots)))\\ &\quad = G(\overline{G}(y_1, z_1), G(\overline{G}(y_1, z_2), \ldots, G(\overline{G}(y_k, z_{t-1}), \overline{G}(y_k, z_t))\ldots)).\end{aligned} \tag{7.3.3}$$

Note that this hyperidentity contains one binary functional variable, hence by Corollary 7.2.4 it is a consequence of the hyperidentities (1)–(7).

And also notice that (7.3.3) can be written concisely in the following form:

$$\overline{G}(\overline{G_{i\in\{1,\ldots,k\}}\{\overline{y}_i\}}, \overline{G_{j\in\{1,\ldots,t\}}\{\overline{z}_j\}}) = G_{i\in\{1,\ldots,k\}, j\in\{1,\ldots,t\}}\{\overline{G(y_i, z_j)}\}.$$

If $S = \{s_1, ..., s_m\}$, $S' = \{s'_1, ..., s'_l\}$ are complete sets, then we have

$$\begin{aligned}\overline{G}(\overline{f_S(G)}, \overline{f_{S'}(G)}) &= \overline{G}(\overline{G}(\overline{G_{s_1}\{x_i\}}, G(\overline{G_{s_2}\{x_i\}}, \ldots,\\ &\quad G(\overline{G_{s_{m-1}}\{x_i\}}, \overline{G_{s_m}\{x_i\}}) \ldots)),\\ &\quad \overline{G}(\overline{G_{s'_1}\{x_i\}}, G(\overline{G_{s'_2}\{x_i\}}, \ldots,\\ &\quad G(\overline{G_{s'_{l-1}}\{x_i\}}, \overline{G_{s'_l}\{x_i\}}) \ldots)))\\ &\overset{(7.3.3)}{=} G(\overline{G_{s_1\cup s'_1}\{x_i\}}, G(\overline{G_{s_1\cup s'_2}\{x_i\}}, \ldots,\\ &\quad G(\overline{G_{s_m\cup s'_{l-1}}\{x_i\}}, \overline{G_{s_m\cup s'_l}\{x_i\}}) \ldots)) = G_{s\in\widetilde{S}}\{\overline{G_s\{x_i\}}\},\end{aligned}$$

where $\widetilde{S} = \{s_1 \cup s'_1, s_1 \cup s'_2, \ldots, s_m \cup s'_l\} = \{a \cup b : a \in S, b \in S'\}$. We claim that $\widetilde{S} = S \cap S'$ (and therefore $\widetilde{S}$ is a complete set). For, if $s \in S \cap S'$, then $s \in S$ and $s \in S'$, hence $s = s \cup s \in \widetilde{S}$. Conversely, if $s \in \widetilde{S}$, then there exist $a \in S, b \in S'$ such that $s = a \cup b$. But then $a \subseteq s \subseteq I$, $b \subseteq s \subseteq I$ and so $s \in S$, $s \in S'$ (as S and S' are complete sets). So we conclude that

$$\overline{G}(\overline{f_S(G)}, \overline{f_{S'}(G)}) = f_{S\cap S'}(G), \tag{**}$$

Now using the hyperidentity

$$F(x, y) = F(G(x, y), \overline{G}(\overline{x}, \overline{y})), \tag{7.3.4}$$

we have

$$\begin{aligned}F(f_{S_1}(G), f_{S_2}(G)) = F(G(f_{S_1}(G), f_{S_2}(G)), \overline{G}(\overline{f_{S_1}(G)}, \overline{f_{S_2}(G)})) \overset{(**)}{=}\\ F(f_{S_1\cup S_2}(G), f_{S_1\cap S_2}(G)).\end{aligned}$$

Now simplify the word $F(f_{S_3}, F(f_{S_2}, f_{S_1}))$ using the hyperidentity:

$$F(x, F(y, z)) = F(G(x, G(y, z)), \overline{G}(\overline{x}, G(\overline{y}, \overline{z}))). \tag{7.3.5}$$

We have

$$\begin{aligned}F(f_{S_3}, F(f_{S_2}, f_{S_1})) &\overset{(7.3.5)}{=} F(G(f_{S_3}, G(f_{S_2}, f_{S_1})), \overline{G}(\overline{f_{S_3}}, G(\overline{f_{S_2}}, \overline{f_{S_1}})))\\ &\overset{(**)}{=} F(f_{S_1\cup S_2\cup S_3}, f_{S_1\cap S_2\cap S_3}).\end{aligned}$$

Note that (7.3.4) is a consequence of the hyperidentity (7.3.5) (in the case $z = y$).

Now we can prove by induction that:

$$F(f_{S_m}(G), F(\ldots F(f_{S_2}(G), f_{S_1}(G)) \ldots)) = F(f_{S_1\cap\ldots\cap S_m}(G), f_{S_1\cup\ldots\cup S_m}(G)).$$

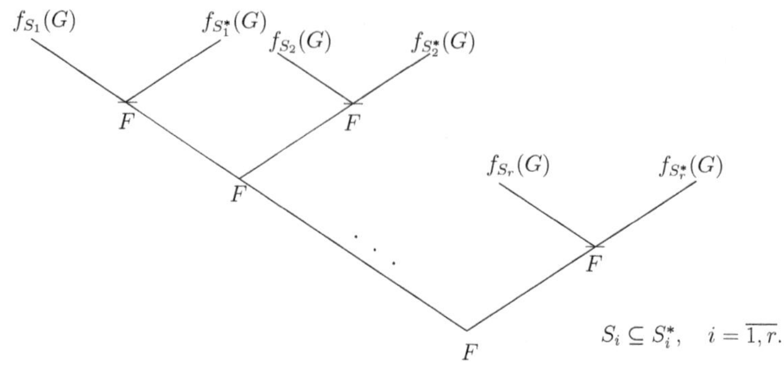

Fig. 7.2.

We saw that the word of the form $\overline{F}(f_{S_m}(G), F(\ldots F(f_{S_2}(G), f_{S_1}(G))\ldots))$, for $m \geq 2$ can be modified to the form $\overline{F}(f_S(G), f_{S^*}(G))$ where $S \subseteq S^*$. Notice that in the case $m = 1$ it can be modified to a similar form. For example, $\overline{f_{S_1}} = \overline{F}(f_{S_1}, f_{S_1})$, for $S_1 \subseteq S_1$. Hence, the original word is modified to the form $\overline{F}(f_{S_1}, f_{S_1^*})$, $S_1 \subseteq S_1^*$ or to a word of the form of Fig. 7.2.

Notice that $G(f_S(G), f_E(G)) = f_S(G)$ as $E \subseteq S$.

Let $f_A(G)$ and $f_B(G)$ be the d_1-canonical forms of the words $\overline{f_E(G)}$ and $\overline{f_U(G)}$ respectively. Then we have

$$\begin{aligned} f_E(G) = f_{B\cap E}(G) &\overset{(**)}{=} \overline{G}(\overline{f_B(G)}, \overline{f_E(G)}) = \overline{G}(f_U(G), \\ f_A(G)) &= \overline{f_{U\cup A}(G)} = \overline{f_U(G)}. \end{aligned}$$

Hence, $\overline{f_E(G)} = f_U(G)$.

Now use the following hyperidentity of the variety of De Morgan algebras:

$$\begin{aligned} &F(\overline{F}(G(x,e), G(x, G(y,e))), \overline{F}(G(z,e), G(z, G(t,e)))) \\ &\quad = F(\overline{F}(G(x, G(z,e)), G(x, G(y, G(z, G(u,e))))), \\ &\qquad \overline{F}(\overline{G}(\overline{G}(x, G(y,e)), \overline{G}(z, G(t,e))), e)). \qquad (7.3.6) \end{aligned}$$

Substituting $x = f_{S_1}$, $y = f_{S_1^*}$, $z = f_{S_2}$, $t = f_{S_2^*}$, $u = f_U$, $e = f_E$ we get

$$\begin{aligned} &F(\overline{F}(f_{S_1}, f_{S_1^*}), \overline{F}(f_{S_2}, f_{S_2^*})) \\ &\quad = F(\overline{F}(G(f_{S_1}, f_E), G(f_{S_1}, G(f_{S_1^*}, f_E))), \overline{F}(G(f_{S_2}, f_E), G(f_{S_2}, G(f_{S_2^*}, f_E)))) \\ &\quad \overset{(7.3.6)}{=} F(\overline{F}(G(f_{S_1}, G(f_{S_2}, f_E)), G(f_{S_1}, G(f_{S_1^*}, G(f_{S_2}, G(f_U, f_E))))), \\ &\quad \overline{F}(\overline{G}(\overline{G}(f_{S_1}, G(f_{S_1^*}, f_E)), \overline{G}(f_{S_2}, G(f_{S_2^*}, f_E))), f_E)) \\ &\quad = F(\overline{F}(f_{S_1\cup S_2}, f_U), \overline{F}(f_{S_1^*\cap S_2^*}, f_E)). \end{aligned}$$

Analogously:

$$\begin{aligned}
&F(F(\overline{F}(f_{S_1}, f_{S_1^*}), \overline{F}(f_{S_2}, f_{S_2^*})), \overline{F}(f_{S_3}, f_{S_3^*}))\\
&\quad = F(F(\overline{F}(f_{S_1 \cup S_2}, f_U), \overline{F}(f_{S_1^* \cap S_2^*}, f_E)), \overline{F}(f_{S_3}, f_{S_3^*}))\\
&\quad = F(\overline{F}(f_{S_1^* \cap S_2^*}, f_E), F(\overline{F}(f_{S_1 \cup S_2}, f_U), \overline{F}(f_{S_3}, f_{S_3^*})))\\
&\quad = F(\overline{F}(f_{S_1^* \cap S_2^*}, f_E), F(\overline{F}(f_{S_1 \cup S_2 \cup S_3}, f_U), \overline{F}(f_{S_3^* \cap U}, f_E)))\\
&\quad = F(\overline{F}(f_{S_1 \cup S_2 \cup S_3}, f_U), F(\overline{F}(f_E, f_{S_1^* \cap S_2^*}), \overline{F}(f_E, f_{S_3^*})))\\
&\quad = F(\overline{F}(f_{S_1 \cup S_2 \cup S_3}, f_U), F(\overline{F}(f_E, f_U), \overline{F}(f_{S_1^* \cap S_2^* \cap S_3^*}, f_E)))\\
&\quad = F(\overline{F}(f_{S_1 \cup S_2 \cup S_3}, f_U), \overline{F}(f_{S_1^* \cap S_2^* \cap S_3^*}, f_E)).
\end{aligned}$$

In the last step we used the hyperidentity

$$F(\overline{F}(e, G(y, G(u, e))), \overline{F}(G(y, e), e)) = \overline{F}(G(y, e), e), \tag{7.3.7}$$

which is a consequence of the hyperidentity (7.3.6) (if we take $x = z = e,\ t = y$).

Now we can proceed by induction:

$$\begin{aligned}
&F(\overline{F}(f_{S_r}, f_{S_r^*}), F(\dots F(\overline{F}(f_{S_2}, f_{S_2^*}), \overline{F}(f_{S_1}, f_{S_1^*}))\dots))\\
&\quad = F(\overline{F}(f_{S_1 \cup \dots \cup S_r}, f_U), \overline{F}(f_{S_1^* \cap \dots \cap S_r^*}, f_E)),
\end{aligned}$$

which is a d_2-canonical word.

We can modify the word $\overline{F}(f_{S_1}, f_{S_1^*})$ to the d_2-canonical form as well. To do so, we use the following consequence of the hyperidentity (7.3.6):

$$\begin{aligned}
&\overline{F}(G(x, e), G(x, G(y, e)))\\
&\quad = F(\overline{F}(G(x, e), G(x, G(y, G(u, e)))), \overline{F}(G(x, G(y, e)), e)). \quad (7.3.8)
\end{aligned}$$

If we take $x = f_{S_1},\ y = f_{S_1^*},\ u = f_U,\ e = f_E$ then we receive

$$\overline{F}(f_{S_1}, f_{S_1^*}) = F(\overline{F}(f_{S_1}, f_U), \overline{F}(f_{S_1^*}, f_E)).$$

Thus in all cases the word ω is modified to the d_2-canonical form.

Now let us prove the uniqueness. Suppose that $f_{(S_1,S_2)}(G, F, \bar{}, x_1, \dots, x_n) = \omega$ and $f_{(S_3,S_4)}(G, F, \bar{}, x_1, \dots, x_n) = \omega$ and $S_1 \neq S_3$. Then the hyperidentity $f_{(S_1,S_2)}(G, F, \bar{}, x_1, \dots, x_n) = f_{(S_3,S_4)}(G, F, \bar{}, x_1, \dots, x_n)$ holds in any De Morgan algebra. But we can show that it is not true in the De Morgan algebra **4**. Take $F = \cdot, G = \cdot$. By Theorem 7.2.2 there are $x_1, \dots, x_n \in \mathbf{4}$ such that $f_{S_1}(G) \neq f_{S_3}(G)$, as $S_1 \neq S_3$. For those values of the variables

$x_1, \ldots, x_n$ described in Theorem 7.2.2, we have $f_E(G, \bar{}, x_1, \ldots, x_n) = 1$ and $f_U(G) = \overline{f_E(G)} = 0$. And we get (for those values):

$$\begin{aligned} f_{(S_1,S_2)}(G, F, \bar{}, x_1, \ldots, x_n) &= F(\overline{F}(f_{S_1}(G), f_E(G)), \overline{F}(f_{S_2}(G), f_U(G))) \\ &= \overline{f_{S_1}(\cdot) \cdot 1} \cdot \overline{f_{S_2}(\cdot) \cdot 0} = \overline{f_{S_1}(\cdot)}, \\ f_{(S_3,S_4)}(G, F, \bar{}, x_1, \ldots, x_n) &= F(\overline{F}(f_{S_3}(G), f_E(G)), \overline{F}(f_{S_4}(G), f_U(G))) \\ &= \overline{f_{S_3}(\cdot) \cdot 1} \cdot \overline{f_{S_4}(\cdot) \cdot 0} = \overline{f_{S_3}(\cdot)}, \end{aligned}$$

i.e., $f_{(S_1,S_2)} \neq f_{(S_3,S_4)}$.

If $S_2 \neq S_4$ then we take $F = +,\ G = \cdot$. □

Corollary 7.3.3. *A hyperidentity $\omega_1 = \omega_2$ with two binary functional variables F and G holds in the variety of De Morgan algebras if and only if the d_2-canonical forms of the hyperterms ω_1 and ω_2 with respect to $\{G, F, x_1, \ldots, x_n,\}$ are the same, where $\{x_1, \ldots, x_n\} = [\omega_1] \cup [\omega_2]$.*

As the hyperidentity (6) (from Theorem 7.2.2) is the consequence of the hyperidentity (9) (from Theorem 7.3.2) in the case $F = G$ we receive the following corollary.

Corollary 7.3.4. *In the variety of De Morgan algebras every hyperidentity with no more than two binary functional variables is a consequence of the hyperidentities (1)–(5), (7), (8)–(11) listed in Theorems 7.2.2 and 7.3.2.*

7.4. Hyperidentities of the variety of De Morgan algebras: The case of m binary functional variables

Define the concept of the d_m-canonical word (form) by induction.

Definition 7.4.1. Let $S_1, \ldots, S_{2^{m-1}}$ be complete sets and $F_1, \ldots, F_m$ be binary functional variables. A hyperterm of the following type:

$$F_m(\overline{F}_m(f_{(S_1,\ldots,S_{2^{m-2}})}, f_E(F_1)), \overline{F}_m(f_{(S_{2^{m-2}+1},\ldots,S_{2^{m-1}})}, f_U(F_1))),$$

where

$$f_{(S_1,\ldots,S_{2^{m-2}})} = f_{(S_1,\ldots,S_{2^{m-2}})}(F_1, \ldots, F_{m-1}, \bar{}, x_1, \ldots, x_n)$$

and

$$f_{(S_{2^{m-2}+1},\ldots,S_{2^{m-1}})} = f_{(S_{2^{m-2}+1},\ldots,S_{2^{m-1}})}(F_1, \ldots, F_{m-1}, \bar{}, x_1, \ldots, x_n)$$

are d_{m-1}-canonical words (forms) with respect to $\{F_1, \ldots, F_{m-1}, x_1, \ldots, x_n\}$, and $f_E(F_1)$, $f_U(F_1)$ are d_1-canonical words with respect to $\{F_1, x_1, \ldots, x_n\}$, is called a d_m-canonical word (form) with respect to $\{F_1, \dots, F_m, x_1, \dots, x_n\}$ and it is denoted by $f_{(S_1,\dots,S_{2^{m-1}})}(F_1, \dots, F_m, \bar{;} x_1, \dots, x_n)$ or briefly $f_{(S_1,\dots,S_{2^{m-1}})}(F_1, \dots, F_m)$ or $f_{(S_1,\dots,S_{2^{m-1}})}$ or $f_{S[1,k]}(F_{[1,m]})$ or $f_{S[1,k]}$, where $k = 2^{m-1}$.

Two d_m-canonical words $f_{(S_1,\dots,S_{2^{m-1}})}$ and $f_{(S'_1,\dots,S'_{2^{m-1}})}$ are said to be equal if $S_i = S'_i$ for all $1 \le i \le 2^{m-1}$.

The analogue of Theorem 7.3.2 holds here as well.

Theorem 7.4.2. *For any De Morgan hyperterm ω, $[\omega] \subseteq \{x_1, \dots, x_n\}$, with at most m binary functional variables $F_1, \dots, F_m$ there exists a unique d_m-canonical word $\widetilde{\omega}$ with respect to $\{F_1, \dots, F_m, x_1, \dots, x_n\}$ such that the hyperidentity $\omega = \widetilde{\omega}$ holds in the variety of De Morgan algebras. The word $\widetilde{\omega}$ is called the d_m-canonical form of the hyperterm ω with respect to $\{F_1, \dots, F_m, x_1, \dots, x_n\}$.*

Moreover, $\widetilde{\omega}$ is obtained from the hyperterm ω with a finite number of transitions realized by the following finite number of hyperidentities of the variety of De Morgan algebras, which do not depend on ω: (1)–(5), (7)–(11) *(listed in Theorems 7.2.2 and 7.3.2) and*

(12) $G(F(\overline{F}(\overline{G}(x,e),\overline{e}),\overline{F}(\overline{G}(y,e),\overline{G}(x,G(y,G(u,e))))),e)$
$= F(\overline{F}(\overline{G}(x,e),\overline{e}),\overline{F}(\overline{G}(y,e),\overline{G}(x,G(y,G(u,e))))),$

(13) $G(F(\overline{F}(G(x,e),e),\overline{F}(G(y,e),G(x,G(y,G(\overline{e},e))))),\overline{e}) = \overline{e},$

(14) $G(F(\overline{F}(G(x,e),e),\overline{F}(G(y,e),G(x,G(y,G(u,e))))),$
$F(\overline{F}(G(z,e),e),\overline{F}(G(t,e),G(z,G(t,G(u,e))))))$
$= F(\overline{F}(\overline{G}(\overline{G}(x,e),\overline{G}(z,e)),e),$
$\overline{F}(\overline{G}(\overline{G}(y,e),\overline{G}(t,e)),G(x,G(y,G(z,G(t,G(u,e))))))),$

(15) $\overline{G}(\overline{F}(\overline{F}(G(x,e),e),\overline{F}(G(y,e),G(x,G(y,G(u,e))))),$
$\overline{F}(\overline{F}(G(z,e),e),\overline{F}(G(t,e),G(z,G(t,G(u,e))))))$
$= F(\overline{F}(G(G(x,e),G(z,e)),e),$
$\overline{F}(G(G(y,e),G(t,e)),G(x,G(y,G(z,G(t,G(u,e))))))).$

Proof. For $m = 1$ the theorem is proved. Suppose that it is proved for n binary functional variables for all positive integers $n < m$, and let us prove for m.

First we replace all unary functional variables in the hyperterm ω by negation and exclude all double negations by taking into account the (hyper)identity $\overline{\overline{x}} = x$.

Then using hyperidentities (7.3.1) and (7.3.2) we can modify ω to a form whose tree contains all the binary functional symbols F_m below all the binary functional symbols $F_1, \dots, F_{m-1}$. Then all the subwords (subtrees) which do

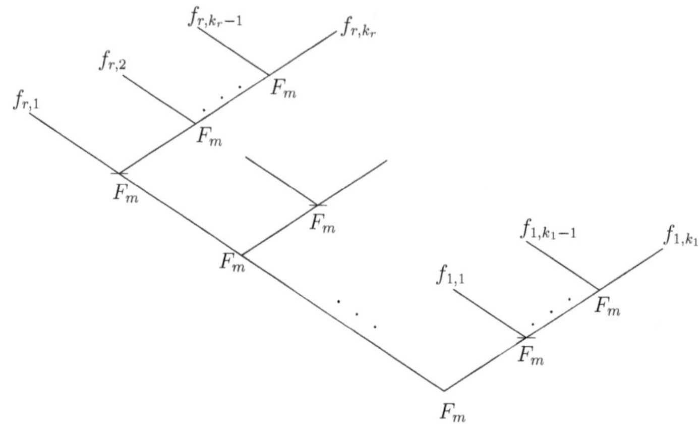

Fig. 7.3.

not include F_m, can be modified to the d_{m-1}-canonical form with respect to $\{F_1, \ldots, F_{m-1}, x_1, \ldots, x_n\}$ by the induction hypothesis. Denote those d_{m-1}-canonical words by $y_1, \ldots, y_p$ and consider this word as a word with one binary functional variable F_m, which can be modified to the d_1-canonical form with respect to $\{F_m, y_1, \ldots, y_p\}$. If subwords of the form $\overline{y}_j$ appear then they can be modified to the d_{m-1}-canonical form as well, because they include no more than $m - 1$ binary functional variables. We get a word of the form of Fig. 7.3. □

Now we need the following properties.

Proposition 7.4.3. *For any d_t-canonical word $f_{S[1,k]}(F_{[1,t]})$ with $k = 2^{t-1}$ and $t \leq m$ the following hyperidentities hold in the variety of De Morgan algebras:*

$$F_1(f_{S[1,k]}, f_E(F_1)) = f_{S[1,k]}, \tag{7.4.1}$$

$$F_1(f_{S[1,k]}, f_U(F_1)) = f_U(F_1). \tag{7.4.2}$$

Proof. We proceed to the proof of both equalities simultaneously by induction on t. For $t = 1$ they are obvious. Suppose (7.4.1) and (7.4.2) are proved for $t - 1$, and let us prove that they hold for t.

Notice that if μ is a word with no more than $t - 1$ binary functional variables, then μ can be modified to d_{t-1}-canonical form as $t - 1 < m$. Hence, from the induction hypothesis we can conclude that $F_1(\mu, f_E(F_1)) = \mu$ and $F_1(\mu, f_U(F_1)) = f_U(F_1)$ for any such word μ. Afterwards we will use this remark many times.

Therefore we have:

$$
\begin{aligned}
&F_t(\overline{F_t}(\overline{F_1}(\overline{f_{S[k/2+1,k]}}, f_E(F_1)), \overline{f_E(F_1)}), \overline{F_t}(\overline{F_1}(\overline{f_{S[1,k/2]}}, f_E(F_1)),\\
&\overline{F_1}(\overline{f_{S[k/2+1,k]}}, F_1(\overline{f_{S[1,k/2]}}, F_1(f_U(F_1), f_E(F_1))))))\\
&\quad = F_t(\overline{F_t}(f_{S[k/2+1,k]}, f_U(F_1)), \overline{F_t}(f_{S[1,k/2]}, f_E(F_1))) = f_{S[1,k]}.
\end{aligned}
$$

Now use the following hyperidentity:

$$
\begin{aligned}
&G(F(\overline{F}(\overline{G}(x,e),\overline{e}), \overline{F}(\overline{G}(y,e), \overline{G}(x, G(y, G(u,e))))), e)\\
&\quad = F(\overline{F}(\overline{G}(x,e),\overline{e}), \overline{F}(\overline{G}(y,e), \overline{G}(x, G(y, G(u,e))))).
\end{aligned}
\tag{7.4.3}
$$

Substituting $x = \overline{f_{S[k/2+1,k]}}$, $y = \overline{f_{S[1,k/2]}}$, $e = f_E(F_1)$, $u = f_U(F_1)$, $F = F_t$, $G = F_1$ we get:

$$
F_1(f_{S[1,k]}, f_E(F_1)) = f_{S[1,k]}.
$$

Further using the hyperidentity

$$
G(F(\overline{F}(G(x,e),e), \overline{F}(G(y,e), G(x, G(y, G(\overline{e}, e))))), \overline{e}) = \overline{e},
\tag{7.4.4}
$$

we get (if we take $x = f_{S[1,k/2]}$, $y = f_{S[k/2+1,k]}$, $e = f_E(F_1)$, $F = F_t$, $G = F_1$):

$$
\begin{aligned}
F_1(f_{S[1,k]}, f_U(F_1)) &\overset{\text{(by induction)}}{=} F_1(F_t(\overline{F_t}(F_1(f_{S[1,k/2]}, f_E(F_1)), f_E(F_1)),\\
&\overline{F_t}(F_1(f_{S[k/2+1,k]}, f_E(F_1)), F_1(f_{S[1,k/2]}, F_1(f_{S[k/2+1,k]},\\
&F_1(\overline{f_E(F_1)}, f_E(F_1)))))), \overline{f_E(F_1)}) = \overline{f_E(F_1)} = f_U(F_1).
\end{aligned}
$$

□

For $S = (S_1, \ldots, S_k)$, $S' = (S'_1, \ldots, S'_k)$ let $S \cup S' = (S_1 \cup S'_1, \ldots, S_k \cup S'_k)$ and $S \cap S' = (S_1 \cap S'_1, \ldots, S_k \cap S'_k)$. Also by $S \subseteq S'$ we mean that $S_i \subseteq S'_i$ for $i = \overline{1,k}$.

Proposition 7.4.4. *If $k = 2^{t-1}$, $t \leq m$, then*

$$
F_1(f_{S[1,k]}, f_{S'[1,k]}) = \begin{cases} f_{S\cup S'[1,k]}, & \text{if } t \text{ is odd,}\\ f_{S\cap S'[1,k]}, & \text{if } t \text{ is even;} \end{cases}
\tag{7.4.5}
$$

$$
\overline{F_1}(\overline{f_{S[1,k]}}, \overline{f_{S'[1,k]}}) = \begin{cases} f_{S\cap S'[1,k]}, & \text{if } t \text{ is odd,}\\ f_{S\cup S'[1,k]}, & \text{if } t \text{ is even.} \end{cases}
\tag{7.4.6}
$$

Proof. We simultaneously prove (7.4.5) and (7.4.6) by induction. For $t = 1$ hyperidentities (7.4.5) and (7.4.6) are obvious. Suppose they are true for $t-1$, and let us prove that they hold for t.

We need the following hyperidentity:

$$\begin{aligned}&G(F(\overline{F}(G(x,e),e),\overline{F}(G(y,e),G(x,G(y,G(u,e))))),\\&\quad = F(\overline{F}(G(z,e),e),\overline{F}(G(t,e),G(z,G(t,G(u,e))))))\\&F(\overline{F}(\overline{G}(\overline{G}(x,e),\overline{G}(z,e)),e),\overline{F}(\overline{G}(\overline{G}(y,e),\\&\quad\overline{G}(t,e)),G(x,G(y,G(z,G(t,G(u,e))))))).\end{aligned} \tag{7.4.7}$$

If t is odd then $t-1$ is even and so:

$$\begin{aligned}&F_1(f_{S[1,k]},f_{S'[1,k]}) \overset{(7.4.1),(7.4.2)}{=} F_1(F_t(\overline{F_t}(F_1(f_{S[1,k/2]},f_E(F_1)),f_E(F_1)),\\&\overline{F_t}(F_1(f_{S[k/2+1,k]},f_E(F_1)),F_1(f_{S[1,k/2]},F_1(f_{S[k/2+1,k]},F_1(f_U(F_1),f_E(F_1)))))),\\&F_t(\overline{F_t}(F_1(f_{S'[1,k/2]},f_E(F_1)),f_E(F_1)),\overline{F_t}(F_1(f_{S'[k/2+1,k]},f_E(F_1)),F_1(f_{S'[1,k/2]},\\&F_1(f_{S'[k/2+1,k]},F_1(f_U(F_1),f_E(F_1)))))))\\&\overset{(7.4.7)}{=} F_t(\overline{F_t}(\overline{F_1}(\overline{F_1}(f_{S[1,k/2]},f_E(F_1)),\overline{F_1}(f_{S'[1,k/2]},f_E(F_1))),f_E(F_1)),\\&\overline{F_t}(\overline{F_1}(\overline{F_1}(f_{S[k/2+1,k]},f_E(F_1)),\overline{F_1}(f_{S'[k/2+1,k]},f_E(F_1))),F_1(f_{S[1,k/2]},\\&F_1(f_{S[k/2+1,k]},F_1(f_{S'[1,k/2]},F_1(f_{S'[k/2+1,k]},F_1(f_U(F_1),f_E(F_1)))))))\\&\overset{(7.4.1),(7.4.2)}{=} F_t(\overline{F_t}(\overline{F_1}(\overline{f_{S[1,k/2]}},\overline{f_{S'[1,k/2]}}),\\&f_E(F_1)),\overline{F_t}(\overline{F_1}(\overline{f_{S[k/2+1,k]}},\overline{f_{S'[k/2+1,k]}}),f_U(F_1)))\\&\overset{\text{(by induction)}}{=} F_t(\overline{F_t}(f_{S\cup S'[1,k/2]},f_E(F_1)),\overline{F_t}(F_1(f_{S\cup S'[k/2+1,k]},f_U(F_1)))\\&= f_{S\cup S'[1,k]}.\end{aligned}$$

The case of even t is considered similarly.

Analogously, (7.4.6) can be proved using the following hyperidentity:

$$\begin{aligned}&\overline{G}(\overline{F}(\overline{F}(G(x,e),e),\overline{F}(G(y,e),G(x,G(y,G(u,e))))),\\&\overline{F}(\overline{F}(G(z,e),e),\overline{F}(G(t,e),G(z,G(t,G(u,e))))))\\&\quad = F(\overline{F}(G(G(x,e),G(z,e)),e),\overline{F}(G(G(y,e),G(t,e)),\\&\quad G(x,G(y,G(z,G(t,G(u,e))))))).\end{aligned} \tag{7.4.8}$$

□

Proposition 7.4.5. *If* $k=2^{t-1}$, $t\le m$, *then*

$$F_t(f_{S[1,k]},f_{S'[1,k]}) = F_t(f_{S\cup S'},f_{S\cap S'}). \tag{7.4.9}$$

Proof. We have:

$$F_t(f_{S[1,k]}, f_{S'[1,k]}) \overset{(7.3.4)}{=} F_t(F_1(f_{S[1,k]}, f_{S'[1,k]}), \overline{F_1}(\overline{f_{S[1,k]}}, \overline{f_{S'[1,k]}})).$$

Now (7.4.9) follows from property (7.4.4). □

Let us continue the proof of the theorem. Using the hyperidentity (7.3.5) we show that

$$F_m(f_{S_1[1,k]}, F_m(f_{S_2[1,k]}, f_{S_3[1,k]}))$$

$$\overset{(7.3.5)}{=} F_m(F_1(f_{S_1[1,k]}, F_1(f_{S_2[1,k]}, f_{S_3[1,k]})), \overline{F_1}(\overline{f_{S_1[1,k]}}, F_1(\overline{f_{S_2[1,k]}}, \overline{f_{S_3[1,k]}})))$$

$$\overset{(7.4.5),\ (7.4.6)}{=} F_m(f_{S_1 \cup S_2 \cup S_3[1,k]}, f_{S_1 \cap S_2 \cap S_3[1,k]}),$$

where $k = 2^{m-2}$, i.e., $f_{S_1[1,k]}, f_{S_2[1,k]}, f_{S_3[1,k]}$ are d_{m-1}-canonical words. Further, we will denote a d_{m-1}-canonical word $f_{S[1,k]}$ concisely by f_S.

Then proceeding by induction (or using the analogue of the hyperidentity (7.3.5) for l objective variables) we prove that:

$$F_m(f_{S_1}, F_m(\ldots F_m(f_{S_{l-1}}, f_{S_l})\ldots)) = F_m(f_{S_1 \cup \ldots \cup S_l}, f_{S_1 \cap \ldots \cap S_l})$$

for any positive integer l.

After that the hyperterm ω is modified to the form of Fig. 7.4, where $S_i \subseteq S_i^*$.

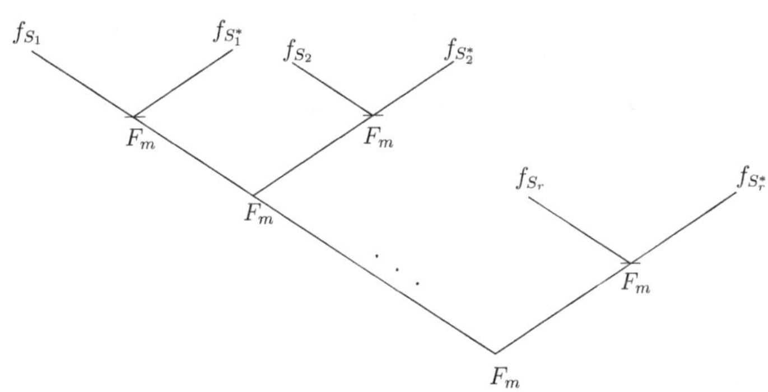

Fig. 7.4.

Further, using property (7.4.3) and hyperidentity (7.3.6), as in the third section we can show that for an even m we have:

$$F_m(\overline{F_m}(f_{S_1}, f_{S_1^*}), \overline{F_m}(f_{S_2}, f_{S_2^*})) = F_m(\overline{F_m}(f_{S_1 \cup S_2}, f_U(F_1)), \overline{F_m}(f_E(F_1), f_{S_1^* \cap S_2^*})).$$

(we use the fact that $m-1$ is odd, f_{S_i} and $f_{S_i^*}$ ($i = 1, 2$) are d_{m-1}-canonical words).

If m is odd, then

$$\begin{aligned}
&F_m(\overline{F_m}(f_{S_1}, f_{S_1^*}), \overline{F_m}(f_{S_2}, f_{S_2^*})) \overset{(7.4.5),(7.4.6),(7.4.1),(7.4.2)}{=} F_m(\overline{F_m}(F_1(f_{S_1^*}, f_E(F_1)),\\
&F_1(f_{S_1}, F_1(f_{S_1^*}, f_E(F_1)))), \overline{F_m}(F_1(f_{S_2^*}, f_E(F_1)), F_1(f_{S_2}, F_1(f_{S_2^*}, f_E(F_1)))))\\
&\overset{(7.3.6)}{=} F_m(\overline{F_m}(F_1(f_{S_1^*}, F_1(f_{S_2^*}, f_E(F_1))),\\
&F_1(f_{S_1^*}, F_1(f_{S_1}, F_1(f_{S_2^*}, F_1(f_U(F_1), f_E(F_1)))))),\\
&\overline{F_m}(\overline{F_1}(\overline{F_1}(f_{S_1^*}, F_1(f_{S_1}, f_E(F_1))), \overline{F_1}(f_{S_2^*}, F_1(f_{S_2}, f_E(F_1)))), f_E(F_1)))\\
&\overset{(7.4.5),(7.4.6),(7.4.1),(7.4.2)}{=} F_m(\overline{F_m}(f_{S_1^* \cap S_2^*}, f_U(F_1)), \overline{F_m}(f_{S_1 \cup S_2}, f_E(F_1))).
\end{aligned}$$

Further, for an odd m we have (as in the third section):

$$\begin{aligned}
&F_m(\overline{F_m}(f_{S_1}, f_{S_1^*}), F_m(\overline{F_m}(f_{S_2}, f_{S_2^*}), \overline{F_m}(f_{S_3}, f_{S_3^*})))\\
&\quad = F_m(\overline{F_m}(f_{S_1^* \cap S_2^* \cap S_3^*}, f_U(F_1)), \overline{F_m}(f_E(F_1), f_{S_1 \cup S_2 \cup S_3})).
\end{aligned}$$

We prove (by induction) that it is true for all $r > 1$, and in this way we modify the hyperterm ω to the d_m-canonical form (for $r > 1$).

The case for an even m is considered analogously.

Now only the case $r = 1$ is left. In this case we have (if m is odd)

$$\overline{F_m}(f_{S_1^*}, f_{S_1}) \overset{(7.3.8)}{=} F_m(\overline{F_m}(f_{S_1^*}, f_U(F_1)), \overline{F_m}(f_{S_1}, f_E(F_1))).$$

And again for an even m everything is analogous.

Thus the first part of the theorem is proved. Now let us prove the second part — the uniqueness. We proceed to the proof by induction. For $m = 1$ the uniqueness was proved in Theorem 7.2.2. Now we assume the uniqueness is true for $m-1$ and prove it for m.

For $\alpha \in \{+, \cdot\}$ denote

$$\alpha^* = \begin{cases} + & \text{if } \alpha = \cdot, \\ \cdot & \text{if } \alpha = +. \end{cases}$$

Suppose that the following hyperidentity holds in the variety of De Morgan algebras: $f_{(S_1,\ldots,S_{2^{m-1}})}(F_1, \ldots, F_m, \bar{}, x_1, \ldots, x_n) = f_{(S'_1,\ldots,S'_{2^{m-1}})}(F_1, \ldots, F_m, \bar{}, x_1, \ldots, x_n)$ and $S_i \neq S'_i$ for some $i \in \{1, \ldots, 2^{m-1}\}$.

Let $i > 2^{m-2}$. Then by the induction hypothesis there exists a De Morgan algebra $\mathfrak{D}$ such that the following hyperidentity does not hold in it: $f_{(S_{2^{m-2}+1},\dots,S_{2^{m-1}})}(F_1,\dots,F_{m-1},\bar{},x_1,\dots,x_n) = f_{(S'_{2^{m-2}+1},\dots,S'_{2^{m-1}})}(F_1,\dots,F_{m-1},\bar{},x_1,\dots,x_n)$.

It means that there are such values of $x_1,\dots,x_n \in \mathfrak{D}$ and $F_1,\dots,F_{m-1} \in \{+,\cdot\}$ for which the equality is not true. Denote these values by $a_1,\dots,a_n,\alpha_1,\dots,\alpha_{m-1}$ respectively. Take $F_m = \alpha_1^*$. Then, using property (7.4.3) we get:

$$\begin{aligned}
&f_{(S_1,\dots,S_{2^{m-1}})}(\alpha_1,\dots,\alpha_{m-1},\alpha_1^*,\bar{},a_1,\dots,a_n)\\
&= \alpha_1^*(\overline{\alpha_1^*}(f_{S[1,k/2]}, f_E(\alpha_1)), \overline{\alpha_1^*}(f_{S[k/2+1,k]}, f_U(\alpha_1)))\\
&= \alpha_1(\overline{\alpha_1}(\overline{f_{S[1,k/2]}}, \overline{f_E(\alpha_1)}), \overline{\alpha_1}(\overline{f_{S[k/2+1,k]}}, \overline{f_U(\alpha_1)}))\\
&= \overline{f_{(S_{2^{m-2}+1},\dots,S_{2^{m-1}})}(\alpha_1,\dots,\alpha_{m-1},\bar{},a_1,\dots,a_n)}\\
&\neq \overline{f_{(S'_{2^{m-2}+1},\dots,S'_{2^{m-1}})}(\alpha_1,\dots,\alpha_{m-1},\bar{},a_1,\dots,a_n)}\\
&= \overline{\alpha}_1(\overline{\alpha_1}(\overline{f_{S'[1,k/2]}}, \overline{f_E(\alpha_1)}), \overline{\alpha_1}(\overline{f_{S'[k/2+1,k]}}, \overline{f_U(\alpha_1)}))\\
&= \alpha_1^*(\overline{\alpha_1^*}(f_{S'[1,k/2]}, f_E(\alpha_1)), \overline{\alpha_1^*}(f_{S'[k/2+1,k]}, f_U(\alpha_1)))\\
&= f_{(S'_1,\dots,S'_{2^{m-1}})}(\alpha_1,\dots,\alpha_{m-1},\alpha_1^*,\bar{},a_1,\dots,a_n).
\end{aligned}$$

Contradiction!

The case $i \leq 2^{m-2}$ is considered analogously (but we take $F_m = \alpha_1$ here). □

Corollary 7.4.6. *In the variety of De Morgan algebras the hyperidentity $\omega_1 = \omega_2$ with binary functional variables $F_1,\dots,F_m$ is satisfied if and only if the d_m-canonical forms of hyperterms ω_1 and ω_2 with respect to $\{F_1,\dots,F_m,x_1,\dots,x_n\}$ are the same, where $\{x_1,\dots,x_n\} = [\omega_1] \cup [\omega_2]$.*

Corollary 7.4.7. *The variety of De Morgan algebras has a decidable hyperequational theory.*

Corollary 7.4.8. *The variety of De Morgan algebras has a finite base of hyperidentities. Namely every hyperidentity of the variety of De Morgan algebras is a consequence of the following hyperidentities: (1)–(5), (7)–(11) (from Corollary 7.3.4).*

Proof. Let H_m be the set of all hyperidentities with at most m binary functional variables. It is clear that $H_1 \subseteq H_2 \subseteq \cdots$. From the proof of Theorem 7.4.2 it follows that the hyperidentities of H_m are consequences of the hyperidentities of $H_{m-1} \cup \cdots \cup H_2 \cup H_1$ for $m \geq 3$. Now if $m \geq 4$ then the same is true for $m-1$, hence the hyperidentities of H_m are the consequences of the hyperidentities of $H_{m-2} \cup \cdots \cup H_2 \cup H_1$. Proceeding in this way, we conclude that the hyperidentities of H_m are consequences of the hyperidentities of H_2. Now we can apply Corollary 7.3.4 of Theorem 7.3.2. □

7.5. On the hyperequational theory of the variety of De Morgan algebras

It is proved in Chapter 5 that the hyperequational theory of the variety of Boolean algebras is one-based. However, for the variety of De Morgan algebras the similar result is not valid.

The following result on ordinary equational theories is proved in [233]. Let Λ be the equational theory based on the lattice axioms, and Ω be the equational theory based on the single equation $x = y$.

Theorem 7.5.1.

(1) *Let θ be a finitely based equational theory. Assume that for some term $\tau(x_1, \ldots, x_k)$ $(k \geq 2)$ θ contains each one of the equations:*

$$x = \tau(y, x, \ldots, x),\ \ x = \tau(x, y, \ldots, x), \ldots, x = \tau(x, x, \ldots, y).$$

Assume also that θ has a base composed solely of equations of the form $x = \sigma$. Under this assumptions θ is one-based.

(2) *Λ is one-based equational theory. Moreover, there does not exist a one-based equational theory Γ with $\Lambda \subseteq \Gamma \subseteq \Omega$ and $\Gamma \neq \Lambda$, $\Gamma \neq \Omega$.*

Using these results we prove the following theorem.

Theorem 7.5.2. *The hyperequational theory of the variety of De Morgan algebras is not one-based (cf. Theorem 5.7.9).*

Proof. We proceed to the proof of the theorem by contradiction. Assume there exists a hyperidentity:

$$w_1 = w_2, \tag{7.5.1}$$

which holds in the variety of De Morgan algebras and it is a base of hyperidentities of this variety. First we claim that (7.5.1) has the form:

$$x = \tau, \tag{7.5.2}$$

where x is an objective variable. To show this suppose that it is not true. So the hyperterms w_1 and w_2 have the form $F(\omega_1, \omega_2)$ or $\overline{\omega}$ where F is a binary functional variable. Consider the algebra: $\mathfrak{F} = (\{0,1\}, \{+, ^-\})$ with $0+0 = 0+1 = 1+0 = 1+1 = 1$, $\overline{0} = \overline{1} = 1$. It is clear that (7.5.1) holds in $\mathfrak{F}$, but the hyperidentity $F(x,x) = x$, which holds in the variety of De Morgan algebras, does not hold in $\mathfrak{F}$. We arrived at a contradiction, therefore (7.5.1) has the form (7.5.2).

Now if we replace all binary functional variables in the word τ by the operations $+$ or $\cdot$, then we will get identities:

$$\begin{cases} x = \tau_1, \\ x = \tau_2, \\ \quad \vdots \\ x = \tau_n. \end{cases} \tag{7.5.3}$$

The hyperidentity (7.5.2) holds in a T-algebra $Q(+, \cdot, ^-)$ $(T = \{1,2\})$ if and only if the identities (7.5.3) hold in it.

Now let $\mathfrak{L} = Q(+, \cdot)$ be an arbitrary distributive lattice. Consider the algebra $\mathfrak{U} = (Q \times Q, \{+, \cdot, ^-\})$, where the operations are defined as follows:

$$\begin{cases} (x_1, x_2) + (y_1, y_2) = (x_1 + y_1, x_2 \cdot y_2), \\ (x_1, x_2) \cdot (y_1, y_2) = (x_1 \cdot y_1, x_2 + y_2), \\ \qquad\qquad \overline{(x_1, x_2)} = (x_2, x_1). \end{cases} \tag{7.5.4}$$

It is easy to verify that $\mathfrak{U}$ is a De Morgan algebra. Hence, the hyperidentity (7.5.1) holds in it, and therefore the identities (7.5.3) also hold in $\mathfrak{U}$. If we replace the objective variables of the words $\tau_1, \ldots, \tau_n$ by the elements of the set $Q \times Q$ and also replace all the subwords of the form $\overline{(u_1, u_2)}$ by (u_2, u_1), then we will get the identities:

$$\begin{cases} (x_1, x_2) = (\tau_{11}, \tau_{12}), \\ (x_1, x_2) = (\tau_{21}, \tau_{22}), \\ \qquad \vdots \\ (x_1, x_2) = (\tau_{n1}, \tau_{n2}), \end{cases} \tag{7.5.5}$$

where τ_{i1}, τ_{i2} are the lattice words, i.e. the words with the operations $+$ and $\cdot$. So the identities:

$$\begin{cases} x_1 = \tau_{11}, \\ x_2 = \tau_{12}, \\ \quad\vdots \\ x_1 = \tau_{n1}, \\ x_2 = \tau_{n2} \end{cases} \tag{7.5.6}$$

hold in $\mathfrak{L}$. As $\mathfrak{L}$ is an arbitrary distributive lattice, we conclude that the identities (7.5.6) hold in the variety of distributive lattices.

Now we prove that the identity of distributivity is the consequence of identities (7.5.6), i.e., if an algebra $\mathfrak{A} = A(+,\cdot)$ satisfies (7.5.6), then it also satisfies the identity:

$$x(y+z) = xy + xz. \tag{7.5.7}$$

To show this consider the algebra $\mathfrak{W} = (A \times A, \{+,\cdot,^{-}\})$, where the operations are defined according to (7.5.4). The algebra $\mathfrak{A} = A(+,\cdot)$ satisfies (7.5.6), therefore $\mathfrak{W}$ satisfies (7.5.3); hence, the hyperidentity (7.5.2) holds in it. Therefore, all hyperidentities of the variety of De Morgan algebras hold in $\mathfrak{W}$. In particular, $\mathfrak{W}$ satisfies the hyperidentity of distributivity. Hence, $(x_1,x_2)((y_1,y_2)+(z_1,z_2)) = (x_1,x_2)(y_1,y_2)+(x_1,x_2)(z_1,z_2)$ for any $x_1,x_2,y_1,y_2,z_1,z_2 \in A$. So $x_1(y_1+z_1) = x_1y_1 + x_1z_1$ for all $x_1,y_1,z_1 \in A$, i.e., $\mathfrak{A}$ satisfies (7.5.7).

According to Theorem 7.5.1 there exists an identity

$$x = \sigma \tag{7.5.8}$$

of the variety of lattices, which is a base for the equational theory of that variety. Therefore, identities (7.5.6) and (7.5.8) form a base for the equational theory of the variety of distributive lattices. Hence, by Theorem 7.5.1 the equational theory of the variety of distributive lattices is one-based (we can take $\tau(x_1,x_2,x_3) = x_1x_2 + x_2x_3 + x_3x_1$). But it follows from the second part of the Theorem 7.5.1 that equational theory of the variety of distributive lattices is not one-based. Contradiction! Hence, the hyperequational theory of the variety of De Morgan algebras is not one-based. □

Chapter 8

Functional Completeness Theorem for De Morgan Functions

8.1. Boolean functions case

In this section we introduce the notion of complete set of Boolean functions (see [34,58,59,68,72,316,333,334,443]).

A *Boolean function* in the variables $x_1, \ldots, x_n$ is a map $f : \{0,1\}^n \to \{0,1\}$. The collection of all n-ary Boolean functions is denoted by $\mathcal{B}_n$; clearly $|\mathcal{B}_n| = 2^{2^n}$.

Some examples of Boolean functions include:

- the 0-ary constant functions 0 and 1;
- the unary function $\neg$ (negation);
- the binary functions $\vee$ (OR), $\wedge$ (AND), $\oplus$ (EXCLUSIVE-OR, where $x \oplus y$ is defined as $(x \wedge \neg y) \vee (\neg x \wedge y)$), $\to$ (implication, where $x \to y$ is defined by $\neg x \vee y$), and $\equiv$ (equivalence, also called biimplication).

A set $\{f_1, \ldots, f_s\}$ of Boolean functions is called *complete* for the class of boolean functions if every Boolean function f can be obtained as the superposition (see the next section for definition) of these functions. A complete set is called a *basis* if in addition it is minimal, i.e., no proper subset is complete.[a] The following are complete sets of Boolean functions:

- $\{\vee, \wedge, \neg, 0, 1\}$,
- $\{\vee, \neg, 0, 1\}$,
- $\{\wedge, \neg, 0, 1\}$,
- $\{\oplus, \wedge, 0, 1\}$,

[a]By an abuse of terminology, a non-minimal complete set of connectives is often called a basis as well.

- $\{|, 0, 1\}$ where $|$ denotes NOR; i.e., $x|y = \neg x$ if $x = y$, and $= 0$ otherwise.

In this section, we characterize the complete sets of Boolean functions. To this end, we distinguish the following characteristic classes of boolean functions:

- T_n^0: The class of all Boolean functions $f \in \mathcal{B}_n$ satisfying $f(0^n) = 0$.
- T_n^1: The class of all Boolean functions $f \in \mathcal{B}_n$ satisfying $f(1^n) = 1$.
- D_n (*Self-dual*): The class of all Boolean functions $f \in \mathcal{B}_n$ satisfying

$$f(x_1, \dots, x_n) = \neg f(\neg x_1, \dots, \neg x_n).$$

- M_n (*Monotone*): The class of all Boolean functions $f \in \mathcal{B}_n$ satisfying

$$x \leqslant y \Rightarrow f(x) \leq f(y),$$

where for n tuples $x = (x_1, \dots, x_n)$, $y = (y_1, \dots, y_n)$ we define $x \leqslant y$ if and only if $(\forall i \leqslant n)(x_i \leqslant y_i)$.[b] If $u = (u_1, \dots, u_n)$, $v = (v_1, \dots, v_n) \in B^n$ then we write $u \preceq v$ if there exists k $(1 \leq k \leq n)$ such that $u_i = v_i$ for all $i \neq k$ and $u_k = 0$, $v_k = 1$. Clearly, the following lemma is valid.

Lemma 8.1.1. *A Boolean function $f : B^n \to B$ is monotone if and only if*

$$x \preceq y \Rightarrow f(x) \leq f(y),$$

where $x, y \in B^n$.

- L_n (*Linear*): The class of all Boolean functions $f \in \mathcal{B}_n$ of the form

$$f(x) = \sum_{i=1}^{k} b_i x_i \bmod 2,$$

where $b_i \in \{0, 1\}$, for $i = 1, \dots, k$.

Now we can formulate the following classical Post's functional completeness theorem for Boolean functions (Post [333,334]).

Theorem 8.1.2. *A class $\mathcal{C} \subseteq \mathcal{B}_n$ of Boolean functions is complete if and only if $\mathcal{C}$ is not a subset of any of the classes T_n^0, T_n^1, D_n, M_n, L_n.*

---

[b]Throughout we write $(\forall i \leqslant n)$ rather than the more cumbersome, but correct form $(\forall 1 \leqslant i \leqslant n)$ or $(\forall i \in \{1, \dots, n\})$

The condition $\mathcal{C} \nsubseteq T_n^0, T_n^1, D_n, M_n, L_n$ is clearly necessary. To prove sufficiency choose Boolean functions

$$f_T^i \in \mathcal{C} \backslash T_n^i, \quad f_D \in \mathcal{C} \backslash D_n, \quad f_M \in \mathcal{C} \backslash M_n, \quad f_L \in \mathcal{C} \backslash L_n.$$

Since

$$\{f_T^0, f_T^1, f_D, f_M, f_L\} \subseteq \mathcal{C}$$

it is enough to show that the set $\{f_T^0, f_T^1, f_D, f_M, f_L\}$ is complete. In turn this will follow from the fact that these functions can generate the constants 0, 1 and the functions $\neg x$, $x \wedge y$.

8.2. Closed and complete classes of De Morgan functions

In this section, we define the concepts of closed and complete classes of De Morgan functions analogously to Boolean functions.

First, we define essential and inessential variables for functions mapping D^n into D as in the case of Boolean functions.

Definition 8.2.1. The variable x_k is called *essential* for the function $f(x_1, \dots, x_n)$ if there exist such values $\alpha_1, \dots, \alpha_{k-1}, \alpha_{k+1}, \dots, \alpha_n, \beta, \gamma \in D$ that

$$f(\alpha_1, \dots, \alpha_{k-1}, \beta, \alpha_{k+1}, \dots, \alpha_n) \neq f(\alpha_1, \dots, \alpha_{k-1}, \gamma, \alpha_{k+1}, \dots, \alpha_n).$$

In other words we can say that x_k is called essential for the function $f(x_1, \dots, x_n)$ if there exist such values $\alpha_1, \dots, \alpha_{k-1}, \alpha_{k+1}, \dots, \alpha_n \in D$ of the variables $x_1, \dots, x_{k-1}, x_{k+1}, \dots, x_n$, that the function $g(x) = f(\alpha_1, \dots, \alpha_{k-1}, x, \alpha_{k+1}, \dots, \alpha_n)$ is not constant function. In this case, we say that the function f essentially depends on the variable x_k. A variable x_k is called *inessential* for the function f if it is not an essential variable for f.

Afterwards we will consider De Morgan functions up to inessential variables. This allows us to consider a function of n variables as a function of more or fewer variables.

Denote by $\mathfrak{D}$ the set of all De Morgan functions, i.e.,

$$\mathfrak{D} = \bigcup_n \mathcal{D}_n.$$

Let $f(x_1, \dots, x_m), f_1, \dots, f_m$ be De Morgan functions. Clearly we can suppose that $f_1, \dots, f_m$ are functions of the same variables $y_1, \dots, y_p$, i.e., $f_i = f_i(y_1, \ldots, y_p)$, $i = 1, \ldots, m$. Then the function

$$g(y_1, \ldots, y_p) = f(f_1(y_1, \ldots, y_p), \ldots, f_m(y_1, \ldots, y_p))$$

is a mapping from D^p into D (we will denote it by $f(f_1, \ldots, f_m))$). We claim that g is a De Morgan function too. This follows from the fact that De Morgan functions are precisely the term functions of the algebra **4** (we can also deduce this argument from the definition of De Morgan functions as functions preserving the relations σ and ρ).

Let us fix the alphabet of variables $U = \{x, x_1, x_2, \ldots, x_n, \ldots, y, y_1, y_2, \ldots, y_n, \ldots, z, z_1, z_2, \ldots, z_n, \ldots\}$. Let $\mathfrak{N} \subseteq \mathfrak{D}$ be a non-empty set of De Morgan functions. Below by induction we define the concept of *formula with respect to* $\mathfrak{N}$. Also we will correspond a De Morgan function to each formula, which is said to be *realized by* that formula.

If $f \in \mathfrak{N}$ is an n-ary function and $x_1, \ldots, x_n$ are variables then the expression $f(x_1, \ldots, x_n)$ is a formula with respect to $\mathfrak{N}$. We correspond the function f to the formula $f(x_1, \ldots, x_n)$.

Let $g \in \mathfrak{N}$ be an m-ary function and $A_1, \ldots, A_m$ be variables or formulas with respect to $\mathfrak{N}$. Then the expression $g(A_1, \ldots, A_m)$ is a formula with respect to $\mathfrak{N}$. Suppose the functions f_i are realized by those formulas A_i which are not variables. If $A_j = x_{p_j}$ is a variable then take f_j to be the projection map on the p_jth coordinate. Now we correspond the function $g(f_1, \ldots, f_m)$ to the formula $g(A_1, \ldots, A_m)$.

There are no other formulas with respect to $\mathfrak{N}$.

Note that the concepts of a formula with respect to $\mathfrak{N}$ and a function realized by a formula are essentially the same as the concepts of a term of algebra $(D; \mathfrak{N})$ and a term operation. The only difference between terms and formulas lies in the fact that variable symbols are terms but they are not formulas. To avoid this problem we can just consider only those operations that are induced by terms with at least one occurrence of a functional symbol.

If the function f is realized by a formula which consists only of the functional symbols $f_1, \ldots, f_n$ and also variable symbols then we say that f is the superposition of the functions $f_1, \ldots, f_n$. As we saw above the superposition of De Morgan functions is also a De Morgan function, i.e., the set of all De Morgan functions is closed under superposition.

Definition 8.2.2. Let $\mathfrak{N} \subseteq \mathfrak{D}$ be a set of De Morgan functions. The set of all functions that can be realized by formulas with respect to $\mathfrak{N}$, is called the *closure* of $\mathfrak{N}$ and it is denoted by $[\mathfrak{N}]$.

In other words $[\mathfrak{N}]$ is the set of all functions which can be presented as superpositions of functions from the class $\mathfrak{N}$.

Here it is useful to compare the set of all functions that can be realized by formulas with respect to $\mathfrak{N}$ to the clone generated by $\mathfrak{N}$. The only difference between those two objects is in the fact that projections should belong to the clone, but they are not necessarily present in $[\mathfrak{N}]$. In particular, if $\mathfrak{N}$ contains the projection functions then $[\mathfrak{N}]$ is the clone generated by $\mathfrak{N}$.

Definition 8.2.3. The class $\mathfrak{N} \subseteq \mathfrak{D}$ is called *(functionally) closed* (with respect to superpositions) if $[\mathfrak{N}] = \mathfrak{N}$. The class $\mathfrak{N} \subseteq \mathfrak{D}$ is called *(functionally) complete* or *complete in* $\mathfrak{D}$ if $[\mathfrak{N}] = \mathfrak{D}$.

Note that if a closed class contains the projection functions then it is a clone.

For example, the class of all De Morgan functions is closed and also complete. The class $\{0, 1, \overline{x}, x \cdot y, x + y\}$ is complete. This follows from the fact that every non-constant De Morgan function can be presented in DNF (CNF). And as $x + y = \overline{\overline{x} \cdot \overline{y}}$, $x \cdot y = \overline{\overline{x} + \overline{y}}$, $1 = \overline{0}$, we conclude that the classes $\{0, \overline{x}, x + y\}$ and $\{0, \overline{x}, x \cdot y\}$ are complete too.

In the next section, we consider some important closed classes of De Morgan functions and establish a criterion to determine whether the given class of De Morgan functions is complete.

Note that if the identical function $\varepsilon(x) = x$ belongs to a class $\mathfrak{N}$ then to show that $\mathfrak{N}$ is closed it is sufficient to show that if $f, f_1, \ldots, f_n \in \mathfrak{N}$, where f is an n-ary function, then $f(f_1, \ldots, f_n) \in \mathfrak{N}$.

Now let us study some further properties of De Morgan functions which will be used in the next section.

First notice that if $f(x_1, \ldots, x_n)$ is a non-constant De Morgan function, then it can be presented in DNF and so $f(a, \ldots, a) = a$, $f(b, \ldots, b) = b$.

Lemma 8.2.4. *Let f be a De Morgan function. The following statements hold:*

(1) (i) *if we switch the values of some variables of f from 0 to a or from b to 1 or 0 and if the value of function is not b then it can remain the same or switch to a;*

(ii) *if we switch the values of some variables of f from 0 to b or from a to 1 or 0 and if the value of function is not a then it can remain the same or switch to b;*

(2) (i) *if the values of all variables are a's, 1's or 0's then the value of f is not b;*

(ii) *if the values of all variables are b's, 1's or 0's then the value of f is not a;*

(3) (i) *if the values of all variables are 0's, a's or 1's and the value of f is 0 or 1 then if we switch some a's to 0's or 1's then the value of f remains the same;*

(ii) *if the values of all variables are 0's, b's or 1's and the value of f is 0 or 1 then if we switch some b's to 0's or 1's then the value of f remains the same.*

Proof. The first statement follows immediately from the fact that De Morgan functions preserve the relations ρ and ρ^{-1}.

The second statement follows from the fact that De Morgan functions are the term functions of the algebra **4**.

The third statement follows from the second statement and from the fact that f preserves the order relations ρ and ρ^{-1}. □

8.3. Some important closed classes of De Morgan functions

In this section, we define some classes of De Morgan functions and prove that they are closed. Those classes are important in formulation of the main theorem in the next section.

For $u \in D$ denote by $\mathfrak{T}_u$ the set of all De Morgan functions which preserve the value u, i.e.,

$$f \in \mathfrak{T}_u \Leftrightarrow f(u, u, \dots, u) = u.$$

Thus $f \in \mathfrak{T}_u$ if and only if f preserves the unary relation $\{u\}$, and hence the classes $\mathfrak{T}_u$ are closed under superposition. Taking into account that the identical function $\varepsilon(x) = x$ belongs to $\mathfrak{T}_u$ we conclude that the class $\mathfrak{T}_u$ for all $u \in D$ is closed.

For example, $x \in \mathfrak{T}_u$ for all $u \in D$, i.e., x preserves all values. The function $x \cdot \overline{x}$ preserves 0, a and b, but does not preserve 1. The function $x + \overline{y}$ preserves 1, a and b and does not preserve 0.

As we saw in the previous section, $\mathfrak{T}_a$ and $\mathfrak{T}_b$ consist exactly of all non-constant functions, i.e., $\mathfrak{T}_a = \mathfrak{T}_b$. Denote

$$\mathfrak{T} = \mathfrak{T}_a = \mathfrak{T}_b = \mathfrak{D} \setminus \{0, 1\}.$$

The next class is the class of monotone De Morgan functions. The definition of monotone De Morgan functions is analogous to definition of monotone Boolean functions (Definition 4.1.4), only we should by "$\leq$" mean the order of De Morgan algebra **4** instead of the order of Boolean algebra **2**. So a De Morgan function is *monotone* if and only if it preserves the order $\{(0,0),(0,a),(0,b),(0,1),(a,a),(a,1),(b,b),(b,1),(1,1)\} \subseteq D^2$.

For example, the functions $0,\ 1,\ x,\ x+y,\ x\cdot y$ are monotone De Morgan functions and the De Morgan functions $\overline{x},\ \overline{x}+y$ are not.

Denote the class of all monotone De Morgan functions by $\mathfrak{M}$. Obviously $\mathfrak{M}$ is a closed class.

It is easy to see that if f and g are monotone De Morgan functions then $f\cdot g$ and $f+g$ are monotone too. Hence all De Morgan functions of the following form are monotone De Morgan functions:

$$\sum_{s\in S}\prod_{i\in s} x_i, \tag{8.3.1}$$

where $S \subseteq 2^{\{1,\dots,n\}}$ is an antichain. We will prove that all monotone De Morgan functions can be presented in the form (8.3.1).

Theorem 8.3.1. *A De Morgan function $f(x_1,\dots,x_n)$ is monotone if and only if the corresponding monotone Boolean function $\varphi_f(t_1,\dots,t_{2n})$ does not essentially depend on variables $t_{n+1},\dots,t_{2n}$.*

Proof. Let φ_f does not essentially depend on last n variables. Then we have that $\varphi_f(t_1,\dots,t_n) = \sum_{s\in S}\prod_{i\in s} t_i$ for some antichain $S\subseteq 2^{\{1,\dots,n\}}$. Therefore, we can conclude that f has the form (8.3.1). Hence f is monotone.

Now suppose f is monotone. Let us prove that $\varphi_f(t_1,\dots,t_{2n})$ does not essentially depend on variables $t_{n+1},\dots,t_{2n}$. Suppose this is not true. Then there exists an integer k with $n+1\leq k\leq 2n$ such that t_k is an essential variable for φ_f. This means that there are $\alpha_1,\dots,\alpha_{k-1},\alpha_{k+1},\dots,\alpha_{2n}\in\{0,1\}$ such that

$$\varphi_f(\alpha_1,\dots,\alpha_{k-1},0,\alpha_{k+1},\dots,\alpha_{2n})=0,\quad \varphi_f(\alpha_1,\dots,\alpha_{k-1},1,\alpha_{k+1},\dots,\alpha_{2n})=1.$$

Denote $\tau_i = (\alpha_i,\overline{\alpha_{n+i}})\in D$ for $i=1,\dots,n,\ i\neq k-n,\ \mu=(\alpha_{k-n},0)\in D,\ \nu=(\alpha_{k-n},1)\in D$. And let $\xi=(\tau_1,\dots,\tau_{k-n-1},\mu,\tau_{k-n+1},\dots,\tau_n)\in D^n,\ \zeta=(\tau_1,\dots,\tau_{k-n-1},\nu,\tau_{k-n+1},\dots,\tau_n)\in D^n$. Obviously $\xi\leq\zeta$. Now we have:

$$f(\xi) = (\varphi_f(\alpha_1,\dots,\alpha_{k-1},1,\alpha_{k+1},\dots,\alpha_{2n}),\lambda_1) = (1,\lambda_1),$$
$$f(\zeta) = (\varphi_f(\alpha_1,\dots,\alpha_{k-1},0,\alpha_{k+1},\dots,\alpha_{2n}),\lambda_2) = (0,\lambda_2).$$

Here we denoted by λ_1 and λ_2 the corresponding second components (we did not evaluate them as we do not need them). As f is monotone and $\xi \leq \zeta$, we should have $(1, \lambda_1) \leq (0, \lambda_2)$. But this is a contradiction. $\square$

Corollary 8.3.2. *A De Morgan function $f : D^n \to D$ is monotone if and only if f can be presented in the form (8.3.1).*

In other words, this means that $[\{0, 1, x \cdot y, x + y\}] = \mathfrak{M}$.

Corollary 8.3.3. *There are exactly m_n monotone De Morgan functions of n variables.*

Thus for monotone De Morgan functions we get the same form as for monotone Boolean functions. If $\mathfrak{M}_n$ is the set of all monotone De Morgan functions of n variables, then the algebra $(\mathfrak{M}_n; \{+, \cdot\})$ is a distributive lattice. Moreover, $(\mathfrak{M}_n; \{+, \cdot\})$ is the free distributive lattice with system of free generators $\Delta = \{\delta_n^1, \ldots, \delta_n^n\}$. Hence, the distributive lattice of monotone De Morgan functions of n variables is isomorphic to the distributive lattice of monotone Boolean functions of n variables.

Recall that a closed class $\mathfrak{R} \neq \mathfrak{D}$ of De Morgan functions is called a *maximal* closed class in the class $\mathfrak{D}$ of all De Morgan functions or simply a *maximal* closed class, if there does not exist a closed class $\mathfrak{P} \subseteq \mathfrak{D}$ with $\mathfrak{R} \subseteq \mathfrak{P}$ and $\mathfrak{R} \neq \mathfrak{P}$, $\mathfrak{P} \neq \mathfrak{D}$. The class $\mathfrak{M}$ of all monotone De Morgan functions is not a maximal closed class. To prove this we define a larger class of De Morgan functions, which we call quasimonotone functions.

Definition 8.3.4. A De Morgan function $f : D^n \to D$ is called *quasi-monotone* if $f|_{\{0,1\}^n}$ is a monotone Boolean function, i.e., if for any two n-tuples $\alpha = (\alpha_1, \ldots, \alpha_n)$, $\beta = (\beta_1, \ldots, \beta_n) \in \{0, 1\}^n$ with $\alpha \preceq \beta$ we have $f(\alpha) \leq f(\beta)$.

Denote the set of all quasimonotone De Morgan functions by $\mathfrak{QM}$. Hence $f \in \mathfrak{QM}$ if and only if f preserves the order relation $\{(0, 0), (0, 1), (1, 1)\} \subseteq D^2$.

For example, the De Morgan functions 0, 1, x, $x+y$, $x\cdot y$, $x\cdot y+z+y\cdot\overline{y}$ are quasimonotone and the De Morgan functions $x\cdot\overline{y}$, $\overline{x}$, $\overline{x}+y+\overline{z}$, $x+\overline{y}+z\cdot\overline{z}$ are not.

Lemma 8.3.5. *If f is a nonquasimonotone De Morgan function, i.e., $f \notin \mathfrak{QM}$, then replacing its variables by functions $0, 1$ or x, we can obtain the nonquasimonotone function $\overline{x}$.*

Proof. As f is a non-quasimonotone De Morgan function, there exist n-tuples $\alpha = (\alpha_1, \dots, \alpha_n)$, $\beta = (\beta_1, \dots, \beta_n) \in \{0,1\}^n$ with $\alpha \preceq \beta$ such that $f(\alpha) > f(\beta)$. But $\alpha, \beta \in \{0,1\}^n$, hence $f(\alpha), f(\beta) \in \{0,1\}$ and so

$$f(\alpha) = 1, \ f(\beta) = 0.$$

As $\alpha \preceq \beta$ there exists k $(1 \le k \le n)$ such that $\alpha_i = \beta_i$ for all $i \ne k$ and $\alpha_k = 0$, $\beta_k = 1$.

Consider the function $\varphi(x) = f(\alpha_1, \dots, \alpha_{k-1}, x, \alpha_{i+1}, \dots, \alpha_n)$. We have

$$\varphi(1) = f(\beta) = 0, \ \varphi(0) = f(\alpha) = 1.$$

Therefore, φ is a non-constant De Morgan function and so

$$\varphi(a) = a, \ \varphi(b) = b.$$

Hence $\varphi(x) = \overline{x}$. □

Theorem 8.3.6. *The class $\mathfrak{QM}$ of all quasimonotone De Morgan functions is closed.*

Proof. The proof follows immediately from the facts that quasimonotone functions can be defined as functions that preserve a relation and that the identity function $\varepsilon(x) = x$ is a quasimonotone De Morgan function. □

Now let us give some notations which will be used in the definition of the next important class of De Morgan functions.

For an n-tuple $\gamma = (\gamma_1, \dots, \gamma_n) \in D^n$ and for $u \in D$ denote $S_\gamma(u) = \{i : 1 \le i \le n, \ \gamma_i = u\}$ and $|\gamma_u| = |S_\gamma(u)|$ (here for a finite set A we denote by $|A|$ its cardinality, i.e., the number of its elements). Now let $S_\gamma(a) = \{i_1, \dots, i_m\}$, $S_\gamma(b) = \{j_1, \dots, j_k\}$. Also let $\alpha = (\alpha_1, \dots, \alpha_m) \in \{0,1\}^m$, $\beta = (\beta_1, \dots, \beta_k) \in \{0,1\}^k$ and $i_1 < i_2 < \dots < i_m$, $j_1 < j_2 < \dots < j_k$. For $i = 1, 2, \dots, n$ denote

$$\delta_i(a) = \begin{cases} \gamma_i, & \text{if } i \notin S_\gamma(a), \\ \alpha_l, & \text{if } i = i_l \in S_\gamma(a), \end{cases}$$

and

$$\delta_i(b) = \begin{cases} \gamma_i, & \text{if } i \notin S_\gamma(b), \\ \beta_l, & \text{if } i = i_l \in S_\gamma(b). \end{cases}$$

And finally denote $\gamma_a(\alpha) = (\delta_1(a), \dots, \delta_n(a))$, $\gamma_b(\beta) = (\delta_1(b), \dots, \delta_n(b))$. In other words if we move along the components of γ from left to right and replace every component a by corresponding α_i (i.e., i is the number of a's from the beginning of γ to current position including that position too) then we get $\gamma_a(\alpha)$. Analogously $\gamma_b(\beta)$ can be described. So if we write $\gamma_a(\alpha)$, then we suppose that the number of components of α is equal to the number of a's among the components of γ. Also if that number is zero then we mean $\gamma_a(\alpha) = \gamma$. The same is valid for $\gamma_b(\beta)$ too. To avoid complex notations we will write $\gamma_{a,b}(\alpha, \beta)$ instead of $(\gamma_a(\alpha))_b(\beta)$ (i.e., $\gamma_{a,b}(\alpha, \beta)$ is obtained from γ if we replace all a's by corresponding α_i's and all b's by corresponding β_i's).

Definition 8.3.7. A De Morgan function $f : D^n \to D$ is called a *zigzag* function if for every n-tuple $\gamma = (\gamma_1, \dots, \gamma_n) \in D^n$ with $f(\gamma) \in \{0, 1\}$ the following condition holds:

for any $\alpha, \alpha' \in \{0,1\}^{|\gamma_a|}$, $\beta, \beta' \in \{0,1\}^{|\gamma_b|}$ with

$$f(\gamma_a(\alpha)) \neq f(\gamma_a(\alpha')),\ f(\gamma_b(\beta)) \neq f(\gamma_b(\beta'))$$

we have

$$f(\gamma_{a,b}(\alpha, \beta)) = f(\gamma_{a,b}(\alpha', \beta)) = f(\gamma_{a,b}(\alpha, \beta')) = f(\gamma_{a,b}(\alpha', \beta')).$$

Denote by $\mathfrak{Z}$ the class of all zigzag De Morgan functions.

For example, the functions 0, 1, x, $\overline{x}$, $x+\overline{x}$, $x\cdot\overline{x}$, $x\cdot\overline{y}+y\cdot\overline{y}$, $x\cdot\overline{x}\cdot y\cdot\overline{z}$ are zigzag De Morgan functions and $x + y$, $x \cdot y$, $x + \overline{y}$, $x \cdot y \cdot \overline{z} + x \cdot \overline{x} \cdot z$ are not. For instance, let us prove that the function $x \cdot y$ is not a zigzag De Morgan function. To see this, notice that:

- $a \cdot b = 0$;
- $a \cdot 0 = 0 \neq a = a \cdot 1$;
- $0 \cdot b = 0 \neq b = 1 \cdot b$;
- $0 \cdot 0 = 0 \neq 1 = 1 \cdot 1$.

So, the function $x \cdot y$ does not satisfy the condition of Definition 8.3.7 and hence it is not a zigzag De Morgan function.

Lemma 8.3.8. *If f is a non-zigzag De Morgan function then replacing its variables by functions* 0, 1, x, $\overline{x}$, y *or* $\overline{y}$, *we can obtain the non-zigzag De Morgan function* $x \cdot y$ *or* $x + y$.

Proof. Suppose $f(x_1, \dots, x_n)$ is a non-zigzag De Morgan function. Then from Definition 8.3.7 we conclude that there exist an n-tuple γ and α, $\alpha' \in \{0,1\}^{|\gamma_a|}$ and β, $\beta' \in \{0,1\}^{|\gamma_b|}$ such that the following four conditions are satisfied:

- $f(\gamma) \in \{0, 1\}$,
- $f(\gamma_a(\alpha)) \neq f(\gamma_a(\alpha'))$,
- $f(\gamma_b(\beta)) \neq f(\gamma_b(\beta'))$,

- not all of the values $f(\gamma_{a,b}(\alpha,\beta))$, $f(\gamma_{a,b}(\alpha',\beta))$, $f(\gamma_{a,b}(\alpha,\beta'))$, $f(\gamma_{a,b}(\alpha',\beta'))$ are equal.

Denote $\omega = f(\gamma) \in \{0,1\}$. As f preserves the relation ρ we conclude that $f(\gamma_a(\alpha)), f(\gamma_a(\alpha')) \in \{\omega, b\}$ and $f(\gamma_b(\beta)), f(\gamma_b(\beta')) \in \{\omega, a\}$. Without loss of generality we can suppose that $f(\gamma_a(\alpha)) = f(\gamma_b(\beta)) = \omega$, $f(\gamma_a(\alpha')) = b$, $f(\gamma_b(\beta')) = a$.

Now we make the following replacements:

- if $\gamma_i \in \{0,1\}$, i.e., $i \in S_\gamma(0) \cup S_\gamma(1)$, then replace x_i by γ_i,
- if $i = i_j \in S_\gamma(a)$ then

  if $\alpha_j = \alpha'_j$ replace x_i by α_j,
  if $\alpha_j = \omega$, $\alpha'_j \neq \omega$ replace x_i by x,
  if $\alpha'_j = \omega$, $\alpha_j \neq \omega$ replace x_i by $\overline{x}$,

- if $i = i_j \in S_\gamma(b)$ then

  if $\beta_j = \beta'_j$ replace x_i by β_j,
  if $\beta_j = \omega$, $\beta'_j \neq \omega$ replace x_i by y,
  if $\beta'_j = \omega$, $\beta_j \neq \omega$ replace x_i by $\overline{y}$.

After this we get a function $g(x,y)$. Below we show that $g(x,y) = x \cdot y$ or $g(x,y) = x + y$.

It is easy to see that:

$$\omega = f(\gamma_a(\alpha)) = g(\omega, b),\ b = f(\gamma_a(\alpha')) = g(\overline{\omega}, b) \tag{8.3.2}$$

and

$$\omega = f(\gamma_b(\beta)) = g(a, \omega),\ a = f(\gamma_b(\beta')) = g(a, \overline{\omega}). \tag{8.3.3}$$

Using the fact that f preserves the relation ρ we get $g(a,b) \in \{\omega, a\}$ and $g(a,b) \in \{\omega, b\}$. Therefore

$$g(a,b) = \omega. \tag{8.3.4}$$

Further, using the second condition of the definition of De Morgan functions we conclude that:

$$g(b,a) = \omega,\ g(\omega,a) = \omega,\ g(\overline{\omega},a) = a,\ g(b,\omega) = \omega,\ g(b,\overline{\omega}) = b. \tag{8.3.5}$$

Now using Lemma 8.2.4 from (8.3.2) and (8.3.3) we get

$$g(\omega,\omega) = g(\omega,\overline{\omega}) = g(\overline{\omega},\omega) = \omega. \tag{8.3.6}$$

Notice that:

$$f(\gamma_{a,b}(\alpha,\beta)) = g(\omega,\omega),\ \ f(\gamma_{a,b}(\alpha',\beta)) = g(\overline{\omega},\omega),$$
$$f(\gamma_{a,b}(\alpha,\beta')) = g(\omega,\overline{\omega}),\ \ f(\gamma_{a,b}(\alpha',\beta')) = g(\overline{\omega},\overline{\omega}).$$

Therefore two of the values $g(\omega,\omega), g(\omega,\overline{\omega}), g(\overline{\omega},\omega), g(\overline{\omega},\overline{\omega})$ are not equal.

But we saw that $g(\omega,\omega) = g(\omega,\overline{\omega}) = g(\overline{\omega},\omega) = \omega$, hence $g(\overline{\omega},\overline{\omega}) \neq \omega$, i.e.,

$$g(\overline{\omega},\overline{\omega}) = \overline{\omega}. \tag{8.3.7}$$

And as g is a non-constant De Morgan function

$$g(a,a) = a,\ \ g(b,b) = b \tag{8.3.8}$$

(as we mentioned in Sec. 8.4, before Lemma 8.2.4).

Now from (8.3.2)–(8.3.8) we conclude that

$$g(x,y) = \begin{cases} x \cdot y & \text{if } \omega = 0, \\ x + y & \text{if } \omega = 1. \end{cases}$$

□

Now we can show that the class of all zigzag De Morgan functions is closed.

Theorem 8.3.9. *The class $\mathfrak{Z}$ of all zigzag De Morgan functions is closed.*

Proof. We should show that if $h(x_1,\dots,x_n), h_1(y_1,\dots,y_p),\dots,h_n(y_1,\dots,y_p)$ are zigzag De Morgan functions then $h(h_1,\dots,h_n)$ is a zigzag De Morgan function too (as the identical function x is a zigzag function). First we notice that this is true if $h_1,\dots,h_n$ are constant functions (i.e., 0 or 1) or functions of the form $q(x) = x$ or $q(x) = \overline{x}$ where x is a variable. In other words, if we replace some of the variables of a zigzag function by functions 0, 1 or functions of the form x or $\overline{x}$ then the result will also be a zigzag function. This argument follows immediately from the definition of zigzag functions. Further, notice that it is enough to show that if h and h_1 are zigzag functions then $h(h_1(y_1,\dots,y_p),x_2,\dots,x_n)$ is also a zigzag function. Indeed, assume this is true. We prove by induction on k that if the functions $h,h_1,\dots,h_k$ are zigzag functions then the function $h(h_1,\dots,h_k,x_{k+1},\dots,x_n)$ is a zigzag function too ($k \leq n$). For $k = 1$ this is our assumption. Suppose this is true for k, let us prove for $k+1$. Replace the variable x_{k+1} by a variable $z \notin \{x_1,\dots,x_n,y_1,\dots,y_p\}$ in the function $h(x_1,\dots,x_n)$ and get the function $h(x_1,\dots,x_k,z,x_{k+2},\dots,x_n)$. Now if we replace the variable z

in the function $h(h_1, \ldots, h_k, z, x_{k+2}, \ldots, x_n)$ (which is a zigzag function by induction hypothesis) by function h_{k+1}, then by our assumption we will get a zigzag function, i.e., $h(h_1, h_2, \ldots, h_{k+1}, x_{k+2}, \ldots, x_n)$ is a zigzag function too.

Now suppose for some zigzag functions $h(x_1, \ldots, x_n)$ and $h_1(y_1, \ldots, y_p)$ the function $\varphi(t_1, \ldots, t_m) = h(h_1(y_1, \ldots, y_p), x_2, \ldots, x_n)$ is not a zigzag function (here $\{t_1, \ldots, t_m\} = \{x_2, \ldots, x_n, y_1, \ldots, y_p\}$). Then from Lemma 8.3.8 it follows that we can replace the variables $t_1, \ldots, t_m$ by functions 0, 1, x, $\overline{x}$, y, $\overline{y}$ and in this way we get the non-zigzag De Morgan function $x \cdot y$ or $x + y$. After replacing variables $x_2, \ldots, x_n$ in the function h we get the function $f(x_1, x, y)$, and replacing the variables $y_1, \ldots, y_p$ in h_1 we get the function $g(x, y)$. From the argument stated above (at the beginning of the proof) it follows that f and g are zigzag De Morgan functions. It is obvious that the corresponding replacement of the variables $t_1, \ldots, t_m$ in φ results the function $\psi(x, y) = f(g(x, y), x, y)$. Thus $\psi(x, y) = x \cdot y$ or $\psi(x, y) = x + y$. Without loss of generality suppose $\psi(x, y) = x \cdot y$. Thus we have

$$f(g(x, y), x, y) = x \cdot y. \tag{8.3.9}$$

Now let us consider cases.

Case 1: $g(a, b) \in \{a, b\}$. Without loss of generality suppose

$$g(a, b) = a. \tag{8.3.10}$$

Substituting $x = a,\ y = b$ in (8.3.9) we get

$$f(a, a, b) = 0.$$

Applying Lemma 8.2.4 to the De Morgan function g, from (8.3.10) we get:

$$g(a, 0) = g(a, 1) = a.$$

Hence

$$f(a, a, 0) = 0,\ f(a, a, 1) = a. \tag{8.3.11}$$

Also we have $f(g(1, 1), 1, 1) = 1$. Therefore

$$f(0, 1, 1) = 1 \quad \text{and} \quad f(0, 1, b) \neq 0$$

or

$$f(1, 1, 1) = 1 \quad \text{and} \quad f(1, 1, b) \neq 0.$$

Note that $g(0, b) \neq a$ according to Lemma 8.2.4.

If $g(0,b) = b$ then from (8.3.9) we have $f(b,0,b) = 0$ and so

$$f(0,0,b) = 0 \quad \text{and} \quad f(0,0,0) = 0.$$

If $g(0,b) = 0$ then

$$f(0,0,b) = 0 \quad \text{and} \quad f(0,0,0) = 0.$$

If $g(0,b) = 1$ then

$$f(1,0,b) = 0 \quad \text{and} \quad f(1,0,0) = 0.$$

In all cases we found $\lambda_1, \lambda_2, \lambda_1', \lambda_2' \in \{0,1\}$ such that

$$f(\lambda_1, \lambda_2, b) \neq f(\lambda_1', \lambda_2', b),$$

and not all of

$$f(\lambda_1, \lambda_2, 0), f(\lambda_1', \lambda_2', 0), f(\lambda_1, \lambda_2, 1), f(\lambda_1', \lambda_2', 1)$$

are equal. This with (8.3.11) shows that f is not a zigzag De Morgan function, which is a contradiction.

Case 2: $g(a,b) \in \{0,1\}$. Without loss of generality suppose

$$g(a,b) = 0.$$

From (8.3.9) we conclude that

$$f(0,a,b) = 0. \tag{8.3.12}$$

As g is a zigzag function, we should have

$$g(a,0) = g(a,1)$$

or

$$g(0,b) = g(1,b)$$

or

$$g(a,0) \neq g(a,1), \ g(0,b) \neq g(1,b),$$
$$g(0,0) = g(0,1) = g(1,0) = g(1,1).$$

As $g(a,b) = 0$, we have $g(a,0), g(a,1) \in \{0,a\}$. Suppose

$$g(a,0) = g(a,1) = 0.$$

Then $g(0,0) = g(0,1) = g(1,0) = g(1,1) = 0$. Therefore:

$$f(0,a,b) = 0,\ f(0,a,0) = 0,\ f(0,a,1) = a,$$
$$f(0,0,0) = f(0,0,1) = f(0,1,0) = 0,\ f(0,1,1) = 1.$$

Applying Lemma 8.2.4 to the De Morgan function f we get $f(0,1,b) = f(0,0,b) \in \{0,b\}$. If $f(0,1,b) = 0$ then $f(0,1,1) = 0$, which is a contradiction. So

$$f(0,1,b) = f(0,0,b) = b,\ f(b,0,b) = b. \tag{8.3.13}$$

The latter equality is obtained from the former using Lemma 8.2.4. From $g(a,b) = 0$ we conclude that $g(0,b) = 0$ or $g(0,b) = b$. Substituting $x = 0,\ y = b$ in (8.3.9) we get $f(0,0,b) = 0$ or $f(b,0,b) = 0$. But this contradicts (8.3.13).

If $g(a,0) = g(a,1) = a$ then

$$f(a,a,0) = 0,\ f(a,a,1) = a. \tag{8.3.14}$$

From the former equality it follows that $f(a,a,b) \in \{0,b\}$. But from (8.3.12) and from Lemma 8.2.4 we have $f(a,a,b) \in \{0,a\}$. And we conclude that

$$f(a,a,b) = 0.$$

Continuing as in Case 1 we get a contradiction.

The case $g(0,b) = g(1,b)$ is considered similarly.

Now suppose

$$g(a,0) \neq g(a,1),\ g(0,b) \neq g(1,b),$$
$$g(0,0) = g(0,1) = g(1,0) = g(1,1).$$

As $g(a,b) = 0$, we have $g(a,0), g(a,1) \in \{0,a\}$ and $g(b,0), g(b,1) \in \{0,b\}$ (we apply Lemma 8.2.4). First let

$$g(a,0) = g(0,b) = 0,\ g(a,1) = a,\ g(1,b) = b. \tag{8.3.15}$$

As above we have

$$f(0,0,0) = f(0,0,1) = f(0,1,0) = 0,\ f(0,1,1) = 1. \tag{8.3.16}$$

From (8.3.9) and (8.3.15) we get

$$f(0,a,0) = 0,\ f(0,0,b) = 0.$$

If $f(0,a,1) = 0$ or $f(0,1,b) = 0$ then $f(0,1,1) = 0$, which contradicts (8.3.16). Hence

$$f(0,a,1) \neq f(0,a,0), \;\; f(0,1,b) \neq f(0,0,b).$$

And we conclude (taking into account (8.3.16)) that f is not a zigzag De Morgan function. But this is a contradiction.

Now suppose

$$g(a,1) = g(1,b) = 0, \;\; g(a,0) = a, \;\; g(0,b) = b.$$

Then

$$f(a,a,0) = f(b,0,b) = 0, \;\; f(0,a,1) = a, \;\; f(0,1,b) = b.$$

Applying Lemma 8.2.4 to f we conclude that the first three equalities of (8.3.16) are true for this case too and also $f(0,a,0) = 0$, $f(0,0,b) = 0$. Then from Lemma 8.2.4 and from $g(a,1) = 0$ it follows that $g(1,1) = 0$. Now (8.3.9) implies $f(0,1,1) = 1$. Hence (8.3.16) is true for this case too. Now we have $f(0,1,1) = 1$, therefore $f(0,a,1) \in \{1,a\}$ and $f(0,1,b) \in \{1,b\}$ (again we apply Lemma 8.2.4). So $f(0,a,1) \neq f(0,a,0)$ and $f(0,1,b) \neq f(0,0,b)$, hence f is not a zigzag De Morgan function, which is a contradiction.

And finally suppose

$$g(a,0) = g(1,b) = 0, \;\; g(a,1) = a, \;\; g(0,b) = b.$$

In this case we have

$$f(0,a,0) = f(b,0,b) = 0, \;\; f(a,a,1) = a, \;\; f(0,1,b) = b. \tag{8.3.17}$$

And as $f(0,a,b) = 0$ therefore $f(0,a,1) \in \{0,a\}$ and $f(0,0,b) \in \{0,b\}$ (we apply Lemma 8.2.4). But $f(0,a,1) = 0$ contradicts (8.3.16) (which is true for this case too), and $f(0,0,b) = b$ contradicts (8.3.17). Hence $f(0,a,1) = a$ and $f(0,0,b) = 0$. But this with (8.3.16) and (8.3.17) contradicts the fact that f is a zigzag De Morgan function.

The case

$$g(a,1) = g(0,b) = 0, \;\; g(a,0) = a, \;\; g(1,b) = b$$

can be considered in an analogous way.

So we finished the proof of the theorem. □

Note that it is not known yet how one can define the zigzag De Morgan functions as functions preserving some relations on the set D.

Note that all the closed classes $\mathfrak{D}, \mathfrak{T}_0, \mathfrak{T}_1, \mathfrak{T}, \mathfrak{M}, \mathfrak{QM}, \mathfrak{Z}$ are clones as they contain the projection functions.

It is easy to verify that none of the closed classes $\mathfrak{T}_0, \mathfrak{T}_1, \mathfrak{T}, \mathfrak{QM}, \mathfrak{Z}$ is contained in another class. This follows from the following table, where the sign $+$ indicates that the function belongs to the corresponding class and the sign $-$ means the contrary:

	$\mathfrak{T}_0$	$\mathfrak{T}_1$	$\mathfrak{T}$	$\mathfrak{QM}$	$\mathfrak{Z}$
1	$-$	$+$	$-$	$+$	$+$
0	$+$	$-$	$-$	$+$	$+$
$\overline{x}$	$-$	$-$	$+$	$-$	$+$
$x \cdot \overline{y}$	$+$	$-$	$+$	$-$	$-$
$x + \overline{y}$	$-$	$+$	$+$	$-$	$-$
$x \cdot y$	$+$	$+$	$+$	$+$	$-$

8.4. Functional completeness theorem for De Morgan functions

Here we use the results of the previous section to prove the functional completeness theorem for De Morgan functions.

Theorem 8.4.1 (Functional completeness theorem for De Morgan functions). *A class $\mathfrak{B}$ of De Morgan functions is functionally complete if and only if none of the five closed classes $\mathfrak{T}_0, \mathfrak{T}_1, \mathfrak{T}, \mathfrak{QM}, \mathfrak{Z}$ contains $\mathfrak{B}$.*

Proof. The necessity is evident. If $\mathfrak{B}$ were contained in a closed class $\mathfrak{N}$ (different from $\mathfrak{D}$) then obviously $[\mathfrak{B}] \subseteq [\mathfrak{N}] = \mathfrak{N} \neq \mathfrak{D}$ and so $\mathfrak{B}$ can not be complete.

Let us prove the sufficiency. Let $f_1, f_2, f_3, f_4, f_5 \in \mathfrak{B}$ such that $f_1 \notin \mathfrak{T}_0$, $f_2 \notin \mathfrak{T}_1$, $f_3 \notin \mathfrak{T}$, $f_4 \notin \mathfrak{QM}$, $f_5 \notin \mathfrak{Z}$. As $\mathfrak{T}$ consists of all non-constant functions, f_3 must be a constant function, say 0. Then using the function f_1 we obtain the constant De Morgan function 1, as

$$f_1(0, \ldots, 0) = 1.$$

Now from constant functions and non-quasimonotone function f_4 we construct the function $\overline{x}$ (Lemma 8.3.5). And finally using the functions $0, 1, \overline{x}$ and f_5 (which is a non-zigzag function), we construct the non-zigzag De Morgan function $x \cdot y$ or $x + y$ (Lemma 8.3.8). Having those functions we can construct all De Morgan functions (by formula (12.4.1)). And so $\mathfrak{B}$ is complete. □

Corollary 8.4.2. *If $\mathfrak{B}$ is a complete class, then there exists a complete subclass of $\mathfrak{B}$ which contains not more than four De Morgan functions.*

Proof. From the functional completeness theorem it follows that there are functions $f_1, f_2, f_3, f_4, f_5 \in \mathfrak{B}$ such that $f_1 \notin \mathfrak{T}_0$, $f_2 \notin \mathfrak{T}_1$, $f_3 \notin \mathfrak{T}$, $f_4 \notin \mathfrak{QM}$, $f_5 \notin \mathfrak{Z}$. But $f_3 \notin \mathfrak{T}$ implies that f_3 is a constant function. If $f_3 = 0$ then $f_3 \notin \mathfrak{T}_1$ and if $f_3 = 1$ then $f_3 \notin \mathfrak{T}_0$. In the former case, the class $\{f_1, f_3, f_4, f_5\}$ is complete and in the latter case $\{f_2, f_3, f_4, f_5\}$ is complete (due to the functional completeness theorem). □

Following to [191] we can introduce the following concept.

Definition 8.4.3. A class $\mathfrak{N}$ of De Morgan functions is called *precomplete* if it is not complete and for any De Morgan function $f \notin \mathfrak{N}$ the class $\mathfrak{N} \cup \{f\}$ is complete.

It is obvious that precomplete class is closed. And also note that precomplete classes are the same as maximal closed classes defined in the previous section.

Corollary 8.4.4. *Every closed class of De Morgan functions is contained in one of the five classes of De Morgan functions: $\mathfrak{T}_0, \mathfrak{T}_1, \mathfrak{T}, \mathfrak{QM}$ and $\mathfrak{Z}$.*

Corollary 8.4.5. *There are exactly five precomplete classes of De Morgan functions: $\mathfrak{T}_0, \mathfrak{T}_1, \mathfrak{T}, \mathfrak{QM}$ and $\mathfrak{Z}$.*

In other words, those five precomplete classes are maximal subclones of the clone $\mathfrak{D}$.

Finally note that there is an algorithm which, given a De Morgan function f, determines whether f belongs to the given precomplete class. Hence we conclude that there is an algorithm, which for a given finite set of De Morgan functions determines whether it is complete. The decidability of completeness also follows from the fact that $\mathfrak{D}$ is finitely generated, hence it is sufficient to test whether each of the generators can be obtained from the given finite set of functions.

Chapter 9

Bilattices

9.1. Preliminaries

In [21,22] Belnap introduced a logic, which is based on the algebraic structure called *FOUR*, having four truth values. Bilattices as algebras with two lattice structures were introduced by M. Ginsberg and M. Fitting, in 1986–90 (see [107–109,128]). They have found wide applications in studies of logic programming, multi-valued logic and artificial intelligence (also see [162,239,240]). These bilattices we called Ginsberg's bilattices. Bilattices are natural generalization of Belnap's *FOUR*.

In this chapter, we prove that the description of Ginsberg's bilattices with conditions of interlacement and boundness stated in papers of various authors keeps on remaining true without the second condition, and a weaker form of interlacement called *weak interlacement* here. In particular, we prove that every weakly interlaced bilattice is isomorphic to the superproduct of two lattices, while every weakly interlaced Ginsberg bilattice is isomorphic to the Ginsberg's superproduct of two equal lattices.

The lattices with third additional binary operation have been considered since the 1950s (see [17,158,159,175]); in these works, the initial lattices are distributive, or the third operation connected with the lattice operation by the identity of distributivity. However, often we meet algebras with two lattice structures.

An algebra $(L; \wedge, \vee, \cdot, +)$ with four binary operations is called bilattice if the reducts $L_1 = (L; \wedge, \vee)$ and $L_2 = (L; \cdot, +)$ are lattices. It is called *bounded*, if both reducts L_1 and L_2 are bounded lattices.

The bilattice $(L; \wedge, \vee, \cdot, +)$ is called *modular* (or *distributive*) if the reducts L_1 and L_2 are modular (or distributive) lattices.

It is well known that the automorphism φ of a given algebra is called *involution*, if φ^2 is an identical mapping. A mapping $\varphi : L \to L$ is called *antiautomorphism* of the lattice $(L; \wedge, \vee)$ if the following DeMorgan identities are valid:

$$\varphi(x \vee y) = \varphi x \wedge \varphi y, \quad \varphi(x \wedge y) = \varphi x \vee \varphi y.$$

And if φ^2 is an identity mapping, then φ is called *coinvolution*. Following to the tradition, here the unary operations are denoted by $(')$. If $(L; \wedge, \vee)$ is a lattice and a unary operation $'$ is an involution (coinvolution), then the algebra $(L; \wedge, \vee, ')$ with two binary and one unary operations is called *a lattice with involution (coinvolution)*.

An algebra $(L; \wedge, \vee, \cdot, +, ')$ with the additional unary operation $'$ is called *bilattice with involution (coinvolution)*, if the reducts $L^1 = (L; \wedge, \vee, ')$ and $L^2 = (L; \cdot, +, ')$ are lattices with involution (coinvolution).

An algebra $(L; \wedge, \vee, \cdot, +, ')$ is called *Ginsberg bilattice*, if one of the reducts L^1 and L^2 is a lattice with involution, while the other reduct is a lattice with coinvolution. Without loss of generality, we can consider L^1 is a lattice with coinvolution, and L^2 is a lattice with involution.

An algebra $(L; \wedge, \vee, \cdot, +, ')$ with four binary and one unary operation $'$ is called *a Boolean bilattice (De Morgan bilattice)*, if the reducts $L^1 = (L; \wedge, \vee, ')$ and $L^2 = (L; \cdot, +, ')$ are Boolean lattices (De Morgan lattices).

A bilattice $(L; \wedge, \vee, \cdot, +)$ is called *interlaced* if any of the four binary operations preserve the both lattice orders $\leqslant_1$ in L_1 and $\leqslant_2$ in L_2, i.e., the following implications are valid:

$$x \leqslant_1 y,\ u \leqslant_1 v \longrightarrow x \cdot u \leqslant_1 y \cdot v,\ x + u \leqslant_1 y + v, \tag{9.1.1}$$

and

$$x \leqslant_2 y,\ u \leqslant_2 v \longrightarrow x \wedge u \leqslant_2 y \wedge v,\ x \vee u \leqslant_2 y \vee v. \tag{9.1.2}$$

This means that interlaced bilattices form a quasivariety. Let us recall that lattice operations preserve the lattice order in a given lattice. So we can consider every lattice $(L; \wedge, \vee)$ as an interlaced bilattice $(L; \wedge, \vee, \wedge, \vee)$.

A bilattice $(L; \wedge, \vee, \cdot, +)$ is called *weakly interlaced*, if the condition (9.1.2) and one of the following conditions are valid:

$$x \leqslant_1 y,\ u \leqslant_1 v \longrightarrow x \cdot u \leqslant_1 y \cdot v,$$

or

$$x \leqslant_1 y,\ u \leqslant_1 v \longrightarrow x + u \leqslant_1 y + v.$$

An algebra $(L; \wedge, \vee, \cdot, +, ')$ with the additional unary operation $'$ is called interlaced (weakly interlaced, bounded, modular, distributive) bilattice if $(L; \wedge, \vee, \cdot, +)$ is an interlaced (weakly interlaced, bounded, modular, distributive) bilattice.

A characterization of interlaced bounded Ginsberg bilattices, and also bounded bilattices (and bounded Ginsberg bilattices) with identities of distributivity for every pair of its binary operations is given in [107,128,347]. The characterization of bounded interlaced bilattice is given in [18] (also see [162,239,240,347]). Paper [299] is dedicated to characterization of interlaced bilattices without bounds, and characterization of Boolean bilattices. One of the main result of this chapter is the characterization of Ginsberg bilattices with the condition of a weakly interlacement (with the help of many-sorted algebras, which has become a useful tool in cryptography and automation theory).

Recall that the superproduct $L_1 \bowtie L_2$ of the two lattices $(L_1; \wedge_1, \vee_1)$ and $(L_2; \wedge_2, \vee_2)$ is an algebra $(L_1 \times L_2; \wedge, \vee, \cdot, +)$ with four binary operations:

$$\wedge = (\wedge_1, \vee_2),\ \vee = (\vee_1, \wedge_2),$$
$$\cdot = (\wedge_1, \wedge_2),\ \ + = (\vee_1, \vee_2),$$

which operate component-wise.

It is easy to prove that the superproduct of two lattices is an interlaced bilattice. Thus, the superproduct of two modular (distributive, Boolean, DeMorgan) lattices is an interlaced modular (distributive, Boolean, DeMorgan) bilattice. The superproduct of two lattices with involution (coinvolution) is an interlaced bilattice with involution (coinvolution).

The proofs of the following results see in previous chapters.

Theorem 9.1.1. *The variety of lattices satisfies the following hyperidentities:*

$$X(x,x) = x, \tag{9.1.3}$$
$$X(x,y) = X(y,x), \tag{9.1.4}$$
$$X(x, X(y,z)) = X(X(x,y),z), \tag{9.1.5}$$
$$X(Y(X(x,y),z), Y(y,z)) = Y(X(x,y),z). \tag{9.1.6}$$

And conversely, every hyperidentity of the variety of lattices is a consequence of the hyperidentities (9.1.3), (9.1.4), (9.1.5), (9.1.6) (see Chapter 2).

Theorem 9.1.2. *The variety of modular lattices satisfies the following hyperidentities: (9.1.3)–(9.1.6) and*

$$X(Y(x,X(y,z)),Y(y,z)) = Y(X(x,Y(y,z)),X(y,z)). \tag{9.1.7}$$

And conversely, every hyperidentity of the variety of modular lattices is a consequence of the hyperidentities (9.1.3)–(9.1.7) (see Chapter 2).

Theorem 9.1.3. *The variety of distributive lattices satisfies the following hyperidentities: (9.1.3)–(9.1.5) and*

$$X(x,Y(y,z)) = Y(X(x,y),X(x,z)). \tag{9.1.8}$$

And conversely, every hyperidentity of the variety of distributive lattices is a consequence of the hyperidentities (9.1.3)–(9.1.8) (see Chapter 2).

Theorem 9.1.4. *The variety of Boolean algebras satisfies the following hyperidentities: (9.1.3)–(9.1.5), (9.1.8) and*

$$F(F(x)) = x, \tag{9.1.9}$$
$$X(F(x),y) = X(F(X(x,y)),y), \tag{9.1.10}$$
$$F(X(F(X(x,y)),F(X(x,F(y))))) = x. \tag{9.1.11}$$

And conversely, every hyperidentity of the variety of Boolean algebras is a consequence of the hyperidentities (9.1.3)–(9.1.5), (9.1.8)–(9.1.11).

All hyperidentities of the variety of Boolean algebras are consequences of one of its hyperidentities (*see Chapter* 5).

Theorem 9.1.5. *The variety of De Morgan algebras satisfies the following hyperidentities: (9.1.3)–(9.1.5), (9.1.8), (9.1.9) and*

$$F(x) = G(x), \tag{9.1.12}$$
$$F(Y(F(X(x,y)),z)) = X(F(Y(F(x),z)),F(Y(F(y),z))), \tag{9.1.13}$$
$$X(F(X(x,y)),F(x)) = F(x), \tag{9.1.14}$$
$$X(x,X(y,z)) = X(Y(x,Y(y,z)),F(Y(F(x),Y(F(y),F(z))))), \tag{9.1.15}$$
$$\begin{aligned} &X(F(X(Y(x,e),Y(x,Y(y,e)))),F(X(Y(z,e),Y(z,Y(t,e))))) \\ &= X(F(X(Y(x,Y(z,e))),Y(x,Y(y,Y(z,Y(u,e)))))), \\ &F(X(F(Y(F(Y(x,Y(y,e))),F(Y(z,Y(t,e))))),e)))). \end{aligned} \tag{9.1.16}$$

And conversely, every hyperidentity of the variety of De Morgan algebras is a consequence of the hyperidentities: (9.1.3)–(9.1.5), (9.1.8), (9.1.9), (9.1.12)–(9.1.16).

The hyperequational theory of the variety of De Morgan algebras is not one-based (see Chapter 7).

9.2. Interlaced semilattice operations

Let $(L; \cap, \cup)$ be a lattice and let $(L; *)$ be a semilattice with the corresponding orders $\leqslant_\cap$ and $\leqslant_*$:

$$x \leqslant_\cap y \longleftrightarrow x \cap y = x \longleftrightarrow x \cup y = y,$$
$$x \leqslant_* y \longleftrightarrow x * y = x.$$

The smallest equivalence relation on L is denoted by $\mathbf{0}$, and the largest equivalence relation on the set L is denoted by $\mathbf{1}$.

The semilattice operation $*$ is called interlaced with the operations $\cap$, $\cup$ if the order $\leqslant_\cap$ is preserved by the operation $*$ and the order $\leqslant_*$ is preserved by the operations $\cap$, $\cup$, i.e.,

$$a \leqslant_\cap b \,\&\, c \leqslant_\cap d \longrightarrow a * c \leqslant_\cap b * d,$$
$$a \leqslant_* b \,\&\, c \leqslant_* d \longrightarrow a \cap c \leqslant_* b \cap d \,\&\, a \cup c \leqslant_* b \cup d$$

(cf. [111,178]).

Lemma 9.2.1. *The operation $*$ of the semilattice $(L; *)$ is interlaced with the operations of the lattice $(L; \cap, \cup)$ if and only if for every operation $A, B \in \{\cap, \cup, *\}$ the following identity is valid:*

$$A(B(A(x, y), z), B(y, z)) = B(A(x, y), z), \tag{9.2.1}$$

i.e., in the algebra $(L; \cap, \cup, *)$ the hyperidentity (9.1.6) is satisfied.

Proof. From the condition

$$x \leqslant_\cap y \longrightarrow x * z \leqslant_\cap y * z$$

follows the equality:

$$((x \cap y) * z) \cap (y * z) = (x \cap y) * z,$$

and conversely. Indeed:

$$x \cap y \leqslant_\cap y \longrightarrow (x \cap y) * z \leqslant_\cap y * z,$$

and

$$((x \cap y) * z) \cap (y * z) = (x \cap y) * z.$$

If $x \leqslant_{\cap} y$, then $x \cap y = x$ and

$$(x * z) \cap (y * z) = ((x \cap y) * z) \cap (y * z) = (x \cap y) * z = x * z.$$

Hence, $x * z \leqslant_{\cap} y * z$.

Similarly, from the condition

$$x \leqslant_{*} y \longrightarrow x \cap z \leqslant_{*} y \cap z$$

it follows the identity: $((x * y) \cap z) * (y \cap z) = (x * y) \cap z$, and conversely.

From the condition

$$x \leqslant_{*} y \longrightarrow x \cup z \leqslant_{*} y \cup z$$

it follows the identity:

$$((x * y) \cup z) * (x \cup z) = (x * y) \cup z,$$

and conversely. In addition, let us recall that any lattice $(L; \cap, \cup)$ satisfies the hyperidentity (9.1.6) (see Theorem 9.1.1). □

Corollary 9.2.2. *A bilattice is interlaced if and only if it satisfies hyperidentity (9.1.6). The superproduct of two lattices is an interlaced bilattice. Interlaced (modular, distributive, DeMorgan, Boolean) bilattices form a variety. Interlaced bilattices with involution (coinvolution) and interlaced Ginsberg bilattices form varieties.*

Lemma 9.2.3. *If $(L; \cap, \cup)$ is a lattice and $(L; *)$ is a semilattice, the operation of which is interlaced with the operations $\cap$, $\cup$ then for every $x, y \in L$ the following inequalities are valid:*

$$x \cap y \leqslant_{\cap} x * y \leqslant_{\cap} x \cup y,$$
$$x * y \leqslant_{*} x \cap y,$$
$$x * y \leqslant_{*} x \cup y.$$

Proof. Since $x \cap y \leqslant_{\cap} x$, $x \cap y \leqslant_{\cap} y$, we have $(x \cap y) * (x \cap y) \leqslant_{\cap} x * y$ and $x \cap y \leqslant_{\cap} x * y$. Similarly, since $x \leqslant_{\cap} x \cup y$, $y \leqslant_{\cap} x \cup y$, we have $x * y \leqslant_{\cap} x \cup y$, i.e., the first inequality is proved. Further, from $x * y \leqslant_{*} x$, $x * y \leqslant_{*} y$ it follows: $x * y \leqslant_{*} x \cap y$ and $x * y \leqslant_{*} x \cup y$. □

Lemma 9.2.4. *If $(L; \cap, \cup)$ is a lattice and $(L; *)$ is a semilattice, the operation of which is interlaced with the operations, $\cap$, $\cup$, then the following implications are valid:*

$$a \leqslant_{\cap} x \leqslant_{\cap} b \,\&\, a \leqslant_{*} b \longrightarrow a \leqslant_{*} x \leqslant_{*} b,$$
$$a \leqslant_{\cap} x \leqslant_{\cap} b \,\&\, b \leqslant_{*} a \longrightarrow b \leqslant_{*} x \leqslant_{*} a.$$

Proof. From the condition $a \leqslant_* b$ it follows that $a \cap x \leqslant_* b \cap x$ and $a \cup x \leqslant_* b \cup x$. Besides, from the condition $a \leqslant_\cap x \leqslant_\cap b$ it follows that $a \cap x = a$, $b \cap x = x$, $a \cup x = x$, $b \cup x = b$. Hence, $a \leqslant_* x$, $x \leqslant_* b$. The proof of the second implication is similar. □

Lemma 9.2.5. *If $(L; \cap, \cup)$ is a lattice and $(L; *)$ is a semilattice, the operation of which is interlaced with the operations $\cap$, $\cup$, then the following implications are valid:*

$$u \leqslant_\cap x \,\&\, u \leqslant_\cap y \,\&\, u \leqslant_* x \,\&\, u \leqslant_* y \longrightarrow x \cap y = x * y\,,$$
$$x \leqslant_\cap u \,\&\, y \leqslant_\cap u \,\&\, u \leqslant_* x \,\&\, u \leqslant_* y \longrightarrow x \cup y = x * y\,.$$

Proof. From the condition $u \leqslant_\cap x$, $u \leqslant_\cap y$ it follows that $u \leqslant_\cap x \cap y \leqslant_\cap x$ and since $u \leqslant_* x$, we have $u \leqslant_* x \cap y \leqslant_* x$ by Lemma 9.2.3. Analogously: $u \leqslant_* x \cap y \leqslant_* y$. Thus, $x \cap y \leqslant_* x$, $x \cap y \leqslant_* y$, and hence, $x \cap y \leqslant_* x * y$. On the other hand $x * y \leqslant_* x \cap y$ by Lemma 9.2.3. So, $x \cap y = x * y$. □

Lemma 9.2.6. *If $(L; \cap, \cup)$ is a lattice, $(L; *)$ is a semilattice, the operation of which is interlaced with the operations, $\cap$, $\cup$, and there exists a binary operation $\triangle$ on L such that $(L; *, \triangle)$ is a lattice, then*

$$a \leqslant_\cap b \longrightarrow a \leqslant_\cap a \triangle b \leqslant_\cap b.$$

Proof. According to Lemma 9.2.3, we have

$$a = a * (a \triangle b) \leqslant_* a \cup (a \triangle b).$$

Hence, from the condition, $a \leqslant_\cap b$, we also have

$$a = b \cap a \leqslant_* b \cap (a \cup (a \triangle b)).$$

From the conditions $b \leqslant_* a \triangle b$ and $a \leqslant_\cap b$ it follows that

$$b = a \cup b \leqslant_* a \cup (a \triangle b),$$

so

$$b = b \cap b \leqslant_* b \cap (a \cup (a \triangle b)),$$

i.e.,

$$b \leqslant_* b \cap (a \cup (a \triangle b)).$$

Thus, $b \cap (a \cup (a \triangle b))$ is an upper bound for the elements a, b with respect to the order $\leqslant_*$. Hence,

$$a \triangle b \leqslant_* b \cap (a \cup (a \triangle b)).$$

Further, from the condition $a \leqslant_* a\triangle b$ and according to the interlacement, we have

$$a \cup (a\triangle b) \leqslant_* a\triangle b,$$
$$a \cap (a\triangle b) \leqslant_* a\triangle b.$$

Similarly, from the condition $b \leqslant_* a\triangle b$ and according to the interlacement, we have

$$b \cup (a\triangle b) \leqslant_* a\triangle b,$$
$$a \cap (a\triangle b) \leqslant_* a\triangle b.$$

We see that

$$a\triangle b \leqslant_* b \cap (a \cup (a\triangle b)) \leqslant_* b \cap (a\triangle b) \leqslant_* a\triangle b,$$

i.e.,

$$b \cap (a\triangle b) = a\triangle b,$$

and $a\triangle b \leqslant_\cap b$.

The proof of the property $a \leqslant_\cap a\triangle b$ is similar. □

9.3. The congruences θ_1 and θ_2

Let us define the binary relations θ_1 and θ_2 in the algebra $(L; \cap, \cup, *)$ by the rule:

$$a\,\theta_1\,b \longleftrightarrow a * b = a \cup b,$$
$$a\,\theta_2\,b \longleftrightarrow a * b = a \cap b.$$

Lemma 9.3.1. *If $(L; \cap, \cup)$ is a lattice and $(L; *)$ is a semilattice, the operation of which is interlaced with the operations $\cap, \cup$, then the relations θ_1 and θ_2 are congruences of the algebra $(L; \cap, \cup, *)$.*

Proof. First, we proof that θ_1 is a equivalence relation.

(1) *Reflexivity:*

$$a * a = a = a \cup a, \quad \text{i.e.,} \ a\,\theta_1\,a;$$

(2) *Symmetry:*

$$a\,\theta_1\,b \longrightarrow b * a = a * b = a \cup b = b \cup a \longrightarrow b\,\theta_1\,a;$$

(3) *Transitivity.* If $a\,\theta_1\,b$ and $b\,\theta_1\,c$, then $a*b = a\cup b$ and $b*c = b\cup c$. Thus, $a\cup b \leqslant_* a$, $a\cup b \leqslant_* b$, $b\cup c \leqslant_* b$, $b\cup c \leqslant_* c$. According to the interlacement of the operation $*$ we have $a\cup b\cup c \leqslant_* b\cup c$, $a\cup b\cup c \leqslant_* a\cup b$. On the other hand, $b\cup c \leqslant_\cap a\cup b\cup c$, $a\cup b \leqslant_\cap a\cup b\cup c$. So,

$$a\cup b\cup c = (b\cup c)\cup(a\cup b) = (b\cup c)*(a\cup b) = a*b*c,$$

by Lemma 9.2.5, i.e., $a\cup b\cup c = a*b*c$. Hence,

$$a \leqslant_\cap a*b*c, \quad c \leqslant_\cap a*b*c,$$

and since $a*b*c \leqslant_* a$, $a*b*c \leqslant_* c$, we have $a\cup c = a*c$ by Lemma 9.2.5, i.e., $a\,\theta_1\,c$.

We now check that θ_1 is a congruence of the lattice $(L;\cap,\cup)$.

If $a\,\theta_1\,b$, then $a*b = a\cup b$. Consequently, $a\cup b \leqslant_* a$, $a\cup b \leqslant_* b$ and according to the interlacement of the operation $*$, we have

$$a\cup b\cup c \leqslant_* a\cup c, \quad a\cup b\cup c \leqslant_* b\cup c.$$

Note that $a\cup c \leqslant_\cap a\cup b\cup c$, $b\cup c \leqslant_\cap a\cup b\cup c$ and

$$(a\cup c)\cup(b\cup c) = (a\cup c)*(b\cup c),$$

by Lemma 9.2.5, i.e., $a\cup c\,\theta_1\,b\cup c$. Let us prove that $a\cap c\,\theta_1\,b\cap c$. Indeed, if $a\,\theta_1\,b$, then

$$(a\cup b)\cap c \leqslant_* a\cap c, \quad (a\cup b)\cap c \leqslant_* b\cap c.$$

Since $a\cap c \leqslant_\cap (a\cup b)\cap c$, $b\cap c \leqslant_\cap (a\cup b)\cap c$, we have

$$(a\cap c)\cup(b\cap c) = (a\cap c)*(b\cap c),$$

by Lemma 9.2.5, which means that $a\cap c\,\theta_1\,b\cap c$.

It is remained to prove that $a*c\,\theta_1\,b*c$, if $a\,\theta_1\,b$. Namely, if $a\,\theta_1\,b$, then $a*b = a\cup b$. Thus, $a \leqslant_\cap a*b$, $b \leqslant_\cap a*b$ and $a*c \leqslant_\cap a*b*c$, $b*c \leqslant_\cap a*b*c$. Since $a*b*c \leqslant_* a*c$, $a*b*c \leqslant_* b*c$ and $(a*c)\cup(b*c) = (a*c)*(b*c)$ by Lemma 9.2.5, $a*c\,\theta_1\,b*c$.

The second part of the assertion has a similar proof. □

It is evident that $\theta_1\cap\theta_2 = \mathbf{0}$ (null congruence). Indeed,if $a\,\theta_1\cap\theta_2\,b$, then $a\,\theta_1\,b$ and $a\,\theta_2\,b$, hence, $a*b = a\cup b$, $a*b = a\cap b$, i.e. $a\cup b = a\cap b$ and $a = b$.

Lemma 9.3.2. *By the condition of the preceding lemma, we have:*

(1) $a\cap b\,\theta_1\,a*b$, $a*b\,\theta_2\,a\cup b$ *for every* $a,b\in L$;
(2) $a \leqslant_\cap b \longrightarrow a\,\theta_1\theta_2\,b$.

Proof. (1) By Lemma 9.2.3, we have

$$(a \cap b) * (a * b) = a * b = (a \cap b) \cup (a * b),$$

i.e., $a \cap b \, \theta_1 \, a * b$. Similarly,

$$(a \cup b) * (a * b) = a * b = (a \cup b) \cap (a * b),$$

i.e., $a * b \, \theta_2 \, a \cup b$.

(2) If $a \leqslant_\cap b$, then $a \cap b = a$, $a \cup b = b$. By property (1) we have: $a \cap b \, \theta_1 \, a * b$, $a * b \, \theta_2 \, a \cup b$, i.e., $a \cap b \, \theta_1 \theta_2 \, a \cup b$ and $a \, \theta_1 \theta_2 \, b$. □

Lemma 9.3.3. *If $(L; \cap, \cup)$ is a lattice and $(L; *)$ is a semilattice, the operation of which is an interlaced with the operations $\cap$, $\cup$ and there exists a binary operation $\triangle$ on L such that $(L; *, \triangle)$ is a lattice, then*

$$a \leqslant_\cap b \longrightarrow a \, \theta_2 \theta_1 \, b.$$

Proof. By Lemma 9.2.6 we have

$$a \leqslant_\cap b \longrightarrow a \leqslant_\cap a \triangle b \leqslant_\cap b.$$

Thus,

$$a * (a \triangle b) = a = a \cap (a \triangle b),$$
$$b * (b \triangle a) = b = b \cup (a \triangle b).$$

Hence, $a \, \theta_2 \, a \triangle b$, $b \, \theta_1 \, a \triangle b$ and $a \, \theta_2 \theta_1 \, b$. □

Since every lattice $(L; \cap, \cup)$ satisfies the hyperidentity (9.1.6), we have

$$X(Y(X(x, y), z), Y(y, z)) \, \theta_i \, Y(X(x, y), z),$$

where $X, Y \in \{\cap, \cup\}$, $x, y, z \in L$, $i = 1, 2$.

9.4. Weakly interlaced bilattices. The structural results

Proposition 9.4.1. *If $(L; \cap, \cup)$ is a lattice and $(L; *)$ is a semilattice with the operation $*$ interlaced with the operations $\cap$, $\cup$, then there exists a pair (θ_1, θ_2) of the congruences of the lattice $(L; \cap, \cup)$ such that:*

(1) $\theta_1 \cap \theta_2 = \mathbf{0}$ *(null congruence)*,
(2) $a \leqslant_\cap b \longrightarrow a \, \theta_1 \theta_2 \, b$,
(3) $X(Y(X(x, y), z), Y(y, z)) \, \theta_i \, Y(X(x, y), z)$,

for all $X, Y \in \{\cap, \cup\}$, $x, y, z \in L$, $i = 1, 2$.

Conversely, to the every pair (θ_1, θ_2) *of congruences of the lattice* $(L; \cap, \cup)$ *with conditions* (1) – (3), *corresponds a semilattice* $(L; *)$, *the operation of which is interlaced with the operations* $\cap$, $\cup$.

Proof. Condition (2) follows from Lemma 9.3.2 (as it is mentioned above, condition (1) is evident, and condition (3) follows from hyperidentity (9.1.6)). Let us prove the converse.

Let θ_1 and θ_2 be congruences satisfying the condition of the proposition. Let us define the following operation $*$ on L by the rule:

$$a * b = d \longleftrightarrow d\,\theta_1\, a \cap b,\ d\,\theta_2\, a \cup b.$$

Since $a \cap b \leqslant_\cap a \cup b$, we have $a \cap b\,\theta_1\theta_2\, a \cup b$, by condition (2), i.e., there exists an element $d \in L$ such that $a \cap b\,\theta_1\, d$ and $d\,\theta_2\, a \cup b$.

Let us show the uniqueness of $d \in L$. If there exist the elements $d_1, d_2 \in L$, such that $d_1\,\theta_1\, a \cap b$, $d_1\,\theta_2\, a \cup b$, $d_2\,\theta_1\, a \cap b$, $d_2\,\theta_2\, a \cup b$, then $d_1\,\theta_1\, d_2$ and $d_1\,\theta_2\, d_2$, hence, $d_1\,(\theta_1 \cap \theta_2)\, d_2$ and $d_1 = d_2$, because, $\theta_1 \cap \theta_2 = \mathbf{0}$, by condition (1).

Let us proof the idempotency, commutativity, associativity of the operation $*$ and its interlacement with the operations $\cap$, $\cup$ (using Lemma 9.2.1). To prove the hyperidentity (9.1.6), let us check the following identity:

$$((x * y) \cap z) * (y \cap z) = (x * y) \cap z$$

(the other identities have similar proofs).

Let $u_1 = (x * y) \cap z$ and $u_2 = ((x * y) \cap z) * (y \cap z)$. From the definition of the operation $*$ we have $x * y\,\theta_1\, x \cap y$, $x * y\,\theta_2\, x \cup y$. Thus, $(x * y) \cap z\,\theta_1\, x \cap y \cap z$, $(x * y) \cap z\,\theta_2\,(x \cup y) \cap z$, $x \cap y \cap z\,\theta_1 u_1 \theta_2\,(x \cup y) \cap z$ and

$$((x * y) \cap z) * (y \cap z)\,\theta_1\,(x \cap y \cap z) * (y \cap z)\,\theta_1\,(x \cap y \cap z) \cap (y \cap z).$$

Hence, $u_2 \theta_1\, x \cap y \cap z$. Similarly, from the condition $(x * y) \cap z\,\theta_2\,(x \cup y) \cap z$ it follows that

$$((x * y) \cap z) \cup (y \cap z)\,\theta_2\,((x \cup y) \cap z) \cup (y \cap z)$$

and

$$((x*y)\cap z)*(y\cap z)\,\theta_2\,((x*y)\cap z)\cup(y\cap z)\,\theta_2\,((x\cup y)\cap z)\cup(y\cap z)\,\theta_2\,(x\cup y)\cap z.$$

by condition (3). So, $x \cap y \cap z\,\theta_1\, u_2\,\theta_2\,(x \cup y) \cap z$.

Thus, $u_1\,\theta_1\, u_2$, $u_1\,\theta_2\, u_2$, i.e. $u_1\,\theta_1 \cap \theta_2\, u_2$ and $u_1 = u_2$ by condition (1).

Idempotency of the operation $*$. If $a * a = d$, then $d\,\theta_1\, a \cap a = a$ and $d\,\theta_2\, a \cup a = a$. Hence, $d\,\theta_1 \cap \theta_2\, a$ and $d = a$ by condition (1).

Commutativity of the operation $*$. If $a * b = d$, then $d\,\theta_1\, a \cap b = b \cap a$ and $d\,\theta_2\, a \cup b = b \cup a$. Hence, $d = b * a$ and $a * b = b * a$.

Associativity of the operation $*$. If $d_1 = (a * b) * c$ and $d_2 = a * (b * c)$, then $d_1 \theta_1\, a \cap b \cap c$, $d_1 \theta_2\, a \cup b \cup c$, $d_2 \theta_1\, a \cap b \cap c$, $d_2 \theta_2\, a \cup b \cup c$, and $d_1\, \theta_1\, d_2$, $d_1\, \theta_2\, d_2$ and $d_1\, (\theta_1 \cap \theta_2)\, d_2$, hence, $d_1 = d_2$ by condition (1). □

Proposition 9.4.2. *Let $(L; \cap, \cup)$ be a lattice. There exists a bijective correspondence between the set of semilattice operations $*$ on L, that are interlaced with the operations $\cap$, $\cup$, and the set of representations of the lattice L as a subdirect product of two lattices, such that if (a, b), (a', b') are the elements of this subdirect product and $(a, b) \leqslant_\cap (a', b')$, then (a, b') is also an element of this subdirect product. An operation $*$ on L corresponding to such a subdirect representation $\varphi : L \to A \times B$ is determined by the following rule:*

$$x * y = \varphi^{-1}(a \cap a', b \cup b'),$$

where $\varphi(x) = (a, b)$ and $\varphi(y) = (a', b')$, $a, a' \in A$, $b, b' \in B$.

Proof. If $(L; *)$ is a semilattice, the operation of which is interlaced with the operations $\cap$, $\cup$, then according to Proposition 9.4.1, there exist congruences θ_1, θ_2 of the algebra $(L; \cap, \cup)$ such that $\theta_1 \cap \theta_2 = \mathbf{0}$. Hence, the lattice $(L; \cap, \cup)$ is isomorphic to the subdirect product of the factor-lattices, L/θ_1 and L/θ_2, i.e., there exists a monomorphism, $\varphi : L \to L/\theta_1 \times L/\theta_2$, such that $\varphi(x) = ([x]_{\theta_1}, [x]_{\theta_2})$, where $x \in L$.

If $(a, b), (a', b') \in \varphi(L)$ and $(a, b) \leqslant_\cap (a', b')$, then there exist elements, $u, v \in L$, such that $\varphi(u) = (a, b) \leqslant_\cap (a', b') = \varphi(v)$. Hence, $u \leqslant_\cap v$ and, according to Proposition 9.4.1, we have $u\, \theta_1 \theta_2\, v$. So there exists an element, $t \in L$, such that $u\, \theta_1\, t\, \theta_2\, v$. Besides, $\varphi(t) = ([t]_{\theta_1}, [t]_{\theta_2})$, $\varphi(u) = ([u]_{\theta_1}, [u]_{\theta_2})$ and $a = [u]_{\theta_1}$. Since $u\, \theta_1\, t$, then $[u]_{\theta_1} = [t]_{\theta_1}$, i.e. $a = [t]_{\theta_1}$. Similarly, since $\varphi(v) = ([v]_{\theta_1}, [v]_{\theta_2}) = (a', b')$, then $b' = [v]_{\theta_2}$, and since $t\, \theta_2\, v$ we have $[t]_{\theta_2} = [v]_{\theta_2}$ and $b' = [t]_{\theta_2}$. Thus, $\varphi(t) = ([t]_{\theta_1}, [t]_{\theta_2}) = (a, b') \in \varphi(L)$.

Conversely, if $\varphi : L \to A \times B$ is a subdirect product of the lattice L with the corresponding congruences θ_1 and θ_2, then $\theta_1 \cap \theta_2 = \mathbf{0}$ and $A \simeq L/\theta_1$, $B \simeq L/\theta_2$. Since the lattices L/θ_1 and L/θ_2 satisfy hyperidentity (9.1.6), then the congruences, θ_1 and θ_2 satisfy condition (3) of Proposition 9.4.1.

If $a \leqslant_\cap b$, where $a, b \in L$, then $\varphi(a) \leqslant_\cap \varphi(b)$, where $\varphi(a) = ([a]_{\theta_1}, [a]_{\theta_2})$, $\varphi(b) = ([b]_{\theta_1}, [b]_{\theta_2})$, $([a]_{\theta_1}, [b]_{\theta_2}) \in \varphi(L)$. Hence, there exists an element $t \in L$ such that $\varphi(t) = ([a]_{\theta_1}, [b]_{\theta_2}) = ([t]_{\theta_1}, [t]_{\theta_2})$. So, $[a]_{\theta_1} = [t]_{\theta_1}$, $[b]_{\theta_2} = [t]_{\theta_2}$, i.e., $a\,\theta_1\,t$, $t\,\theta_2\,b$ and $a\,\theta_1\theta_2\,b$. Thus, the congruences, θ_1, θ_2, of the lattice, $(L;\cap,\cup)$, satisfy the conditions of Proposition 9.4.1. Hence, we have a semilattice operation $*$ on L, which is interlaced with the operation $\cap$, $\cup$, and $x*y\,\theta_1\,x\cap y$, $x*y\,\theta_2\,x\cup y$. □

Now we can calculate $x*y$, where $x,y\in L$.

If $\varphi(x)=(a,b)$, $\varphi(y)=(a',b')$, i.e., $a=[x]_{\theta_1}$, $b=[x]_{\theta_2}$, $a'=[y]_{\theta_1}$, $b'=[y]_{\theta_2}$, then

$$[x*y]_{\theta_1}=[x\cap y]_{\theta_1}=[x]_{\theta_1}\cap[y]_{\theta_1}=a\cap a',$$
$$[x*y]_{\theta_2}=[x\cup y]_{\theta_2}=[x]_{\theta_2}\cup[y]_{\theta_2}=b\cup b'.$$

Hence,

$$\varphi(x*y)=([x*y]_{\theta_1},[x*y]_{\theta_2})=(a\cap a',b\cup b').$$

Proposition 9.4.3. (a) *A semilattice $(L;*)$ from Proposition 9.4.1 becomes a lattice $(L;*,\triangle)$ for some binary operation $\triangle$ on L (uniquely) if and only if the congruences θ_1 and θ_2 is permutable;*

(b) *A semilattice $(L;*)$ from Proposition 9.4.2 becomes a lattice $(L;*,\triangle)$ for some binary operation $\triangle$ on L (uniquely) if and only if the corresponding subdirect representation becomes a direct representation.*

Proof. (a) If there exists a binary operation $\triangle$ on L such that $(L;*,\triangle)$ is a lattice, then according to Lemmas 9.3.2, 9.3.3 we have $\theta_1\theta_2=\theta_2\theta_1$.

If $\theta_1\theta_2=\theta_2\theta_1$, then according to Proposition 9.4.1, there exists a binary operation $\triangle$ on L, corresponding to the pair of congruences (θ_2,θ_1) such that $\triangle$ is interlaced with the operations $\cap$, $\cup$, and

$$a\triangle b=d\longleftrightarrow d\,\theta_2\,a\cap b,\quad d\,\theta_1\,a\cup b.$$

Since $a\cap b\,\theta_2\,a\triangle b\,\theta_1\,a\cup b$, then, according to property (1) of Lemma 9.3.2, we have:

$$(a\triangle b)*a\,\theta_2(a\cap b)*a\,\theta_2\,(a\cap b)\cup a=a,$$
$$(a\triangle b)*a\,\theta_1(a\cup b)*a\,\theta_1\,(a\cup b)\cap a=a;$$

So, $(a\triangle b)*a\,\theta_1\cap\theta_2\,a$ and $a=(a\triangle b)*a$ by property (1) of Proposition 9.4.1. Similarly, $a=(a*b)\triangle a$, i.e., $(L;*,\triangle)$ is a lattice. Besides, if $(L;*,\triangle_1)$ and $(L;*,\triangle_2)$ are lattices, then $\triangle_1=\triangle_2$ [143], i.e., in the lattice $(L;*,\triangle)$ the operation $\triangle$ is determined in a unique way.

(b) We have $\theta_1\cap\theta_2=\mathbf{0}$. Let us proof that $\theta_1\cup\theta_2=\mathbf{1}$. If $a,b\in L$, then, according to property (1) of Lemma 9.3.2, we have $a\cap b\,\theta_1\,a*b\,\theta_2\,a\cup b$ and, according to Lemma 9.2.3, we have $a \cap b \leqslant_\cap a * b \leqslant_\cap a \cup b$. The rest follows from [143].

Theorem 9.4.4. *Every weakly interlaced bilattice is isomorphic to the superproduct of two lattices* (cf. [300]).

Proof. Let $(L; \cap, \cup, *, \triangle)$ be a weakly interlaced bilattice. If $L_1 = (L; \cap, \cup)$ and $L_2 = (L; *, \triangle)$ are lattices and the operation $*$ is interlaced with the operations $\cap$, $\cup$, then, according to Propositions 9.4.2 and 9.4.3, there exist the lattices A and B such that L_1 is isomorphic to the direct product $A \times B$, i.e., there exists a bijection $\varphi : L \to A \times B$, such that

$$\varphi(x \cap y) = \varphi(x) \cap \varphi(y) = (a \cap a', b \cap b'),$$

$$\varphi(x \cup y) = \varphi(x) \cup \varphi(y) = (a \cup a', b \cup b'),$$

where $\varphi(x) = (a, b)$ and $\varphi(y) = (a', b')$.

Now we have to prove only the following equalities: $\varphi(x * y) = (a \cap a', b \cup b')$ and $\varphi(x \triangle y) = (a \cup a', b \cap b')$. The first equality follows from Proposition 9.4.1.

Let us prove the second one. We have

$$x \cap y \, \theta_2 \, x \triangle y \, \theta_1 \, x \cup y,$$

i.e.,

$$[x \triangle y]_{\theta_2} = [x \cap y]_{\theta_2} = [x]_{\theta_2} \cap [y]_{\theta_2},$$
$$[x \triangle y]_{\theta_1} = [x \cup y]_{\theta_1} = [x]_{\theta_1} \cup [y]_{\theta_1}.$$

Since $\varphi(x) = (a, b)$, $\varphi(y) = (a', b')$, then $a = [x]_{\theta_1}$, $a' = [y]_{\theta_1}$, $b = [x]_{\theta_2}$, $b' = [y]_{\theta_2}$, then

$$[x \triangle y]_{\theta_1} = [x]_{\theta_1} \cup [y]_{\theta_1} = a \cup a',$$
$$[x \triangle y]_{\theta_2} = [x]_{\theta_2} \cap [y]_{\theta_2} = b \cap b'.$$

Hence,

$$\varphi(x \triangle y) = ([x \triangle y]_{\theta_1}, [x \triangle y]_{\theta_2}) = (a \cup a', b \cap b').$$

Thus, the bilattice $(L; \cap, \cup, *, \triangle)$ is isomorphic to the superproduct $A \bowtie B$ of the two lattices A and B. □

Corollary 9.4.5. *Every bounded (modular, distributive) weakly interlaced bilattice is isomorphic to the superproduct of two bounded (modular, distributive) lattices.*

Corollary 9.4.6. *Every weakly interlaced (modular, distributive) bilattice satisfies the hyperidentities of the variety of (modular, distributive) lattices (see Theorems 9.1.1–9.1.3).*

9.5. Weakly interlaced bilattices with involution: The structural results

An algebra $(L; *, ')$ with one binary and one unary operations is called semilattice with involution, if $(L; *)$ is a semilattice and the unary operation $'$ is its second order automorphism, i.e.,

$$(x * y)' = x' * y',$$
$$(x')' = x.$$

An algebra $(L; \cap, \cup, ')$ with two binary and one unary operations is called lattice with involution, if $(L; \cap, \cup)$ is a lattice, and the unary operation $'$ is its second order automorphism, i.e.,

$$(x \cap y)' = x' \cap y',$$
$$(x \cup y)' = x' \cup y',$$
$$(x')' = x.$$

An algebra $(L; \cap, \cup, *, \triangle, ')$ with four binary and one unary operations is called bilattice with involution if $(L; \cap, \cup, ')$ and $(L; *, \triangle, ')$ are lattices with involution.

It is evident that every bilattice is a bilattice with an identical involution.

Proposition 9.5.1. *If $(L; \cap, \cup, ')$ is a lattice with involution, $(L; *, ')$ is a semilattice with involution and the operation $*$ is interlaced with the operations $\cap$ and $\cup$, then there exists a pair of congruences (θ_1, θ_2) of the algebra $(L; \cap, \cup, ')$ that satisfies conditions (1)–(3) of Proposition 9.4.1.*

Conversely, to every pair (θ_1, θ_2) of congruences of the lattice with involution $(L; \cap, \cup, ')$ that satisfies conditions (1)–(3) of Proposition 9.4.1, there corresponds a semilattice with involution $(L; *, ')$ whose binary operation is interlaced with the operations $\cap$ and $\cup$.

Proof. One should repeat the proof of Proposition 9.4.1, simultaneously checking that θ_1, θ_2 also preserve the unary operation . To prove the second part, we only add the identity:

$$(x * y)' = x' * y'.$$

If $x' * y' = d$, then $d\,\theta_1\, x' \cap y'$ and $d\,\theta_2\, x' \cup y'$, i.e., $d\,\theta_1\,(x \cap y)'$, $d\,\theta_2\,(x \cup y)'$. Hence, $d'\,\theta_1\, x \cap y$, $d'\,\theta_2\, x \cup y$ and $d' = x * y$. So, $d = (x * y)'$ and $(x * y)' = x' * y'$. □

Proposition 9.5.2. *Let $(L; \cap, \cup, ')$ be a lattice with involution. Then there exists a bijective correspondence between the set of semilattice operations $*$ that define a semilattice with involution $(L; *, ')$, and are interlaced with* *the operations* $\cap$ *and* $\cup$ *and the set of representations of* $(L; \cap, \cup, ')$ *as a subdirect product of two lattices such that if* (a, b), (a', b') *are the elements of this subdirect product and* $(a, b) \leqslant_{\cap} (a', b')$, *then* (a, b') *is also an element of this subdirect product. An operation* $*$ *on* L *corresponding to such a subdirect representation* $\varphi : L \to A \times B$ *is determined by the following rule:*

$$x * y = \varphi^{-1}(a \cap a', b \cup b'),$$

where $\varphi(x) = (a, b)$ *and* $\varphi(y) = (a', b')$, $a, a' \in A$, $b, b' \in B$.

Proof. The proof of Proposition 9.4.2 is repeated with the help of Proposition 9.5.1. □

Proposition 9.5.3. (a) *A semilattice with involution* $(L; *, ')$ *from Proposition 9.5.1 becomes a lattice with involution* $(L; *, \triangle, ')$ *for some binary operation* $\triangle$ *on* L *(uniquely) if and only if the congruences* θ_1 *and* θ_2 *are permutable.*

(b) *A semilattice with involution* $(L; *, ')$ *from Proposition 9.5.2 becomes a lattice with the involution* $(L; *, \triangle, ')$ *for some operation* $\triangle$ *on* L *(uniquely) if and only if the corresponding subdirect representation becomes a direct representation.*

Proof. The proof of Proposition 9.4.3 is repeated with the help of Propositions 9.5.1 and 9.5.2. □

Theorem 9.5.4. *Every weakly interlaced bilattice with an involution is isomorphic to the superproduct of two lattices with involution.*

Proof. The proof of Theorem 9.4.4 is repeated with the help of Propositions 9.5.1, 9.5.2 and 9.5.3. □

Corollary 9.5.5. *Every weakly interlaced bounded (modular, distributive) bilattice with involution is isomorphic to the superproduct of two bounded (modular, distributive) lattices with involution.*

Corollary 9.5.6. *Every weakly interlaced (modular, distributive) bilattice with involution satisfies the hyperidentities of the variety of (modular, distributive) lattices with involution.*

9.6. Weakly interlaced bilattices with coinvolution: The structural results

Let us recall that an algebra $(L; \cap, \cup, ')$ with two binary operations $\cap$, $\cup$ and one unary operation $'$ is called a lattice with coinvolution, if $(L; \cap, \cup)$ is a lattice and unary operation $'$ is its second order antiautomorphism, i.e.,

$$\begin{aligned}(x \cap y)' &= x' \cup y',\\ (x \cup y)' &= x' \cap y',\\ (x')' &= x.\end{aligned}$$

For example, the De Morgan algebras, Boolean algebras, ortholattices are lattices with coinvolution. An algebra $(L; \cap, \cup, ', 0, 1)$ is called ortholattice, if $(L; \cap, \cup, ')$ is a lattice with coinvolution, $(L; \cap, \cup, 0, 1)$ bounded lattice and the following identities are valid:

$$\begin{aligned}x \cup x' &= 1,\\ x \cap x' &= 0.\end{aligned}$$

Proposition 9.6.1. *Let $(L; \cap, \cup, ')$ be a lattice with coinvolution. If the semilattice operation $*$ on L is interlaced wit the operations $\cap$, $\cup$ and there exists a lattice with coinvolution $(L; *, \triangle, ')$ (for some binary operation $\triangle$ on L), then there exists a pair of congruences (θ_1, θ_2) of algebra $(L; \cap, \cup, ')$ that satisfies the conditions (1)–(3) of Proposition 9.4.1 and the condition*

(4) $\theta_1\theta_2 = \theta_2\theta_1$.

Conversely, to every pair (θ_1, θ_2) of congruences of the lattice wit coinvolution $(L; \cap, \cup, ')$ with conditions (1)–(4), there corresponds a lattice with coinvolution $(L; *, \triangle, ')$, whose operation $*$ is interlaced with the operations $\cap$, $\cup$.

Proof. If the relations θ_1, θ_2 are defined as above, then, as is known, they are congruences of $(L; \cap, \cup)$ and satisfy conditions (1)–(4). It remains to add the following conditions:

$$\begin{aligned}a\,\theta_1\,b &\longrightarrow a'\,\theta_1\,b',\\ a\,\theta_2\,b &\longrightarrow a'\,\theta_2\,b'.\end{aligned}$$

Namely, if $a\,\theta_1\,b$, then $a * b = a \cup b$, $(a * b)' = (a \cup b)'$, $a'\triangle b' = a' \cap b'$ and $a'\triangle b'\,\theta_1\,a' \cap b'$. In addition $a'\triangle b'\,\theta_1\,a' \cup b'$; hence, $a' \cap b'\,\theta_1\,a' \cup b'$ and $a'\,\theta_1\,b'$.

The second condition has the similar proof. □

Conversely, for every pair (θ_1, θ_2) of congruences of the algebra $(L; \cap, \cup, ')$, we determine the operations $*$ and $\triangle$ by the following rule:

$$\begin{aligned}a * b = d &\longleftrightarrow d\,\theta_1\,a \cap b,\; d\,\theta_2\,a \cup b,\\ a\triangle b = c &\longleftrightarrow c\,\theta_1\,a \cup b,\; c\,\theta_2\,a \cap b.\end{aligned}$$

As in the case of $*$, it is ease to proof the correctness of the operation $\triangle$, as well as its idempotency, commutativity, associativity, the interlacement of $*$ with the operations $\cap$, $\cup$, and the absorption low.

Now we only have to prove the De Morgan law. If $x' * y' = d$, then $x' \cap y' \,\theta_1\, d$ and $x' \cup y' \,\theta_2\, d$, i.e. $(x \cup y)' \,\theta_1\, d$, $d \,\theta_2\, (x \cap y)'$. Hence, $x \cup y \,\theta_1\, d'$, $d' \,\theta_2\, x \cap y$, i.e., $d' = x \triangle y$ and $d = (x \triangle y)'$. So

$$(x \triangle y)' = x' * y'.$$

Proposition 9.6.2. *Let $(L; \cap, \cup, ')$ be a lattice wit coinvolution. Then there exists a bijection between set of lattices wit coinvolution $(L; *, \triangle, ')$ whose operation $*$ is interlaced with the operations $\cap$, $\cup$, and the set of isomorphisms from the algebra $(L; \cap, \cup, ')$ to the direct products of two lattices with coinvolution. If $\varphi : L \to A \times B$ is such an isomorphism, then the operations $*$ and $\triangle$ on L are determined by the rule:*

$$x * y = \varphi^{-1}(a \cap a', b \cup b'),$$
$$x \triangle y = \varphi^{-1}(a \cup a', b \cap b'),$$

where $\varphi(x) = (a, b)$ and $\varphi(y) = (a', b')$, $a, a' \in A$, $b, b' \in B$.

Proof. See the proofs of the Propositions 9.4.2 and 9.4.3 with the help of Propositions 9.6.1. □

Theorem 9.6.3. *Every weakly interlaced bilattice with coinvolution is isomorphic to the superproduct of two lattices with coinvolution.*

Proof. The proof of the Theorem 9.4.4 is repeated with the help of the Propositions 9.6.1 and 9.6.2. □

Corollary 9.6.4. *Every weakly interlaced bounded bilattice with coinvolution is isomorphic to the superproduct of two bounded lattices with coinvolution.*

Corollary 9.6.5. *Every weakly interlaced De Morgan (Boolean) bilattice is isomorphic to the superproduct of two De Morgan (Boolean) lattices.*

Corollary 9.6.6. *Every weakly interlaced bilattice with coinvolution satisfies the hyperidentities of the variety of lattices with coinvolution.*

Corollary 9.6.7. *Every weakly interlaced De Morgan bilattice satisfies the hyperidentities of the variety of De Morgan lattices (see Theorem 9.1.5).*

Corollary 9.6.8. *Every weakly interlaced Boolean bilattice satisfies the hyperidentities of the variety of Boolean algebras (see Theorem 9.1.4).*

Theorem 9.6.9 ([299]). *If bilattice* $(L; \wedge_1, \vee_1, \wedge_2, \vee_2, ')$ *with additional unary operation* $'$ *satisfies the hyperidentities of the variety of Boolean algebras, then it is an interlaced Boolean bilattice.*

Proof. It is enough to mention that in the variety of Boolean algebras satisfied the following hyperidentities:

$$(x')' = x \quad \text{and} \quad X(x, X(y, y')') = x.$$

Hence, the bilattice $(L; \wedge_1, \vee_1, \wedge_2, \vee_2, ')$, which satisfies these hyperidentities, also satisfies the following identities too:

$$\begin{aligned} x \wedge_1 (y \wedge_1 y')' &= x, \\ x \vee_1 (y \vee_1 y')' &= x, \\ x \wedge_2 (y \wedge_2 y')' &= x, \\ x \vee_2 (y \vee_2 y')' &= x. \end{aligned}$$

From these identities it follow that the reducts $L^1 = (L; \wedge_1, \vee_1, ')$ and $L^2 = (L; , \wedge_2, \vee_2, ')$ are bounded lattices with the following largest and smallest elements:

$$\begin{aligned} \top_1 &= (y \wedge_1 y')', \bot_1 = (y \vee_1 y')', \\ \top_2 &= (y \wedge_2 y')', \bot_2 = (y \vee_2 y')', \end{aligned}$$

i.e.,

$$\begin{aligned} \top_1 &= y \vee_1 y', \bot_1 = y \wedge_1 y', \\ \top_2 &= y \vee_2 y', \bot_2 = y \wedge_2 y'. \end{aligned}$$

Thus the reducts L^1 and L^2 are complemented lattices. From the hyperidentity (9.1.6) follows that $(L; \wedge_1, \vee_1, \wedge_2, \vee_2, ')$ is an interlaced bilattice (Lemma 9.2.1), and from the hyperidentity of distributivity (9.1.8) it follows that L^1 and L^2 are Boolean algebras. □

For a Boolean bilattice, a reduct that contains the unary operation and one or two of its binary operation is called a Boolean reduct .

Corollary 9.6.10. *A Boolean reduct of the weakly interlaced Boolean bilattice has a decidable elementary theory (see Chapter 6).*

9.7. Weakly interlaced Ginsberg bilattices: The structural results

For various characterizations of weakly interlaced Ginsberg bilattices without the boundedness condition, we consider the lattices with coinvolution $(L;\cap,\cup,\ ')$ and the semilattices with involution $(L;*,\ ')$ as two-sorted algebras $(L,L;\cap,\cup,\ ')$ and $(L,L;*,\ ')$, where $' : L \to L$ is a mapping from the first L to the second one. Then we consider theirs homomorphisms, kernels of homomorphisms, congruences and subalgebras in the category of many-sorted algebras [41,151,329]. For example, the congruence of algebra $(L,L;\cap,\cup,\ ')$ is a pair (α_1,α_2), where α_1 and α_2 are congruences of algebra $(L;\cap,\cup)$, and

$$x\,\alpha_1\,y \longrightarrow x'\,\alpha_2\,y',$$

where $x,y \in L$. The congruence of algebra $(L,L;*,\ ')$ has the similar definition.

Lemma 9.7.1. *If $(L;\cap,\cup,\ ')$ is a lattice with coinvolution and $(L;*,\ ')$ is a semilattice with involution whose binary operation is interlaced with the operations $\cap$ and $\cup$, then the pairs (θ_1,θ_2) and (θ_2,θ_1) are congruences of two-sorted algebra $(L,L;\cap,\cup,\ ')$, where θ_1 and θ_2 were defined in Sec. 9.3.*

Proof. Indeed,

$$\begin{aligned}
&x\,\theta_1\,y \longrightarrow x*y = x\cup y \longrightarrow (x*y)' = (x\cup y)'\\
&\quad\longrightarrow x'*y' = x'\cap y' \longrightarrow x'\,\theta_2\,y',\\
&x\,\theta_2\,y \longrightarrow x*y = x\cap y \longrightarrow (x*y)' = (x\cap y)' \longrightarrow x'*y' = x'\cup y'\\
&\quad\longrightarrow x'\,\theta_1\,y'.
\end{aligned}$$

□

Lemma 9.7.2. *If $(L;\cap,\cup,\ ')$ is a lattice with coinvolution and $(L;*,\ ')$ is a semilattice with involution whose binary operation is interlaced with the operations $\cap$ and $\cup$, then the pairs (θ_1,θ_2) and (θ_2,θ_1) are congruences of two-sorted algebra $(L,L;*,\ ')$.*

Proposition 9.7.3. *Let $(L;\cap,\cup,\ ')$ be lattice with coinvolution. If $(L;*,\ ')$ is a semilattice with involution whose binary operation is interlaced with the operations $\cap$ and $\cup$, then there exist congruences (θ_1,θ_2) and (θ_2,θ_1) two-sorted algebra $(L,L;\cap,\cup,\ ')$ that satisfy conditions (1)–(3) of Proposition 9.4.1.*

Conversely, to every pair of congruence $((\theta_1, \theta_2), (\theta_2, \theta_1))$ *of two-sorted algebra* $(L, L; \cap, \cup, ')$ *with conditions (1)–(3) of Proposition 9.4.1, there corresponds a semilattice with involution* $(L; *, ')$ *whose binary operation is interlaced with the operations* $\cap$ *and* $\cup$.

Proof. The first part of this assertion follows from Lemma 9.7.1 with the help of Proposition 9.4.1. For the second part of the assertion, we add the following identity to the proof of Proposition 9.4.1:

$$(x * y)' = x' * y'.$$

If $x' * y' = d$, then $d\,\theta_1\,x' \cap y'$, $d\,\theta_2\,x' \cup y'$, i.e., $d\,\theta_1\,(x \cup y)'$ and $d\,\theta_2\,(x \cap y)'$. Hence, $d'\,\theta_2\,x \cup y$ and $d'\,\theta_1\,x \cap y$. So, $d' = x * y$ and $d = (x * y)'$, i.e., $(x * y)' = x' * y'$. □

We considered two-sorted algebras of the form $L = (L_1, L_2; \cap, \cup, ')$, where the binary operation $\cap$ and $\cup$ are defined on the set L_1, and the unary operation is defined as follows: $' : L_1 \to L_2$, i.e., the binary operations have the type $(1,1;1)$, and the unary operation has the type $(1;2)$ in the sense of monograph [329].

For two-sorted algebras $A = (A_1, A_2; \cap_1, \cup_1, ')$ and $B = (B_1, B_2; \cap_2, \cup_2, ^\star)$, let us define the following two-sorted algebra:

$$A \boxdot B = (A_1 \times B_1, B_2 \times A_2; \cap, \cup, ^-)$$

where

$$(a_1, b_1) \cap (a_2, b_2) = (a_1 \cap_1 a_2, b_1 \cap_2 b_2),$$
$$(a_1, b_1) \cup (a_2, b_2) = (a_1 \cup_1 a_2, b_1 \cup_2 b_2),$$

for $a_1, a_2 \in A_1$ and $b_1, b_2 \in B_1$, i.e., $\cup, \cap : (A_1 \times B_1) \times (A_1 \times B_1) \longrightarrow A_1 \times B_1$, and

$$\overline{(a, b)} = (b^\star, a'),$$

for $a \in A_1$, $b \in B_1$, $a' \in A_2$ and $b^\star \in B_2$, i.e. $- : A_1 \times B_1 \to B_2 \times A_2$.

We call the algebra $A \boxdot B$ the Ginsberg product of the algebras A and B. The Ginsberg product has the advantage, that in the case of $A = (A_1, A_2; \cap_1, \cup_1, ')$ and $B = (A_2, A_1; \cap_2, \cup_2, ^\star)$ the product $A \boxdot B$ is an ordinary algebra, i.e., all of its operation are defined on the same set $A_1 \times A_2$.

Two-sorted algebras $A = (A_1, A_2; \cap_1, \cup_1, \,')$ and $B = (A_2, A_1; \cap_2, \cup_2, ^\star)$ are called lattice-connected, if $(A_1; \cap_1, \cup_1)$ and $(A_2; \cap_2, \cup_2)$ are lattices and

$$\begin{aligned}
&(x \cup_1 y)' = x' \cup_2 y',\\
&(x \cap_1 y)' = x' \cap_2 y',\\
&(u \cup_2 v)^\star = u^\star \cup_1 v^\star,\\
&(u \cap_2 v)^\star = u^\star \cap_1 v^\star,\\
&(x')^\star = x, \quad (u^\star)' = u,
\end{aligned}$$

i.e., $'$ is an isomorphism from $(A_1; \cap_1, \cup_1)$ to $(A_2; \cap_2, \cup_2)$, and * is an isomorphism from $(A_2; \cap_2, \cup_2)$ to $(A_1; \cap_1, \cup_1)$, and the mappings $'$ and * are inverse mappings (with respect to each other). Clearly, it suffices to require that only one of the mappings should be an isomorphism.

If

$$\begin{aligned}
&e_1 : A_1 \times B_1 \to A_1,\ \tilde{e}_1 : B_2 \times A_2 \to A_2,\\
&e_2 : A_1 \times B_1 \to B_1,\ \tilde{e}_2 : B_2 \times A_2 \to B_2
\end{aligned}$$

are projections, then the pairs $(e_1, \tilde{e}_1)$ and $(e_2, \tilde{e}_2)$ are homomorphisms from algebra $A \boxdot B$ to A and to B. Let us denoted $\mathrm{Ker}\,(e_i, \tilde{e}_i) = (\mathrm{Ker}(e_i), \mathrm{Ker}\,(\tilde{e}_i))$ and considered the quotient-algebra of algebra $A \boxdot B$ modulo $\mathrm{Ker}\,(e_i, \tilde{e}_i)$:

$$A \boxdot B/\mathrm{Ker}\,(e_i, \tilde{e}_i) = \left(A_1 \times B_1/\mathrm{Ker}\,(e_i)\,, B_2 \times A_2/\mathrm{Ker}\,(\tilde{e}_i)\,; \cap, \cup, -\right).$$

In the case $i = 1$, we have isomorphism

$$A \boxdot B/\mathrm{Ker}\,(e_1, \tilde{e}_1) \simeq A,$$

and in the case $i = 2$, we have

$$A \boxdot B/\mathrm{Ker}\,(e_2, \tilde{e}_2) \simeq B.$$

Lemma 9.7.4. *Let $L = (L, L; \cap, \cup, \,')$, $A = (A_1, A_2; \cap, \cup, \,')$ and $B = (B_1, B_2; \cap, \cup, ^\star)$ be two-sorted algebras. If there exists an isomorphism $(\varphi_1, \varphi_2) : L \to A \boxdot B$, then there exist congruences $\left(\theta_1, \tilde{\theta}_1\right)$ and $\left(\theta_2, \tilde{\theta}_2\right)$ of the algebra L such that*

$1'$. *$L/\left(\theta_1, \tilde{\theta}_1\right) \simeq A$ and $L/\left(\theta_2, \tilde{\theta}_2\right) \simeq B$,*
$2'$. *$\theta_1 \vee \theta_2 = \mathbf{1}$ and $\tilde{\theta}_1 \vee \tilde{\theta}_2 = \mathbf{1}$,*
$3'$. *$\theta_1 \cap \theta_2 = \mathbf{0}$ and $\tilde{\theta}_1 \cap \tilde{\theta}_2 = \mathbf{0}$,*
$4'$. *$\theta_1\theta_2 = \theta_2\theta_1$ and $\tilde{\theta}_1\tilde{\theta}_2 = \tilde{\theta}_2\tilde{\theta}_1$.*

Corollary 9.7.5. *Let* $L = (L, L; \cap, \cup, ')$, $A = (A_1, A_2; \cap, \cup, ')$ *and* $B = (A_2, A_1; \cap, \cup, ^{\star})$ *be two-sorted algebras. If there exists an isomorphism* $(\varphi_1, \varphi_2) : L \to A \boxdot B$, *then there exist congruences* (θ_1, θ_2) *and* (θ_2, θ_1) *of the algebra* L *such that*

$1''$. $L/(\theta_1, \theta_2) \simeq A$ *and* $L/(\theta_2, \theta_1) \simeq B$,
$2''$. $\theta_1 \vee \theta_2 = \mathbf{1}$,
$3''$. $\theta_1 \cap \theta_2 = \mathbf{0}$,
$4''$. $\theta_1\theta_2 = \theta_2\theta_1$.

A subalgebra $(M_1, M_2; \cap, \cup, ^{-})$ of algebra $A \boxdot B$ such that $e_1(M_1) = A_1$, $e_2(M_1) = B_1$, $\tilde{e}_1(M_2) = A_2$ and $\tilde{e}_2(M_2) = B_2$ is called the Ginsberg subproduct of algebras A and B.

Lemma 9.7.6. *Let* $L = (L, L; \cap, \cup, ')$, $A = (A_1, A_2; \cap, \cup, ')$ *and* $B = (B_1, B_2; \cap, \cup, ^{\star})$ *be two-sorted algebras. If there exists an isomorphism from* L *to Ginsberg subproduct of algebras* A *and* B, *then there exist the congruences* $\left(\theta_1, \tilde{\theta}_1\right)$ *and* $\left(\theta_2, \tilde{\theta}_2\right)$ *of the algebra* L *such that*

$1'''$. $L/\left(\theta_1, \tilde{\theta}_1\right) \simeq A$ *and* $^{L}/\left(\theta_2, \tilde{\theta}_2\right) \simeq B$,
$2'''$. $\theta_1 \cap \theta_2 = \mathbf{0}$ *and* $\tilde{\theta}_1 \cap \tilde{\theta}_2 = \mathbf{0}$.

Corollary 9.7.7. *Let* $L = (L, L; \cap, \cup, ')$, $A = (A_1, A_2; \cap, \cup, ')$ *and* $B = (A_2, A_1; \cap, \cup, ^{\star})$ *be two-sorted algebras. If there exists an isomorphism from* L *to Ginsberg subproduct of algebras* A, B, *then there exist the congruences* (θ_1, θ_2) *and* (θ_2, θ_1) *of algebra* L *such that*

1^{IV}. $^{L}/(\theta_1, \theta_2) \simeq A$ *and* $^{L}/(\theta_2, \theta_1) \simeq B$,
2^{IV}. $\theta_1 \cap \theta_2 = \mathbf{0}$.

Lemma 9.7.8. *Let* $\left(\theta_1, \tilde{\theta}_1\right)$ *and* $\left(\theta_2, \tilde{\theta}_2\right)$ *be the congruences of algebra* $L = (L, L; \cap, \cup, ')$ *such that*

1^{V}. $\theta_1 \vee \theta_2 = \mathbf{1}$ *and* $\tilde{\theta}_1 \vee \tilde{\theta}_2 = \mathbf{1}$;
2^{V}. $\theta_1 \cap \theta_2 = \mathbf{0}$ *and* $\tilde{\theta}_1 \cap \tilde{\theta}_2 = \mathbf{0}$;
3^{V}. $\theta_1\theta_2 = \theta_2\theta_1$ *and* $\tilde{\theta}_1\tilde{\theta}_2 = \tilde{\theta}_2\tilde{\theta}_1$.

Then there exists an isomorphism $(\varphi_1, \varphi_2) : L \to L/(\theta_1, \tilde{\theta}_1) \boxdot L/(\theta_2, \tilde{\theta}_2)$ *such that* $\varphi_1(a) = ([a]_{\theta_1}, [a]_{\theta_2})$, $\varphi_2(a) = ([a]_{\tilde{\theta}_2}, [a]_{\tilde{\theta}_1})$, *where* $a \in L$.

Corollary 9.7.9. *If the congruences* $(\theta_1, \tilde{\theta}_1)$ *and* $(\theta_2, \tilde{\theta}_2)$ *of the algebra* $L = (L, L; \cap, \cup, ')$ *satisfy condition* 2^{V}, *then there exists an isomorphism* (φ_1, φ_2) *from algebra* L *to the Ginsberg subproduct of algebras* $L/(\theta_1, \tilde{\theta}_1)$

and $L/(\theta_2,\tilde{\theta}_2)$, *such that* $\varphi_1(a) = ([a]_{\theta_1},[a]_{\theta_2})$ *and* $\varphi_2(a) = ([a]_{\tilde{\theta}_2},[a]_{\tilde{\theta}_1})$, *where* $a \in L$. *In the case of congruences* (θ_1,θ_2) *and* (θ_2,θ_1), *we have the equality* $\varphi_1 = \varphi_2$, *i.e., we get an isomorphism of the ordinary algebras.*

Proposition 9.7.10. *Let* $(L;\cap,\cup,{}')$ *be a lattice with coinvolution. Then there exists a bijective correspondence between the set of semilattice operations* $*$ *that define a semilattice with involution* $(L;*,{}')$ *and are interlaced with the operations* $\cap$, $\cup$, *and the set of representations of* $L = (L,L;\cap,\cup,{}')$ *as a Ginsberg subproduct of two lattice-connected algebras such that if* (a,b), (a',b') *are elements of this Ginsberg subproduct and* $(a,b) \leqslant (a',b')$, *then* (a,b') *is also an element of this Ginsberg subproduct. An operation* $*$ *on* L *corresponding to such a representation* $(\varphi_1,\varphi_2): L \to A \boxdot B$ *is determined by the rule*

$$x * y = \varphi_1^{-1}(a \cap a', b \cup b'),$$

where $\varphi_1(x) = (a,b)$ *and* $\varphi_1(y) = (a',b')$, $a,a' \in A_1$, $b,b' \in A_2$.

Proof. If $(L;*,{}')$ is a semilattice with involution whose binary operation is interlaced with the operations $\cap$ and $\cup$, then, according to Proposition 9.7.3, to the binary operation $*$ there correspond congruences (θ_1,θ_2) and (θ_2,θ_1) of the algebra $L = (L,L;\cap,\cup,{}')$ with conditions (1)–(3) of Proposition 9.4.1. According to Corollary 9.7.9, the algebra L is isomorphic to the Ginsberg subproduct of the algebras $L/(\theta_1,\theta_2)$ and $L/(\theta_2,\theta_1)$, i.e., there exists a monomorphism $(\varphi_1,\varphi_2): L \to L/(\theta_1,\theta_2) \boxdot L/(\theta_2,\theta_1)$ such that $(\varphi_1(L),\varphi_2(L);\cap,\cup,\bar{})$ is a Ginsberg subproduct. It is evident that $L/(\theta_1,\theta_2$ and $L/(\theta_2,\theta_1)$ lattice-connected algebras. □

Corollary 9.7.11. (a) *The semilattice with involution* $(L;*,{}')$ *from Proposition 9.7.3 becomes a lattice with involution* $(L;*,\triangle,{}')$ *for some binary operation* $\triangle$ *on* L *(in a unique way) if and only if the congruences* (θ_1,θ_2) *and* (θ_2,θ_1) *is permutable.*

(b) *The semilattice with involution* $(L;*,{}')$ *from Proposition 9.7.10 becomes a lattice with involution* $(L;*,\triangle,{}')$ *for some binary operation* $\triangle$ *on* L *(in a unique way) if and only if the corresponding representation of* $(L,L;\cap,\cup,{}')$ *as Ginsberg subproduct becomes a representation of* $(L,L;\cap,\cup,{}')$ *as Ginsberg product.*

Let $A = (A_1,A_2;\cap_1,\cup_1,{}')$ and $B = (B_1,B_2;\cap_2,\cup_2,{}^*)$ be two-sorted algebras. An algebra

$$A \boxtimes B = (A_1 \times B_1, B_2 \times A_2; (\cap_1,\cup_2),(\cup_1,\cap_2),(\cap_1,\cap_2),(\cup_1,\cup_2),\bar{}),$$

is called Ginsberg superproduct of algebras A and B, where

$$\overline{(a,b)} = (b^\star, a'), \qquad a \in A_1, \; b \in B_1,$$

and all binary operations operate component-wise, i.e., the reduct $(A_1 \times B_1; (\cap_1, \cup_2), (\cup_1, \cap_2), (\cap_1, \cap_2), (\cup_1, \cup_2))$ is a superproduct of algebras $(A_1; \cap_1, \cup_1)$ and $(B_1; \cap_2, \cup_2)$.

Lemma 9.7.12. *The Ginsberg superproduct of two lattice-connected algebras is interlaced Ginsberg bilattice.*

Proof. The binary reduct of Ginsberg superproduct of two lattice-connected algebras is a superproduct of two lattices, and superproduct of two lattices preserves the hyperidentities of the variety of lattices. Hence, hyperidentity (9.1.6) satisfied in the Ginsberg superproduct of two lattice-connected algebras. Thus, according to Corollary 9.2.2, the Ginsberg superproduct of two lattice-connected algebras is an interlaced bilattice. The rest of the assertion has the easy proof. Namely, if

$$\begin{aligned} \wedge = (\cap_1, \cup_2), \vee = (\cup_1, \cap_2), \\ \cdot = (\cap_1, \cap_2), + = (\cup_1, \cup_2), \end{aligned}$$

then

$$\begin{aligned} \overline{(a_1,b_1) \wedge (a_2,b_2)} &= \overline{(a_1 \cap_1 a_2, b_1 \cup_2 b_2)} = ((b_1 \cup_2 b_2)^\star, (a_1 \cap_1 a_2)') \\ &= (b_1^\star \cup_1 b_2^\star, a_1' \cap_2 a_2') = (b_1^\star, a_1') \vee (b_2^\star, a_2') \\ &= \overline{(a_1,b_1)} \vee \overline{(a_2,b_2)}. \end{aligned}$$

Similarly,

$$\begin{aligned} \overline{(a_1,b_1) \vee (a_2,b_2)} &= \overline{(a_1,b_1)} \wedge \overline{(a_2,b_2)}, \\ \overline{(a_1,b_1) + (a_2,b_2)} &= \overline{(a_1,b_1)} + \overline{(a_2,b_2)}, \\ \overline{(a_1,b_1) \cdot (a_2,b_2)} &= \overline{(a_1,b_1)} \cdot \overline{(a_2,b_2)}, \\ \overline{\overline{(a,b)}} &= (a,b). \end{aligned}$$

□

Theorem 9.7.13. *Let $(L; \cap, \cup, ')$ is a lattice with coinvolution and $(L; *, \triangle, ')$ is a lattice with involution. If the operation $*$ or the operation $\triangle$ is interlaced with the operations $\cap$ and $\cup$, then the algebra $(L; \cap, \cup, *, \triangle, ')$ is isomorphic to the Ginsberg superproduct of two lattice-connected algebras, i.e., every weakly interlaced Ginsberg bilattice is isomorphic to the superproduct of two lattice-connected algebras.*

Proof. According to Proposition 9.7.10 and Corollary 9.7.11, there exist two lattice-connected algebras $L/(\theta_1, \theta_2) = (A_1, A_2; \cap_1, \cup_1, ')$ and $L/(\theta_2, \theta_1) = (A_2, A_1; \cap_2, \cup_2, ^\star)$ such that $(L, L; \cap, \cup, ')$ is isomorphic to their Ginsberg product. In addition, according to Corollary 9.7.9, in this isomorphism (φ_1, φ_2) we have $\varphi_1 = \varphi_2$, i.e., we have an isomorphism of the ordinary algebras. We also see that the mapping $\varphi_1 = \varphi_2$ is an isomorphism between the algebras: $(L; \cap, \cup, *, \triangle, ')$ and Ginsberg superproduct of algebras $L/(\theta_2, \theta_1)$ and $L/(\theta_1, \theta_2)$ (see the proof of Theorem 9.4.4). □

If in the lattice-connected algebras A and B we have $A_1 = A_2$, and the unary operations are $' = {}^\star = \varepsilon$, where ε is an identical mapping, then $A = B = (A_1; \cap_1, \cup_2, \varepsilon)$ is a lattice with identical involution, and $\overline{(x, y)} = (y, x)$, and in this case we denote

$$A \boxtimes A = G(A).$$

Theorem 9.7.14. *Every weakly interlaced Ginsberg bilattice is isomorphic to the bilattice $G(A)$ for some lattice $A = (A; \cap, \cup, \varepsilon)$ with identical involution.*

Proof. By previous theorem every Ginsberg bilattice $(L; \cap, \cup, *, \triangle, ')$ is isomorphic to the Ginsberg superproduct $A \boxtimes B$ of two lattice-connected algebras $A = (A_1, A_2; \cap_1, \cup_1, ')$ and $B = (A_2, A_1; \cap_2, \cup_2, ^\star)$. Let us denote $C = A_1$ and proof there existence of an isomorphism $\varphi : A \boxtimes B \to G(C)$. If $x \in A_1$, $y \in A_2$ and $\varphi((x, y)) = (x, y^\star)$, then φ is a bijection and preserves the bilattice operations. □

Corollary 9.7.15. *Every bounded (modular, distributive) weakly interlaced Ginsberg bilattice is isomorphic to the bilattice $G(A)$ for some (modular, distributive) lattice $A = (A; \cap, \cup, \varepsilon)$ with identical involution.*

Problem 9.7.16. *Characterize the hyperidentities of the variety of (modular, distributive) lattices with involution, and describe the algebras with these hyperidentities.*

Problem 9.7.17. *Characterize the hyperidentities of the variety of (modular) lattices with coinvolution, and describe the algebras with these hyperidentities.*

Problem 9.7.18. *Characterize the hyperidentities of the variety of (modular, distributive) Ginsberg bilattices, and describe the algebras with these hyperidentities.*

Chapter 10

Super-Boolean Functions and Free Super-Boolean Algebras

10.1. Preliminaries

Definition 10.1.1 ([256,269]). An algebra $\mathfrak{A} = (Q; \Sigma)$ is called a super-Boolean algebra if it satisfies all hyperidentities of the variety of Boolean algebras (see Chapter 5).

For a set $T \subseteq \mathbb{N} = \{0, 1, 2, \ldots\}$ a super-Boolean T-algebra is a super-Boolean algebra with arithmetic type T. The following result is proved in Sec. 5.5.

Lemma 10.1.2. *If a super-Boolean algebra $\mathfrak{A} = (Q; \Sigma)$ is subdirectly irreducible then there are at most two binary operations in the set Σ.*

In Chapter 5 a structural theorem for super-Boolean algebras with two binary and one unary operations is proved. Let us recall the definition of Boolean sum introduced in [256] to formulate that structural result.

Definition 10.1.3. Let $\mathfrak{A} = (Q; \Omega \cup \{F\})$ be an algebra with a unary operation F. Let $(Q_i; \Omega)$, $i \in I$ be subalgebras of the algebra $\mathfrak{A}$, and $\mathfrak{A}_i = (Q_i; \Omega \cup \{F_i\})$ be algebras with a unary operation F_i. The algebra $\mathfrak{A}$ is called *Boolean sum* of algebras $\mathfrak{A}_i$, if the following conditions hold true:

(1) $Q_i \cap Q_j = \emptyset$ for all $i, j \in I, i \neq j$.
(2) $Q = \bigcup_{i \in I} Q_i$.
(3) There exist two elements $0, 1 \in I$ and two binary operations $+, \cdot$ and a unary operation $'$ can be defined on I such that $(I; \{+, \cdot, ', 0, 1\})$ is a Boolean algebra.

(4) If $i, j \in I$ and $i \leq j$ (here "$\leq$" is the order of the lattice $(I; \{+, \cdot\})$), then there exists an (bi)isomorphism

$$(\varphi_{i,j}, \tilde{\varepsilon}) : \mathfrak{A}_i \to \mathfrak{A}_j,$$

where $\tilde{\varepsilon}(F_i) = F_j$, $\tilde{\varepsilon}(A) = A$ for any $A \in \Omega$. Moreover, $\varphi_{i,i}$ is the identical mapping of the set Q_i, and for all $i \leq j \leq k$ we have $\varphi_{i,j} \cdot \varphi_{j,k} = \varphi_{i,k}$.

(5) For every $i \in I$ there exists an (bi)isomorphism

$$(h_{i,i'}, \tilde{\varepsilon}) : \mathfrak{A}_i \to \mathfrak{A}_{i'}$$

such that $h_{i,i'}^{-1} = h_{i',i}$ and $h_{i,i'} \cdot \varphi_{i',1} = \varphi_{i,1}$.

(6) For any n-ary operation $A \in \Omega$ $(n \geq 2)$ and for any $x_1, \ldots, x_n \in Q$ we have

$$A(x_1, \ldots, x_n) = A(\varphi_{i_1,i_0}(x_1), \ldots, \varphi_{i_n,i_0}(x_n)),$$

where $x_j \in Q_{i_j}$, $i_j \in I$, $j = 1, \ldots, n$, $i_0 = i_1 + \cdots + i_n$.

(7) For any $x \in Q$ we have

$$F(x) = h_{i,i'}(F_i(x)),$$

where $x \in Q_i$.

Th following structural result is proved in Chapter 5.

Theorem 10.1.4. *An algebra $\mathfrak{A} = (Q, \{+, \cdot, {}'\})$ with two binary operations $+, \cdot$ and one unary operation ${}'$ is a super-Boolean algebra if and only if it is a Boolean lattice or Boolean sum of Boolean lattices.*

In this chapter, we give a functional representation of the free n-generated super-Boolean algebra with two binary, one unary and two nullary operations. Namely, we define the concept of super-Boolean function and prove that the free super-Boolean algebra with two binary, one unary and two nullary operations on n free generators is isomorphic to the super-Boolean algebra of super-Boolean functions of n variables.

10.2. Auxiliary results

The T-hypervariety of super-Boolean T-algebras, where $T = \{1, 2\}$, is denoted by $\mathfrak{QB}$. From the results of Chapter 5 follows that this hypervariety can be axiomatized by finite number of hyperidentities.

Note that from the hyperidentity $X(x) = Y(x)$ of Boolean algebras it follows that any super-Boolean algebra has a unique unary operation, which we will denote by ${}'$.

Denote by $\mathfrak{QB}(2,2,1)$ the variety of all super-Boolean algebras with two binary and one unary operations. So the variety $\mathfrak{QB}(2,2,1)$ is a hypervariety.

We will consider the super-Boolean algebras with two binary, one unary and two nullary operations (i.e., constants) that satisfy some identities (as given below). We will denote the variety of such super-Boolean algebras by $\mathfrak{QB}(2,2,1,0,0)$. Namely, an algebra $\mathfrak{A} = (Q; \{\vee, \cdot, ', 0, 1\})$ belongs to $\mathfrak{QB}(2,2,1,0,0)$ if and only if it satisfies all the hyperidentities of the variety of Boolean algebras and also the following two identities:

$$x \cdot 1 = x,$$
$$1' = 0.$$

The free algebras of the variety $\mathfrak{QB}(2,2,1,0,0)$ are called the free super-Boolean algebras with two binary, one unary and two nullary operations. Our main result is the characterization of the finitely generated free super-Boolean algebras with two binary, one unary and two nullary operations.

Similarly, we define the variety $\mathfrak{QB}(2,1,0)$ to be the class of all super-Boolean algebras with one binary, one unary and one nullary operations, where the constant (the nullary operation) is the unity of the binary operation.

Let $B = \{0,1\}$. Define the operations $+, \cdot, '$, on B by the following way: $0+0=0,\ 0+1=1+0=1+1=1,\ 0\cdot 1=0\cdot 0=1\cdot 0=0,\ 1\cdot 1=1,\ 0'=1,\ 1'=0$. We get the Boolean algebra $(B; \{+, \cdot, ', 0, 1\})$. Denote $\mathbf{2} = (B; \{+, \cdot, ', 0, 1\})$, $\mathbf{2}^+ = (B; \{+, ', 0\})$, $\mathbf{2}^\bullet = (B; \{\cdot, ', 1\})$. Obviously, $\mathbf{2}^+$ and $\mathbf{2}^\bullet$ are isomorphic, namely, the mapping $x \mapsto x'$ is an isomorphism. Clearly, $\mathbf{2}^+, \mathbf{2}^\bullet \in \mathfrak{QB}(2,1,0)$.

Theorem 10.2.1 ([281]). *The super-Boolean algebras* $\mathbf{2}$ *and* $\mathbf{2}^\bullet$ *are subdirectly irreducible, and those algebras are the only non-trivial subdirectly irreducible super-Boolean* T*-algebras up to isomorphism, where* $T = \{2,1,0\}$. *Hence* $\mathbf{2}^\bullet$ *is the only subdirectly irreducible algebra of the variety* $\mathfrak{QB}(2,1,0)$.

Corollary 10.2.2. *Heq* $\mathfrak{QB} =$ *Heq* $\mathbf{2}$.

Denote $D = \{0, a, b, 1\}$. Let us define the operations $+, \cdot, \vee, \wedge, '$ on the set D as follows: $0+x=x+0=x,\ 1+x=x+1=1,\ x+x=x,\ 0\cdot x=x\cdot 0=0,\ 1\cdot x=x\cdot 1=x,\ x\cdot x=x,\ a\vee x=x\vee a=a,\ b\vee x=x\vee b=x,\ a\wedge x=x\wedge a=x,\ b\wedge x=x\wedge b=b,\ x\vee x=x,\ x\wedge x=x$ for all $x \in D$, and $a+b=b+a=1,\ a\cdot b=b\cdot a=0,\ 0\vee 1=1\vee 0=a,\ 0\wedge 1=1\wedge 0=b,\ 1'=0,\ 0'=1,\ a'=b,\ b'=a$. Clearly, $(D; \{+, \cdot, ', 0, 1\})$ is a Boolean algebra, while $(D; \{+, \vee, ', 0, 1\})$, $(D; \{+, \wedge, ', 0, 1\})$, $(D; \{\vee, \cdot, ', 0, 1\})$, $(D; \{\wedge, \cdot, ', 0, 1\})$ are super-Boolean algebras and are not Boolean algebras. Clearly, these four algebras are isomorphic. Denote $\mathfrak{QB}_4 = (D; \{\vee, \cdot, \ ', 0, 1\})$. Thus $\mathfrak{QB}_4 \in \mathfrak{QB}(2,2,1,0,0)$. Notice that $\mathfrak{QB}_4$ is isomorphic to the superproduct $\mathbf{2} \bowtie \mathbf{2}^{\bullet}$.

Let us prove the following lemma.

Lemma 10.2.3. *$Eq\ \mathfrak{QB}(2,2,1,0,0) = Eq\ \mathfrak{QB}_4$.*
Moreover, an identity $\tau_1(x_1,\ldots,x_n) = \tau_2(x_2,\ldots,x_n)$ with binary functional symbols $\vee, \cdot$, unary functional symbol $'$ and nullary functional symbols $0, 1$ holds in any super-Boolean algebra with two binary, one unary and two nullary operations if the equality $\tau_1 = \tau_2$ holds for every $x_1,\ldots,x_n \in B \subseteq D$.

Proof. Obviously $Eq\ \mathfrak{QB}(2,2,1,0,0) \subseteq Eq\ \mathfrak{QB}_4$. We shall prove that any identity $\tau_1 = \tau_2$ of the super-Boolean algebra $\mathfrak{QB}_4$ is true in all super-Boolean algebras with two binary, one unary and two nullary operations. First suppose that the terms τ_1 and τ_2 do not contain any constants.

As we mentioned above, $\mathfrak{QB}_4$ is isomorphic to the superproduct $\mathbf{2} \bowtie \mathbf{2}^{\bullet}$. Hence the identity $\tau_1 = \tau_2$ holds in the algebra $\mathbf{2} \bowtie \mathbf{2}^{\bullet}$ too. Denote by τ_1^1 and τ_2^1 the terms that are obtained from the terms τ_1 and τ_2 if we replace all operations $\vee$ by $+$. Similarly, let τ_1^2 and τ_2^2 be the terms that are obtained from the terms τ_1 and τ_2 by replacement of all operations $\vee$ by $\cdot$. And by ω_1, ω_2 denote the second-order terms that are obtained from the terms τ_1 and τ_2 if we replace all occurrences of the operation $\vee$ by a binary functional symbol F, all occurrences of the operation $\cdot$ by a binary functional symbol G and all occurrences of the operation $'$ by a unary functional symbol X. Clearly, the identities $\tau_1^1 = \tau_2^1$ and $\tau_1^2 = \tau_2^2$ hold in the algebra $\mathbf{2}$. From the duality principle, we conclude that the hyperidentity $\omega_1 = \omega_2$ holds in $\mathbf{2}$. According to Corollary 10.2.2 that hyperidentity is valid in all super-Boolean algebras. Therefore, the identity $\tau_1 = \tau_2$ is true in all super-Boolean algebras with two binary, one unary and two nullary operations.

Now let us consider the case when τ_1 and τ_2 can contain constants 1 and 0. From the hyperidentity $F(x, X(x)) = F(y, X(y))$ of Boolean algebras we conclude that the identity $(x \cdot x')' = (y \cdot y')' = (1 \cdot 1')' = (1')' = 1$ holds in any super-Boolean algebra from $\mathfrak{QB}(2,2,1,0,0)$. And therefore the identity $0 = x \cdot x'$ also holds in any super-Boolean algebra from $\mathfrak{QB}(2,2,1,0,0)$. Now in the terms τ_1 and τ_2 we can replace all constants 1 and 0 by the terms $(x \cdot x')'$ and $x \cdot x'$ respectively, and we will arrive at the first case.

The second part is proved by the same way, we only need to notice that in this case too the identities $\tau_1^1 = \tau_2^1$ and $\tau_1^2 = \tau_2^2$ hold in the algebra $\mathbf{2}$. □

For a set X denote the set of all its subsets by 2^X or $\mathcal{P}(X)$. If we consider subsets of a given set X then for a subset $s \subseteq X$ we denote $\overline{s} = X \setminus s$.

A function $f : B^n \to B$ is called a *Boolean function* of n variables. The following result is well known.

Theorem 10.2.4 ([72]). *For every Boolean function $f : B^n \to B$ there exists a unique set $S \subseteq 2^{\{1,\dots,n\}}$ such that*

$$f(x_1, \dots, x_n) = \sum_{s \in S} \left(\prod_{i \in s} x_i \cdot \prod_{i \in \overline{s}} x'_i \right),$$

where the operations on the right-hand side are the operations of Boolean algebra **2**.

10.3. Super-Boolean functions

Below everywhere by a super-Boolean algebra (with one or two binary operations) we mean a super-Boolean algebra with nullary operations.

Recall that $B = \{0, 1\}$, $D = \{0, 1, a, b\}$. Let us construct a one-to-one correspondence between the sets D and $B \times B$ as follows:

$$0 \leftrightarrow (0, 0), \ \ a \leftrightarrow (1, 0), \ \ b \leftrightarrow (0, 1), \ \ 1 \leftrightarrow (1, 1).$$

We define the operations $+, \cdot, \vee, \wedge, ^-, \, '$ on the set $B \times B$ as follows:

$$(u, v)' = (u', v'),$$
$$(u_1, \varphi_1) + (u_2, \varphi_2) = (u_1 + u_2, \varphi_1 + \varphi_2), \ (u_1, \varphi_1) \cdot (u_2, \varphi_2) = (u_1 \cdot u_2, \varphi_1 \cdot \varphi_2),$$
$$(u_1, \varphi_1) \vee (u_2, \varphi_2) = (u_1 + u_2, \varphi_1 \cdot \varphi_2), \ (u_1, \varphi_1) \wedge (u_2, \varphi_2) = (u_1 \cdot u_2, \varphi_1 + \varphi_2),$$

where the operations on the right-hand side are the operations of the Boolean algebra **2**. These operations are isomorphic to the corresponding operations on the set D (the one-to-one correspondence described above is an isomorphism).

However, if the tuple $(y, z) \in B \times B$ corresponds to $x \in D$ then we will write $x = (y, z)$ (this causes no confusion).

Definition 10.3.1. A function $f : B^n \to D$ is called a *super-Boolean function* of n variables.

We will consider the set B as a subset of D. And so all Boolean functions are super-Boolean functions.

Lemma 10.3.2. *For any super-Boolean function $f : B^n \to D$ there exist two Boolean functions $f_1, f_2 : B^n \to B$ such that for all $x_1, \dots, x_n \in B$ the following equality holds:*

$$f(x_1, \dots, x_n) = (f_1(x_1, \dots, x_n), f_2(x_1, \dots, x_n)). \tag{10.3.1}$$

These Boolean functions f_1, f_2 *are uniquely determined by the super-Boolean function* f.

Proof. As we said above we will not distinguish the sets $B \times B$ and D. So we can consider the projection functions f_1 and f_2 such that for any $x_1, \dots, x_n \in B$ $f_i(x_1, \dots, x_n)$ is the ith coordinate of $f(x_1, \dots, x_n)$ in the set $B \times B$, where $i = 1, 2$. Then f_1 and f_2 map B^n into B, i.e., they are Boolean functions and obviously the equality (10.3.1) holds. The uniqueness of such functions is obvious. □

Note that here for a Boolean function $f : B^n \to B$ we have $f_1 = f_2 = f$.

Theorem 10.3.3. *For any super-Boolean function* $f : B^n \to D$ *there exist two Boolean functions* $f_1, f_2 : B^n \to B$ *with*

$$f(x_1, \dots, x_n) = (a \cdot f_1(x_1, \dots, x_n)) + (b \cdot f_2(x_1, \dots, x_n)), \tag{10.3.2}$$

where the operations are the operations on the set D *defined above. These Boolean functions* f_1, f_2 *are uniquely determined by the super-Boolean function* f.

Proof. From equality (10.3.1) of Lemma 10.3.2 we have (we omit the variables):

$$f = (f_1, f_2) = ((1,0) \cdot (f_1, f_1)) + ((0,1) \cdot (f_2, f_2)) = (a \cdot f_1) + (b \cdot f_2).$$

The uniqueness follows from Lemma 10.3.2. □

Taking into account the equalities in D:

$$a \cdot x = 0 \vee x,\ b \cdot y = 0 \wedge y,\ (x \vee y)' = x' \wedge y',$$
$$(x \wedge y)' = x' \vee y',\ x + y = (1 \wedge (x \vee y)) \vee (0 \wedge x \wedge y),\ x, y \in D,$$

we get

$$(a \cdot x) + (b \cdot y) = (0 \vee x) + (0 \wedge y) = (1 \wedge (0 \vee x \vee (0 \wedge y)))$$
$$\times \vee (0 \wedge y \wedge (0 \vee x)) = (1 \wedge x) \vee (0 \wedge y) = (0 \vee x')' \vee (1 \vee y')'.$$

Thus we conclude that any super-Boolean function f can be represented in the following form:

$$f = (0 \vee f_1')' \vee (1 \vee f_2')',$$

where f_1 and f_2 are the corresponding Boolean functions with the Boolean operations $\vee$ and $'$. Note that this form of super-Boolean functions is the analogue of the b_2-canonical form of terms with two binary functional variables introduced in [253].

Now, from Theorem 10.2.4 we conclude that any super-Boolean function can be represented in the following form:

$$f(x_1,\dots,x_n)=\left(0\vee\prod_{s\in S_1}\left(\prod_{i\in s}x_i\cdot\prod_{i\in\overline{s}}x_i'\right)'\right)'\vee\left(1\vee\prod_{s\in S_2}\left(\prod_{i\in s}x_i\cdot\prod_{i\in\overline{s}}x_i'\right)'\right)', \tag{10.3.3}$$

for uniquely determined sets $S_1,S_2\subseteq 2^{\{1,\dots,n\}}$ (here the domain of variables is the set B).

This form is called the disjunctive normal form (DNF) of the super-Boolean function f (and it is unique for a given super-Boolean function). Note that here if $S_1=S_2=\emptyset$ then $f=0$ and if $S_1=S_2=2^{\{1,\dots,n\}}$ then $f=1$.

Denote by $\mathcal{SB}_n$ the set of all super-Boolean functions of n variables. For any two functions $f,g\in\mathcal{SB}_n$ define $f+g$, $f\cdot g$, $f\vee g$, $f\wedge g$, f' by the standard way, i.e., $(f\lambda g)(x)=f(x)\lambda g(x)$, for all $\lambda\in\{+,\cdot,\vee,\wedge\}$, and $f'(x)=(f(x))'$, $x\in B^n$, where the operations on the right-hand side are the operations on the set D defined above. Thus we get the algebras $(\mathcal{SB}_n;\{+,\vee,\ ',0,1\})$, $(\mathcal{SB}_n;\{+,\wedge,\ ',0,1\})$, $(\mathcal{SB}_n;\{\vee,\cdot,\ ',0,1\})$, $(\mathcal{SB}_n;\{\wedge,\cdot,\ ',0,1\})$ (here the nullary operations are the constant functions $0(x)=0$ and $1(x)=1$ for all $x\in B^n$), which are super-Boolean algebras (and they are isomorphic). Define $\mathfrak{Q}\mathfrak{B}_n=(\mathcal{SB}_n;\{\vee,\cdot,\ ',0,1\})$.

Consider the projection functions

$$\delta_n^i(x_1,\dots,x_n)=x_i,\ i=1,\dots,n,$$

as super-Boolean functions $B^n\to D$. Clearly, (10.3.3) implies

$$f=\left(0\vee\prod_{s\in S_1}\left(\prod_{i\in s}\delta_n^i\cdot\prod_{i\in\overline{s}}(\delta_n^i)'\right)'\right)'\vee\left(1\vee\prod_{s\in S_2}\left(\prod_{i\in s}\delta_n^i\cdot\prod_{i\in\overline{s}}(\delta_n^i)'\right)'\right)'.$$

Now let us prove the following result.

Theorem 10.3.4 (Functional representation theorem for free super-Boolean algebras ([279]). *The algebra $\mathfrak{Q}\mathfrak{B}_n$ is the free super-Boolean algebra with the system of free generators $\Delta=\{\delta_n^1,\dots,\delta_n^n\}$. Hence every free n-generated super-Boolean algebra with two binary, one unary and two nullary operations is isomorphic to the super-Boolean algebra $\mathfrak{Q}\mathfrak{B}_n$.*

Proof. Let $\mathfrak{F}=(Q;\{\vee,\cdot,\ ',0,1\})$ be a super-Boolean algebra and $\mu:\Delta\to Q$ be a mapping. We prove that there exists a unique homomorphism

$\nu : \mathcal{SB}_n \to \mathfrak{F}$ with $\nu|_\Delta = \mu$. As we saw above, any element $f \in \mathcal{SB}_n$ can be presented in the form

$$f = \left(0 \vee \prod_{s \in S_1} \left(\prod_{i \in s} \delta_n^i \cdot \prod_{i \in \overline{s}} (\delta_n^i)'\right)'\right)' \vee \left(1 \vee \prod_{s \in S_2} \left(\prod_{i \in s} \delta_n^i \cdot \prod_{i \in \overline{s}} (\delta_n^i)'\right)'\right)'$$

for the uniquely determined sets $S_1, S_2 \subseteq 2^{\{1,\dots,n\}}$.

Now set

$$\nu(f) = \left(0 \vee \prod_{s \in S_1} \left(\prod_{i \in s} \mu(\delta_n^i) \cdot \prod_{i \in \overline{s}} (\mu(\delta_n^i))'\right)'\right)' \vee$$
$$\left(1 \vee \prod_{s \in S_2} \left(\prod_{i \in s} \mu(\delta_n^i) \cdot \prod_{i \in \overline{s}} \mu((\delta_n^i))'\right)'\right)'.$$

Obviously, $\nu(\delta_n^i) = \mu(\delta_n^i)$, $i = 1, \dots, n$. Let us prove that ν is a homomorphism. First notice that $\nu(0) = 0$, $\nu(1) = 1$. Now let $f_1, f_2 \in \mathcal{SB}_n$. Let the terms w_1 and w_2 be the disjunctive normal forms of super-Boolean functions f_1 and f_2, respectively; and let w be the disjunctive normal form of the super-Boolean function $f_1 \vee f_2$. Then the equality $w = w_1 \vee w_2$ holds when the variables take values from the set $B \subseteq D$. From Lemma 10.2.3, it follows that the identity $w = w_1 \vee w_2$ holds in the super-Boolean algebra $\mathfrak{F}$. Hence $\nu(f_1 \vee f_2) = \nu(f_1) \vee \nu(f_2)$. Analogously, we prove that $\nu(f_1 \cdot f_2) = \nu(f_1) \cdot \nu(f_2)$, $\nu(f') = (\nu(f))'$. Therefore, ν is a homomorphism. The uniqueness of ν is evident. □

Corollary 10.3.5. *Every finitely generated free super-Boolean algebra with two binary, one unary and two nullary operations is finite.*

Corollary 10.3.6. *Every finitely generated super-Boolean algebra with two binary, one unary and two nullary operations is finite, i.e., the variety of super-Boolean algebras with two binary, one unary and two nullary operations is a Burnside variety.*

Problem 10.3.7. *Prove the Post type functional completeness theorem for super-Boolean functions (see [193,276,333,334]).*

Problem 10.3.8. *Develop the super-Boolean analogue of the theory of Boolean functions.*

10.4. Another approach

In this section, we show that instead of super-Boolean functions we can define the concept of hyper-Boolean functions (from D^n into D), so that the functional representation theorem remains true, if we replace super-Boolean functions by hyper-Boolean functions.

For $\alpha \in D$ denote by α^1 its first coordinate from $B \times B$ and by α^2 its second coordinate from $B \times B$ (according to the one-to-one correspondence described above), i.e., $\alpha = (\alpha^1, \alpha^2)$.

Definition 10.4.1. A *hyper-Boolean function* of n variables is a function mapping D^n into D that satisfies the following condition:

$$f(x_1, \ldots, x_n) = ((f(x_1^1, \ldots, x_n^1))^1, (f(x_1^2, \ldots, x_n^2))^2), \tag{10.4.1}$$

for all $x_1, \ldots, x_n \in D$.

This condition means that f is uniquely determined by its restriction on the set B^n (which is a super-Boolean function). This condition is equivalent to the following:

$$f(x_1, \ldots, x_n) = (a \cdot f(x_1^1, \ldots, x_n^1)) + (b \cdot f(x_1^2, \ldots, x_n^2)).$$

Let us define the Boolean functions $\varphi_1, \varphi_2 : B^n \to B$ by the following way:

$$\varphi_1(u_1, \ldots, u_n) = (f(u_1, \ldots, u_n))^1,$$
$$\varphi_2(u_1, \ldots, u_n) = (f(u_1, \ldots, u_n))^2,$$

for all $u_1, \ldots, u_n \in B$. Then we can write the equality (10.4.1) in the following way:

$$\begin{aligned} &f(x_1, \ldots, x_n) \\ &\quad = (1,0) \cdot (\varphi_1(x_1^1, \ldots, x_n^1), \varphi_1(x_1^2, \ldots, x_n^2)) + (0,1) \cdot (\varphi_2(x_1^2, \ldots, x_n^2), \\ &\quad \varphi_2(x_1^2, \ldots, x_n^2)) = (a \cdot g_1(x_1, \ldots, x_n)) + (b \cdot g_2(x_1, \ldots, x_n)), \end{aligned}$$

where g_1 and g_2 are functions from D^n into D defined by the following equalities:

$$g_1(x_1, \ldots, x_n) = (\varphi_1(x_1^1, \ldots, x_n^1), \varphi_1(x_1^2, \ldots, x_n^2)),$$
$$g_2(x_1, \ldots, x_n) = (\varphi_2(x_1^2, \ldots, x_n^2), \varphi_2(x_1^2, \ldots, x_n^2)).$$

One can prove that the functions g_1 and g_2 are term functions of the Boolean algebra $(D; \{+, \cdot, {}', 0, 1\})$ and can be represented in the form

$$g_k(x_1, \ldots, x_n) = \sum_{s \in S_k} \left(\prod_{i \in s} x_i \cdot \prod_{i \in \overline{s}} x_i' \right), \quad k = 1, 2,$$

for uniquely determined sets $S_1, S_2 \subseteq 2^{\{1,\ldots,n\}}$, where the operations on the right-hand side are the operations of the Boolean algebra $(D; \{+, \cdot, \ ', 0, 1\})$.

Now as in Sec. 10.3 we can show that $f = (0 \vee g_1')' \vee (1 \vee g_2')'$. And the representation of the hyper-Boolean function f in the form (10.3.3) is true too, only the domain of variables is the set D now (instead of B). Thus every hyper-Boolean function can be uniquely represented in the disjunctive normal form. This means that hyper-Boolean functions are term functions of the super-Boolean algebra $\mathfrak{QB}_4$.

For two hyper-Boolean functions f_1 and f_2 we can define $f_1 + f_2$, $f_2 \cdot f_2$, $f_1 \vee f_2$, $f_1 \wedge f_2$, f_1' by standard way and we can show that these new functions are also hyper-Boolean functions. And again we get the super-Boolean algebra $(\mathcal{HB}_n; \{\vee, \cdot, \ ', 0, 1\})$ where $\mathcal{HB}_n$ is the set of all hyper-Boolean functions of n variables.

Now we prove that all term functions of the super-Boolean algebra QB_4 are hyper-Boolean functions. We induct on the structure of terms. (For a term ω we will denote by $\widehat{\omega}$ the term function of the algebra QB_4 corresponding to that term.) If $\omega = x_i$ is a variable, then the corresponding term function $\widehat{\omega}$ is a projection function, which is a hyper-Boolean function. If $\omega = 0$ or $\omega = 1$ then $\widehat{\omega}$ is either the constant hyper-Boolean function 0 or 1. Now suppose $\omega = \omega_1 \vee \omega_2$ or $\omega = \omega_1 \cdot \omega_2$, where $\widehat{\omega}_1$ and $\widehat{\omega}_2$ are hyper-Boolean functions. In the first case $\widehat{\omega} = \widehat{\omega_1 \vee \omega_2} = \widehat{\omega}_1 \vee \widehat{\omega}_2$, in the second case $\widehat{\omega} = \widehat{\omega_1 \cdot \omega_2} = \widehat{\omega}_1 \cdot \widehat{\omega}_2$. Hence $\widehat{\omega}$ is a hyper-Boolean function. Finally, if $\omega = \lambda'$ where $\widehat{\lambda}$ is a hyper-Boolean function then $\widehat{\omega} = \widehat{(\lambda')} = (\widehat{\lambda})'$ which is a hyper-Boolean function.

Thus hyper-Boolean functions are precisely the term functions (operations) of the algebra QB_4 (and this is the advantage of this approach), i.e., the set of all hyper-Boolean functions is the clone of term functions of the algebra QB_4.

Now the functional representation theorem can be proved by the same way as in the previous section, i.e., the algebra $(\mathcal{HB}_n; \{\vee, \cdot, \ ', 0, 1\})$ is the free n-generated super-Boolean algebra freely generated by the projections. Hence, the algebra $(\mathcal{HB}_n; \{\vee, \cdot, \ ', 0, 1\})$ is isomorphic to the algebra $(\mathcal{SB}_n; \{\vee, \cdot, \ ', 0, 1\})$.

Problem 10.4.2. *Characterize finitely generated free super-Boolean algebras with two binary and one unary operations via Boolean sum (see Theorem 10.1.4).*

Chapter 11

Super-De Morgan Functions and Free Super-De Morgan Algebras

11.1. Preliminaries

In this chapter, we give a functional representation of the free n-generated algebras with two binary and one unary operations with hyperidentities of the variety of De Morgan algebras. Namely, we define the concept of super-De Morgan function and prove that the free super-De Morgan algebras with two binary and one unary operations on n free generators is isomorphic to the super-De Morgan algebras of super-De Morgan functions of n variables.

Definition 11.1.1. An algebra $\mathfrak{A} = (Q; \Sigma)$ is called a *super-De Morgan algebra* if it satisfies all hyperidentities of the variety of De Morgan algebras (see [269]).

For example, the superproduct of any two De Morgan algebras is a super-De Morgan algebra with four binary operations.

Note that from the hyperidentity $F(x) = G(x)$ of De Morgan algebras it follows that any super-De Morgan algebra has a unique unary operation, which is usually denoted by $^{-}$.

For a set $T \subseteq \mathbb{N} = \{0, 1, 2, \dots\}$ a super-De Morgan T-algebra is a super-De Morgan algebra with arithmetic type T.

Lemma 11.1.2 ([274]). *If a super-De Morgan algebra $\mathfrak{A} = (Q; \Sigma)$ is subdirectly irreducible, then there are at most two binary operations in the set Σ (cf. Sec. 5.5).*

In [274] a structural theorem for super-De Morgan algebras with two binary and one unary operations is proved. Let us give the following definition of De Morgan sum to formulate that structural result (cf. Sec. 5.5).

Definition 11.1.3. Let $\mathfrak{A} = (Q; \Omega \cup \{F\})$ be an algebra with one unary operation F. Let (Q_i, Ω), $i \in I$ be subalgebras of the algebra $\mathfrak{A}$, and $\mathfrak{A}_i = (Q_i; \Omega \cup \{F_i\})$ be algebras with one unary operation F_i. The algebra $\mathfrak{A}$ is called *De Morgan sum* of algebras $\mathfrak{A}_i$, if the following conditions hold:

(1) $Q_i \cap Q_j = \emptyset$ for all $i, j \in I, i \neq j$.
(2) $Q = \bigcup_{i \in I} Q_i$.
(3) Two binary operations $+, \cdot$ and a unary operation $^-$ can be defined on I such that $I(+, \cdot, ^-)$ is a De Morgan lattice.
(4) If $i, j \in I$ and $i \leq j$ (here "$\leq$" is the order of the lattice $I(+, \cdot)$), then there exists a (bi)isomorphism

$$(\varphi_{i,j}, \tilde{\varepsilon}) : \mathfrak{A}_i \to \mathfrak{A}_j,$$

where $\tilde{\varepsilon}(F_i) = F_j$, $\tilde{\varepsilon}(A) = A$ for any $A \in \Omega$. Moreover, $\varphi_{i,i}$ is the identical mapping of the set Q_i, and for all $i \leq j \leq k$ we have $\varphi_{i,j} \cdot \varphi_{j,k} = \varphi_{i,k}$.
(5) For every $i \in I$, there exists a (bi)isomorphism

$$(h_{i,\bar{i}}, \tilde{\varepsilon}) : \mathfrak{A}_i \to \mathfrak{A}_{\bar{i}},$$

such that $h^{-1}_{i,\bar{i}} = h_{\bar{i},i}$ and $h_{i,\bar{i}} \cdot \varphi_{\bar{i},k} = \varphi_{i,k}$ for all $k \geq i + \bar{i}$, $k \in I$.
(6) For any n-ary operation $A \in \Omega$ $(n \geq 2)$ and for any $x_1, \ldots, x_n \in Q$ we have

$$A(x_1, \ldots, x_n) = A(\varphi_{i_1,i_0}(x_1), \ldots, \varphi_{i_n,i_0}(x_n)),$$

where $x_j \in Q_{i_j}$, $i_j \in I$, $j = \overline{1, n}$, $i_0 = i_1 + \cdots + i_n$;
(7) For any $x \in Q$ we have

$$F(x) = h_{i,\bar{i}}(F_i(x)),$$

where $x \in Q_i$ (cf. [256]).

Theorem 11.1.4 ([274]). *An algebra $\mathfrak{A} = (Q; \{+, \cdot, ^-\})$ with two binary operations $+, \cdot$ and one unary operation $^-$ is a super-De Morgan algebra if and only if it is a De Morgan lattice or De Morgan sum of De Morgan lattices (cf. Sec. 5.5).*

Note that there exists a super-De Morgan algebra with two binary and one unary operations which is not a De Morgan lattice. For example, let $\mathfrak{A} = (Q; +, \cdot, ^-)$ be a (non-trivial) De Morgan algebra and take the superproduct $\mathfrak{A} \bowtie \mathfrak{A}$. Now we consider its reduct $\mathfrak{B} = (Q \times Q; \{\vee, \cdot, ^-\})$, where $\vee = (+, \cdot)$, $\cdot = (\cdot, \cdot)$, $^- = (^-, ^-)$. Clearly, this is a super-De Morgan algebra with two binary and one unary operations. But it is not a De Morgan lattice since the absorption law is not valid in that algebra.

11.2. Auxiliary results

The T-hypervariety of super-De Morgan T-algebras, where $T = \{1, 2\}$, is denoted by $\mathfrak{Q}\mathfrak{D}$. From the results of Chapter 7 follows that this hypervariety can be axiomatized by finite number of hyperidentities.

As it was noted above, from the hyperidentity $X(x) = Y(x)$ of De Morgan algebras it follows that any super-De Morgan algebra has a unique unary operation, which we will denote by $^-$.

Denote by $\mathfrak{Q}\mathfrak{D}(2,2,1)$ the variety of all super-De Morgan algebras with two binary and one unary operations. So the variety $\mathfrak{Q}\mathfrak{D}(2,2,1)$ is a hypervariety.

Free algebras of the variety $\mathfrak{Q}\mathfrak{D}(2,2,1)$ are called free super-De Morgan algebras with two binary and one unary operations. Our main result is the characterization of finitely generated free super-De Morgan algebras with two binary and one unary operations.

Similarly, we define the variety $\mathfrak{Q}\mathfrak{D}(2,1)$ to be the class of all super-De Morgan algebras with one binary and one unary operations.

Let $D = \{0, a, b, 1\}$. Let us define the operations $+, \cdot, ^-$ on the set D as follows: $0+x = x+0 = x$, $1+x = x+1 = 1$, $x+x = x$, $0\cdot x = x\cdot 0 = 0$, $1\cdot x = x\cdot 1 = x$, $x\cdot x = x$ for all $x \in D$, and $a+b = b+a = 1$, $a\cdot b = b\cdot a = 0$, $\overline{1} = 0$, $\overline{0} = 1$, $\overline{a} = a$, $\overline{b} = b$. Clearly, $\mathbf{4} = (D; \{+, \cdot, ^-\})$ is a De Morgan lattice. Let $\mathbf{2} = (\{0,1\}; \{+, \cdot, ^-\})$ and $\mathbf{3} = (\{0, a, 1\}; \{+, \cdot, ^-\})$ be its subalgebras. And let $\mathbf{2}^\bullet = (\{0,1\}; \{\cdot, ^-\})$, $\mathbf{3}^\bullet = (\{0, a, 1\}; \{\cdot, ^-\})$, $\mathbf{4}^\bullet = (\{0, a, b, 1\}; \{\cdot, ^-\})$ be their reducts.

Note that the De Morgan lattice $\mathbf{4}$ is the reduct of the four-element diamond De Morgan algebra, which generates the variety of all De Morgan algebras.

Obviously, $\mathbf{2}, \mathbf{3}, \mathbf{4} \in \mathfrak{Q}\mathfrak{D}(2,2,1)$, $\mathbf{2}^\bullet, \mathbf{3}^\bullet, \mathbf{4}^\bullet \in \mathfrak{Q}\mathfrak{D}(2,1)$.

Let us consider the superproduct $QD_4 = \mathbf{4} \bowtie \mathbf{4}^\bullet = (D \times D; \{\vee, \cdot, ^-\})$, where $\vee = (+, \cdot)$, $\cdot = (\cdot, \cdot)$, $^- = (^-, ^-)$. Clearly $QD_4 \in \mathfrak{Q}\mathfrak{D}(2,2,1)$. Also note that QD_4 is not a De Morgan lattice (see the example after Theorem 11.1.4).

Theorem 11.2.1 ([275]). *The super-De Morgan algebras* $\mathbf{2}$, $\mathbf{3}$, $\mathbf{4}$, $\mathbf{2}^\bullet$, $\mathbf{3}^\bullet$, $\mathbf{4}^\bullet$ *are subdirectly irreducible, and those algebras are the only non-trivial subdirectly irreducible super-De Morgan* T*-algebras up to isomorphism, where*

$T = \{1, 2\}$. Hence $\mathbf{2}^\bullet$, $\mathbf{3}^\bullet$, $\mathbf{4}^\bullet$ are the only subdirectly irreducible algebras of the variety $\mathfrak{QD}(2, 1)$.

Corollary 11.2.2. *Heq $\mathfrak{QD}$ = Heq* **4**.

The set of all identities of the given class V of algebras is called the equational theory of the class V and it is denoted by $Eq\ V$. If the class V consists of one algebra $\mathfrak{A}$ then instead of $Eq\{\mathfrak{A}\}$ we write simply $Eq\ \mathfrak{A}$.

Define $E = D \times D$. We will identify the diagonal subset $\Delta = \{(0,0), (1,1), (a,a), (b,b)\}$ with the set D (we identify (x, x) with x). And thus we will consider the set D as a subset of E.

Let us prove the following lemma.

Lemma 11.2.3. *$Eq\ \mathfrak{QD}(2, 2, 1) = Eq\ QD_4$. Moreover, an identity $\tau_1(x_1, \ldots, x_n) = \tau_2(x_1, \ldots, x_n)$ with binary functional symbols $\vee, \cdot$ and unary functional symbol $^-$ holds in any super-De Morgan algebras with two binary and one unary operations if the equality $\tau_1 = \tau_2$ holds for every $x_1, \ldots, x_n \in D \subseteq E$.*

Proof. Obviously $Eq\ \mathfrak{QD}(2, 2, 1) \subseteq Eq\ QD_4$. We shall prove that any identity $\tau_1 = \tau_2$ of the super-De Morgan algebra QD_4 is true in all super-De Morgan algebras with two binary and one unary operations.

The algebra QD_4 is equal to the superproduct $\mathbf{4} \bowtie \mathbf{4}^\bullet$. Hence the identity $\tau_1 = \tau_2$ holds in the algebra $\mathbf{4} \bowtie \mathbf{4}^\bullet$ too. Denote by τ_1^1 and τ_2^1 the terms that are obtained from the terms τ_1 and τ_2 if we replace all occurrences of the operations $\vee$ by $+$. Similarly, let τ_1^2 and τ_2^2 be the terms that are obtained from the terms τ_1 and τ_2 by replacement of all operations $\vee$ by $\cdot$. And by ω_1, ω_2 denote the second-order terms that are obtained from the terms τ_1 and τ_2 if we replace all occurrences of the operation $\vee$ by a binary functional variable X, all occurrences of the operation $\cdot$ by a binary functional variable Y and all occurrences of the operation $^-$ by a unary functional variable F. Clearly, the identities $\tau_1^1 = \tau_2^1$ and $\tau_1^2 = \tau_2^2$ hold in the algebra **4**. From the duality principle we conclude that the hyperidentity $\omega_1 = \omega_2$ holds in **4**. According to Corollary 11.2.2 that hyperidentity is valid in all super-De Morgan algebras. Therefore the identity $\tau_1 = \tau_2$ is true in all super-De Morgan algebras with two binary and one unary operations.

The second part is proved in the same way, we only need to notice that in this case also the identities $\tau_1^1 = \tau_2^1$ and $\tau_1^2 = \tau_2^2$ hold in the algebra **4**. □

11.3. Super-De Morgan functions

Let us start with the definition of De Morgan functions.

Recall that $D = \{0, a, b, 1\}$ and the algebra $\mathbf{4} = (D; \{+, \cdot, ^{-}\})$ is a De Morgan lattice.

For $\delta \in D$ let

$$\delta^* = \begin{cases} \delta \text{ if } \delta \in \{0,1\}, \\ a \quad \text{if } \delta = b, \\ b \quad \text{if } \delta = a. \end{cases}$$

For $\alpha = (\alpha_1, \ldots, \alpha_n)$, $\beta = (\beta_1, \ldots, \beta_n) \in D^n$ we say that β is a *permitted modification* of α if $\beta_i = \alpha_i$ for all $1 \le i \le n$ with the possible exception of some k $(1 \le k \le n)$ such that

$$\beta_k = \begin{cases} a \text{ if } \alpha_k = 0, \\ 1 \text{ if } \alpha_k = b. \end{cases}$$

Definition 11.3.1. A function $f : D^n \to D$ is called a *De Morgan function* if the following conditions hold (see Chapters 3 and 4):

(1) if $x_i \in \{0,1\}$, $i = 1, \ldots, n$, then $f(x_1, \ldots, x_n) \in \{0,1\}$;
(2) for any $x_i \in D$, $i = 1, \ldots, n$, we have $f(x_1^*, \ldots, x_n^*) = (f(x_1, \ldots, x_n))^*$;
(3) if $x, y \in D^n$ with $f(x) \neq b$ and y is a permitted modification of x then $f(y) \in \{f(x), a\}$.

For example, the constant functions $f = 1$ and $f = 0$ are De Morgan functions, but the constant functions $f = a$ and $f = b$ are not. Further examples of De Morgan functions are $f(x) = x$, $g(x) = \overline{x}$, $h(x, y) = x \cdot y$, $q(x, y) = x + y$, where the operations on the right-hand side are the operations of the De Morgan algebra $\mathbf{4}$.

If $\varrho \subseteq D \times D$ is a binary relation on D, then we say that a function $f : D^n \to D$ preserves ϱ if the following implication holds for any $x_i, y_i \in D$:

$$(x_i, y_i) \in \varrho, \ i = 1, \ldots, n \Rightarrow (f(x_1, \ldots, x_n), f(y_1, \ldots, y_n)) \in \varrho.$$

Now note that De Morgan functions can also be defined as functions mapping D^n into D that preserve the following binary relations on the set D: $\sigma = \{(0,0), (a,b), (b,a), (1,1)\} \subseteq D^2$, which is the graph of the automorphism $x \mapsto x^*$, and the order relation $\rho = \{(b,b), (b,0), (b,1), (b,a), (0,0), (0,a), (1,1), (1,a), (a,a)\} \subseteq D^2$. To see this, first notice that condition (3) means that if we switch the value of one variable of f from 0 to a or from b to 1 and if the value of function is not b then it can remain the same or switch to a. Clearly, this is also true in the case when we switch the values of some variables of f from 0 to a or from b to 1. From condition (2) it follows that the same argument will also be true if we replace a by b and b by a. We mean that if we switch the values of some variables from 0 to b or from a to 1 and if the value of function is not a then it can remain the same or switch to b. And this argument implies that if we switch the values of some variables from b to 0 or from 1 to a and if the value of function is not b then it can remain the same or switch to a (we can easily prove this by contradiction). For instance, let us prove that if we switch the values of some variables from b to 0 and if the value of function is not b then it can remain the same or switch to a. Obviously, it is sufficient to prove this for the case when we switch only the value of one variable. We proceed to the proof by contradiction. Suppose α, $\beta \in D^n$ and β is obtained from α by switching a component of α from b to 0 and also $f(\alpha) \neq b$ and $f(\beta) \notin \{a, f(\alpha)\}$. Then α is obtained from β by switching one of the components of β from 0 to b. And as $f(\beta) \neq a$ then $f(\alpha) = f(\beta)$ or $f(\alpha) = b$ (as we stated above). But $f(\alpha) \neq f(\beta)$ and so $f(\alpha) = b$.

And finally if we switch the values of some variables from b to a and if the value of function is not b then it can remain the same or switch to a. To see this we can first switch the values of variables from b to 1 and then switch them from 1 to a.

Now all mentioned facts show that a De Morgan function f preserves the order relation ρ.

Let $(P; \leq)$ be a partially ordered set. A subset $A \subseteq P$ is called an antichain [143] if any two different elements of A are not comparable with respect to the order $\leq$.

We can define the componentwise order of inclusion $\subseteq$ on the set $2^{\{1,\dots,n\}} \times 2^{\{1,\dots,n\}}$:

$$(a_1, b_1) \subseteq (a_2, b_2) \Leftrightarrow a_1 \subseteq a_2 \text{ and } b_1 \subseteq b_2.$$

And thus we turn that set into a partially ordered set. Below we will use antichains of that poset.

De Morgan functions can be represented in disjunctive normal form. Namely, for any De Morgan function f of n variables there exists a unique antichain $S \subseteq 2^{\{1,\dots,n\}} \times 2^{\{1,\dots,n\}}$ such that

$$f(x_1, \dots, x_n) = \sum_{s=(s_1,s_2)\in S} \left(\prod_{i\in s_1} x_i \cdot \prod_{i\in s_2} \overline{x}_i \right). \tag{11.3.1}$$

Here for $S = \emptyset$ we set $f(x_1, \dots, x_n) = 0$ and for $S = \{(\emptyset, \emptyset)\}$ we set $f(x_1, \dots, x_n) = 1$ (the constant De Morgan functions). As a consequence we get that there are m_{2n} De Morgan functions of n variables, where we have denoted by m_n the number of all antichains of the partially ordered set $(2^{\{1,\dots,n\}};\subseteq)$ (or equivalently, m_n is the number of all monotone Boolean functions in n variables).

And finally note that De Morgan functions are precisely the term functions of the De Morgan algebra **4**.

Recall that $E = D \times D$. We define the operations $+, \cdot, \vee, \wedge, ^-$ on the set E as follows. For $u, v, u_1, \varphi_1, u_2, \varphi_2 \in D$ we set

$$\overline{(u,v)} = (\overline{u}, \overline{v}),$$

$$(u_1,\varphi_1)+(u_2,\varphi_2)=(u_1+u_2,\varphi_1+\varphi_2),\ \ (u_1,\varphi_1)\cdot(u_2,\varphi_2)=(u_1\cdot u_2,\varphi_1\cdot\varphi_2),$$
$$(u_1,\varphi_1)\vee(u_2,\varphi_2)+(u_1+u_2,\varphi_1\cdot\varphi_2),\ \ (u_1,\varphi_1)\wedge(u_2,\varphi_2)=(u_1\cdot u_2,\varphi_1+\varphi_2),$$

where the operations on the right-hand side are the operations of the De Morgan algebra **4**.

For an element $\alpha \in E$ we denote by α^1 its first coordinate in D and by α^2 the second coordinate in D, i.e., $\alpha = (\alpha^1, \alpha^2)$. Obviously, $\alpha \in D$ if and only if $\alpha^1 = \alpha^2$.

For a function $f : D^n \to E$ we define its projection functions $f^1, f^2 : D^n \to D$ as follows:

$$f^i(x_1,\dots,x_n) = (f(x_1,\dots,x_n))^i, \quad i = 1, 2.$$

So every function from D^n to E can be uniquely represented as a tuple of functions from D^n into D: $f = (f^1, f^2)$ (for simplicity we omit the variables).

Definition 11.3.2. A function $f : D^n \to E$ is called a *super-De Morgan function* of n variables, if its projection functions f^1, f^2 are non-constant De Morgan functions.

Clearly, all non-constant De Morgan functions are super-De Morgan functions (as we mentioned above, we consider the set D as a subset of E). And for a non-constant De Morgan function f we have $f^1 = f^2 = f$.

If we define $c = (1,0), d = (0,1) \in E$ then we have $f = (f^1, f^2) = (1,0)\cdot(f^1,f^1) + (0,1)\cdot(f^2,f^2) = c\cdot f^1 + d\cdot f^2$ (we assume that $\cdot$ has precedence over $+$).

It follows immediately from Definition 11.3.2 that there are $(m_{2n}-2)^2$ super-De Morgan functions of n variables.

Denote by $\mathcal{SD}_n$ the set of all super-De Morgan functions of n variables. For any two functions $f, g \in \mathcal{SD}_n$ define $f+g$, $f\cdot g$, $f\vee g$, $f\wedge g$, $\overline{f}$ in the standard way, i.e., $(f\lambda g)(x) = f(x)\lambda g(x)$, for all $\lambda \in \{+,\cdot,\vee,\wedge\}$, and $\overline{f}(x) = \overline{f(x)}$, $x \in D^n$, where the operations on the right-hand side are the operations on the set E defined above. Clearly, the set $\mathcal{SD}_n$ is closed under these operations. Thus, we get the algebras $(\mathcal{SD}_n; \{+, \vee, ^-\})$, $(\mathcal{SD}_n; \{+, \wedge, ^-\})$, $(\mathcal{SD}_n; \{\vee, \cdot, ^-\})$, $(\mathcal{SD}_n; \{\wedge, \cdot, ^-\})$, which are super-De Morgan algebras (and they are isomorphic). Define $\mathfrak{Q}\mathfrak{D}_n = (\mathcal{SD}_n; \{\vee, \cdot, ^-\})$.

Define $I = \{(\{1, \ldots, n\}, \{1, \ldots, n\})\}$, $U = \{(\{i\}, \emptyset), (\emptyset, \{i\}) : 1 \leq i \leq n\}$. Clearly, these sets are antichains. We denote the De Morgan functions corresponding to these antichains by f_I and f_U, respectively, i.e.,

$$f_I(x_1, \ldots, x_n) = x_1 \cdot \ldots \cdot x_n \cdot \overline{x}_1 \cdot \ldots \cdot \overline{x}_n,$$
$$f_U(x_1, \ldots, x_n) = x_1 + \cdots + x_n + \overline{x}_1 + \cdots + \overline{x}_n = \overline{f_I(x_1, \ldots, x_n)}.$$

It is easy to see that for any non-constant De Morgan function f the following equalities hold:

$$f + f_I = f, \ f \cdot f_I = f_I, \ f + f_U = f_U, \ f \cdot f_U = f.$$

And so we conclude that for any super-De Morgan function f the following equality is true:

$$\begin{aligned} f &= (f^1, f^2) = (f^1, f_U) \vee (f_I, f^2) = \overline{(\overline{f^1}, f_I)} \vee \overline{(f_U, \overline{f^2})} \\ &= \overline{(f_I, f_I) \vee \overline{(f^1, f^1)}} \vee \overline{(f_U, f_U) \vee \overline{(f^2, f^2)}} = \overline{f_I \vee \overline{f^1}} \vee \overline{f_U \vee \overline{f^2}}. \end{aligned}$$

Now from the representation of De Morgan functions in DNF we conclude that for any super-De Morgan function $f : D^n \to E$ there exist two antichains $S_1, S_2 \subseteq 2^{\{1,\ldots,n\}} \times 2^{\{1,\ldots,n\}}$ (that are uniquely determined by the function f) with $S_1, S_2 \neq \emptyset$ and $S_1, S_2 \neq \{(\emptyset, \emptyset)\}$ such that

$$\begin{aligned} &f(x_1, \ldots, x_n) \\ &= \overline{\prod_{i=1}^n x_i \cdot \prod_{i=1}^n \overline{x}_i \vee \prod_{(s_1,s_2) \in S_1} \overline{\prod_{i \in s_1} x_i \cdot \prod_{i \in s_2} \overline{x}_i}} \\ &\quad \vee \overline{\overline{\prod_{i=1}^n x_i \cdot \prod_{i=1}^n \overline{x}_i} \vee \prod_{(s_1,s_2) \in S_2} \overline{\prod_{i \in s_1} x_i \cdot \prod_{i \in s_2} \overline{x}_i}}. \end{aligned} \tag{11.3.2}$$

Here the domain of the variables is the set D. This form is called the disjunctive normal form (DNF) of the super-De Morgan function f. DNF is unique for a given super-De Morgan function.

Consider the projection functions

$$\delta_n^i(x_1,\dots,x_n) = x_i,\ i = 1,\dots,n,$$

as super-De Morgan functions $D^n \to E$. Clearly, (11.3.2) implies

$$f = \overline{\prod_{i=1}^{n}\delta_n^i \cdot \prod_{i=1}^{n}\overline{\delta_n^i} \vee \prod_{(s_1,s_2)\in S_1}\overline{\prod_{i\in s_1}\delta_n^i \cdot \prod_{i\in s_2}\overline{\delta_n^i}}}$$

$$\vee\, \overline{\overline{\prod_{i=1}^{n}\delta_n^i \cdot \prod_{i=1}^{n}\overline{\delta_n^i}} \vee \prod_{(s_1,s_2)\in S_2}\overline{\prod_{i\in s_1}\delta_n^i \cdot \prod_{i\in s_2}\overline{\delta_n^i}}}.$$

Now let us prove the following result.

Theorem 11.3.3 (Functional representation theorem for free super-De Morgan algebras ([280]). *The algebra $\mathfrak{Q}\mathfrak{D}_n$ is the free super-De Morgan algebra with the system of free generators $\Delta = \{\delta_n^1,\dots,\delta_n^n\}$. Hence, every free n-generated super-De Morgan algebra with two binary and one unary operations is isomorphic to the super-De Morgan algebra $\mathfrak{Q}\mathfrak{D}_n$.*

Proof. Let $\mathfrak{F} = (Q;\{\vee,\cdot,\bar{\ },0,1\})$ be a super-De Morgan algebra and $\mu : \Delta \to Q$ be a mapping. We prove that there exists a unique homomorphism $\nu : \mathcal{SD}_n \to \mathfrak{F}$ with $\nu|_\Delta = \mu$. As we saw above, any element $f \in \mathcal{SD}_n$ can be presented in the form

$$f = \overline{\prod_{i=1}^{n}\delta_n^i \cdot \prod_{i=1}^{n}\overline{\delta_n^i} \vee \prod_{(s_1,s_2)\in S_1}\overline{\prod_{i\in s_1}\delta_n^i \cdot \prod_{i\in s_2}\overline{\delta_n^i}}}$$

$$\vee\, \overline{\overline{\prod_{i=1}^{n}\delta_n^i \cdot \prod_{i=1}^{n}\overline{\delta_n^i}} \vee \prod_{(s_1,s_2)\in S_2}\overline{\prod_{i\in s_1}\delta_n^i \cdot \prod_{i\in s_2}\overline{\delta_n^i}}}$$

for uniquely determined antichains $S_1, S_2 \subseteq 2^{\{1,\dots,n\}} \times 2^{\{1,\dots,n\}}$.

Now set

$$\nu(f) = \overline{\prod_{i=1}^{n}\mu(\delta_n^i) \cdot \prod_{i=1}^{n}\overline{\mu(\delta_n^i)} \vee \prod_{(s_1,s_2)\in S_1}\overline{\prod_{i\in s_1}\mu(\delta_n^i) \cdot \prod_{i\in s_2}\overline{\mu(\delta_n^i)}}}$$

$$\vee\overline{\overline{\prod_{i=1}^{n}\mu(\delta_n^i) \cdot \prod_{i=1}^{n}\mu(\overline{\delta_n^i})} \vee \prod_{(s_1,s_2)\in S_2}\overline{\prod_{i\in s_1}\mu(\delta_n^i) \cdot \prod_{i\in s_2}\overline{\mu(\delta_n^i)}}}.$$

Obviously, $\nu(\delta_n^i) = \mu(\delta_n^i),\ i = 1,\dots,n$. Let us prove that ν is a homomorphism. Suppose $f_1, f_2 \in \mathcal{SD}_n$ and let the terms w_1 and w_2 be the disjunctive normal forms of super-De Morgan functions f_1 and f_2 respectively; and let w be the disjunctive normal form of the super-De Morgan function $f_1 \vee f_2$. Then the equality $w = w_1 \vee w_2$ holds when the variables take values from the set $D \subseteq E$. From Lemma 11.2.3 it follows that the identity $w = w_1 \vee w_2$ holds in the super-De Morgan algebra $\mathfrak{F}$. Hence $\nu(f_1 \vee f_2) = \nu(f_1) \vee \nu(f_2)$. Analogously we prove that $\nu(f_1 \cdot f_2) = \nu(f_1) \cdot \nu(f_2)$, $\nu(\overline{f}) = \overline{\nu(f)}$. Therefore, ν is a homomorphism. The uniqueness of ν is evident. □

Corollary 11.3.4. *Every finitely generated free super-De Morgan algebra with two binary and one unary operations is finite.*

Corollary 11.3.5. *Every finitely generated super-De Morgan algebra with two binary and one unary operations is finite, i.e. the variety of super-De Morgan algebras with two binary and one unary operations is a Burnside variety.*

Problem 11.3.6. *Prove the Post type functional completeness theorem for super-De Morgan functions (see [193,276,333,334]).*

Problem 11.3.7. *Develop the super-De Morgan analogue of the theory of Boolean functions.*

11.4. Another approach

In this section, we show that instead of super-De Morgan functions we can define the concept of hyper-De Morgan functions (from E^n into E), so that the main result (the functional representation theorem) remains true, if we replace super-De Morgan functions by hyper-De Morgan functions. The advantage of this approach is that the hyper-De Morgan functions are precisely the term functions (operations) of the algebra QD_4, i.e., the set of all hyper-De Morgan functions is the clone of term functions of the algebra QD_4.

As in the previous section, for $\alpha \in E$ denote by α^1 its first coordinate from D and by α^2 its second coordinate from D, i.e., $\alpha = (\alpha^1, \alpha^2)$.

For a function $f : E^n \to E$ define the functions $\varphi_f^1, \varphi_f^2 : D^n \to D$ in the following way:

$$\varphi_f^1(u_1, \dots, u_n) = (f(u_1, \dots, u_n))^1,$$
$$\varphi_f^2(u_1, \dots, u_n) = (f(u_1, \dots, u_n))^2,$$

where $u_1, \dots, u_n \in D$ (again we consider D as a subset of E).

Definition 11.4.1. A *hyper-De Morgan function* of n variables is a function f mapping E^n into E for which φ_f^1, φ_f^2 are nonconstant De Morgan functions and

$$f(x_1, \dots, x_n) = (\varphi_f^1(x_1^1, \dots, x_n^1), \varphi_f^2(x_1^2, \dots, x_n^2)), \tag{11.4.1}$$

for all $x_1, \dots, x_n \in E$.

This condition means that f is uniquely determined by its restriction on the set D^n (which is a super-De Morgan function). Equality (11.4.1) is equivalent to the following:

$$f(x_1, \dots, x_n) = (c \cdot f(x_1^1, \dots, x_n^1)) + (d \cdot f(x_1^2, \dots, x_n^2)).$$

Now we can write the equality (11.4.1) in the following way:

$$\begin{aligned} &f(x_1, \dots, x_n) \\ &\quad = (1,0) \cdot (\varphi_f^1(x_1^1, \dots, x_n^1), \varphi_f^1(x_1^2, \dots, x_n^2)) \\ &\qquad +(0,1) \cdot (\varphi_f^2(x_1^2, \dots, x_n^2), \varphi_f^2(x_1^2, \dots, x_n^2)) \\ &\quad = c \cdot g_1(x_1, \dots, x_n) + d \cdot g_2(x_1, \dots, x_n), \end{aligned}$$

where g_1 and g_2 are functions from E^n into E defined by the following equalities:

$$\begin{aligned} g_1(x_1, \dots, x_n) &= (\varphi_f^1(x_1^1, \dots, x_n^1), \varphi_f^1(x_1^2, \dots, x_n^2)), \\ g_2(x_1, \dots, x_n) &= (\varphi_f^2(x_1^2, \dots, x_n^2), \varphi_f^2(x_1^2, \dots, x_n^2)). \end{aligned}$$

One can prove that the functions g_1 and g_2 are term functions of the De Morgan algebra $(E; \{+, \cdot, {}^-\})$ and can be represented in the form

$$g_k(x_1, \dots, x_n) = \sum_{(s_1, s_2) \in S_k} \left(\prod_{i \in s_1} x_i \cdot \prod_{i \in s_2} \overline{x}_i \right),$$

for uniquely determined antichains $S_1, S_2 \subseteq 2^{\{1,\dots,n\}} \times 2^{\{1,\dots,n\}}$.

Now as in previous section we can show that $f = \overline{f_I \vee \overline{g_1} \vee \overline{f_U \vee \overline{g_2}}}$. And the representation of the hyper-De Morgan function f in the form (11.3.2) is true too, only the domain of variables is the set E now (instead of D). Thus every hyper-De Morgan function can be uniquely represented in the disjunctive normal form. This means that hyper-De Morgan functions are term functions of the super-De Morgan algebra QD_4.

For two hyper-De Morgan functions f_1 and f_2 we can define $f_1 + f_2$, $f_2 \cdot f_2$, $f_1 \vee f_2$, $f_1 \wedge f_2$, $\overline{f}_1$ in the standard way and we can show that these new functions are also hyper-De Morgan functions. And again we get the super-De Morgan algebra $(\mathcal{HD}_n; \{\vee, \cdot, ^{-}\})$, where $\mathcal{HD}_n$ is the set of all hyper-De Morgan functions of n variables.

Now we prove that all term functions of the super-De Morgan algebra QD_4 are hyper-De Morgan functions. We proceed by induction on the structure of terms. (For a term ω we will denote by $\widehat{\omega}$ the term function of the algebra QD_4 corresponding to that term.) If $\omega = x_i$ is a variable, then the corresponding term function $\widehat{\omega}$ is a projection function, which is a hyper-De Morgan function. Now suppose $\omega = \omega_1 \vee \omega_2$ or $\omega = \omega_1 \cdot \omega_2$, where $\widehat{\omega}_1$ and $\widehat{\omega}_2$ are hyper-De Morgan functions. In the first case $\widehat{\omega} = \widehat{\omega_1 \vee \omega_2} = \widehat{\omega}_1 \vee \widehat{\omega}_2$, and in the second case $\widehat{\omega} = \widehat{\omega_1 \cdot \omega_2} = \widehat{\omega}_1 \cdot \widehat{\omega}_2$. Hence, $\widehat{\omega}$ is a hyper-De Morgan function. Finally, if $\omega = \overline{\lambda}$ where $\widehat{\lambda}$ is a hyper-De Morgan function, then $\widehat{\omega} = \widehat{(\overline{\lambda})} = \overline{\widehat{\lambda}}$, which is a hyper-De Morgan function.

Thus, hyper-De Morgan functions are precisely the term functions (operations) of the algebra QD_4, i.e., the set of all hyper-De Morgan functions is the clone of term functions of the algebra QD_4.

Now the functional representation theorem can be proved in the same way as in the previous section, i.e., the algebra $(\mathcal{HD}_n; \{\vee, \cdot, ^{-}\})$ is the free n-generated super-De Morgan algebra freely generated by the projections. Hence, the algebra $(\mathcal{HD}_n; \{\vee, \cdot, ^{-}\})$ is isomorphic to the algebra $(\mathcal{SD}_n; \{\vee, \cdot, ^{-}\})$.

Problem 11.4.2. *Characterize finitely generated free super-De Morgan algebras with two binary and one unary operations via De Morgan sum (see Theorem 11.1.4).*

Chapter 12

Bi-De Morgan Functions and Free Distributive Bilattices

12.1. Introduction

In this chapter, we introduce the concept of bi-De Morgan function and the main result is formulated through this idea: bi-De Morgan functions of n variables form a free n-generated distributive bilattice. Note that another characterization of free n-generated distributive bilattices via specification of the (super)product representation of distributive bilattices is given in [63]. Advantage of our approach is that it is applicable for various varieties.

Let us recall the definitions of some preliminary concepts.

An algebra $(Q;\{+,\cdot,\ ',0,1\})$ with two binary, one unary and two nullary operations is called a *Boolean algebra* if $(Q;\{+,\cdot,0,1\})$ is a bounded distributive lattice with least element 0 and greatest element 1 and the algebra $(Q;\{+,\cdot,\ ',0,1\})$ satisfies the following identities:

$$x + x' = 1,$$
$$x \cdot x' = 0.$$

An algebra $(Q;\{+,\cdot,\bar{\ },0,1\})$ with two binary, one unary and two nullary operations is called a *De Morgan algebra* if $(Q;\{+,\cdot,0,1\})$ is a bounded distributive lattice with least element 0 and greatest element 1 and the algebra $(Q;\{+,\cdot,\bar{\ }\})$ satisfies the following identities:

$$\overline{x+y} = \overline{x}\cdot\overline{y},$$
$$\overline{\overline{x}} = x,$$

where $\overline{\overline{x}} = \overline{(\overline{x})}$ (see [19,35,38,54–56,104,167,241,260,261,263,273,277]). For example, the standard fuzzy algebra $F = ([0,1];\max(x,y),\min(x,y),1-x,0,1)$ is a De Morgan algebra.

An algebra $(Q; \{+, \cdot, ^-, ', 0, 1\})$ with two binary, two unary and two nullary operations is called a *Boole-De Morgan* algebra if $(Q; \{+, \cdot, ^-, 0, 1\})$ is a De Morgan algebra and $(Q; \{+, \cdot, ', 0, 1\})$ is a Boolean algebra and the two unary operations commute, i.e., $(\overline{x})' = \overline{(x')}$ [260,261,268] (see Chapter 3).

In [21,22] Belnap introduced a logic, which is based on the algebraic structure called *FOUR*, having four truth values (see also [16]). In papers [125–128] M.L. Ginsberg proposed algebraic structures called bilattices that naturally generalize Belnap's *FOUR*.

We need the following definitions of bilattices and distributive bilattices in this chapter. An algebra $\mathfrak{B} = (Q; \{+, \cdot, \vee, \wedge, ^-, 0_1, 1_1, 0_2, 1_2\})$ with four binary, one unary and four nullary operations is called a *bilattice* if $(Q; \{+, \cdot, 0_1, 1_1\})$ and $(Q; \{\vee, \wedge, 0_2, 1_2\})$ are bounded lattices and $\mathfrak{B}$ satisfies the following identities:

$$\begin{aligned}
\overline{\overline{x}} &= x, \\
\overline{x + y} &= \overline{x} \cdot \overline{y}, \\
\overline{x \cdot y} &= \overline{x} + \overline{y}, \\
\overline{x \vee y} &= \overline{x} \vee \overline{y}, \\
\overline{x \wedge y} &= \overline{x} \wedge \overline{y}.
\end{aligned}$$

A bilattice is said to be *distributive* if any two of its binary operations distribute over each other. Note that in this case the algebra $(Q; \{+, \cdot, ^-, 0_1, 1_1)$ is a De Morgan algebra.

For the applications and characterizations of bilattices in various varieties see [18,71,75,108,109,166,239,240,262,264,286,300,335,337,343,347].

For a Boole-De Morgan algebra $\mathfrak{A} = (Q; \{+, \cdot, ^-, ', 0, 1\})$ we define one more unary operation * by the following way: $x^* = (\overline{x})' = \overline{(x')}$. It is easy to see that $(x+y)^* = x^* + y^*$, $(x \cdot y)^* = x^* \cdot y^*$, $\overline{(x^*)} = (\overline{x})^*$, $(x^*)' = (x')^*$. Thus the mapping $x \to x^*$ is an automorphism of the Boole-De Morgan algebra $\mathfrak{A}$. Also it is easy to see that $x^* = x$ if and only if $x' = \overline{x}$.

Let $B = \{0, 1\}$. Define the operations $+, \cdot, '$, on B by the following way: $0+0 = 0$, $0+1 = 1+0 = 1+1 = 1$, $0 \cdot 1 = 0 \cdot 0 = 1 \cdot 0 = 0$, $1 \cdot 1 = 1$, $0' = 1$, $1' = 0$. We get the Boolean algebra $\mathbf{2} = (B; \{+, \cdot, ', 0, 1\})$.

Every Boolean algebra can be considered as a Boole-De Morgan algebra with two equal unary operations. In particular if $\mathbf{2} = (B; \{+, \cdot, ', 0, 1\})$ is the two-element Boolean algebra and $\overline{x} = x'$ (i.e., the unary operations $^-$ and $'$ are equal) then the algebra $(B; \{+, \cdot, ^-, ', 0, 1\})$ is a Boole-De Morgan algebra and we have denoted it by BM_2.

Further, denote $D = \{0, a, b, 1\}$. Defining $0{+}x = x{+}0 = x$, $0{\cdot}x = x{\cdot}0 = 0$ and $1 \cdot x = x \cdot 1 = x$, $1 + x = x + 1 = 1$ for all $x \in D$, and $a + b = b + a = 1$, $a \cdot b = b \cdot a = 0$, $\overline{0} = 1$, $\overline{1} = 0$, $\overline{a} = a$, $\overline{b} = b$, $1' = 0$, $0' = 1$, $a' = b$, $b' = a$ we get the Boole-De Morgan algebra $BM_4 = (D; \{+, \cdot, ^-, {}', 0, 1\})$. If we define the binary operations $\vee$ and $\wedge$ on D by setting $a \vee x = a$, $b \vee x = x$ and $a \wedge x = x$, $b \wedge x = b$ for all $x \in D$, and $0 \vee 1 = a$, $0 \wedge 1 = b$ then we get the distributive bilattice $B_4 = (D; \{+, \cdot, \vee, \wedge, ^-, 0, 1, b, a\})$ (see [21,22]) and the algebra $BMB_4 = (D; \{+, \cdot, \vee, \wedge, ^-, {}', 0, 1, b, a\})$.

For the definition of free algebras of the given variety also see [38,61,143, 329]. The following classical results are well known.

Theorem 12.1.1 (R. Dedekind, 1897). *The free Boolean algebra on n free generators is isomorphic to the Boolean algebra of Boolean functions of n variables [38,72,84,143,329] (see Chapter 4).*

Theorem 12.1.2 (R. Dedekind, 1897). *The free bounded distributive lattice on n free generators is isomorphic to the bounded distributive lattice of monotone Boolean functions of n variables [38,72,84,143,180,329] (see Chapter 4).*

A problem posed by B.I. Plotkin in 1970s has required finding the varieties of algebras with analogous functional representations of free finitely generated algebras. The importance of this problem is that the generated functions are alternative to Boolean functions with their respective applications which are comparable to the applications of Boolean functions.

In Chapter 4, the concept of De Morgan function was introduced and it is proved that the free De Morgan algebra on n free generators is isomorphic to the De Morgan algebra of De Morgan functions of n variables. Note that another characterization of free De Morgan algebras is in [19].

In Chapter 8, a functional completeness criterion for De Morgan functions was established.

In Chapter 10, the concept of a super-Boolean function is introduced and it is proved that the free super-Boolean algebra on n free generators is isomorphic to the super-Boolean algebra of super-Boolean functions of n variables. An algebra is called a super-Boolean algebra if it satisfies all hyperidentities of the variety of Boolean algebras [253,256,257].

A super-De Morgan algebra is an algebra satisfying hyperidentities of the variety of De Morgan algebras (lattices) [269]. In Chapter 11, we give a functional representation of the free n-generated super-De Morgan algebras with two binary and one unary operations. Namely, we define the concept of super-De Morgan function and prove that the free super-De Morgan algebra with two binary and one unary operations on n free generators is isomorphic to the super-De Morgan algebra of super-De Morgan functions of n variables.

In Chapter 3, the concept of quasi-De Morgan function was also introduced and it is proved that the free Boole-De Morgan algebra on n free generators is isomorphic to the Boole-De Morgan algebra of quasi-De Morgan functions of n variables.

All these functions are more general than the Boolean functions and, in contrast to the other generalized Boolean functions discovered and investigated so far (see [227,387,409], etc.), these have clearly normal forms, which are importance for the applications of these functions in pure and applied mathematics, especially in discrete mathematics, quantum computation, quantum information theory, quantum logic and the theory of quantum computers.

In this chapter, we define the concepts of bi-De Morgan function and prove that the free distributive bilattice on n free generators is isomorphic to the distributive bilattice of bi-De Morgan functions of n variables.

12.2. Some preliminary results

Theorem 12.2.1. *The Boole-De Morgan algebras BM_2 and BM_4 are subdirectly irreducible, and those algebras are the only nontrivial subdirectly irreducible Boole-De Morgan algebras up to isomorphism (see Theorem 3.2.12).*

Lemma 12.2.2 ([347]). *Let $\mathfrak{B} = (Q; \{+, \cdot, \vee, \wedge, ^-, 0_1, 1_1, 0_2, 1_2\})$ be a distributive bilattice. A relation $\theta \subseteq Q \times Q$ is a congruence relation of the bilattice $\mathfrak{B}$ if and only if it is a congruence relation of the reduct $(Q; \{+, \cdot, ^-\})$.*

Corollary 12.2.3. *The distributive bilattice B_4 is subdirectly irreducible, and it is the only subdirectly irreducible distributive bilattice up to isomorphism.*

Now we can say that any distributive bilattice is isomorphic to a subalgebra of a direct power of algebra B_4.

For a variety of algebras $\mathcal{V}$ denote by $Eq(\mathcal{V})$ the set of all identities that are satisfied in all algebras of $\mathcal{V}$ (the equational theory of the variety $\mathcal{V}$). Also for $\mathfrak{M} \in \mathcal{V}$ by $Eq(\mathfrak{M})$ denote the set of all identities of that algebra (the equational theory of that algebra). By $\mathcal{B}$ denote the variety of all distributive bilattices.

Corollary 12.2.4. $Eq(\mathcal{B}) = Eq(B_4)$.

Proof. Obviously $Eq(\mathcal{B}) \subseteq Eq(B_4)$. Now if $\mathfrak{M} \in \mathcal{B}$ then $\mathfrak{M}$ is isomorphic to a subalgebra of direct power of algebra B_4. But any identity of the algebra B_4 is valid also in a direct power of that algebra, and hence, in $\mathfrak{M}$. Therefore $Eq(\mathcal{B}) \supseteq Eq(B_4)$. □

Corollary 12.2.5. *The following identities hold in any distributive bilattice:*

$$\begin{aligned} x \wedge y &= 1_2 \cdot xy + 0_2 \cdot (x+y), \\ x \vee y &= 0_2 \cdot xy + 1_2 \cdot (x+y), \\ x + y &= (1_1 \wedge (x \vee y)) \vee (0_1 \wedge x \wedge y), \\ x \cdot y &= (0_1 \wedge (x \vee y)) \vee (1_1 \wedge x \wedge y). \end{aligned}$$

Proof. It follows from the previous corollary and from the fact that the following identities are valid in the distributive bilattice B_4:

$$\begin{aligned} x \wedge y &= a \cdot xy + b \cdot (x+y), \\ x \vee y &= a \cdot (x+y) + b \cdot xy, \\ x + y &= (1 \wedge (x \vee y)) \vee (0 \wedge x \wedge y), \\ x \cdot y &= (0 \wedge (x \vee y)) \vee (1 \wedge x \wedge y). \end{aligned}$$

□

Corollary 12.2.6. *The free n-generated distributive bilattice is isomorphic to the subalgebra of $B_4^{B_4^n}$ generated by the projections (cf. [61]).*

12.3. Functions on the set D

Recall that $D = \{0, a, b, 1\}$ and $B = \{0, 1\}$. Let us consider a one-to-one correspondence between the sets D and $B \times B$ as follows:

$$0 \leftrightarrow (0,0), \;\; a \leftrightarrow (1,0), \;\; b \leftrightarrow (0,1), \;\; 1 \leftrightarrow (1,1).$$

We define the operations $+, \cdot, \vee, \wedge, {}^-, {}'$ on the set $B \times B$ as follows:

$$\begin{gathered} \overline{(u,v)} = (v', u'), \;\; (u,v)' = (u', v'), \\ (u_1, v_1) + (u_2, v_2) = (u_1 + u_2, v_1 + v_2), \;\; (u_1, v_1) \cdot (u_2, v_2) = (u_1 \cdot u_2, v_1 \cdot v_2), \\ (u_1, v_1) \vee (u_2, v_2) = (u_1 + u_2, v_1 \cdot v_2), \;\; (u_1, v_1) \wedge (u_2, v_2) = (u_1 \cdot u_2, v_1 + v_2), \end{gathered}$$

where the operations on the right-hand side are the operations of the Boole-De Morgan algebra BM_2. However, if the couple $(y, z) \in B \times B$ corresponds to $x \in D$ then we will write $x = (y, z)$ (this causes no confusion).

For $\alpha \in D$ let

$$\alpha^* = \begin{cases} \alpha & \text{if } \alpha = 0, 1, \\ a & \text{if } \alpha = b, \\ b & \text{if } \alpha = a. \end{cases}$$

The unary operation * can also be defined on $B \times B$ taking into account the isomorphism described above. In result we get $(u, v)^* = (v, u) = (v^*, u^*)$, $u, v \in B$. It is clear that $x^* = (\overline{x})' = \overline{x'}$ (which agrees with the notation from the previous section) and $(x^*)^* = x$.

Definition 12.3.1. A function $f : D^n \to D$ is called a *bi-De Morgan function* if it preserves the order relation $\rho = \{(b, b), (b, 0), (b, 1), (b, a), (0, 0), (0, a), (1, 1), (1, a), (a, a)\} \subseteq D^2$.

Note that the order relation ρ is the order of the lattice $(D; \{\vee, \wedge\})$ (the operations $\vee$ and $\wedge$ are defined in the first section) and if $x \rho y$ then $y^* \rho x^*$. Thus bi-De Morgan functions are those functions mapping D^n into D that are monotone with respect to the order of the lattice $(D; \{\vee, \wedge\})$, i.e., an n-ary function f is a bi-De Morgan function if

$$u_i, v_i \in D,\ u_i \rho v_i,\ i = 1, \ldots, n \Rightarrow f(u_1, \ldots, u_n) \rho f(v_1, \ldots, v_n).$$

A function $f : D^n \to D$ is called a *quasi-De Morgan function* if the following conditions hold:

(1) if $x_i \in \{0, 1\}$, $i = 1, \ldots, n$, then $f(x_1, \ldots, x_n) \in \{0, 1\}$,
(2) if $x_i \in D$, $i = 1, \ldots, n$, then $f(x_1^*, \ldots, x_n^*) = (f(x_1, \ldots, x_n))^*$.

Condition (1) means that the function f preserves the unary relation $\theta = \{0, 1\} \subseteq D$, and condition (2) means that f preserves the binary relation $\sigma = \{(0, 0), (a, b), (b, a), (1, 1)\} \subseteq D^2$, which is the graph of the automorphism $x \mapsto x^*$.

Notice that condition (1) is a consequence of condition (2), but however it is more convenient to write it as a separate condition.

A quasi-De Morgan function $f : D^n \to D$ is called a *De Morgan function* if it preserves the order relation ρ.

In other words, a De Morgan function is a bi-De Morgan function that preserves the relations θ and σ.

Let us give some examples. The constant functions $f = 1$ and $f = 0$ are quasi-De Morgan (and also De Morgan) functions, but the constant functions $f = a$ and $f = b$ are not. Instead, the latter two functions are bi-De Morgan functions. The functions $f(x) = x$, $g(x) = \overline{x}$, $h(x, y) = x \cdot y$, $q(x, y) = x + y$,

$p(x) = x'$, where the operations on the right-hand side are the operations of the Boole-De Morgan algebra BM_4, are quasi-De Morgan functions. Also note that the function p is an example of quasi-De Morgan function which is not a De Morgan function. The functions $a \cdot x$, $x \vee y$, $x \wedge y$ are bi-De Morgan functions, and $b \cdot x'$, $x' \wedge y$ are not bi-De Morgan functions.

Now let us mention some properties of quasi-De Morgan and De Morgan functions.

Below, for $x_i \in D$ we denote by (y_i, z_i) the pair from $B \times B$ which corresponds to x_i, i.e., $x_i = (y_i, z_i)$.

A function $f : D^n \to D$ is a quasi-De Morgan function if and only if there exists a Boolean function $\varphi : B^{2n} \to B$ such that

$$f(x_1, \ldots, x_n) = (\varphi(y_1, \ldots, y_n, z_1', \ldots, z_n'), \varphi(z_1, \ldots, z_n, y_1', \ldots, y_n')), \tag{12.3.1}$$

for all $x_1, \ldots, x_n \in D$ (see Theorem 3.3.3).

The Boolean function φ is determined uniquely by the quasi-De Morgan function f, and we will denote it by φ_f.

There are exactly 2^{4^n} quasi-De Morgan functions of n variables.

Recall that a Boolean function is called monotone if it preserves the order of the lattice $(\{0, 1\}; \{+, \cdot\})$.

A quasi-De Morgan function f is a De Morgan function if and only if the corresponding Boolean function φ_f is monotone (see Theorem 4.3.4).

Denote by m_n the number of monotone Boolean functions of n variables.

There are m_{2n} De Morgan functions of n variables.

For any quasi-De Morgan function f of n variables there exists a unique set $S \subseteq 2^{\{1,\ldots,n\}} \times 2^{\{1,\ldots,n\}}$ such that

$$f(x_1, \ldots, x_n) = \sum_{(s_1,s_2)\in S} \left(\prod_{i\in s_1} x_i \cdot \prod_{i\in \overline{s}_1} x_i' \cdot \prod_{i\in s_2} \overline{x}_i \cdot \prod_{i\in \overline{s}_2} x_i^* \right), \tag{12.3.2}$$

where the operations on the right-hand side are the operations of the algebra BM_4.

To formulate the next result let us recall the definition of antichain. If $(L; \leq)$ is a partially ordered set then $A \subseteq L$ is called an *antichain* if any two different elements of the set A are incomparable (for antichain see [84,143,180,386]). Defining the order relation $\subseteq$ on the set $2^{\{1,\ldots,n\}} \times 2^{\{1,\ldots,n\}}$ componentwisely (i.e., $(a_1, a_2) \subseteq (b_1, b_2)$, if $a_1 \subseteq b_1$ and $a_2 \subseteq b_2$) we get a partially ordered set $2^{\{1,\ldots,n\}} \times 2^{\{1,\ldots,n\}}(\subseteq)$. Below we will use antichains of that poset.

For any De Morgan function f of n variables, there exists a unique antichain $S \subseteq 2^{\{1,\dots,n\}} \times 2^{\{1,\dots,n\}}$ such that

$$f(x_1,\dots,x_n) = \sum_{s=(s_1,s_2)\in S} \left(\prod_{i\in s_1} x_i \cdot \prod_{i\in s_2} \overline{x}_i \right). \tag{12.3.3}$$

The terms in the right-hand side of Eqs. (12.3.2) and (12.3.3) are called disjunctive normal forms (DNF) for quasi-De Morgan and De Morgan functions, respectively. Thus every quasi-De Morgan (De Morgan) function can be uniquely represented in the disjunctive normal form.

Denote the set of all bi-De Morgan functions, quasi-De Morgan functions and De Morgan functions of n variables by $\mathcal{B}_n$, $\mathcal{BM}_n$ and $\mathcal{D}_n$, respectively. For any two functions $f, g : D^n \to D$ define $f+g$, $f\cdot g$, $f\vee g$, $f\wedge g$, $\overline{f}$ and f' by the standard way, i.e., $(f\lambda g)(x) = f(x)\lambda g(x)$, for all $\lambda \in \{+,\cdot,\vee,\wedge\}$, and $\overline{f}(x) = \overline{f(x)}$, $f'(x) = (f(x))'$, $x \in D^n$.

Let us prove that the set $\mathcal{B}_n$ is closed under the operations $+,\cdot,\vee,\wedge,^{-}$. This set consists of all functions from D^n into D that preserve the order of the lattice $(D;\{\vee,\wedge\})$. So the functions $x\vee y$ and $x\wedge y$ are bi-De Morgan functions, i.e., belong to $\mathcal{B}_n$. Hence if $f,g\in\mathcal{B}_n$, i.e., f and g preserve the order ρ, then $f\vee g, f\wedge g$ also preserve the order ρ, and therefore $f\vee g, f\wedge g \in \mathcal{B}_n$. Also note that all the constant functions $0,1,a,b$ are bi-De Morgan functions. Now from the last two identities of Corollary 12.2.5 it follows that the functions $x+y$ and $x\cdot y$ are also bi-De Morgan functions. Hence $f+g$, $f\cdot g\in\mathcal{B}_n$ for any functions $f,g\in\mathcal{B}_n$.

Analogously we can prove the similar results for the sets $\mathcal{BM}_n$ and $\mathcal{D}_n$. Namely that the set $\mathcal{BM}_n$ is closed under the operations $+,\cdot,^{-},{}'$ and the set $\mathcal{D}_n$ is closed under the operations $+,\cdot,^{-}$.

Thus, we get the following algebras:

$$\mathfrak{B}_n = (\mathcal{B}_n, \{+,\cdot,\vee,\wedge,^{-}0,1,b,a\}),$$
$$\mathfrak{BM}_n = (\mathcal{BM}_n, \{+,\cdot,^{-}{}',0,1\}),$$
$$\mathfrak{D}_n = (\mathcal{D}_n, \{+,\cdot,^{-}0,1\}).$$

Evidently, $\mathfrak{B}_n$ is a distributive bilattice, $\mathfrak{BM}_n$ is a Boole-De Morgan algebra and $\mathfrak{D}_n$ is a De Morgan algebra.

Let us consider the projection functions

$$\delta_n^i(x_1,\dots,x_n) = x_i, \ i=1,\dots,n,$$

as functions $D^n \to D$.

The algebra $\mathfrak{BM}_n$ is a free Boole-De Morgan algebra with the system of free generators $\Delta = \{\delta_n^1, \ldots, \delta_n^n\}$. Hence every free n-generated Boole-De Morgan algebra is isomorphic to the Boole-De Morgan algebra $\mathfrak{BM}_n$ (see Theorem 3.3.8).

The algebra $\mathfrak{D}_n$ is a free De Morgan algebra with the system of free generators $\Delta = \{\delta_n^1, \ldots, \delta_n^n\}$. Hence every free n-generated De Morgan algebra is isomorphic to the De Morgan algebra $\mathfrak{D}_n$ (see Theorem 4.3.7).

12.4. Disjunctive normal forms of bi-De Morgan functions

Lemma 12.4.1. *For any function $f : D^n \to D$ there exists a unique pair (φ_1, φ_2) of Boolean functions of $2n$ variables such that*

$$f(x_1, \ldots, x_n) = (\varphi_1(y_1, \ldots, y_n, z_1', \ldots, z_n'), \varphi_2(z_1, \ldots, z_n, y_1', \ldots, y_n')), \tag{12.4.1}$$

for all $x_1, \ldots, x_n \in D$.

Proof. There are 2^{2^n} Boolean functions of n variables, hence there are $(2^{2^{2n}})^2 = 4^{4^n}$ pairs (φ_1, φ_2) of Boolean functions of $2n$ variables. And also there are 4^{4^n} functions mapping D^n into D. Now the statement of the theorem follows immediately. □

Note that if here the Boolean functions φ_1 and φ_2 are equal then the function f is a quasi-De Morgan function.

Theorem 12.4.2. *Any function $f : D^n \to D$ can be uniquely represented in the form*

$$f = (a \cdot f_1) + (b \cdot f_2), \tag{12.4.2}$$

where f_1 and f_2 are quasi-De Morgan functions of n variables (here the operations on the right-hand side are the operations of the algebra BMB_4).

Proof. For simplicity we will write equality (12.4.1) omitting the variables. Thus we have

$$f = (\varphi_1, \varphi_2) = ((1,0) \cdot (\varphi_1, \varphi_1)) + ((0,1) \cdot (\varphi_2, \varphi_2)) = (a \cdot f_1) + (b \cdot f_2),$$

where $f_1 = (\varphi_1, \varphi_1)$ and $f_2 = (\varphi_2, \varphi_2)$ are quasi-De Morgan functions by Theorem 3.3.3. The uniqueness follows from Lemma 12.4.1. □

Corollary 12.4.3. *For every function $f : D^n \to D$ of n variables there exists a unique pair of sets (S_1, S_2) $(S_1, S_2 \subseteq 2^{\{1,\ldots,n\}} \times 2^{\{1,\ldots,n\}})$ such that*

$$\begin{aligned} f(x_1, \ldots, x_n) \\ &= a \cdot \sum_{(s_1,s_2)\in S_1} \left(\prod_{i\in s_1} x_i \cdot \prod_{i\in \overline{s}_1} x_i' \cdot \prod_{i\in s_2} \overline{x}_i \cdot \prod_{i\in \overline{s}_2} x_i^*\right) \\ &\quad + b \cdot \sum_{(s_1,s_2)\in S_2} \left(\prod_{i\in s_1} x_i \cdot \prod_{i\in \overline{s}_1} x_i' \cdot \prod_{i\in s_2} \overline{x}_i \cdot \prod_{i\in \overline{s}_2} x_i^*\right), \end{aligned} \tag{12.4.3}$$

where the operations on the right-hand side are the operations of the algebra BMB_4.

This form is unique for a given function $f : D^n \to D$ and is called the disjunctive normal form of that function.

Theorem 12.4.4. *A function $f : D^n \to D$ is a bi-De Morgan function if and only if f can be represented in the form*

$$f = (a \cdot f_1) + (b \cdot f_2), \tag{12.4.4}$$

where f_1 and f_2 are De Morgan functions of n variables (here the operations on the right hand side are the operations of the algebra B_4). And this representation is unique.

Proof. Let f can be represented in the form (12.4.4). The functions f_1 and f_2 are De Morgan functions, hence they preserve the order ρ. And so the function $(a\cdot f_1)+(b\cdot f_2)$ preserves that order as well (from the arguments stated above). Thus the sufficiency is proved. Let us prove the necessity.

The function f can be uniquely represented in the form (12.4.4) with quasi-De Morgan functions f_1 and f_2. We must prove that f_1 and f_2 are De Morgan functions, i.e., preserve the order ρ. Suppose f preserves the order ρ. Then the function $a \cdot f = a \cdot f_1$ preserves ρ too. We claim that $b \cdot f_1$ also preserves ρ. Let $\alpha_i \rho\, \beta_i$, $i = 1, \ldots, n$ and let $\alpha = (\alpha_1, \ldots, \alpha_n) \in D^n$, $\beta = (\beta_1, \ldots, \beta_n) \in D^n$. Define the unary operation * on n-tuples from D^n componentwisely. Then $\beta_i^* \rho\, \alpha_i^*$, hence $a \cdot f_1(\beta^*)\, \rho\, a \cdot f_1(\alpha^*)$. As f_1 is a quasi-De Morgan function, we conclude that $(b \cdot f_1(\beta))^* = a \cdot (f_1(\beta))^*\ \rho\ a \cdot f_1(\alpha^*) = (b \cdot f_1(\alpha))^*$. Therefore $b \cdot f_1(\alpha)\ \rho\ b \cdot f_1(\beta)$. Now we conclude that $f_1 = (a \cdot f_1) + (b \cdot f_1)$ preserves the order ρ. By the same way, we show that f_2 also preserves ρ. And hence f_1 and f_2 are De Morgan function. This completes the proof of the theorem. □

This theorem makes clear why such functions are called bi-De Morgan functions. Now we can state that for a bi-De Morgan function the corresponding Boolean functions in the representation (12.4.1) are monotone.

Corollary 12.4.5. *There are m_{2n}^2 bi-De Morgan functions of n variables.*

Corollary 12.4.6. *For every bi-De Morgan function f of n variables there exists a unique pair of antichains (S_1, S_2) $(S_1, S_2 \subseteq 2^{\{1,\dots,n\}} \times 2^{\{1,\dots,n\}})$ such that*

$$f(x_1, \dots, x_n) = a \cdot \sum_{(s_1,s_2)\in S_1} \left(\prod_{i\in s_1} x_i \cdot \prod_{i\in s_2} \overline{x}_i\right) + b \cdot \sum_{(s_1,s_2)\in S_2} \left(\prod_{i\in s_1} x_i \cdot \prod_{i\in s_2} \overline{x}_i\right), \tag{12.4.5}$$

where the operations on the right-hand side are the operations of the algebra B_4.

The form (12.4.5) is called the disjunctive normal form of bi-De Morgan function f.

12.5. Finitely generated free distributive bilattices

In this section, we characterize finitely-generated free distributive bilattices by means of bi-De Morgan functions.

Theorem 12.5.1 ([271]). *The algebra $\mathfrak{B}_n$ is a free distributive bilattice with the system of free generators $\Delta = \{\delta_n^1, \dots, \delta_n^n\}$. Hence every free n-generated distributive bilattice is isomorphic to the distributive bilattice $\mathfrak{B}_n$.*

Proof. Let $\mathfrak{B} = (Q; \{+, \cdot, \vee, \wedge, {}^-, 0_1, 1_1, 0_2, 1_2\})$ be a distributive bilattice and $\mu : \Delta \to Q$ be a mapping. We should prove that there exists a unique homomorphism $\nu : \mathcal{B}_n \to Q$ which extends μ, i.e., $\nu|_\Delta = \mu$. The algebra $(Q; \{+, \cdot, {}^-, 0_1, 1_1\})$ is a De Morgan algebra and from Theorem 4.3.7 we conclude that there exists a mapping $\xi : \mathcal{D}_n \to B$ which extends μ and is a homomorphism from the algebra $\mathfrak{D}_n$ into the algebra $(Q; \{+, \cdot, {}^-, 0_1, 1_1\})$. From Theorem 12.4.4, it follows that any function $f \in \mathcal{B}_n$ can be uniquely represented in the form $f = (a \cdot f_1) + (b \cdot f_2)$, where f_1 and f_2 are De Morgan functions. Define $\nu(f) = (1_2 \cdot \xi(f_1)) + (0_2 \cdot \xi(f_2))$. Obviously ν is a homomorphism. The uniqueness is also evident. □

Corollary 12.5.2. *Every finitely generated free distributive bilattice is finite.*

Corollary 12.5.3. *Every finitely generated distributive bilattice is finite, i.e. the variety of distributive bilattices is a Burnside variety.*

Problem 12.5.4. *Prove the Post type functional completeness theorem for bi-De Morgan functions (see [193,276,333,334]).*

Problem 12.5.5. *Develop the bi-De Morgan analogue of the theory of Boolean functions.*

Chapter 13

Weakly Idempotent Lattices and Bilattices

13.1. Preliminaries

In this chapter, we study weakly idempotent lattices with an additional interlaced operation. A representation theorem for interlaced weakly idempotent bilattices is proved. We obtain a representation theorem for interlaced weakly idempotent pre-bilattices, too, as a corollary. In the second part of the chapter we introduce the concept of non-idempotent Plonka function and the weakly Plonka sum and characterize the subdirectly irreducible weakly idempotent quasilattices, using these functions. As a consequence, we characterize the hyperidentities of the variety of weakly idempotent lattices.

There exist various extensions of the concept of lattice. For example, in [112,113] weakly associative lattices are introduced and in [17,158,159,175] the lattices with a third operation were studied. In [238] an algebra with a system of identities was introduced, which we call weakly idempotent lattices (also see [135,325]).

Definition 13.1.1. The algebra $(L; \wedge)$ with one binary operation is called *weakly idempotent semilattice*, if it satisfies the following identities:

$$a \wedge b = b \wedge a \quad (commutativity) \tag{13.1.1}$$

$$(a \wedge b) \wedge c = a \wedge (b \wedge c) \quad (associativity) \tag{13.1.2}$$

$$a \wedge (b \wedge b) = a \wedge b \quad (weakly\ idempotency) \tag{13.1.3}$$

The operation $\wedge$ is called product. Adding the idempotent identity $a \wedge a = a$ to it, we obtain a semilattice. The element $a \in L$ is called an idempotent of the *weakly idempotent semilattice* $(L; \wedge)$, if $a \wedge a = a$. The set of the idempotent elements of each weakly idempotent semilattice form a semilattice, i.e., the product of any two idempotent elements in the weakly idempotent semilattice is an idempotent element.

Definition 13.1.2. The algebra $(L; \wedge, \vee)$ with two binary operations is called *weakly idempotent lattice*, if the reducts $(L; \wedge)$ and $(L; \vee)$ are weakly idempotent semilattices and the following identities are valid:

$$a \wedge (b \vee a) = a \wedge a, a \vee (b \wedge a) = a \vee a \quad (\textit{weakly absorption}) \tag{13.1.4}$$

$$a \wedge a = a \vee a \quad (\textit{equalization}) \tag{13.1.5}$$

For example, the following algebra $(Z \setminus \{0\}; \wedge, \vee)$, where $x \wedge y = (|x|, |y|)$ and $x \vee y = [|x|, |y|]$, for which $(|x|, |y|)$ and $[|x|, |y|]$ are the greatest common division (gcd) and the least common multiple (lcm) of $|x|$ and $|y|$, is a weakly idempotent lattice, which is not a lattice, since $x \wedge x \neq x$ and $x \vee x \neq x$ for negative x.

The operation $\vee$ is called sum. The element $a \in L$ is called an idempotent of the weakly idempotent lattice $(L; \wedge, \vee)$ if $a \wedge a = a$ and $a \vee a = a$. Note that the product (sum) of two elements of a weakly idempotent lattice $(L; \wedge, \vee)$ is an idempotent element:

$$\begin{aligned}
(x \wedge y) \vee (x \wedge y) &\overset{(13.1.5)}{=} (x \wedge y) \wedge (x \wedge y) \\
&\overset{(13.1.2)}{=} ((x \wedge y) \wedge x) \wedge y \overset{(13.1.1)}{=} ((y \wedge x) \wedge x) \wedge y \\
&\overset{(13.1.2)}{=} (y \wedge (x \wedge x)) \wedge y \overset{(13.1.3)}{=} (y \wedge x) \wedge y \\
&\overset{(13.1.1)}{=} (x \wedge y) \wedge y \overset{(13.1.2)}{=} x \wedge (y \wedge y) \\
&\overset{(13.1.3)}{=} x \wedge y.
\end{aligned}$$

The other condition is proved similarly. So, the set of all idempotent elements of a weakly idempotent lattice is a lattice.

The weakly idempotent lattice $(L; \wedge, \vee)$ is called distributive, if it satisfies the following identities of distributivity:

$$x \wedge (y \vee z) = (x \wedge y) \vee (x \wedge z),$$

$$x \vee (y \wedge z) = (x \vee y) \wedge (x \vee z).$$

The weakly idempotent lattice $(L; \wedge, \vee)$ is called modular, if it satisfies the following identity of modularity:

$$(x \wedge (y \vee z)) \vee (y \wedge z) = (x \vee (y \wedge z)) \wedge (y \vee z).$$

For each weakly idempotent semilattice $(L; *)$ there is a corresponding quasiorder Θ (i.e., a reflexive and transitive relation), defined in the following manner:

$$a\Theta b \iff a * b = a * a. \tag{13.1.6}$$

For each weakly idempotent lattice $(L; \wedge, \vee)$, we have

$$a\Theta b \iff a \wedge b = a \wedge b \iff a \vee b = b \vee b. \tag{13.1.7}$$

Definition 13.1.3. We say that the operation $*$ of the weakly idempotent semilattice $(L; *)$ is interlaced with the operations $\wedge$ and $\vee$ of the weakly idempotent lattice $(L; \wedge, \vee)$, if the weakly idempotent semilattice operation $*$ preserves the weakly idempotent lattice quasiorder, and the operations $\wedge, \vee$ preserve the weakly idempotent semilattice quasiorder.

Definition 13.1.4. An algebra $(L; \wedge, \vee, *, \Delta)$ with four binary operations is called a *weakly idempotent pre-bilattice* if the reducts $L_1 = (L; \wedge, \vee)$ and $L_2 = (L; *, \Delta)$ are weakly idempotent lattices and the following identity is valid: $a * a = a \wedge a$. If the reducts $L_1 = (L; \wedge, \vee)$ and $L_2 = (L; *, \Delta)$ are lattices, then the algebra $(L; \wedge, \vee, *, \Delta)$ is called a pre-bilattice.

Definition 13.1.5. An algebra $(L; \wedge, \vee, *, \Delta, ')$ with four binary operations and one unary operation of negation is called a *weakly idempotent bilattice* if $(L; \wedge, \vee, *, \Delta)$ is a weakly idempotent pre-bilattice and the following additional identities are valid:

$$(a \wedge b)' = a' \vee b', \quad (a \vee b)' = a' \wedge b',$$
$$(a * b)' = a' * b', (a\Delta b)' = a'\Delta b', (a')' = a.$$

If $(L; \wedge, \vee, *, \Delta)$ is a pre-bilattice, then $(L; \wedge, \vee, *, \Delta, ')$ is called a bilattice.

The weakly idempotent (pre-)bilattice $(L; \wedge, \vee, *, \Delta, ')$ ($(L; \wedge, \vee, *, \Delta)$) is called distributive (modular) if each pair of the operations from the set $\{\wedge, \vee, *, \Delta\}$ satisfies the identities of distributivity (modularity).

The weakly idempotent (pre-)bilattices are extension of the concept of (pre-) bilattice. Bilattices were introduced by Ginsberg [125–128] as a general and uniform framework for a diversity of applications in artificial intelligence. In a series of papers it was shown that these structures may serve as foundation for many areas, such as logic programming [107–109], artificial intelligence [128], truth theory [110] and others. For applications and characterization of (pre-)bilattices in various varieties see also [18,45,46,75,117,166,262,264,299,300,335,337,347].

Since every weakly idempotent(pre-)bilattice has structure of two weakly idempotent lattices, then two quasiorders correspond to every weakly idempotent (pre-) bilattice.

Definition 13.1.6. A weakly idempotent (pre-)bilattice is called *interlaced*, if all the basic weakly idempotent (pre-)bilattice operations are quasiorder preserving with respect to both quasiorders.

Every distributive weakly idempotent (pre-)bilattices is interlaced.

Note that the operations of the weakly idempotent lattice $(L; \wedge, \vee)$ preserve the corresponding quasiorder, hence the weakly idempotent pre-bilattice $(L; \wedge, \vee, \wedge, \vee)$ is interlaced. The weakly idempotent pre-bilattice $L = (L; \wedge, \vee, *, \triangle)$ is distributive if and only if it satisfies the following hyperidentity:

$$X(Y(x, y), z) = Y(X(x, z), X(y, z)). \tag{13.1.8}$$

For the categorical definition of the hyperidentity in [249] the bihomomorphisms between two algebras $(Q; F)$ and $(Q'; F')$ are defined as the pair $(\varphi; \tilde{\psi})$ of the mappings:

$$\varphi : Q \to Q', \tilde{\psi} : F \to F', |A| = |\tilde{\psi} A|,$$

with the following condition:

$$\varphi(a_1, \ldots, a_n) = (\tilde{\psi} A)(\varphi a_1, \ldots, \varphi a_n)$$

for any $A \in F, a_1, \ldots, a_n \in Q, |A| = n$. More about the application of such morphisms in the cryptography can be found in [15].

Algebras with their bihomomorphisms $(\varphi; \tilde{\psi})$ (as morphisms) form a category. The product in this category is called a superproduct of algebras and is denoted by $Q \bowtie Q'$ for the two algebras Q and Q'. For example, the superproduct of the two weakly idempotent lattices $(Q; +, \cdot)$ and $(Q'; +, \cdot)$ is the binary algebra $(Q \times Q'; (+, \cdot), (\cdot, +), (+, +), (\cdot, \cdot))$, with four binary operations, where the pairs of the operations operate component-wise, i.e.,

$$(A, B)((x, y), (u, v)) = (A(x, u), B(y, v)),$$

and $Q \bowtie Q'$ is an interlaced weakly idempotent pre-bilattice.

Indeed: denote the quasiorders of Q and Q' by $\leq_1$ and $\leq_2$, correspondingly, the quasiorders of $Q \bowtie Q'$ – by $\leq_I$ and $\leq_{II}$, which are defined in the following way: $(a, b) \leq_I (c, d) \iff a \leq_1 c, b \leq_2 d$ and $(a, b) \leq_{II} (c, d) \iff a \leq_1 c, d \leq_2 b$. Take $(a, b) \leq_I (c, d)$ and $(e, f) \in Q \times Q'$. Show that $(a, b)(+, \cdot)(e, f) \leq_I (c, d)(+, \cdot)(e, f)$. From $(a, b) \leq_I (c, d)$ we have $a \leq_1$

$c, b \leq_2 d$, hence for $e \in Q$ and $f \in Q'$ we get: $a+e \leq_1 c+e$ and $b \cdot f \leq_2 d \cdot f$, thus $(a+e, b \cdot f) \leq_I (c+e, d \cdot f)$ and $(a,b)(+,\cdot)(e,f) \leq_I (c,d)(+,\cdot)(e,f)$. The rest of the conditions are proved similarly.

13.2. Some properties of weakly idempotent lattices

Let $(L; \wedge, \vee)$ be a weakly idempotent lattice and $(L; *)$ be a weakly idempotent semilattice. Denote the quasiorder corresponding to $(L; \wedge, \vee)$ by $\leq_\wedge$ and the quasiorder corresponding to $(L; *)$ — by $\leq_*$ for $(L; *)$, which are defined by (13.1.6) and (13.1.7).

The following lemmas are proved, using the definition of weakly idempotent lattices and weakly idempotent semilattices.

Lemma 13.2.1. *Let $(L; \wedge, \vee)$ be a weakly idempotent lattice and $(L, *)$ be a weakly idempotent semilattice. Then:*

$$a \leq_* b \leq_* a \to a * a = b * b,$$

$$a \leq_\wedge b \leq_\wedge a \to a \wedge a = b \wedge b.$$

Lemma 13.2.2. *Let $(L; \wedge, \vee)$ be a weakly idempotent lattice. Then:*

$$x \wedge y \leq_\wedge x \leq_\wedge x \vee y,$$

$$x \wedge y \leq_\wedge y \leq_\wedge x \vee y,$$

where $x, y \in L$.

Theorem 13.2.3. *The operation $*$ of the weakly idempotent semilattice $(L; *)$ is interlaced with the operations of the weakly idempotent lattice $(L; \wedge, \vee)$ if and only if the algebra $(L; \wedge, \vee, *)$ satisfies the following hyperidentity:*

$$X(Y(X(x,y),z), Y(y,z)) = X(Y(X(x,y),z), Y(X(x,y),z)).$$

In particular, the class of such algebras $(L; \wedge, \vee, *)$ forms a variety.

Proof. Let us show, for example, that if the operation $*$ is interlaced with the operations $\wedge, \vee$, then the following identity is valid:

$$((x \wedge y) * z) \wedge (y * z) = ((x \wedge y) * z) \wedge ((x \wedge y) * z). \qquad (13.2.1)$$

From Lemma 13.2.2 we have $x \wedge y \leq_\wedge z$, then $((x \wedge y) * z) \leq_\wedge y * z$. Hence, $((x \wedge y) * z) \wedge (y * z) = ((x \wedge y) * z) \wedge ((x \wedge y) * z)$. Conversely, let in $(L; \wedge, \vee, *)$ the equality (13.2.1) be valid, then $((x \wedge x) * z) \vee (x * z) = ((x \vee x) * z) \vee (x * z) = ((x \vee x) * z) \vee ((x \vee x) * z) = ((x \vee x) * z) \wedge ((x \vee x) * z)$, hence, $x * z \leq_\wedge (x \wedge x) * z \leq_\wedge x * z$, thus by Lemma 13.2.1 we get: $((x \wedge x) * z) \wedge ((x \wedge x) * z) = (x * z) \wedge (x * z)$. Suppose that $x \leq_\wedge y$, then $(x*z)\wedge(y*z) = ((x*z)\wedge(y*z))\wedge(y*z) = ((x\wedge x)*z)\wedge((x\wedge x)*z)\wedge(y*z) = ((x \wedge y) * z) \wedge ((x \wedge y) * z) = ((x \wedge x) * z) \wedge ((x \wedge x) * z) = (x * z) \wedge (x * z)$.

So, if $x \leq_\wedge y$, then $x * z \leq_\wedge y * z$. □

Corollary 13.2.4. *Let $(L; \wedge, \vee)$ be a weakly idempotent lattice and $(L, *)$ be a weakly idempotent semilattice and the following identity $a \wedge a = a \vee a$ is valid. Then the operation $*$ is interlaced with the operations $\wedge, \vee$ if and only if the algebra $(L; \wedge, \vee, *)$ satisfies the following hyperidentity:*

$$X(Y(X(x, y), z), Y(y, z)) = Y(X(x, y), z). \tag{13.2.2}$$

Corollary 13.2.5. *A weakly idempotent (pre-)bilattice $(L; \wedge, \vee, *, \Delta, ')$ $((L; \wedge, \vee, *, \Delta))$ is interlaced if and only if the hyperidentity (13.2.2) is valid in this (pre-)bilattice.*

In Lemmas 13.2.6–13.2.17 we suppose that $(L; \wedge, \vee)$ is a weakly idempotent lattice, $(L; *)$ is a weakly idempotent semilattice and the operation $*$ is interlaced with the operations $\wedge, \vee$ and the following identity $a \wedge a = a * a$ is valid.

Lemma 13.2.6. *For any $x, y \in L$, the following inequalities are valid:*

$$x \wedge y \leq_\wedge x * y \leq_\wedge x \vee y,$$
$$x * y \leq_* x \wedge y, x * y \leq_* x \vee y.$$

Lemma 13.2.7. *For $a, x, b \in L$, we have*

$$a \leq_\wedge x \leq_\wedge b \text{ and } a \leq_* b \rightarrow a \leq_* x \leq_* b;$$
$$a \leq_\wedge x \leq_\wedge b \text{ and } b \leq_* a \rightarrow b \leq_* x \leq_* a.$$

Proof. Let us proof the first statement. From $a \leq_* b$ we obtain that $a \wedge x \leq_* b \wedge x$ and $a \vee x \leq_* b \vee x$. Using the condition: $a \leq_\wedge x \leq_\wedge b$, we get: $a \wedge x = x \wedge x$ and $a \vee z = x \vee x$, then $a \wedge a \leq_* x \wedge x$ and $x \wedge x \leq_* b \wedge b$, so $a \leq_* x \leq_* b$. □

Lemma 13.2.8. *For $u, x, y \in L$, we have*

$$u \leq_\wedge x \text{ and } u \leq_\wedge y \text{ and } u \leq_* x \text{ and } u \leq_* y \rightarrow x \wedge y = x * y;$$
$$x \leq_\wedge u \text{ and } y \leq_\wedge u \text{ and } u \leq_* x \text{ and } u \leq_* y \rightarrow x \vee y = x * y.$$

Proof. From $u \leq_\wedge x$ and $u \leq_\wedge y$ by Lemma 13.2.6 it follows that $u \wedge u \leq_\wedge x \wedge y \leq_\wedge x * y$. From the conditions $u \leq_* x$ and $u \leq_* y$ and $x * x = x \wedge x$ we have $u \wedge u = u * u \leq_* x * y$, then by Lemma 13.2.7

we get $u \wedge u \leq_* x \wedge y \leq_* x * y$. On the other hand, $x * y \leq_* x \wedge y$ (see Lemma 13.2.2), then $x * y \leq_* x \wedge y \leq_* x * y$, hence, by Lemma 13.2.1: $x \wedge y = (x \wedge y) * (x \wedge y) = (x * y) * (x * y) = x * y$.

□

Lemma 13.2.9. *$a \leq_\wedge b \to a \leq_\wedge a\Delta b \leq_\wedge b$, where $a, b \in L$.*

Proof. By Lemma 13.2.6 $a \wedge a = a * a = a * (a\Delta b) \leq_* a \vee (a\Delta b)$, then from $a \leq_\wedge b$ we get $a * a = a \wedge a = b \wedge a = b \wedge (a \wedge a) \leq_* b \wedge [a \vee (a\Delta b)]$, hence,

$$a * a \leq_* b \wedge [a \vee (a\Delta b)]. \tag{13.2.3}$$

From $b \leq_* a\Delta b$ and $a \leq_\wedge b$ we obtain $b \wedge b = a \vee b \leq_* a \vee (a\Delta b)$, hence, $b * b = b \wedge b = (b \wedge b) \wedge b \leq_* b \wedge [a \vee (a\Delta b)]$ so,

$$b * b \leq_* b \wedge [a \vee (a\Delta b)]. \tag{13.2.4}$$

From (13.2.3) and (13.2.4) it follows that

$$(a * a)\Delta(b * b) \leq_* (b \wedge [a \vee (a\Delta b)])\Delta(b \wedge [a \vee (a\Delta b)])$$
$$(b \wedge [a \vee (a\Delta b)]) \wedge (b \wedge [a \vee (a\Delta b)]) = b \wedge [a \vee (a\Delta b)].$$

Then $a\Delta b \leq_* b \wedge [a \vee (a\Delta b)]$.

Further, from $a \leq_* a\Delta b$ it follows that

$$a \vee (a\Delta b) \leq_* (a\Delta b) \vee (a\Delta b) = a\Delta b, a \wedge (a\Delta b) \leq_* (a\Delta b) \wedge (a\Delta b) = a\Delta b.$$

From $b \leq_* a\Delta b$ we deduce that $b \vee (a\Delta b) \leq_* a\Delta b, b \wedge (a\Delta b) \leq_* a\Delta b$. So, $a\Delta b \leq_* b \wedge [a \vee (a\Delta b)] \leq_* b \wedge (a\Delta b) \leq_* a\Delta b$, hence,

$$[b \wedge (a\Delta b)] = (a\Delta b) * (a\Delta b).$$
$$[b \wedge (a\Delta b)] * [b \wedge (a\Delta b)] = [b \wedge (a\Delta b)] \wedge [b \wedge (a\Delta b)] = b \wedge (a\Delta b),$$
$$(a\Delta b) * (a\Delta b) = (a\Delta b) \wedge (a\Delta b),$$

hence $b \wedge (a\Delta b) = (a\Delta b) \wedge (a\Delta b)$. So, $a\Delta b \leq_\wedge b$. The second part of the inequality is proved in same way. □

Let us define the relations θ_1 and θ_2 on the algebra $(L; \wedge, \vee, *)$ in the following manner:

$$a\theta_1 b \iff a * b = a \vee b,$$
$$a\theta_2 b \iff a * b = a \wedge b.$$

Lemma 13.2.10. *θ_1, θ_2 are congruences of $(L; \wedge, \vee)$.*

Proof. Reflexivity and symmetricity are obvious, so let us show the transitivity. Let $a\theta_1 b$ and $b\theta_1 c$, then $a \vee b = a * b$ and $b \vee c = b * c$; hence by Lemma 13.2.6 $a \vee b \leq_* b$ and $b \vee c \leq_* b$. So, $a \vee b \vee c \leq_* b \vee c$ and $a \vee b \vee c \leq_* a \vee b$. On the other hand, by Lemma 13.2.2 we have $b \vee c \leq_\wedge a \vee b \vee c, a \vee b \leq_\wedge a \vee b \vee c$. Using Lemma 13.2.8 we get $a \vee b \vee c = (a \vee b) \vee (b \vee c) = (a \vee b) * (b \vee c) = (a * b) \vee (b * c) = ((a * b) * b) * c = (a * b) * c$. From here it follows that $a \leq_\wedge a * b * c, b \leq_\wedge a * b * c$. From the inequalities $a * b * c \leq_* a, a * b * c \leq_* c$, applying Lemma 13.2.8, we get: $a \vee c = a * c$; hence $a\theta_1 c$.

Congruence: Let $a\theta_1 b$, then $a * b = a \vee b$, hence $a \vee b \leq_* a$ and $a \vee b \leq_* b$; then $a \vee b \vee c \leq_* a \vee c, a \vee b \vee c \leq_* b \vee c$ for any $c \in L$. On the other hand, by Lemma 13.2.2 we have: $a \vee c \leq_\wedge a \vee b \vee c$ and $b \vee c \leq_\wedge a \vee b \vee c$, then using Lemma 13.2.8 we get: $(a \vee c) \vee (b \vee c) = (a \vee c) * (b \vee c)$, so $a \vee c\theta_1 b \vee c$. □

Lemma 13.2.11. *θ_1, θ_2 are congruences on $(L; *)$.*

Proof. Let $a\theta_1 b$, i.e., $a * b = a \vee b$, then $a \leq_\wedge a * b$ and $b \leq_\wedge a * b$, hence $a * c \leq_\wedge a * b * c$ and $b * c \leq_\wedge a * b * c$. Further, by Lemma 13.2.2: $a * b * c \leq_* a * c$ and $a * b * c \leq_* b * c$. Then, by Lemma 13.2.8 we have $(a * c) * (b * c) = (a * c) \vee (b * c)$, hence $a * c\theta_1 b * c$. Similarly we prove that $a * c\theta_2 b * c$, if $a\theta_2 b$. □

Remark 13.2.12. $a \in [a \wedge a]_{\theta_1} = [a \vee a]_{\theta_1}$, $a \in [a \wedge a]_{\theta_2} = [a \vee a]_{\theta_2}$, for all $a \in L$.

Lemma 13.2.13. *For all $a, b \in L$ we have $a(\theta_1 \cap \theta_2)b \iff a \wedge a = b \wedge b$.*

Proof. $a(\theta_1 \cap \theta_2)b \iff a\theta_1 b$ and $a\theta_2 b \iff a * b = a \vee b$ and $a * b = a \wedge b \iff a \vee b = a \wedge b \iff (a \vee b) \wedge a = (a \wedge b) \wedge a$ and $(a \vee b) \wedge b = (a \wedge b) \wedge b \iff a \wedge a = a \wedge b$ and $b \wedge b = a \wedge b \iff a \wedge a = b \wedge b$. □

Lemma 13.2.14. *For all $a, b \in L$ we have $(a \wedge b)\theta_1(a * b)\theta_2(a \vee b)$.*

Proof. By Lemma 13.2.6 we have $(a \wedge b) * (a * b) = a * b = (a \wedge b) \vee (a * b)$ hence, $a \wedge b\theta_1 a * b$. Similarly we get $(a \vee b) * (a * b) = a * b = (a \vee b) \wedge (a * b)$ hence, $a \vee b\theta_2 a * b$. Hence, we have that $a * b\theta_2 a \vee b$. □

Lemma 13.2.15. *$a \leq_\wedge b \to a\theta_1\theta_2 b$, where $a, b \in L$.*

Proof. From Lemma 13.2.15 it follows that $a \wedge b\theta_1 a * b$ and $a * b\theta_2 a \vee b$, then $a \wedge b\theta_1\theta_2 a \vee b$, hence $a \wedge a\theta_1\theta_2 b \wedge b$ and by Remark 13.2.12 we get $a\theta_1\theta_2 b$. □

Lemma 13.2.16. *Let $(L; *, \Delta)$ be a weakly idempotent lattice, then $a \leq_\wedge b \to a\theta_2\theta_1 b$, where $a, b \in L$.*

Proof. By Lemma 13.2.9 from $a \leq_\wedge b$, then $a \leq_\wedge a\Delta b \leq_\wedge b$. Then $a * (a\Delta b) = a * a = a \wedge a = a \wedge (a\Delta b)$ and $b * (a\Delta b) = b * b = b \wedge b = b \vee (a\Delta b)$, hence $a\theta_2 a\Delta b$ and $b\theta_1 a\Delta b$. So, $a\theta_2\theta_1 b$. □

Lemma 13.2.17. *The algebras* $(L/\theta_1; \wedge, \vee)$ *and* $(L/\theta_2; \wedge, \vee)$ *are lattices.*

Proof. Every element of algebra $(L/\theta_1; \wedge, \vee)$ is idempotent. Indeed, $[a]_{\theta_1} \wedge [a]_{\theta_1} = [a \wedge a]_{\theta_1} = [a]_{\theta_1}$ (by Remark 13.2.12). □

Lemma 13.2.18. *The congruences* Θ, Φ *of the weakly idempotent lattice* $(L; \wedge, \vee)$*, which satisfy the following conditions:* $a\Theta(a \wedge a)$ *and* $a\Psi(a \wedge a)$*, commute if and only if the following conditions are equivalent:* $a\Theta\Phi b$ *and* $a\Phi\Theta b$*, where* $a \leq_\wedge b$*.*

Proof. Its necessity is clear. Let us show its sufficiency. Let $x, y \in L$ and $x\Theta z$, $z\Theta y$. Then $x \wedge y \wedge z\Theta x \wedge z\Theta x \wedge x$, hence, there exists $t \in L$ such that $x \wedge y \wedge z\Theta t\Phi x \wedge x$, so $y \vee y\Phi y \vee t$. Further, $x \wedge y \wedge z\Theta y \wedge z\Phi y \wedge y$ hence, $y \wedge z\Phi\Theta y \vee t, y \wedge z\Theta\Phi y \vee t$ and $t\Theta y \wedge z$, so $t\Theta\Phi y \vee t, x \wedge x\phi\Theta t\Phi\Theta y \vee t\Theta y \vee y$, then $x \wedge x\Phi\Theta y \wedge y$. Hence, $x\Phi\Theta y$. it means that $\Theta\Phi \leq \Phi\Theta$. So, $\Theta\Phi = \Phi\Theta$ □

Consider the subset $[a, b] = \{x \in L | a \leq_\wedge x \leq_\wedge b\}$ of $(L; \wedge, \vee)$. It is obvious that $[a, b]$ is closed under the operations of the weakly idempotent lattice $(L; \wedge, \vee)$.

Remark 13.2.19. Let $(L; \wedge, \vee)$ be a weakly idempotent lattice. If $b, c \in [a, d]$ and $a\theta d$, then $b\theta c$.

Lemma 13.2.20. *The reflexive binary relation* θ *for the weakly idempotent lattice* $(L; \wedge, \vee)$*, which satisfies the condition* $a\theta(a \wedge a)$*, is a congruence if and only if the following conditions are valid:*

(1) $x\theta y \iff x \wedge y\theta x \vee y$;
(2) $x \leq_\wedge y \leq_\wedge z, x\theta y, y\theta z \to x\theta z$;
(3) $x \leq_\wedge y, x\theta y \to x \wedge t\theta y \wedge t, x \vee t\theta y \vee t$.

Proof. Its necessity is clear, let us show its sufficiency.

Let us proof that θ is transitive. Let $x\theta y$ and $y\theta z$, then we have: $x \wedge y\theta x \vee y$. From condition (3) we obtain $y \vee z = (y \vee z) \vee (y \wedge x)\theta(y \vee z) \vee (y \vee x) = x \vee y \vee z$. Similarly, we show that $x \wedge y \wedge z\theta y \wedge z$. Hence, we obtain $x \wedge y \wedge z\theta y \wedge z\theta y \vee z\theta x \vee y \vee z$, and by Lemma 13.2.2 we have $x \wedge y \wedge z \leq_\wedge y \wedge z \leq_\wedge y \vee z \leq_\wedge x \vee y \vee z$. Applying twice condition (2), we get: $x \wedge y \wedge z\theta x \vee y \vee z$. Now let us make the following designations: $a = x \wedge y \wedge z$, $b = x$, $c = z$, $d = x \vee y \vee z$, then $x\theta z$ (by Remark 13.2.19).

Let $x\theta y$, show that $x \vee t\theta y \vee t$. Indeed, from $x\theta y$ we get that $x \wedge y\theta x \vee y$; hence, $(x\wedge y)\vee t\theta x\vee y\vee t$. Take $a = (x\wedge y)\vee t$, $d = x\vee y\vee t$, $b = x\vee t$, $c = y\vee t$, then we have: $x \vee t\theta y \vee t$ (by Remark 13.2.19). Let us show that if $x_0\theta y_0$ and $x_1\theta y_1$, then $x_\theta a\vee x_1\theta y_0\vee y_1$. Since $x_0\theta y_0$ and $x_1\theta y_1$, then $x_0\vee x_1\theta x_0\vee y_1\theta y_0\vee y_1$, hence $x_0 \vee x_1\theta y_0 \vee y_1$. Similarly we show that $x_0 \wedge x_1\theta y_0 \vee y_1$. □

Lemma 13.2.21. *Let Θ and Φ be congruencies for a weakly idempotent lattice $(L; \wedge, \vee)$ such that $a\Theta(a \wedge a)$ and $a\Phi(a \wedge a)$ for any $a \in L$. Then union of this congruencies can be described in the following manner:*

$$x(\Theta \cup \Phi)y \iff \text{there exists a consequence } z_0 = x \wedge y,$$
$$z_1, \ldots, z_{(n-1)} = x \vee y \text{ such that}$$
$$z_0 = x \wedge y \leq_\wedge z_1 \leq_\wedge \ldots \leq_\wedge z_{(n-1)} = x \vee y,$$

where $z_i\Theta z_{(i+1)}$ or $z_i\Phi z_{(i+1)}$, for all $i = 0, \ldots, n-1$.

13.3. Interlaced weakly idempotent (pre-)bilattices

Theorem 13.3.1. *Let $(L; \wedge, \vee)$ be a weakly idempotent lattice, $(L; *)$ be a weakly idempotent semilattice, having the operation interlaced with the operations $\wedge, \vee$ and the following identity is valid: $a \wedge a = a * a$. Then there exists a pair of congruencies (θ_1, θ_2) for the weakly idempotent lattice $(L; \wedge, \vee)$, which satisfies the following conditions:*

(1) $a(\theta_1 \cap \theta_2)b \iff a \wedge a = b \wedge b$;
(2) $a \leq_\wedge b \to a\theta_1\theta_2 b$;
(3) $X(Y(X(x, y), z), Y(y, z))\theta_i Y(X(x, y), z)$

for all $X, Y \in \{\wedge, \vee\}, x, y, z \in L, i = 1, 2$.

Conversely, each pair of congruencies (θ_1, θ_2) for the weakly idempotent lattice $(L; \wedge, \vee)$, which satisfies conditions $(1) - (3)$, corresponds the weakly idempotent semilattice $(L; *)$, with the operation interlaced with the operations $\wedge, \vee$ and the following identity is valid: $a \wedge a = a * a$.

Proof. Define the relations θ_1 and θ_2 on $(L; \wedge, \vee, *)$, as above, i.e.,

$$a\theta_1 b \iff a * b = a \vee b, \quad a\theta_2 b \iff a * b = a \wedge b.$$

From Lemmas 13.2.10, 13.2.13 and 13.2.15 it follows that θ_1 and θ_2 are congruences for $(L; \wedge, \vee)$ satisfying conditions (1) and (2). Condition (3) is valid, since any weakly idempotent lattice is interlaced. Conversely, let θ_1 and θ_2 are congruences satisfying the conditions of this theorem. Define the operation $*$ by the following rule: $a * b = d \wedge d \iff d\theta_1 a \wedge b$ and $d\theta_2 a \vee b$.

Existence of such d follows from condition (2). Obviously, the operation $*$ is commutative, and the following identities are true: $a * (b * b) = a * b$, $a*a = a\wedge a$. The elements $d_1 = a*(b*c), d_2 = (a*b)*c$ satisfy $d_1 \cap d_2 \theta_1 a \wedge b \wedge c$ and $d_1 \cap d_2 \theta_2 a \wedge b \wedge c$, $d_1 \cap d_2 \theta_1 a \vee b \vee c$, $d_1 \cap d_2 \theta_2 a \vee b \vee c$ consequently, $d_1 \theta_1 d_2$ and $d_1 \theta_2 d_2$; hence, $a * (b * c) = (a * b) * c$. In the similar way, using Corollary 13.2.4 we show that the operation $*$ is interlaced with the operations $\wedge, \vee$. □

Theorem 13.3.2. *Let $(L; \wedge, \vee)$ be a weakly idempotent lattice. There exists a bijective correspondence between the weakly idempotent semilattice operations $*$ on L (which are interlaced with the operations $\wedge, \vee$ and satisfy the identity $a \wedge a = a * a$) and the epimorphisms ϕ, from $(L; \wedge, \vee)$ to the subdirect product of two lattices, satisfying $\phi(x) = \phi(y) \iff x \wedge x = y \wedge y$. Moreover, if (a, b) and (a_1, b_1) are elements of this subdirect product and $(a, b) \leq_\wedge (a_1, a_2)$, then (a, b_1) belongs to this subdirect product too, and if $\phi(x) = (a, b), \phi(y) = (a_1, b_1)$, then $\phi(x * y) = (a \wedge a_1, b \vee b_1)$.*

Proof. Let $(L; *)$ be a weakly idempotent semilattice satisfying the theorems conditions, and θ_1 and θ_2 be the congruence relations from Theorem 13.3.1. Then the weakly idempotent semilattice $(L; *)$ is epimorphically mapped to the subdirect product of the two lattices L/θ_1 and L/θ_2, $\phi(x) = ([x]_{\theta_1}, [x]_{\theta_2})$ such that $\phi(x) = \phi(y) \iff x \wedge x = y \wedge y$. Let (a, b) and (a_1, b_1) belong to this subdirect product and $(a, b) \leq_\wedge (a_1, a_2)$. Then there exist $u, v \in L$ such that $\phi(u) = (a, b)$ and $\phi(v) = (a_1, b_1)$. Then by Theorem 13.3.1, there exists $t \in L$ with the property $u\theta_1 t\theta_2 v$; hence, $\phi(t) = (a, b_1)$.

Conversely, let ϕ be an epimorphism between L and the subdirect product of lattices A and B satisfy the theorems conditions. Then there exist congruence relations θ_1 and θ_2 such that $a(\theta_1 \cap \theta_2)b \iff a \wedge a = b \wedge b$ and L/θ_1, L/θ_2 are isomorphic to A and B, respectively. Hence, L/θ_i satisfy the condition (1).

If $a, b \in L$ and $a \leq_\wedge b$, then $\phi(a) \leq_\wedge \phi(b)$; hence, by our assumption we get that $([a]_{\theta_1}, [b]_{\theta_2})$ belong to the subdirect product, hence there exists $t \in L$ such that $\phi(t) = ([a]_{\theta_1}, [b]_{\theta_2})$, then $a\theta_1 t$, $t\theta_2 b$, yield $a\theta_1\theta_2 b$. Hence, the pair of congruences (θ_1, θ_2) in L satisfies the conditions of Theorem 13.3.1, and it follows that there exists the weakly idempotent semilattice operation $*$ which is interlaced with the operations $\wedge$ and $\vee$ and satisfies the identity $a \wedge a = a * a$. The last statement of the Theorem can be proved with help of the relation $a \wedge b\theta_1 a * b\theta_2 a \vee b$. □

Theorem 13.3.3. (a) *The weakly idempotent semilattice* $(L; *)$ *from the Theorem 13.3.1 can be transformed into the weakly idempotent lattice* $(L; *, \Delta)$ *(moreover, in the unique way) if and only if the corresponding congruences* θ_1 *and* θ_2 *commute.*
(b) *The weakly idempotent semilattice* $(L; *)$ *of the Theorem 13.3.2 can be transformed into the weakly idempotent lattice* $(L; *, \Delta)$ *iff the corresponding subdirect product is a direct product.*

Proof.

(a) If there is a weakly idempotent lattice, then $\theta_1\theta_2 = \theta_2\theta_1$, by Lemmas 13.2.15, 13.2.16, 13.2.18. Conversely, let $\theta_1\theta_2 = \theta_2\theta_1$. By Theorem 13.3.1 there exists a weakly idempotent semilattice operation Δ on L, corresponding to the pair (θ_2, θ_1), which is interlaced with the operations $\wedge$ and $\vee$ and satisfies the identities $a\Delta a = a \wedge a$, $a \wedge b\theta_2 a\Delta b\Theta_1$. Hence, $(a\Delta b)*a\theta_2(a\wedge b)*a\theta_2(a\vee b)\wedge a = a\vee a$ and $(a\Delta b)*a\theta_1(a\vee b)*a\theta_1(a\vee b)\wedge a = a \wedge a$, so $a * a = (a\Delta b) * a$. Similarly, we get $a\Delta a = (a * b)\Delta a$. Hence, $(L; *, \Delta)$ is a weakly idempotent lattice.

(b) Let $a, b \in L$ then by Lemmas 13.2.21, 13.2.14 $a \wedge b\theta_1 a * b\theta_2 a \vee b$, and $a \wedge b \leq_\wedge a * b \leq_\wedge a \vee b$, then $\theta_1 \cup \theta_2 = i$ [142,143]. Hence, the subdirect product is a direct product, if and only if $\theta_1\theta_2 = \theta_2\theta_1$ [142,143]. By a), this is equivalent to the following condition: $(L; *, \Delta)$ forms a weakly idempotent lattice.

□

Theorem 13.3.4. *Let* $(L; \wedge, \vee)$ *and* $(L; *, \Delta)$ *be weakly idempotent lattices. If the operation* $*$ *is interlaced with the operations* $\wedge$ *and* $\vee$ *and satisfies the identity* $a \wedge b = a * a$*; then the operation* Δ *is interlaced with the operations* $\wedge$ *and* $\vee$*, too.*

Proof. The proof follows from Theorems 13.3.1 and 13.3.3. □

Lemma 13.3.5. *Let* $(L; \wedge, \vee, *, \Delta, ')$ *be a weakly idempotent bilattice. Then:*

$$a\theta_1 b \iff a'\theta_2 b', \; a\theta_2 b \iff a'\theta_1 b'.$$

Lemma 13.3.6. *Let* $(L; \wedge, \vee, *, \Delta, ')$ *be a weakly idempotent bilattice and the operation* $\wedge$ *is interlaced with the operations* $*, \Delta$*. Then there exists an isomorphism between the following lattices:* $(L/\theta_1; *, \Delta)$ *and* $(L/\theta_2; *, \Delta)$.

Proof. The mapping $h : (L/\theta_1; *, \Delta) \mapsto (L/\theta_2; *, \Delta)$, which is defined in the following way: $h : [x]_{\theta_1} \mapsto [x']_{\theta_2}$, is an isomorphism. □

Let $L = (L; +, \cdot)$ be weakly idempotent lattice, then on the superproduct $L \bowtie L$ is defined the operation of negation $'$ in the following way:

$$(a, b)' = (b, a).$$

The obtained algebra $(L \times L; (+, \cdot), (\cdot, +), (+, +).(\cdot, \cdot), ')$ is called a superproduct with negation and is denoted by $L\tilde{\bowtie}L$.

Theorem 13.3.7. *Let $L = (L; \wedge, \vee .*, \Delta, ')$ be a weakly idempotent bilattice. The operation $\wedge$ is interlaced with the operations $\wedge, \vee$ if and only if there exists an epimorphism φ between the weakly idempotent bilattice L and the superproduct with negation $A\tilde{\bowtie}A$, where A is a lattice; Moreover, this epimorphism satisfies the following condition: $\varphi(x) = \varphi(y) \iff x * x = y * y$. Hence, this epimorphism is an isomorphism on the bilattice of idempotent elements of L.*

Proof. By Theorems 13.3.2 and 13.3.3 there exists an epimorphism $\phi : L \mapsto A \times B$ between the weakly idempotent lattice $(L; *, \Delta)$ and the subdirect product of the two lattices $A = (L/\theta_1; *, \Delta)$ and $B = (L/\theta_2; *, \Delta)$, which satisfies the condition $\phi(x) = \phi(y) \iff x * x = y * y$. The map ϕ can be continued to the epimorphism between the weakly idempotent pre-bilattice $(L; \wedge, \vee, *, \Delta)$ and the superproduct $A \bowtie B$ in the following manner:

$$\phi(x \wedge y) = (a * a_1, b\Delta b_1), \phi(x \vee y) = (a\Delta a_1, b * b_1)$$

where $\phi(x) = (a, b)$, $\phi(y) = (a_1, b_1)$.

By Lemma 13.3.6 we know that there exists an isomorphism h between $(L/\theta_1; *, \Delta)$ and $(L/\theta_2; *, \Delta)$. Hence, $\phi \circ h : L \mapsto L/\theta_1 \bowtie L/\theta_2$ is an epimorphism between weakly idempotent pre-bilattice $(L; \wedge, \vee, *, \Delta)$ and the algebra $L/\theta_1 \bowtie L/\theta_1$, which is defined in the following way:

$$\psi(x) = ([x]_{\theta_1}, [x']_{\theta_1}).$$

Let us show that this map can be continued until epimorphism between the bilattice L and the superproduct with negation $L/\theta_1\tilde{\bowtie}L/\theta_1$. Thus, we need to show that $(\psi(x))' = \psi(x')$. Indeed,

$$(\psi(x))' = ([x]_{\theta_1}, [x']_{\theta_1})' = ([x']_{\theta_1}, [x]_{\theta_1}) = ([x']_{\theta_1}, [(x')']_{\theta_1}) = \psi(x'). \quad \square$$

Corollary 13.3.8. *Let $L = (L; \wedge, \vee, *, \Delta)$ be a weakly idempotent pre-bilattice. The operation $*$ is interlaced with the operations $\wedge, \vee$ if and only if there exists an epimorphism φ between the weakly idempotent pre-bilattice L and the superproduct of two lattices. Moreover, this epimorphism satisfies the* *following condition:* $\varphi(x) = \varphi(y) \iff x{*}x = y{*}y$. *Hence, this epimorphism is an isomorphism on the pre-bilattice of idempotent elements of* L.

Note that if $L = (L; \wedge, \vee, *, \Delta)$ be a weakly idempotent pre-bilattice, then $(L/\theta_i; *, \Delta) = (L/\theta_i; \wedge, \vee)$, since $(a \wedge b)\theta_1(a * b)$ and $(a \vee b)\theta_2(a\Delta b)$.

Corollary 13.3.9. *The weakly idempotent bilattice* $L = (L; \wedge, \vee.*, \Delta, ')$ *is distributive if and only if there exists an epimorphism* φ *between the weakly idempotent bilattice* L *and the superproduct with negation* $A\tilde{\bowtie}A$, *where* A *is a distributive lattice. Moreover, this epimorphism satisfies the following condition:* $\varphi(x) = \varphi(y) \iff x * x = y * y$. *Hence, this epimorphism is an isomorphism on the bilattice of the idempotent elements of* L.

The similar result is valid for interlaced modular weakly idempotent bilattices.

13.4. Non-idempotent Plonka functions and weakly Plonka sums

Definition 13.4.1. The algebra $\mathfrak{U} = (U, \Sigma)$ is called weakly Plonka sum of its subalgebras $(U_i; \Sigma)$, where $i \in I$, if the following conditions are valid (cf. [323,327,346]):

(i) $U_i \cap U_j = \emptyset$, for all $i, j \in I, i \neq j$.
(ii) $U = \bigcup_{i \in I} U_i$.
(iii) On the set of indexes I there exists a relation "$\leq$" such that $(I; \leq)$ is an upper semilattice with the following properties.
(iv) If $i \leq j$, then there exists a homomorphism $\varphi_{i,j} : (U_i, \Sigma) \mapsto (U_j, \Sigma)$, where $\varphi_{i,j} \cdot \varphi_{j,k} = \varphi_{i,k}$, for $i \leq j \leq k$ and $\varphi_{i,i}(x) = F(x, \ldots, x)$ for any $F \in \Sigma$ and any $x \in U_i$.
(v) For all $A \in \Sigma$ and for all $x_1, \ldots, x_n \in Q$ the following equality is valid:

$$A(x_1, \ldots, x_n) = A(\varphi_{i_1,i_0}(x_1), \ldots, \varphi_{i_n,i_0}(x_n)),$$

where the arity $|A| = n$, $x_1 \in U_{i_1}, \ldots, x_n \in U_{i_n}, i_1, \ldots, i_n \in I, i_0 = \sup\{i_1, \ldots, i_n\}$.

Let $T_{\mathfrak{U}} = \{|A||A \in \Sigma\}$ be an arithmetic type of an algebra $\mathfrak{U}$. From the conditions (iv) of the definition it follows that if the algebra $\mathfrak{U} = (U, \Sigma)$

is a weakly Plonka sum of its subalgebras, then $\mathfrak{U}$ satisfies the following hyperidentity:

$$X(\underbrace{x,\ldots,x}_{n}) = Y(\underbrace{x,\ldots,x}_{m}), \tag{13.4.1}$$

where $m, n \in T_{\mathfrak{U}}$.

Definition 13.4.2. A binary function $f : U \times U \to U$ is called a non-idempotent Plonka function of $\mathfrak{U}$, if it satisfies the following identities (cf. [323,327,346]):

(1) $f(f(x,y),z) = f(x,f(y,z))$;
(2) $f(x,x) = F_t(x,\ldots,x)$, for any operation $F_t \in \Sigma$;
(3) $f(x,f(y,z)) = f(x,f(z,y))$;
(4) $f(F_t(x_1,\ldots,x_{n(t)}),y) = F_t(f(x_1,y),\ldots f(x_{n(t)},y))$, for any operation $F_t \in \Sigma$;
(5) $f(y,F_t(x_1,\ldots,x_{n(t)})) = f(y,F_t(f(y,x_1),\ldots f(y,x_{n(t)})))$, for any operation $F_t \in \Sigma$;
(6) $f(F_t(x_1,\ldots,x_{n(t)}),x_i) = F_t(x_1,\ldots,x_{n(t)})$, for any $1 \leq i \leq n(t)$ and for any operation $F_t \in \Sigma$;
(7) $f(F_t(x_1,\ldots,x_{n(t)}),F_t(x_1,\ldots,x_{n(t)})) = F_t(x_1,\ldots,x_{n(t)})$, for any operation $F_t \in \Sigma$;
(8) $f(x,f(x,y)) = f(x,y)$.

From condition (2) of Definition 13.4.2 it follows that the algebra $\mathfrak{U} = (U;\Sigma)$, possessing a non-idempotent Plonka function satisfies the hyperidentity (13.4.1).

Theorem 13.4.3. *To every non-idempotent Plonka function of algebra $\mathfrak{U} = (U;\Sigma)$ corresponds a representation of $\mathfrak{U}$ as a weakly Plonka sum of its subalgebras.*

Proof. Define on the set U a relation $\alpha \subseteq U \times U$ in the following way:

$$a\alpha b \leftrightarrow f(a,b) = f(a,a), f(b,a) = f(b,b),$$

where f is a non-idempotent Plonka function for $\mathfrak{U}$.
Let us show that α is an equivalence on U. Indeed, reflexivity and symmetricity are immediately follow from the definition. Show transitivity: let $a\alpha b$ and

$b\alpha c$, then $f(a,b) = f(a,a), f(b,a) = f(b,b), f(b,c) = f(b,b), f(c,b) = f(c,c)$. Hence:

$$\begin{aligned} f(a,c) &\stackrel{8}{=} f(a, f(a,c)) \stackrel{1}{=} f(f(a,a),c) = f(f(a,b),c) \stackrel{1}{=} f(a, f(b,c)) \\ &= f(a, f(b,b)) = f(a, f(b,a)) \stackrel{3}{=} f(a, f(a,b)) \stackrel{1}{=} f(a,b) = f(a,a); \\ f(c,a) &\stackrel{8}{=} f(c, f(c,a)) \stackrel{1}{=} f(f(c,c),a) \\ &= f(f(c,b),a) \stackrel{1}{=} f(c, f(b,a)) = f(c, f(b,b)) \stackrel{1}{=} f(f(c,b),b) \\ &= f(f(c,c),b) \stackrel{1}{=} f(c, f(c,b)) \stackrel{8}{=} f(c,b) = f(c,c). \end{aligned}$$

Thus, $a\alpha c$. Denote the corresponding equivalence classes by $U_i, i \in I$. So we obtain a partition of U: $\{U_i \subseteq U, i \in I\}$.

Let us prove that U_i are subalgebras. Indeed, let $a_1, \ldots, a_{n(t)} \in U_i, i \in I$, then for any $F_t \in \Sigma, (|F_t| = t)$ we get

$$\begin{aligned} f(F_t(a_1, \ldots, a_{n(t)}), a_1) &\stackrel{6}{=} F_t(a_1, \ldots a_{n(t)}) \\ &\stackrel{7}{=} f(F_t(a_1, \ldots a_{n(t)}), F_t(a_1, \ldots, a_{n(t)})); \\ f(a_1, F_t(a_1, \ldots, a_{n(t)})) &\stackrel{5}{=} f(a_1, F_t(f(a_1,a_1), \ldots f(a_1, a_{n(t)}))) \\ &\stackrel{2}{=} f(a_1, F_t(F_t(a_1, \ldots, a_1), \ldots, F_t(a_1, \ldots, a_1))) \\ &\stackrel{2,7}{=} f(a_1, F_t(a_1, \ldots, a_1)) \\ &= F_t(a_1, \ldots, a_1) \stackrel{2}{=} f(a_1, a_1), \\ &\quad \text{i.e. } F_t(a_1, \ldots, a_{n(t)}), a_1 \in U_i. \end{aligned}$$

On the set of indices I define an order "$\leq$" in the following manner: $i_1 \leq i_2$ if and only if there exist $a \in U_{i_1}$, $b \in U_{i_2}$ such that $f(b,a) = f(b,b)$. This order gives I a structure of semilattice. Indeed, reflexivity immediately follows from the definition. Let us show that "$\leq$" is antisymmetric:

Let $i_1 \leq i_2$ and $i_2 \leq i_1$, then there exist $a, a' \in U_{i_1}, b, b' \in U_{i_2}$ such that $f(b,a) = f(b,b)$ and $f(a',b') = f(a',a')$. Hence,

$$\begin{aligned} & f(F_t(a', \ldots, a'), F_t(b, \ldots, b)) \\ &\stackrel{5}{=} f(f(a',a'), F_t(f(F_t(a', \ldots, a'), b), \ldots, f(F_t(a', \ldots, a'), b))) \\ &= f(f(a',a'), F_t(f(f(a',b'),b), \ldots, f(f(a',b'),b))) \\ &\stackrel{1}{=} f(f(a',a'), F_t(f(a', f(b',b)), \ldots, f(a', f(b',b)))) \end{aligned}$$

$$
\begin{aligned}
&= f(f(a',a'), F_t(f(a', F_t(b',\dots,b')),\dots, f(a', F_t(b',\dots,b'))))\\
&\stackrel{5}{=} f(f(a',a'), F_t(f(a', F_t(f(a',b'),\dots, f(a',b'))),\dots,\\
&\quad f(a', F_t(f(a',b'),\dots, f(a',b')))))\\
&= f(f(a',a'), F_t(f(a', (F_t(F_t(a',\dots,a'),\dots, F_t(a',\dots,a')))),\dots,\\
&\quad f(a', (F_t(F_t(a',\dots,a'),\dots, F_t(a',\dots,a'))))))\\
&\stackrel{7}{=} f(f(a',a'), F_t(f(a', F_t(a',\dots,a')),\dots f(a', F_t(a',\dots,a')))\\
&\stackrel{7}{=} f(f(a',a'), f(a', F_t(a',\dots,a'))) = f(F_t(a',\dots,a'), F_t(a',\dots,a'));
\end{aligned}
$$

and

$$
\begin{aligned}
f(f(b,b), F_t(a',\dots,a')) &\stackrel{5}{=} f(f(b,b), F_t(f(f(b,b),a'),\dots, f(f(b,b),a')))\\
&= f(f(b,b), F_t(f(f(b,a),a'),\dots, f(f(b,a),a')))\\
&\stackrel{1}{=} f(f(b,b), F_t(f(b, f(a,a')),\dots, f(b, f(a,a'))))\\
&\stackrel{5}{=} f(f(b,b), F_t(f(b, F_t(a,\dots,a)),\dots,\\
&\qquad f(b, F_t(a,\dots,a))))\\
&= f(f(b,b), F_t(f(b, F_t(f(b,a),\dots, f(b,a))),\dots,\\
&\qquad f(b, F_t(f(b,a),\dots, f(b,a)))))\\
&= f(f(b,b), F_t(f(b, F_t(F_t(b,\dots,b),\dots,\\
&\qquad F_t(b,\dots,b))),\dots,\\
&\qquad f(b, F_t(F_t(b,\dots,b),\dots, F_t(b,\dots,b)))))\\
&\stackrel{7}{=} f(f(b,b), F_t(f(b, F_t(b,\dots,b)),\dots f(b, F_t(b,\dots,b)))\\
&\stackrel{7}{=} f(f(b,b), f(b, F_t(b,\dots,b)))\\
&= f(F_t(b,\dots,b), F_t(b,\dots,b))
\end{aligned}
$$

Thus, we get that: $F_t(b,\dots,b) \in U_{i_1} \cap U_{i_2}$, i.e., $U_{i_1} \cap U_{i_2} \neq \emptyset$, so $U_{i_1} = U_{i_2}$, hence, $i_1 = i_2$.

Let $i_1 \leq i_2$ and $i_2 \leq i_3$. Then there exist $a \in U_{i_1}, b, c \in U_{i_2}, d \in A_{i_3}$, such that $f(b,a) = f(b,b)$, $f(d,c) = f(d,d)$. So

$$
\begin{aligned}
f(d,b) &\stackrel{8}{=} f(d, f(d,b)) \stackrel{1}{=} f(f(d,d),b) = f(f(d,c),b)\\
&\stackrel{1}{=} f(d, f(c,b)) = f(d, F_t(c\dots,c))
\end{aligned}
$$

$$
\begin{aligned}
&\stackrel{5}{=} f(d, F_t(f(d,c), \ldots, f(d,c))) \\
&= f(d, F_t(F_t(d, \ldots, d), \ldots, F_t(d, \ldots, d))) \\
&= F_t(d, \ldots, d) = f(d,d);
\end{aligned}
$$

and

$$
\begin{aligned}
f(d,a) &= f(d, f(d,a)) = f(f(d,d),a) = f(f(d,b),a) = f(d, f(b,a)) \\
&= f(f(d,d), F_t(b, \ldots, b)) \\
&= f(f(d,d), F_t(f(f(d,d),b), \ldots, f(f(d,d),b))) \\
&= f(f(d,d), F_t(f(d,d), \ldots, f(d,d))) \stackrel{7}{=} f(F_t(d, \ldots, d), F_t(d, \ldots, d)),
\end{aligned}
$$

hence, $i_1 \le i_3$. Thus, $(I; \le)$ is an ordered set. Let us prove that for any $i, j \in I$ there exists the supremum. From the following equalities: $f(f(a,b),a) = f(a, f(b,a)) = f(a, f(a,b)) = f(f(a,a),b) = f(F_t(a, \ldots, a), b) = F_t(f(a,b), \ldots, f(a,b))$ and $f(a, f(f(a,b),b)) = F_t(f(a,b), \ldots, f(a,b))$ (for all $a, b \in U$) holds that for all $i, j \in I$ there exists an upper bound and it is such that the index $k \in I$ and that $f(a,b) \in U_k$ for some $a \in U_i, b \in U_j$. Let us show that this $k \in I$ is the supremum of i, j. Suppose that there exists $l \in I$ such that $i \le l$ and $j \le l$, i.e., there exist $a \in U_i, b \in U_l$ such that $f(b,a) = f(b,b)$ and there exist $c \in U_j, d \in U_l$ such that $f(d,c) = f(d,d)$. Let us prove that $k \le l$, where $k \in I$ is an index, satisfying the condition: $f(a,c) \in U_k$. Hence, we have

$$
\begin{aligned}
f(f(b,d), f(a,c)) &= f(b, f(d, f(a,c))) = f(b, f(d, f(c,a))) \\
&= f(b, f(f(d,c),a)) \\
&= f(b, f(F_t(d, \ldots, d), a) = f(b, f(a, F_t(d, \ldots, d))) \\
&= f(f(b,a), F_t(d, \ldots, d)) \\
&= f(F_t(b, \ldots, b), F_t(d, \ldots, d)) = f(f(b,b), f(d,d)) \\
&= f(b, f(b, f(d,d))) = f(b, f(f(b,d),d)) = f(f(b, f(b,d)), d) \\
&= f(f(b, f(d,b)), d) = f(f(f(b,d),b), d) = f(f(b,d), f(b,d)),
\end{aligned}
$$

where $f(b,d) \in U_k, f(a,c) \in U_l$. Thus, $k \le l$.

Define mappings $\varphi_{i_1,i_2} : U_{i_1} \mapsto U_{i_2}$ for $i_1 \le i_2$ in the following way:

$$
\varphi_{i_1,i_2}(a) = f(a,b),
$$

where $a \in U_{i_1}, b \in U_{i_2}$.

First of all, let us show that $f(a,b) \in U_{i_2}$ for all $a \in U_{i_1}, b \in U_{i_2}$. Since $i_1 \leq i_2$, then there exist $c \in A_{i1}, d \in A_{i_2}$ such that $f(d,c) = f(d,d)$. Thus, we obtain

$$\begin{aligned} f(f(a,d),d) &\overset{1}{=} f(a,f(d,d)) \overset{2}{=} f(a,F_t(d,\dots,d)) = f(a,f(d,c)) \\ &\overset{3}{=} f(a,f(c,d)) \overset{1}{=} f(f(a,c),d) = f(F_t(a,\dots,a),d) \\ &\overset{4}{=} F_t(f(a,d),\dots,f(a,d)) \overset{2}{=} f(f(a,d),f(a,d)); \\ f(d,f(a,d)) &\overset{3}{=} f(d,f(d,a)) \overset{1}{=} f(f(d,d),a) \\ &\overset{2}{=} f(F_t(d,\dots,d),a) = f(f(d,c),a) \\ &\overset{1}{=} f(d,f(c,a)) = f(d,F_t(c,\dots,c)) \\ &\overset{5}{=} f(d,F_t(f(d,c),\dots,f(d,c))) \\ &= f(d,F_t(F_t(d,\dots,d),\dots,F_t(d,\dots,d))) \\ &\overset{7}{=} F_t(d,\dots,d) \overset{2}{=} f(d,d). \end{aligned}$$

Hence, $f(a,d) \in U_{i_2}$. Further:

$$\begin{aligned} f((a,b),f(a,d)) &\overset{1}{=} f(a,f(b,f(a,d))) = f(a,f(F_t(b,\dots,b),a)) \\ &= f(a,f(f(b,b),a)) = f(a,f(b,f(b,a))) \\ &= f(a,f(b,f(a,b))) = f(f(a,b),f(a,b)); \\ f(f(a,d),f(a,b)) &= f(a,f(d,f(a,b))) = f(a,f(d,f(b,a))) \\ &= f(a,f(f(d,b),a)) \\ &= f(a,f(F_t(d,\dots,d),a)) = f(a,f(f(d,d),a)) \\ &= f(a,f(d,f(d,a))) \\ &= f(a,f(d,f(a,d))) = f(f(a,d),f(a,d)). \end{aligned}$$

Hence, we get that $f(a,b) \in U_{i_2}$ for all $a \in U_{i_1}, b \in U_{i_2}$.

The definition of mappings φ_{i_1,i_2} is consistent, i.e., it is independent from the choosing of element $b \in U_{i_2}$. Indeed, let $f(a,b_1), f(a,b_2)$ be arbitrary elements from U_{i_2}, then:
$f(f(a,b_1),f(a,b_2)) = f(f(a,b_1),f(a,b_1))$. On the other hand,

$$\begin{aligned} f(f(a,b_1),f(a,b_2)) &\overset{3}{=} f(f(a,b_1),f(b_2,a)) \overset{1}{=} f(f(a,f(b_1,b_2)),a)\\ &\overset{3}{=} f(f(a,f(b_2,b_1)),a) = f(f(a,f(b_2,b_2)),a)\\ &\overset{1}{=} f(f(f(a,b_2),b_2),a) \overset{1}{=} f(f(a,b_2),f(b_2,a))\\ &\overset{3}{=} f(f(a,b_2),f(a,b_2)). \end{aligned}$$

Hence, $f(f(a,b_1),f(a,b_1)) = f(f(a,b_2),f(a,b_2))$.

Transform the left and right sides of the obtained equality:

$$\begin{aligned} f(f(a,b_1),f(a,b_1)) &\overset{2,4}{=} f(F_t(a,a),b_1) \overset{2}{=} f(f(a,a),b_1)\\ &\overset{2}{=} f(a,f(a,b_1)) \overset{8}{=} f(a,b_1). \end{aligned}$$

In the similar way, we get $f(f(a,b_2),f(a,b_2)) = f(a,b_2)$.

Thus, we have $f(a,b_1) = f(a,b_2)$.

Let us proof that the mappings φ_{i_1,i_2} are homomorphisms. Let $a_1,\ldots,a_{n(t)} \in U_{i_1}$ and for the arbitrary $c \in U_{i_2}$ we have $\varphi_{i_1,i_2}(a_j) = f(a_j,c)$ (where $j \in \{1,\ldots,n(t)\}$), then for $F_t \in \Sigma$, $|F_t| = t$:

$$\begin{aligned} F_t(\varphi_{i_1,i_2}(a_1),\ldots,\varphi_{i_1,i_2}(a_{n(t)})) &= F_t(f(a_1,c),\ldots,f(a_{n(t)},c))\\ &= f(F_t(a_1,\ldots,a_{n(t)}),c)\\ &= \varphi_{i_1,i_2}(F_t(a_1,\ldots,a_{n(t)})). \end{aligned}$$

Finally, let us prove that: $F_t(x_1,\ldots,x_{n(t)}) = F_t(\varphi_{i_1,i_0}(x_1),\ldots,\varphi_{i_1,i_{n(t)}}(x_{n(t)}))$, where $|A| = n, x_1 \in U_{i_1},\ldots,x_n \in U_{i_n}, i_1,\ldots,i_n \in I, i_0 = \sup\{i_1,\ldots,i_n\}$.

It is sufficient to proof that $F_t(x_1,\ldots,x_{n(t)}) \in U_{i_0}$. Let there exists such $j \in I$ that $F_t(x_1,\ldots,x_{n(t)}) \in U_j$.

By induction we get, that $y = f(\ldots f(f(x_1,x_2),x_3),\ldots,x_n) \in U_{i_0}$ and $f(y,x_i) = F_t(y,\ldots,y)$ for any $x_1 \in U_{i_1},\ldots,x_n \in U_{i_{n(t)}}, i \in \{1,\ldots,n(t)\}$. So, $f(y,F_t(x_1,\ldots,x_n)) = f(y,F_t(f(y,x_1),\ldots,f(y,x_{n(t)}))) = f(y,F_t(F_t(y,\ldots,y),\ldots,F_t(y,\ldots,y))) = F_t(y,\ldots,y)$.

Hence, $j \le i_0$. On the other hand, $F_t(x_1,\ldots,x_{n(t)}) = f(F_t(x_1,\ldots,x_{n(t)}),x_k), k \in \{1,\ldots,n(t)\}$, then $i_0 \le j$. Thus we get that $i_0 = j$ and $F_t(x_1,\ldots,x_{n(t)}) \in U_{i_0}$. Further: $F_t(\varphi_{i_1,i_0}(x_1),\ldots,\varphi_{i_{n(t)},i_0}(x_{n(t)})) = F_t(f(x_1,x),\ldots,f(x_{n(t)},x)) = f(F_t(x_1,\ldots,x_{n(t)}),x) = F_t(F_t(x_1,\ldots,x_{n(t)}),\ldots,F_t(x_1,\ldots,x_{n(t)})) = f(F_t(x_1,\ldots,x_{n(t)}),F_t(x_1,\ldots,x_{n(t)}))$, where $x \in U_{i_0}$ and $i_0 = \sup\{i_1,\ldots,i_{n(t)}\}$. □

Definition 13.4.4. A binary algebra $\mathfrak{U} = (U, \Sigma)$ is called a weakly idempotent quasilattice, if it satisfies the following hyperidentities:

$$X(x,x) = Y(x,x), \tag{13.4.2}$$
$$X(x,y) = X(y,x), \tag{13.4.3}$$
$$X(x,X(y,z)) = X(X(x,y),z), \tag{13.4.4}$$
$$X(x,X(y,y)) = X(x,y), \tag{13.4.5}$$
$$X(Y(X(x,y),z),Y(x,z)) = Y(X(x,y),z). \tag{13.4.6}$$

Note that each weakly idempotent semilattice, weakly idempotent lattice and the superproduct of weakly idempotent lattices satisfy the above hyperidentities.

If $L = (L;\wedge,\vee)$ is a weakly idempotent lattice, then the superproduct $L \bowtie L$ satisfies all hyperidentities of the variety of weakly idempotent lattices. The reduct $(L \times L; (\wedge,\wedge),(\wedge,\vee))$ also satisfies the hyperidentities of the variety of weakly idempotent lattices, but it does not satisfy the law of weak absorption: $a \wedge (a \vee b) = a \wedge a$, $a \vee (a \wedge b) = a \vee a$, so it is not a weakly idempotent lattice.

The following hyperidentities are the consequences of the hyperidentities (13.4.2)–(13.4.6), we need for proving the following theorem.

$$X(Y(X(Y(z,y),x),X(y,x)),Y(x,X(y,x)) = Y(X(Y(z,y),x),X(y,x)), \tag{13.4.7}$$
$$Y(X(Y(z,y),x),X(y,x)) = X(Y(X(Y(z,y),x),X(y,x)),Y(y,x)), \tag{13.4.8}$$
$$X(x,X(Y(x,y),Y(Y(x,y),z))) = X(x,Y(z,Y(x,y))), \tag{13.4.9}$$
$$X(y,Y(y,z)) = Y(y,X(y,z)), \tag{13.4.10}$$
$$Y(Y(x,X(z,Y(y,z))),y) = Y(x,Y(y,z)), \tag{13.4.11}$$
$$X(x,Y(x,X(y,Y(y,z))) = Y(x,Y(y,z)), \tag{13.4.12}$$
$$X(x,Y(x,X(z,Y(y,z))) = Y(x,Y(y,z)). \tag{13.4.13}$$

Theorem 13.4.5. *Every weakly idempotent quasilattice $(Q;A,B)$ with two binary operations is a weakly idempotent lattice or a weakly Plonka sum of its subalgebras which are weakly idempotent lattices.*

Proof. Define the mapping $f : Q \times Q \to Q$ in the following way:

$$f(x,y) = A(x,B(x,y)) = B(x,A(x,y)).$$

The function f is a non-idempotent Plonka function. The consistence of f follows from the hyperidentity (13.4.10). Let us check the conditions of Definition 13.4.2.

$$
\begin{aligned}
(1)\quad f(f(x,y),z) &= f(A(x,B(x,y)),z)\\
&= A(A(x,B(x,y)),B(A(x,B(x,y)),z))\\
&\overset{(13.4.6)}{=} A(A(x,B(x,y)),A(B(A(x,B(x,y)),z)),\\
&\qquad B(B(x,y),z))\\
&\overset{(13.4.11),(13.4.3)}{=} A(A(x,B(x,y)),A(B(A(x,B(x,y)),z),\\
&\qquad B(B(z,A(x,B(x,y))),z)))\\
&\overset{(13.4.9)}{=} A(A(x,B(x,y)),B(B(z,A(x,B(x,y))),y))\\
&\overset{(13.4.11)}{=} A(A(x,B(x,y)),B(B(x,y),z))\\
&\overset{(13.4.4)}{=} A(x,A(B(x,y),B(B(x,y),z)))\\
&\overset{(13.4.9)}{=} A(x,B(B(x,y),z)).\\
f(x,f(y,z)) &= f(x,A(y,B(y,z)))\\
&= A(x,B(x,A(y,Y(y,z))))\\
&\overset{(13.4.12)}{=} A(x,B(x,B(y,z)))).
\end{aligned}
$$

(2) $f(x,x)=A(x,B(x,x)) \overset{(13.4.2)}{=} A(x,A(x,x)) \overset{(13.4.5)}{=} A(x,x)$.

(3) $f(x,f(y,z))=f(x,A(y,B(y,z)))=A(x,B(x,A(y,B(y,z))))$;
$f(x,f(z,y))=f(x,A(z,B(z,y)))=A(x,B(x,A(z,B(z,y))))$.
From the hyperidentity (13.4.11) follows that $f(x,f(y,z)) = f(x,f(z,y))$.

Further, without loss of generality, we will suppose that $F_t=A$.

$$
\begin{aligned}
(4)\quad f(A(x_1,x_2),y) &= A(A(x_1,x_2),B(A(x_1,x_2),y))\\
&\overset{(13.4.6)}{=} A(A(x_1,x_2),A(B(x_1,x_2),y),B(x_1,y))\\
&\overset{(13.4.7)}{=} A(A(x_1,x_2),B(x_1,y)).\\
A(f(x_1,y),f(x_2,y)) &= A(A(x_1,Y(x_1,y)),A(x_2,Y(x_2,y)))\\
&\overset{(13.4.4)}{=} A(x_1,A(B(x_1,y),A(x_2,B(x_2,y))))
\end{aligned}
$$

$$\stackrel{(13.4.2)}{=} A(x_1, A(x_2, A(B(x_1, y), B(x_2, y))))$$
$$\stackrel{(13.4.7)}{=} A(x_1, A(x_2, B(x_1, y)))$$
$$\stackrel{(13.4.4)}{=} A(A(x_1, x_2), B(x_1, y)).$$

(5)

$$\begin{aligned}
f(y, A(x_1, x_2)) &= A(y, B(y, A(x_1, x_2))) \\
&= B(y, A(y, A(x_1, x_2))). \\
f(y, A(f(y, x_1), f(y, x_2))) &= f(y, A(A(y, B(y, x_1)), \\
&\quad A(y, B(y, x_2)))) \\
&= A(y, B(y, A(A(y, B(y, x_1)), \\
&\quad A(y, (y, x_2))))) \\
&\stackrel{(13.4.10)}{=} B(y, A(y, A(A(y, B(y, x_1)), \\
&\quad A(y, B(y, x_2))))) \\
&\stackrel{(13.4.4),(13.4.5)}{=} B(y, A(y, A(B(y, x_1), \\
&\quad B(y, x_2)))) \\
&\stackrel{(13.4.13)}{=} B(y, A(y, A(A(B(y, x_1), \\
&\quad A(y, x_2)), B(B(x_1, x_2), y)))) \\
&\stackrel{(13.4.4)}{=} B(y, A(B(y, x_1), A(y, A(B(y, x_2), \\
&\quad B(B(x_1, x_2), y)))))) \\
&\stackrel{(13.4.8)}{=} B(y, A(B(y, x_1), \\
&\quad A(y, B(Y(x_1, x_2), y)))) \\
&= B(y, A(y, B(B(x_1, x_2), y))) \\
&\stackrel{(13.4.10)}{=} A(y, B(y, B(B(x_1, x_2), y))) \\
&\stackrel{(13.4.5),(13.4.4)}{=} A(y, A(B(x_1, x_2), y)) \\
&\stackrel{(13.4.11)}{=} B(y, A(A(x_1, x_2), y)).
\end{aligned}$$

$$\begin{aligned}
(6)\qquad f(A(x_1,x_2),x_i) &= A(A(x_1,x_2),B(A(x_1,x_2),x_i)) \\
&\overset{(13.4.11)}{=} B(A(x_1,x_2),A(A(x_1,x_2),x_i)) \\
&\overset{(13.4.4),(13.4.5)}{=} B(A(x_1,x_2),A(x_1,x_2)) \\
&= A(A(x_1,x_2),A(x_1,x_2)) = A(x_1,x_2).
\end{aligned}$$

$$\begin{aligned}
(7)\ f(A(x_1,x_2),A(x_1,x_2)) &= A(B(x_1,x_2),B(A(x_1,x_2),A(x_1,x_2))) \\
&\overset{(13.4.2)}{=} A(A(x_1,x_2),A(A(x_1,x_2),X(x_1,x_2))) \\
&\overset{(13.4.4),(13.4.5)}{=} A(A(x_1,x_2),A(x_1,x_2)) \\
&\overset{(13.4.4),(13.4.5)}{=} A(x_1,x_2).
\end{aligned}$$

$$\begin{aligned}
(8)\ f(x,f(x,y)) &= f(x,A(x,B(x,y))) = A(x,B(x,A(x,B(x,y)))) \\
&\overset{(13.4.10)}{=} A(x,B(x,B(x,A(x,Y)))) \\
&\overset{(13.4.4),(13.4.5)}{=} A(x,B(x,A(x,y))) \\
&\overset{(13.4.10)}{=} B(x,A(x,A(x,y))) \\
&\overset{(13.4.4),(13.4.5)}{=} B(x,A(x,y)) \\
&\overset{(13.4.10)}{=} A(x,B(x,y)) = f(x,y).
\end{aligned}$$

Applying Theorem 13.4.3, we obtain that $(Q; A, B)$ is a weakly idempotent lattice or is a weakly Plonka sum of its subalgebras, which are weakly idempotent lattices. The law of weak absorption for subalgebras U_i follows from the conditions $f(x,y) = f(x,x)$ and $f(y,x) = f(y,y)$. □

Corollary 13.4.6. *For subdirectly irreducible weakly idempotent quasilattice $\mathfrak{U} = (U;\Sigma)$ we have $|\Sigma| \leq 2$.*

Proof. Let us show that if $|\Sigma| \geq 3$, then $\mathfrak{U}$ is not subdirectly irreducible. From $|\Sigma| \geq 3$ follows that there exist pairwise distinct operations $A_1, A_2, A_3 \in \Sigma$. Define a function $f_{i,j}$ in the following way:

$$f_{ij}(x,y) = A_i(x, A_j(x,y)),$$

and relations $\widetilde{\theta}_{i,j}$ on U by the following rule: $x\widetilde{\theta}_{i,j}y \leftrightarrow f_{i,j}(x,y) = x, f_{i,j}(y,x) = y$.

Then $\theta_{i,j} = \widetilde{\theta}_{i,j} \cup \{x = x\}$ are non-trivial congruences on $\mathfrak{U}$, having the trivial intersection: $\theta_{1,2} \cap \theta_{1,3} \cap \theta_{2,3} = \mathbf{0}$. $\square$

Corollary 13.4.7 ([82,287]). *Every hyperidentity of the variety of weakly idempotent lattices is a consequence of the following hyperidentities: (13.4.2)–(13.4.6).*

Corollary 13.4.8 ([82,287]). *Every hyperidentity of the variety of weakly idempotent distributive lattices is a consequence of the hyperidentities (13.4.2)–(13.4.5) and the hyperidentity (13.1.8).*

Chapter 14

A Set-Theoretical Representation for Weakly Idempotent Lattices and Interlaced Weakly Idempotent Bilattices

14.1. Preliminaries

In this chapter, we give set-theoretical characterizations both for weakly idempotent lattices and interlaced weakly idempotent bilattices. In particular, we obtain a set-theoretical representation for interlaced bilattices and distributive bilattices (without bounds), too. In [40], arbitrary lattices were characterized by sets, which generalize the well-known classical Birkhoff–Stone set-theoretical representation theorem for distributive lattices and Boolean algebras [39,390] (for another generalization, see [43,54,55,103,263]). A set-theoretical representation for bounded distributive bilattices is well known and were given in [117]. However, the same problem is still opened for interlaced bilattices, which have applications in multi-valued logic, artificial intelligence, logic programming and other directions of applied mathematics and computer science.

We already mentioned that there exist various extensions of the concept of a lattice. For example, in [112,113] the weakly associative lattices were introduced. In [238] an algebra with a system of identities was introduced, which we call a weakly idempotent lattice (see also [135,320,325]). The variety of weakly idempotent lattices is a nilpotent shift of the variety of lattices [238]. In the previous chapter, we studied weakly idempotent lattices with an additional interlaced operation. In this chapter, we give set-theoretical representations both for arbitrary weakly idempotent lattices and for interlaced weakly idempotent bilattices. In particular, we obtain a set-theoretical representation for interlaced bilattices, too. This problem was left open in the paper [117].

In Sec. 14.2, we begin at the definitions of a weakly idempotent semilattice, a weakly idempotent lattice, a quasiordered set, an inf-quasiordered set, a sup-quasiordered set and an $inf\,sup$-quasiordered set. We establish a connection among these weakly idempotent structures and the corresponding quasiorders (Lemmas 14.2.6–14.2.9, Corollaries 14.2.15, 14.2.16). In Sec. 14.3, we prove inequalities valid for weakly idempotent lattices. Further, we give definitions of a filter and ideal of a weakly idempotent lattice, then we characterize the filter and the ideal of a weakly idempotent lattice that are generated by a non-empty subset of the weakly idempotent lattice. On a subset of the set of all filters (ideals) of a weakly idempotent lattice, we define binary operations $\cap^*$ and $\cup^*$. In Sec. 14.4 we prove a set-theoretical representation theorem for arbitrary weakly idempotent lattices (Theorem 14.4.1). Then we define the prime filter and the prime ideal of a weakly idempotent lattice. We redefine the binary operations $\cap^*$ and $\cup^*$ on a subset of the set of all filters (ideals) of a weakly idempotent lattice by the new operations $\cap^{**}$ and $\cup^{**}$. We get as a corollary from Theorem 14.4.1 set-theoretical representations for distributive weakly idempotent lattices and for arbitrary lattices (Corollaries 14.4.6, 14.4.7). In Sec. 14.5, we define a weakly idempotent bilattice, an interlaced weakly idempotent bilattice, a distributive weakly idempotent bilattice, a hyperidentity and superproduct, a filter and an ideal of the weakly idempotent bilattice. Then we prove certain inequalities valid on a weakly idempotent bilattice. Furthermore, we characterize the filter (ideal) of a weakly idempotent bilattice that is generated by a non-empty subset of the weakly idempotent bilattice (Lemmas 14.5.10 and 14.5.11). Also we define the binary operations $\cap^*$ and $\cup^*$ on a subset of the set of all filters (ideals) of a weakly idempotent bilattice. In Theorem 14.5.16, we give a set-theoretical representation for interlaced weakly idempotent bilattices. Then we define a prime filter and prime ideal of weakly idempotent bilattices, also we redefine the binary operations $\cap^*$ and $\cup^*$ on a subset of the set of all filters (ideals) of a weakly idempotent bilattice by the operations $\cap^{**}$ and $\cup^{**}$. Finally, we get as corollaries the set-theoretical representations for distributive weakly idempotent bilattices, interlaced bilattices and distributive bilattices without bounds (Corollaries 14.5.19–14.5.21).

14.2. Preliminary concepts

Let us recall the concepts of weakly idempotent semilattice and weakly idempotent lattice.

Definition 14.2.1. An algebra $(L; \wedge)$ with one binary operation is called a weakly idempotent semilattice, if it satisfies the following identities:

$$a \wedge b = b \wedge a \ (commutativity); \tag{14.2.1}$$

$$(a \wedge b) \wedge c = a \wedge (b \wedge c); \ (associativity); \tag{14.2.2}$$

$$a \wedge (b \wedge b) = a \wedge b \ (weakley\ idempotency). \tag{14.2.3}$$

The operation $\wedge$ is called a product. Adding the idempotence identity $a \wedge a = a$ we obtain a semilattice. The element $a \in L$ is called an idempotent of the weakly idempotent semilattice $(L; \wedge)$ if $a \wedge a = a$. The set of the idempotent elements of each weakly idempotent semilattice form a semilattice, i.e., the product of any two idempotent elements in the weakly idempotent semilattice is an idempotent element. Namely:

$$\begin{aligned}(x \wedge y) \wedge (x \wedge y) &\overset{(14.2.2)}{=} x \wedge (y \wedge (x \wedge y)) \overset{(14.2.1)}{=} x \wedge (y \wedge (y \wedge x)) \\ &\overset{(14.2.2)}{=} x \wedge ((y \wedge y) \wedge x) \\ &\overset{(14.2.1),(14.2.3)}{=} x \wedge (x \wedge y) \\ &\overset{(14.2.2)}{=} (x \wedge x) \wedge y \\ &\overset{(14.2.1),(14.2.3)}{=} y \wedge x = x \wedge y.\end{aligned}$$

Definition 14.2.2. The relation $\theta \subseteq L \times L$ is called a quasiorder if it is reflexive and transitive.

Example 14.2.3. The cover of a set L is a family $P = \{X_i\}_{i \in I}$ of subsets of L such that $\bigcup_{i \in I} X_i = L$. A relation Q, defined on the set of all covers of the set L:

$$P_1 Q P_2 \iff \forall X \in P_1 \exists Y \in P_2 (X \subseteq Y)$$

is a quasiorder. It is not an order, because there exist such different covers P_1 and P_2 that $P_1 Q P_2$ and $P_2 Q P_1$.

Example 14.2.4. The classical relation of divisibility on $\mathbf{Z}$ is a quasiorder. So, the results of this chapter are related to the number theory too.

Every quasiorder generates an order as follows.
Let θ be a quasiorder on the set $L \neq \emptyset$; then $E_\theta = \theta \cap \theta^{-1} \subseteq L \times L$ is an equivalence. For any element $x \in L$ let us denote by $[x]$ the class of the relation E_θ which contains the element x. Let $\leq_\theta$ be a relation induced on the set L/E_θ from θ in the following manner: for $[a], [b] \in L/E_\theta$

$$[a] \leq_\theta [b] \iff a\theta b.$$

A straightforward argument show that this definition is correct and that it is an order.

The function $K : L/E_\theta \mapsto L$ is called a choice function, if $K([a]) \in [a]$ for each $[a] \in L/E_\theta$.

Definition 14.2.5. The pair $(L;\theta)$ is called an inf-quasiordered (a *sup*-quasiordered) set, if for any two classes of equivalences $[a], [b] \in L/E_\theta$ there exists $\inf([a],[b]) = [a] \wedge [b]$ (dually $\sup([a],[b]) = [a] \vee [b]$), i.e., if $(L/E_\theta; \leq_\theta)$ is a lower (an upper) semilattice.

Lemma 14.2.6. *Let $(L;\theta)$ be an inf-quasiordered set and let $K : L/E_\theta \to L$ be an arbitrary choice function. If for $x, y \in L$, $x \wedge y = K(\inf([x],[y])) = K([x] \wedge [y])$, then the algebra $(L;\wedge)$ is a weakly idempotent semilattice, which we call a lower weakly idempotent semilattice.*

Proof. It is straightforward to check that $(L;\wedge)$ satisfies the identities (14.2.1)–(14.2.3). □

Lemma 14.2.7. *Let $(L;\wedge)$ be a weakly idempotent semilattice. Then the relation $a\theta b \leftrightarrow a \wedge b = a \wedge a$ is a quasiorder on the set L, the mapping $K : L/E_\theta \mapsto L$, $[a] \mapsto a \wedge a$ is a choice function, and the pair $(L;\theta)$ is an inf-quasiordered set with $inf([a],[b]) = [a \wedge b]$, and $x \wedge y = K(inf([x],[y]))$.*

Proof. The first part is obvious. Let us show that $\inf([a],[b]) = [a \wedge b]$. First note that $[a \wedge b] \leq_\theta [a]$ and $[a \wedge b] \leq_\theta [b]$. Indeed $(a \wedge \ b) \wedge a = a \wedge b = (a \wedge b) \wedge (a \wedge b)$, then $[a \wedge b] \leq_\theta [a]$. Similarly for $[a \wedge b] \leq_\theta [b]$. Consider $[c] \in L/E_\theta$ a lower bound for $[a], [b] \in L/E_\theta$, then $c\theta a$ and $c\theta b$, thus $c \wedge (a \wedge b) = (c \wedge a) \wedge b = (c \wedge c) \wedge b = c \wedge (c \wedge c) = c \wedge c$, yield $[c] \leq_\theta [a \wedge b]$ and $\inf([a],[b]) = [a \wedge b]$. □

The following two lemmas are dual to Lemma 14.2.6 and Lemma 14.2.7, respectively.

Lemma 14.2.8. *Let $(L;\theta)$ be a sup-quasiordered set and let $K : L/E_\theta \mapsto L$ be an arbitrary choice function. If for each two elements, $x, y \in L$:*

$$x \vee y = K(\sup([x],[y])) = K([x] \vee [y]),$$

then the algebra $(L;\vee)$ is a weakly idempotent semilattice, which we call an upper weakly idempotent semilattice.

Lemma 14.2.9. *Let $(L;\vee)$ be a weakly idempotent semilattice. Then the relation $a\theta b \leftrightarrow a \vee b = b \vee b$ is a quasiorder on the set L, the mapping $K : L/E_\theta \mapsto L$, $[a] \mapsto a \vee a$ is a choice function, and the pair $(L;\theta)$ is a sup-quasiordered set with $sup([a],[b]) = [a \vee b]$, and $x \vee y = K(sup([x],[y]))$.*

Definition 14.2.10. An algebra $(L;\wedge,\vee)$ with two binary operations is called a *weakly idempotent lattice*, if the reducts $(L;\wedge)$ and $(L;\vee)$ are weakly idempotent semilattices and the following identities are valid:

$$a \wedge (b \vee a) = a \wedge a, a \vee (b \wedge a) = a \vee a, \ (\textit{weak absorption}) \tag{14.2.4}$$

$$a \wedge a = a \vee a. \ (\textit{equalization}) \tag{14.2.5}$$

The operation $\vee$ is called a sum.

Example 14.2.11. Let us consider the relation of divisibility on the set $\mathbf{Z} \setminus \{0\}$, which is a quasiorder on $\mathbf{Z} \setminus \{0\}$. The corresponding equivalence classes are the following sets $\{x, -x\}$. Define a choice function as $K([x]) = |x|$. Then we have: $x \wedge_1 y = \gcd(|x|,|y|)$ and $x \vee_1 y = \mathrm{lcm}(|x|,|y|)$. Thus, the algebra $(Z \setminus \{0\};\wedge_1,\vee_1)$ is a weakly idempotent lattice, which is not a lattice, since $x \wedge_1 x \neq x$ for negative x.

Example 14.2.12. If we define the choice function on $\mathbf{Z} \setminus \{0\}$ as follows $K([x]) = -|x|$, then we have: $x \wedge_2 y = -\gcd(|x|,|y|)$ and $x \vee_2 y = -\mathrm{lcm}[|x|,|y|]$. Hence, the algebra $(Z\setminus\{0\};\wedge_2,\vee_2)$ also is a weakly idempotent lattice.

The element $a \in L$ is called an idempotent of a weakly idempotent lattice $(L;\wedge,\vee)$ if $a \wedge a = a$ and $a \vee a = a$.

Remark 14.2.13. We know that the product (sum) of any two elements of a weakly idempotent lattice $(L;\wedge,\vee)$ is an idempotent element. So, the set of all idempotent elements of a weakly idempotent lattice is a lattice.

Definition 14.2.14. A pair $(L;\theta)$ is called an *inf sup-quasiordered set* if for any two classes of equivalences $[a],[b] \in L/E_\theta$ there exist both $\inf([a],[b]) = [a] \wedge [b]$ and $\sup([a],[b]) = [a] \vee [b]$, i.e., if $(L/E_\theta;\leq_\theta)$ is a lattice.

The following two corollaries immediately follow from Lemmas 14.2.6–14.2.9.

Corollary 14.2.15. *Let* $(L;\theta)$ *be an* $inf sup$*-quasiordered set and let* $K : L/E_\theta \mapsto L$ *be an arbitrary choice function. If for any two elements* $x, y \in L$*:*

$$x \wedge y = K(\inf([x],[y])) = K([x] \wedge [y]),$$
$$x \vee y = K(\sup([x],[y])) = K([x] \vee [y]),$$

then the algebra $(L;\wedge,\vee)$ *is a weakly idempotent lattice.*

Corollary 14.2.16. *Let* $(L;\wedge,\vee)$ *be a weakly idempotent lattice. Then the relation* $a\theta b \leftrightarrow a \wedge b = a \wedge a \leftrightarrow a \vee b = b \vee b$ *is a quasiorder on the set* L*, the mapping* $K : L/E_\theta \mapsto L$*;* $[a] \mapsto a \vee a$ *is a choice function, and the pair* $(L;\theta)$ *is an* $inf sup$*-quasiordered set with* $inf([a],[b]) = [a \wedge b]$*,* $sup([a],[b]) = [a \vee b]$ *and*

$$x \wedge y = K(\inf([x],[y])), \quad x \vee y = K(\sup([x],[y])). \tag{14.2.6}$$

Definition 14.2.17. A weakly idempotent lattice $(L;\wedge,\vee)$ is called *bounded* if the corresponding lattice $(L/E_\theta;\leq_\theta)$ is bounded.

14.3. Filters and ideals of weakly idempotent lattices

Let $L = (L;\wedge,\vee)$ be a weakly idempotent lattice. By Corollary 14.2.16, each weakly idempotent lattice defines the quasiorder θ (which is denoted by $\leq$ in this section) and it is defined by the following rule:

$$x \leq y \leftrightarrow x \wedge y = x \wedge x. \tag{14.3.1}$$

Remark 14.3.1. Note that the operations of the weakly idempotent lattice preserve the corresponding quasiorder. Indeed, let $a \leq b$ and $c \leq d$, i.e., $a \vee b = b \vee b$ and $c \vee d = d \vee d$, then $(a \vee c) \vee (b \vee d) = (a \vee b) \vee (c \vee d) = (b \vee b) \vee (d \vee d) = b \vee d$, thus $a \vee c \leq b \vee d$. In the same way, we show that $a \wedge c \leq b \wedge d$.

Lemma 14.3.2. *Let* $L = (L;\wedge,\vee)$ *be a weakly idempotent lattice. Then*

$$a \leq b \leq a \leftrightarrow a \wedge a = b \wedge b.$$

Proof. Let $a \leq b \leq a$, then by (14.3.1) $a \wedge b = a \wedge a$ and $a \wedge b = b \wedge b$, hence, $a \wedge a = b \wedge b$. Conversely, let $a \wedge a = b \wedge b$, then $a \wedge b \overset{(14.2.3)}{=} a \wedge (b \wedge b) = a \wedge (a \wedge a) \overset{(14.2.3)}{=} a \wedge a$ and $a \vee b \overset{(14.2.3)}{=} a \vee (b \vee b) = a \vee (a \vee a) \overset{(14.2.3)}{=} a \vee a$, thus, by (14.3.1) $a \leq b$ and $b \leq a$. □

Lemma 14.3.3. *Let $(L; \wedge, \vee)$ be a weakly idempotent lattice. Then the following inequality is valid on L:*

$$x \wedge y \leq x \leq x \vee y.$$

Proof. From the identities (14.2.1)–(14.2.3) and Remark 14.2.13 we have $(x \wedge y) \wedge x \overset{(14.2.1)}{=} (y \wedge x) \wedge x \overset{(14.2.2)}{=} y \wedge (x \wedge x) \overset{(14.2.3)}{=} y \wedge x \overset{(14.2.1)}{=} x \wedge y = (x \wedge y) \wedge (x \wedge y)$. Hence, by (14.3.1) we obtain $x \wedge y \leq x$. The second side of the inequality proved in a similar way. □

We say that a non-empty subset A of a weakly idempotent lattice $(L; \wedge, \vee)$ is a weakly idempotent sublattice if it is closed under the weakly idempotent lattice operations, i.e., if the following condition is valid:

$$x, y \in A \rightarrow x \wedge y, x \vee y \in A. \tag{14.3.2}$$

Definition 14.3.4. The weakly idempotent sublattice F of a weakly idempotent lattice $L = (L; \wedge, \vee)$ is called a *filter of* L if the following condition is satisfied: $x \in F, y \in L \rightarrow x \vee y \in F$.

A filter is called proper if it is different from L. The set of all the filters of the weakly idempotent lattice L is denoted by $F(L)$.

Definition 14.3.5. A weakly idempotent sublattice I of a weakly idempotent lattice $L = (L; \wedge, \vee)$ is called an *ideal of* L if the following condition is satisfied: $x \in I, y \in L \rightarrow x \wedge y \in I$.

An ideal is called proper if it is different from L. The set of all the ideals of the weakly idempotent lattice L is denoted by $I(L)$.

Lemma 14.3.6. *Let $L = (L; \wedge, \vee)$ be a weakly idempotent lattice. The following statements are equivalent for L:*

(1) *F is a filter of L.*
(2) *If $x, y \in F$, then $x \wedge y \in F$ and if $x \in F, y \in L, x \leq y$, then $y \vee y \in F$.*

Proof. (1)→(2) Let F be a filter on L; then, for any $x, y \in F$, $x \wedge y \in F$ (by (14.3.2)) and for any $x \in F$, $y \in L$ $x \vee y \in F$. Assume that $x \leq y$, (i.e., $x \vee y = y \vee y$), then $y \vee y \in F$.
(2)→(1) For $x \in F$, $y \in L$, using Lemma 14.3.3, we get $(x \vee y) \vee (x \vee y) \in F$, since $x \leq x \vee y$. From Remark 14.2.13, we obtain that $(x \vee y) \vee (x \vee y) = x \vee y$. Thus, $x \vee y \in F$ for each $x \in F, y \in L$. □

Lemma 14.3.7. *Let $L = (L; \wedge, \vee)$ be a weakly idempotent lattice. The following statements are equivalent for L:*

(1) *I is an ideal of L.*
(2) *If $x, y \in I$, then $x \vee y \in F$ and if $x \in F, y \in L, x \leq y$, then $x \wedge x \in I$.*

Lemma 14.3.8. *Let $L = (L; \wedge, \vee)$ be a weakly idempotent lattice and $X \subseteq L$. Then*

$$[X) = X \cup \{x \in L | \exists x_1, \ldots, x_n \in X, x_1 \wedge \ldots \wedge x_n \leq x \ \& \ x = K([x])\}$$

is the smallest filter that contains $X \subseteq L$.

Proof. Let us show that $[X)$ is a filter. Take $x, y \in [X)$, then consider the following cases:

- $x, y \in [X) \setminus X$, then there exist $x_1, \ldots, x_n, y_1, \ldots, y_m \in X$ such that $x_1 \wedge \ldots \wedge x_n \leq x = K([x])$ and $y_1 \wedge \cdots \wedge y_m \leq y = K([y])$. By Remark 14.3.1 and the equality (14.2.6) we get: $x_1 \wedge \cdots \wedge x_n \wedge y_1 \wedge \cdots \wedge y_m \leq x \wedge y = K(\inf([x], [y]))$; hence, $x \wedge y \in [X)$.

The following cases are proved similarly:

- $x \in [X) \setminus X, y \in X$;
- $y \in [X) \setminus X, x \in X$;
- $x, y \in X$.

Thus, for each $x, y \in [X)$ we obtain that $x \wedge y \in [X)$. Similarly, we prove the second condition of Lemma 14.3.6; hence, $[X)$ is a filter.

Suppose $F \neq [X)$ is a filter of the weakly idempotent lattice L, which contains X. Let $x \in [X) \setminus X$, then there exist $x_1, \ldots, x_n \in X$ such that $x_1 \wedge \cdots \wedge x_n \leq x = K([x])$. Since F is a filter which contains X, we obtain that $x_1 \wedge \cdots \wedge x_n \in F$ and $x \wedge x \in F$. By equality (14.2.6), we have $x \wedge x = K(inf([x], [x])) = K([x]) = x$, and $x \in F$. Hence, $[X) \subseteq F$ and $[X)$ is the smallest filter that contains X. □

Lemma 14.3.9. *Let $L = (L; \wedge, \vee)$ be a weakly idempotent lattice and $X \subseteq L$. Then*

$$(X] = X \cup \{x \in L | \exists x_1, \ldots, x_n \in X, x \leq x_1 \wedge \ldots \wedge x_n \ \& \ x = K([x])\}$$

is the smallest ideal that contains $X \subseteq L$.

Corollary 14.3.10. *Let $L = (L; \wedge, \vee)$ be a weakly idempotent lattice and $a \in L$. Then*

$$[a) = \{a\} \cup \{x \in L | a \leq x \ \& \ x = K([x])\}$$

is the smallest filter that contains $a \in L$.

Proof. Take $X = \{a\}$ in Lemma 14.3.8. $\square$

Corollary 14.3.11. *Let $L = (L; \wedge, \vee)$ be a weakly idempotent lattice and $a \in L$. Then*

$$(a] = \{a\} \cup \{x \in L | x \leq a \ \& \ x = K([x])\}$$

is the smallest ideal that contains $a \in L$.

Proof. Take $X = \{a\}$ in Lemma 14.3.9. $\square$

Lemma 14.3.12. *Let $L = (L; \wedge, \vee)$ be a weakly idempotent lattice. Then the following assertions are valid on L:*

$$[a \wedge a) \cup [b \wedge b) \subseteq [u \wedge b), \quad [a \vee b) = [a \vee a) \cap [b \vee b),$$

$$(a \wedge b] = (a \wedge a] \cap (b \wedge b], \quad (a \vee a] \cup (b \vee b] \subseteq (a \vee b].$$

Proof. Let us show that $[a \wedge a) \cup [b \wedge b) \subseteq [a \wedge b)$. Suppose $y \in [a \wedge a) \cup [b \wedge b)$, then let us consider the following possible cases:

(1) In the first case, $y = a \wedge a$; from Lemma 14.3.3 and the equality (14.2.6) we obtain $a \wedge b \leq a \leq a \wedge a = K(inf([a],[a])) = K([a])$. Hence, $y = a \wedge a \in (a \wedge b]$.
(2) In the second case, $a \leq y = K([y])$; by Lemma 14.3.3 we obtain $a \wedge b \leq a \leq y$, i.e. $y \in [a \wedge b)$.
The following two cases are proved similarly:
(3) $y = b \wedge b$,
(4) $b \leq y = K([y])$.

Thus, $[a \wedge a) \cup [b \wedge b) \subseteq [a \wedge b)$. $\square$

For each $a \in L$, let $f(a) = \{F \in F(L) | a \in F\}$ and $i(a) = \{I \in I(L) | a \in I\}$. Define on the sets $f(L) = \{f(a) | a \in L\}$ and $i(L) = \{i(a) | a \in L\}$ the operations $\cap^*$ and $\cup^*$, by the following rules:

$$f(a) \cap^* f(b) = \{F \in F(L) | \exists X \in f(a \wedge a), \exists Y \in f(b \wedge b), X \cup Y \subseteq F\};$$

$$f(a) \cup^* f(b) = \{F \in F(L) | \exists X \in f(a \wedge a), \exists Y \in f(b \wedge b), X \cap Y \subseteq F\};$$

$$i(a) \cap^* i(b) = \{I \in I(L) | \exists X \in i(a \wedge a), \exists Y \in i(b \wedge b), X \cup Y \subseteq I\};$$

$$i(a) \cup^* i(b) = \{I \in I(L) | \exists X \in i(a \wedge a), \exists Y \in i(b \wedge b), X \cap Y \subseteq I\}.$$

Lemma 14.3.13. *Let $(L; \wedge, \vee)$ be a weakly idempotent lattice. Then the following identities are valid:*

$$f(a \wedge b) = f(a) \cap^* f(b); \tag{14.3.3}$$

$$f(a \vee b) = f(a) \cup^* f(b); \tag{14.3.4}$$

$$i(a \wedge b) = i(a) \cup^* i(b); \tag{14.3.5}$$

$$i(a \vee b) = i(a) \cap^* i(b). \tag{14.3.6}$$

Proof. Let us prove the first equality. Suppose $F \in f(a \wedge b)$, then $a \wedge b \in F$. If $c \in [a \wedge b)$, then $c = a \wedge b$ and hence, $c \in F$, or $a \wedge b \leq c = K([c])$. Thus, since F is a filter, we obtain that $c = K([c]) = K(\inf([c],[c])) = c \wedge c \in F$. Hence, $[a \wedge b) \subset F$, and by Lemma 14.3.12, we have $[a \wedge a) \cup [b \wedge b) \subseteq [a \wedge b)$. Thus, there exist $X = [a \wedge a)$ and $Y = [b \wedge b)$ such that $X \cup Y \subseteq F$, i.e., $F \in f(a) \cap^* f(b)$.

Conversely, if $F \in f(a) \cap^* f(b)$, then there exist $X \in f(a \wedge a)$ and $Y \in f(b \wedge b)$, such that $X \cup Y \subseteq F$ hence, $a \wedge a \in X$ and $b \wedge b \in Y$. Then $a \wedge a, b \wedge b \in X \cup Y \subseteq F$; hence, since F is a filter, we get $a \wedge b = (a \wedge a) \wedge (b \wedge b) \in F$. Thus, $F \in f(a \wedge b)$.

The rest of the identities are proved similarly. □

14.4. A set-theoretical representation for arbitrary weakly idempotent lattices

From Lemma 14.3.13 it immediately follows that the algebras $f(L) = (f(L); \cap^*, \cup^*)$ and $i(L) = (i(L); \cup^*, \cap^*)$ are weakly idempotent lattices.

Theorem 14.4.1. *Every weakly idempotent lattice $L = (L; \wedge, \vee)$ is isomorphic to the weakly idempotent lattice $f(L) = (f(L); \cap^*, \cup^*)$, i.e.,*

$$L \cong f(L).$$

Proof. The mapping $\varphi : L \to (f(L); \cap^*, \cup^*)$, which is defined by the following rule:

$$\varphi : x \mapsto f(x)$$

is an isomorphism. Indeed:

(1) *Injectivity.* Suppose $\varphi(a) = \varphi(b)$, i.e. $f(a) = f(b)$. Then, since $[a) \in f(a)$ and $[b) \in f(b)$, we obtain that $b \in [a)$ and $a \in [b)$. By Corollary 14.3.10 we have two cases: $b = a$ or $a \leq b = K([b])$ and $b \leq a = K([a])$. In the second case, by Lemma 14.3.2, we have $a \wedge a = b \wedge b$; and by

equality (14.2.6), we obtain that $a \wedge a = K(\inf([a],[a])) = K([a]) = a$ and $b \wedge b = K(inf([b],[b])) = K([b]) = b$. Hence, $a = b$.

(2) The proof of the surjectivity of φ is obvious.

(3) The fact that φ is a homomorphism follows directly from the identities (14.3.3) and (14.3.4):

$$\varphi(x \wedge y) = f(x \wedge y) = f(x) \cap^* f(y) = \varphi(x) \cap^* \varphi(y);$$
$$\varphi(x \vee y) = f(x \vee y) = f(x) \cup^* f(y) = \varphi(x) \cup^* \varphi(y). \qquad \square$$

Corollary 14.4.2. *For every weakly idempotent lattice $L = (L; \wedge, \vee)$ we have*

$$L \cong i(L).$$

Proof. The mapping $\varphi_1 : L \to (i(L); \cup^*, \cap^*)$, which is defined by the following rule:

$$\varphi_1 : x \mapsto i(x)$$

is an isomorphism.
The proof of the injectivity of φ_1 is similar to Theorem 14.4.1, and the fact that φ_1 is a homomorphism follows immediately from equalities (14.3.5) and (14.3.6). $\square$

Note that the operation $\cup^*$, defined on $f(L)$, coincides with the operation $\cup^{**}$, defined in the following way:

$$f(a) \cup^{**} f(b) = f(a \wedge a) \cup f(b \wedge b) \cup \{F \in F(L) | \exists X \in f(a \wedge a), \exists Y \in f(b \wedge b), X \cap Y \subseteq F\}.$$

Indeed, let $F' \in f(a \wedge a) \cup f(b \wedge b)$, then $a \wedge a \in F'$ or $b \wedge b \in F'$. Hence $a \vee b \in F'$. From Lemma 14.3.12 we have $[a \vee b) = [a \wedge a) \cap [b \wedge b)$. Hence, there exist $X = [a \wedge a)$ and $Y = [b \wedge b)$ such that $X \cap Y \subseteq F'$, yield $F' \in \{F \in F(L) | \exists X \in f(a \wedge a), \exists Y \in f(b \wedge b), X \cap Y \subseteq F\}$ and thus $f(a \wedge a) \cup f(b \wedge b) \subseteq \{F \in F(L) | \exists X \in f(a \wedge a), \exists Y \in f(b \wedge b), X \cap Y \subseteq F\}$. So, $f(a) \cup^* f(b) = f(a) \cup^{**} f(b)$ on $f(L)$.

Also note that the operation $\cap^*$ defined on $f(L)$ can be replaced with the operation $\cap^{**}$, defined in the following way:

$$f(a) \cap^{**} f(b) = f(a \wedge a) \cap f(b \wedge b) \cap \{F \in F(L) | \exists X \in f(a \wedge a), \exists Y \in f(b \wedge b), X \cup Y \subseteq F\}.$$

Indeed, let $F \in \{F \in F(L) | \exists X \in f(a \wedge a), \exists Y \in f(b \wedge b), X \cup Y \subseteq F\}$; then there exist filters X and Y of L such that $a \wedge a \in X$, $b \wedge b \in Y$ and $X \cup Y \subseteq F$.

So $F \in f(a \wedge a), F \in f(b \wedge b)$, hence $\{F \in F(L) | \exists X \in f(a \wedge a), \exists Y \in f(b \wedge b), X \cup Y \subseteq F\} \subseteq f(a \wedge a) \cap f(b \wedge b)$. Hence $f(a) \cap^* f(b) = f(a) \cap^{**} f(b)$.

Moreover, $f(a) \cap^{**} f(b) = f(a \wedge a) \cap f(b \wedge b)$.

In the same way, we can show that the operations $\cap^*$ and $\cup^*$ on the set $i(L)$ coincide with the operations $\cap^{**}$ and $\cup^{**}$ defined as follows:

$$i(a) \cup^{**} i(b) = i(a \wedge a) \cup i(b \wedge b) \cup \{I \in I(L) | \exists X \in i(a \wedge a),\ \exists Y \in i(b \wedge b), X \cap Y \subseteq I\},$$

$$i(a) \cap^{**} i(b) = i(a \wedge a) \cap i(b \wedge b) \cap \{I \in I(L) | \exists X \in i(a \wedge a),\ \exists Y \in i(b \wedge b), X \cup Y \subseteq I\}.$$

Moreover:

$$i(a) \cap^{**} i(b) = i(a \wedge a) \cap i(b \wedge b).$$

Definition 14.4.3. A filter F of a weakly idempotent lattice $(L; \wedge, \vee)$ is called prime if F is a proper filter and

$$x \vee y \in F \to x \in F \text{ or } y \in F,$$

for $x, y \in L$.

Definition 14.4.4. An ideal I of a weakly idempotent lattice $(L; \wedge, \vee)$ is called prime if I is a proper ideal and

$$x \wedge y \in I \to x \in I \text{ or } y \in I,$$

for $x, y \in L$.

Definition 14.4.5. The weakly idempotent lattice $(L; \wedge, \vee)$ is called distributive if it satisfies the identities of distributivity:

$$x \wedge (y \vee z) = (x \wedge y) \vee (x \wedge z),$$

$$x \vee (y \wedge z) = (x \vee y) \wedge (x \vee z).$$

Denote the set of all prime filters of a weakly idempotent lattice L by $pF(L)$, and the set of all prime ideals of a weakly idempotent lattice L by $pI(L)$. Let us make the following designations: $pf(a) = \{X \in pF(L) | a \in X\}$ and $pi(a) = \{X \in pI(L) | a \in X\}$, $pf(L) = \{pf(a) | a \in L\}$ and $pi(L) = \{pi(a) | a \in L\}$.

The operations $\cap^{**}$ and $\cup^{**}$ can be defined on $pf(L)$ and $pi(L)$ and then identities similar to those in Lemma 14.3.13 can be proven for prime filters and prime ideals too. Thus, the sets $pF(L)$ and $pI(L)$ are closed under the operations $\cap^{**}$ and $\cup^{**}$.

Corollary 14.4.6. *Each distributive weakly idempotent lattice* $L = (L; \wedge, \vee)$ *is isomorphic to the distributive weakly idempotent lattice* $pf(L) = (pf(L); \cap^{**}, \cup)$, *i.e.,*

$$L \cong pf(L).$$

Proof. Let us show that on the set $pf(L)$ the operation $\cup^{**}$ converts to the set-theoretical union. Namely, let us consider $Z \in pf(x) \cup^{**} pf(y)$. From Lemma 14.3.13 we have $x \vee y \in Z$. Since Z is a prime filter, then $x \in Z$ or $y \in Z$, hence $Z \in pf(x \wedge x) \cup pf(y \wedge y)$. Thus, $pf(x) \cup^{**} pf(y) \subseteq pf(x \wedge x) \cup pf(y \wedge y)$.

The converse, namely, $pf(x \wedge x) \cup pf(y \wedge y) \subseteq pf(x) \cup^{**} pf(y)$ is trivial. Using the identities of Lemma 14.3.13 for the case of $pf(L)$ we immediately get that $pf(L)$ is a weakly idempotent lattice. On the set $pf(L)$, we redefine in the following way the operation $\cap^{**}$: $pf(a) \cap^{**} pf(b) = pf(a \wedge a) \cap pf(b \wedge b)$. Hence, the weakly idempotent lattice $pf(L)$ is distributive. $\square$

Let us make some designations: the set of all prime filters of the lattice L is denoted by $\tilde{F}(L)$, $\tilde{f}(a) = \{X \in \tilde{F}(L) | a \in X\}$, $\tilde{f}(L) = \{\tilde{f}(a) | a \in L\}$.

Corollary 14.4.7 ([40]). *Every lattice* $L = (L; \wedge, \vee)$ *is isomorphic to the lattice* $\tilde{f}(L) = (\tilde{f}(L); \cap, \cup^*)$, *i.e.,*

$$L \cong \tilde{f}(L).$$

Proof. Since L is a lattice, the operation $\cap^*$ is converted into the set-theoretical intersection and the operation $\cup^*$ is defined as follows:

$$\tilde{f}(a) \cup^* \tilde{f}(b) = \{F \in \tilde{F}(L) | (\exists X \in \tilde{f}(a))(\exists Y \in \tilde{f}(b)) X \cap Y \subseteq F\}.$$

It is obvious that $(\tilde{f}(L); \cap, \cup^*)$ is a lattice. Indeed: $\tilde{f}(a) \cap \tilde{f}(a) = \tilde{f}(a \wedge a) = \tilde{f}(a)$, $\tilde{f}(a) \cup^* \tilde{f}(a) = \tilde{f}(a \vee a) = \tilde{f}(a)$. $\square$

14.5. A set-theoretical representation for interlaced weakly idempotent bilattices

Bilattices are algebras with two separate lattice structures and weakly idempotent bilattices are algebras with two separate weakly idempotent lattice structures. Bilattices have been used as the basis for a denotational semantics for systems of inference that arise in artificial intelligence and knowledge-based logic programming (see [108,128]). In particular, they have been used to provide a general framework for an efficient procedural semantics of logic programming languages that can deal with incomplete as well as contradictory information.

Originally, Ginsberg [128] suggested to use bilattices as the underlying framework for various AI inference systems including those based on default logics, truth maintenance systems, probabilistic logics and others. These ideas were later pursued in [108,109,239], in the context of logic programming semantics. Studies of the algebraic properties of bilattices have provided insight into their internal structures, and have led to practical results, especially in reducing the computational complexity of bilattice-based multi-valued logic programs.

Ginsberg's original definition of bilattice postulated a connection of two lattice structures through a negation operation. However, the terminology for bilattices is not uniform. For this reason, we often speak of a bilattice and weakly idempotent bilattice without negation for emphasis.

Definition 14.5.1. The algebra $L = (L; \wedge, \vee, *, \triangle)$ with four binary operations is called a *weakly idempotent bilattice*, if the reducts $L_1 = (L; \wedge, \vee)$ and $L_2 = (L; *, \triangle)$ are weakly idempotent lattices and it also satisfies the following identity:

$$a \wedge a = a * a. \tag{14.5.1}$$

If $L_1 = (L; \wedge, \vee)$ and $L_2 = (L; *, \triangle)$ are lattices, then the algebra $L = (L; \wedge, \vee, *, \triangle)$ is called a bilattice.

Every bilattice is a weakly idempotent bilattice. A bilattice L is called bounded if the reducts L_1 and L_2 are bounded lattices. A weakly idempotent bilattice L is called bounded if the reducts L_1 and L_2 are bounded weakly idempotent lattices.

Definition 14.5.2. A (weakly idempotent) bilattice $(L; \wedge, \vee, *, \triangle)$ is called distributive if each pair of the operations from the set $\{\wedge, \vee, *, \triangle\}$ satisfies the identities of distributivity.

Let us denote the quasiorder of the first reduct $L_1 = (L; \wedge, \vee)$ by the following: $\leq_\wedge$ and that of the second reduct $L_2 = (L; *, \triangle)$ by: $\leq_*$ (see Corollary 14.2.16), i.e.,

$$x \leq_\wedge y \leftrightarrow x \wedge y = x \wedge x,$$
$$x \leq_* y \leftrightarrow x * y = x * x.$$

Definition 14.5.3. The weakly idempotent bilattice $(L; \wedge, \vee, *, \triangle)$ is called interlaced if all the basic weakly idempotent bilattice operations are quasiorder-preserving with respect to the both quasiorders, i.e.,

$$\begin{aligned} a \leq_\wedge b, c \leq_\wedge d &\to a * c \leq_\wedge b * d, a\triangle c \leq_\wedge b\triangle d, \\ a \leq_* b, c \leq_* d &\to a \wedge c \leq_* b \wedge d, a \vee c \leq_* b \vee d. \end{aligned}$$

Example 14.5.4. Every distributive weakly idempotent bilattice $(L; \wedge, \vee, *, \triangle)$ is interlaced.

Namely, if $x \leq_\wedge y$, then $(z * x) \wedge (z * y) = z * (x \wedge y) = z * (x \wedge x) = (z*x) \wedge (z*x)$ and $(z\triangle x) \wedge (z\triangle y) = z\triangle(x \wedge y) = z\triangle(x \wedge x) = (z\triangle x) \wedge (z\triangle x)$. So, $z * x \leq_\wedge z * y$ and $z\triangle x \leq_\wedge z\triangle y$ for any $z \in L$. If $x \leq_* y$, then $(z \wedge x) * (z \wedge y) = z \wedge (x * y) = z \wedge (x * x) = (z \wedge x) * (z \wedge x)$, hence $z \wedge x \leq_* z \wedge y$ and $(z \vee x) * (z \vee y) = z \vee (x * y) = z \vee (x * x) = (z \vee x) * (z \vee x)$. Thus, $z \vee x \leq_* z \vee y$ and $z \wedge x \leq_* z \wedge y$ for any $z \in L$.

Example 14.5.5. If $(L; \wedge, \vee)$ is a weakly idempotent lattice, then $(L; \wedge, \vee, \wedge, \vee)$ is an interlaced weakly idempotent bilattice. If $(L; \wedge, \vee)$ is a distributive weakly idempotent lattice, then $(L; \wedge, \vee, \wedge, \vee)$ is a distributive weakly idempotent bilattice. If $(L; \wedge, \vee)$ is a lattice, then $(L; \wedge, \vee, \wedge, \vee)$ is an interlaced bilattice. If $(L; \wedge, \vee)$ is a distributive lattice, then $(L; \wedge, \vee, \wedge, \vee)$ is a distributive bilattice.

For the applications and characterizations of bilattices in various varieties see [18,46,45,62,63,75,107–110,117,166,239,240,262,299,300,335,337,347]. An important open problem in the study of varieties of bilattices is to characterize the finitely generated free algebras. Reference [357] provides further context for the importance of this problem in pure and applied algebra.

Note that until 2000 bilattices with bounds were investigated, but since 2004 bilattices and weakly idempotent bilattices without bounds have been studied (see [299,300]). In current book, we also considered bilattices and weakly idempotent bilattices without bounds (i.e., the general case).

On the characterization of hyperidentities of varieties of lattices, modular lattices, distributive lattices, Boolean and De Morgan algebras see previous chapters.

A weakly idempotent bilattice $L = (L; \wedge, \vee, *, \triangle)$ is distributive if and only if it satisfies the following hyperidentity:

$$X(Y(x, y), z) = Y(X(x, z), X(y, z));$$

One can prove that a weakly idempotent bilattice $L = (L; \wedge, \vee, *, \triangle)$ is interlaced iff it satisfies the following hyperidentity:

$$X(Y(X(x,y),z),Y(y,z)) = Y(X(x,y),z).$$

Example 14.5.6. The superproduct $Q \bowtie Q'$ is an interlaced weakly idempotent bilattice for every two weakly idempotent lattices $(Q;+,\cdot)$ and $(Q';+,\cdot)$. The superproduct $Q \bowtie Q'$ is a distributive weakly idempotent bilattice for every two weakly idempotent distributive lattices $(Q;+,\cdot)$ and $(Q';+,\cdot)$. The superproduct $Q \bowtie Q'$ is a distributive bilattice for every two distributive lattices $(Q;+,\cdot)$ and $(Q';+,\cdot)$. The superproduct $Q \bowtie Q'$ is an interlaced bilattice for every two lattices $(Q;+,\cdot)$ and $(Q';+,\cdot)$.

Lemma 14.5.7. *Let $L = (L; \wedge, \vee, *, \triangle)$ be an interlaced weakly idempotent bilattice. Then for any $x, y \in L$, the following inequalities are valid:*

$$x \wedge y \leq_\wedge x * y \leq_\wedge x \vee y, x \wedge y \leq_\wedge x \triangle y \leq_\wedge x \vee y,$$
$$x * y \leq_* x \wedge y \leq_* x \triangle y, x * y \leq_* x \vee y \leq_* x \triangle y.$$

Proof. From Lemma 14.3.3 we have: $x \wedge y \leq_\wedge x$ and $x \wedge y \leq_\wedge y$; then, since L is an interlaced weakly idempotent bilattice, we have $(x \wedge y) * (x \wedge y) \leq_\wedge x * y$. From the following identity: $x * x = x \wedge x$, of a weakly idempotent bilattice we obtain $(x \wedge y) * (x \wedge y) = (x \wedge y) \wedge (x \wedge y) = x \wedge y$, i.e. $x \wedge y \leq_\wedge x * y$. Similarly, from Lemma 14.3.3, we have $x \leq_\wedge x \vee y$ and $y \leq_\wedge x \vee y$, then, since L is an interlaced weakly idempotent bilattice, we obtain $x * y \leq_\wedge x \vee y$. Thus, $x \wedge y \leq_\wedge x * y \leq_\wedge x \vee y$. □

Definition 14.5.8. The non-empty subset $F \subseteq L$ is called a filter of a weakly idempotent bilattice $L = (L; \wedge, \vee, *, \triangle)$ if F is a filter for both reducts: $L_1 = (L; \wedge, \vee)$ and $L_2 = (L; *, \triangle)$, i.e., if it satisfies the following conditions:

(ff1) if $x, y \in F$, then $x \wedge y \in F$;
(ff2) if $x, y \in F$, then $x * y \in F$;
(ff3) if $x \in F, y \in L$ and $x \leq_\wedge y$, then $y \vee y \in F$;
(ff4) if $x \in F, y \in L$ and $x \leq_* y$, then $y \triangle y \in F$.

Denote the set of all the filters of a weakly idempotent bilattice L by $FF(L)$.

Definition 14.5.9. The non-empty subset $I \subseteq L$ is called an ideal of a weakly idempotent bilattice $L = (L; \wedge, \vee, *, \triangle)$, if I is a filter for the reduct $L_1 = (L; \wedge, \vee)$ and the ideal for the reduct $L_2 = (L; *, \triangle)$, i.e., if it satisfies the following conditions:

(fi1) if $x, y \in I$, then $x \wedge y \in I$;
(fi2) if $x, y \in I$, then $x \triangle y \in I$;
(fi3) if $x \in I, y \in L$ and $x \leq_\wedge y$, then $y \vee y \in I$;
(fi4) if $y \in I, x \in L$ and $x \leq_* y$, then $x * x \in I$.

Denote the set of all the ideals of a weakly idempotent bilattice L by $FI(L)$.

Lemma 14.5.10. *Let X be a non-empty subset of the interlaced weakly idempotent bilattice $L = (L; \wedge, \vee, *, \triangle)$. Then*

$$[X]_f = X \cup \{x \in L | \exists x_1, \ldots, x_n \in X, \exists y \in L, x_1 \wedge \ldots \wedge x_n \leq_\wedge y \leq_* x \;\&\; x = K([x])\}$$

is the smallest filter of the interlaced weakly idempotent bilattice L that contains X.

Proof. First of all it is straightforward to prove that

$$[X]_f = X \cup \{x \in L | \exists x_1, \ldots, x_n \in X, \exists y \in L, x_1 * \ldots * x_n \leq_* y \leq_\wedge x \;\&\; x = K([x])\}.$$

Then, similar to Lemma 14.3.8, we show that $[X]_f$ is a filter; moreover, it is the smallest filter that contains X. □

Lemma 14.5.11. *Let X be a non-empty subset of the interlaced weakly idempotent bilattice $L = (L; \wedge, \vee, *, \triangle)$. Then*

$$[X]_i = X \cup \{x \in L | \exists x_1, \ldots, x_n \in X, \exists y \in L, x_1 \wedge \ldots \wedge x_n \leq_\wedge y \geq_* x \;\&\; x = K([x])\}$$

is the smallest ideal of the interlaced weakly idempotent bilattice L that contains X.

Corollary 14.5.12. *Let $L = (L; \wedge, \vee, *, \triangle)$ be an interlaced weakly idempotent bilattice. Then*

$$[a]_f = \{a\} \cup \{x \in L | \exists y \in L, a \leq_\wedge y \leq_* x \;\&\; x = K([x])\}$$

is the smallest filter of the interlaced weakly idempotent bilattice L that contains $a \in L$.

Proof. Take $X = \{a\}$ in Lemma 14.5.10. □

Corollary 14.5.13. *Let $L = (L; \wedge, \vee, *, \triangle)$ be an interlaced weakly idempotent bilattice. Then*

$$[a]_i = \{a\} \cup \{x \in L | \exists y \in L, a \leq_\wedge y \geq_* x \;\&\; x = K([x])\}$$

is the smallest ideal of the interlaced weakly idempotent bilattice L that contains $a \in L$.

Proof. Take $X = \{a\}$ in Lemma 14.5.11. □

Lemma 14.5.14. *Let $L = (L; \wedge, \vee, *, \triangle)$ be an interlaced weakly idempotent bilattice. The following assertions are valid on L:*

$$[a \wedge a]_f \cup [b \wedge b]_f \subseteq [a \wedge b]_f; \quad [a \vee b]_f = [a \wedge a]_f \cap [b \wedge b]_f;$$
$$[a \wedge a]_f \cup [b \wedge b]_f \subseteq [a * b]_f; \quad [a \triangle b]_f = [a \wedge a]_f \cap [b \wedge b]_f;$$
$$[a \wedge a]_i \cup [b \wedge b]_i \subseteq [a \wedge b]_i; \quad [a \vee b]_i = [a \wedge a]_i \cap [b \wedge b]_i;$$
$$[a \wedge a]_i \cup [b \wedge b]_i \subseteq [a \triangle b]_i; \quad [a * b]_i = [a \wedge a]_i \cap [b \wedge b]_i.$$

Proof. Let us prove the equality: $[a \vee b]_f = [a \wedge a]_f \cap [b \wedge b]_f$. If $x \in [a \vee b]_f$, then, by Corollary 14.5.12 we have: $a \vee b \leq_\wedge c \leq_* x = K([x])$, then $a \leq_\wedge c \leq_* x$ and $b \leq_\wedge c \leq_* x$; thus, $a \vee a \leq_\wedge a \leq_\wedge c \leq_* x$ and $a \vee b \leq_\wedge b \leq_\wedge c \leq_* x$. Thus, $x \in [a \vee a]_f$ and $x \in [b \vee b]_f$, i.e., $x \in [a \vee a]_f \cap [b \wedge b]_f$.

Conversely, let $x \in [a \vee a]_f \cap [b \vee b]_f$, then: $x \in [a \vee a]_f$ and $x \in [b \vee b]_f$. From $x \in [a \wedge a]_f$ by Corollary 14.5.12 follows that there exists $c \in L$ such that: $a \vee a \leq_\wedge c \leq_* x$ and from Lemma 14.3.3 we have: $a \vee b \leq_\wedge a \leq_\wedge a \vee a$ and it follows that $a \vee b \leq_\wedge c \leq_* x$, i.e. $x \in [a \vee b]_f$. If $x \in [b \wedge b]_f$, then we similarly obtain that $x \in [a \vee b]_f$. Thus, $[a \vee b]_f = [a \vee a]_f \cup [b \vee b]_f$. The rest of the assertions are proved similarly. □

For every $a \in L$, we denote: $Bf(a) = \{X \in FF(L) | a \in X\}$ and $Bi(a) = \{X \in FI(L) | a \in X\}$.
Let us define on the sets $Bf(L) = \{Bf(a) | a \in L\}$ and $Bi(L) = \{Bi(a) | a \in L\}$ the binary operations $\cap^*$ and $\cup^*$ in the following manner:

$$Bf(a) \cap^* Bf(b) = \{F \in FF(L) | (\exists X \in Bf(a \wedge a))(\exists Y \in Bf(b \wedge b)),$$
$$X \cup Y \subseteq F\};$$
$$Bf(a) \cup^* Bf(b) = \{F \in FF(L) | (\exists X \in Bf(a \wedge a))(\exists Y \in Bf(b \wedge b)),$$
$$X \cap Y \subseteq F\};$$
$$Bi(a) \cap^* Bi(b) = \{I \in FI(L) | (\exists X \in Bi(a \wedge a))(\exists Y \in Bi(b \wedge b)),$$
$$X \cup Y \subseteq I\};$$

$$Bi(a) \cup^* Bi(b) = \{I \in FI(L) | (\exists X \in Bi(a \wedge a))(\exists Y \in Bi(b \wedge b)), \\ X \cap Y \subseteq I\}.$$

Lemma 14.5.15. *Let $L = (L; \wedge, \vee, *, \triangle)$ be an interlaced weakly idempotent bilattice. Then the following equalities are valid on $(Bf(L); \cap^*, \cup^*)$ and $(Bi(L); \cap^*, \cup^*)$:*

$$Bf(a \wedge b) = Bf(a) \cap^* Bf(b);\ \ Bf(a \vee b) = Bf(a) \cup^* Bf(b);$$
$$Bf(a * b) = Bf(a) \cap^* Bf(b);\ \ Bf(a \triangle b) = Bf(a) \cup^* Bf(b);$$
$$Bi(a \wedge b) = Bi(a) \cap^* Bi(b);\ \ Bi(a \vee b) = Bi(a) \cup^* Bi(b);$$
$$Bi(a * b) = Bi(a) \cup^* Bi(b); Bi(a \triangle b) \ \ = Bi(a) \cap^* Bi(b).$$

Proof. Let us show the first equality. Suppose that $F \in Bf(a \wedge b)$, then $a \wedge b \in F$. Let $c \in [a \wedge b]_f$, then by Corollary 14.5.12, there exists $c \in L$ such that: $a \wedge b \leq_\wedge z \leq_* c = K([c])$, so $[a \wedge b]_f \subseteq F$. By Lemma 14.5.14 we have $[a \wedge a]_f \cup [b \wedge b]_f \subseteq [a \wedge b]_f$. Thus, there exist $X = [a \wedge a]_f$ and $Y = [b \wedge b]_f$ such that $X \cap Y \in F$, i.e., $F \in Bf(a) \cap^* Bf(b)$. Conversely, let $F \in Bf(a) \cap^* Bf(b)$, then there exist $X \in Bf(a \wedge a)$ and $Y \in Bf(b \wedge b)$ such that $X \cup Y \in F$; hence, $a \wedge a \in X$ and $b \wedge b \in Y$. Then we see that $a \wedge a, b \wedge b \in X \cup Y \in F$ and $a \wedge b = (a \wedge a) \wedge (b \wedge b) \in F$. Thus, $F \in Bf(a \wedge b)$ and $Bf(a \wedge b) = Bf(a) \cap^* Bf(b)$. The rest of the equalities are proved similarly. □

From Lemma 14.5.15, it immediately follows that $Bf(L) = (Bf(L); \cap^*, \cup^*)$ and $Bi(L) = (Bi(L); \cap^*, \cup^*)$ satisfy the identities (14.2.1)–(14.2.5). Hence, $Bf(L)$ and $Bi(L)$ are weakly idempotent lattices.

Theorem 14.5.16 ([288]). *Every interlaced weakly idempotent bilattice $L = (L; \wedge, \vee, *, \triangle)$ is isomorphic to the superproduct of the two weakly idempotent lattices $Bf(L) = (Bf(L); \cap^*, \cup^*)$ and $Bi(L) = (Bi(L); \cap^*, \cup^*)$, i.e.,*

$$L \cong Bf(L) \bowtie Bi(L).$$

Proof. Let us show that the mapping:

$$\varphi : L \to (Bf(L); \cap^*, \cup^*) \bowtie (Bi(L); \cap^*, \cup^*)$$

which is defined by:

$$\varphi : x \mapsto (Bf(x), Bi(x)) \tag{14.5.2}$$

is an isomorphism. The proof of the injectivity of φ is similar to the proof of Theorem 14.4.1, while the surjectivity is obvious.
The fact that φ is a homomorphism follows from Lemma 14.5.15:

$$\begin{aligned}\varphi(x \wedge y) &= (Bf(x \wedge y), Bi(x \wedge y)) = (Bf(x) \cap^* Bf(y), Bi(x) \cap^* Bi(y))\\ &= (Bf(x), Bi(x))(\cap^*, \cap^*)(Bf(y), Bi(y)) = \varphi(x)(\cap^*, \cap^*)\varphi(y);\\ \varphi(x \vee y) &= (Bf(x \vee y), Bi(x \vee y)) = (Bf(x) \cup^* Bf(y), Bi(x) \cup^* Bi(y))\\ &= (Bf(x), Bi(x))(\cup^*, \cup^*)(Bf(y), Bi(y)) = \varphi(x)(\cup^*, \cup^*)\varphi(y);\\ \varphi(x * y) &= (Bf(x * y), Bi(x * y)) = (Bf(x) \cap^* Bf(y), Bi(x) \cup^* Bi(y))\\ &= (Bf(x), Bi(x))(\cap^*, \cup^*)(Bf(y), Bi(y)) = \varphi(x)(\cap^*, \cup^*)\varphi(y);\\ \varphi(x \triangle y) &= (Bf(x \triangle y), Bi(x \triangle y)) = (Bf(x) \cup^* Bf(y), Bi(x) \cap^* Bi(y))\\ &= (Bf(x), Bi(x))(\cup^*, \cap^*)(Bf(y), Bi(y)) = \varphi(x)(\cup^*, \cap^*)\varphi(y).\end{aligned}$$

□

Note that the operations $\cup^*$ and $\cap^*$ on the sets $Bf(L)$ and $Bi(L)$ coincide with the operations $\cup^{**}$ and $\cap^{**}$ respectively, which are defined in the following way:

$$\begin{aligned}Bf(a) \cup^{**} Bf(b) &= Bf(a \wedge a) \cup Bf(b \wedge b) \cup \{F \in FF(L) | \exists X \in Bf(a \wedge a),\\ &\qquad \exists Y \in Bf(b \wedge b), X \cap Y \subseteq F\},\\ Bf(a) \cap^{**} Bf(b) &= Bf(a \wedge a) \cap Bf(b \wedge b) \cap \{F \in FF(L) | \exists X \in Bf(a \wedge a),\\ &\qquad \exists Y \in Bf(b \wedge b), X \cup Y \subseteq F\},\\ Bi(a) \cup^{**} Bi(b) &= Bi(a \wedge a) \cup Bi(b \wedge b) \cup \{I \in FI(L) | \exists X \in i(a \wedge a),\\ &\qquad \exists Y \in i(b \wedge b), X \cap Y \subseteq I\},\\ Bi(a) \cap^{**} Bi(b) &= Bi(a \wedge a) \cap Bi(b \wedge b) \cap \{I \in FI(L) | \exists X \in i(a \wedge a),\\ &\qquad \exists Y \in i(b \wedge b), X \cup Y \subseteq I\}.\end{aligned}$$

Moreover, $Bf(a) \cap^{**} Bf(b) = Bf(a \wedge a) \cap Bf(b \wedge b)$ and $Bi(a) \cap^{**} Bi(b) = Bi(a \wedge a) \cap Bi(b \wedge b)$.

Now, let us introduce the concepts of prime filters and prime ideals for a weakly idempotent bilattice.

Definition 14.5.17. The filter F of a weakly idempotent bilattice $L = (L; \wedge, \vee, *, \triangle)$ is called prime, if it is prime filter for both reducts $L_1 = (L; \wedge, \vee)$ and $L_2 = (L; *, \triangle)$.

Definition 14.5.18. The ideal I of a weakly idempotent bilattice $L = (L; \wedge, \vee, *, \triangle)$ is called prime, if it is a prime filter for the reduct $L_1 = (L; \wedge, \vee)$ and prime ideal for the reduct $L_2 = (L; *, \triangle)$.

Denote the set of all prime filters of a weakly idempotent bilattice L by $pBF(L)$, and the set of all prime ideals of a weakly idempotent bilattice L by $pBI(L)$. Let us make the following designations: $pBf(a) = \{X \in pBF(L)|a \in X\}$ and $pBi(a) = \{X \in pBI(L)|a \in X\}$, $pBf(L) = \{pBf(a)|a \in L\}$ and $pBi(L) = \{pBi(a)|a \in L\}$.

The operations $\Cap^{**}$ and $\Cup^{**}$ can be defined on $pBf(L)$ and $pBi(L)$ and an analogue of Lemma 14.5.15 can be proven. So, the sets $pBF(L)$ and $pBI(L)$ are closed under the operations $\Cap^{**}$ and $\Cup^{**}$.

Corollary 14.5.19. *Every distributive weakly idempotent bilattice $L = (L; \wedge, \vee, *, \triangle)$ is isomorphic to the superproduct of the two distributive weakly idempotent lattices: $pBf(L) = (pBf(L); \Cap^{**}, \Cup)$ and $pBi(L) = (pBi(L); \Cap^{**}, \Cup)$, i.e.,*

$$L \cong pBf(L) \bowtie pBi(L).$$

Proof. Similarly to the proof of Corollary 14.4.6, we show that on the sets $pBf(L)$ and $pBi(L)$ the operations $\Cup^{**}$ and $\Cup^{**}$ convert to the set-theoretical union. Using the identities of Lemma 14.5.15 for prime filters and prime ideals we get that $pBf(L)$ and $pBi(L)$ are weakly idempotent lattices. We redefine the operation $pBf(L)$ on the set $\Cap^{**}$, in the following way: $pBf(a) \Cap^{**} pBf(b) = pBf(a \wedge a) \cap pBf(b \wedge b)$ and $pBi(a) \Cap^{**} pBi(b) = pBi(a \wedge a) \cap pBi(b \wedge b)$. Hence, the weakly idempotent lattices $pBf(L)$ and $pBi(L)$ are distributive. □

Denote the set of all filters of the bilattice L by $\tilde{BF}(L)$ and the set of all ideals by $\tilde{BI}(L)$. Let us make the following designations: $\tilde{Bf}(a) = \{X \in \tilde{BF}(L)|a \in X\}$, $\tilde{Bi}(a) = \{X \in \tilde{Bi}(L)|a \in X\}$, $\tilde{Bf}(L) = \{\tilde{Bf}(a)|a \in L\}$, $\tilde{Bi}(L) = \{\tilde{Bi}(a)|a \in L\}$.

Corollary 14.5.20. *Every interlaced bilattice $L = (L; \wedge, \vee, *, \triangle)$ is isomorphic to the superproduct of the two lattices $\tilde{Bf}(L) = (\tilde{Bf}(L); \Cap, \Cup^{**})$ and $\tilde{Bi}(L) = (\tilde{Bi}(L); \Cap, \Cup^{**})$, i.e.,*

$$L \cong \tilde{Bf}(L) \bowtie \tilde{Bi}(L).$$

Proof. Since L is a bilattice, the operation $\cap^{**}$ is converted into the set-theoretical intersection, and the operation $\cup^{**}$ is defined as follows:

$$\tilde{f}(a)\cup^{**}\tilde{f}(b) = \tilde{f}(a)\cup\tilde{f}(b)\cup\{F \in \tilde{F}(L)|(\exists X \in \tilde{f}(a))(\exists Y \in \tilde{f}(b))X\cap Y \subseteq F\},$$

$$\tilde{i}(a)\cup^{**}\tilde{i}(b) = \tilde{i}(a)\cup\tilde{i}(b)\{F \in \tilde{F}(L)|(\exists X \in \tilde{i}(a))(\exists Y \in \tilde{i}(b))X\cap Y \subseteq F\}.$$

It is obvious that $\tilde{Bf}(L)$ and $\tilde{Bi}(L)$ are lattices. Indeed: $\tilde{Bf}(a)\cap\tilde{Bf}(a) = \tilde{Bf}(a\wedge a) = \tilde{Bf}(a)$, $\tilde{Bi}(a)\cup^{**}\tilde{Bi}(a) = \tilde{Bi}(a\vee a) = \tilde{Bi}(a)$. □

Denote the set of all prime filters of a bilattice L by $p\tilde{BF}(L)$ and the set of all prime ideals of L by $p\tilde{BI}(L)$. Let us make the following designations: $p\tilde{Bf}(a) = \{X \in p\tilde{BF}(L)|a \in X\}$, $p\tilde{Bi}(a) = \{X \in p\tilde{BI}(L)|a \in X\}$, $p\tilde{Bf}(L) = \{p\tilde{Bf}(a)|a \in L\}$, $p\tilde{Bi}(L) = \{p\tilde{Bi}(a)|a \in L\}$

Corollary 14.5.21. *Every distributive bilattice* $L = (L;\wedge,\vee,*,\Delta)$ *is isomorphic to the superproduct of the two distributive lattices* $p\tilde{Bf}(L) = (p\tilde{Bf}(L);\cap,\cup)$ *and* $p\tilde{Bi}(L) = (p\tilde{Bi}(L);\cap,\cup)$, *i.e.,*

$$L \cong pB\tilde{f}(L) \bowtie pB\tilde{i}(L).$$

Proof. Similarly to the proof of Corollary 14.4.6, we get that the operations $\cup^{**}$ and $\cup^{**}$ convert into the set-theoretical union on the sets $p\tilde{Bf}(L)$ and $p\tilde{Bi}(L)$, respectively. □

For the bounded case last result was proved by Gargov [117].

Chapter 15

Bigroups and Grätzer Algebras

15.1. Preliminary concepts and results

By analogy of bilattices [108,117,125–128,240,264,300,347] we introduce the concepts of a bisemigroup, a bimonoid, a De Morgan bisemigroup and a bigroup.

Let Q be an arbitrary non-empty set; let $O_p^{(n)}Q$ be a set of all n-ary operations on Q, and

$$O_pQ = \bigcup_n O_p^{(n)}Q;$$

(For every non-empty subset $\Sigma \subseteq O_pQ$, the pair $(Q; \Sigma)$ is an algebra.)

A bisemigroup is an algebra $Q(\cdot, \circ)$ equipped with two binary associative operations $\cdot$ and $\circ$. If both of these operations have an identity element, then the bisemigroup is called a bimonoid. A commutative bisemigroup is a bisemigroup in which both operations are commutative. A bisemilattice is a commutative bisemigroup in which both operations are idempotent. In any bisemilattice $Q(\cdot, \circ)$, the binary operations determine two partial orders $\leqslant_1$ and $\leqslant_2$. A bisemilattice is a bilattice, if the partial orders $\leqslant_1$ and $\leqslant_2$ are lattice orders. Since every lattice order is characterized by two binary operations, every bilattice is a binary algebra with four operations and corresponding identities. A De Morgan bisemigroup (see [54,303]) is an algebra $Q(\cdot, \circ, ^-, 0, 1)$ such that $Q(\cdot, \circ)$ is a bimonoid with identity elements 0 (for operation $\cdot$), 1 (for operation $\circ$) and such that the following De Morgan identities

$$\overline{\overline{x}} = x,$$
$$\overline{x \cdot y} = \overline{x} \circ \overline{y},$$
$$\overline{x \circ y} = \overline{x} \cdot \overline{y},$$

and the identities

$0 \circ x = 0,$

$1 \cdot x = 1$

hold. A De Morgan bisemigroup $Q(\cdot, \circ, ^{-}, 0, 1)$ is a De Morgan algebra if $Q(\cdot, \circ)$ is a distributive lattice.

A bimonoid $Q(\cdot, \circ)$ with identity elements 0 (for the operation $\cdot$) and 1 (for the operation $\circ$) is called a bigroup, if for every $x \in Q$:

$0 \circ x = 0,$

$1 \cdot x = 1,$

and the following conditions are valid:

(a) $Q \setminus \{1\}$ is a group with an identity element 0 under the multiplication $\cdot$;
(b) $Q \setminus \{0\}$ is a group with an identity element 1 under the multiplication $\circ$;

A bigroup of order > 3 is called non-trivial.

A bimonoid $Q(\cdot, \circ)$ with identity elements 0 (for operation $\cdot$) and 1 (for operation $\circ$) is called a half-bigroup, if for every $x \in Q$:

$0 \circ x = 0,$

$1 \cdot x = 1,$

and at least one of the two conditions a), b) is valid.

An element $a \in Q$ is called invertible in the bimonoid $Q(\cdot, \circ)$, if a is invertible in both semigroups $Q(\cdot)$ and $Q(\circ)$. We denote the inverse element of a in semigroup $Q(\cdot)$ by a^{-1}, and that in the semigroup $Q(\circ)$ we denote by ${}^{-1}a$. An invertible element of bimonoid $Q(\cdot, \circ)$ is also called an invertible element of De Morgan bisemigroup $Q(\cdot, \circ, ^{-}, 0, 1)$.

The set $O_p^{(2)}Q$ of all binary operations on Q is a bimonoid under the following operations:

$$f \cdot g(x, y) = f(x, g(x, y)), \tag{15.1.1}$$

$$f \circ g(x, y) = f(g(x, y), y), \tag{15.1.2}$$

in which the identity elements are the identical operations δ_2^2 and δ_2^1, where $\delta_2^1(x, y) = x$, and $\delta_2^2(x, y) = y$ for all $x, y \in Q$. Sometimes we define: $\delta_2^2 = E$ and $\delta_2^1 = F$.

The Mal'tsev-type characterization of bisemigroup $O_p^{(2)}Q$ is considered in [422]. For applications of the mentioned binary superpositions, see [25,29,30,43,122,221,251,257,263,295,296,298,311,315,336,388,413].

Any subset $S \subseteq O_p^{(2)}Q$ which is closed under these two operations is called a bisemigroup of the operations (on the set Q). The bisemigroup of operations (on the set Q) is called a bimonoid of operations (on set Q) if it contains the identical operations δ_2^1 and δ_2^2. Besides, the dual operation $f^*(x, y) = f(y, x)$ defines the unary operation $^- : f \to f^*$, which is an antiautomorphism of the bimonoid $O_p^{(2)}Q$, since the following De Morgan identities are valid:

$$(f^*)^* = f,$$
$$(f \cdot g)^* = f^* \circ g^*,$$
$$(f \circ g)^* = f^* \cdot g^*.$$

So the set $O_p^{(2)}Q$ of all binary operations on Q is a De Morgan bisemigroup. Every subalgebra of a De Morgan bisemigroup $O_p^{(2)}Q$ is called a De Morgan bisemigroup of operations (on set Q).

The bimonoid S of operations (on the set Q) is a bigroup, if both of the following conditions are valid (cf. [25]):

(c) $S \setminus \{\delta_2^1\}$ is a group with an identity element δ_2^2 under the multiplication (15.1.1);
(d) $S \setminus \{\delta_2^2\}$ is a group with an identity element δ_2^1 under the multiplication (15.1.2);

This bigroup is called a bigroup of operations (on the set Q).

Example 15.1.1. Let $Q(+, \cdot)$ be a field and for every $a \in Q$:

$$A_a(x, y) = (1 - a)x + ay.$$

If $\Sigma = \{A_a | \, a \in Q\}$ then the algebra $(Q; \Sigma)$ is a bigroup of operation (on the set Q).

The bimonoid S of operations (on the set Q) is a half-bigroup, if at least one of the two conditions (c), (d) is valid. This half-bigroup is called a half-bigroup of operations (on the set Q).

The binary operation A on Q is called a left quasigroup operation, if the equation $A(x, a) = b$ has a unique solution $x \in Q$ for any $a, b \in Q$. If A is a left-quasigroup operation on Q, then $Q(A)$ is called a left-quasigroup. The definitions of a right quasigroup operation and a right quasigroup are analogously. A left quasigroup with a left identity element is called a left loop. A right-quasigroup with a right identity element is called a right-loop. The binary operation A on Q is called a quasigroup operation, if it is a left and a right quasigroup operation simultaneously. If A is a quasigroup operation on Q, then $Q(A)$ is called a quasigroup (see [12,13,27,220,249,370, 382]). A quasigroup with an identity element is called a loop.

On applications of right (left) quasigroups in geometry and topology (knot theory) see [228,312].

The operation $A \in O_p^{(2)}Q$ is called right invertible, if it is invertible in the semigroup $O_p^{(2)}Q(\cdot)$; this inverse is denoted by A^{-1}, which is called the right inverse of A. The operation $A \in O_p^{(2)}Q$ is called left invertible, if it is invertible in the semigroup $O_p^{(2)}Q(\circ)$; this inverse is denoted by ${}^{-1}A$, which is called the left inverse of A.

The following propositions are consequences of the above definitions.

Proposition 15.1.2 ([263]). *The binary operation A on Q is a right-quasigroup operation if and only if A is right invertible. The binary operation A on Q is a left-quasigroup operation if and only if A is left invertible. A groupoid $Q(A)$ is quasigroup if and only if the operation $A \in O_p^{(2)}Q$ is right and left invertible simultaneously. Hence, in a bigroup S of operations, any operation $A \in S, A \neq \delta_2^1, \delta_2^2$ is a quasigroup operation.*

Proposition 15.1.3. (1) *The operation $A \in O_p^{(2)}Q$ is right invertible iff*

$$A(x, A^{-1}(x, y)) = y, \quad A^{-1}(x, A(x, y)) = y$$

for every $x, y \in Q$.

(2) *The operation $A \in O_p^{(2)}Q$ is left invertible if and only if*

$$A({}^{-1}A(y, x), x) = y, \quad {}^{-1}A(A(y, x), x) = y$$

for every $x, y \in Q$.

Proposition 15.1.4. (1) *For every right invertible operation $A \in O_p^{(2)}Q$ and for every operations $B, C \in O_p^{(2)}Q$:*

$$C(A, B) = (C^* \circ (B \cdot A^{-1})) \cdot A,$$

where $C(A, B)(x, y) = C(A(x, y), B(x, y))$ for every $x, y \in Q$;

(2) *For every left invertible operation $A \in O_p^{(2)}Q$ and for every operations $B, C \in O_p^{(2)}Q$:*

$$C(A, B) = (C \cdot (B \circ {}^{-1} A)) \circ A.$$

(3) *For every right invertible operation* $B \in O_p^{(2)}Q$ *and for every operations* $A, C \in O_p^{(2)}Q$:

$$C(A, B) = (C \circ (A \cdot B^{-1})) \cdot B.$$

(4) *For every left invertible operation* $B \in O_p^{(2)}Q$ *and for every operations* $A, C \in O_p^{(2)}Q$:

$$C(A, B) = (C^* \cdot (A \circ^{-1} B)) \circ B.$$

Proof. (1) We have

$$\begin{aligned}
(C^* \circ (B \cdot A^{-1})) \cdot A(x, y) &= C^* \circ (B \cdot A^{-1})(x, A(x, y)) \\
&= C^*(B \cdot A^{-1}(x, A(x, y)), A(x, y)) = C^*(B(x, A^{-1}(x, A(x, y))), A(x, y)) \\
&= C^*(B(x, y), A(x, y)) = C(A(x, y), B(x, y)) = C(A, B)(x, y).
\end{aligned}$$

(2) We have:

$$\begin{aligned}
(C \cdot (B \circ^{-1} A)) \circ A(x, y) &= C \cdot (B \circ^{-1} A)(A(x, y), y) \\
&= C(A(x, y), B \circ^{-1} A(A(x, y), y)) = C(A(x, y), B(^{-1}A(A(x, y), y), y)) \\
&= C(A(x, y), B(x, y)) = C(A, B)(x, y).
\end{aligned}$$

The proof of (3) and (4) are analogously. □

Lemma 15.1.5. *If* A *and* B *are left-quasigroup operations on* Q, *then* $A \circ B$ *is a left-quasigroup operation on* Q. *If* A *and* B *are right-quasigroup operations on* Q, *then* $A \cdot B$ *is a right-quasigroup operation on* Q. *Hence, the set of all left-quasigroup operations on* Q *is a subgroup of semigroup* $O_p^{(2)}Q(\circ)$, *and the set of all right-quasigroup operations on* Q *is a subgroup of semigroup* $O_p^{(2)}Q(\cdot)$.

Example 15.1.6. If $Q = \{1, 2\}$, there are four right-quasigroup operations on Q formed a Klein four-group under multiplication (15.1.1).

Lemma 15.1.7. *Every bigroup of operations is a De Morgan bisemigroup of operations.*

Two binary operations A, B on Q are called orthogonal (see [61,105]), if the system of equations

$$A(x,\, y) = a,$$
$$B(x,\, y) = b$$

has a unique solution $(x,\, y) \in Q^2$ for any $a, b \in Q$. In this case, we also say that the groupoids $Q(A)$ and $Q(B)$ are orthogonal.

It is obvious that two commutative operations defined on the same non-trivial set can not be orthogonal. In particular, two non-trivial commutative quasigroups cannot be orthogonal. Two non-trivial groupoids with identity elements cannot be orthogonal. In particular, two non-trivial loops defined on the same set cannot be orthogonal.

The following lemma is the generalization of the corresponding result in [221,388].

Lemma 15.1.8. *If A is a left-quasigroup operation on Q and B is a right-quasigroup operation on Q, then operations B and $A \cdot B$ are orthogonal. In particular, the quasigroup operations A and $A \cdot A$ are orthogonal. Hence, if $Q(A)$ is a loop, then $Q(A \cdot A)$ cannot be a loop. If A is a right-quasigroup operation on Q and B is a left-quasigroup operation on Q, then the operations B and $A \circ B$ are orthogonal. In particular, the quasigroup operations A and $A \circ A$ are orthogonal. Hence, if $Q(A)$ is a loop, then $Q(A \circ A)$ cannot be a loop.*

Lemma 15.1.9. *In a half-bigroup S of operations (on the set Q) every operation $A \in S$ is orthogonal to any quasigroup operation $B \in S$. In particular, any two quasigroup operations in a half-bigroup of operations (on the set Q) are orthogonal. Hence, any two operations in a bigroup of operations (on the set Q) are orthogonal.*

Theorem 15.1.10. *A half-bigroup of operations $S(\cdot, \circ)$ (on set Q) has a commutative quasigroup operation if and only if it has a quasigroup operation $H \in S$ with condition $H \cdot H = E$ or $H \circ H = F$. In particular, a finite half-bigroup of operations (on the set Q) with a commutative quasigroup operation has an odd order.*

Proof. Let $S(\cdot, \circ)$ be a half-bigroup of operations, and let $S \setminus \{\delta_2^1\}$ be a group with an identity element δ_2^2 under the multiplication (15.1.1). If A is a commutative quasigroup operation, $A \in S$ and $H = ^{-1}A$, then $H \cdot H = E$:

$$A^{-1} = (A^*)^{-1} = (^{-1}A)^* = ((^{-1}A \cdot^{-1} A) \circ A) \cdot A^{-1},$$
$$(^{-1}A \cdot^{-1} A) \circ A = E,$$
$$^{-1}A \cdot^{-1} A = E,$$

where $B^*(x, y) = B(y, x)$ for all $x, y \in Q$.

If $S\backslash\{\delta_2^2\}$ is a group with an identity element δ_2^1 under the multiplication (15.1.2), $A \in S$ is a commutative quasigroup operation and $H = A^{-1}$, then $H \circ H = F$. It has a similar proof. The remaining proof of the theorem is clear.

A bimonoid S of operations (on set Q) is called a (non-trivial) local bigroup, if for every quasigroup operation $A \in S$ there exists some (non-trivial) bigroup of operations (on set Q), which includes A. □

Proposition 15.1.11. *If A and B are idempotent binary operations on Q, then A^*, $A \cdot B$ and $A \circ B$ are also idempotent operations. So the set of all idempotent binary operations on Q is a De Morgan bisemigroup of operations on Q. For every quasigroup operation A the following operations are idempotent:*

$$(A \cdot A) \circ^{-1} A,$$
$$^{-1}A \cdot A^{-1},$$
$$(A \circ A) \cdot A^{-1},$$
$$A \cdot (A^{-1} \circ^{-1} A)^* \cdot A,$$
$$A^{-1} \circ^{-1} A,$$
$$A \circ (^{-1}A \cdot A^{-1})^* \circ A.$$

Example 15.1.12. If $Q = \{1, 2\}$, then the De Morgan bisemigroup of idempotent operations (on set Q) is a Boole–De Morgan algebra of order 4.

If S is a bigroup of binary operations (on the set Q) and $A \in S$, then $A^* \in S$, where $A^*(x, y) = A(y, x)$ for every $x, y \in Q$. Indeed, if A is a quasigroup operation, then:

$$(^{-1}(A^{-1}))^{-1} = ^{-1}((^{-1}A)^{-1}) = A^*.$$

This fact follows from the following result too.

Lemma 15.1.13. *For any operation $B \in O_p^{(2)}Q$ and for any quasigroup operation $C \in O_p^{(2)}Q$ we have*

$$B^* = ((B \circ C^{-1}) \cdot C) \circ^{-1} C,$$
$$B^* = ((B \cdot^{-1} C) \circ C) \cdot C^{-1}.$$

Proof. Indeed:

$$(((B \circ C^{-1}) \cdot C) \circ^{-1} C)(x, y) = (B \circ C^{-1}) \cdot C(^{-1}C(x, y), y)$$
$$= B \circ C^{-1}(^{-1}C(x, y), C(^{-1}C(x, y), y)))$$
$$= B \circ C^{-1}(^{-1}C(x, y), x) = B(C^{-1}(^{-1}C(x, y), x), x) = B(y, x) = B^*(x, y),$$

since:

$$C^{-1}(^{-1}C(x, y), x) = y,$$
$$C(^{-1}C(x, y), y) = x.$$

The second equality is proved analogically. □

15.2. Idempotency of quasigroup operations in bigroups of operations

Theorem 15.2.1. *Every quasigroup operation A of a non-trivial bigroup S of binary operations (on the set Q) is idempotent* ($|S| > 3$).

Proof. Let S_0 be the subset of binary idempotent operation of the set S. We have $E, F \in S_0$. Besides, it is evident, that $S_0 \backslash \{F\}$ is a subgroup of the group $S \backslash \{F\}$ under the right multiplication of binary operations, and $S_0 \backslash \{E\}$ is a subgroup of the group $S \backslash \{E\}$ under the left multiplication of binary operations. According to Proposition 15.1.11, the operations

$$S_A = (A \circ A) \cdot A^{-1}, \quad T_A =^{-1} A \cdot A^{-1}$$

are idempotent. Besides, $T_A \neq F$, since from the condition $^{-1}A \cdot A^{-1} = F$ the equality: $^{-1}A = F$ and $A = F$ follows, which is contradiction, since A is a quasigroup operation. According to Proposition 15.1.11, $U_A = A \circ T_A^* \circ A \in S_0$. Let us consider the operation $V_A = A \circ T_A^*$. If $V_A = F$, then

$A =^{-1} (T_A^*) \in S_0$. If $V_A \neq F$, then

$$V_A \circ V_A = A \circ T_A^* \circ A \circ T_A^* = U_A \circ T_A^* \in S_0.$$

On the other hand:

$$S_{V_A} \cdot V_A = (V_A \circ V_A) \cdot V_A^{-1} \cdot V_A = V_A \circ V_A.$$

Let us consider the following two cases:

(1) The group $S \setminus \{E\}$ has not a second order element. In this case we have $S_{V_A} \neq F$, since if $S_{V_A} = F$, the

$$F \cdot V_A = V_A \circ V_A,$$
$$F = V_A \circ V_A,$$

and V_A has the second order. Hence, there exists $S_{V_A}^{-1}$ and since $V_A \neq F$, there exists V_A^{-1} too. We have:

$$S_{V_A} = (V_A \circ V_A) \cdot V_A^{-1} \in S_0.$$

Indeed:

$$\begin{aligned} S_{V_A}(x,x) &= (V_A \circ V_A) \cdot V_A^{-1}(x,x) = V_A \circ V_A(x, V_A^{-1}(x,x)) \\ &= V_A(V_A(x, V_A^{-1}(x,x)), V_A^{-1}(x,x)) = V_A(x, V_A^{-1}(x,x)) = x. \end{aligned}$$

Now,

$$V_A = S_{V_A}^{-1} \cdot (V_A \circ V_A) \in S_0,$$

and

$$A \circ T_A^* \in S_0.$$

Further, $T_A \neq F$ and $T_A^* \neq E$. Hence, there exists ${}^{-1}(T_A^*)$, therefore:

$$(A \circ T_A^*) \circ^{-1} (T_A^*) \in S_0,$$

and $A \in S_0$;

(2) In the group $S \setminus \{E\}$ there exists a second order element (quasigroup operation), moreover this operation is unique. Indeed, if $C \circ C = F$ and $D \circ D = F$, then the operations C^{-1} and D^{-1} are commutative operations, which contradicts orthogonality of operations C^{-1} and D^{-1}. In fact,

if $H = C^{-1}$, then: $H^{-1} \circ H^{-1} = (C^{-1})^{-1} \circ (C^{-1})^{-1} = C \circ C = F$ and $(H^{-1} \circ H^{-1}) \cdot H = F \cdot H = F$. From here:

$$^{-1}H = ((H^{-1} \circ H^{-1}) \cdot H) \circ^{-1} H$$

and according to Lemma 15.1.13, $^{-1}H = (H^{-1})^*$ or $^{-1}H =^{-1} (H^*)$, and $H = H^*$. Hence, the operation $H = C^{-1}$ is a commutative quasigroup.

Let C be a second order quasigroup operation in the group $S \setminus \{E\}$. Then the operation C is contained in the centre of the group $S \setminus \{E\}$, since the element $X \circ C \circ X^{-1}$ also has the second order and therefore $X \circ C \circ X^{-1} = C$. Hence, $X \circ C = C \circ X$ for every $X \in S \setminus \{E\}$.

If $S_{V_A} \neq F$, then, as above, we obtain $A \in S_0$.

Let A be an operation such that $S_{V_A} = F$. Then $V_A \circ V_A = F$, and $V_A = C$, according to uniqueness of the operation C. On the other hand, we have:

$$F = V_A \circ V_A = V_A \circ A \circ T_A^* = A \circ V_A \circ T_A^* = A \circ A \circ T_A^* \circ T_A^*.$$

Hence,

$$A \circ A =^{-1} (T_A^*) \circ^{-1} (T_A^*) \in S_0.$$

If $A \neq C$, then $A \circ A \neq F$ and

$$A^{-1} = (A \circ A)^{-1} \cdot S_A \in S_0.$$

Thus $A^{-1} \in S_0$ and $A \in S_0$.

Now we only need to prove that the operation C is idempotent, too.

We can assume that $|S| \geq 5$, i.e., the bigroup S contains at least three quasigroup operations. Hence, there exists a quasigroup operation $C_1 \in S$, $C_1 \neq C$, such that the operation $C_2 = C_1 \circ C$ differs from C and C_1. Since C_1 and C_2 differ from C, then, as is proved above, they are idempotent. Therefore, $C =^{-1} C_1 \circ C_2 \in S_0$, i.e., the operation C is idempotent, too. □

For finite Q the result follows from [25].

15.3. Characterization of bigroups of operations by Grätzer algebras

Now we characterize a bigroup of operations (on the set Q) through special algebras introduced by G. Grätzer [140] for characterization of minimal doubly transitive groups of permutations. Such algebras are called Grätzer algebras. The group G of permutations of set Q is called minimally double transitive, if for every (a, b) and (c, d), where $a \neq b$ and $c \neq d$, there exists a unique permutation $\alpha \in G$ such that $\alpha(a) = c$ and $\alpha(b) = d$, where $a, b, c, d \in Q$ [148,165,171,174,196,229,230,404–406,418,425,430]. We use here the concept of Grätzer algebra (G-field) for characterization of $A(x, y)$, in which $x \neq y$ and A is a quasigroup operation in the bigroup of operations.

The algebra $Q(-, \cdot)$ with two binary operations is called a (left) Grätzer algebra or (left) G-field ([26,116,140]), if the following conditions are valid:

(G1) $Q(\cdot)$ is a semigroup with the zero element 0, i.e., $x \cdot 0 = 0 \cdot x = 0$ for any $x \in Q$;
(G2) $|Q| \geq 2$ and $Q'(\cdot)$ is a group, where $Q' = Q\backslash\{0\}$ (the identity element of this group is denoted by 1);
(G3) $x - 0 = x$ for every element $x \in Q$;
(G4) $x(y - z) = xy - xz$ for every elements $x, y, z \in Q$;
(G5) $x - (x - y) = y$, $x - (y - z) = (x - y) - (x - y)(y - x)^{-1}z$, $x \neq y$,

for every $x, y, z \in Q$.

Here a^{-1} is the inverse element of $a \neq 0$ in the group $Q'(\cdot)$. Note that $x - y = 0$ if and only if $x = y$. Indeed $x - x = x - (x - 0) = 0$. And conversely, from the equality $x - y = 0$ it follows:

$$y = x - (x - y) = x - 0 = x.$$

Besides, $0 \neq 1$ (since if $0 = 1$ then $x = x \cdot 1 = x \cdot 0 = 0$, which contradicts to the condition: $|Q| \geq 2$).

Let us denote $-x = 0 - x$. We obtain $-(-x) = x$ and $x(-y) = -(xy)$ for every $x, y \in Q$.

Proposition 15.3.1. *In the Grätzer algebra $Q(-, \cdot)$ the groupoid $Q(-)$ is a right quasigroup, i.e., the equation $a - x = b$ has a unique solution $x \in Q$ for every $a, b \in Q$.*

Proof. It is evident that $x = a - b$ is a solution of the equation $a - x = b$. Indeed, as according to the definition of Grätzer algebra we have $a - (a - b) = b$. If $a - x_1 = b$ and $a - x_2 = b$, then $a - x_1 = a - x_2$ and

$$x_1 = a - (a - x_1) = a - (a - x_2) = x_2.$$

□

If $Q(-)$ is a quasigroup, then the Grätzer algebra $Q(-, \cdot)$ is called a Grätzer q-algebra , and $Q(-)$ is called a Grätzer quasigroup.

The algebra $Q(+, \cdot)$ with two binary operations is called (left) near-field (see [147]), if the following conditions are valid:

(NF1) $Q(+)$ is a non-trivial abelian group with an identity element 0;

(NF2) $Q(\cdot)$ is a semigroup and $Q' = Q\backslash\{0\}$ is a group under multiplication $\cdot$ with the identity element $1 \in Q$;

(NF3) In $Q(+,\cdot)$ the identity of left distributivity is valid.

Near-fields were first considered in 1905 by Dickson [97]. A complete classification of finite near-fields was obtained in 1936 by Zassenhaus [424]. In particular, they have an order p^n, where p is a prime.

Example 15.3.2. (1) If we define the following multiplication on abelian group $Z_2(+)$:

$$0 \cdot 0 = 1 \cdot 0 = 0, 0 \cdot 1 = 1 \cdot 1 = 1,$$

we obtain a near-field, which is not a field. If a near-field contains at least 3 elements, then $x \cdot 0 = 0 \cdot x = 0$ for any x. Finite near-fields of order $3, 4, 5$ are fields.

(2) If the near-field $Q(+,\cdot)$ contains at least 3 elements, then $Q(-,\cdot)$ is a Grätzer q-algebra, where $x - y = x + (-y)$. The obtained Grätzer q-algebra $Q(-,\cdot)$ is called a derivative of the near-field $Q(+,\cdot)$. There exists a Grätzer algebra of order 3, which is not a Grätzer q-algebra. For example, if $Q = \{0, 1, 2\}$, $Q(\cdot)$ is a semigroup with the zero element 0, $Q' = Q\backslash\{0\}$ is a two-element group under multiplication $\cdot$, with identity element 1, and $1 - 2 = 2, 2 - 1 = 1, x - 0 = 0 - x = x, x - x = 0$, then $Q(-)$ is a right but not a left quasigroup and $Q(-,\cdot)$ is a Grätzer algebra. Consequently there exists a finite Grätzer algebra which is not a derivative of a finite near-field. However, the multiplicative groups of finite Grätzer algebras and finite near-fields are the same (see [269]).

Lemma 15.3.3. *Every bigroup S of binary operations (on the set Q) form a Grätzer algebra under the following operations*:

$$A \odot B = B \cdot A,$$

$$A - B = \begin{cases} (B \cdot A^{-1})^* \cdot A & \text{if } A \neq F, \\ B & \text{if } A = F, \end{cases}$$

where $A, B \in S$.

Proof. The axioms of (G1) and (G2) are evident. Let us prove the condition (G3): if $A = F$, then $A - F = F$. If $A \neq F$, then

$$A - F = (F \cdot A^{-1})^* \cdot A = F^* \cdot A = E \cdot A = A.$$

Let us check (G4):

$$A \odot (B - C) = (A \odot B) - (A \odot C).$$

If $A = F$, then the equation is valid. Let $A \neq F$. Then:

$$A \odot (B - C) = (B - C) \cdot A = \begin{cases} C \cdot A, & \text{if } B = F, \\ (C \cdot B^{-1})^* BA, & \text{if } B \neq F, \end{cases}$$

$$(A \odot B) - (A \odot C) = BA - CA$$

$$= \begin{cases} F - CA = CA & \text{if } B = F, \\ (CA(BA)^{-1})^* BA = (CB^{-1})^* BA & \text{if } B \neq F. \end{cases}$$

Now we can check the first identity of $G5$:

$$F - (F - C) = F - C = C.$$

If $A \neq F$, then

$$\begin{aligned} A - (A - C) &= A - (CA^{-1})^* A = ((CA^{-1})^* A \cdot A^{-1})^* A \\ &= ((CA^{-1})^*)^* A = CA^{-1} \cdot A = C \cdot A^{-1} A = C \cdot E = C. \end{aligned}$$

Now about the validity of the second identity of (G5). Denote:

$$L_1 = A - (B - C),$$
$$L_2 = (A - B) - (A - B) \odot (B - A)^{-1} \odot C,$$

and consider the following three cases:

(1) $A = F$ (and $B \neq F$),
(2) $B = F$ (and $A \neq F$),
(3) $A \neq F$, $B \neq F$ (and $A \neq B$).

In the first case, we have

$$L_1 = L_2 = B - C.$$

In the second case,

$$L_1 = L_2 = A - C.$$

For the third case, we find:

$$\begin{aligned} L_1 &= A - (CB^{-1})^* B = ((CB^{-1})^* BA^{-1})^* A, \\ L_2 &= (A - B) - C(B - A)^{-1}(A - B). \end{aligned}$$

Define:

$$T = (B - A)^{-1}(A - B).$$

We obtain:

$$\begin{aligned} T &= ((AB^{-1})^* B)^{-1}(BA^{-1})^* A = B^{-1}((AB^{-1})^*)^{-1}(BA^{-1})^* A \\ &= B^{-1}U(BA^{-1})^* A, \end{aligned}$$

where

$$U = ((AB^{-1})^*)^{-1}.$$

Using the equality $(X \circ Y)^* = X^* \cdot Y^*$, we find

$$\begin{aligned} L_2 &= (A - B) - CT = (BA^{-1})^* A - CT = (CT((BA^{-1})^* A)^{-1})^*(BA^{-1})^* A \\ &= (CB^{-1}U(BA^{-1})^* A((BA^{-1})^* \cdot A)^{-1})^*(BA^{-1})^* A \\ &= (CB^{-1}U)^*(BA^{-1})^* A = ((CB^{-1}U) \circ (BA^{-1}))^* A. \end{aligned}$$

On the other hand, if D is a quasigroup operation, then

$$D^* = ({}^{-1}(D^{-1}))^{-1}.$$

Further,

$$AB^{-1} = (U^{-1})^*, \quad BA^{-1} = ((U^{-1})^*)^{-1} =^{-1} U.$$

Therefore:

$$L_2 = ((CB^{-1} \cdot U) \circ^{-1} U)^* A.$$

However, according to Lemma 15.1.13:

$$W({}^{-1}V) \circ V = W^* V,$$

and by $W = CB^{-1}$, $V =^{-1} U$, we find

$$(CB^{-1} \cdot U) \circ^{-1} U = (CB^{-1})^*(^{-1}U).$$

Hence,

$$L_2 = ((CB^{-1})^*(^{-1}U))^*A = ((CB^{-1})^*BA^{-1})^*A = L_1.$$

□

We call the following result a local characterization of the bigroups of binary operations (see [260,268]).

Theorem 15.3.4. *Let S be a non-trivial bigroup of operations (on set Q). For every two different elements $a, b \in Q$ there exists a Grätzer algebra $H_{a,b}(-,\cdot)$, $H_{a,b} \subseteq Q$, such that $a, b \in H_{a,b}$ and for every $x, y \in H_{a,b}$, for every operation $C \in S$:*

$$C(x, y) = x - (x - y)c,$$

where $c \in H_{a,b}$. Moreover, the Grätzer algebras $H_{a,b}(-,\cdot)$ and $H_{c,d}(-,\cdot)$ are isomorphic for every $a, b, c, d \in Q$, where $a \neq b$ $c \neq d$.

Proof. Let us define:

$$H_{a,b} = \{A(a, b)|A \in S\} \subseteq Q.$$

From Proposition 15.1.4 it follows that $C(A, B) \in S$ for every $A, B, C \in S$, where

$$C(A, B)(u, v) = C(A(u, v), B(u, v)).$$

Therefore, the subset $H_{a,b}$ is a subalgebra of the algebra $(Q; S)$. It is evident, $F(a, b) = a \in H_{a,b}$ and $E(a, b) = b \in H_{a,b}$.

For every element $d \in H_{a,b}$ there exists a unique operation $A \in S$ with the equality: $d = A(a, b)$. Indeed, let $d \neq a$ and $d = B(a, b)$, where $B \in S$

and $B \neq A$. Since $B \neq F$, there exists B^{-1}. Consider the operation: $U = B^{-1} \cdot A \in S$. Then:

$$U(a,b) = B^{-1} \cdot A(a,b) = B^{-1}(a, A(a,b)) = B^{-1}(a,d) = b.$$

According to the previous theorem, the operation $U \in S$ is idempotent. We have:

$$\begin{cases} U(a,b) = b, \\ E(a,b) = b, \end{cases} \qquad \begin{cases} U(b,b) = b, \\ E(b,b) = b, \end{cases}$$

that is the system of equations

$$\begin{cases} U(x,y) = b, \\ E(x,y) = b, \end{cases}$$

has two solutions: (a,b) and (b,b), which contradicts the orthogonality of operations U and E. Now we have to consider the following case only: $d = a$. In this case $F(a,b) = a$ and if $A(a,b) = a$, then we obtain a contradiction, according to idempotency of the operation $A \in S$.

Thus, there exists a bijective mapping φ between the sets S and $H_{a,b}$ for every $a, b \in Q$, where $a \neq b$. Namely: $\varphi(A) = x$, where $x = A(a,b)$.

Let us define the following two operations on the set $H_{a,b}$:

$$x \cdot y = B \cdot A(a,b),$$

$$x - y = \begin{cases} (B \cdot A^{-1})^* A(a,b) & \text{if } x \neq a, \\ y & \text{if } x = a, \end{cases}$$

where $x = A(a,b)$, $y = B(a,b)$, i.e., $x = \varphi A$, $y = \varphi B$. According to Lemma 15.3.3, $S(-, \odot)$ is a bigroup. Moreover,

$$\varphi(A \odot B) = \varphi(B \cdot A) = x \cdot y = \varphi(A) \cdot \varphi(B),$$

$$\varphi(A - B) = x - y = \varphi(A) - \varphi(B),$$

that is the mapping $\varphi : S \to H_{a,b}$ is an isomorphism from Grätzer algebra $S(-, \odot)$ to the algebra $H_{a,b}(-, \cdot)$. Hence, the algebra $H_{a,b}(-, \cdot)$ is also a Grätzer algebra.

Evidently, $a = \varphi(F)$, $b = \varphi(E)$. If $d = \varphi(A)$, then $d^{-1} = \varphi(A^{-1})$ $b - d = \varphi(A^*)$:

$$b - d = (A \cdot E^{-1})^* \cdot E(a, b) = A^*(a, b).$$

Now we have only to prove that:

$$C(x, y) = x - (x - y)c,$$

where $x = A(a, b)$, $y = B(a, b)$, $c = C(a, b)$, $C, B, A \in S$. We have

$$C(x, y) = C(A(a, b), B(a, b)) = C(A, B)(a, b).$$

If $x = a$, then

$$C(a, y) = C(a, B(a, b)) = C \cdot B(a, b) = y \cdot c,$$

i.e.,

$$C(a, y) = y \cdot c = a - (a - y)c.$$

If $x \neq a$, then

$$x - y = (B \cdot A^{-1})^* A(a, b),$$
$$(x - y)c = C(B \cdot A^{-1})^* A(a, b),$$

i.e.,

$$\begin{aligned} x - (x - y)c &= (C(B \cdot A^{-1})^* A \cdot A^{-1})^* A(a, b) \\ &= (C(BA^{-1})^*)^* A(a, b) \\ &= (C^* \circ (BA^{-1}))A(a, b) = C(A, B)(a, b), \end{aligned}$$

according to Proposition 15.1.4. □

Problem 15.3.5. *For which bigroups of operations is Theorem 15.3.4 valid for Grätzer q-algebras?*

Problem 15.3.6. *For which bigroups of operations is Theorem 15.3.4 valid for near-fields?*

Problem 15.3.7. *For which bigroups of operations is Theorem 15.3.4 valid for fields?*

Problem 15.3.8. *Characterize the De Morgan bisemigroup of idempotent operations on the set Q.*

Problem 15.3.9. *Which loops are Grätzer quasigroups isotopic?*

Problem 15.3.10. *Develop the bigroup analogue of the group theory.*

15.4. On Steiner, Stein, and Belousov quasigroups

We now apply the results of the previous sections to quasigroups.

The quasigroup $Q(\circ)$ is called TS-quasigroup (see [27,51,319]) if the identities

$$x \circ (x \circ y) = y,$$

$$x \circ y = y \circ x$$

are valid. In general case, a TS-quasigroup is not idempotent.

An idempotent TS-quasigroup is called a Steiner quasigroup ([27,51,65, 173,319,338,389]). In a non-trivial Steiner quasigroup $Q(\circ)$, for any $a \neq b$ in Q the set $\{a, b, a \circ b\}$ is a three-element subquasigroup, which is isomorphic to a three element quasigroup with the following multiplication table:

	0	1	2
0	0	2	1
1	2	1	0
2	1	0	2

The variety of all Steiner quasigroups has permutable, uniform and regular congruences. Any finite simple Steiner quasigroup with at least four elements is functionally complete (see [338]).

Lemma 15.4.1. *If $Q(A)$ is a TS-quasigroup, then $A \cdot A = \delta_2^2$ and $A \circ A = \delta_2^1$. So if $Q(A)$ is a non-trivial TS-quasigroup, then the set $\{\delta_2^1, \delta_2^2, A\}$ is a bigroup of operations.*

Corollary 15.4.2. *If $Q(\circ)$ is a non trivial Steiner quasigroup, then for every $u, v \in Q, u \neq v$ there exists a three-element field $H_{u,v}(+, \cdot), H_{u,v} \subseteq Q$ such that $u, v \in H_{u,v}$ and for every $x, y \in H_{u,v}$:*

$$x \circ y = (y - x)a + x, a \in H_{u,v}.$$

Example 15.4.3. Let $Q = \{0, 1, 2\}$. For the abelian group $Q(+)$ of order 3 and for the idempotent TS-quasigroup operation $A(x, y) = x + x + y + y$ on Q we have $A(x, y) = (y - x)2 + x$, where $Q(+, \cdot)$ is a field of order 3. Here $H = Q$.

The quasigroup $Q(\circ)$ is called a Stein quasigroup (see [25,65,388]) if the identities

$$x \circ (x \circ y) = y \circ x,$$

$$x \circ (y \circ x) = y$$

are valid. Every Stein quasigroup is idempotent. A non-trivial Stein quasigroup is not commutative. In a non-trivial Stein quasigroup $Q(\circ)$, for any $a \neq b$ in Q, the set $\{a, b, a \circ b, b \circ a\}$ is a four-element subquasigroup, which is isomorphic to the four-element quasigroup with the following multiplication table:

	0	1	2	3
0	0	2	3	1
1	3	1	0	2
2	1	3	2	0
3	2	0	1	3

Lemma 15.4.4. *If $Q(A)$ is a Stein quasigroup, then $A \cdot A = A \circ A = A^* = A^{-1} = {}^{-1}A$ and $A^* \cdot A^* = A = A^* \circ A^*$. So if $Q(A)$ is a non-trivial Stein quasigroup, then the set $\{\delta_2^1, \delta_2^2, A, A^*\}$ is a bigroup of operations.*

Corollary 15.4.5. *If $Q(\circ)$ is a non-trivial Stein quasigroup, then for every $u, v \in Q, u \neq v$ there exists a four-element field $H_{u,v}(+, \cdot), H_{u,v} \subseteq Q$ such that $u, v \in H_{u,v}$ and for every $x, y \in H_{u,v}$:*

$$x \circ y = (y - x)a + x, a \in H_{u,v};$$

In order to move to the next step we present the concept of the following variety of quasigroups.

The quasigroup $Q(\circ)$ is called a Belousov quasigroup, if the identities

$$x \circ (x \circ y) = y \circ x,$$

$$(x \circ y) \circ y = x,$$

$$x \circ (y \circ x) = (y \circ x) \circ y$$

are valid. A non-trivial Belousov quasigroup is not a Stein quasigroup and not commutative.

Lemma 15.4.6. *If $Q(A)$ is a non-trivial Belousov quasigroup, then it is idempotent and $A \cdot A = A^*$, $A \cdot A^* = A \circ A^*$, $A \circ A = \delta_2^1$, $A^* \cdot A^* = \delta_2^2, A^* \circ A^* = A$. So if $Q(A)$ is a non-trivial Belousov quasigroup, then the set $\{\delta_2^1, \delta_2^2, A, A^*, A \cdot A^* = A \circ A^*\}$ is a bigroup of operations.*

Lemma 15.4.7. *In every Belousov quasigroup $Q(\circ)$ the identities $(x \circ y) \circ (y \circ x) = y, (x \circ y) \circ (x \circ (y \circ x)) = y \circ x, (y \circ x) \circ (x \circ (y \circ x)) = x \circ y$ are valid. In a non-trivial Belousov quasigroup $Q(\circ)$, for any $a \neq b$ in Q the set $\{a, b, a \circ b, b \circ a, a \circ (b \circ a)\}$ is a five-element subquasigroup, which is isomorphic to the five-element quasigroup with the following multiplication table:*

	0	1	2	3	4
0	0	2	4	1	3
1	4	1	3	0	2
2	3	0	2	4	1
3	2	4	1	3	0
4	1	3	0	2	4

If we take such subquasigroups as blocks, we obtain a block design on Q.

Corollary 15.4.8. *If $Q(\circ)$ is a non trivial Belousov quasigroup, then for every $u, v \in Q, u \neq v$ there exists a five-element field $H_{u,v}(+,\cdot), H_{u,v} \subseteq Q$ such that $u, v \in H_{u,v}$ and for every $x, y \in H_{u,v}$:*

$$x \circ y = (y - x)a + x, a \in H_{u,v};$$

It follows from Corollary 15.4.8 (or Lemma 15.4.7) that the non-trivial Belousov quasigroup has at least five elements. The variety of Belousov quasigroups is called a Belousov variety, which is a subvariety of the Mikado variety (see [116]). Hence, the Belousov variety has a solvable word problem and is congruence-permutable. Every Belousov quasigroup of prime order is a simple algebra .

The applications of similar quasigroups in cellular automata see in [242].

Problem 15.4.9. *Which loops are Belousov quasigroups isotopic?*

Chapter 16

Abelian Algebras

16.1. Introduction

The n-ary operation A on the set Q is called idempotent if it satisfies the identity $A(x,\dots,x) = x$. An algebra $(Q;\Sigma)$ is called idempotent if every operation $A \in \Sigma$ is idempotent.

Let A be n-ary and B be m-ary operations on the set Q. The pair (A,B) is called abelian (possibly medial, entropic, bisymmetric, bicommutative, etc.) if it satisfies the identity:

$$\begin{aligned} A\,(B\,(x_{11},\dots,x_{1m}),\,\dots,B\,(x_{n1},\dots,x_{nm})) \\ = B\,(A\,(x_{11},\dots,x_{n1}),\dots,A\,(x_{1m},\dots,x_{nm})). \end{aligned}$$

In the case of the binary operations A, B we have the identity:

$$A\,(B\,(x,y),\,B\,(u,v)) = B\,(A\,(x,u),\,A\,(y,v)).$$

An operation A on the set Q is called abelian (medial) if the pair (A,A) is abelian. In the case of one binary operation we have [7,8]:

$$A\,(A\,(x,y),\,A\,(u,v)) = A\,(A\,(x,u),\,A\,(y,v)).$$

An algebra $(Q;\Sigma)$ is called abelian (medial) if the pair (A,B) is abelian for every $A,B \in \Sigma$ [106,190] (for algebras without nullary operations). In other words, an algebra $(Q;\Sigma)$ is called abelian (medial) if $(Q;\Sigma)$ satisfies the hyperidentity of abelity (mediality):

$$\begin{aligned} X\,(Y\,(x_{11},\dots,x_{1m}),\,\dots,Y\,(x_{n1},\dots,x_{nm})) \\ = Y\,(X\,(x_{11},\dots,x_{n1}),\dots,X\,(x_{1m},\dots,x_{nm})) \end{aligned}$$

for any $m,n \in T_{(Q;\Sigma)}$.

An idempotent and abelian algebra is called a mode [345,346].

The set of all binary operations defined on the set Q is denoted by $\mathcal{F}_Q^2$ in this chapter, and we consider the following two operations (binary superpositions) on this set introduced in previous chapters:

$$A \cdot B(x,y) = A(x, B(x,y)),$$
$$A \circ B(x,y) = A(B(x,y, y),$$

where $A, B \in \mathcal{F}_Q^2$, $x, y \in Q$. These operations $(\cdot)$ and $(\circ)$ are called the right and left multiplications of binary operations (see Chapter 15).

The set $\mathcal{F}_Q^2$ forms a monoid under the right (and left) multiplication of binary operations. These two semigroups are isomorphic. The identity element of the semigroup $\mathcal{F}_Q^2(\cdot)$ is $E \in \mathcal{F}_Q^2$, and it is defined by the rule: $E(x,y) = y$; and the identity element of the semigroup $\mathcal{F}_Q^2(\circ)$ is $F \in \mathcal{F}_Q^2$, and it is defined by the rule: $F(x,y) = x$. The mapping $A \to A^*$ is the isomorphism of these two semigroups, where $A^*(x,y) = A(y,x)$, according to the following De Morgan identities:

$$(A^*)^* = f,$$
$$(A \cdot B)^* = A^* \circ B^*,$$
$$(A \circ B)^* = A^* \cdot B^*.$$

The set of idempotent binary operations on Q is a subsemigroup in the semigroups $\mathcal{F}_Q^2(\cdot)$ and $\mathcal{F}_Q^2(\circ)$.

The binary operation $A \in \mathcal{F}_Q^2$ is the right (left) invertible one if the equation $A(a,x) = b$ ($A(y,a) = b$) has a unique solution $x \in Q$ ($y \in Q$) for every $a, b \in Q$. The unique solutions $x, y \in Q$ are usually denoted by $x = A^{-1}(a,b)$ and $y =^{-1} A(b,a)$. Hence,

$$A \cdot A^{-1} = A^{-1} \cdot A = E,$$

for the right invertible operation A, and we have

$$^{-1}A \circ A = A \circ^{-1} A = F,$$

for the left invertible operation A. The operation A^{-1} (or ^{-1}A) is the right (or left) inverse one for the right (left) invertible operation $A \in \mathcal{F}_Q^2$. It is evident that A^{-1} (or ${}^{-1}A$) is right (or left) invertible, and

$$\left(A^{-1}\right)^{-1} = A =^{-1} \left({}^{-1}A\right), \quad \left(A^{-1}\right)^{*} =^{-1} (A^{*}), \quad \left({}^{-1}A\right)^{*} = (A^{*})^{-1}.$$

The binary operation $A \in \mathcal{F}_Q^2$ is invertible if it is right and left invertible. In this case,

$$\left({}^{-1}(A^{-1})\right)^{-1} =^{-1} \left(({}^{-1}A)^{-1}\right) = A^{*}.$$

For example,

$$\left({}^{-1}\left(A^{-1}\right)\right)^{-1}(x,y) = z \leftrightarrow^{-1} \left(A^{-1}\right)(x,z) = y \leftrightarrow A^{-1}(y,z) = x$$
$$\leftrightarrow A(y,x) = z \leftrightarrow A^{*}(x,y) = z.$$

On the applications of right (left) invertible operations in axiomatic characterization of geometric structures and equivalency problem in knot theory see [228,297,312].

The set of all right (left) binary invertible operations on the set Q is denoted by $\mathcal{F}_Q^r$ (and $\mathcal{F}_Q^\ell$).

The set $\mathcal{F}_Q^r$ is a group under the right multiplication of binary operations. The set $\mathcal{F}_Q^\ell$ is a group under the left multiplication of binary operations. These two groups are isomorphic according to the above De Morgan identities.

The groups $\mathcal{F}_Q^r(\cdot)$ and $\mathcal{F}_Q^\ell(\circ)$ are called the right and left groups of binary operations, respectively.

The binary algebra $(Q;\Sigma)$ is called (right, left) invertible if every operation $A \in \Sigma$ is (right, left) invertible.

For any right (left) invertible algebra $(Q;\Sigma)$ we define the algebra $(Q;\Sigma^{-1})$ $((Q;^{-1}\Sigma))$, where

$$\Sigma^{-1} = \{A^{-1} | \, A \in \Sigma\},$$
$${}^{-1}\Sigma = \{{}^{-1}A | \, A \in \Sigma\}.$$

So, for any invertible algebra $(Q;\Sigma)$ we have the algebras: $(Q;\Sigma^{-1})$, $(Q;^{-1}\Sigma)$, $(Q;^{-1}(\Sigma^{-1}))$, $\left(Q;({}^{-1}\Sigma)^{-1}\right)$ and $(Q;\Sigma^{*})$, where

$${}^{-1}\left(\Sigma^{-1}\right) = \left\{{}^{-1}\left(A^{-1}\right) | \, A \in \Sigma\right\},$$
$$\left({}^{-1}\Sigma\right)^{-1} = \left\{\left({}^{-1}A\right)^{-1} | \, A \in \Sigma\right\},$$
$$\Sigma^{*} = \{A^{*} | \, A \in \Sigma\}.$$

However, the algebra $(Q; \Sigma^*)$ is meaningful for every algebra $(Q; \Sigma)$.

In this chapter, the connection between the right (left) invertible abelian algebras and the right (left) group of binary operations is studied (cf. [251, 257]).

16.2. Preliminary concepts and results

Let the operation A be defined on the set Q and let the operation B be defined on the set Q'.

The (binary) operations A and B are called isotopic [12] if there exist tree bijections $\alpha, \beta, \gamma : Q \to Q'$ such that the following equality is valid:

$$\alpha A(x, y) = B(\beta x, \gamma y),$$

or

$$A(x, y) = \alpha^{-1} B(\beta x, \gamma y)$$

for every $x, y \in A$, and the triple, $T = (\alpha, \beta, \gamma)$, is called isotopy and is denoted by $(\alpha, \beta, \gamma) : A \to B$. We denoted $A \sim B$ if A and B are isotopic, i.e., if there exists an isotopy $(\alpha, \beta, \gamma) : A \to B$. It is easy to prove the following tree conditions of equivalence relation:

$$A \sim A,$$
$$A \sim B \longrightarrow B \sim A,$$
$$A \sim B,\ B \sim C \longrightarrow A \sim C.$$

If $T_1 = (\alpha_1, \beta_1, \gamma_1)$ and $T_2 = (\alpha_2, \beta_2, \gamma_2)$ be isotopies, then the product $T_1 \cdot T_2 = (\alpha_1 \cdot \alpha_2, \beta_1 \cdot \beta_2, \gamma_1 \cdot \gamma_2)$ is also an isotopy:

$$\begin{aligned}\alpha_1 \cdot \alpha_2 A(x, y) &= \alpha_2(\alpha_1 A(x, y)) \\ &= \alpha_2(B(\beta_1 x, \gamma_1 y)) = C(\beta_2(\beta_1 x), \gamma_2(\gamma_1 y)) = C((\beta_1 \cdot \beta_2)x, (\gamma_1 \cdot \gamma_2)y).\end{aligned}$$

If the operations A and B are isotopic then the groupoids $Q(A)$ and $Q'(B)$ also are called isotopic. So, we have the definition of isotopic semigroups, quasigroups and groups. The definition of isotopic rings is similar to this [189]. On isotopic algebras see in monographs [249,252].

An isotopy is said to be principal if its third component is the identity map. Note that any isotopy (α, β, γ) is factorized as a product of the principal isotopy with an isomorphism:

$$(\alpha,\beta,\gamma) = (\alpha\gamma^{-1},\beta\gamma^{-1},\varepsilon)(\gamma,\gamma,\gamma) = (\gamma,\gamma,\gamma)(\gamma^{-1}\alpha,\gamma^{-1}\beta,\varepsilon),$$

where ε is the identity map.

Any groupoid $Q(A)$ isotopic to the quasigroup $Q(B)$ is also a quasigroup.

Proposition 16.2.1 ([12]). *Let $Q(\cdot)$ be a quasigroup with the element $e \in Q$. Then $Q(\cdot)$ is the principal isotopic to a loop with the identity element e.*

Proposition 16.2.2 ([12,27,52,189]). *If the loop $Q(\cdot)$ is isotopic to the group $Q'(\circ)$, then they are isomorphic. In particular, isotopic groups are isomorphic.*

Proposition 16.2.3 ([52,189]). *If the groupoid $Q(\cdot)$ with the identity element is isotopic to the semigroup $Q'(\circ)$, then they are isomorphic.*

Proposition 16.2.4 ([249,252]). *If every operation of the algebra $(Q;\Sigma)$ has an identity element and $(Q;\Sigma)$ is isotopic to the algebra with semigroup operations, then these algebras are isomorphic.*

The following result is known as the Belousov theorem for four quasigroup operations.

An algebra $(Q;\Sigma)$ is called isotopic to the groupoid $Q(\cdot)$ if $Q(A)$ isotopic to the $Q(\cdot)$ for any $A \in \Sigma$ [269].

Theorem 16.2.5 ([24,25]). *If the four quasigroup operations A_1, A_2, A_3, A_4 (defined on the same non-empty set Q) satisfy the following general associativity equation:*

$$A_1(A_2(x,y),z) = A_3(x,A_4(y,z)), \tag{16.2.1}$$

then the binary invertible algebra $(Q;A_1,A_2,A_3,A_4)$ is isotopic to a certain group $Q(\cdot)$. Moreover, there exist a group $Q(\cdot)$ and bijections $\alpha,\beta,\gamma,\delta,\varphi : Q \to Q$ such that

$$\begin{aligned}
A_1(x,y) &= \alpha x \cdot \beta y,\\
A_2(x,y) &= \alpha^{-1}(\gamma x \cdot \delta y),\\
A_3(x,y) &= \gamma x \cdot \varphi y,\\
A_4(x,y) &= \varphi^{-1}(\delta x \cdot \beta y)
\end{aligned}$$

for every $x, y \in Q$.

The group $Q(\cdot)$ is unique up to isomorphism.

Proof. Let $a \in Q$ and

$$L_i(x) = A_i(a, x), \quad R_i(x) = A_i(x, a).$$

If we put $x = z = a$ in (16.2.1), we have

$$A_1(A_2(a, y), a) = A_3(a, A_4(y, a)),$$

i.e., $R_1(L_2 y) = L_3(R_4 y)$ and $L_2 R_1 = R_4 L_3$. If we put $x = a$, $y = a$, $z = a$ in (16.2.1) accordingly we get

$$A_1(L_2 y, z) = L_3 A_4(y, z),$$
$$A_1(R_2 x, z) = A_3(x, L_4 z),$$
$$R_1 A_2(x, y) = A_3(x, R_4 y).$$

Let us consider the following isotopy of the operation A_1:

$$x \cdot y = A_1\left(R_1^{-1} x, L_4^{-1}\left(L_3^{-1} y\right)\right).$$

From the obtained equalities it follows that every operation A_i is isotopic to A_1 and the operation A_1 is isotopic to the operation $(\cdot)$. Thus,

$$A_1(x, y) = R_1 x \cdot (L_4 L_3) y,$$
$$A_2(x, y) = R_1^{-1}\left((R_2 R_1) x \cdot (R_4 L_3) y\right),$$
$$A_3(x, y) = (R_2 R_1) x \cdot L_3 y,$$
$$A_4(x, y) = L_3^{-1}\left((L_2 R_1) x \cdot (L_4 L_3) y\right).$$

Now by the substitutions $A_i(x, y)$ in equation (16.2.1) we obtain

$$((R_2 R_1) x \cdot (R_4 L_3) y) \cdot (L_4 L_3) z = (R_2 R_1) x \cdot ((L_2 R_1) y \cdot (L_4 L_3) z).$$

Denoting $(R_2 R_1) x = u$, $(L_4 L_3) z = w$ and $(R_4 L_3) y = (L_2 R_1) y = v$, we get

$$(u \cdot v) \cdot w = u \cdot (v \cdot w).$$

So $Q(\cdot)$ is a group. □

Theorem 16.2.6 ([7,25]). *If the six quasigroup operations A_1, A_2, A_3, A_4, A_5, A_6 (defined on the same non-empty set Q) satisfy the general mediality equation:*

$$A_1(A_2(x, y), A_3(u, v)) = A_4(A_5(x, u), A_6(y, v)), \tag{16.2.2}$$

then the binary invertible algebra $(Q; A_1, A_2, A_3, A_4, A_5, A_6)$ *is isotopic to a certain abelian group* $Q(\circ)$. *Moreover, there exist an abelian group* $Q(\circ)$ *and bijections* $\alpha, \beta, \gamma, \delta, \varphi, \psi, \sigma, \lambda : Q \to Q$ *such that*

$$\begin{aligned}
A_1(x, y) &= \alpha x \circ \varphi y, \\
A_2(x, y) &= \alpha^{-1}(\gamma x \circ \delta y), \\
A_3(x, y) &= \varphi^{-1}(\lambda x \circ \beta y), \\
A_4(x, y) &= \psi x \circ \sigma y, \\
A_5(x, y) &= \psi^{-1}(\gamma x \circ \lambda y), \\
A_6(x, y) &= \sigma^{-1}(\delta x \circ \beta y),
\end{aligned}$$

for every $x, y \in Q$.

The abelian group $Q(\circ)$ *is unique up to isomorphism.*

Proof. If we put $u = a \in Q$ in (16.2.2) we have an equation of the form (16.2.1):

$$A_1'(A_2(x, y), v) = A_4'(a, A_6(y, v)),$$

where

$$\begin{aligned}
A_1'(x, y) &= A_1(x, A_3(a, y)) = A_1(x, L_3(y)), \\
A_4'(x, y) &= A_4(A_5(x, a), y) = A_4(R_5(x), y),
\end{aligned}$$

and $L_i(x) = A_i(a, x)$, $R_i(x) = A_i(x, a)$, $i = 1, 2, 3, 4, 5, 6$.

According to Theorem 16.2.5 there exist a group $Q(\circ)$ and bijections $\alpha, \beta, \gamma, \delta, \sigma : Q \to Q$ such that

$$\begin{aligned}
A_1'(x, y) &= \alpha x \circ \beta y, \\
A_2(x, y) &= \alpha^{-1}(\gamma x \circ \delta y), \\
A_4'(x, y) &= \gamma x \circ \sigma y, \\
A_6(x, y) &= \sigma^{-1}(\delta x \circ \beta y).
\end{aligned}$$

Thus,

$$\begin{aligned}
A_1(x, y) &= A_1'\left(x, L_3^{-1}(y)\right) \\
&= \alpha x \circ \beta\left(L_3^{-1} y\right) = \alpha x \circ \varphi y, \\
A_4(x, y) &= A_4'\left(R_5^{-1}(x), y\right) \\
&= \gamma\left(R_5^{-1} x\right) \circ \sigma y = \psi x \circ \sigma y,
\end{aligned}$$

where

$$\psi x = \gamma\left(R_5^{-1}x)\right) = \left(R_5^{-1}\gamma\right)x,$$
$$\varphi y = \beta\left(L_3^{-1}y\right) = \left(L_3^{-1}\beta\right)y,$$

So, $\psi = R_5^{-1}\gamma$ and $\varphi = L_3^{-1}\beta$. Hence, φ and ψ also are bijections.

Now by the substitutions $A_i(x, y)$ in Eq. (16.2.2) we obtain

$$\gamma x \circ \delta y \circ \varphi A_3(u, v) = \psi A_5(x, u) \circ \delta y \circ \beta v.$$

From this equality by $x = y = a$ it follows that

$$A_3(u, v) = \varphi^{-1}(\lambda u \circ \beta v),$$

where $\lambda : Q \to Q$ also is a bijection. Hence, from Eq. (16.2.2) we have

$$\gamma x \circ \delta y \circ \lambda u \circ \beta v = \psi A_5(x, u) \circ \delta y \circ \beta v.$$

Now if we put $y = \delta^{-1}(e)$ (e is the unit of the group $Q(\circ)$, then

$$A_5(x, y) = \psi^{-1}(\gamma x \circ \lambda u).$$

The substitution $A_5(x, y)$ in the previous equality gives the abelity of the group $Q(\circ)$:

$$\delta y \circ \lambda u = \lambda u \circ \delta y.$$

□

The bijection $\alpha : Q \to Q$ is called autotopy [52] for the operation $A : Q^2 \to Q$ if there exist bijections $\beta, \gamma : Q \to Q$ such that

$$\alpha A(x, y) = A(\beta x, \gamma y),$$

for every $x, y \in Q$, i.e., if there exists the isotopy $(\alpha, \beta, \gamma) : A \to A$. The set of all autotopies of the operation A is a group under superposition of maps: $(\alpha \cdot \alpha')x = \alpha'(\alpha x)$.

It is evident that any automorphism for the groupoid $Q(A)$ is an autotopy for the operation A. If $Q(\cdot)$ is a group, then the left and right translations L_a and R_a are autotopies for $(\cdot)$. Indeed

$$R_a(x) = a \cdot x,$$
$$L_a(x) = x \cdot a,$$

hence,

$$L_a(x \cdot y) = L_a(x) \cdot y,$$

and

$$R_a(x \cdot y) = x \cdot R_a(y).$$

Definition 16.2.7. The bijection $\alpha : Q \to Q$ is called *holomorphism* for the group $Q(\cdot)$ (see [24,202,249]) if

$$\alpha(x \cdot y^{-1} \cdot z) = \alpha x \cdot (\alpha y)^{-1} \cdot \alpha z, \tag{16.2.3}$$

for every $x, y, z \in Q$.

Substituting here $y = e$ (where e is the identity element of $Q(\cdot)$) we obtain:

$$\alpha(x \cdot z) = \alpha x \cdot (\alpha e)^{-1} \cdot \alpha z, \tag{16.2.4}$$

for every $x, z \in Q$. The converse also is valid, i.e., the bijection $\alpha : Q \to Q$ with property (16.2.4) is a holomorphism for the group $Q(\cdot)$.

Indeed, from equality (16.2.4) it follows:

$$\alpha(x^{-1}) = \alpha e \cdot (\alpha x)^{-1} \cdot \alpha e$$

and

$$\begin{aligned}
\alpha(x \cdot y^{-1} \cdot z) &= \alpha\left(x \cdot (y^{-1} \cdot z)\right) \\
&= \alpha x \cdot (\alpha e)^{-1} \cdot \alpha(y^{-1} \cdot z) \\
&= \alpha x \cdot (\alpha e)^{-1} \cdot \alpha(y^{-1}) \cdot (\alpha e)^{-1} \cdot \alpha z \\
&= \alpha x \cdot (\alpha e)^{-1} \cdot \alpha e \cdot (\alpha y)^{-1} \cdot \alpha e \cdot (\alpha e)^{-1} \cdot \alpha z \\
&= \alpha x \cdot (\alpha y)^{-1} \cdot \alpha z.
\end{aligned}$$

So α is the holomorphism for the group $Q(\cdot)$.

The set of all holomorphisms of the group $Q(\cdot)$ is a group under the superposition of mappings. This group is called holomorph of $Q(\cdot)$ and is denoted by $\mathrm{Hol}\, Q(\cdot)$ [170]. For example, if $\alpha, \beta \in \mathrm{Hol}\, Q(\cdot)$ then

$$\begin{aligned}
\alpha \cdot \beta(x \cdot y) &= \beta(\alpha\,(x \cdot y)) \\
&= \beta\left(\alpha x \cdot (\alpha e)^{-1} \cdot \alpha y\right) \\
&= \beta(\alpha x) \cdot (\beta e)^{-1} \cdot \beta\left(\alpha e\right)^{-1} \cdot \alpha y) \\
&= \beta(\alpha x) \cdot (\beta e)^{-1} \cdot \beta\left((\alpha e)^{-1}\right) \cdot (\beta e)^{-1} \cdot \beta(\alpha y),
\end{aligned}$$

and for $x = y = e$ we have:

$$\begin{aligned}(\alpha \cdot \beta)e &= \beta(\alpha e) \cdot (\beta e)^{-1} \cdot \beta\left((\alpha e)^{-1}\right) \cdot (\beta e)^{-1} \cdot \beta(\alpha e),\\ \beta\left(\alpha e\right) &= \beta e\left(\beta\left((\alpha e)^{-1}\right)\right)^{-1} \cdot \beta e,\\ \left(\beta\left(\alpha e\right)\right)^{-1} &= (\beta e)^{-1} \cdot \beta\left(\left((\alpha e)^{-1}\right)\right)^{-1} \cdot (\beta e)^{-1},\end{aligned}$$

hence,

$$\alpha \cdot \beta(x \cdot y) = (\alpha \cdot \beta)x \cdot ((\alpha \cdot \beta)e)^{-1} \cdot (\alpha \cdot \beta)y.$$

Let us consider the following simple properties for holomorphisms of groups.

Proposition 16.2.8. *Any automorphism of the group $Q(\cdot)$ is its holomorphism. The holomorphism α is an automorphism if and only if $\alpha(e) = e$.*

Proposition 16.2.9. *If $\alpha : Q \to Q$ is an automorphism of the group $Q(\cdot)$ and $k \in Q$, then the bijection*

$$\beta(x) = \alpha(x) \cdot k$$

is a holomorphism of $Q(\cdot)$.

Proof. Indeed, $\beta(e) = \alpha(e) \cdot k = e \cdot k = k$ and:

$$\begin{aligned}\beta(x \cdot y) &= \alpha(x \cdot y) \cdot k = \alpha x \cdot \alpha y \cdot k = (\alpha x \cdot k) \cdot k^{-1} \cdot (\alpha y \cdot k)\\ &= \beta(x) \cdot (\beta e)^{-1} \cdot \beta(y).\end{aligned}$$

□

Proposition 16.2.10. *If $\alpha : Q \to Q$ is an automorphism of the group $Q(\cdot)$ and $k \in Q$, then the bijection*

$$\gamma(x) = k \cdot \alpha(x)$$

is a holomorphism of $Q(\cdot)$.

Proposition 16.2.11. *Every holomorphism $\beta : Q \to Q$ of the group $Q(\cdot)$ has the representation:*

$$\beta(x) = \alpha(x) \cdot k,$$

where $\alpha \in \operatorname{Aut} Q(\cdot)$ and $k \in Q$.

Proof. Indeed, let $\beta(e) = k$ and let $\alpha(x) = \beta(x) \cdot k^{-1}$. Let us prove that $\alpha \in \operatorname{Aut} Q(\cdot)$:

$$\begin{aligned}\alpha(x \cdot y) &= \beta(x \cdot y) \cdot k^{-1} \\ &= \beta(x) \cdot (\beta e)^{-1} \cdot \beta(y) \cdot k^{-1} \\ &= \big(\beta(x) \cdot k^{-1}\big) \cdot \big(\beta(y) \cdot k^{-1}\big) \\ &= \alpha(x) \cdot \alpha(y).\end{aligned}$$

So, $\beta(x) = \alpha(x) \cdot k$, where $\alpha \in \operatorname{Aut} Q(\cdot)$ and $k \in Q$. □

Proposition 16.2.12. *Every holomorphism $\gamma : Q \to Q$ of the group $Q(\cdot)$ has the representation:*

$$\gamma(x) = k \cdot \alpha(x),$$

where $\alpha \in \operatorname{Aut} Q(\cdot)$ and $k \in Q$.

Proposition 16.2.13. *If $\alpha : Q \to Q$ is a holomorphism of the group $Q(\cdot)$ and $k \in Q$, then the bijections*

$$\beta(x) = \alpha(x) \cdot k$$

and

$$\gamma(x) = k \cdot \alpha(x)$$

are also holomorphisms of the group $Q(\cdot)$.

Proof. Indeed, $\alpha(x) = \alpha_0(x) \cdot t$, where $\alpha_0 \in \operatorname{Aut} Q(\cdot)$ and $t \in Q$, by Proposition 16.2.11. Hence,

$$\beta(x) = \alpha(x) \cdot k = (\alpha_0(x) \cdot t) \cdot k = \alpha_0(x) \cdot (t \cdot k) = \alpha_0(x) \cdot \ell$$

and β is a holomorphism of $Q(\cdot)$ by Proposition 16.2.9. □

Proposition 16.2.14. *If $\alpha : Q \to Q$ is a holomorphism of the group $Q(\cdot)$, then the bijections $\alpha_1(x) = \alpha(x) \cdot (\alpha e)^{-1}$ and $\alpha_2(x) = (\alpha e)^{-1} \cdot \alpha(x)$ are automorphisms of the group $Q(\cdot)$.*

Proof. We have: $\alpha_1, \alpha_2 \in \operatorname{Hol} Q(\cdot)$ by Proposition 16.2.13 and $\alpha_1(e) = e$, $\alpha_2(e) = e$. Now the assertion follows from Proposition 16.2.8. □

Proposition 16.2.15. *Every autotopy $\alpha : Q \to Q$ of the group $Q(\cdot)$ is its holomorphism.*

Proof. Indeed, there exist the bijections $\beta : Q \to Q$ and $\gamma : Q \to Q$ such that

$$\alpha(x \cdot y) = \beta(x) \cdot \gamma(y)$$

for every $x, y \in Q$. If $x = e$, $y = e$ or $x = y = e$ we have

$$\begin{aligned}
\alpha(y) &= \beta(e) \cdot \gamma(y), \\
\gamma(y) &= (\beta e)^{-1} \cdot \alpha(y), \\
\alpha(x) &= \beta(x) \cdot \gamma(e), \\
\beta(x) &= \alpha(x) \cdot (\gamma e)^{-1}, \\
\alpha(e) &= \beta(e) \cdot \gamma(e).
\end{aligned}$$

Thus

$$\begin{aligned}
\alpha(x \cdot y) &= \alpha(x) \cdot (\gamma e)^{-1} \cdot (\beta e)^{-1} \cdot \alpha(y) \\
&= \alpha(x) \cdot (\alpha e)^{-1} \cdot \alpha(y),
\end{aligned}$$

i.e., α is a holomorphism of the group $Q(\cdot)$. □

Moreover, the bijections β and γ are also holomorphisms of the group $Q(\cdot)$ because

$$\begin{aligned}
\beta(x) &= \alpha(x) \cdot (\gamma e)^{-1} = \alpha(x) \cdot k, \\
\gamma(x) &= (\beta e)^{-1} \cdot \alpha(x) = \ell \cdot \alpha(x).
\end{aligned}$$

Proposition 16.2.16. *If in the group $Q(\cdot)$, the identity*

$$\alpha(x \cdot y) = \beta(x) \cdot \gamma(y),$$

is valid for every $x, y \in Q$, where the mappings $\alpha, \beta, \gamma : Q \to Q$ are bijections, then all these mappings are holomorphisms of the group $Q(\cdot)$. □

Proposition 16.2.17. *If in the group $Q(\cdot)$ the identity,*

$$\beta\left(\alpha(x \cdot y) \cdot z\right) = \gamma x \cdot \delta(\sigma y \cdot \tau z), \tag{16.2.5}$$

is valid for every $x, y, z \in Q$, where the mappings $\alpha, \beta, \gamma, \delta, \sigma, \tau : Q \to Q$ are bijections, and all these mappings are holomorphisms of the group $Q(\cdot)$.

Proof. Substituting $y = e$ in Eq. (16.2.5) we have

$$\beta(\alpha x \cdot z) = \gamma x \cdot \delta(\sigma e \cdot \tau z)$$

and

$$\beta(x \cdot z) = \gamma(\alpha^{-1}x) \cdot \lambda z,$$

where $\lambda z = \delta(\sigma e \cdot \tau z)$. Hence, β is an autotopy of $Q(\cdot)$, and β is a holomorphism, by Proposition 16.2.15. Putting $z = e$ in Eq. (16.2.5) we have

$$\beta(\alpha(x \cdot y)) = \gamma x \cdot \delta(\sigma y \cdot \tau e)$$

and

$$(\alpha \cdot \beta)(x \cdot y) = \gamma x \cdot \mu y,$$

where

$$\mu y = \delta(\sigma y \cdot \tau e).$$

Thus, $\theta = \alpha \cdot \beta$, γ and μ are the holomorphisms of $Q(\cdot)$. Hence, $\alpha = \theta \cdot \beta^{-1} \in \operatorname{Hol} Q(\cdot)$.

If $x = e$ in Eq. (16.2.5) we have

$$\beta(\alpha y \cdot z) = \gamma e \cdot \delta(\sigma y \cdot \tau z)$$

and

$$(\gamma e)^{-1} \cdot \beta\left(\alpha(\sigma^{-1}y) \cdot \tau^{-1}z\right) = \delta(y \cdot z),$$
$$(\gamma e)^{-1} \cdot \beta\left(\alpha(\sigma^{-1}y)\right) \cdot (\beta e)^{-1} \cdot \beta(\tau^{-1}z) = \delta(y \cdot z),$$
$$\rho(y) \cdot \beta\left(\tau^{-1}z\right) = \delta(y \cdot z),$$

where

$$\rho(y) = (\gamma e)^{-1} \cdot \beta\left(\alpha(\sigma^{-1}y)\right)(\beta e)^{-1}.$$

Hence, ρ, δ and $\theta' = \tau^{-1} \cdot \beta$ are the holomorphism of $Q(\cdot)$ by Proposition 16.2.16. It follows that:

$$\tau = \beta \cdot (\theta')^{-1} \in \operatorname{Hol} Q(\cdot).$$

Now, from the equality $\mu y = \delta(\sigma y \cdot \tau e)$, we have

$$\sigma y = \delta^{-1}(\mu y) \cdot (\tau e)^{-1} = (\mu \delta^{-1})y \cdot (\tau e)^{-1}$$

and $\sigma \in \operatorname{Hol} Q(\cdot)$ since $\mu \cdot \delta^{-1} \in \operatorname{Hol} Q$. □

For a more general results on holomorphisms of loops, see [249].

Example 16.2.18. Let $S_{|Q|}$ be the group of permutations of the set Q. We have the following isomorphism:

$$\mathrm{Hol}\,(Z_2 \times Z_2) \simeq S_4.$$

Indeed, let us note first that $\mathrm{Aut}\,(Z_2 \times Z_2) \simeq S_3$, because if $Z_2 \times Z_2 = \{e, a, b, c\}$, then every permutation of the elements a, b, c determines an automorphism of the group $Z_2 \times Z_2$ and conversely, any automorphism of $Z_2 \times Z_2$ is given by a permutation of a, b, c. So, we have $6 = 3!$ different automorphisms and $24 = 3! \cdot 4$ different holomorphisms of $Z_2 \times Z_2$ by Proposition 16.2.9. Hence,

$$\mathrm{Hol}\,(Z_2 \times Z_2) \simeq S_4.$$

The first well-known result on medial algebras concerns medial quasigroups and is called Toyoda theorem [51,410]: *For every medial quasigroup $Q(\cdot)$ there exists an abelian group $Q(+)$ such that the operation $(\cdot)$ is determined by the rule:*

$$x \cdot y = \varphi x + c + \psi y,$$

where $\varphi, \psi \in Aut\, Q(+)$ and $c \in Q$.

The next result is more general [258]:

Theorem 16.2.19. *If the pair (A, B) of binary quasigroup operations on the set Q is abelian then there is an abelian group $Q(+)$ such that the operations A, B are determined by the rules:*

$$A(x, y) = \varphi_1 x + c_1 + \psi_1 y,$$
$$B(x, y) = \varphi_2 x + c_2 + \psi_2 y,$$

where $c_1, c_2 \in Q$ and $\varphi_1, \varphi_2, \psi_1, \psi_2 \in Aut\, Q(+)$.

Proof. Indeed, if the quasigroup operations $A_1, A_2, A_3, A_4, A_5, A_6$ on the set Q satisfy the identity:

$$A_1(A_2(x,y), A_3(u,v)) = A_4(A_5(x,u), A_6(y,v)),$$

then there exists an abelian group $Q(+)$ such that

$$\begin{aligned} A_1(x,y) &= \alpha x + \tau y, & A_4(x,y) &= \mu x + \sigma y,\\ A_2(x,y) &= \alpha^{-1}(\gamma x + \delta y), & A_5(x,y) &= \mu^{-1}(\gamma x + \lambda y),\\ A_3(x,y) &= \tau^{-1}(\lambda x + \beta y), & A_6(x,y) &= \sigma^{-1}(\delta x + \beta y), \end{aligned}$$

for every $x, y \in Q$. We have: $A_1 = A_5 = A_6 = A$ and $A_2 = A_3 = A_4 = B$. So, $A_1(x,y) = A_5(x,y)$ and $A_1(x,y) = A_6(x,y)$, i.e.,

$$\begin{aligned} \alpha x + \tau y &= \mu^{-1}(\gamma x + \lambda y),\\ \alpha x + \tau y &= \sigma^{-1}(\delta x + \beta y), \end{aligned}$$

for every $x, y \in Q$. Thus:

$$\begin{aligned} \mu(\alpha x + \tau y) &= \gamma x + \lambda y,\\ \sigma(\alpha x + \tau y) &= \delta x + \beta y, \end{aligned}$$

i.e.,

$$\begin{aligned} \mu(x+y) &= \gamma(\alpha^{-1}x) + \lambda(\tau^{-1}y),\\ \sigma(x+y) &= \delta(\alpha^{-1}x) + \beta(\tau^{-1}y). \end{aligned}$$

It follows from these equalities:

$$\begin{aligned} \gamma\left(\alpha^{-1}x\right) &= \mu(x) + \left(-\lambda\left(\tau^{-1}0\right)\right),\\ \lambda\left(\tau^{-1}y\right) &= \left(-\gamma\left(\alpha^{-1}0\right)\right) + \mu(y),\\ \delta\left(\alpha^{-1}x\right) &= \sigma(x) + \left(-\beta\left(\tau^{-1}0\right)\right),\\ \beta\left(\tau^{-1}y\right) &= \left(-\delta\left(\alpha^{-1}0\right)\right) + \sigma(y). \end{aligned}$$

Hence,

$$\mu(x+y) = \mu(x) + (-\mu 0) + \mu(y)$$

and

$$\sigma(x+y) = \sigma(x) + (-\sigma 0) + \sigma(y)$$

for every $x, y \in Q$, where $0 \in Q$ is the identity element of the group $Q(+)$, i.e., the bijections $\mu, \sigma : Q \to Q$ are holomorphisms of the group $Q(+)$. So, the mappings $\varphi_2(x) = \mu(x) + (-\mu 0)$ and $\psi_2(x) = (-\sigma 0) + \sigma(x)$ are automorphisms of the group $Q(+)$. Hence, $\mu(x) = \varphi_2(x) + \mu 0$ and $\sigma(x) = \sigma 0 + \psi_2(x)$, where $\varphi_2, \psi_2 \in Aut\, Q(+)$. So,

$$A_4(x,y) = \mu x + \sigma y = \varphi_2(x) + \mu 0 + \sigma 0 + \psi_2(y)$$
$$= \varphi_2(x) + c_2 + \psi_2(y) = B(x,y),$$

where $c_2 = \mu 0 + \sigma 0$. For $A(x,y)$ we have the similar proof taking into account the equalities $A_2 = A_4$ and $A_3 = A_4$. □

As a consequence we get the following characterization (cf. [392]).

Corollary 16.2.20. *If $(Q; \Sigma)$ is a binary invertible medial algebra then there exists an abelian group $Q(+)$ such that every operation $A_i \in \Sigma$ is determined by the rule:*

$$A_i(x,y) = \varphi_i x + c_i + \psi_i y,$$

where $c_i \in Q$ and $\varphi_i, \psi_i \in Aut\, Q(+)$. Moreover, if the algebra $(Q; \Sigma)$ also is idempotent (is a mode), then

$$A_i(x,y) = \varphi_i x + \psi_i y,$$

where $\varphi_i, \psi_i \in Aut\, Q(+)$ and $\varphi_i, \psi_i \in Aut\,(Q; \Sigma)$.

The abelian group $Q(+)$ is unique up to isomorphism.

Proof. Namely, if $A_0 \in \Sigma$ is a fixed operation, then by Toyoda theorem, A_0 is principally isotopic to the abelian group operation $+$ on Q. If $B \in \Sigma$ is any operation, then the pair (A_0, B) is medial, hence, A_0 and B are principally isotopic to the abelian group operation $*$ on Q. Thus, any operation B is principally isotopic to the same abelian group operation $+$ by transitivity of isotopy. So

$$A_0(x,y) = \alpha x + \tau y = A_1(x,y) = A_5(x,y) = A_6(x,y),$$

and for any operation, $B \in \Sigma$, we have

$$B(x,y) = \mu x + \sigma y = A_2(x,y) = A_3(x,y) = A_4(x,y).$$

Hence, for every $x, y \in Q$ we have

$$B(x,y) = \varphi x + t + \psi y,$$

where $\varphi, \psi \in Aut\, Q(+)$ and $t \in Q$, according to the proof of the previous theorem. For the idempotent case we have $t = 0$ and $\varphi x = B(x, 0)$,

$\psi y = B(0, y)$. Hence, $\varphi, \psi \in Aut\,(Q; \Sigma)$. For instance:

$$\begin{aligned}\varphi A(x, y) &= B(A(x, y), 0) = B(A(x, y), A(0, 0)) \\ &= A(B(x, 0), B(y, 0)) = A(\varphi x, \varphi y).\end{aligned}$$

□

From the proof of Theorem 16.2.19 also follows the result:

Corollary 16.2.21. *If the binary quasigroup operations A, B_1, B_2 and C on the set Q satisfy the identity:*

$$A(B_1(x, y), B_2(u, v)) = C(A(x, u), A(y, v)),$$

then there exists an abelian group $Q(+)$ such that the operation C is determined by the rule:

$$C(x, y) = \varphi x + c + \psi y,$$

where $\varphi, \psi \in Aut\, Q(+)$ and $c \in Q$.

16.3. Auxiliary results

The subgroup of the group $\mathcal{F}_Q^r(\cdot)$ generated by the subset $\Sigma \subseteq \mathcal{F}_Q^r$ is denoted by $(\Sigma)_r$. Similarly, the subgroup of the group $\mathcal{F}_Q^\ell(\circ)$ generated by the subset $\Sigma \subseteq \mathcal{F}_Q^\ell$ is denoted by $(\Sigma)_\ell$.

Lemma 16.3.1. *If a right invertible algebra $(Q; \Sigma)$ is abelian (mode), then the algebra $\left(Q; \Sigma \cup \Sigma^{-1}\right)$ also is abelian (mode).*

Proof. For every $A, B \in \Sigma$ and $x, y, u, v \in Q$ there exists an element $v' \in Q$ such that

$$A^{-1}\left(B(x, y), B(u, v')\right) = B\left(A^{-1}(x, u), A^{-1}(y, v)\right).$$

From this equality, we have

$$A\left(B(x, y), B\left(A^{-1}(x, u), A^{-1}(y, v)\right)\right) = B(u, v'),$$

and by the mediality:

$$B\left(A\left(x, A^{-1}(x, u)\right), A\left(y, A^{-1}(y, v)\right)\right) = B(u, v'),$$

i.e., $B(u, v) = B(u, v')$ and $v = v'$. So,

$$A^{-1}\left(B(x, y), B(u, v)\right) = B\left(A^{-1}(x, u), A^{-1}(y, v)\right).$$

Hence

$$A^{-1}\left(B^{-1}(x,y),B^{-1}(u,v)\right)=B^{-1}\left(A^{-1}(x,u),A^{-1}(y,v)\right).$$

□

Lemma 16.3.2. *If the left invertible algebra $(Q;\Sigma)$ is abelian (is a mode), then the algebra $\left(Q;\Sigma\cup^{-1}\Sigma\right)$ also is abelian (is a mode).*

Corollary 16.3.3. *If the invertible algebra $(Q;\Sigma)$ is abelian (is a mode), then the algebra $(Q;\Sigma\cup\Sigma^{-1}\cup^{-1}\Sigma\cup^{-1}(\Sigma^{-1})\cup(^{-1}\Sigma)^{-1}\cup\Sigma^*)$ also is abelian (is a mode).*

Lemma 16.3.4. *If a right invertible algebra $(Q;\Sigma)$ is abelian (is a mode), then the extended algebra $(Q;(\Sigma)_r)$ also is abelian (is a mode).*

Proof. Let us proof the equality:

$$S\left(P(x,y),\ P(u,v)\right)=P\left(S(x,u),\ S(y,v)\right)$$

for every $S,P\in(\Sigma)_r$ by induction. Let

$$\begin{aligned}S&=A_1^{\varepsilon_1}\cdots A_m^{\varepsilon_m},\quad \varepsilon_i=\pm1,\\ P&=B_1^{\delta_1}\cdots B_k^{\delta_k},\quad \delta_j=\pm1,\end{aligned}\tag{16.3.1}$$

where $A_i,B_j\in\Sigma$. The length $\partial(S)$ of the operation $S\in(\Sigma)$ is the number of the operations $A_i\in\Sigma$ in some representation (16.3.1). If $\partial(S)=\partial(P)=1$, then $S,P\in\Sigma\cup^{-1}\Sigma$ and the assertion is followed from Lemma 16.3.1. Suppose the assertion be true for $\partial(S),\ \partial(P)\leqslant n-1$. Consider the following three possibilities:

(1) $\partial(S)=n$, $\partial(P)<n$. Then $S=S_1\cdot S_2$, where $\partial(S_1)<n$ and $\partial(S_2)<n$, and

$$\begin{aligned}S\left(P(x,y),\ P(u,v)\right)&=S_1\cdot S_2\left(P(x,y),\ P(u,v)\right)\\&=S_1\left(P(x,y),\ S_2\left(P(x,y),\ P(u,v)\right)\right)\\&=S_1\left(P(x,y),\ P\left(S_2(x,u),\ S_2(y,v)\right)\right)\\&=P\left(S_1\left(x,S_2(x,u)\right),\ S_1\left(y,S_2(y,v)\right)\right)\\&=P\left(S_1\cdot S_2(x,u),\ S_1\cdot S_2(y,v)\right)\\&=P\left(S(x,u),\ S(y,v)\right),\end{aligned}$$

by assumption;

(2) $\partial(S)<n$, $\partial(P)=n$. So, $P=P_1\cdot P_2$, where $\partial(P_1)<n$ and $\partial(P_2)<n$, and this case followed from case (1);

(3) $\partial(S) = \partial(P) = n$. So, $S = S_1 \cdot S_2$, where $\partial(S_1) < n$ and $\partial(S_2) < n$ and this case followed from case (2).

□

Lemma 16.3.5. *If a left invertible algebra $(Q; \Sigma)$ is abelian (is a mode), then the extended algebra $(Q; (\Sigma)_\ell)$ also is abelian (is a mode).*

Example 16.3.6. If $|Q| = 2$, then the right invertible algebra $\left(Q; \mathcal{F}_Q^r\right)$ is medial.

Following the papers [251,257], we call a binary algebra $(Q; \Sigma)$, with the property $E, F \in \Sigma$, transitive if

(a) $|Q| \geqslant 2$;
(b) for every $a, b, c \subset Q$, where $b \neq a$, there exists an operation $A \subset \Sigma$ such that $A(a, b) = c$;
(c) the algebra $(Q; \Sigma \setminus \{F\})$ is right invertible.

Example 16.3.7. If $|Q| = 2$ and $\Sigma = \{E, F\}$, then $(Q; \Sigma)$ is a transitive mode.

Example 16.3.8. Let $Q(\cdot)$ be a non-trivial abelian group and

$$\Sigma^0 = \{A_q | \, A_q(x, y) = y \cdot q, \; q \in Q, \; x, y \in Q\}.$$

If $\Sigma = \Sigma^0 \cup \{F\}$, then $(Q; \Sigma)$ is a medial and transitive algebra. However, $(Q; \Sigma)$ is not a mode.

Example 16.3.9. If $Q(+, \cdot)$ is a field with the identity $e \in Q$ and

$$\Sigma = \{A_q | \, A_q(x, y) = qx + (e - q)y, \; q \in Q, \; x, y \in Q\},$$

then $(Q; \Sigma)$ is a transitive mode, where $(Q; \Sigma \setminus \{E, F\})$ is an invertible algebra.

If $Q = \{1, 2\}$ and $\Sigma^0 = \mathcal{F}_Q^r$, $\Sigma = \Sigma^0 \cup \{F\}$, then $(Q; \Sigma)$ is a transitive and medial algebra and there exists the operation $A \neq E$ such that $A(2, 1) = 1$. However, the following result is valid for transitive modes.

Lemma 16.3.10. *Let $(Q; \Sigma)$ be a transitive mode and $A \in \Sigma$. If the equation $A(x, a) = a$ has a solution $x \neq a \in Q$, then $A = E$.*

Proof. If $|Q| = 2$ then the idempotent and right invertible operation A is unique: $A = E$. Let $|Q| \geqslant 3$ and the equation $A(x, a) = a$ has a solution $x_0 \neq a$. Let us choose an element $b \in Q$, where $b \neq a$, x_0. There exists an operation $B \in \Sigma \setminus \{F\}$ such that $B(a,b) = x_0$ according to the condition b) of transitivity. So,

$$\begin{aligned} B(a,a) &= B\left(A(x_0,a), A(x_0,a)\right) = A\left(B(x_0,x_0), B(a,a)\right) \\ &= A\left(x_0, B(a,a)\right) = A\left(B(a,b), B(a,a)\right) \\ &= B\left(A(a,a), A(b,a)\right) = B\left(a, A(b,a)\right). \end{aligned}$$

Hence, $a = A(b,a)$ for every $b \neq x_0, a$. However, this equality is also valid for $b = a, x_0$ (because, $A(x_0,a) = a$ and $A(a,a) = a$). Let $c \in Q$ and fix the element $d \neq a, c$. Then there exists an operation $C \in \Sigma \setminus \{F\}$ such that $C(d,a) = c$. Thus

$$\begin{aligned} c = C(d,a) &= C\left(d, A(b,a)\right) = C\left(A(d,d), A(b,a)\right) \\ &= A\left(C(d,b), C(d,a)\right) = A\left(C(d,b), c\right), \end{aligned}$$

for every element $b \in Q$. Hence, $A(u,c) = c$ for every $u, c \in Q$. □

Corollary 16.3.11. *If $(Q;\Sigma)$ is a transitive mode then for every $a,b,c \in Q$, where $b \neq a$ there exists a unique operation $A \in \Sigma$ such that $A(a,b) = c$.*

Proof. Let there exist the two operations $A, B \in \Sigma \setminus \{F\}$ such that $A(a,b) = c$ and $B(a,b) = c$. Consider the group $(\Sigma \setminus \{F\})_r$. The algebra $(Q;(\Sigma \setminus \{F\})_r)$ is medial (and idempotent) by Lemma 16.3.4. Hence, the algebra $(Q;(\Sigma \setminus \{F\})_r \cup \{F\})$ is a transitive mode and we can apply Lemma 16.3.10. Thus, we have $A(a,b) = B(a,b)$ and

$$A^{-1}(a, A(a,b)) = A^{-1}(a, B(a,b)),$$

i.e., $A^{-1} \cdot A(a,b) = A^{-1} \cdot B(a,b)$ and $b = A^{-1} \cdot B(a,b)$. Hence, the equation $A^{-1} \cdot B(x,b) = b$ has a solution $x \neq b$ and it follows that $A^{-1} \cdot B = E$ and $B = A$ by the previous lemma.

If $A = F$ but $B \neq F$ then $B \neq E$ and $B(a,b) = F(a,b) = a$. Hence,

$$B^{-1}(a, B(a,b)) = B^{-1}(a,a) = a$$

and $B^{-1} \cdot B(a,b) = a$, i.e., $E(a,b) = a$ and $b = a$. Contradiction! □

Corollary 16.3.12. *If $(Q;\Sigma)$ is a transitive mode and $A(a,b) = B(a,b)$, where $A, B \in \Sigma$, $a, b \in Q$ and $a \neq b$, then $A = B$.*

Corollary 16.3.13. *If $(Q;\Sigma)$ is a transitive mode, then $(Q;\Sigma)$ is maximal in the following sense: there does not exist a transitive mode $(Q;\Sigma')$, where $\Sigma \subseteq \Sigma'$, $\Sigma \neq \Sigma'$. Hence, $(\Sigma \backslash \{F\})^{-1} \subseteq \Sigma \backslash \{F\}$.*

Proof. If $|Q| = 2$, then $\Sigma = \{E, F\}$, and assertion is valid. Let $|Q| \geqslant 3$ and $(Q; \Sigma')$ also is a transitive mode, where $\Sigma \subseteq \Sigma$ and $\Sigma \neq \Sigma'$. So, there exists an operation, $A \in \Sigma'$, such that $A \notin \Sigma$. Hence, $A \neq E, F$. Let us prove that there exists a triple of pair-vise distinct elements $a, b, c \in Q$ such that $A(a, b) = c$. Indeed:

(1) if for all pairs of distinct elements $a \neq b$ we have: $A(a, b) = a$, then $A = F$;
(2) if for all pairs of distinct elements $a \neq b$ we have: $A(a, b) = b$, then $A = E$;
(3) if for some pairs of distinct elements $a \neq b$ we have $A(a, b) = b$ and for other pairs of distinct elements $c \neq d$ we have $A(c, d) = c$, then $A = E$, by Lemma 16.3.10.

Besides, there exists an operation $B \in \Sigma$ such that $B(a, b) = c$ by transitivity of $(Q; \Sigma)$. The algebra $(Q; \Sigma \cup \{A\})$ is also a transitive mode. Thus, there exists a pair of elements $a \neq b$ in Q and the pair of operations $A, B \in \Sigma_1 \cup \{A\}$ such that $A(a, b) = B(a, b)$, where $\Sigma_1 = \Sigma \setminus \{F\}$. Hence, $A = B$ by Corollary 16.3.12, and $A \in \Sigma$. Contradiction! Now the second part of the assertion follows from Lemma 16.3.1. □

Corollary 16.3.14. *If $(Q; \Sigma)$ is a transitive mode, then $\Sigma \backslash \{F\}$ is an abelian group under the right multiplication of binary functions.*

Proof. The set $\Sigma \setminus \{F\}$ is a group under the right multiplication of binary functions, because $\Sigma \setminus \{F\} = (\Sigma \setminus \{F\})_r$ by maximality of $(Q; \Sigma)$. This group is abelian:

$$\begin{aligned} A \cdot B(x, y) &= A(x, B(x, y)) = A(B(x, x), B(x, y)) \\ &= B(A(x, x), A(x, y)) = B(x, A(x, y)) = B \cdot A(x, y), \end{aligned}$$

for every $A, B \in \Sigma \setminus \{F\}$. □

Theorem 16.3.15. *Let $|Q| \geqslant 3$ and $(Q; \Sigma)$ be a transitive mode. Then every operation, $A \in \Sigma \backslash \{F\}$, $A \neq E$, is left invertible. Hence,*

$$\begin{aligned} &{}^{-1}(\Sigma \backslash \{E\}) \subseteq \Sigma \backslash \{E\}, \quad \left({}^{-1}(\Sigma \backslash \{E, F\})\right)^{-1} \subseteq \Sigma \backslash \{E, F\}, \\ &{}^{-1}\left((\Sigma \backslash \{E, F\})^{-1}\right) \subseteq \Sigma \backslash \{E, F\}, \quad (\Sigma \backslash \{E, F\})^{*} \subseteq \Sigma \backslash \{E, F\}. \end{aligned}$$

The set $\Sigma \backslash \{E\}$ is closed under the left multiplication of binary functions.

Proof. Since $A \neq E, F$ then there are pair-wise distinct elements $a, b, c \in Q$ such that $A(a,b) = c$. Hence, the equation $A(x,b) = c$ has a solution $x = a$.

Consider the equation $A(x,b) = p$, where $p \neq c, b$ (since in the cases $p = c, b$ a solution exists). There is an operation $B \in \Sigma\backslash\{F\}$ such that $B(b,c) = p$. So,

$$B(b,c) = B\left(A(b,b),\ A(a,b)\right) = A\left(B(b,a),\ B(b,b)\right) = A\left(B(b,a), b\right) = p,$$

and the solution of the equation $A(x,b) = p$ is an element $B(b,a) \in Q$.

Consider the equation $A(x,q) = p$, where $q \neq b, p$. Let $d \in Q$ and $d \neq b, q$. There is an operation $C \in \Sigma\backslash\{F\}$ such that $C(d,b) = q$. Thus:

$$\begin{aligned} C(d,c) &= C\left(d, A(a,b)\right) = C\left(A(d,d),\ A(a,b)\right) \\ &= A\left(C(d,a),\ C(d,b)\right) = A\left(C(d,a), q\right). \end{aligned}$$

Thus, if $p = C(d,c)$ then the equation $A(x,q) = p$ has a solution $x = C(d,a)$. If $p \neq C(d,c)$ and since $q \neq C(d,c)$ there is an operation $D \in \Sigma\backslash\{F\}$ such that $D\left(q, C(d,c)\right) = p$. Hence,

$$\begin{aligned} p &= D\left(q, C(d,c)\right) = D\left(q, A(C(d,a), q)\right) = D\left(A(q,q),\ A(C(d,a),\ q)\right) \\ &= A\left(D\left(q, C(d,a)\right),\ D(q,q)\right) = A\left(D\left(q, C(d,a)\right),\ q\right), \end{aligned}$$

and the equation $A(x,q) = p$ also has a solution in the case $p \neq C(d,c)$.

From Lemma 16.3.10 it follows that the equation $A(x,q) = p$ has a unique solution, where $A \neq E$. Namely, if $q = p$ then the uniqueness of the solution $x = q = p$ follows from Lemma 16.3.10. If $p \neq q$ and the equation $A(x,q) = p$ has solutions $x_0 \neq y_0$, then $q \neq y_0, x_0$ and there is an operation $T \in \Sigma \setminus \{F\}$ such that $T(q, x_0) = y_0$. Hence, $T \neq E$ and

$$\begin{aligned} p = A(y_0, q) &= A\left(T(q, x_0),\ T(q,q)\right) = T\left(A(q,q),\ A(x_0, q)\right) \\ &= T\left(q, A(x_0, q)\right) = T(q, p). \end{aligned}$$

It follows from Lemma 16.3.10 that $T = E$. Contradiction!

The last part of the assertion is valid according to Corollary 16.3.3, Lemmas: 16.3.4, 16.3.5, and Corollary 16.3.13 about maximality of $(Q; \Sigma)$. □

16.4. The structural results

Let $(Q; \Sigma)$ be a transitive mode and let $|Q| \geqslant 3$. Then $\Sigma\backslash\{E, F\} \neq \emptyset$. Consider $B \in \Sigma\backslash\{E, F\}$ and $0 \in Q$. According to Theorem 16.3.15, the operation B is invertible, medial, idempotent and:

$$A\left(B(x,y),\, B(u,v)\right) = B\left(A(x,u),\, A(y,v)\right),$$

for every $A \in \Sigma$. According to Corollary 16.2.20 we have $B(x,y) = \varphi x + \psi y$, where

$$x + y = B\left({}^{-1}B(x,0),\, B^{-1}(0,y)\right), \quad \varphi x = B(x,0), \quad \psi y = B(0,y),$$

and

$$A(\varphi x + \psi y, \varphi u + \psi v) = \varphi A(x,u) + \psi A(y,v) = A(\varphi x, \varphi u) + A(\psi y, \psi v),$$

i.e.,

$$A(x + y, u + v) = A(x,u) + A(y,v).$$

Theorem 16.4.1. *If $(Q;\Sigma)$ is a transitive mode, then there exists a field $Q(+,\cdot)$ such that every operation $A \in \Sigma$ is defined by the rule:*

$$A(x,y) = (e - a)x + ay,$$

where e is the identity element of the field and $a \in Q$ (and depends on A). The field $Q(+,\cdot)$ is unique up to isomorphism.

Proof. If $|Q| = 2$ then the assertion is evident, because in this case $\Sigma = \{E, F\}$. Let $|Q| \geqslant 3$. Let us define the addition as above, while we define the multiplication by the following way. Let e be a fixed element of Q and let $e \neq 0$. For every element $x \in Q$ there exists an operation $A_x \in \Sigma$ such that $x = A_x(0,e)$ (if $x = 0, e$, we have $A_x = F, E$). The mapping $\Phi : x \to A_x$ is a bijection from Q to Σ by Corollary 16.3.12. We define $x \cdot y = A_x(0,y)$, where $A_x = \Phi(x)$. Then we prove the field axioms. For example

$$x \cdot y = A_{x\cdot y}(0,e), \quad x \cdot y = A_x(0,y) = A_x\left(0, A_y(0,e)\right) = A_x \cdot A_y(0,e),$$

and according to Corollary 16.3.12, we have $A_x \cdot A_y = A_{x\cdot y}$; then $\Phi(x \cdot y) = \Phi(x) \cdot \Phi(y)$, hence, $x \cdot y = A_x(0,y)$ is associative and commutative. For every operation $A = A_a \in \Sigma$ we have:

$$A(x,y) = A(x + 0, 0 + y) = A(x,0) + A(0,y),$$

$$A(x,x) = A(x,0) + A(0,x) = x;$$

So, $A(x,0) = x - A(0,x)$, hence:

$$A(x,y) = A(x,0) + A(0,y) = x - A(0,x) + A(0,y) = x - ax + ay = (e - a)x + ay,$$

where $a = A(0,e)$.

Let us prove the second part of the theorem. Let there exist a second field $Q(\oplus, \circ)$ satisfying the condition of the assertion. If $0' \in Q$ is the zero element, $e' \in Q$ is the identity element, and $\ominus$ is the subtraction of the new field $Q(\oplus, \circ)$ then for every element $a \in Q$ there exists an element $a' \in Q$ such that

$$A(x,y) = (e-a)x + ay = ((e' \ominus a') \circ x) \oplus (a' \circ y),$$

for every $x, y \in Q$. It is easy to prove that the mapping $\varphi : Q \to Q$, where $\varphi(a) = a'$ is a bijection. If we put $x = 0$ and $y = 0$ in the last equality, we have respectively:

$$\begin{aligned} ay &= ((e' \ominus \varphi a) \circ 0) \oplus (\varphi a \circ y), \\ (e-a)x &= ((e' \ominus \varphi a) \circ x) \oplus (\varphi a \circ 0). \end{aligned}$$

Thus,

$$\begin{aligned} (e-a)x \oplus ay &= ay \oplus (e-a)x \\ &= ((e' \ominus \varphi a) \circ 0)(\varphi a \circ y) \oplus ((e' \ominus \varphi a) \circ x) \oplus (\varphi a \circ 0) \\ &= ((e' \ominus \varphi a) \circ x) \oplus (\varphi a \circ y) \oplus ((e' \ominus \varphi a) \circ 0) \oplus (\varphi a \circ 0) \\ &= ((e' \ominus \varphi a) \circ x) \oplus (\varphi a \circ y) \oplus (e' \circ 0) \\ &= ((e' - \varphi a) \circ x) \oplus (\varphi a \circ y) \oplus 0 \\ &= (((e-a)x) + ay) \oplus 0. \end{aligned}$$

i.e.,

$$u \oplus v = (u+v) \oplus 0,$$

for every $u, v \in Q$. So, with $v = 0'$ we have: $u = (u + 0') \oplus 0$. Hence,

$$u \oplus v = ((u + v - 0') + 0') \oplus 0 = u + v - 0'.$$

From this equality, we also have

$$\begin{aligned} 0' &= (\ominus v) \oplus v = (\ominus v) \oplus v - 0' \\ \ominus v &= 0' - v + 0' \end{aligned}$$

and

$$u \ominus v = u \oplus (\ominus v) = u + (\ominus v) - 0' = u + 0' - v + 0' - 0' = u - v + 0'.$$

Taking into account the values $u \oplus v$ and $u \ominus v$, we have

$$\begin{aligned} a(y-x)+x = A(x,y) &= (\varphi a \circ (y \ominus x)) \oplus x = \varphi a \circ (y - x + 0') \oplus x \\ &= (\varphi a \circ (y - x + 0')) + x - 0', \end{aligned}$$

i.e.,

$$a(y-x) = (\varphi a \circ (y - x + 0')) - 0',$$

and

$$\varphi a \circ t = a(t - 0') + 0',$$

where $t = y - x + 0'$. Hence, if $t = e'$ we have

$$\varphi a = a(e' - 0') + 0' = ka + 0',$$

where $k = e' - 0'$ (and φ is a linear mapping). It remains now to note:

$$\varphi(a+b) = \varphi a \oplus \varphi b,$$

and $\varphi(ab) = \varphi a \circ \varphi b$, for every $a, b \in Q$. For example

$$\begin{aligned} \varphi a \oplus \varphi b &= (ka + 0') \oplus (kb + 0') \\ &= ka + 0' + kb + 0' - 0' = k(a+b) + 0' = \varphi(a+b). \end{aligned}$$

□

Corollary 16.4.2. *If $(Q;\Sigma)$ is a transitive mode, then there exists a field $Q(+,\cdot)$ such that every operations $A \in \Sigma$ is determined by the rule:*

$$A(x,y) = ax + (e-a)y,$$

where e is the identity element of the field and $a \in Q$ (and depends on A). The field $Q(+,\cdot)$ is unique up to isomorphism.

An algebra $(Q;\Sigma)$ is called three-abelian algebra if the subalgebra of the algebra $(Q;\Sigma)$ generated by any three elements $a, b, c \in Q$ is abelian.

Problem 16.4.3. *Characterize abelian transitive algebras.*

Problem 16.4.4. *Characterize three-abelian invertible algebras.*

Problem 16.4.5. *Characterize three-abelian transitive algebras.*

Problem 16.4.6. *Characterize three-abelian, idempotent and transitive algebras.*

16.5. Pre-quasigroups and pre-invertible algebras

In this section, we present a certain useful generalizations of the previous results on abelian algebras via concept of pre-quasigroup (cf. [81]).

Let $(Q;\cdot)$ be a groupoid and $a \in Q$. Denote by L_a (R_a) the map of Q to Q such that $L_a(x) = ax$ $(R_a(x) = xa)$ for all $x \in Q$. They are called left and right translations of the $(Q;\cdot)$. In detail we denote $L_{(\cdot),a}$ $(R_{(\cdot),a})$ for L_a (R_a).

A groupoid $(Q;\cdot)$ is said to be a division (cancellation) groupoid if L_a and R_a are surjective (injective) for every $a \in Q$. If the groupoid $(Q;\cdot)$ is a division (cancellation) groupoid then the operation $(\cdot)$ is called division (cancellation) operation, too. A quasigroup is a cancellation and division groupoid. A loop is a quasigroup $(Q;\cdot)$ with the unit e such that $ex = xe = x$ for any $x \in Q$.

A binary algebra $(Q;\Sigma)$ is called division (cancellation) algebra if $(Q;A)$ is a division (cancellation) groupoid for any operation $A \in \Sigma$. Division and cancellation algebra is called invertible algebra.

A groupoid $(Q;\cdot)$ is called left pre-cancellative if

$$ca = cb \to R_a = R_b,$$

where $a, b, c \in Q$. Right pre-cancellative groupoids are defined dually. A groupoid is called pre-cancellative if it is both left and right pre-cancellative. If groupoid $Q(A)$ is pre-cancellative then the operation A is also called pre-cancellative. A binary algebra $(Q;\Sigma)$ is called pre-cancellative if the groupoid $(Q;A)$ is pre-cancellative for any operation $A \in \Sigma$.

Every quasigroup is a pre-cancellative and division algebra.

Any division and pre-cancellative groupoid is called a pre-qusigroup. A binary algebra $(Q;\Sigma)$ is called pre-invertible algebra if $Q(A)$ is a pre-quasigroup for any operation $A \in \Sigma$.

Example 16.5.1. An example $(Q;\cdot)$ of a pre-cancellative groupoid (which is not a quasigroup) is given by the Cayley table of the operation $(\cdot)$ below:

$\cdot$	1	2	3
1	2	1	1
2	3	2	2
3	1	3	3

Example 16.5.2. Let $(\mathbb{C}^*; \cdot)$ be the multiplicative group of complex numbers and

$$x \circ y = x^2 \cdot y$$

for any $x, y \in \mathbb{C}^*$. Then $\mathbb{C}^*(\circ)$ is a pre-quasigroup, which is a non-quasigroup.

Example 16.5.3. Let $(\mathbb{C}^*; \cdot)$ be the multiplicative group of complex numbers and

$$x \circ y = x \cdot y^2$$

for any $x, y \in \mathbb{C}^*$. Then $\mathbb{C}^*(\circ)$ is a pre-quasigroup, which is a non-quasigroup.

Example 16.5.4. Let $(\mathbb{C}^*; \cdot)$ be the multiplicative group of complex numbers and

$$x \circ y = x^2 \cdot y^2$$

for any $x, y \in \mathbb{C}^*$. Then $\mathbb{C}^*(\circ)$ is a pre-quasigroup, which is a non-quasigroup.

The groupoid $(Q; \cdot)$ is called homotopic to the groupoid $(Q; \circ)$ if there exist three maps α, β, γ of Q to Q such that $\gamma(x \cdot y) = \alpha x \circ \beta y$ for all $x, y \in Q$. The homotopy of the form $T = (\alpha, \beta, id_Q)$ (where id_Q is the identity map of Q) is called principal homotopy. We say that groupoid $(Q; \cdot)$ is epitopic to the groupoid $(Q; \circ)$ if the mappings α, β, γ are surjections. So we have the concept of principal epitopicity as well. Binary algebra $(Q; \Sigma)$ is called epitopic to a groupoid (group, semigroup, loop) $(Q; \circ)$ if $(Q; A)$ is epitopic to $(Q; \circ)$ for any operation $A \in \Sigma$.

The following lemmas are general versions of the corresponding Albert theorems for quasigroups [12,13].

Lemma 16.5.5. *If the groupoid $(Q; \circ)$ with the identity element is principally homotopic to the group $(Q; \cdot)$, then they are isomorphic. Hence, if the loop $(Q; \circ)$ is principally homotopic to the group $(Q; \cdot)$, then they are isomorphic.*

Proof. If $(Q; \circ)$ is a groupoid with the identity element $e \in Q$ and $x \circ y = \alpha x \cdot \beta y$, then, by $y = e$ and $x = e$, we have respectively:

$$x = \alpha x \cdot \beta e,$$

$$y = \alpha e \cdot \beta y.$$

Hence, $\alpha x = x \cdot t^{-1}$ and $\beta y = s^{-1} \cdot y$, where $t = \beta e$, $s = \alpha e$. Thus, $x \circ y = x \cdot t^{-1} \cdot s^{-1} \cdot y = x \cdot w \cdot y$ and

$$\varphi(x \circ y) = \varphi(x) \cdot \varphi(y),$$

where $w = t^{-1} \cdot s^{-1}$ and $\varphi(z) = z \cdot w$. □

Lemma 16.5.6. *Every pre-quasigroup $(Q; \cdot)$ is principally epitopic to a certain loop $(Q; +)$.*

Proof. Let $(Q; \cdot)$ be a pre-cancellation and division groupoid and a, b be its two elements. Take the corresponding mappings h and k of Q into itself such that $hR_a = id_Q$, $kL_b = id_Q$, i.e., $h(x) \cdot a = x$ and $b \cdot k(x) = x$ for every $x \in Q$. Hence, $h(xa) \cdot a = xa$ and $b \cdot k(by) = by$ for every $x, y \in Q$. Put $0 = ba$ and define $x + y = h(x) \cdot k(y)$ for all $x, y \in Q$. Since $(Q; \cdot)$ is a pre-cancellative groupoid, we get $xa + by = h(xa) \cdot k(by) = x \cdot k(by) = xy$. Thus:

$$x \cdot y = R_a(x) + L_b(y)$$

for all $x, y \in Q$.

Let us prove that $(Q; +)$ is a loop.

If $x, y, z \in Q$ and $x + y = x + z$, then $h(x) \cdot k(y) = h(x) \cdot k(z)$; since $(Q; \cdot)$ is a pre-cancellative groupoid, $b \cdot k(y) = b \cdot k(z)$, i.e., $(kL_b)y = (kL_b)z$, $id_Q(y) = id_Q(z)$ and $y = z$. Similarly, $y = z$, whenever $y + x = z + x$. We have proved that $(Q; +)$ is a cancellation groupoid.

Let $x, y \in Q$. There exists $v \in Q$ with $h(x) \cdot v = y$. From the equality $b \cdot k(x) = x$ we have: $b \cdot k(bv) = bv$. If $z = b \cdot v$, then $x + z = x + bv = h(x) \cdot k(bv) = h(x) \cdot v = y$, since $(Q; \cdot)$ is a pre-cancellative groupoid. Similarly, there exists an element u with $u + x = y$, hence $(Q; +)$ is a quasigroup.

We have $b \cdot k(ba) = ba$, hence, by the pre-cancellative property, we obtain $x + 0 = h(x) \cdot k(ba) = h(x) \cdot a = x$ and, since $h(ba) \cdot a = ba$, then, from the pre-cancellative property, we get $0 + x = h(ba) \cdot k(x) = b \cdot k(x) = x$ for all $x \in Q$. Hence, $(Q; +)$ is a loop. □

Theorem 16.5.7. *The pre-quasigroup $(Q; \cdot)$ is principally epitopic to a certain group $(Q; +)$ if and only if that pre-quasigroup satisfies the following Mal'tsev-Reidemeister quasi-identity*

$$x_1y_2 = x_2y_1,\ x_1y_4 = x_2y_3,\ x_3y_2 = x_4y_1 \to x_3y_4 = x_4y_3 \tag{16.5.1}$$

(for Mal'tsev-Reidemeister quasi-identity see [211,341]).

Proof. Any pre-quasigroup $(Q;\cdot)$ is principally epitopic to certain loop $(Q;+)$ by previous lemma: $x \cdot y = \beta x + \gamma y$. Let us note that if $(Q;\cdot)$ satisfies the quasi-identity (16.5.1) then $(Q;+)$ also satisfies the quasi-identity (16.5.1). Namely, if

$$u_1 + v_2 = u_2 + v_1,\ u_1 + v_4 = u_2 + v_3,\ u_3 + v_2 = u_4 + v_1$$

and $u_1 = \beta x_1$, $u_2 = \beta x_2$, $u_3 = \beta x_3$, $u_4 = \beta x_4$, $v_1 = \gamma y_1$, $v_2 = \gamma y_2$, $v_3 = \gamma y_3$, $v_4 = \gamma y_4$, then

$$x_1 \cdot y_2 = x_2 \cdot y_1,\ x_1 \cdot y_4 = x_2 \cdot y_3,\ x_3 \cdot y_2 = x_4 \cdot y_1,$$

hence $x_3 \cdot y_4 = x_4 \cdot y_3$ and $\beta x_3 + \gamma y_4 = \beta x_4 + \gamma y_3$, i.e., $u_3 + v_4 = u_4 + v_3$.

Let $0 \in Q$ be the identity element of loop $(Q;+)$. For any $u_4, v_1, v_4 \in Q$ let us denote $v_1 + v_4 = v_3$ and $u_4 + v_1 = u_3$. We have

$$0 + v_1 = v_1 + 0,\ v_1 + v_4 = v_3 = 0 + v_3,\ u_4 + v_1 = u_3 = u_3 + 0.$$

Hence, $u_3 + v_4 = u_4 + v_3$ and

$$u_3 + v_4 = (u_4 + v_1) + v_4,\quad u_4 + v_3 = u_4 + (v_1 + v_4),$$

i.e., $(u_4 + v_1) + v_4 = u_4 + (v_1 + v_4)$ and $(Q;+)$ is a group.

(*Necessity*). If $(Q;+)$ is a group and

$$x_1 + y_2 = x_2 + y_1,\ x_1 + y_4 = x_2 + y_3,\ x_3 + y_2 = x_4 + y_1,$$

then

$$\begin{aligned} x_3 + y_4 &= (x_4 + y_1 + (-y_2)) + ((-x_1) + x_2 + y_3) \\ &= x_4 + (y_1 + (-y_2) + (-x_1) + x_2) + y_3 = x_4 + y_3. \end{aligned}$$

□

Theorem 16.5.8. *The pre-quasigroup $(Q;\cdot)$ is principally epitopic to a certain abelian group $(Q;+)$ if and only if that pre-quasigroup satisfies the following Thomsen quasi-identity*

$$x_1y_2 = x_2y_1,\ x_1y_3 = x_3y_1 \rightarrow x_2y_3 = x_3y_2 \tag{16.5.2}$$

(for Thomsen quasi-identity see [319]).

The notion of quasiautomorphism of a group [24] is equivalent to the notion of holomorphism of a group [202] (for the holomorphism of Moufang loops, rings and monoids see [249]). Now we introduce the general concept of quasiendomorphism of groups. The mapping $\varphi : Q \to Q$ is called quasiendomorphism of the group $(Q; \cdot)$ if

$$\varphi(x \cdot y) = \varphi x \cdot (\varphi e)^{-1} \cdot \varphi y,$$

for every $x, y \in Q$, where $e \in Q$ is the identity element of the group $(Q; \cdot)$. The well-known properties of quasiautomorphisms of groups are reformulated for quasiendomorphisms. For example we need the following results (see Sec. 16.2).

Lemma 16.5.9. *If $Q(\cdot)$ is a group and for mappings $\alpha, \beta, \gamma : Q \to Q$ the following identity is valid*

$$\alpha(x \cdot y) = \beta x \cdot \gamma y,$$

then α is a quasiendomorphism of $(Q; \cdot)$.

Lemma 16.5.10. *Any quasiendomorphism γ of the group $(Q; \cdot)$ has the form:*

$$\gamma(x) = \gamma_0(x) \cdot s, \tag{16.5.3}$$

and

$$\gamma(x) = s \cdot \gamma_0(x), \tag{16.5.4}$$

where γ_0 is an endomorphism of the group $(Q; \cdot)$, $s \in Q$ and, conversely: the map γ defined by the equality (16.5.3) or (16.5.4) is a quasiendomorphism of the group $(Q; \cdot)$.

Theorem 16.5.11. *Let four operations A, B, C, D be division operations on Q and let either $(Q; A)$ or $(Q; C)$ be a pre-quasigroup. If these operations satisfy the following identity:*

$$A(x, B(y, z)) = C(D(x, y), z), \tag{16.5.5}$$

then there is a group $(Q; \cdot)$ such that the binary algebra $(Q; A, B, C, D)$ is epitopic to the group $(Q; \cdot)$. Moreover, there are surjections $A_1, A_2, B_1, B_2, C_1, C_2, D_1, D_2 : Q \to Q$ such that

$$\begin{cases} A(x, y) = A_1x \cdot A_2y, \\ A_2B(x, y) = A_2B_1x \cdot A_2B_2y, \\ C(x, y) = C_1x \cdot C_2y, \\ C_1D(x, y) = C_1D_1x \cdot C_1D_2y, \end{cases} \tag{16.5.6}$$

for all $x, y \in Q$ and

$$\begin{cases} A_1 = C_1 D_1, \\ A_2 B_1 = C_1 D_2, \\ A_2 B_2 = C_2. \end{cases} \tag{16.5.7}$$

The group $(Q; \cdot)$ is unique up to isomorphism.

Proof. Let $p, q, r \in Q$, $b = B(q, r)$, $d = D(p, q)$ and $e = A(p, b)$. Let also:

$$\begin{cases} A_1 x = A(x, b) = R_{A,b} x, & A_2 x = A(p, x) = L_{A,p} x, \\ B_1 x = B(x, r) = R_{B,r} x, & B_2 x = B(q, x) = L_{B,q} x, \\ C_1 x = C(x, r) = R_{C,r} x, & C_2 x = C(d, x) = L_{C,d} x, \\ D_1 x = D(x, q) = R_{D,q} x, & D_2 x = D(p, x) = L_{D,p} x. \end{cases} \tag{16.5.8}$$

Since A, B, C, D are division operations on Q, then A_1, A_2, B_1, B_2, C_1, C_2, D_1 and D_2 defined by (16.5.8) are surjections.

We obtain

$$A(x, B_1 y) = C_1 D(x, y),$$
$$A_2 B(y, z) = C(D_2 y, z),$$
$$A(x, B_2 z) = C(D_1 x, z).$$

From this identities we obtain the equalities (16.5.7).

Since A_1 and A_2 are surjections, there exist mappings $h_1, h_2 : Q \to Q$ such that $h_1 A_1 = id_Q$ and $h_2 A_2 = id_Q$.

Let A be a pre-cancellative operation. Let us define an operation on the set Q by

$$x \cdot y = A(h_1 x, h_2 y).$$

According to proof of Lemma 16.5.6, the groupoid $(Q; \cdot)$ is a loop and

$$A(x, y) = A_1 x \cdot A_2 y.$$

There exists a mapping $k_1 : Q \to Q$ such that $k_1 D_1 = id_Q$. Hence, we get

$$\begin{aligned} C(x, y) &= C(D_1 k_1 x, y) = A(k_1 x, B_2 y) = A_1 k_1 x \cdot A_2 B_2 y \\ &= C_1 D_1 k_1 x \cdot C_2 y = C_1 x \cdot C_2 y, \\ A_2 B(x, y) &= A(p, B(x, y)) = C(D(p, x), y) = C_1 D(p, x) \cdot C_2 y \\ &= C_1 D_2 x \cdot C_2 y = A_2 B_1 x \cdot A_2 B_2 y, \end{aligned}$$

$$C_1D(x,y) = C(D(x,y),r) = A(x,B(y,r)) = A_1x \cdot A_2B(y,r)$$
$$= A_1x \cdot A_2B_1y = C_1D_1x \cdot C_1D_2y.$$

Thus, we obtained (16.5.6).

Let us prove that $(Q;\cdot)$ is a group. Indeed, from (16.5.5) we have

$$A_1x \cdot A_2B(y,z) = C_1D(x,y) \cdot C_2z,$$
$$A_1x \cdot (A_2B_1y \cdot A_2B_2z) = (C_1D_1x \cdot C_1D_2y) \cdot C_2z.$$

According to (16.5.7) we get

$$A_1x \cdot (A_2B_1y \cdot A_2B_2z) = (A_1x \cdot A_2B_1y) \cdot A_2B_2z,$$

and since A_1, A_2B_1 and A_2B_2 are surjections, we obtain associativity of $(\cdot)$.

For the case of pre-cancellativity of the operation C, the proof is similar.

The last statement of the theorem follows from Lemma 16.5.5. □

Theorem 16.5.12. *Let the set Q form a division groupoid under the six operations A_i (for $i = 1,\dots,6$) and either $(Q;A_1)$ or $(Q;A_4)$ is a pre-quasigroup. If these operations satisfy the following identity:*

$$A_1(A_2(x,y),A_3(u,v)) = A_4(A_5(x,u),A_6(y,v)), \tag{16.5.9}$$

then there exists an abelian group $(Q;\cdot)$ such that the binary algebra $(Q;A_1,A_2,A_3,A_4,A_5,A_6)$ is epitopic to the abelian group $(Q;\cdot)$. Moreover, there exist eight surjective mappings: $\alpha,\beta,\gamma,\delta,\lambda,\sigma,\varphi,\psi$ of Q onto itself such that

$$A_1(x,y) = \alpha x \cdot \varphi y,$$
$$\alpha A_2(x,y) = \gamma x \cdot \delta y,$$
$$\varphi A_3(x,y) = \lambda x \cdot \beta y,$$
$$A_4(x,y) = \psi x \cdot \sigma y,$$
$$\psi A_5(x,y) = \gamma x \cdot \lambda y,$$
$$\sigma A_6(x,y) = \delta x \cdot \beta y.$$

The abelian group $(Q;\cdot)$ is unique up to isomorphism.

Proof. If we put $u = a \in Q$ in (16.5.9), we have an equation of associativity (16.5.5):

$$A'_1(A_2(x,y),v) = A'_4(x, A_6(y,v)),$$

where

$$A'_1(x,y) = A_1(x, A_3(a,y)) = A_1(x, L_3 y),$$
$$A'_4(x,y) = A_4(A_5(x,a),y) = A_4(R_5 x, y),$$

and $L_i(x) = A_i(a,x)$, $R_i(x) = A_i(x,a)$, $i = 1,\dots,6$. It is obvious that the operations A'_1 and A'_4 are division operations, and if A_1 (A_4) is pre-cancellative, then A'_1 (A'_4) also is pre-cancellative.

According to Theorem 16.5.11, there exists a group $(Q;\cdot)$ and surjections $\alpha, \beta, \gamma, \delta, \sigma : Q \to Q$ such that

$$A'_1(x,y) = \alpha x \cdot \beta y,$$
$$\alpha A_2(x,y) = \gamma x \cdot \delta y,$$
$$A'_4(x,y) = \gamma x \cdot \sigma y,$$
$$\sigma A_6(x,y) = \delta x \cdot \beta y.$$

Since L_3 is a surjection, there is a mapping $h_3 : Q \to Q$ such that $h_3 L_3 = id_Q$. Thus

$$A_1(x,y) = A'_1(x, h_3 y) = \alpha x \cdot \beta h_3 y = \alpha x \cdot \varphi y,$$

where $\varphi = h_3\beta$. Let us prove that φ is a surjection. If we take in the last equality $x = a$, we have $A_1(a,y) = \alpha a \cdot \varphi y$, or $L_1 y = \tilde{L}_{\alpha a}\varphi y$, where $\tilde{L}_{\alpha a}(x) = \alpha a \cdot x$ is a left translation of the group $(Q;\cdot)$. Thus, $\varphi = L_1 \tilde{L}_{\alpha a}^{-1}$ and φ is a surjection.

It can be shown that

$$A_4(x,y) = \psi x \cdot \sigma y,$$

where $\psi : Q \to Q$ is a surjection.

Now, by the substitutions $A_i(x,y)$ in Eq. (16.5.9), we obtain

$$\gamma x \cdot \delta y \cdot \varphi A_3(u,v) = \psi A_5(x,u) \cdot \delta y \cdot \beta v.$$

Now, if we choose the elements x, y such that $\gamma x = e$, $\delta y = e$, where e is the identity element of the group $(Q;\cdot)$, we get

$$\varphi A_3(u,v) = \lambda u \cdot \beta v,$$

where $\lambda = \psi L_5$ is a surjection from Q to Q. Hence, from the previous equality we have

$$\gamma x \cdot \delta y \cdot \lambda u \cdot \beta v = \psi A_5(x, u) \cdot \delta y \cdot \beta v.$$

Choose the elements y, v such that $\delta y = e$, $\beta v = e$, then

$$\psi A_5(x, u) = \gamma x \cdot \lambda u.$$

The substitution $\psi A_5(x, y)$ in the previous equality gives the abelity of the group $(Q; \cdot)$: $\delta y \cdot \lambda u = \lambda u \cdot \delta y$, where δ, λ are surjections. □

Theorem 16.5.13. *Let the set Q form a division groupoid under the six operations A_i (for $i = 1, \ldots, 6$) and either $(Q; A_1)$ or $(Q; A_4)$ is a pre-quasigroup. If these operations satisfy the following paramedial identity:*

$$A_1(A_2(x, y), A_3(u, v)) = A_4(A_5(v, y), A_6(u, x)), \tag{16.5.10}$$

then there exists an abelian group $(Q; \cdot)$ such that the binary algebra $(Q; A_1, A_2, A_3, A_4, A_5, A_6)$ is epitopic to the abelian group $(Q; \cdot)$. Moreover, there exist eight mappings: $\alpha, \beta, \gamma, \delta, \lambda, \sigma, \varphi, \psi$ of Q onto itself such that

$$A_1(x, y) = \alpha x \cdot \varphi y,$$
$$\alpha A_2(x, y) = \gamma x \cdot \delta y,$$
$$\varphi A_3(x, y) = \lambda x \cdot \beta y,$$
$$A_4(x, y) = \sigma x \cdot \psi y,$$
$$\sigma A_5(x, y) = \beta x \cdot \delta y,$$
$$\psi A_6(x, y) = \lambda x \cdot \gamma y.$$

The abelian group $(Q; \cdot)$ is unique up to isomorphism.

Proof. Let $A_i^*(x, y) = A_i(y, x)$, then using the equality (16.5.10), we have

$$A_1(A_2(x, y), A_3(u, v)) = A_4^*(A_6^*(x, u), A_5^*(y, v)),$$

where A_4^*, A_5^*, A_6^* are also division operations and $(Q; A_4^*)$ is a pre-quasigroup if $(Q; A_4)$ is a pre-quasigroup. Now, the statement of the theorem follows from Theorem 16.5.12 and Lemma 16.5.5. □

Let us consider the identity, which is defined by Kolmogoroff [3] equality: $(xy)(yz) = xz$ (see also [252]). A pre-quasigroup with this identity is called a Kolmogoroff pre-quasigroup.

Theorem 16.5.14. *Let the set Q form a division groupoid under the four binary operations A, B, C, D and either $(Q; A)$ or $(Q; D)$ is a pre-quasigroup. If these operations satisfy the following identity:*

$$A(B(x, y), C(y, z)) = D(x, z), \tag{16.5.11}$$

then there exists a group $(Q; \cdot)$ such that the binary algebra $(Q; A, B, C, D)$ is epitopic to the group $(Q; \cdot)$. Moreover, there exist the surjective mappings A_1, A_2, B_1, B_2, C_1, C_2, D_1, D_2 of Q onto itself such that

$$\begin{cases} A(x, y) = A_1x \cdot A_2y, \\ A_1B(x, y) = B_1x \cdot B_2y, \\ A_2C(x, y) = C_1x \cdot C_2y, \\ D(x, y) = D_1x \cdot D_2y. \end{cases} \tag{16.5.12}$$

The group $(Q; \cdot)$ is unique up to isomorphism.

Proof. Suppose that A is a pre-cancellative operation. Then, by Lemma 16.5.6, there exists a loop $(Q; \cdot)$ such that

$$A(x, y) = A_1x \cdot A_2y,$$

where A_1 and A_2 are surjective mapping of Q to Q. From equality (16.5.11) we have:

$$A_1B(x, y) \cdot A_2C(y, z) = D(x, z).$$

If we take $y = t \in Q$ in the last equality, we obtain

$$D(x, z) = D_1x \cdot D_2z,$$

where D_1 and D_2 are surjective mappings of Q to Q. Now, from (16.5.11) we get

$$A_1B(x, y) \cdot A_2C(y, z) = D_1x \cdot D_2z. \tag{16.5.13}$$

Let us prove that $(Q; \cdot)$ is a group. Choose an element z, in the last equality, such that $D_2z = e$ (e is a unit of the loop $Q(\cdot)$); then $A_1B(x, y) \cdot \alpha y = D_1x$, where α is a surjection of Q to Q. According to the last equality, from (16.5.13), we obtain

$$A_1B(x, y) \cdot A_2C(y, z) = (A_1B(x, y) \cdot \alpha y) \cdot D_2z. \tag{16.5.14}$$

For every y there exists x such that $A_1B(x, y) = e$; therefore for every $y, z \in Q$, we have

$$A_2C(y, z) = \alpha y \cdot D_2 z.$$

Substituting the last equality into (16.5.14), we obtain

$$A_1B(x, y) \cdot (\alpha y \cdot D_2 z) = (A_1B(x, y) \cdot \alpha y) \cdot D_2 z.$$

Hence, $(Q; \cdot)$ is a group. Now, if in the equality (16.5.13) we take $z = t \in Q$, we have

$$A_1B(x, y) \cdot A_2C(y, t) = B_1 x,$$

where $B_1 : Q \to Q$ is a surjective. Let $I : Q \to Q$ defined by $Ix = x^{-1}$, where x^{-1} is an inverse of x in the group $(Q; \cdot)$. Then, from the last equality, we get

$$A_1B(x, y) = B_1 x \cdot (A_2C(y, t))^{-1} = B_1 x \cdot IA_2R_{C,t} y = B_1 x \cdot B_2 y,$$

where $B_2 : Q \to Q$ is a surjection.

Analogously, we can prove the equality: $A_2C(x, y) = C_1 x \cdot C_2 y$. □

Corollary 16.5.15. *Any Kolmogoroff pre-quasigroup $(Q; \cdot)$ is principally epitopic to a certain group $(Q; \circ)$.*

Let us consider the identity which is defined by the left Euclidean equation: $(xy)(zy) = xz$. A pre-quasigroup with this identity is called a left Euclidean pre-quasigroup.

Theorem 16.5.16. *Let the set Q form a division groupoid under the four binary operations A, B, C, D and either $(Q; A)$ or $(Q; D)$ is a pre-quasigroup. If these operations satisfy the following identity:*

$$A(B(x, y), C(z, y)) = D(x, z), \tag{16.5.15}$$

then there exists a group $(Q; \cdot)$ such that the binary algebra $(Q; A, B, C, D)$ is epitopic to the group $(Q; \cdot)$. Moreover, there exist the surjective mappings

A_1, A_2, B_1, B_2, C_1, C_2, D_1, D_2 *of* Q *onto itself such that*

$$\begin{cases} A(x,y) = A_1x \cdot A_2y, \\ A_1B(x,y) = B_1x \cdot B_2y, \\ IA_2C(x,y) = C_1x \cdot C_2y, \\ D(x,y) = D_1x \cdot D_2y, \end{cases} \tag{16.5.16}$$

where $I : Q \to Q$ *defined by* $Ix = x^{-1}$ *and* x^{-1} *is an inverse of* x *in the the group* $(Q;\cdot)$.

The group $(Q;\cdot)$ *is unique up to isomorphism.*

Corollary 16.5.17. *Any left Euclidean pre-quasigroup* $(Q;\cdot)$ *is principally epitopic to a certain group* $(Q;\circ)$.

Now, we consider the identity which is defined by the right Euclidean equation: $(xy)(xz) = yz$. A pre-quasigroup with this identity is called a right Euclidean pre-quasigroup.

Theorem 16.5.18. *Let the set* Q *form a division groupoid under the four binary operations* A, B, C, D *and either* $(Q;A)$ *or* $(Q;D)$ *is a pre-quasigroup. If these operations satisfy the following identity:*

$$A(B(x,y), C(x,z)) = D(y,z), \tag{16.5.17}$$

then there exists a group $(Q;\cdot)$ *such that the binary algebra* $(Q;A,B,C,D)$ *is epitopic to the group* $(Q;\cdot)$. *Moreover, there exist the surjective mappings* A_1, A_2, B_1, B_2, C_1, C_2, D_1, D_2 *of* Q *onto itself such that*

$$\begin{cases} A(x,y) = A_1x \cdot A_2y, \\ IA_1B(x,y) = B_1x \cdot B_2y, \\ A_2C(x,y) = C_1x \cdot C_2y, \\ D(x,y) = D_1x \cdot D_2y, \end{cases} \tag{16.5.18}$$

where $I : Q \to Q$ *defined by* $Ix = x^{-1}$ *and* x^{-1} *is an inverse of* x *in the group* $(Q;\cdot)$.

The group $(Q;\cdot)$ *is unique up to isomorphism.*

Corollary 16.5.19. *Any right Euclidean pre-quasigroup* $(Q;\cdot)$ *is principally epitopic to a certain group* $(Q;\circ)$.

Theorem 16.5.20. *If the pair* (A,B) *of binary division operations on the set* Q *is medial (abelian) and either* $(Q;A)$ *or* $(Q;B)$ *is a pre-quasigroup, then* *there is an abelian group $(Q;\cdot)$ such that the operations A, B are determined by the rules:*

$$A(x,y) = \psi_1 x \cdot c_1 \cdot \sigma_1 y,$$
$$B(x,y) = \psi_2 x \cdot c_2 \cdot \sigma_2 y,$$

where $c_1, c_2 \in Q$ and $\psi_1, \psi_2, \sigma_1, \sigma_2$ are surjective endomorphisms of the group $(Q;\cdot)$ such that

$$\psi_1\sigma_2 = \sigma_2\psi_1, \psi_2\sigma_1 = \sigma_1\psi_2, \psi_1\psi_2 = \psi_2\psi_1, \sigma_1\sigma_2 = \sigma_2\sigma_1.$$

The abelian group $(Q;\cdot)$ is unique up to isomorphism.

Proof. According to Theorem 16.5.12, if the division operations A_i ($i = 1,\ldots,6$) on the set Q satisfy the identity

$$A_1(A_2(x,y), A_3(u,v)) = A_4(A_5(x,u), A_6(y,v)),$$

then there exists an abelian group $(Q;\cdot)$ such that

$$A_1(x,y) = \alpha x \cdot \varphi y,$$
$$\alpha A_2(x,y) = \gamma x \cdot \delta y,$$
$$\varphi A_3(x,y) = \lambda x \cdot \beta y,$$
$$A_4(x,y) = \psi x \cdot \sigma y,$$
$$\psi A_5(x,y) = \gamma x \cdot \lambda y,$$
$$\sigma A_6(x,y) = \delta x \cdot \beta y$$

for every $x, y \in Q$. We have $A_1 = A_5 = A_6 = A$ and $A_2 = A_3 = A_4 = B$. Hence, $A_1(x,y) = A_5(x,y)$ and $A_1(x,y) = A_6(x,y)$, i.e.,

$$\psi(\alpha x \cdot \varphi y) = \gamma x \cdot \lambda y,$$
$$\sigma(\alpha x \cdot \varphi y) = \delta x \cdot \beta y,$$

for every $x, y \in Q$. Thus

$$\psi(x \cdot y) = \gamma h_\alpha x \cdot \lambda h_\varphi y,$$
$$\sigma(x \cdot y) = \delta h_\alpha x \cdot \beta h_\varphi y,$$

where h_α and h_φ are mappings from Q to Q such that $h_\alpha \alpha = id_Q$ and $h_\varphi \varphi = id_Q$. Therefore, ψ and σ are quasiendomorphisms of the group $(Q;\cdot)$ and, by Lemma 16.5.10, we have $\psi = \tilde{R}_s \psi_2$ and $\sigma = \tilde{R}_t \sigma_2$, where $\tilde{R}_s$ and $\tilde{R}_t$

are right translations of the group $(Q;\cdot)$ and ψ_2,σ_2 are endomorphism of the group $Q(\cdot)$. Hence, we get

$$B(x,y) = A_4(x,y) = \psi_2 x \cdot c_2 \cdot \sigma_2 y,$$

where $c_2 \in Q$. It is obvious that ψ_2 and σ_2 are surjective endomorphisms of the group $(Q;\cdot)$. For $A(x,y)$ we have a similar proof, if we take into account the equalities $A_2 = A_4$ and $A_3 = A_4$.

The last statement of the theorem follows from Lemma 16.5.5. □

Corollary 16.5.21. *If $(Q;\Sigma)$ is a binary pre-invertible abelian algebra, then there exists an abelian group $(Q;+)$ such that every operation $A_i \in \Sigma$ is determined by the rule:*

$$A_i(x,y) = \varphi_i x + c_i + \psi_i y,$$

where $c_i \in Q$ and φ_i, ψ_i are surjective endomorphisms of the group $(Q;+)$.

The abelian group $(Q;+)$ is unique up to isomorphism.

Let A and B be binary operations on the set Q. The pair (A,B) is called paramedial [81,100] if it satisfies the identity:

$$A(B(x,y),B(u,v)) = B(A(v,y),A(u,x)).$$

A binary algebra $(Q;\Sigma)$ is called paramedial if the pair (A,B) is paramedial for every $A,B \in \Sigma$. In other words, a binary algebra $(Q;\Sigma)$ is called paramedial if $(Q;\Sigma)$ satisfies the hyperidentity of paramediality:

$$X(Y(x,y),Y(u,v)) = Y(X(v,y),X(u,x)).$$

Theorem 16.5.22. *If the pair (A,B) of binary division operations on the set Q is paramedial and either $(Q;A)$ or $(Q;B)$ is a pre-quasigroup, then there is an abelian group $(Q;\cdot)$ such that the operations A,B are determined by the rules:*

$$A(x,y) = \psi_1 x \cdot c_1 \cdot \sigma_1 y,$$

$$B(x,y) = \psi_2 x \cdot c_2 \cdot \sigma_2 y,$$

where $c_1, c_2 \in Q$ and $\psi_1,\psi_2,\sigma_1,\sigma_2$ are surjective endomorphisms of the group $(Q;\cdot)$ such that:

$$\psi_1\psi_2 = \sigma_2\sigma_1, \psi_2\psi_1 = \sigma_1\sigma_2, \psi_1\sigma_2 = \psi_2\sigma_1, \sigma_1\psi_2 = \sigma_2\psi_1.$$

The abelian group $(Q;\cdot)$ is unique up to isomorphism.

Proof. The proof of this theorem is similar to the proof of Theorem 16.5.20. □

Corollary 16.5.23. *If $(Q;\Sigma)$ is a binary paramedial pre-invertible algebra, then there exists an abelian group $(Q;+)$ such that every operation $A_i \in \Sigma$ is determined by the rule:*

$$A_i(x, y) = \varphi_i x + c_i + \psi_i y,$$

where $c_i \in Q$ and φ_i, ψ_i are surjective endomorphisms of the group $(Q;+)$.

The abelian group $(Q;+)$ is unique up to isomorphism.

Problem 16.5.24. *Characterize three-abelian pre-invertible algebras.*

Chapter 17

Non-trivial Associative and Distributive Hyperidentities in q-Algebras

17.1. Classification of associative hyperidentities

A hyperidentity $w_1 = w_2$ is said to be an associative (Abelian, commutative, left distributive, right distributive, ...) hyperidentity if, ignoring the functional variables, it becomes the associative (Abelian, commutative, left distributive, right distributive, ...) law; that is, it becomes $((x, y), z) = (x, (y, z))$ (or, respectively, $((x, y), (u, v)) = ((x, u), (y, v))$, $(x, y) = (y, x)$, $(x, (y, z)) = ((x, y), (x, z))$, $((x, y), z) = ((x, z), (y, z)), \ldots$).

A binary algebra $(Q; \Sigma)$ is said to be a q-algebra if Σ contains an invertible operation, that is, $Q(A)$ is a quasigroup for some A in Σ. A binary algebra $(Q; \Sigma)$ is called an e-algebra if Σ contains an operation having an identity element.

A binary algebra $(Q; \Sigma)$ is called functionally non-trivial, if $|\Sigma| > 1$.

Definition 17.1.1. A functional variable X is said to be singular in a hyperidentity $w_1 = w_2$ if X occurs just once in it and at least one of the following conditions holds:

(a) a subword $w = X(\omega_1, \omega_2)$ involves object variables $x \in [\omega_1]$ and $y \in [\omega_2]$, each of which occurs just once in w;
(b) a subword $w = X(\omega_1, \omega_2)$ has the form $X(\omega_1, x)$ or $X(x, \omega_2)$, and there exists an object variable $y \in [w]$ different from x and occurring just once in w, where $[w]$ is the set of object variables occurring in w.

A functional variable X is said to be strictly singular in a hyperidentity $w_1 = w_2$ if X occurs just once in it and a subword $X(\omega_1, \omega_2)$ of w_i ($i = 1$ or 2) contains object variables $x \in [\omega_1]$ and $y \in [\omega_2]$ occurring just once in w_i.

Lemma 17.1.2 ([249]). *No functionally non-trivial q-algebra can satisfy a hyperidentity with a singular functional variable. No functionally non-trivial e-algebra can satisfy a hyperidentity with a strictly singular functional variable.*

Proof. Assume not. Let $\mathfrak{A} = \langle Q; \Sigma \rangle$ be a functionally non-trivial q-algebra such that Σ contains an invertible operation A, and suppose that $\mathfrak{A}$ satisfies the hyperidentity

$$w_1 = w_2 \tag{17.1.1}$$

having a singular functional variable X. We give X two different values A_1 and A_2 from Σ, and for each such substitution we give the remaining functional variables the same value A from Σ, where of course $Q(A)$ is a quasigroup. This gives two identities:

$$w_1' = w_2', \qquad w_1'' = w_2''.$$

Suppose for definiteness that X occurs in w_1. Then the expressions for w_2' and w_2'' are identical, so that $w_2' = w_2''$ and thus $w_1' = w_1''$. All the operations in w_1' and w_1'' are equal to A except for A_1 and A_2. Since $Q(A)$ is a quasigroup, after all cancellations we get the identity

$$A_1(\omega_1^0, \omega_2^0) = A_2(\omega_1^0, \omega_2^0).$$

Let $X(\omega_1, \omega_2)$ be a subword satisfying condition (a) of Definition 17.1.1. Apart from x and y, we give each object variable an arbitrary but fixed value in Q. Then ω_1^0 becomes $\lambda(x)$ and ω_2^0 becomes $\mu(y)$, where λ and μ are permutations of Q, since they are products of translations in the quasigroup $Q(A)$. Thus, we have

$$A_1(\lambda x, \mu y) = A_2(\lambda x, \mu y).$$

Since λ and μ are permutations, we deduce from this that $A_1 = A_2$, contradicting the choice of these operations. Case (b) is dealt with in a similar fashion.

To prove the second part of the lemma, all object variables other than x and y are replaced by the identity element. □

Corollary 17.1.3. *Every non-trivial associative hyperidentity satisfied in a functionally non-trivial q-algebra must be of functional rank* 2 *and take one* *of the following forms:*

$$X(x, Y(y, z)) = Y(X(x, y), z), \qquad ((ass)_1)$$

$$X(x, Y(y, z)) = X(Y(x, y), z), \qquad ((ass)_2)$$

$$X(x, X(y, z)) = Y(Y(x, y), z). \qquad ((ass)_3)$$

Moreover, $(ass)_2$ *follows from* $(ass)_3$ *and* $(ass)_1$ *follows from* $(ass)_2$ *in the class of q-algebras. A q-algebra satisfying* $(ass)_1$ *satisfies* $(ass)_2$ *if and only if it satisfies the hyperidentity*

$$X(x, Y(y, x)) = X(Y(x, y), x);$$

a q-algebra satisfying $(ass)_2$ *satisfies* $(ass)_3$ *if and only if it satisfies the coindentity*

$$X(a, X(b, c)) = Y(Y(a, b), c).$$

Finally, a q-algebra with a non-trivial associative hyperidentity must be invertible.

Proof. If the associative hyperidentity has functional rank 4 or 3, it must involve singular functionally variables; thus, by Lemma 17.1.2, it cannot be satisfied in a functionally non-trivial q-algebra. If the functional rank is 2, only the three cases enumerated can occur.

If a q-algebra $(Q; \Sigma)$ satisfies one of $(ass)_1$, $(ass)_2$ and $(ass)_3$, it must also satisfy the trivial associative hyperidentity

$$X(x, X(y, z)) = X(X(x, y), z),$$

so that all operations in Σ are semigroup operations and $Q(\cdot)$ is a group for some $(\cdot)$ in Σ. We claim that if a q-algebra $(Q; \Sigma)$ satisfies $(ass)_1$, then every operation A_i in Σ is determined by the rule $A_i(x, y) = x \cdot t_i \cdot y$, where t_i is some element of Q depending on A_i. To see this, in $(ass)_1$ we put $X = A_i$, $Y = (\cdot)$ and $y = e$, the identity element of the group $Q(\cdot)$. What we get is $A_i(x, z) = A_i(x, e) \cdot z = \lambda_i(x) \cdot z$, where $\lambda_i(x) = A_i(x, e)$. On putting $X = (\cdot)$ and $Y = A_i$ in $(ass)_1$, we get $x \cdot (\lambda_i y \cdot z) = \lambda_i(x \cdot y) \cdot z$, $x \cdot \lambda_i y = \lambda_i(x \cdot y)$, $x \cdot \lambda_i e = \lambda_i(x)$, $x \cdot t_i = \lambda_i(x)$, where $t_i = \lambda_i e$. Thus $A_i(x, y) = x \cdot t_i \cdot y$. By straightforward verification, we can convince ourselves that the converse is true: if each operation A_i in Σ is determined by a rule $A_i(x, y) = x \cdot t_i \cdot y$, where $Q(\cdot)$ is a group and $t_i \in Q$, then $(ass)_1$ is satisfied in $(Q; \Sigma)$.

Next we consider a q-algebra $(Q; \Sigma)$ satisfying $(ass)_2$. Just as in the preceding case, we put $X = (\cdot) \in \Sigma$, $Y = A_i \in \Sigma$ and $z = e$, giving

$A_i(x,y) = x\cdot\lambda_i(y)$, where $\lambda_i(y) = A_i(y,e)$. Further, we find that $\lambda_i(y) = y\cdot t_i$, and hence that $A_i(x,y) = x\cdot y\cdot t_i$ and $x\cdot y\cdot z\cdot t_i = x\cdot y\cdot t_i\cdot z$, that is, t_i lies in the centre of the group $Q(\cdot)$. Easy verification confirms the converse, namely that $(ass)_2$ is indeed satisfied in such an algebra.

It can be proved in the same way that a q-algebra $(Q;\Sigma)$ satisfies $(ass)_3$ if and only if there exists a group $Q(\cdot)$ such that every operation A_i in Σ is determined by the rule $A_i(x,y) = x\cdot t_i\cdot y$, where t_i is an involution ($t_i^2 = e$) in the centre of $Q(\cdot)$. □

Corollary 17.1.4. *Any non-trivial associative hyperidentity satisfied in a functionally non-trivial e-algebra must have functional rank* 2 *and be one of* $(ass)_1$, $(ass)_2$ *and* $(ass)_3$. *Moreover, in the class of e-algebras,* $(ass)_2$ *follows from* $(ass)_3$ *and* $(ass)_1$ *from* $(ass)_2$.

The proof of Corollary 17.1.4 is similar to that of Corollary 17.1.3.

Note without proof that the following result, which corresponds to a classical result of Bruck and Hughes (see [51]) and Albert [12].

Theorem 17.1.5 ([252]). *If an e-algebra is isotopic to a binary algebra satisfying a non-trivial associative hyperidentity, then it is isomorphic to it.*

Associative hyperidentities in invertible algebras were first discussed in [25]. See also [352] and [74].

17.2. Classification of distributive hyperidentities

Example 17.2.1. Let $Q(+,\cdot)$ be a nilpotent associative ring of index of nilpotency 5, suppose that $A_a(x,y) = x\cdot a\cdot y$ for $a\in Q$, and put $\Sigma = \{A_a \mid a\in Q\}$. The following non-trivial associative, left and right distributive hyperidentities hold in $(Q;\Sigma)$:

$$X(x,Y(y,z)) = U(V(x,y),z),$$

$$X(x,Y(y,z)) = U(V(x,y),Z(x,z)),$$

$$X(Y(x,y),z) = U(V(x,z),Z(y,z)).$$

We come now to a classification of non-trivial left and right distributive hyperidentities satisfied in non-trivial q-algebras.

Theorem 17.2.2. *Every non-trivial left distributive hyperidentity satisfied in a functionally non-trivial q-algebra satisfying the trivial right distributive hyperidentity*

$$X(X(x,y),z) = X(X(x,z),X(y,z)) \qquad (17.2.1)$$

must be of functional rank 2 and of the form

$$X(x, Y(y, z)) = Y(X(x, y), X(x, z)). \tag{d_1}$$

Every non-trivial right-distributive hyperidentity satisfied in a functionally non-trivial q-algebra satisfying the trivial left distributive hyperidentity

$$X(x, X(y, z)) = X(X(x, y), X(x, z)) \tag{17.2.2}$$

must be of functional rank 2 *and of the form*

$$X(Y(x, y), z) = Y(X(x, z), X(y, z)). \tag{d_2}$$

Proof. Since there are five places in all for functional variables in a left or right distributive hyperidentity, Lemma 17.1.2 tells us that the rank of a non-trivial hyper-identity of one of these forms must be 2, otherwise it would involve singular functional variables. Thus, the functional rank must be two, and one of the functional variables occurs twice and the other one three times. In addition to (d_1), there are nine left distributive hyperidentities like this:

$$X(x, X(y, z)) = Y(Y(x, y), Y(x, z)), \tag{17.2.3}$$

$$X(x, X(y, z)) = Y(X(x, y), Y(x, z)), \tag{17.2.4}$$

$$X(x, X(y, z)) = Y(Y(x, y), X(x, z)), \tag{17.2.5}$$

$$X(x, X(y, z)) = X(Y(x, y), Y(x, z)), \tag{17.2.6}$$

$$X(x, Y(y, z)) = X(Y(x, y), Y(x, z)), \tag{17.2.7}$$

$$X(x, Y(y, z)) = X(Y(x, y), X(x, z)), \tag{17.2.8}$$

$$X(x, Y(y, z)) = X(X(x, y), Y(x, z)), \tag{17.2.9}$$

$$X(x, Y(y, z)) = Y(Y(x, y), X(x, z)), \tag{17.2.10}$$

$$X(x, Y(y, z)) = Y(X(x, y), Y(x, z)). \tag{17.2.11}$$

We have to prove that none of (17.2.3)–(17.2.11) can hold in any non-trivial q-algebra satisfying the trivial right distributive hyperidentity (17.2.1).

Suppose that (17.2.3) is satisfied in a non-trivial q-algebra $(Q; \Sigma)$ satisfying (17.2.1). By the definition of q-algebra, there exists an operation A in Σ for which $Q(A)$ is a quasigroup; this $Q(A)$ is therefore a distributive quasigroup, which means that the following equalities hold for all x, y, z in Q:

$$A(x, A(y, z)) = A(A(x, y), A(x, z)),$$
$$A(A(x, y), z) = A(A(x, z), A(y, z)).$$

Putting $y = x$ in the first of these and cancelling, we get $A(x, x) = x$ for all x in Q, that is, A is idempotent.

We define a new operation $(\circ)$ on Q:

$$x \circ y = A(R_e^{-1}x, L_e^{-1}x), \qquad e \in Q,$$

where $R_e(x) = A(x, e)$, $L_e(x) = A(e, x)$. What we get is a loop $Q(\circ)$ with identity element $e = A(e, e)$. Thus,

$$A(x, y) = R_e(x) \circ L_e(x),$$

where $R_e, L_e \in \operatorname{Aut} Q(\circ)$. For,

$$\begin{aligned} R_e(x \circ y) &= R_e A(R_e^{-1}x, L_e^{-1}y) = A(A(R_e^{-1}x, L_e^{-1}y), e) \\ &= A(A(R_e^{-1}x, e), A(L_e^{-1}y, e)) = A(R_e(R_e^{-1}x), R_e(L_e^{-1}y)) \\ &= A(R_e^{-1}(R_e x), L_e^{-1}(R_e y)) = R_e(x) \circ R_e(y), \end{aligned}$$

so that $R_e L_e = L_e R_e$, since

$$R_e(L_e x) = A(A(e, x), e) = A(A(e, e), A(x, e)) = A(e, A(x, e)) = L_e(R_e x).$$

In the same way, we can prove that L_e is an automorphism of $Q(\circ)$. Putting $R_e = \varphi$ and $L_e = \psi$, we get

$$A(x, y) = \varphi x \circ \psi y, \qquad \varphi, \psi \in \operatorname{Aut} Q(\circ),$$

where $A \in \Sigma$ and $Q(A)$ is a quasigroup.

Putting $X = A_i \in \Sigma$ and $Y = A$ in (17.2.3), we get

$$\varphi(\varphi x \circ \psi y) \circ \psi(\varphi x \circ \psi z) = A_i(x, A_i(y, z)),$$

which for $x = e$ gives

$$\varphi(\psi y) \circ \psi^2(z) = \mu_i A_i(y, z),$$

where $\mu_i(x) = A_i(e, x)$. Further,

$$\mu_i A_i(e, z) = \psi^2(z),$$
$$\mu_i^2(z) = \psi^2(z).$$

From the last of these equalities, it follows that μ_i^2 is a permutation and thus that μ_i is a permutation, and $Q(A_i)$ is a quasigroup isotopic to the loop $Q(\circ)$ for all $A_i \in \Sigma$. In other words, $(Q;\Sigma)$ is an invertible algebra. It is therefore an idempotent algebra, that is, the idempotent hyperidentity $X(x,x) = x$ holds in it.

Let A_1 and A_2 be different elements of Σ. Putting $X = A_1$, $Y = A_2$ and $y = z$ in (17.2.3), we get the following contradiction:

$$\begin{aligned} A_1(x, A_1(y,y)) &= A_2(A_2(x,y), A_2(x,y)), \\ A_1(x,y) &= A_2(x,y), \\ A_1 &= A_2. \end{aligned}$$

The hyperidentities (17.2.4)–(17.2.10) are treated in the same way, and similar arguments establish the second part of the theorem. □

Corollary 17.2.3 ([6,25]). *A non-trivial left (right, respectively) distributive hyperidentity satisfied in a functionally non-trivial invertible algebra must have functional rank* 2 *and take the form* (d_1) *(*(d_2)*, respectively).*

Proof. Idempotency of an invertible algebra with a left or right distributive hyper-identity follows from the cancellation in invertible algebras. Further, in cancellative idempotent algebras each of the hyperidentities (17.2.3)–(17.2.10) has the hyperidentity $X(x,y) = Y(x,y)$ as a consequence. For example, for the hyperidentity (17.2.4) we have:

$$\begin{aligned} X(x, X(y,y)) &= Y(X(x,y), Y(x,y)), \\ X(x,y) &= Y(X(x,y), Y(x,y)), \\ Y(X(x,y), X(x,y)) &= Y(X(x,y), Y(x,y)), \\ X(x,y) &= Y(x,y). \end{aligned}$$

□

Corollary 17.2.4 ([249,252]). *A non-trivial left and right hyperidentity of distributivity satisfied in a functionally non-trivial q-algebra must be of functional rank* 2 *and be* (d_1) *in the left distributive case and* (d_2) *in the right distributive case.*

See also [73,388] on distributive hyperidentities.

17.3. Invertible algebras with hyperidentities of distributivity

In this section, the solution of the problem of Belousov (1965) on characterization of invertible algebras with the non-trivial hyperidentities of distributivity (d_1), (d_2) is presented.

Theorem 17.3.1. *An invertible algebra $(Q;\Sigma)$ satisfies hyperidentities (d_1) and (17.2.1) if and only if there exists a commutative Moufang loop $Q(\circ)$ such that every operation A in Σ is defined as follows:*

$$A(x,y) = \varphi_A(x) \circ \psi_A(y),$$

where $\varphi_A, \psi_A \in \operatorname{Aut} Q(\circ)$, $\varphi_A \in \operatorname{Aut} Q(A)$, $(\psi_A, \tilde{\varepsilon}) \in \operatorname{Aut}(Q;\Sigma)$, $\varphi_A(x) \circ \psi_A(x) = x$, $x \circ \varphi_A(x) \in K_Q$ (the kernel[a] of the loop $Q(\circ)$) for all $A \in \Sigma$ and all $x, y \in Q$.

Proof (*Sufficiency*). Since $x \circ \varphi_B(x) \in K_Q$, on replacing x by $\varphi_B^{-1}(x)$ here we get

$$\varphi_B^{-1}(x) \circ x = x \circ \varphi_B^{-1}(x) \in K_Q.$$

On the other hand, since Moufang loops are biassociative,

$$x \circ \varphi_B^{-1}(x) = x^3 \circ (x^{-2} \circ \varphi_B^{-1}(x)),$$

whence we deduce, since $x^3 \in K_Q$ in commutative Moufang loops [52], that

$$x^{-2} \circ \varphi_B^{-1}(x) \in K_Q,$$
$$x^2 \circ (\varphi_B^{-1}(x))^{-1} \in K_Q$$

for all $x \in Q$ and $B \in \Sigma$.

By assumption, $Q(\circ)$ is a commutative Moufang loop, and so it satisfies the identity $x^2 \circ (y \circ z) = (x \circ y) \circ (x \circ z)$. However, since $x^2 \circ (\varphi_B^{-1}(x))^{-1} \in K_Q$,

[a]The left (right, middle) kernel of a loop $Q(\cdot)$ is the set of all a in Q such that $ax \cdot y = a \cdot xy$ ($xy \cdot a = x \cdot ya$, $xa \cdot y = x \cdot ay$, respectively) for all $x, y \in Q$. The kernel is the intersection of these three sets. The three kernels coincide in Moufang loops; and the same is true in loops with the inverse property (the so-called IP-loops). A loop $Q(\cdot)$ is a Moufang loop if it satisfies the identity $x(yz \cdot x) = xy \cdot zx$. Commutative Moufang loops are characterized by the single identity $x^2 \cdot yz = xy \cdot xz$.

we have

$$\begin{aligned}
&(\varphi_B^{-1}(x) \circ (y \circ z)) \circ (x^2 \circ (\varphi_B^{-1}(x))^{-1}) \\
&\quad = (x^2 \circ (\varphi_B^{-1}(x))^{-1}) \circ (\varphi_B^{-1}(x) \circ (y \circ z)) \\
&\quad = ((x^2 \circ (\varphi_B^{-1}(x))^{-1})(\varphi_B^{-1}(x)) \circ (y \circ z) \\
&\quad = x^2 \circ (y \circ z)
\end{aligned}$$

and similarly

$$(x \circ y) \circ (x \circ z) = ((x \circ y) \circ ((x^{-1} \circ \varphi_B^{-1}(x)) \circ z)) \circ (x^2 \circ (\varphi_B^{-1}(x))^{-1}).$$

Therefore,

$$\begin{aligned}
\varphi_B^{-1}(x) \circ (y \circ z) &= (x \circ y) \circ ((x^{-1} \circ \varphi_B^{-1}(x)) \circ z), \\
\varphi_B(\varphi_B^{-1}(x) \circ (y \circ z)) &= \varphi_B((x \circ y) \circ ((x^{-1} \circ \varphi_B^{-1}(x)) \circ z)), \\
x \circ \varphi_B(y \circ z) &= \varphi_B(x \circ y) \circ ((\varphi_B(x^{-1}) \circ x) \circ \varphi_B(z)), \qquad (17.3.1) \\
x \circ (\varphi_B(y) \circ \varphi_B(z)) &= (\varphi_B x \circ \varphi_B y) \circ (\psi_B x \circ \varphi_B z), \\
x \circ (y \circ z) &= (\varphi_B(x) \circ y) \circ (\psi(x) \circ z)
\end{aligned}$$

for all $x, y, z \in Q$ and all $B \in \Sigma$. Thus, for all $B \in \Sigma$ we get

$$\begin{aligned}
x \circ B(y, z) = x \circ (\varphi_B y \circ \psi_B z) &\overset{(17.3.1)}{=} (\varphi_B x \circ \varphi_B y) \circ (\psi_B x \circ \psi_B z) \\
&= \varphi_B(x \circ y) \circ \psi_B(x \circ z) = B(x \circ y, x \circ z).
\end{aligned}$$

Hyperidentities (d_1) and (17.2.1) are now easily verified:

$$\begin{aligned}
B(A(x, y), A(x, z)) &= B(\varphi_A x \circ \psi_A y, \varphi_A x \circ \psi_A z) \\
&= \varphi_A x \circ B(\psi_A y, \psi_A z) = \varphi_A x \circ \psi_A B(y, z) \\
&= A(x, B(y, z)); \\
A(A(x, z), A(y, z)) &= A(\varphi_A x \circ \psi_A z, \varphi_A y \circ \psi_A z) \\
&= A(\psi_A z \circ \varphi_A x, \psi_A z \circ \varphi_A y) = \psi_A z \circ A(\varphi_A x, \varphi_A y) \\
&= \psi_A z \circ \varphi_A A(x, y) = \varphi_A A(x, y) \circ \psi_A z = A(A(x, y), z).
\end{aligned}$$

(*Necessity*). Let $(Q; \Sigma)$ be an invertible algebra satisfying (d_1) and (17.2.1), and let e be any fixed element of Q. Clearly, $(Q; \Sigma)$ is idempotent. For each

$A \in \Sigma$ we consider the bijections

$$R_A : x \to A(x,e), \qquad x \in Q,$$
$$L_A : x \to A(e,x), \qquad x \in Q,$$

and define a new operation $\circ_A$ as follows:

$$u \circ_A v = A(R_A^{-1}(u), L_A^{-1}(v)), \qquad u, v \in Q.$$

It is easy to see that $Q(\circ_A)$ is a loop with identity element $e = A(e,e)$, for every A in Σ. Since $A(u,v) = R_A(u) \circ_A L_A(v)$ we have $u = A(u,u) = R_A(u) \circ_A L_A(u)$. Further,

$$\begin{aligned} R_A(L_B(u)) &= R_A B(e,u) = A(B(e,u),e) \\ &= A(B(e,u),B(e,e)) = B(e,A(u,e)) = L_B(R_A(u)), \\ L_A(L_B(u)) &= L_A B(e,u) = A(e,B(e,u)) \\ &= B(A(e,e),A(e,u)) = B(e,A(e,u)) = L_B(L_A(u)). \end{aligned}$$

Thus, we get for all A and B in Σ:

$$R_A L_B = L_B R_A, \qquad L_A L_B = L_B L_A.$$

Moreover, it follows from (d_1) that L_A is an automorphism of $(Q;\Sigma)$, and (17.2.1) implies that R_A is an automorphism of the quasigroup $(Q;A)$, for each $A \in \Sigma$:

$$L_A B(x,v) = A(e,B(x,y)) = B(A(e,x),A(e,y)) = B(L_A(x),L_A(y)),$$
$$R_A A(x,y) = A(A(x,y),e) = A(A(x,e),A(y,e)) = A(R_A(x),R_A(y)).$$

Starting from what was said above, we get

$$\begin{aligned} L_A(x \circ_B y) &= L_A B(R_B^{-1}x, L_B^{-1}y) \\ &= B(L_A(R_B^{-1}x), L_A(L_B^{-1}y)) = B(R_B^{-1}(L_A x), L_B^{-1}(L_A y)) \\ &= R_B(R_B^{-1}(L_A x)) \circ_B L_B(L_B^{-1}(L_A y) = L_A x \circ_B L_A y \end{aligned}$$

and similarly

$$R_A(x \circ_A y) = R_A(x) \circ_A R_A(y),$$

for all $A, B \in \Sigma$.

Putting $\varphi_A = R_A$ and $\psi_A = L_A$, we arrive at the equality

$$A(x, y) = \varphi_A(x) \circ_A \psi_A(y).$$

Let $(\circ_A) = (\circ)$, $\varphi_A = \varphi$, $\psi_A = \psi$. Then, starting from the left distributive law

$$A(x, A(y, z)) = A(A(x, y), A(x, z)),$$

we get

$$\varphi x \circ \psi(\varphi y \circ \psi z) = \varphi(\varphi x \circ \psi y) \circ \psi(\varphi x \circ \psi z),$$
$$\varphi x \circ (\psi(\varphi y) \circ \psi^2 z) = (\varphi^2 x \circ \varphi(\psi y)) \circ (\psi(\varphi x) \circ \psi^2 z).$$

We make the substitution $x \to \varphi^{-1}x$, $y \to \varphi^{-1}(\psi^{-1}y)$, $z \to \psi^{-2}z$ in this last equality; since $\varphi\psi = \psi\varphi$, we therefore deduce that

$$x \circ (y \circ z) = (\varphi x \circ y) \circ (\psi x \circ z). \tag{17.3.2}$$

Let $y = e$, so that

$$x \circ z = \varphi x \circ (\psi x \circ z);$$

on replacing z by $y \circ (\psi x)^{-1}$, where s^{-1} denotes the right inverse of s in the loop $Q(\circ)$, we get

$$x \circ (y \circ (\psi x)^{-1}) = \varphi x \circ (\psi x \circ (y \circ (\psi x)^{-1}))$$

or, by (17.3.2),

$$(\varphi x \circ y) \circ \psi x \circ (\psi x)^{-1}) = \varphi x \circ (\psi x \circ (y \circ (\psi x)^{-1})),$$
$$\varphi x \circ y = \varphi x \circ (\psi x \circ (y \circ (\psi x)^{-1}),$$
$$y = \psi x \circ (y \circ (\psi x)^{-1}).$$

Replacing x by $\psi^{-1}(x)$ in this equality gives the identity

$$x \circ (y \circ x^{-1}) = y. \tag{17.3.3}$$

We consider the left distributive law for the operation A in Σ, and transform it as follows:

$$\begin{aligned}
A(x, A(y, z)) &= A(A(x, y), A(x, z)),\\
A(x, A(y, z)) &= A(A(x, A(x, z)), A(y, A(x, z))),\\
A(x, A(x, z)) &= {}^{-1}A(A(x, A(y, z)), A(y, A(x, z))),\\
A(x, A(x, z)) &= {}^{-1}A(A(x, A(y, z)), A(A(y, x), A(y, z))),\\
A(x, A(x, z)) &= A({}^{-1}A(x, A(y, x)), A(y, z)).
\end{aligned}$$

Let $x = e$ and $\alpha(y) = {}^{-1}A(e, y)$; the last of these equalities then becomes

$$\begin{aligned}
\psi_A^2 = A(\alpha(\varphi(y)), A(y, z)) &= \varphi(\alpha(\varphi y)) \circ \psi(\varphi y \circ \psi z)\\
&= \varphi(\alpha(\varphi y)) \circ (\psi(\varphi y) \circ \psi^2(z)),
\end{aligned}$$

that is, $z = \varphi(\alpha(\psi^{-1}(y))) \circ (y \circ z) = I(y) \circ (y \circ z)$, where $I : Q \to Q$ is a permutation, and clearly $I(y) = y^{-1}$. Thus, we get the identity

$$x^{-1} \circ (x \circ y) = y, \tag{17.3.4}$$

and

$$x \circ y = (x^{-1})^{-1} \circ (x^{-1} \circ (x \circ y)) = (x^{-1})^{-1} \circ y.$$

So, $(x^{-1})^{-1} = x$.

Comparing identities (17.3.3) and (17.3.4), we find that $y \circ x^{-1} = x^{-1} \circ y$, that is, $Q(\circ)$ is a commutative inverse loop.

Replacing x by $\varphi^{-1}x$ in identity (17.3.2) and using the equality

$$\varphi^{-1}x = A(\varphi^{-1}x, \varphi^{-1}x) = \varphi(\varphi^{-1}x) \circ \psi(\varphi^{-1}x) = x \circ \psi(\varphi^{-1}x),$$

we get

$$(x \circ \psi(\varphi^{-1}x)) \circ (y \circ z) = (x \circ y) \circ (\psi(\varphi^{-1}x) \circ z),$$

that is,

$$(x \circ f(x)) \circ (y \circ z) = (x \circ y) \circ (f(x) \circ z), \tag{17.3.5}$$

where $f(x) = \psi(\varphi^{-1}x)$. Thus,

$$(x \circ y) \circ (f(x) \circ z) = (x \circ f(x)) \circ (z \circ y) = (x \circ z) \circ (f(x) \circ y),$$

that is,

$$\begin{aligned}(x\circ y)\circ L_{f(x)}(z) &= (x\circ z)\circ L_{f(x)}(y),\\ (x\circ L^{-1}_{f(x)}(y))\circ z &= (x\circ L^{-1}_{f(x)}(z))\circ y,\\ L_x(L^{-1}_{f(x)}(y))\circ z &= L_x(L^{-1}_{f(x)}(z))\circ y,\end{aligned}$$

where $L_a(z) = a\circ z$, $L_a(L_a^{-1}(z)) = a\circ L_a^{-1}(z)$, $z = a\circ L_a^{-1}(z), a^{-1}\circ z = a^{-1}(a\circ L^{-1}(z))$, $a^{-1}\circ z = L_a^{-1}(z)$. Consequently,

$$\begin{aligned}(u\circ L_x(L^{-1}_{f(x)}(e)))\circ v &= (e\circ L^{-1}_{f(x)}(u)))\circ v\\ &= L_x(L^{-1}_{f(x)}(u))\circ v = u\circ L_x(L^{-1}_{f(x)}(v))\\ &= u\circ(e\circ L_x(L^{-1}_{f(x)}v)) = u\circ(L_x(L^{-1}_{f(x)}e)\circ v),\end{aligned}$$

that is,

$$L_x(L^{-1}_{f(x)}e) = x\circ(f(x)^{-1}\circ e) = x\circ f(x)^{-1}\in K_Q \tag{17.3.6}$$

(the three kernels coincide in inverse loops).

Starting from the inclusion just proved and using (17.3.5), it is easy to check that the commutative loop $Q(\circ)$ is Moufang:

$$\begin{aligned}x^2\circ(y\circ z) &= (y\circ z)\circ x^2\\ &= (y\circ z)\circ((x\circ x)\circ(f(x)\circ f(x)^{-1}))\\ &= (y\circ z)\circ((x\circ f(x))\circ(x\circ f(x)^{-1}))\\ &= ((y\circ z)\circ(x\circ f(x)))\circ(x\circ f(x)^{-1})\\ &= (x\circ f(x))\circ(y\circ z)\circ(x\circ f(x)^{-1})\\ &= ((x\circ y)\circ(f(x)\circ z))\circ(x\circ f(x)^{-1})\\ &= (x\circ y)\circ((f(x)\circ z)\circ(x\circ f(x)^{-1}))\\ &= (x\circ y)\circ((z\circ f(x))\circ(x\circ f(x)^{-1}))\\ &= (x\circ y)\circ(z\circ(f(x)\circ(f(x)^{-1}\circ x)))\\ &= (x\circ y)\circ(z\circ x) = (x\circ y)\circ(x\circ z).\end{aligned}$$

We claim that $x \circ \varphi(x) \in K_Q$. Starting from (17.3.2), we get

$$\begin{aligned}(x \circ (x^{-1} \circ \varphi^{-1}(x))) \circ (y \circ z)\varphi^{-1}x \circ (y \circ z) &= (x \circ y) \circ (\psi(\varphi^{-1}x) \circ z) \\ &= (x \circ y) \circ ((\varphi^{-1}x \circ x^{-1}) \circ z) \\ &= (x \circ y) \circ ((x^{-1} \circ \varphi^{-1}x) \circ z),\end{aligned}$$

since $\psi(x) = x \circ \varphi(x^{-1})$. Therefore, we have arrived at an equality of the form (17.3.5) for the function $f(x) = x^{-1} \circ \varphi^{-1}(x)$. By (17.3.6),

$$x^{-1} \circ f(x) \in K_Q,$$
$$x^{-1} \circ (x^{-1} \circ \varphi^{-1}(x)) = x^{-2} \circ \varphi^{-1}(x) \in K_Q.$$

However, $x^3 \in K_Q$, and it is clear that K_Q is invariant under the automorphism φ, so that $x^3 \circ (x^{-2} \circ \varphi^{-1}x) \in K_Q$, $x \circ \varphi^{-1}(x) \in K_Q$, $\varphi(x) \circ x \in K_Q$.

All that remains to complete the proof of Theorem 17.3.1 is this: in an invertible algebra $(Q; \Sigma)$ satisfying (d_1) and (17.2.1) we have $\circ_A = \circ_B$ for all $A, B \in \Sigma$. □

Lemma 17.3.2. *If $(Q; \Sigma)$ is an invertible algebra with (d_1), the following condition holds for all $A, B \in \Sigma$:*

$$x \circ_A B(y, z) = B(x \circ_A y, x \circ_A z). \qquad (17.3.7)$$

Proof. By (d_1) we have

$$Ax, B(y, z) = B(A(x, y), A(x, z)),$$

so that

$$R_A x \circ_A L_A B(y, z) = B(R_A x \circ_A L_A y, R_A x \circ_A L_A z).$$

Since L_A is an automorphism of the algebra $(Q; \Sigma)$ and of the loop $Q(\circ_A)$, so is L_A^{-1}:

$$L_A^{-1}(R_A x \circ_A L_A B(y, z)) = L_A^{-1} B(R_A x \circ_A L_A y, R_A x \circ_A L_A z),$$
$$L_A^{-1}(R_A x) \circ_A B(y, z) = B(L_A^{-1}(R_A x) \circ_A y, L_A^{-1}(R_A x) \circ_A z),$$

that is,

$$x \circ_A B(y, z) = B(x \circ_A y, x \circ_A z).$$

□

Lemma 17.3.3. *Let $(Q;\Sigma)$ be an invertible algebra satisfying hyperidentities (d_1) and (17.2.1), and let $(\circ_A) = (\circ)$ for an element of A of Σ. For arbitrary x, y in Q and B in Σ, the subloop of the loop $Q(\circ)$ generated by x, y and $L_B(x)$ is a group.*

Proof. By (17.3.7),

$$x \circ B(y,z) = B(x \circ y, x \circ z),$$

whence, with $x = z^{-1}$ (inverse in $Q(\circ)$), we get

$$\begin{aligned} z^{-1} \circ B(y,z) &= B(z^{-1} \circ y, e) = R_B(y \circ z^{-1}), \\ B(y,z) &= R_B(y \circ z^{-1}) \circ z. \end{aligned}$$

Similarly, with $x = y^{-1}$ we get

$$B(y,z) = L_B(z \circ y^{-1}) \circ y,$$

so that

$$R_B(y \circ z^{-1}) \circ z = L_B(z \circ y^{-1}) \circ y.$$

Putting $y = e$ here and replacing z by $z \circ y^{-1} = x$, we get

$$\begin{aligned} R_B(y \circ z^{-1}) \circ (z \circ y^{-1}) &= L_B(z \circ y^{-1}), \\ R_B(y \circ z^{-1}) &= L_B(z \circ y^{-1}) \circ (z \circ y^{-1})^{-1}, \\ R_B(y \circ z^{-1}) \circ z &= (L_B(z \circ y^{-1}) \circ (z \circ y^{-1})^{-1}) \circ z, \\ L_B(z \circ y^{-1}) \circ y &= (L_B(z \circ y^{-1}) \circ (z \circ y^{-1})^{-1}) \circ z, \\ L_B(x) \circ (x^{-1} \circ z) &= (L_B(x) \circ x^{-1}) \circ z, \end{aligned}$$

that is, the associative law holds for $L_B(x)$, x^{-1} and z. Moufang's theorem [243] says that three elements a, b and c of a Moufang loop such that $a \circ (b \circ c) = (a \circ b) \circ c$ generate an associative subloop, that is, a group, so that $L_B(x)$, x and z generate an associative subloop, as required. □

Corollary 17.3.4. *If $(Q;\Sigma)$ is an invertible algebra satisfying (d_1) and (17.2.1), then $(\circ_A) = (\circ_B)$ for all A and B in Σ.*

Proof. Let $(\circ_B) = (\oplus)$, $(\circ_A) = (\circ)$. Since L_B is an automorphism of the loop $Q(\circ)$, we have

$$B(y,z) = R_B y \oplus L_B z = L_B(z \circ y^{-1}) \circ y = (L_B(z) \circ (L_B y)^{-1}) \circ y,$$

whence we get $R_B y = (L_B y)^{-1} \circ y$ on putting $z = e$. Applying Lemma 17.3.3 to the elements y, $L_B y$ and $L_B z$, we get

$$\begin{aligned} R_B y \oplus L_B z &= (L_B z \circ (L_B y)^{-1}) \circ y = L_B z \circ ((L_B y)^{-1} \circ y) \\ &= L_B z \circ R_B y = R_B y \circ L_B z. \end{aligned}$$

This completes the proof of Theorem 17.3.1. □

We have the similar proof for the following result.

Theorem 17.3.1′. *An invertible algebra $(Q;\Sigma)$ satisfies the hyperidentities (d_2) and (17.2.2) if and only if there is a commutative Moufang loop $Q(\circ)$ such that every operation A in Σ is defined by the rule*

$$A(x,y) = \varphi_A(x) \circ \psi_A(y),$$

in which $\varphi_A, \psi_A \in \operatorname{Aut} Q(\circ)$, $\psi_A \in \operatorname{Aut} Q(A)$, $(\varphi_A, \tilde{\varepsilon}) \in \operatorname{Aut}(Q;\Sigma)$, $\varphi_A(x) \circ \psi_A(x) = x$, $x \circ \varphi_A(x) \in K_Q$ for all $A \in \Sigma$, $x, y \in Q$.

Corollary 17.3.5. *An invertible algebra $(Q;\Sigma)$ satisfies the hyperidentities (d_1) and (d_2) if and only if there is a commutative Moufang loop $Q(\circ)$ such that every operation A in Σ is defined by the rule*

$$A(x,y) = \varphi_A(x) \circ \psi_A(y),$$

where $\varphi_A, \psi_A \in \operatorname{Aut} Q(\circ)$, $(\varphi_A, \tilde{\varepsilon})$, $(\psi_A, \tilde{\varepsilon}) \in \operatorname{Aut}(Q;\Sigma)$, $\varphi_A(x) \circ \psi_A(x) = x$, $x \circ \varphi_A(x) \in K_Q$ for all $A \in \Sigma$, $x, y \in Q$.

The problem of characterizing q-algebras with (d_1) and (17.2.1), or with (d_2) and (17.2.2), is still open (see [252] for the situation in which (d_1) and (d_2) are both satisfied).

In conclusion, we state some results that follow from Theorems 17.3.1 and 17.3.1′ and Corollary 17.3.5.

Corollary 17.3.6. *If the hyperidentities (d_1) and (17.2.1) (or (d_2) and (17.2.2), or (d_1) and (d_2)) are satisfied in an invertible algebra $(Q;\Sigma)$, the same is true in the algebra $(Q; \Sigma \cup \Sigma^{-1} \cup {}^{-1}\Sigma)$.*

Corollary 17.3.7. *If the hyperidentities (d_1) and (17.2.1) (or (d_2) and (17.2.2), or (d_1) and (d_2)) are satisfied in an invertible algebra $(Q;\Sigma)$, the same is true in the algebra $(Q; \Sigma \cup \Sigma^{-1} \cup {}^{-1}\Sigma \cup {}^{-1}(\Sigma^{-1}) \cup ({}^{-1}\Sigma)^{-1} \cup \Sigma^*)$*

Corollary 17.3.8. *An invertible algebra satisfying hyperidentities* (d_1) *and* (17.2.1) *is Abelian if and only if it satisfies* (d_2) *and the corresponding Moufang loop (as in Theorem* 17.3.1*) is a commutative group.*

Corollary 17.3.8′. *An invertible algebra satisfying hyperidentities* (d_2) *and* (17.2.2) *is Abelian if and only if it satisfies* (d_1) *and the corresponding Moufang loop (as in Theorem* 17.2.10′*) is associative.*

Corollary 17.3.9. *An invertible algebra* $(Q;\Sigma)$ *satisfying hyperidentities* (d_1) *and* (d_2) *is Abelian if and only if it satisfies the identity*

$$A(B(x,y),B(z,u)) = B(A(x,z),A(y,u)$$

for some A *and* B *in* Σ.

Corollary 17.3.10. *Let* $(Q;\Sigma)$ *be an invertible algebra satisfying hyperidentities* (d_1) *and* (d_2) *and also the relation*

$$A(B(a,b),B(c,d) = B(A(a,c),A(b,d))$$

for some $a,b,c,d \in Q$ *and* $A,B \in \Sigma$. *Then the invertible subalgebra generated by* $\{a,b,c,d\}$ *is Abelian.*

Corollary 17.3.11. *Every three-generator subalgebra of an invertible algebra satisfying* (d_1) *and* (d_2) *is Abelian.*

Chapter 18

Essentially Non-trivial Associative and Distributive Hyperidentities in Semigroups

18.1. Preliminary concepts and results

Proposition 18.1.1. *In a non-singleton binary algebra $(Q;\Sigma)$ such that Σ contains δ_2^1 and δ_2^2 (and therefore in the termal algebra of every non-singleton algebra), we have:*

(1) *no non-trivial associative hyperidentity is satisfied;*

If the algebra $(Q;\Sigma)$ satisfies:

(2) *a non-trivial left distributive hyperidentity, then the hyperidentity has functional rank 2 and is of the form (d_1);*
(3) *a non-trivial right distributive hyperidentity, then the hyperidentity has functional rank 2 and is of the form (d_2),*

where $\delta_2^1(x,y) = x$ and $\delta_2^2(x,y) = y$ for any $x, y \in Q$.

The proof of this proposition is by straightforward verification.

Let $Q(\cdot)$ be a semigroup. The following function is said to be its *binary polynomial* (term):

$$F(x,y) = z_1^{\varepsilon_1} z_2^{\varepsilon_2} \dots z_n^{\varepsilon_n}, \tag{18.1.1}$$

where $n \in \mathbb{N}, \varepsilon_1, \varepsilon_2, \dots, \varepsilon_n \in \mathbb{N}$, $z_1, z_2, \dots, z_n \in \{x, y\}$ and $z_i \neq z_{i+1}$. The number n is called the length of this representation of the polynomial $F(x,y)$. However, due to the identities in the semigroup $Q(\cdot)$, the same polynomial $F(x,y)$ can have different representations of the form (18.1.1). The ternary polynomial $F(x,y,z)$ is defined similarly.

For example, if semigroup $Q(\cdot)$ satisfies the identities $x^2 = x^3$ and $x^2 = y^2$, then it has the following binary polynomials only: x, y, xy, yx, xyx, yxy, x^2.

By Q^2_{pol} we denote the collection of all binary polynomials of the semigroup $Q(\cdot)$.

We say that the hyperidentity $(*)$ is *polynomially satisfied* (valid) in the semigroup $Q(\cdot)$ if this hyperidentity is satisfied in the binary algebra $\left(Q; Q^2_{\text{pol}}\right)$.

For instance:

(1) The semigroup $Q(\cdot)$ polynomially satisfies the trivial hyperidentity of left distributivity

$$X(x, X(y, z)) = X(X(x, y), X(x, z))$$

if and only if $Q(\cdot)$ satisfies the following identities: $x^2 = x^3$, $xyz = xyxz$, $xyz = xzyz$;

(2) The semigroup $Q(\cdot)$ polynomially satisfies the hyperidentity (d_1) if and only if $Q(\cdot)$ satisfies the following identities: $x^2 = x$, $xyz = xyxz$, $xyz = xzyz$.

In [88], it is proved that the class of all semigroups polynomially satisfying the trivial associative hyperidentity:

$$X(x, X(y, z)) = X(X(x, y), z), \qquad (*,*)$$

forms a finitely based variety of semigroups, and the basis of this variety contains about 1000 identities. These semigroups are called *hyperassociative*. For example, any two-element semigroup is hyperassociative [88]. In [330] (see also [331]), a basis of the identities of the same variety is given, which contains the following four identities:

$$x^4 = x^2,$$

$$xyxzxyx = xyzyx,$$

$$xy^2z^2 = xyz^2yz^2,$$

$$x^2y^2z = x^2yx^2yz,$$

i.e., a semigroup is hyperassociative if and only if it satisfies these four identities (see also [93,179,290,313,314]).

Moreover, the semigroup $Q(\cdot)$ polynomially satisfies the following hyperidentity:

$$X(x, X(x, y)) = X(X(x, x), y)$$

if and only if $Q(\cdot)$ satisfies the following four identities:

$$x^4 = x^2,$$
$$x^3yx^3 = x^2yx^2,$$
$$x^2y^2x^2y^2 = x^2y^2,$$
$$yx^3y = yxyxyxy.$$

We need the following definitions.

Definition 18.1.2. The binary polynomial $F(x, y)$ depends on the variable x essentially in the semigroup $Q(\cdot)$ if there are elements $x_1, x_2, y \in Q$ such that $F(x_1, y) \neq F(x_2, y)$. In the same way, the essentially dependence of the polynomial $F(x, y)$ on the variable y is defined.

Definition 18.1.3. The binary polynomial $F(x, y)$ is called essential if it depends on both variables x and y essentially.

We also let Q^2_{epol} be the collection of all binary essential polynomials of the semigroup $Q(\cdot)$.

We say that the hyperidentity $(*)$ is *essentially satisfied* (valid) or is *satisfied for essential polynomials* in the semigroup $Q(\cdot)$ if this hyperidentity is satisfied in the binary algebra $\left(Q; Q^2_{\text{epol}}\right)$. In this case, the hyperidentity $(*)$ is called *essential hyperidentity in semigroup.*

Definition 18.1.4. We say that two hyperidentities are equivalent in semigroups (written as $\Leftrightarrow$) if they simultaneously are either polynomially satisfied or none of them is polynomially satisfied in any semigroup $Q(\cdot)$. It is said that the hyperidentity (h_1) implies the hyperidentity (h_2), written as $(h_1) \Rightarrow (h_2)$, if in all semigroups where the hyperidentity (h_1) is satisfied polynomially, the hyperidentity (h_2) is also satisfied polynomially.

For example, the hyperidentities (d_1) and (d_2) are equivalent in semigroups. The following two hyperidentities are equivalent in semigroups too:

$$X(x, X(x, y)) = X(X(x, x), y),$$
$$X(x, X(y, y)) = X(X(x, y), y).$$

Indeed, a binary polynomial F of the semigroup $Q(\cdot)$ satisfies to the identity

$$F(F(x,x),y) = F(x,F(x,y))$$

if and only if the binary polynomial $F^*(x,y) = F(y,x)$ satisfies to the identity

$$F^*(F^*(y,x),x) = F^*(y,F^*(x,x)),$$

and $F = (F^*)^*$.

Likewise, we also define the following definition.

Definition 18.1.5. We say that two hyperidentities are essentially equivalent in semigroups (written as $\Leftrightarrow_e$) if they simultaneously are either essentially satisfied or none of them is essentially satisfied in any semigroup $Q(\cdot)$. It is said that the hyperidentity (h_1) essentially implies the hyperidentity (h_2), written as $(h_1) \Rightarrow_e (h_2)$, if in all semigroups where the hyperidentity (h_1) is satisfied essentially, the hyperidentity (h_2) is also satisfied essentially.

Consider the following system of identities:

$$\begin{cases} xyz = zxy, \\ x^2yz = xy^2z, \\ x^3yz = x^2yz. \end{cases} \tag{18.1.2}$$

Lemma 18.1.6. *Any essential polynomial $X(x,y)$ in a semigroup $Q(\cdot)$ which satisfies the identities* (18.1.2) *equals one of the polynomials from the following set: $P = \{xy, yx, x^2y, xy^2, x^2y^2\}$.*

Proof. From Eq. (18.1.2) implies that any essential polynomial $F(x,y)$ is equal to a polynomial of the form x^my^n or yx for some positive integers m,n.

Towards the proof of the lemma it remains to show that any polynomial of the form x^my^n is equal to one in P.

If we take $z = y$ in the third identity of (18.1.2), we get $x^3y^2 = x^2y^2$. On the other hand, by the first identity of (18.1.2), we have $x^2y^3 = y^3x^2$. Therefore, $x^2y^3 = y^3x^2 = y^2x^2 = x^2y^2 = x^3y^2$. Hence,

$$x^2y^n = x^2y^2, \quad n \geqslant 3,$$

$$x^ny^2 = x^2y^2, \quad n \geqslant 3,$$

$$xy^3 = xy^2y = x^2y^2,$$

$$x^3y = xx^2y = yxx^2 = yx^3 = y^2x^2 = x^2y^2,$$
$$xy^n = (xy^3)y^{n-3} = x^2y^2y^{n-3} = x^2y^{n-1} = x^2y^2, \quad n > 3,$$
$$x^ny = x^{n-3}(x^3y) = x^{n-3}x^2y^2 = x^{n-1}y^2 = x^2y^2, \quad n > 3.$$

Let $X(x,y) = x^my^n$ be a polynomial. Now, if $m > 2$, $n > 3$, we have $X(x,y) = x^my^n = x^{m-2}x^2y^3y^{n-3} = x^{m-2}x^2y^2y^{n-3} = x^my^{n-1} = \cdots = x^my^3 = x^{m-2}(x^2y^3) = x^{m-2}x^2y^2 = x^my^2 = x^2y^2$.

By using these procedures, we can reduce the exponents to be less than or equal to 2 and, thus, $X(x,y)$ is equal to a polynomial in P. □

Corollary 18.1.7. *Any polynomial $X(x,y) \neq x^n, y^m$ in a semigroup $Q(\cdot)$ which satisfies the identities* (18.1.2) *equals one of the polynomials from the following set: $P = \{xy, yx, x^2y, xy^2, x^2y^2\}$*

18.2. Essentially non-trivial associative hyperidentities

Theorem 18.2.1. *The hyperidentity*

$$X(X(x,y),z) = X(x,Y(y,z)) \tag{18.2.1}$$

is essentially satisfied in the semigroup $Q(\cdot)$ if and only if that semigroup satisfies one of the following systems of identities:

$$\begin{cases} xyz = xzy, \\ xyz = zxy, \\ x^2y = x^3, \end{cases} \tag{18.2.2}$$

$$\begin{cases} xyz = xzy, \\ xyz = zxy, \\ x^2y = y^3, \end{cases} \tag{18.2.3}$$

$$\begin{cases} xyz = zxy, \\ x^2yz = xyz. \end{cases} \tag{18.2.4}$$

Proof. First, we prove the necessity.

In the case, when the polynomial xy is not essential then we have either $xy = x^2$ or $xy = y^2$ and they are no essential polynomial.

Now consider the case when xy is essential. Then so is the polynomial yx.

By taking $X(x,y) = xy$ and $Y(x,y) = yx$, we get $xyz = xzy$.

By taking $X(x,y) = yx$ and $Y(x,y) = xy$, we get $zyx = yzx$ which with the previous identity implies $xyz = xzy = zxy$.

Consider the case when x^2y is not essential. This implies that $x^2y = x^3$ or $x^2y = y^3$ and, therefore, one of the systems (18.2.2) and (18.2.3) of identities is satisfied.

Now assume the polynomial x^2y is also essential. Then so is the polynomial y^2x.

Take $X(x,y) = yx$ and $Y(x,y) = y^2x$ in the hyperidentity (18.2.1), and get $z^2yx = zyx$ and so $x^2yz = xyz$. Then the semigroup $Q(\cdot)$ satisfies the system (18.2.4) of identities.

This completes the proof of the necessity of the theorem.

Towards the proof of the sufficiency of the theorem, we consider the following cases:

(1) If (18.2.2) is satisfied in $Q(\cdot)$, then the only possible essential polynomials are xy, yx. For these polynomials, the hyperidentity is satisfied due to the first two identities of (18.2.2).

(2) If (18.2.3) is satisfied in $Q(\cdot)$, then the only possible essential polynomials are xy, yx. For these polynomials, the hyperidentity is satisfied due to the first two identities of (18.2.3).

(3) Now, assume (18.2.4) is satisfied in $Q(\cdot)$. Let us first prove that this semigroup satisfies the conditions of Lemma 18.1.6, that is, $Q(\cdot)$ satisfies (18.1.2).

The first identity of (18.1.2) is the same as the first identity of (18.2.4).

For the second identity of (18.1.2), note that $x^2yz = xyz = zxy = yzx = y^2zx = xy^2z$.

For the third identity of (18.1.2), note that $x^3yz = x(x^2yz) = xxyz = xyz$.

Now, by Lemma 18.1.6, the only possible essential polynomials are xy, yx, x^2y, xy^2, x^2y^2. Moreover, we have $x^2y^2 = xxy^2 = y^2xx = yxx = xxy = x^2y$ and $xy^2 = xyy = x^2yy = x^2y^2 = x^2y$. Thus, the possible different essential polynomials are xy, yx, x^2y.

Let us consider those polynomials separately.

(a) If $X(x,y) = xy$, then we should prove $xY(y,z) = xyz$:

(i) $Y(y,z) = yz \Rightarrow xyz = xyz$,

(ii) $Y(y,z) = zy \Rightarrow xzy = x^2zy = (xxz)y = (zxx)y = zx^2y = x^2yz = xyz$,

(iii) $Y(y,z) = y^2z \Rightarrow xy^2z = y^2zx = yzx = xyz$.

(b) If $X(x,y) = yx$, then we should prove $Y(y,z)x = zyx$:

  (i) $Y(y,z) = yz \Rightarrow yzx = xyz = xzy = zyx$, by the previous case,
  (ii) $Y(y,z) = zy \Rightarrow zyx = zyx$,
  (iii) $Y(y,z) = y^2z \Rightarrow y^2zx = y^2zx = yzx = xyz$.

(c) If $X(x,y) = x^2y$, then we should prove $x^2Y(y,z) = x^2yx^2yz$. By the previous cases we have $x^2Y(y,z) = x^2yz = xyz$ and $x^2yx^2yz = (xyx)yz = x^2y^2z = xy^2z = y^2zx = yzx = xyz$.

This finishes the proof of the theorem. □

Theorem 18.2.2. *The hyperidentity*

$$X\left(X\left(x,y\right),z\right) = Y\left(x,X\left(y,z\right)\right) \tag{18.2.5}$$

is essentially satisfied in the semigroup $Q(\cdot)$ if and only if that semigroup satisfies one of the following systems of identities: (18.2.4) and

$$\begin{cases} xyz = zxy, \\ x^2y = x^3, \end{cases} \tag{18.2.6}$$

$$\begin{cases} xyz = zxy, \\ x^2y = y^3. \end{cases} \tag{18.2.7}$$

Proof. First, we prove the necessity.

In the case, when the polynomial xy is not essential then we have either $xy = x^2$ or $xy = y^2$ and they are no essential polynomial.

Now consider the case when xy is essential. Then so is the polynomial yx.

By taking $X(x,y) = xy$ and $Y(x,y) = yx$, we get $xyz = yzx$ and thus $xyz = yzx = zxy$.

Consider the case when x^2y is not essential. This implies that $x^2y = x^3$ or $x^2y = y^3$ and, therefore, one of the systems (18.2.6) and (18.2.7) of identities is satisfied.

Now assume the polynomial x^2y is also essential. Then so is the polynomial y^2x.

Take $X(x,y) = xy$ and $Y(x,y) = x^2y$ in the hyperidentity (18.2.5), and get $x^2yx = xyz$. Then the semigroup $Q(\cdot)$ satisfies the system (18.2.4) of identities.

This completes the proof of the necessity of the theorem.

Towards the proof of the sufficiency of the theorem, we consider the following cases:

(1) If (18.2.6) is satisfied in $Q(\cdot)$, then the only possible essential polynomials are xy, yx. For these polynomials, the hyperidentity is satisfied due to the first identity of (18.2.6).
(2) If (18.2.7) is satisfied in $Q(\cdot)$, then the only possible essential polynomials are xy, yx. For these polynomials, the hyperidentity is satisfied due to the first identity of (18.2.7).
(3) Now, assume (18.2.4) is satisfied in $Q(\cdot)$. As in the proof of the previous theorem, $Q(\cdot)$ satisfies (18.1.2).
   Now, by Lemma 18.1.6, the only possible essential polynomials are xy, yx, x^2y, xy^2, x^2y^2. Moreover, we have $x^2y^2 = xxy^2 = y^2xx = yxx = xxy = x^2y$ and $xy^2 = xyy = x^2yy = x^2y^2 = x^2y$. Thus, the possible different essential polynomials are xy, yx, x^2y.
   Let us consider those polynomials separately.
   (a) If $X(x,y) = xy$, then we should prove $Y(x, yz) = xyz$:
      (i) $Y(x,y) = xy \Rightarrow xyz = xyz$,
      (ii) $Y(x,y) = yx \Rightarrow yzx = xyz$,
      (iii) $Y(x,y) = x^2y \Rightarrow x^2yz = xyz$.
   (b) If $X(x,y) = yx$, then we should prove $Y(x, zy) = zyx$:
      By the previous case, we have $Y(x, zy) = xzy = zyx$.
   (c) If $X(x,y) = x^2y$, then we should prove $Y(x, y^2z) = x^2yx^2yz$.
      By the first cases we have $Y(x, y^2z) = xy^2z = y^2zx = yzx = xyz$ and $x^2yx^2yz = (xyx)yz = x^2y^2z = xy^2z = y^2zx = yzx = xyz$.

This finishes the proof of the theorem. □

Theorem 18.2.3. *The hyperidentity*

$$X\left(X\left(x,y\right),z\right) = Y\left(x, Y\left(y,z\right)\right) \tag{18.2.8}$$

is essentially satisfied in the semigroup $Q(\cdot)$ if and only if that semigroup satisfies one of the following systems of identities: (18.2.4) and

$$\begin{cases} xyz = zyx, \\ xyx = x^3, \\ x^2y = x^3, \end{cases} \tag{18.2.9}$$

$$\begin{cases} xyz = zyx, \\ xyx = y^3, \\ x^2y = x^3, \end{cases} \tag{18.2.10}$$

$$\begin{cases} xyz = zyx, \\ xyx = x^3, \\ x^2y = y^3, \end{cases} \tag{18.2.11}$$

$$\begin{cases} xyz = zyx, \\ xyx = y^3, \\ x^2y = y^3. \end{cases} \tag{18.2.12}$$

Proof. First, we prove the necessity.

In the case, when the polynomial xy is not essential then we have either $xy = x^2$ or $xy = y^2$ and they are no essential polynomial.

Now consider the case when xy is essential. Then so is the polynomial yx.

By taking $X(x, y) = xy$ and $Y(x, y) = yx$, we get $xyz = zyx$.

If the polynomial xyx is essential, then so is the polynomial yxy and so by taking $X(x, y) = yxy$ and $Y(x, y) = xy$, we get $zyxyz = xyz$ and so $xyz = (z(yx)y)z = y(yx)zz = y^2xz^2 = (yyx)z^2 = xy^2z^2 \Rightarrow xyz = xy^2z^2 = x(y^2zz) = xz^2y^2 = xzy$. This implies that $x^2y = xxy = xyx$ is also essential.

On the other hand, if the polynomial x^2y is essential, then so is the polynomial y^2x. Take $X(x, y) = xy^2$ and $Y(x, y) = xy$ in the hyperidentity (18.2.8), and get $xy^2z^2 = xyz$ and so $xyz = xy^2z^2 == x(yyz^2) = xz^2y^2 = xzy \Rightarrow xyz = zyx = zxy$. Thus, $xyx = xxy = x^2y$ is also essential polynomial.

As a result, the polynomials xyx and x^2y are essential at the same time.

Therefore, there are the following cases:

(1) x^2y and xyx are not essential, so
   (a) $xyx = x^3$ and $x^2y = x^3$, which is (18.2.9).
   (b) $xyx = y^3$ and $x^2y = x^3$, which is (18.2.10).
   (c) $xyx = x^3$ and $x^2y = y^3$, which is (18.2.11).
   (d) $xyx = y^3$ and $x^2y = y^3$, which is (18.2.12).

(2) x^2y and xyx are essential, so by the results above, (18.2.4) is satisfied.

This completes the proof of the necessity of the theorem.

Towards the proof of the sufficiency of the theorem, we consider the following cases:

(1) If (18.2.9) is satisfied in $Q(\cdot)$, then the only possible essential polynomials are xy, yx. For these polynomials, the hyperidentity is satisfied due to the first two identities of (18.2.2).

(2) If (18.2.10) is satisfied in $Q(\cdot)$, then the only possible essential polynomials are xy, yx. For these polynomials, the hyperidentity is satisfied due to the first two identities of (18.2.2).

(3) If (18.2.11) is satisfied in $Q(\cdot)$, then the only possible essential polynomials are xy, yx. For these polynomials, the hyperidentity is satisfied due to the first two identities of (18.2.2).

(4) If (18.2.12) is satisfied in $Q(\cdot)$, then the only possible essential polynomials are xy, yx. For these polynomials, the hyperidentity is satisfied due to the first two identities of (18.2.2).

(5) Now, assume (18.2.4) is satisfied in $Q(\cdot)$. As in the proof of the previous theorem, $Q(\cdot)$ satisfies (18.1.2).

Now, by Lemma 18.1.6, the only possible essential polynomials are xy, yx, x^2y, xy^2, x^2y^2. Moreover, we have $x^2y^2 = xxy^2 = y^2xx = yxx = xxy = x^2y$ and $xy^2 = xyy = x^2yy = x^2y^2 = x^2y$. Thus, the possible different essential polynomials are xy, yx, x^2y.

Let us consider those polynomials separately.

(a) If $X(x,y) = xy$, then we should prove $Y(x, Y(y,z)) = xyz$:

  (i) $Y(x,y) = xy \Rightarrow xyz = xyz$,
  (ii) $Y(x,y) = yx \Rightarrow zyx = zxy = xyz$,
  (iii) $Y(x,y) = x^2y \Rightarrow x^2y^2z = y^2zx = yzx = xyz$.

(b) If $X(x,y) = yx$, then we should prove $Y(x, Y(y,z)) = zyx$:
By the previous case, we have $Y(x, Y(y,z)) = xyz = zyx$.

(c) If $X(x,y) = x^2y$, then we should prove $Y(x, Y(y,z)) = x^2yx^2yz$:
By the first case, we have $Y(x, Y(y,z)) = xyz$ and $x^2yx^2yz = (xyx)yz = x^2y^2z = xy^2z = y^2zx = yzx = xyz$.

This finishes the proof of the theorem. □

Theorem 18.2.4. *The hyperidentity*

$$X\left(Y\left(x,y\right),z\right) = X\left(x, X\left(y,z\right)\right) \tag{18.2.13}$$

is essentially satisfied in the semigroup $Q(\cdot)$ if and only if that semigroup satisfies one of the following systems of identities: (18.2.2), (18.2.3) and (18.2.4).

Proof. We prove that this hyperidentity is equivalent to the hyperidentity (18.2.1).

Note that a polynomial $X(x,y)$ is essential if and only if so is the polynomial $Z(x,y) = X(y,x)$. Now, if we replace $X(x,y)$ with $X(y,x)$ and $Y(x,y)$ with $Y(y,x)$ in the hyperidentity (18.2.1), we will get $X(z,X(y,x)) = X(Y(z,y),x)$ and therefore if we also replace z with x and x with z, we will get exactly the hyperidentity (18.2.13).

This implies that the necessary and sufficient conditions for the semigroup $Q(\cdot)$ to satisfy the hyperidentity (18.2.1) or (18.2.13) are the same.

Thus, this finishes the proof of the theorem. □

Theorem 18.2.5. *The hyperidentity*

$$X\left(Y\left(x,y\right),z\right) = X\left(x,Y\left(y,z\right)\right) \tag{18.2.14}$$

is essentially satisfied in the semigroup $Q(\cdot)$ if and only if that semigroup satisfies one of the following systems of identities: (18.2.6), (18.2.7) and

$$\begin{cases} xyz = zxy, \\ x^2yz = xy^2z, \\ x^3yz = x^2yz; \end{cases} \tag{18.2.15}$$

Proof. First, we prove the necessity.

In the case, when the polynomial xy is not essential then we have either $xy = x^2$ or $xy = y^2$ and they are no essential polynomial.

Now consider the case when xy is essential. Then so is the polynomial yx.

By taking $X(x,y) = xy$ and $Y(x,y) = yx$, we get $yxz = xzy$ and so $xyz = yzx = zxy$.

Consider the case when x^2y is not essential. This implies that $x^2y = x^3$ or $x^2y = y^3$ and, therefore, one of the systems (18.2.6) and (18.2.7) of identities is satisfied.

Now assume the polynomial x^2y is also essential. Then so is the polynomial y^2x.

Take $X(x,y) = xy$ and $Y(x,y) = x^2y$ in hyperidentity (18.2.14), and get $x^2yz = xy^2z$.

Take $X(x,y) = x^2y$ and $Y(x,y) = xy$ in hyperidentity (18.2.14), and get $xyxyz = x^2yz$ so $x^2yz = (x(yx)y)z = y^2x^2z = (y^2xx)z = (yx^2x)z = xx^2yz = x^3yz$.

Therefore, the semigroup $Q(\cdot)$ satisfies the system (18.2.15) of identities. This completes the proof of the necessity of the theorem.

Towards the proof of the sufficiency of the theorem, we consider the following cases:

(1) If (18.2.6) is satisfied in $Q(\cdot)$, then the only possible essential polynomials are xy, yx. For these polynomials, the hyperidentity is satisfied due to the first two identities of (18.2.6).
(2) If (18.2.7) is satisfied in $Q(\cdot)$, then the only possible essential polynomials are xy, yx. For these polynomials, the hyperidentity is satisfied due to the first two identities of (18.2.7).
(3) Now, assume (18.2.15) is satisfied in $Q(\cdot)$.
 Now, by Lemma 18.1.6, the only possible essential polynomials are xy, yx, x^2y, xy^2, x^2y^2.
 Let us consider those polynomials separately.

 (a) If $X(x,y) = xy$, then we should prove $Y(x,y)z = xY(y,z)$:

  (i) $Y(x,y) = xy \Rightarrow xyz = xyz$,
  (ii) $Y(x,y) = yx \Rightarrow yxz = xzy$,
  (iii) $Y(x,y) = x^2y \Rightarrow x^2yz = xy^2z$,
  (iv) $Y(x,y) = xy^2 \Rightarrow xy^2z = xyz^2$, as $xy^2z = y^2zx = yz^2x = xyz^2$,
  (v) $Y(x,y) = x^2y^2 \Rightarrow x^2y^2z = xy^2z^2$, as $x^2y^2z = zx^2y^2 = z^2xy^2 = xy^2z^2$.

 (b) If $X(x,y) = yx$, then we should prove $zY(x,y) = Y(y,z)x$:
 Note that as $Y(x,y)$ is essential, there are at least two variables participating in Y and, therefore, by the first case we have $zY(x,y) = Y(x,y)z = xY(y,z) = Y(y,z)x$.
 (c) If $X(x,y) = x^2y$, then we should prove $Y^2(x,y)z = x^2Y(y,z)$.

  (i) $Y(x,y) = xy \Rightarrow x^2y^2z = x^2yz \Leftrightarrow x^3yz = x^2yz$, as $xyxy = x^2y^2$,
  (ii) $Y(x,y) = yx \Rightarrow y^2x^2z = x^2zy \Leftrightarrow x^3yz = x^2yz$,
  (iii) $Y(x,y) = x^2y \Rightarrow x^4y^2z = x^2y^2z$,
  (iv) $Y(x,y) = xy^2 \Rightarrow x^2y^4z = x^2yz^2$,
  (v) $Y(x,y) = x^2y^2 \Rightarrow x^4y^4z = x^2y^2z^2$.

 (d) If $X(x,y) = xy^2$, then we should prove $Y(x,y)z^2 = xY^2(y,z)$.

  (i) $Y(x,y) = xy \Rightarrow xyz^2 = xy^2z^2 \Leftrightarrow xyz^2 = xyz^3$,
  (ii) $Y(x,y) = yx \Rightarrow yxz^2 = xz^2y^2$,
  (iii) $Y(x,y) = x^2y \Rightarrow x^2yz^2 = xy^4z^2$,
  (iv) $Y(x,y) = xy^2 \Rightarrow xy^2z^2 = xy^2z^4$,
  (v) $Y(x,y) = x^2y^2 \Rightarrow x^2y^2z^2 = xy^4z^4$.

 (e) If $X(x,y) = x^2y^2$, then we should prove $Y^2(x,y)z^2 = x^2Y^2(y,z)$.

Note that by the previous cases we have $Y^2(x,y)z^2 = (Y^2(x,y)z)z = x^2Y(y,z)z = x(xY(y,z))z = x(Y(x,y)z)z = x(Y(x,y)z^2) = x(xY^2(y,z)) = x^2Y^2(y,z)$.

This finishes the proof of the theorem. □

Theorem 18.2.6. *The hyperidentity*

$$X\left(Y\left(x,y\right),z\right) = Y\left(x, X\left(y,z\right)\right) \tag{18.2.16}$$

is essentially satisfied in the semigroup $Q(\cdot)$ if and only if that semigroup satisfies one of the following systems of identities: (18.2.2), (18.2.3) and

$$\begin{cases} xyz = xzy, \\ xyz = zxy, \\ x^2yz = xy^2z, \\ x^3yz = x^2yz. \end{cases} \tag{18.2.17}$$

Proof. First, we prove the necessity.

In the case, when the polynomial xy is not essential then we have either $xy = x^2$ or $xy = y^2$ and they are no essential polynomial.

Now consider the case when xy is essential. Then so is the polynomial yx.

By taking $X(x,y) = xy$ and $Y(x,y) = yx$, we get $xyz = yzx$ and so $xyz = zxy$.

By taking $X(x,y) = yx$ and $Y(x,y) = xy$, we get $zxy = xzy$ which with the previous identity implies $xyz = yzx = zyx = xzy$.

Consider the case when x^2y is not essential. This implies that $x^2y = x^3$ or $x^2y = y^3$ and, therefore, one of the systems (18.2.2) and (18.2.3) of identities is satisfied.

Now assume the polynomial x^2y is also essential. Then so is the polynomial y^2x.

Take $X(x,y) = xy$ and $Y(x,y) = y^2x$ in hyperidentity (18.2.16), and get $y^2xz = yzyzx = y^2z^2x$ and so $x^2yz = x^2y^2z = xy^2z$.

Moreover, $x^3yz = x(x^2yz) = x(xy^2z) = x^2y^2z = x^2yz$.

Then the semigroup $Q(\cdot)$ satisfies the system (18.2.17) of identities.

This completes the proof of the necessity of the theorem.

Towards the proof of the sufficiency of the theorem, we consider the following cases:

(1) If (18.2.2) is satisfied in $Q(\cdot)$, then the only possible essential polynomials are xy, yx. For these polynomials, the hyperidentity is satisfied due to the first two identities of (18.2.2).

(2) If (18.2.3) is satisfied in $Q(\cdot)$, then the only possible essential polynomials are xy, yx. For these polynomials, the hyperidentity is satisfied due to the first two identities of (18.2.2).

(3) Now, assume (18.2.17) is satisfied in $Q(\cdot)$.

Now, by Lemma 18.1.6, the only possible essential polynomials are xy, yx, x^2y, xy^2, x^2y^2.

Let us consider those polynomials separately.

(a) If $X(x, y) = xy$, then we should prove $Y(x, y)z = Y(x, yz)$:

(i) $Y(x, y) = xy \Rightarrow xyz = xyz$,
(ii) $Y(x, y) = yx \Rightarrow yxz = yzx$,
(iii) $Y(x, y) = x^2y \Rightarrow x^2yz = x^2yz$,
(iv) $Y(x, y) = xy^2 \Rightarrow xy^2z = xyzyz$, as $xyzyz = xy^2z^2 = xy^4z = xy^3zy = xy^2zy = xy^2z$,
(v) $Y(x, y) = x^2y^2 \Rightarrow x^2y^2z = x^2yzyz$, by the previous case.

(b) If $X(x, y) = yx$, then we should prove $zY(x, y) = Y(x, zy)$:

(i) $Y(x, y) = xy \Rightarrow zxy = xzy$,
(ii) $Y(x, y) = yx \Rightarrow zyx = zyx$,
(iii) $Y(x, y) = x^2y \Rightarrow zx^2y = x^2zy$,
(iv) $Y(x, y) = xy^2 \Rightarrow zxy^2 = xzyzy$, by one of the above cases, $xzyxy = xz^2y = zxy^2$,
(v) $Y(x, y) = x^2y^2 \Rightarrow zx^2y^2 = x^2zyzy$, by the previous case.

(c) If $X(x, y) = x^2y$, then we should prove $Y^2(x, y)z = Y(x, y^2z)$.

(i) $Y(x, y) = xy \Rightarrow xyxyz = xy^2z$,
(ii) $Y(x, y) = yx \Rightarrow yxyxz = y^2zx$,
(iii) $Y(x, y) = x^2y \Rightarrow x^2yx^2yz = x^2y^2z$,
(iv) $Y(x, y) = xy^2 \Rightarrow xy^2xy^2z = xy^2zy^2z$, as $xy^2xy^2z = x^2y^4z = x^2yz$ and $xy^2zy^2z = xy^4z^2 = xyz^2 = x^2yz$,
(v) $Y(x, y) = x^2y^2 \Rightarrow x^2y^2x^2y^2z = x^2y^2zy^2z$, by the previous case.

(d) If $X(x, y) = xy^2$, then we should prove $Y(x, y)z^2 = Y(x, yz^2)$. This is true, by the fact that $Y(x, y)z = Y(x, yz)$.

(e) If $X(x, y) = x^2y^2$, then we should prove $Y^2(x, y)z^2 = Y(x, y^2z^2)$. This is true, by the fact that $Y^2(x, y)z = Y(x, y^2z)$.

This finishes the proof of the theorem. □

Theorem 18.2.7. *The hyperidentity*

$$X\left(Y\left(x,y\right),z\right)=Y\left(x,Y\left(y,z\right)\right) \tag{18.2.18}$$

is essentially satisfied in the semigroup $Q(\cdot)$ if and only if that semigroup satisfies one of the following systems of identities: (18.2.4), (18.2.6) and (18.2.7).

Proof. We prove that this hyperidentity is equivalent to the hyperidentity (18.2.5).

Note that a polynomial $X(x,y)$ is essential if and only if so is the polynomial $Z(x,y)=X(y,x)$. Now, if we replace $X(x,y)$ with $X(y,x)$ and $Y(x,y)$ with $Y(y,x)$ in the hyperidentity (18.2.5), we will get $X(z,X(y,x)) = Y(X(z,y),x)$ and therefore if we also replace z with x, x with z, X with Y and Y with X, we will get exactly the hyperidentity (18.2.18).

This implies that the necessary and sufficient conditions for the semigroup $Q(\cdot)$ to satisfy the hyperidentity (18.2.5) or (18.2.18) are the same.

Thus, this finishes the proof of the theorem. □

18.3. Essentially associative hyperidentities with higher functional rank

In this section, we consider associative hyperidentities with functional rank 3 and 4, which are the following:

$$X(X(x,y),z)=Y(x,Z(y,z)), \tag{18.3.1}$$

$$X(Y(x,y),z)=X(x,Z(y,z)), \tag{18.3.2}$$

$$X(Y(x,y),z)=Y(x,Z(y,z)), \tag{18.3.3}$$

$$X(Y(x,y),z)=Z(x,X(y,z)), \tag{18.3.4}$$

$$X(Y(x,y),z)=Z(x,Y(y,z)), \tag{18.3.5}$$

$$X(Y(x,y),z)=Z(x,Z(y,z)), \tag{18.3.6}$$

$$X(Y(x,y),z)=Z(x,T(y,z)). \tag{18.3.7}$$

Theorem 18.3.1. *The hyperidentity (18.3.1) is essentially satisfied in the semigroup $Q(\cdot)$ if and only if that semigroup satisfies one of the following systems of identities: (18.2.2), (18.2.3) and (18.2.4).*

Proof. We prove that this hyperidentity is equivalent to the hyperidentity (18.2.1).

If we take $Y(x,y) = X(x,y)$ and $Z(x,y) = Y(x,y)$ then we will get the hyperidentity (18.2.1). Thus, (18.3.1) $\Rightarrow$ (18.2.1).

On the other hand, if we assume that the hyperidentity (18.2.1) is satisfied, then we have $X(X(x,y),z) = Y(x,Y(y,z))$ as (18.2.1) $\Rightarrow$ (18.2.8). Likewise, we have $X(X(x,y),z) = Y(x,Y(y,z)) = Y(Y(x,y),z) = Y(x,Z(y,z))$.

Thus, this finishes the proof of the theorem. □

Theorem 18.3.2. *The hyperidentity (18.3.2) is essentially satisfied in the semigroup $Q(\cdot)$ if and only if that semigroup satisfies one of the following systems of identities: (18.2.2), (18.2.3) and (18.2.4).*

Proof. We prove that this hyperidentity is equivalent to the hyperidentity (18.2.1).

If we take $Y(x,y) = X(x,y)$ and $Z(x,y) = Y(x,y)$ then we will get the hyperidentity (18.2.1). Thus, (18.3.2) $\Rightarrow$ (18.2.1).

On the other hand, if we assume that the hyperidentity (18.2.1) is satisfied, then we have $X(Y(x,y),z) = X(x,X(y,z))$ as (18.2.1) $\Rightarrow$ (18.2.13). Likewise, we have $X(Y(x,y),z) = X(x,X(y,z)) = X(X(x,y),z) = X(x,Z(y,z))$.

Thus, this finishes the proof of the theorem. □

Theorem 18.3.3. *The hyperidentity (18.3.3) is essentially satisfied in the semigroup $Q(\cdot)$ if and only if that semigroup satisfies one of the following systems of identities: (18.2.2), (18.2.3) and (18.2.4).*

Proof. We prove that this hyperidentity is equivalent to the hyperidentity (18.2.1).

If we take $Y(x,y) = X(x,y)$ and $Z(x,y) = Y(x,y)$ then we will get the hyperidentity (18.2.1). Thus, (18.3.3) $\Rightarrow$ (18.2.1).

On the other hand, if we assume that the hyperidentity (18.2.1) is satisfied, then we have $X(Y(x,y),z) = Y(x,Y(y,z))$ as (18.2.1) $\Rightarrow$ (18.2.18). Likewise, we have $X(Y(x,y),z) = Y(x,Y(y,z)) = Y(Y(x,y),z) = Y(x,Z(y,z))$.

Thus, this finishes the proof of the theorem. □

Theorem 18.3.4. *The hyperidentity (18.3.4) is essentially satisfied in the semigroup $Q(\cdot)$ if and only if that semigroup satisfies one of the following systems of identities: (18.2.2), (18.2.3) and (18.2.4).*

Proof. We prove that this hyperidentity is equivalent to the hyperidentity (18.2.13).

If we take $Z(x,y) = X(x,y)$ then we will get the hyperidentity (18.2.13). Thus, (18.3.4) $\Rightarrow$ (18.2.13).

On the other hand, if we assume that the hyperidentity (18.2.13) is satisfied, then we have $X(Y(x,y),z) = X(x,X(y,z))$. Now, as (18.2.13) $\Rightarrow$ (18.2.16), we have $X(Y(x,y),z) = X(x,X(y,z)) = X(X(x,y),z) = Z(x,X(y,z))$.

Thus, this finishes the proof of the theorem. □

Theorem 18.3.5. *The hyperidentity (18.3.5) is essentially satisfied in the semigroup $Q(\cdot)$ if and only if that semigroup satisfies one of the following systems of identities: (18.2.4), (18.2.6) and (18.2.7).*

Proof. We prove that this hyperidentity is equivalent to the hyperidentity (18.2.18).

If we take $Z(x,y) = Y(x,y)$ then we will get the hyperidentity (18.2.18). Thus, (18.3.5) $\Rightarrow$ (18.2.18).

On the other hand, if we assume that the hyperidentity (18.2.18) is satisfied, then we have $X(Y(x,y),z) = Y(x,Y(y,z))$. Now as (18.2.18) $\Rightarrow$ (18.2.5), we have $X(Y(x,y),z) = Y(x,Y(y,z)) = Y(Y(x,y),z) = Z(x,Y(y,z))$.

Thus, this finishes the proof of the theorem. □

Theorem 18.3.6. *The hyperidentity (18.3.6) is essentially satisfied in the semigroup $Q(\cdot)$ if and only if that semigroup satisfies one of the following systems of identities: (18.2.2), (18.2.3) and (18.2.4).*

Proof. We prove that this hyperidentity is equivalent to the hyperidentity (18.2.13).

If we take $Z(x,y) = X(x,y)$ then we will get the hyperidentity (18.2.13). Thus, (18.3.6) $\Rightarrow$ (18.2.13).

On the other hand, if we assume that the hyperidentity (18.2.13) is satisfied, then we have $X(Y(x,y),z) = Y(x,Y(y,z))$ as (18.2.1) $\Rightarrow$ (18.2.18). Likewise, we have $X(Y(x,y),z) = Y(x,Y(y,z)) = Y(Y(x,y),z) = Z(x,Z(y,z))$.

Thus, this finishes the proof of the theorem. □

Theorem 18.3.7. *The hyperidentity (18.3.7) is essentially satisfied in the semigroup $Q(\cdot)$ if and only if that semigroup satisfies one of the following systems of identities: (18.2.2), (18.2.3) and (18.2.4).*

Proof. We prove that this hyperidentity is equivalent to the hyperidentity (18.2.1).

If we take $Y(x,y) = X(x,y)$, $Z(x,y) = X(x,y)$ and $T(x,y) = Y(x,y)$ then we will get the hyperidentity (18.2.1). Thus, (18.3.7) $\Rightarrow$ (18.2.1).

On the other hand, if we assume that the hyperidentity (18.2.1) is satisfied, then we have $X(Y(x,y),z) = Y(x,Y(y,z))$ as (18.2.1) $\Rightarrow$ (18.2.18). Likewise, we have $X(Y(x,y),z) = Y(x,Y(y,z)) = Z(Z(x,y),z) = Z(x,T(y,z))$.

Thus, this finishes the proof of the theorem. □

We can summarize the cases of essential satisfiability of associative hyperidentities as follows.

Theorem 18.3.8. *Any associative hyperidentity is essentially equivalent to one of the following hyperidentities (18.2.1), (18.2.5), (18.2.8), (18.2.14), (18.2.16). Moreover, we have the following implications: (18.2.1)* $\Rightarrow_e$ *(18.2.8), (18.2.1)* $\Rightarrow_e$ *(18.2.5)* $\Rightarrow_e$ *(18.2.16)* $\Rightarrow_e$ *(18.2.14).*

Proof. The proof follows from the Theorems 18.3.1–18.3.7. □

18.4. Essentially distributive hyperidentities

The left distributive hyperidentities of functional rank 1 or 2 are as follows:

$$X(x,X(y,z)) = X(X(x,y),X(x,z)), \tag{18.4.1}$$

$$X(x,X(y,z)) = X(X(x,y),Y(x,z)), \tag{18.4.2}$$

$$X(x,X(y,z)) = X(Y(x,y),X(x,z)), \tag{18.4.3}$$

$$X(x,X(y,z)) = X(Y(x,y),Y(x,z)), \tag{18.4.4}$$

$$X(x,X(y,z)) = Y(X(x,y),X(x,z)), \tag{18.4.5}$$

$$X(x,X(y,z)) = Y(X(x,y),Y(x,z)), \tag{18.4.6}$$

$$X(x,X(y,z)) = Y(Y(x,y),X(x,z)), \tag{18.4.7}$$

$$X(x,X(y,z)) = Y(Y(x,y),Y(x,z)), \tag{18.4.8}$$

$$X(x,Y(y,z)) = X(X(x,y),X(x,z)), \tag{18.4.9}$$

$$X(x,Y(y,z)) = X(X(x,y),Y(x,z)), \tag{18.4.10}$$

$$X(x,Y(y,z)) = X(Y(x,y),X(x,z)), \tag{18.4.11}$$

$$X(x,Y(y,z)) = X(Y(x,y),Y(x,z)), \tag{18.4.12}$$

$$X(x,Y(y,z)) = Y(X(x,y),X(x,z)), \tag{18.4.13}$$

$$X(x,Y(y,z)) = Y(X(x,y),Y(x,z)), \tag{18.4.14}$$

$$X(x, Y(y, z)) = Y(Y(x, y), X(x, z)), \tag{18.4.15}$$

$$X(x, Y(y, z)) = Y(Y(x, y), Y(x, z)). \tag{18.4.16}$$

Define x^0 as the empty word.

Let us consider the following set of identities:

$$\begin{cases} xyz = xyxz, \\ xyz = xzyz. \end{cases} \tag{18.4.17}$$

Lemma 18.4.1. *Any essential polynomial $X(x, y)$ in a semigroup $Q(\cdot)$ which satisfies the identities (18.4.17) equals one of the polynomials from the following set: $P = \{xy, yx, x^2y, yx^2, xyx, yxy\}$.*

Proof. Any essential polynomial $X(x, y)$ can be expressed in one of the following forms:

$X(x, y) = x^{\alpha_1}y^{\beta_1}x^{\alpha_2}y^{\beta_2}\ldots x^{\alpha_n}y^{\beta_n}$,
$X(x, y) = y^{\beta_1}x^{\alpha_1}y^{\beta_2}x^{\alpha_2}\ldots y^{\beta_n}x^{\alpha_n}$,
$X(x, y) = x^{\alpha_1}y^{\beta_1}x^{\alpha_2}y^{\beta_2}\ldots x^{\alpha_n}y^{\beta_n}x^{\alpha_{n+1}}$,
$X(x, y) = y^{\beta_1}x^{\alpha_1}y^{\beta_2}x^{\alpha_2}\ldots y^{\beta_n}x^{\alpha_n}y^{\beta_{n+1}}$,

where $n, \alpha_i, \beta_i \in, i = 1, 2, \ldots, n$.

Let us show that

$$\forall a, b, c, d \in \mathbb{N} \exists e, f \in \mathbb{N}\left(x^a y^b x^c y^d = x^e y^f\right). \tag{18.4.18}$$

Indeed, $x^a y^b x^c y^d = x^{a-1}\left(x\left(y^b x^{c-1}\right) x y^d\right) = x^{a-1}\left(x\left(y^b x^{c-1}\right) y^d\right) = x^a y^b x^{c-1} y^d = \cdots = x^a y^b y^d = x^a y^{b+d}$, so the numbers $e = a, f = b + d$ satisfy (18.4.18).

From (18.4.18) we get that any essential polynomial $X(x, y)$ can be expressed in one of the following forms:

$X(x, y) = x^a y^b$,
$X(x, y) = y^a x^b$,
$X(x, y) = x^a y^b x^c$,
$X(x, y) = y^a x^b y^c$,

where $a, b, c \in \mathbb{N}$.

Let us now prove that

$$\forall a, b \in \mathbb{N}, \exists c \in \mathbb{N}\left(x^a y^b = x^c y\right). \tag{18.4.19}$$

If $b = 1$ then the number $c = a$ satisfies (1.2). If $b \geq 2$ then

$x^a y^b = \left(x^a y^2\right) y^{b-2} = \left(x^a y y\right) y^{b-2} = \left(x^a y x^a y\right) y^{b-2} = \left(x^a x^a y\right) y^{b-2} = x^{2a} y^{b-1} = \cdots = x^c y$.

Therefore,
$\exists c \in \mathbb{N}(y^a x^b = y^c x)$,
$\exists d, e \in \mathbb{N}(x^a y^b x^c = x^a y^d x = x^e yx)$,
$\exists d, e \in \left(y^a x^b y^c = y^a x^d y = y^e xy\right)$.

So any essential polynomial $X(x, y)$ can be expressed in one of the following forms:
$X(x, y) = x^a y$,
$X(x, y) = y^a x$,
$X(x, y) = x^a yx$,
$X(x, y) = y^a xy$,
where $a \in \mathbb{N}$.

Let us show that $x^3y = x^2y$ and $x^2yx = xyx$. Indeed, $x^3y = xxxy = xxy = x^2y$, $x^2yx = xxyx = xyx$.

Therefore if $a \geq 2$ then $x^a yx = x^{a-2}x^2yx = x^{a-2}xyx = x^{a-1}yx = \cdots = xyx$.

So, $y^a xy = yxy$, $a \geq 2$. Hence, $x^a y = x^{a-3}x^3 y = x^{a-3}x^2 y = x^{a-1}y = \cdots = x^2 y$, if $a \geq 3$. So we got that $x^a y = x^2 y$ when $a \geq 2$. In the same way $y^a x = y^2 x = yyx = yxyx = yx^2$ when $a \geq 2$.

Hence any essential polynomial $X(x, y)$ can be expressed in one of the following forms:
$X(x, y) = xy$, $X(x, y) = x^2y$,
$X(x, y) = yx$, $X(x, y) = yx^2$,
$X(x, y) = xyx$, $X(x, y) = yxy$.

The lemma is proved. □

Corollary 18.4.2. *Any polynomial $X(x, y) \neq x^n, y^m$ in a semigroup $Q(\cdot)$ which satisfies the identities (18.4.17) equals one of the polynomials from the following set: $P = \{xy, yx, x^2y, yx^2, xyx, yxy\}$.*

Now let's consider the following set of identities:

$$\begin{cases} xyz = xzy, \\ xyz = yxz, \\ xyz = x^2yz. \end{cases} \tag{18.4.20}$$

Definition 18.4.3. We say the variable x participates in the polynomial $X(x, y, z)$ if it has an entrance with a non-zero degree in that polynomial.

In the same way, the participation of the variables y and z in the polynomial $X\,(x, y, z)$ is defined.

For example, if the polynomials $Y\,(x, y)$, $Z\,(x, y)$, $U\,(x, y)$ are essential polynomials in the semigroup $Q(\cdot)$, then the variables x, y, z participate in the following polynomials:

$X(x, y, z) = Y(Z(x, y), z),$
$X(x, y, z) = Y(Z(x, y), U(y, z)).$

Lemma 18.4.4. *If the variables x, y, z participate in any polynomial $X(x, y, z)$ in a semigroup $Q(\cdot)$ which satisfies the identities (18.4.20), then that polynomial $X(x, y, z)$ is equal to the polynomial xyz and thus it is determined uniquely.*

Proof. Let $Q(\cdot)$ be a semigroup satisfying the identities (1.3) and let $X(x, y, z)$ be an arbitrary polynomial such that the variables x, y, z participate in it. In this case, the polynomial $X(x, y, z)$ can be expressed in the following form:

$X(x, y, z) = x^a y^b z^c$, where $a, b, c \in \mathbb{N}$.

If $a \geq 2$ then $x^a y^b z^c = x^{a-2}\left(x^2 y^b z^c\right) = x^{a-2}\left(x y^b z^c\right) = x^{a-1} y^b z^c = \cdots = x y^b z^c$.

So $x^a y^b z^c = x y^b z^c$, for any $a \in \mathbb{N}$.

In the same way, we get $x y^b z^c = x y z^c = xyz$.

The lemma is proved. □

Theorem 18.4.5. *The hyperidentity (18.4.1) is essentially satisfied in the semigroup $Q(\cdot)$ if and only if that semigroup satisfies the system of identities (18.4.17).*

Proof. Let us first prove the necessity.

Let us consider the case when the polynomial xy is not essential. In that case, $xy = x^2$ or $xy = y^2$ and there is no essential polynomial.

Now assume the polynomial xy is essential. In that case, the polynomial yx is also essential.

Taking $X(x, y) = xy$ in (18.4.1) we get $xyz = xyxz$.

Taking $X(x, y) = yx$ in (18.4.1) we get $zyx = zxyx \Rightarrow xyz = xzyz$.

Thus the necessity is proved.

Suppose that (18.4.17) is satisfied in the semigroup $Q(\cdot)$.

Due to Lemma 18.4.1, any essential polynomial $X(x, y)$ belongs to the following set: $P = \{xy, yx, x^2y, yx^2, xyx, yxy\}$.

Let us consider those cases separately:

(1) $X(x, y) = xy \Rightarrow xyz = xyxz.$
(2) $X(x, y) = yx \Rightarrow zyx = zxyx.$
(3) $X(x, y) = x^2y \Rightarrow x^2y^2zy^2z = x^2yx^2yx^2z,$
$x^2y^2zy^2z = x^2\left(y^2z^2\right) = x^2y^2z,$
$x^2yx^2yx^2z = x^2y\left(x^2yx^2z\right) = x^2yx^2yz = x^2y^2z.$

(4) $X(x,y) = yx^2 \Rightarrow zy^2x^2 = zx^2yx^2yx^2$,
$zx^2yx^2yx^2 = zx^2\left(yx^2yx^2\right) = zx^2y^2x^2 = zy^2x^2$.

(5) $X(x,y) = xyx \Rightarrow xyzyx = xyxxzxxyx$,
$xyxxzxxyx = xy\left(x^2zx^2y\right)x = xyx^2zyx = (xyxx)\,zyx = xyxzyx = xyzyx$.

(6) $X(x,y) = yxy \Rightarrow yzyxyzy = xzxxyxxzx$,
$yzyxyzy = yzxyzy = yzxzy = yxzy$,
$xzxxyxxzx = xz\left(x^2yx^2z\right)x = xzx^2yzx = (xzxx)\,yzx = xzxyzx = xzyzx = xyzx = xyzx = yxyzx = y\,(xy\,(zx)\,y) = yxy\,(zyxy) = yxy\,(z\,(yx)\,zy) = yxy\,(zy)\,x\,(zy) = yxyxzy = yxyzy = yxzy$.

This completes the proof of the theorem. □

Theorem 18.4.6. *The hyperidentity (18.4.2) is essentially satisfied in the semigroup $Q(\cdot)$ if and only if that semigroup satisfies the system of identities (18.4.20).*

Proof. Let us prove the necessity.

Let us consider the case when the polynomial xy is not essential. In that case, $xy = x^2$ or $xy = y^2$ and there is no essential polynomial.

Now let us suppose that the polynomial xy is essential. In that case, the polynomial yx is also essential.

Taking $X(x,y) = xy, Y(x,y) = xy$ in (18.4.2) we get $xyz = xyxz$.

Taking $X(x,y) = yx, Y(x,y) = yx$ in (18.4.2) we get $zyx = zxyx \Rightarrow xyz = xzyz$.

Taking $X(x,y) = xy, Y(x,y) = yx$ in (18.4.2) we get $xyz = xyzx = xyzyx = xzyx = xzy$.

Taking $X(x,y) = yx, Y(x,y) = xy$ in (18.4.2) we get $zyx = xzyx \Rightarrow xyz = zxyz = zyxz = yxz$.

Moreover, we have $xyz = x(yxz) = xxyz = x^2yz$.

The necessity is proved.

Let us suppose that (18.4.20) is satisfied in the semigroup $Q(\cdot)$.

Due to Lemma 18.4.4, for binary essential polynomials we have: $X(x, X(y,z)) = xyz = X(X(x,y), Y(x,z))$.

This completes the proof of the theorem. □

Theorem 18.4.7. *The hyperidentity (18.4.3) is essentially satisfied in the semigroup $Q(\cdot)$ if and only if that semigroup satisfies the system of identities (18.4.20).*

Proof. Let us prove the necessity.

Let us consider the case when the polynomial xy is not essential. In that case $xy = x^2$ or $xy = y^2$ and there is no essential polynomial.

Now let us suppose that the polynomial xy is essential. In that case, the polynomial yx is also essential.

Taking $X(x,y) = xy, Y(x,y) = xy$ in (18.4.3) we get $xyz = xyxz$.

Taking $X(x,y) = yx, Y(x,y) = yx$ in (18.4.3) we get $zyx = zxyx \Rightarrow xyz = xzyz$.

Taking $X(x,y) = xy, Y(x,y) = yx$ in (18.4.3) we get $xyz = zxyz = zyxyz = zyxz = yxz$.

Taking $X(x,y) = yx, Y(x,y) = xy$ in (18.4.3) we get $zyx = zyxz \Rightarrow xyz = xyzx = xzyzx = xzyx = xzy$.

Moreover, we have $xyz = x(yxz) = xxyz = x^2yz$. □

Theorem 18.4.8. *The hyperidentity (18.4.4) is essentially satisfied in the semigroup $Q(\cdot)$ if and only if that semigroup satisfies the system of identities (18.4.20).*

Proof. Let us prove the necessity.

Let us consider the case when the polynomial xy is not essential. In that case $xy = x^2$ or $xy = y^2$ and there is no essential polynomial.

Now let us suppose that the polynomial xy is essential. In that case, the polynomial yx is also essential.

Taking $X(x,y) = xy, Y(x,y) = xy$ in (18.4.4) we get $xyz = xyxz$.

Taking $X(x,y) = yx, Y(x,y) = yx$ in (18.4.4) we get $zyx = zxyx \Rightarrow xyz = xzyz$.

Taking $X(x,y) = xy, Y(x,y) = yx$ in (18.4.4) we get $xyz = zxzy = zxy \Rightarrow xyz = z(xz)y = yz(xz) = yxz$.

Taking $X(x,y) = yx, Y(x,y) = xy$ in (18.4.4) we get $zyx = yzxz \Rightarrow xyz = y(xz)x = xzxy = xzy$.

Moreover, we have $xyz = x(yxz) = xxyz = x^2yz$. □

Theorem 18.4.9. *The hyperidentity (18.4.5) is essentially satisfied in the semigroup $Q(\cdot)$ if and only if that semigroup satisfies the system of identities (18.4.20).*

Proof. Let us prove the necessity.

Let us consider the case when the polynomial xy is not essential. In that case $xy = x^2$ or $xy = y^2$ and there is no essential polynomial.

Now let us suppose that the polynomial xy is essential. In that case, the polynomial yx is also essential.

Taking $X(x,y) = xy, Y(x,y) = xy$ in (18.4.5) we get $xyz = xyxz$.

Taking $X(x,y) = yx, Y(x,y) = yx$ in (18.4.5) we get $zyx = zxyx \Rightarrow xyz = xzyz$.

Taking $X(x,y) = xy, Y(x,y) = yx$ in (18.4.5) we get $xyz = yzxz = yxz$.

Taking $X(x,y) = yx, Y(x,y) = xy$ in (18.4.5) we get $zyx = zxzy \Rightarrow xyz = xzxy = xzy$.

Moreover, we have $xyz = x(yxz) = xxyz = x^2yz$. □

Theorem 18.4.10. *The hyperidentity (18.4.6) is essentially satisfied in the semigroup $Q(\cdot)$ if and only if that semigroup satisfies the system of identities (18.4.20).*

Proof. Let us prove the necessity.

Let us consider the case when the polynomial xy is not essential. In that case, $xy = x^2$ or $xy = y^2$ and there is no essential polynomial.

Now let us suppose that the polynomial xy is essential. In that case, the polynomial yx is also essential.

Taking $X(x,y) = xy, Y(x,y) = xy$ in (18.4.6) we get $xyz = xyxz$.

Taking $X(x,y) = yx, Y(x,y) = yx$ in (18.4.6) we get $zyx = zxyx \Rightarrow xyz = xzyz$.

Taking $X(x,y) = xy, Y(x,y) = yx$ in (18.4.6) we get $xyz = zyxz = zxyxz = zxyz = yxz$.

Taking $X(x,y) = yx, Y(x,y) = xy$ in (18.4.6) we get $zyx = zxyz \Rightarrow xyz = xzyx = xzyzx = xyzx = xzy$.

Moreover, we have $xyz = x(yxz) = xxyz = x^2yz$. □

Theorem 18.4.11. *The hyperidentity (18.4.7) is essentially satisfied in the semigroup $Q(\cdot)$ if and only if that semigroup satisfies the system of identities (18.4.20).*

Proof. Let us prove the necessity.

Let us consider the case when the polynomial xy is not essential. In that case $xy = x^2$ or $xy = y^2$ and there is no essential polynomial.

Now let us suppose that the polynomial xy is essential. In that case, the polynomial yx is also essential.

Taking $X(x,y) = xy, Y(x,y) = xy$ in (18.4.7) we get $xyz = xyxz$.

Taking $X(x,y) = yx, Y(x,y) = yx$ in (18.4.7) we get $zyx = zxyx \Rightarrow xyz = xzyz$.

Taking $X(x,y) = xy, Y(x,y) = yx$ in (18.4.7) we get $xyz = yzzx = (yz)zx = zxx(yz) = (zxxy)z = yzxz = yxz$.

Taking $X(x,y) = yx, Y(x,y) = xy$ in (18.4.7) we get $zyx = xzzy \Rightarrow xyz = yxz = zyyx = xzy$.

Moreover, we have $xyz = x(yxz) = xxyz = x^2yz$. □

Theorem 18.4.12. *The hyperidentity (18.4.8) is essentially satisfied in the semigroup $Q(\cdot)$ if and only if that semigroup satisfies the system of identities (18.4.20).*

Proof. Let us prove the necessity.

Let us consider the case when the polynomial xy is not essential. In that case $xy = x^2$ or $xy = y^2$ and there is no essential polynomial.

Now let us suppose that the polynomial xy is essential. In that case, the polynomial yx is also essential.

Taking $X(x, y) = xy, Y(x, y) = xy$ in (18.4.8) we get $xyz = xyxz$.

Taking $X(x, y) = yx, Y(x, y) = yx$ in (18.4.8) we get $zyx = zxyx \Rightarrow xyz = xzyz$.

Taking $X(x, y) = xy, Y(x, y) = yx$ in (18.4.8) we get $xyz = zyzx = zyx \Rightarrow xyz = zy(zx) = (zx)yz = zyxyz = zyxz = yxz$.

Taking $X(x, y) = yx, Y(x, y) = xy$ in (18.4.8) we get $zyx = xzyz \Rightarrow xyz = zx(yx) = (yx)xz = z(xx)y = xzy$.

Moreover, we have $xyz = x(yxz) = xxyz = x^2yz$. □

Theorem 18.4.13. *The hyperidentity (18.4.9) is essentially satisfied in the semigroup $Q(\cdot)$ if and only if that semigroup satisfies the system of identities (18.4.20).*

Proof. Let us prove the necessity.

Let us consider the case when the polynomial xy is not essential. In that case $xy = x^2$ or $xy = y^2$ and there is no essential polynomial.

Now let us suppose that the polynomial xy is essential. In that case, the polynomial yx is also essential.

Taking $X(x, y) = xy, Y(x, y) = xy$ in (18.4.9) we get $xyz = xyxz$.

Taking $X(x, y) = yx, Y(x, y) = yx$ in (18.4.9) we get $zyx = zxyx \Rightarrow xyz = xzyz$.

Taking $X(x, y) = xy, Y(x, y) = yx$ in (18.4.9) we get $yxz = xzyz = xyz$.

Taking $X(x, y) = yx, Y(x, y) = xy$ in (18.4.9) we get $zxy = zyzx = zyx \Rightarrow xyz = xzy$.

Moreover, we have $xyz = x(yxz) = xxyz = x^2yz$. □

Theorem 18.4.14. *The hyperidentity (18.4.10) is essentially satisfied in the semigroup $Q(\cdot)$ if and only if that semigroup satisfies the system of identities (18.4.20).*

Proof. Let us prove the necessity.

Let us consider the case when the polynomial xy is not essential. In that case $xy = x^2$ or $xy = y^2$ and there is no essential polynomial.

Now let us suppose that the polynomial xy is essential. In that case, the polynomial yx is also essential.

Taking $X(x,y) = xy, Y(x,y) = xy$ in (18.4.10) we get $xyz = xyxz$.

Taking $X(x,y) = yx, Y(x,y) = yx$ in (18.4.10) we get $zyx = zxyx \Rightarrow xyz = xzyz$.

Taking $X(x,y) = xy, Y(x,y) = yx$ in (18.4.10) we get $yxz = xzzy = xzyzy = xyzy = xzy \Rightarrow xyz = x(yx)z = zxyx = zyx = xzy$.

Moreover, we have $xyz = zyx = yxz$ and $xyz = x(yxz) = xxyz = x^2yz$. □

Theorem 18.4.15. *The hyperidentity (18.4.11) is essentially satisfied in the semigroup $Q(\cdot)$ if and only if that semigroup satisfies the system of identities (18.4.20).*

Proof. Let us prove the necessity.

Let us consider the case when the polynomial xy is not essential. In that case $xy = x^2$ or $xy = y^2$ and there is no essential polynomial.

Now let us suppose that the polynomial xy is essential. In that case, the polynomial yx is also essential.

Taking $X(x,y) = xy, Y(x,y) = xy$ in (18.4.11) we get $xyz = xyxz$.

Taking $X(x,y) = yx, Y(x,y) = yx$ in (18.4.11) we get $zyx = zxyx \Rightarrow xyz = xzyz$.

Taking $X(x,y) = xy, Y(x,y) = yx$ in (18.4.11) we get $yxz = zxyz = zxyxz = zyxz = xyz$.

Taking $X(x,y) = yx, Y(x,y) = xy$ in (18.4.11) we get $zxy = zyxz \Rightarrow xyz = xy(zx) = yxzx = yzx = zxy = xzy$.

Moreover, we have $xyz = x(yxz) = xxyz = x^2yz$. □

Theorem 18.4.16. *The hyperidentity (18.4.12) is essentially satisfied in the semigroup $Q(\cdot)$ if and only if that semigroup satisfies the system of identities (18.4.20).*

Proof. Let us prove the necessity.

Let us consider the case when the polynomial xy is not essential. In that case $xy = x^2$ or $xy = y^2$ and there is no essential polynomial.

Now let us suppose that the polynomial xy is essential. In that case, the polynomial yx is also essential.

Taking $X(x,y) = xy, Y(x,y) = xy$ in (18.4.12) we get $xyz = xyxz$.

Taking $X(x,y) = yx, Y(x,y) = yx$ in (18.4.12) we get $zyx = zxyx \Rightarrow xyz = xzyz$.

Taking $X(x,y) = xy, Y(x,y) = yx$ in (18.4.12) we get $yxz = zxzy = zxy \Rightarrow xyz = x(yxz) = xzxy = xzy$.

Moreover, we have $xyz = (xzy)z = yzxz = yxz$ and $xyz = x(yxz) = xxyz = x^2yz$. □

Theorem 18.4.17. *The hyperidentity (18.4.13) is essentially satisfied in the semigroup $Q(\cdot)$ if and only if that semigroup satisfies the system of identities (18.4.17).*

Proof. Let us first prove the necessity.

By taking $Y(x,y) = X(x,y)$ in (18.4.13) we get $X(x, X(y,z)) = X(X(x,y), X(x,z))$, which is the hyperidentity (18.4.1). Therefore, the necessity is proved.

Suppose that (18.4.17) is satisfied in the semigroup $Q(\cdot)$.

Due to Lemma 18.4.1, any essential polynomial $X(x,y)$ belongs to the following set: $P = \{xy, yx, x^2y, yx^2, xyx, yxy\}$.

Let us consider those cases separately:

(1) $X(x,y) = xy \Rightarrow xY(y,z) = Y(xy, xz)$.

   (a) $Y(x,y) = xy \Rightarrow xyz = xyxz$,
   (b) $Y(x,y) = yx \Rightarrow xzy = xzxy$,
   (c) $Y(x,y) = x^2y \Rightarrow xy^2z = xyxyxz, xyxyxz = xyxyz = xy^2z$,
   (d) $Y(x,y) = yx^2 \Rightarrow xzy^2 = xzxyxy, xzxyxy = xzxy^2 = xzy^2$,
   (e) $Y(x,y) = xyx \Rightarrow xyzy = xyxzxy, xyxzxy = xyxzy = xyzy$,
   (f) $Y(x,y) = yxy \Rightarrow xzyz = xzxyxz, xzxyxz = xzxyxz = xzxyz = xzyz$.

(2) $X(x,y) = yx \Rightarrow zY(x,y) = Y(zx, zy)$.
   The proof is in the same way as in the preceding case.
(3) $X(x,y) = x^2y \Rightarrow x^2Y(y,z) = Y(x^2y, x^2z)$, due to step (a).
(4) $X(x,y) = yx^2 \Rightarrow Y(y,z)x^2 = Y(yx^2, zx^2)$, due to step (b).
(5) $X(x,y) = xyx \Rightarrow xY(y,z)x = Y(xyx, xzx)$.
(6) $X(x,y) = yxy \Rightarrow Y(y,z)xY(y,z) = Y(yxy, zxz)$.

   (a) $Y(x,y) = xy \Rightarrow yzxyz = yxyzxz, yzxyz = yzyxyz = yzyxz = yzxz = yxz = yzxz = yzzxz = yzyzxz = yyzxz = yyxz = yxyxz = yxyzxz$,
   (b) $Y(x,y) = yx \Rightarrow zyxzy = zxzyxy, zyxzy = zxyzy = zxzy = zxy = zxy^2 = zxyy = zxyzxy$,
   (c) $Y(x,y) = x^2y \Rightarrow y^2zxy^2z = yxyyxyzxz, y^2zxy^2z = y^2zxz = y^2xz = y^2x^2z = yyxzxz = yyxyzxz = yxyyxyzxz$,
   (d) $Y(x,y) = yx^2 \Rightarrow zy^2xzy^2 = zxzyxyyxy, zy^2xzy^2 = zxzy^2 = zxy^2 = zx^2y^2 = zxzxyy = zxzyxyy = zxzyxyyxy$,

(e) $Y(x,y) = xyx \Rightarrow yzyxyzy = yxyzxzyxy, yzyxyzy = yzyxy = yxzy = yx^2zy = yxzxzy = yxyzxzy = yxyzxzyxy,$
(f) $Y(x,y) = yxy \Rightarrow zyzxzyz = zxzyxyzxz, zyzxzyz = zyzxz = zxyz = zx^2yz = zxyxyz = zxzyxyz = zxzyxyzxz.$

This completes the proof of the theorem. □

Theorem 18.4.18. *The hyperidentity (18.4.14) is essentially satisfied in the semigroup $Q(\cdot)$ if and only if that semigroup satisfies the system of identities (18.4.20).*

Proof. Let us prove the necessity.

Let us consider the case when the polynomial xy is not essential. In that case $xy = x^2$ or $xy = y^2$ and there is no essential polynomial.

Now let us suppose that the polynomial xy is essential. In that case, the polynomial yx is also essential.

Taking $X(x,y) = xy, Y(x,y) = xy$ in (18.4.14) we get $xyz = xyxz$.

Taking $X(x,y) = yx, Y(x,y) = yx$ in (18.4.14) we get $zyx = zxyx \Rightarrow xyz = xzyz$.

Taking $X(x,y) = xy, Y(x,y) = yx$ in (18.4.14) we get $yxz = zxxy = zxzxy = zxzy = zxy \Rightarrow xyz = x(yxz) = xzxy = xzy$.

Moreover, we have $xyz = (xzy)z = yzxz = yxz$ and $xyz = x(yxz) = xxyz = x^2yz$. □

Theorem 18.4.19. *The hyperidentity (18.4.15) is essentially satisfied in the semigroup $Q(\cdot)$ if and only if that semigroup satisfies the system of identities (18.4.20).*

Proof. Let us prove the necessity.

Let us consider the case when the polynomial xy is not essential. In that case $xy = x^2$ or $xy = y^2$ and there is no essential polynomial.

Now let us suppose that the polynomial xy is essential. In that case, the polynomial yx is also essential.

Taking $X(x,y) = xy, Y(x,y) = xy$ in (18.4.15) we get $xyz = xyxz$.

Taking $X(x,y) = yx, Y(x,y) = yx$ in (18.4.15) we get $zyx = zxyx \Rightarrow xyz = xzyz$.

Taking $X(x,y) = xy, Y(x,y) = yx$ in (18.4.15) we get $yxz = xzyx = xyzyx = xzyx = zxy$.

Moreover, we have $xyz = (xzy)z = yzxz = yxz$, $xyz = x(yxz) = xzxy = xzy$ and $xyz = x\,(yxz) = xxyz = x^2yz$. □

Theorem 18.4.20. *The hyperidentity (18.4.16) is essentially satisfied in the semigroup $Q(\cdot)$ if and only if that semigroup satisfies the system of identities (18.4.20).*

Proof. Let us prove the necessity.

Let us consider the case when the polynomial xy is not essential. In that case $xy = x^2$ or $xy = y^2$ and there is no essential polynomial.

Now let us suppose that the polynomial xy is essential. In that case, the polynomial yx is also essential.

Taking $X(x, y) = xy, Y(x, y) = xy$ in (18.4.16) we get $xyz = xyxz$.

Taking $X(x, y) = yx, Y(x, y) = yx$ in (18.4.16) we get $zyx = zxyx \Rightarrow xyz = xzyz$.

Taking $X(x, y) = xy, Y(x, y) = yx$ in (18.4.16) we get $yxz = zxyx = zyx \Rightarrow xyz = xy(xz) = xzxy = xzy$.

Moreover, we have $xyz = xz(yz) = yzxz = yxz$ and $xyz = x\,(yxz) = xxyz = x^2yz$. □

We will denote polynomials (as also functional variables) by letters $X, Y, Z, \ldots$.

Theorem 18.4.21. *Suppose that a left distributive hyperidentity, whose functional rank is greater than 2, is given. In that case that hyperidentity is essentially satisfied in the semigroup $Q(\cdot)$ if and only if the semigroup $Q(\cdot)$ satisfies the system of identities (18.4.20).*

Proof. *Necessity.* Our hyperidentity has the following form:

$$X(x, X_1(y, z)) = X_2(X_3(x, y), X_4(x, z)),$$

where $X_1, X_2, X_3, X_4 \in \{X, Y, Z, \ldots\}$.

As the functional rank of our hyperidentity is at least three, then we can plug in it at least three different polynomials X, Y and Z.

If in our hyperidentity there is exactly one polynomial Y, then by taking the remaining polynomials equal to the polynomial X we will get one of the hyperidentities (18.4.2)–(18.4.12) or (18.4.14)–(18.4.16). And if in our hyperidentity there are more than one polynomial Y, then taking the remaining polynomials (besides the polynomial X) equal to the polynomial Y, we will get one of the hyperidentities (18.4.2)–(18.4.12) or (18.4.14)–(18.4.16) again. So the necessity is proved.

Let us prove the sufficiency now.

Let us suppose that (18.4.20) is satisfied in the semigroup $Q(\cdot)$.

Due to Lemma 18.4.4, for binary essential polynomials we have: $X(x, X_1(y, z)) = xyz = X_2(X_3(x, y), X_4(x, z))$.

This completes the proof of the theorem. □

Theorem 18.4.22. *Any left distributive hyperidentity is essentially equivalent either to the hyperidentity (18.4.1) whose functional rank is one or to the hyperidentity (18.4.2) whose functional rank is two. Moreover, we have (18.4.2)* $\Rightarrow_e$ *(18.4.1).*

Proof. The proof easily follows from Theorems 18.4.5–18.4.21. □

The similar results are valid for right distributive hyperidentities.

Chapter 19

Binary Representations of Semigroups: The Multiplicative Groups of Fields and Hyperidentities

19.1. Preliminary results

We denote by $\mathcal{F}_Q^2$ the set of all binary operations on Q, and define the following operations on it:

$$A \cdot B(x, y) = A[x, B(x, y)],$$
$$A \circ B(x, y) = A[B(x, y), y],$$

where $A, B \in \mathcal{F}_Q^2$, $x, y \in Q$. The operations $(\cdot)$ and $(\circ)$ are called right and left multiplication, respectively.

This set $\mathcal{F}_Q^2$ forms a monoid under left and right multiplication. The two semigroups arising are isomorphic. The identity element of $\mathcal{F}_Q^2(\cdot)$ is the element E defined by the rule $E(x, y) = y$, while that of $\mathcal{F}_Q^2(\circ)$ is determined by $F(x, y) = x$. An isomorphism between the two monoids is established by the map $A \to A^*$, where $A^*(x, y) = A(y, x)$. It is bijective, and $(A \cdot B)^* = A^* \circ B^*$, since

$$(A \cdot B)^*(x, y) = A \cdot B(y, x) = A(y, B(y, x)) = A^*(B^*(x, y), y) = A^* \circ B^*(x, y).$$

A binary operation A defined on a set Q is said to be right (left) invertible if, for all a and b in Q, the equation $A(a, x) = b$ (the equation $A(y, a) = b$, respectively) has a unique solution in Q. We denote the unique solutions respectively by $x = A^{-1}(a, b)$, $y = {}^{-1}A(b, a)$.

Thus, for all a and b in Q we have

$$A(a, A^{-1}(a,b)) = A^{-1}(a, A(a,b)) = b$$

if A is right invertible, and

$$A({}^{-1}A(b,a), a) = {}^{-1}A(A(b,a), a) = b$$

if A is left invertible. The operations A^{-1} and ${}^{-1}A$ are called the right and left inverses of A, respectively. It is easy to see that A^{-1} is right invertible and ${}^{-1}A$ left invertible, and moreover

$$(A^{-1})^{-1} = A = {}^{-1}({}^{-1}A), \quad (A^{-1})^{*} = {}^{-1}(A^{*}), \quad ({}^{-1}A)^{*} = (A^{*})^{-1}.$$

A binary operation A in $\mathcal{F}_Q^2$ is said to be invertible if it is left and right invertible. In this case the left and right inverses for Asatisfy the following rules:

$$({}^{-1}(A^{-1}))^{-1} = {}^{-1}(({}^{-1}A)^{-1}) = A^{*}.$$

A binary algebra $(Q;\Sigma)$ is said to be (right, left) invertible if each of the operations in Σ is (right, left) invertible. An invertible algebra with a single operation is a quasigroup.

Every invertible algebra $(Q;\Sigma)$ spawns five further invertible algebras: $(Q;\Sigma^{-1})$, $(Q;{}^{-1}\Sigma)$, $(Q;{}^{-1}(\Sigma^{-1}))$, $(Q;({}^{-1}\Sigma)^{-1})$ and $(Q;\Sigma^{*})$, where

$$\begin{aligned}
\Sigma^{-1} &= \{A^{-1} \mid A \in \Sigma\},\\
{}^{-1}\Sigma &= \{{}^{-1}A \mid A \in \Sigma\},\\
{}^{-1}(\Sigma^{-1}) &= \{{}^{-1}(A^{-1}) \mid A \in \Sigma\},\\
({}^{-1}\Sigma)^{-1} &= \{({}^{-1}A)^{-1} \mid A \in \Sigma\},\\
\Sigma^{*} &= \{A^{*} \mid A \in \Sigma\}.
\end{aligned}$$

The sets Σ^{-1} and ${}^{-1}\Sigma$ are also interesting in the case of right and left invertible algebras, respectively.

Proposition 19.1.1. *The set of all right (left) invertible operations defined on Q is a group under right (left) multiplication of operations. The two groups are isomorphic.*

Proof. We denote the set of all right (left) invertible operations defined on Q by $\mathcal{F}_Q^r$ (by $\mathcal{F}_Q^l$). If A and B are elements of $\mathcal{F}_Q^r$, then $A \cdot B$ lies in $\mathcal{F}_Q^r$, since the equation $A \cdot B(a, x) = A(a, B(a, x)) = b$ has a unique solution x in Q for all a and b in Q. For every right invertible operation $A \in \mathcal{F}_Q^r$ we have

$$A \cdot A^{-1} = A^{-1} \cdot A = E,$$

where A^{-1} is right inverse to A.

In the same way, the set $\mathcal{F}_Q^l$ of left invertible operations defined on Q is a group under left multiplication. The correspondence $A \to A^*$ yields an isomorphism between the two groups. □

For $\Sigma \subseteq \mathcal{F}_Q^r$ we denote the subgroup of $\mathcal{F}_Q^r(\cdot)$ generated by Σ as $(\Sigma)_r$. When $\Sigma \subseteq \mathcal{F}_Q^l$ we denote the subgroup of$\mathcal{F}_Q^l(\circ)$ generated by Σ as $(\Sigma)_l$.

Proposition 19.1.2. *If $(Q; \Sigma)$ is a right invertible algebra satisfying the hyperidentity (d_1), then the algebra $(Q; \Sigma \cup \Sigma^{-1})$ also satisfies (d_1). If (d_2) is satisfied in a left invertible algebra $(Q; \Sigma)$, then it is also satisfied in $(Q; {}^{-1}\Sigma \cup \Sigma)$.*

Proof. We prove the first part of this proposition, since the second part is so similar. For arbitrary A and B in Σ and x, y, z in Q, we consider the equality

$$A^{-1}(x, B(y, z')) = B(A^{-1}(x, y), A^{-1}(x, z));$$

an element z' like this must exist because all operations are right invertible. Using the definition of A^{-1}, we get

$$\begin{aligned} B(y, y') &= A(x, B(A^{-1}(x, y), A^{-1}(x, z))) \\ &= B(A(x, A^{-1}(x, y)), A(x, A^{-1}(x, z))) = B(y, z), \end{aligned}$$

whence it follows that $z = z'$. Thus, the identity

$$A^{-1}(x, B(y, z)) = B(A^{-1}(x, y), A^{-1}(x, z))$$

holds for all A and B in Σ. Using this identity, we establish the further identity

$$A^{-1}(x, B^{-1}(y, z)) = B^{-1}(A^{-1}(x, y), A^{-1}(x, z)).$$

If

$$A^{-1}(x, B^{-1}(y, z) = B^{-1}(A^{-1}(x, y), A^{-1}(x, z')),$$

then

$$A^{-1}(x,z') = B(A^{-1}(x,y), A^{-1}(x, B^{-1}(y,z)))$$
$$= A^{-1}(x, B(y, B^{-1}(y,z))) = A^{-1}(x,z).$$

It follows from this that $z = z'$. It remains to establish the identity

$$A(x, B^{-1}(y,z)) = B^{-1}(A(x,y), A(x,z))$$

for all A and B in Σ. Suppose that

$$A(x, B^{-1}(y,z')) = B^{-1}(A(x,y), A(x,z));$$

then

$$B^{-1}(y,z') = A^{-1}(x, B^{-1}(A(x,y), A(x,z)))$$
$$= B^{-1}(A^{-1}(x, A(x,y)), A^{-1}(x, A(x,z))) = B^{-1}(y,z)$$

so that $z = z'$. □

Corollary 19.1.3. *If the right (left) invertible algebra $(Q;\Sigma)$ satisfies the hyperidentity (d_1) ((d_2), respectively), it is satisfied also in the extended algebra $(Q;(\Sigma)_r)$ (in $(Q;(\Sigma)_l)$, respectively).*

Proof. Proof of the first part. We use induction to show that the identity

$$S(x, P(y,z)) = P(S(x,y), S(x,z))$$

holds for all S and P in (Σ). Suppose that

$$\begin{aligned} S &= A_1^{\varepsilon_1}\cdots A_m^{\varepsilon_m}, \quad \varepsilon_i = \pm 1, \\ P &= B_1^{\delta_1}\cdots B_k^{\delta_k}, \quad \delta_j = \pm 1, \end{aligned} \tag{19.1.1}$$

where the A_i and B_j lie in Σ. By the length $\partial(S)$ of an operation S in (Σ) we mean the number of operations A_i in some representation of the form (19.1.1). If $\partial(S) = \partial(P) = 1$, then S and P lie in $\Sigma \cup \Sigma^{-1}$, and the assertion follows from Proposition 19.1.2. Suppose that the assertion holds when $\partial(S)$, $\partial(P) \leqslant n-1$. Three cases are possible.

(1) $\partial(S) = n$, $\partial(P) < n$. Then $S = S_1 \cdot S_2$, where $\partial(S_1), \partial(S_2) < n$, so by the induction hypothesis we have

$$\begin{aligned} S(x, P(y, z)) &= S_1 \cdot S_2(x, P(y, z)) = S_1(x, S_2(x, P(y, z))) \\ &= S_1(x, P(S_2(x, y), S_2(x, z))) \\ &= P(S_1(x, S_2(x, y)), S_1(x, S_2(x, z))) \\ &= P(S_1 \cdot S_2(x, y), S_1 \cdot S_2(x, z)) = P(S(x, y), S(x, z)). \end{aligned}$$

(2) $\partial(S) < n$, $\partial(P) = n$. Here we put $P = P_1 \cdot P_2$, where $\partial(P_1), \partial(P_2) < n$, and again apply the induction hypothesis.

(3) $\partial(S) = \partial(P) = n$. Putting $S = S_1 \cdot S_2$, where $\partial(S_1), \partial(S_2) < n$, we revert to case (2).

□

19.2. Binary Cayley theorem for semigroups, idempotent and commutative semigroups

When we speak of $\mathcal{F}_Q^2$ in the sequel, we shall always have in mind the semigroup $\mathcal{F}_Q^2(\cdot)$. We call it the semigroup of binary functions on the set Q.

Suppose that we are given a set Q and a semigroup Γ, which in general has no connection with Q. We say [251] that we have a representation of Γ by binary functions on Q or we have a binary action of Γ on Q if there is a map

$$(*) : \Gamma \times Q^2 \to Q,$$

which associates with each $\gamma \in \Gamma$ and every pair $(x, y) \in Q^2$ an element $\gamma * (x, y)$ of Q such that

$$\gamma_1\gamma_2 * (x, y) = \gamma_1 * (x, \gamma_2 * (x, y))$$

for all $x, y \in Q$ and all $\gamma_1, \gamma_2 \in \Gamma$.

Suppose that we are given such a representation. Then to each γ in Γ there corresponds a binary operation $\varphi(\gamma)$ in $\mathcal{F}_Q^2$:

$$\varphi(\gamma)(x, y) = \gamma * (x, y).$$

We arrive at a map $\varphi : \Gamma \to \mathcal{F}_Q^2$ satisfying the following condition:

$$\begin{aligned}\varphi(\gamma_1 \cdot \gamma_2)(x, y) &= \gamma_1\gamma_2 * (x, y) = \gamma_1 * (x, \gamma_2 * (x, y)) \\ &= \varphi\gamma_1(x, \varphi\gamma_2(x, y)) = (\varphi\gamma_1 \cdot \varphi\gamma_2)(x, y),\end{aligned}$$

that is,

$$\varphi(\gamma_1 \cdot \gamma_2) = \varphi\gamma_1 \cdot \varphi\gamma_2$$

and φ is a homomorphism from the semigroup Γ to the semigroup $\mathcal{F}_Q^2$. Therefore, when we set up a representation of Γ by binary functions on a set Q, in so doing we define a semigroup homomorphism

$$\varphi : \Gamma \to \mathcal{F}_Q^2.$$

The converse also holds. Given a homomorphism

$$\varphi : \Gamma \to \mathcal{F}_Q^2,$$

it is easily seen that putting

$$\gamma * (x, y) = (\varphi\gamma)(x, y)$$

for all $x, y \in Q$ and $\gamma \in \Gamma$ gives a representation of Γ by binary functions on Q. Thus, establishing a representation of a semigroup Γ by binary functions on a set Q is equivalent to providing a homomorphism from Γ into $\mathcal{F}_Q^2$. A representation of Γ of this type is called a binary representation in Q.

Given a binary representation of a semigroup Γ, the elements of Γ act on Q as binary operations:

$$\alpha(x, y) = \alpha * (x, y), \qquad \alpha \in \Gamma, \quad x, y \in Q.$$

Consequently, Q becomes a binary Γ-algebra satisfying the identity

$$\alpha \cdot \beta(x, y) = \alpha(x, \beta(x, y)), \qquad \forall \alpha, \beta \in \Gamma,$$

and is called a binary Γ-polygon.

If the binary Γ-algebra thus obtained satisfies a hyperidentity $W_1 = W_2$ (or formula Φ), we shall say that the original representation satisfies this hyperidentity (or the formula Φ).

If Γ is a monoid with identity element e, we add the condition that $\varphi(e) = E$ to the definition of binary representation. A binary representation is said to be right invertible if the associated binary Γ-polygon Q is right invertible; that is, for arbitrary $\alpha \in \Gamma$ and $a, b \in Q$, the equation $\alpha(a, x) = b$ has a unique solution x in Q. Left invertibility and invertibility are defined in the obvious way. It is easy to see that every binary representation of a group is right invertible.

We say that a binary representation of a semigroup is faithful if the corresponding homomorphism is a monomorphism.

Proposition 19.2.1 (binary Cayley theorem for monoids). *Every monoid Γ has a faithful binary representation satisfying the hyperidentities*

$$X(x, X(y, z)) = X(X(x, y), X(x, z)),$$
$$X(X(x, y), X(u, v)) = X(X(x, u), X(y, v)).$$

Proof. Let $Q = \Gamma$ and take $x, y \in Q$ and $\gamma \in \Gamma$. We put $\gamma(x, y) = \gamma \cdot y$, where $(\cdot)$ is the multiplication in Γ. Here, the condition

$$\gamma_1 \cdot \gamma_2(x, y) = \gamma_1(x, \gamma_2(x, y))$$

follows, since multiplication in Γ is associative. Moreover, $e(x, y) = e \cdot y = y$. The existence of the identity elemente of Γ guarantees that the binary representation is faithful:

$$\gamma_1 \neq \gamma_2 \to \gamma_1(x, e) \neq \gamma_2(x, e).$$

The fact that the displayed hyperidentities are satisfied is clear. □

Assume now that Γ is a semigroup without identity element. In this situation we can adjoin an external identity element to Γ and produce a monoid Γ^*. A faithful binary representation of Γ^* induces one on Γ, and we have the following assertions.

Proposition 19.2.2 (binary Cayley theorem for semigroups). *Every semigroup has a faithful binary representation satisfying the hyperidentities*

$$X(x, X(y, z)) = X(X(x, y), X(x, z)),$$
$$X(X(x, y), X(u, v)) = X(X(x, u), X(y, v)).$$

Proposition 19.2.3 (binary Cayley theorem for idempotent semigroups). *Every idempotent semigroup has a faithful binary representation satisfying the hyperidentities*

$$X(x, X(y, z)) = X(X(x, y), z),$$
$$X(x, X(y, z)) = X(X(x, y), X(x, z)),$$
$$X(X(x, y), X(u, v)) = X(X(x, u), X(y, v)).$$

Proposition 19.2.4 (binary Cayley theorem for commutative semigroups). *Every commutative semigroup has a faithful binary representation satisfying the left distributive hyperidentity* (d_1) *and the Abelian hyperidentity:*

$$X(Y(x,y),Y(u,v)) = Y(X(x,u),X(y,v)).$$

Conversely, if a semigroup has a faithful binary representation with these properties, it is commutative.

Proof. *Necessity* is clear. For *sufficiency*:

$$\begin{aligned}\alpha\cdot\beta(x,y) &= \alpha(x,\beta(x,y)) = \beta(\alpha(x,x),\alpha(x,y))\\ &= \alpha(\beta(x,x),\beta(x,y)) = \beta(x,\alpha(x,y)) = \beta\cdot\alpha(x,y).\end{aligned}$$

□

19.3. Binary Cayley theorem for groups

A binary representation

$$\varphi:\Gamma\to\mathcal{F}_Q^2$$

is said to be transitive if Q is a singleton or else for all a, b, c in Q with $b\neq a\neq c$ there is an element α in Γ with $\alpha(a,b)=c$.

Theorem 19.3.1 (binary Cayley theorem for groups). *A monoid is a group if and only if it has a faithful transitive right invertible binary representation satisfying the left distributive hyperidentity* (d_1).

Proof (*Sufficiency*). Let Γ be a monoid and suppose that $\varphi:\Gamma\to\mathcal{F}_Q^2$ is a faithful transitive right invertible binary representation of Γ satisfying (d_1). When $|Q|=1$, the assertion is trivial. Suppose that $|Q|=2$. In all, there are four right invertible binary operations on Q comprising a Klein four-group under right multiplication of operations. All of its subsemigroups are subgroups. Thus, if Γ is embedded in this group, it must itself be a group. It remains to consider the case where $|Q|\geqslant 3$.

By right invertibility, the map $\alpha_a(x)=\alpha(a,x)$ is a bijection on Q, while it follows from (d_1) that α_a is an automorphism of the binary Γ-polygon Q. If two automorphisms have the same effect on an element b of Q, they will have the same effect on the subpolygon generated by b. From the definition of transitivity, the Γ-polygon Q is generated by any two different elements, and thus when two automorphisms α and β coincide on two different elements of Q, they will coincide on every element x of Q.

Next, we establish the implication

$$\alpha(a, b) = \beta(a, b) \to \alpha = \beta,$$

in which a and b are different elements of Q and α and β lies in Γ. We consider two possibilities.

(1) The binary Γ-polygon Q is idempotent, that is, $\alpha(x, x) = x$ for all $\alpha \in \Gamma$ and all $x \in Q$. In this situation, $\alpha(a, a) = \beta(a, a)$, so that the automorphisms α_a and β_a coincide on two different elements a and b. Thus $\alpha_a(x) = \beta_a(x)$ for every x in Q, so that $\alpha(a, x) = \beta(a, x)$. However, since $|Q| \geqslant 3$ and by transitivity, it follows that for every y in Q there exists c in Q with $y \neq c \neq a$ and δ in Γ such that $y = \delta(c, a) = \delta_c(a)$. Since δ_c is an automorphism of the binary Γ-polygon Q, we have

$$\delta_c\alpha(a, x) = \delta_c\beta(a, x)),$$

$$\alpha(\delta_c(a), \delta_c(x) = \beta(\delta_c(a), \delta_c(x)),$$

$$\alpha(y, u) = \beta(y, u)$$

for all $y, u \in Q$.

(2) Q is not idempotent, that is, there exist $\gamma \in \Gamma$ and $x \in Q$ such that $\gamma(x, x) \neq x$. Then $\gamma(y, y) \neq y$ for every y in Q, as we prove now. Suppose that $\gamma(t, t) = t$ for some $t \in Q$ with $t \neq x$. Again since $|Q| \geqslant 3$ and using transitivity, there exist c in Q with $x \neq c \neq t$ and $\delta \in \Gamma$ such that $x = \delta(c, t) = \delta_c(t)$. Thus,

$$\delta_c(\gamma(t, t)) = \delta_c(t),$$

$$\gamma(\delta_c(t), \delta_c(t)) = \delta_c(t),$$

$$\gamma(x, x) = x,$$

and this contradicts the choice of γ and x. Therefore, the subpolygon generated by the element b of Q contains at least two different elements b and $\gamma(b, b)$, so that it is exactly Q. Thus, the automorphisms α_a and β_a coincide on every x in Q, that is, $\alpha(a, x) = \beta(a, x)$; as in case (1), it now follows that $\alpha = \beta$.

Consider the subgroup (Γ) of $\mathcal{F}_Q^r(\cdot)$ generated by the subset Γ of $\mathcal{F}_Q^r$. By Corollary 19.1.3, hyperidentity (d_1) is satisfied in the (Γ)-algebra Q. Therefore, there exists a transitive faithful right invertible binary representation $\psi : (\Gamma) \to \mathcal{F}_Q^2$ satisfying (d_1) and extending φ. To complete the proof, it is enough to verify that $(\Gamma) = \Gamma$. Take α in (Γ). There exist elements a, b and c with $b \neq a \neq c$ such that $\alpha(a, b) = c$. Indeed, on fixing a in Q and choosing an element b such that $b \neq a$, $b \neq \alpha^{-1}(a,a)$ — which is possible since $|Q| \geqslant 3$ — we get $\alpha(a,b) = c \neq a$. For if $\alpha(a,b) = a$, then

$$\alpha^{-1}(a,\alpha(a,b)) = \alpha^{-1}(a,a), \qquad \alpha^{-1}\cdot\alpha(a,b) = \alpha^{-1}(a,a)$$

and $b = \alpha^{-1}(a,a)$, which contradicts the choice of b. Since the original binary representation φ is transitive, there exists β in Γ such that $\beta(a,b) = c$. Thus $\alpha(a,b) = \beta(a,b)$ with $a \neq b$, and therefore $\alpha = \beta \in \Gamma$.

(*Necessity*). Suppose that Γ is a group. With each element γ of Γ we associate the binary operation defined on $Q = \Gamma$ by the rule $\gamma(x,y) = yx^{-1}\gamma^{-1}x$. Then

$$\begin{aligned} e(x,y) &= y, \\ \gamma_1\cdot\gamma_2(x,y) &= yx^{-1}(\gamma_1\gamma_2)^{-1}x = yx^{-1}\gamma_2^{-1}\gamma_1^{-1}x \\ &= yx^{-1}\gamma_2^{-1}xx^{-1}\gamma_1^{-1}x = \gamma_2(x, yx^{-1}\gamma_2^{-1}x) = \gamma_1 x, \gamma_2(x,y), \end{aligned}$$

that is, we have a binary representation of Γ. It is faithful, transitive and right invertible. It is also immediate to check that it satisfies (d_1):

$$\begin{aligned} \gamma_2(\gamma_1(x,y),\gamma_1(x,z)) &= zx^{-1}\gamma_1^{-1}x(yx^{-1}\gamma_1^{-1}x)^{-1}\gamma_2^{-1}yx^{-1}\gamma_1^{-1}x \\ &= zx^{-1}\gamma_1^{-1}xx^{-1}\gamma_1 xy^{-1}\gamma_2^{-1}yx^{-1}\gamma_1^{-1}x \\ &= zy^{-1}\gamma_2^{-1}yx^{-1}\gamma_1^{-1}x = \gamma_1(x,\gamma_2(y,z)). \end{aligned}$$

□

Problem 19.3.2. *Characterize the right invertible algebras satisfying the left distributive hyperidentity* (d_1)*, and the same thing with right and left interchanged.*

19.4. Characterization of multiplicative groups of fields

In 1960, at the Third All Union Colloquium on General Algebra, A. I. Mal'tsev conjectured that the class of multiplicative groups of fields cannot be given an elementary characterization [102]. This conjecture was proved in [177] (see also [350]).

In this section, the solution of Jacobson–Mal'tsev-Fuchs problem on characterization of multiplicative groups of fields is presented.

The next result shows that, in the language of binary representations, the class of multiplicative groups of fields is distinguished in the class of all groups by a single hyperidentity of idempotency (cf. Theorem 19.3.1). For more general result on characterization of multiplicative groups of division rings (skew fields), see [252].

Theorem 19.4.1 ([251]). *A monoid is the multiplicative group of a field if and only if it has a faithful transitive right invertible binary representation satisfying the left distributive hyperidentity* (d_1) *and the hyperidentity of idempotency* $X(x,x) = x$.

Proof (*Sufficiency*). Let Γ be a monoid having a faithful transitive right invertible binary representation $\varphi : \Gamma \to \mathcal{F}_Q^2$ satisfying (d_1) and an idempotent hyperidentity By Theorem 19.3.1, Γ is a group, which by idempotency is Abelian. This is how that goes:

$$\alpha \cdot \beta(x,y) = \alpha(x, \beta(x,y)) = \alpha(\beta(x,x), \beta(x,y)) = \beta(x, \alpha(x,y)) = \beta \cdot \alpha(x,y).$$

This assertion is clear when $|Q| \leqslant 2$, since the only idempotent right invertible binary operation defined on such a set is that given by $E(x,y) = y$. We shall therefore assume that $|Q| \geqslant 3$, and make the further assumption that Γ is non-trivial.

Since φ is transitive, it follows that the group $\operatorname{Aut} Q$ of all automorphisms of the binary Γ-polygon Q is a minimal doubly transitive group in the sense of [140]; that is, for any two pairs (a,b) and (c,d) of elements of Q such that $a \neq b$ and $c \neq d$, there exists exactly one λ in $\operatorname{Aut} Q$ such that $\lambda a = c$ and $\lambda b = d$. We consider the following possibilities.

(1) **Case $c \neq b$.** In this case, there exists α in Γ such that $\alpha(b,a) = c$ and $\alpha(b,b) = b$, so that $\alpha_b(a) = c$ and $\alpha_b(b) = b$. There exists β in Γ such that $\beta(c,b) = d$, $\beta(c,c) = c$, that is, $\beta_c(b) = d$ and $\beta_c(c) = c$. Thus, the product $\lambda = \alpha_b \cdot \beta_c$ is the unique automorphism with the required property.
(2) **Case $d \neq a$.** The argument here is as in case (1).
(3) **Case $c = b$ and $d = a$.** Since $|Q| \geqslant 3$, we can choose an element s in Q different from a and from b, and then apply case (1) to the pairs (a,b), (s,a) and (s,a), (b,a). The product of the two resulting automorphisms is the automorphism we require. Suppose that $\alpha \in \Gamma \backslash \{e\}$ and that a and b are different elements of Q. Then $\alpha(a,b) \neq b$, so that there exists $f(\alpha)$ in Γ such that $f(\alpha)(b,a) = \alpha(a,b)$. If x and y are different elements of Q, the double transitivity of $\operatorname{Aut} Q$ means that there is an automorphism

λ of Q such that $\lambda a = x$ and $\lambda b = y$. Thus

$$\lambda(f(\alpha)(b, a) = \lambda\alpha(a, b),$$
$$f(\alpha)(\lambda b, \lambda a) = \alpha(\lambda a, \lambda b),$$
$$f(\alpha)(y, x) = \alpha(x, y).$$

However, the last equality is satisfied even when $x = y$. Moreover, if we adjoin an external zero 0 to Γ and define its binary action by $0(x, y) = x$ and $f(0) = e$, the identity thus obtained is also true when $\alpha = 0$. The equality

$$f(f(\alpha) \cdot f(\beta)(x, y) = (\beta \cdot f(\alpha \cdot f(\beta^{-1})))(x, y)$$

holds for all $\alpha \in \Gamma \cup \{0\}$, all $\beta \in \Gamma$ and all $x, y \in Q$. In fact,

$$\begin{aligned} f(f(\alpha)f(\beta)(x, y) &= f(\alpha) \cdot f(\beta)(y, x) \\ &= f(\alpha)(y, f(\beta)(y, x)) = f(\alpha)(y, \beta(x, y)) = \alpha(\beta(x, y), y); \end{aligned}$$

$$\begin{aligned} \beta \cdot f(\alpha \cdot f(\beta^{-1})))(x, y) &= \beta(x, f(\alpha \cdot f(\beta^{-1}))(x, y) \\ &= \beta(x, (\alpha \cdot f(\beta^{-1}))(y, x)) = \beta(x, \alpha(y, f(\beta^{-1}))(y, x)) \\ &= \beta(x, \alpha(y, \beta^{-1}(x, y))) = \alpha(\beta(x, y), \beta(x, \beta^{-1}(x, y))) \\ &= \alpha(\beta(x, y), (\beta \cdot \beta^{-1})(x, y)) = \alpha(\beta(x, y), e(x, y)) \\ &= \alpha(\beta(x, y), y). \end{aligned}$$

The sufficiency proof can now be completed using the following assertion [96]: a commutative group $\Gamma(\cdot)$ with identity element e is the multiplicative group of a field if and only if there exists a map $f : \Gamma \cup \{0\} \to \Gamma \cup \{0\}$ such that $f(0) = e$, $f(f(\alpha)) = \alpha$ and

$$f(f(\alpha) \cdot f(\beta)) = \beta \cdot f(\alpha \cdot f(\beta^{-1}))$$

for all $\alpha \in \Gamma \cup \{0\}$ and $\beta \in \Gamma$.

(*Necessity*). Let Γ be the multiplicative group of a field $Q(+, \cdot)$, so that $\Gamma = Q \setminus \{0\}$. With each element γ of Γ we associate the binary operation defined on Q as follows:

$$\gamma(x, y) = (1 - \gamma)x + \gamma y.$$

Then $1(x, y) = y$ and

$$\begin{aligned}\gamma_1 \cdot \gamma_2(x, y) &= (1 - \gamma_1\gamma_2)x + \gamma_1\gamma_2 y = x - \gamma_1\gamma_2 x + \gamma_1\gamma_2 y \\ &= x - \gamma_1 x + \gamma_1 x - \gamma_1\gamma_2 x + \gamma_1\gamma_2 y \\ &= x - \gamma_1 x + \gamma_1(x - \gamma_2 x + \gamma_2 y) \\ &= (1 - \gamma_1)x + \gamma_1((1 - \gamma_2)x + \gamma_2 y) = \gamma_1(x, \gamma_2(x, y)).\end{aligned}$$

It is easy to prove that the binary operation thus obtained is faithful, transitive and right invertible, and that it satisfies the hyperidentity (d_1) and the hyperidentity of idempotency. This completes the proof of Theorem 19.4.1. □

See [140] for general properties of minimal doubly transitive permutation groups.

Remark 19.4.2. The original proof of Theorem 19.4.1 in the paper [251] does not use the assertion of [96].

Problem 19.4.3. *Characterize the right (respectively, left) invertible algebras satisfying the hyperidentities (d_1) and $X(x, x) = x$ (satisfying (d_2) and $X(x, x) = x$, respectively).*

Chapter 20

Associative Formulae with Functional Variables

20.1. Introduction and preliminaries

We consider the following formulae from the second order languages (logic) [64,66,214,216]: hyperidentities of associativity, $\forall\exists(\forall)$-identities of associativity and $\forall\exists\forall(\forall)$-identities of associativity. For $\forall\exists(\forall)$-identities, $\forall\exists\forall(\forall)$-identities and other preliminary concepts and results, see [249,252, 254,268–270,289].

In this chapter, the Belousov theorem on linearity of invertible algebras with the Schauffler type $\forall\exists(\forall)$-identity of associativity is extended over the other $\forall\exists(\forall)$-identities of associativity. As a consequence, we obtain the equivalency of the considered $\forall\exists(\forall)$-identities of associativity and non-trivial hyperidentities of associativity in systems of groups. For the considered formulas we prove the Schauffler–Belousov type theorems, too.

An algebra $(Q;\Sigma)$ with binary operations is called binary algebra. A binary algebra $(Q;\Sigma)$ is said to be a q-algebra (e-algebra) if there is an operation $A \in \Sigma$ such that $Q(A)$ is a quasigroup (a groupoid with a unit). A binary algebra $(Q;\Sigma)$ is called functionally non-trivial if $|\Sigma| > 1$. It is known [249,252] (see also Sec. 17.1) that if an associative non-trivial hyperidentity is satisfied in a functionally non-trivial q-algebra (e-algebra) then this hyperidentity can only be of the functional rank 2 and of one of the following forms:

$$X(x, Y(y, z)) = Y(X(x, y), z), \qquad (ass)_1$$

$$X(x, Y(y, z)) = X(Y(x, y), z), \qquad (ass)_2$$

$$Y(x, Y(y, z)) = X(X(x, y), z). \qquad (ass)_3$$

Moreover, in the class of q-algebras (e-algebras) the hyperidentity $(ass)_3$ implies the hyperidentity $(ass)_2$ which, in its turn, implies the hyperidentity $(ass)_1$.

A binary algebra $(Q;\Sigma)$ is called hyperassociative, if it satisfies the first hyperidentity of associativity:

$$X(x, Y(y, z)) = Y(X(x, y), z). \qquad (ass)_1$$

Thus, the hyperassociative algebras are algebras with semigroup operations. Hyperassociative algebras under the name of Γ-semigroups, Γ-semirings (gamma-semigroups, gamma-semirings) or doppelsemigroups also were considered by various authors [11,20,149,172,188,197,200,306,318, 342,358,359,364–366,369,433–441].

Example 20.1.1. Let A and B be non-empty sets, Σ be the set of all mappings from B to A and Q be the set of all mappings from A to B. Then we can consider every element $\alpha \in \Sigma$ as a binary operation on Q:

$$\alpha(a, b) = a \cdot \alpha \cdot b,$$

where $a, b \in Q$ and $a \cdot \alpha \cdot b$ is the usual superposition of mappings. So we obtain an algebra $(Q;\Sigma)$, which is hyperassociative.

If $(Q;\Sigma)$ is a functionally non-trivial idempotent hyperassociative algebra then the cardinality of Q is $|Q| \geq 4$. An example of an idempotent functionally non-trivial hyperassociative algebra $Q(+,\cdot)$ is given by the Cayley tables of their operations $+$ and $\cdot$ as shown below:

+	**1**	**2**	**3**	**4**
1	1	1	1	1
2	1	2	1	4
3	3	3	3	3
4	4	4	4	4

·	**1**	**2**	**3**	**4**
1	1	1	1	1
2	1	2	4	4
3	3	3	3	3
4	4	4	4	4

.

Moreover, there exist 24 functionally non-trivial idempotent hyperassociative algebras $Q(+,\cdot)$ with the condition, $|Q| = 4$. Further, the superproduct of two such algebras is an idempotent hyperassociative algebra with four binary operations, etc. [249,252,257,264,268,269,300]. For the structure of idempotent and hyperassociative algebras see [304].

The following second-order formula is called a $\forall\exists(\forall)$-identity:

$$\forall X_1, \ldots, X_k \exists X_{k+1}, \ldots, X_m \forall x_1, \ldots, x_n \; (w_1 = w_2), \qquad (20.1.1)$$

where w_1 and w_2 are the hyperterms in the functional variables $X_1, \ldots, X_m$ and the object variables $x_1, \ldots, x_n$. Moreover, k is called functional rank of the given $\forall\exists(\forall)$-identity (20.1.1). The $\forall\exists\forall(\forall)$-identity has a similar definition.

The satisfiability of these second-order formulae in the algebra $(Q; \Sigma)$ is understood by the quantifiers $(\forall X_i)$ and $(\exists X_j)$ to mean: "for every value $X_i = A \in \Sigma$ of the corresponding arity" and "there exists a value $X_j = A \in \Sigma$ of the corresponding arity". This semantics is compatible with bihomomorphism of the algebras introduced in [249,252], that are defined as the pairs $\left(\varphi, \tilde{\psi}\right)$ of maps with the following condition: A and $\tilde{\psi}(A)$ have the same arity and

$$\varphi\, A(x_1, \ldots, x_n) = \left[\tilde{\psi}(A)\right](\varphi x_1, \ldots, \varphi x_n).$$

For the categorical definition of $\forall\exists(\forall)$-identities, see [244].

A binary algebra $(Q; \Sigma)$ is called isotopic to the groupoid $Q(\cdot)$ if its every operation is isotopic to the groupoid $Q(\cdot)$, i.e., for any operation $A \in \Sigma$ there exist bijections $\alpha_A, \beta_A, \gamma_A : Q \to Q$ such that

$$A(x, y) = \alpha_A^{-1}(\beta_A x \cdot \gamma_A y)$$

for every $x, y \in Q$ [12]. A binary algebra $(Q; \Sigma)$ is called isotopic to a groupoid (group, loop, semigroup) if $(Q; \Sigma)$ isotopic to some groupoid (group, loop, semigroup) $Q(\cdot)$ [269].

A binary algebra $(Q; \Sigma)$ is called left (right)-linear on the groupoid $Q(\cdot)$ if its every operation is left (right)-linear on the groupoid $Q(\cdot)$, i.e., for any operation $A \in \Sigma$ there exist automorphisms φ_A of the groupoid $Q(\cdot)$ and permutation α_A of Q with the equality

$$A(x, y) = \varphi_A x \cdot \alpha_A y$$

$$(A(x, y) = \alpha_A x \cdot \varphi_A y)$$

for every $x, y \in Q$. A binary algebra $(Q; \Sigma)$ is called left (right)-linear on a groupoid (group, loop, semigroup) if $(Q; \Sigma)$ is left (right)-linear on some groupoid (group, loop, semigroup) $Q(\cdot)$.

A binary algebra $(Q; \Sigma)$ is called linear on the groupoid $Q(\cdot)$ if its every operation is linear on the groupoid $Q(\cdot)$, i.e., for any operation $A \in \Sigma$ there exist automorphisms φ_A and ψ_A of the groupoid $Q(\cdot)$ and element $t_A \in Q$ with the equality

$$A(x, y) = (\varphi_A x \cdot t_A) \cdot \psi_A y$$

for every $x, y \in Q$.

A binary algebra $(Q; \Sigma)$ is called linear on a groupoid (group, loop, semigroup) if $(Q; \Sigma)$ is linear on some groupoid (group, loop, semigroup) $Q(\cdot)$. (A linear algebra on the commutative group is often called T-linear algebra too [160,410].)

A binary algebra $(Q; \Sigma)$ is called invertible if its every operation is quasigroup operation, i.e., $Q(A)$ is a quasigroup for any operation $A \in \Sigma$. An invertible algebra $(Q; \Sigma)$ is called system of groups if $Q(A)$ is a group for any operation $A \in \Sigma$.

In the paper [24] it is proved that invertible algebra $(Q; \Sigma)$ with the following $\forall\exists(\forall)$-identity of associativity:

$$\forall X, Y \exists X', Y' \forall x, y, z (X(Y(x, y), z) = X'(x, Y'(y, z))) \tag{20.1.2}$$

is linear on a group. Binary algebra $(Q; \Sigma)$ satisfies the $\forall\exists(\forall)$-identity of associativity (20.1.2) if for every operations $A, B \in \Sigma$ there exist operations $A', B' \in \Sigma$ such that the following identity

$$A(B(x, y), z) = A'(x, B'(y, z))$$

is valid.

In the paper [24] the similar result is formulated for the following $\forall\exists(\forall)$-identity of associativity too:

$$\forall X, Y \exists X', Y' \forall x, y, z (X(x, Y(y, z)) = X'(Y'(x, y), z)). \tag{20.1.3}$$

Proposition 20.1.2.

(1) *The $\forall\exists(\forall)$-identity of associativity* (20.1.2) *is equivalent to the following non-trivial hyperidentity of associativity $(ass)_1$ in systems of groups.*
(2) *The $\forall\exists(\forall)$-identities of associativity* (20.1.3) *and* (20.1.2) *are equivalent in systems of groups.*

The formulas (20.1.2) and (20.1.3) are called Schauffler type $\forall\exists(\forall)$-identities of associativity [362]. He was the first to consider these formulas in the coding theory. During the second World War R. Schauffler worked for the German Cryptography service and developed a method of error detection based on using of identities with functional variables, in which the check digits are calculated through an associative system with quasigroup operations [360–362]. Particularly, he proved the following result [362].

Schauffler theorem. *Let Q be a non-empty set. For any pair of (binary) quasigroups $Q(A)$ and $Q(B)$ there exists a pair of (binary) quasigroups $Q(C)$ and $Q(D)$ with the identity*

$$A(B(x,y),z) = C(x,D(y,z)) \tag{20.1.4}$$

if and only if the cardinality of Q is $|Q| \leqslant 3$.

In other words, if Ω_Q is a set of all (binary) quasigroup operations on the set Q then in the algebra $(Q;\Omega_Q)$ the following $\forall\exists(\forall)$-identity of associativity

$$\forall X,Y\exists Z,U\forall x,y,z(X(Y(x,y),z) = Z(x,U(y,z))) \tag{20.1.5}$$

is satisfied if and only if the cardinality of Q is $|Q| \leqslant 3$.

In the paper [25] another proof of Schauffler's theorem is suggested, while in papers [414–417,423,431] this result is extended for ternary and n-ary quasigroups cases, as well as for topological n-ary quasigroups.

In the paper [183] the following result is proved.

Theorem 20.1.3. *Let Q be a non-empty set. For every pair of (binary) groupoids $Q(A)$ and $Q(B)$ there exists a pair of (binary) groupoids $Q(C)$ and $Q(D)$ with the identity (20.1.4) iff the set Q is infinite or singleton.*

In other words, if G_Q is a set of all (binary) operations on the set Q then in the algebra $(Q;G_Q)$ the second order formula (20.1.5) is satisfied if and only if the set Q is infinite or singleton.

A following more general result is proved in paper [254].

Theorem 20.1.4. *Let Q be a non-empty set. For every pair of (binary) quasigroups $Q(A)$ and $Q(B)$ there exists a pair of (binary) groupoids $Q(C)$ and $Q(D)$ with the identity (20.1.4) if and only if the cardinality of Q is $|Q| \leqslant 3$ or the set Q is infinite.*

Besides, in paper [254] the satisfiability of $\forall\exists(\forall)$-identities of associativity of functional rank 1 is also considered.

In this chapter, we characterize invertible algebras with an $\forall\exists(\forall)$-identity of associativity (20.1.6)–(20.1.9) as linear algebras:

$$\forall X, Y \exists X', Y' \forall x, y, z (X(Y'(x, y), z) = X'(x, Y(y, z))), \tag{20.1.6}$$

$$\forall X, Y \exists X', Y' \forall x, y, z (X(Y'(x, y), z) = Y(x, X'(y, z))), \tag{20.1.7}$$

$$\forall X, Y \exists X', Y' \forall x, y, z (Y'(X(x, y), z) = X'(x, Y(y, z))), \tag{20.1.8}$$

$$\forall X, Y \exists X', Y' \forall x, y, z (Y'(X(x, y), z) = Y(x, X'(y, z))). \tag{20.1.9}$$

In the case of group systems we obtain equivalency of any considered $\forall\exists(\forall)$-identity and a non-trivial hyperidentity of associativity.

Note the following earlier related result, too.

Theorem 20.1.5 ([252]). *In invertible algebras any $\forall\exists\forall(\forall)$-identity of associativity is equivalent to a non-trivial hyperidentity of associativity.*

20.2. Auxiliary results and concepts

Lemma 20.2.1 ([24,25]). *Invertible algebra $Q(A, B, C, D)$ with the following identity*

$$A(x, B(y, z)) = C(D(x, y), z)$$

is isotopic to a group (see Theorem 16.2.5).

In detail, we have the following results.

Lemma 20.2.2 ([269]).

(1) *Invertible algebra $Q(A, B)$ with the following identity*

$$A(x, A(y, z)) = B(B(x, y), z)$$

is linear on a group.

(2) *If the identity*

$$A(A(x, y), z) = B(x, C(y, z))$$

is satisfied in algebra $Q(A, B, C)$ with tree quasigroup operations, then the quasigroup $Q(A)$ is left-linear on a group.

(3) *If the identity*

$$A(x, A(y, z)) = B(C(x, y), z)$$

is satisfied in algebra $Q(A, B, C)$ with tree quasigroup operations, then the quasigroup $Q(A)$ is right-linear on a group.

(4) *In binary algebra $Q(A, B)$ with two binary quasigroup operations the identity*

$$A(A(x, y), z) = A(x, B(y, z))$$

holds if and only if there exist a group $Q(\circ)$ and permutations α of Q such that

$$A(x, y) = x \circ \alpha y,$$
$$\alpha B(x, y) = \alpha x \circ \alpha y.$$

Hence, $Q(B)$ is a group (A. Suschkewitsch [393]) isomorphic to $Q(\circ)$.

(5) *In binary algebra $Q(A, B)$ with two binary quasigroup operations the identity*

$$A(x, A(y, z)) = A(B(x, y), z)$$

holds if and only if there exist a group $Q(\circ)$ and permutation α of Q such that

$$A(x, y) = \alpha x \circ y,$$
$$\alpha B(x, y) = \alpha x \circ \alpha y.$$

Let $Q(\cdot)$ be a group. Bijection $\alpha : Q \to Q$ is called holomorphism (see Sec. 16.2) of group $Q(\cdot)$ if the following equality:

$$\alpha(x \cdot y^{-1} \cdot z) = \alpha x \cdot (\alpha y)^{-1} \cdot \alpha z$$

is valid for any $x, y, z \in Q$. Substitute $y = e$, which is the unit element of $Q(\cdot)$, we obtain

$$\alpha(x \cdot z) = \alpha x \cdot (\alpha e)^{-1} \cdot \alpha z$$

for any $x, z \in Q$. And conversely, if bijection $\alpha : Q \to Q$ satisfies the last equality then α is an holomorphism of $Q(\cdot)$. Namely, $\alpha(x^{-1}) = \alpha e \cdot (\alpha x)^{-1} \cdot \alpha e$ and

$$\begin{aligned}\alpha(x \cdot y^{-1} \cdot z) &= \alpha x \cdot (\alpha e)^{-1} \cdot \alpha(y^{-1} \cdot z)\\ &= \alpha x \cdot (\alpha e)^{-1} \alpha(y^{-1}) \cdot (\alpha e)^{-1} \cdot \alpha z\\ &= \alpha x \cdot (\alpha e)^{-1} \cdot \alpha e \cdot \alpha(y^{-1}) \cdot \alpha e \cdot (\alpha e)^{-1} \cdot \alpha z\\ &= \alpha x \cdot (\alpha y)^{-1} \cdot \alpha z.\end{aligned}$$

The set of all holomorphisms of the group $Q(\cdot)$ forms the group $\mathrm{Hol}\, Q(\cdot)$ (which is called holomorph of $Q(\cdot)$). For holomorphisms of semigroups and rings, see [249].

The following properties of holomorphisms are known from Sec. 16.2.

(1) Any holomorphism α of the group $Q(\cdot)$ has the following representations:

$$\alpha x = \alpha_1 x \cdot k_1 \quad \text{and} \quad \alpha x = k_2 \cdot \alpha_2 x,$$

where α_1 and α_2 are automorphisms of the group $Q(\cdot)$ and $k_1, k_2 \in Q$.

(2) If α is a holomorphism of the group $Q(\cdot)$ and $s_1, s_2 \in Q$ then the following bijections

$$\beta_1 x = \alpha x \cdot s_1 \quad \text{and} \quad \alpha_2 x = (\alpha e)^{-1} \cdot \alpha x$$

are automorphisms of $Q(\cdot)$.

(3) If $Q(\cdot)$ is a group and bijections $\alpha, \beta, \gamma : Q \to Q$ satisfy the following identity

$$\alpha(x \cdot y) = \beta x \cdot \gamma y,$$

then α, β, γ are holomorphisms of $Q(\cdot)$.

(4) If $Q(\cdot)$ is a group and bijections $\alpha, \beta, \gamma, \delta, \sigma, \tau : Q \to Q$ satisfy the following identity

$$\alpha(\beta(x \cdot y) \cdot z) = \gamma x \cdot \delta(\sigma y \cdot \tau z),$$

then all the mentioned mappings are holomorphisms of $Q(\cdot)$.

(5) A binary algebra is linear on the group if and only if it is left-linear on a group and right-linear on a group.

20.3. Main results

We need the following semantics of $\forall\exists(\forall)$-identities. We say that the binary algebra $(Q; \Sigma)$ with quasigroup operations $(*)$-satisfies the $\forall\exists(\forall)$-identity of associativity (20.1.6) if for every quasigroup operations $A, B \in \Sigma$ there exist quasigroup operations $A', B' \in \Omega_Q$ with identity:

$$A(B'(x, y), z) = A'(x, B(y, z));$$

Analogously, we understand the $(*)$-satisfiability of invertible algebra of one of the $\forall\exists(\forall)$-identities of associativity (20.1.7)–(20.1.9).

Lemma 20.3.1. *Invertible algebra $(Q; \Sigma)$ $(*)$-satisfying one of the $\forall\exists(\forall)$-identity of associativity (20.1.6)–(20.1.9) is isotopic to a group.*

Proof. In the case of (20.1.6), for every operation $A, B \in \Sigma$ there exist $A', B' \in \Omega_Q$ such that the following identity:

$$A(B'(x,y),z) = A'(x, B(y,z))$$

is valid. According to Lemma 20.2.1, the algebra $Q(A, B', A', B)$ is isotopic to a group $Q(\cdot)$. Let us fix operations A, B', A', B and $(\cdot)$. Now for any $C \in \Sigma$ there exist operations $A'', B'' \in \Omega_Q$ with the following identity:

$$A(B''(x,y),z) = A''(x, C(y,z)).$$

Again, according to Lemma 20.2.1, the algebra $Q(A, B'', A'', C)$ is isotopic to another group $Q(\circ)$. So C is isotopic to $(\circ)$, where $(\circ)$ is isotopic to A, and A is isotopic to $(\cdot)$. Hence, any operation $C \in \Sigma$ is isotopic to fixed operation $(\cdot)$, by transitivity of isotopy. □

Example 20.3.2. Let $Q(\cdot)$ be a group and define the following binary operation $(*)$ on Q isotopic to $(\cdot)$:

$$x * y = \gamma(\alpha x \cdot \beta y).$$

If $\Sigma = \{*\}$, then algebra $(Q; \Sigma)$ $(*)$-satisfies the $\forall\exists(\forall)$-identity of associativity (20.1.6).

Namely, if $A = B = (*)$, then $A'(x,y) = \gamma(x \cdot \gamma^{-1}y)$, $B'(x,y) = \alpha^{-1}(x \cdot \alpha y)$, where $Q(A')$ and $Q(B')$ are quasigroups isotopic to $Q(\cdot)$.

Let us prove the following stronger results in the case of $\forall\exists(\forall)$-identities.

Theorem 20.3.3. *Invertible algebra $(Q; \Sigma)$ with the $\forall\exists(\forall)$-identity of associativity (20.1.6) is linear on a group.*

Proof. According to Lemma 20.3.1 the algebra $(Q; \Sigma)$ is isotopic to a group $Q(\cdot)$. By the condition, for every operations $A, B \in \Sigma$ there exist $A', B' \in \Sigma$ with the following identity:

$$A(B'(x,y),z) = A'(x, B(y,z)). \tag{20.3.1}$$

Hence,

$$A(x,y) = \alpha^{-1}(\beta x \cdot \gamma y), \quad B(x,y) = \lambda^{-1}(\mu x \cdot \nu y),$$
$$A'(x,y) = \alpha_1^{-1}(\beta_1 x \cdot \gamma_1 y), \quad B'(x,y) = \lambda_1^{-1}(\mu_1 x \cdot \nu_1 y).$$

Substituting these values in the equality (20.3.1) we obtain

$$\alpha^{-1}(\beta(\lambda_1^{-1}(\mu_1 x \cdot \nu_1 y))) \cdot \gamma z = \alpha_1^{-1}(\beta_1 x \cdot \gamma_1(\lambda^{-1}(\mu y \cdot \nu z))),$$
$$\alpha_1\alpha^{-1}(\beta\lambda_1^{-1}(x \cdot y) \cdot z) = \beta_1(\mu_1^{-1}x) \cdot \gamma_1\lambda^{-1}(\mu(\nu_1^{-1}y) \cdot \nu(\gamma^{-1}z)).$$

Thus the following products are holomorphisms of the group $Q(\cdot)$: $\gamma_1\lambda^{-1} = \theta_1$ and $\nu\gamma^{-1} = \theta$. If we fix $A = A' \in \Sigma$ we obtain holomorphism $\nu\gamma_1^{-1} = \theta_2$. Hence, we have $\lambda^{-1} = \gamma_1^{-1}\theta_1$, $\nu = \theta_2\gamma_1$ and

$$\begin{aligned}B(x, y) &= \lambda^{-1}(\mu x \cdot \nu y)y \\ &= \gamma_1^{-1}\theta_1(\mu x \cdot \theta_2\gamma_1 y) \\ &= \gamma_1^{-1}(\theta_1(\mu x) \cdot (\theta_1 e)^{-1} \cdot \theta_1(\theta_2\gamma_1 y)) \\ &= \gamma_1^{-1}(\delta x \cdot \psi(\gamma_1 y)),\end{aligned}$$

where ψ is an automorphism of the group $Q(\cdot)$ and δ is a bijection of Q. Now let us define the new operation $(\circ)$ on the set Q by the following rule:

$$x \circ y = \gamma_1^{-1}(\gamma_1 x \cdot \gamma_1 y).$$

So, $Q(\circ)$ is a group since $Q(\circ)$ is isomorphic to $Q(\cdot)$.

Hence,

$$B(x, y) = \gamma_1^{-1}\delta x \circ \gamma_1^{-1}\psi\gamma_1 y = \sigma x \circ \varphi y,$$

where $\sigma x = \gamma_1^{-1}\delta x$, $\varphi y = \gamma_1^{-1}\psi\gamma_1 y$ and φ is an automorphism of $Q(\circ)$:

$$\begin{aligned}\varphi(x \circ y) &= \gamma_1^{-1}\psi\gamma_1(x \circ y) = \gamma_1^{-1}\psi\gamma_1\gamma_1^{-1}(\gamma_1 x \cdot \gamma_1 y) \\ &= \gamma_1^{-1}(\psi\gamma_1 x \cdot \psi\gamma_1 y) = \gamma_1^{-1}(\gamma_1\varphi_1\varphi x \cdot \gamma_1\varphi y) = \varphi x \circ \varphi y.\end{aligned}$$

So, for any operation $B \in \Sigma$ we have

$$B(x, y) = \sigma x \circ \varphi y,$$

where $Q(\circ)$ is a group, φ is an automorphism of the group $Q(\circ)$ and $\sigma : Q \to Q$ is a bijection. Using this result for A, A', B', B in equality (20.3.1) we have

$$\sigma(\tau' x \circ \lambda' y) \circ \varphi z = \sigma' x \circ \varphi'(\tau y \circ \lambda z),$$

where λ', φ, φ' and λ are automorphisms of $Q(\circ)$.

Substituting $z = f$, where f is the identity element of the group $Q(\circ)$, we obtain

$$\begin{aligned}\sigma(\tau' x \circ \lambda' y) &= \sigma' x \circ \varphi'(\tau y), \\ \sigma(x \circ y) &= \sigma'((\tau')^{-1} x \circ \varphi'(\tau(\lambda')^{-1} y),\end{aligned}$$

and σ is a holomorphism of $Q(\circ)$, i.e.,

$$\sigma(x) = \sigma_0 x \circ r,$$

where σ_0 is an automorphism of $Q(\circ)$ and $r \in Q$. So,

$$B(x, y) = \sigma_0 x \circ r \circ \varphi y,$$

i.e., any operation $B \in \Sigma$ is linear on the group $Q(\circ)$. □

Example 20.3.4. A quasigroup $Q(\cdot)$ satisfies the $\forall\exists(\forall)$-identity of associativity (20.1.6) if and only if $Q(\cdot)$ is a group.

Example 20.3.5. Let $Q(+)$ be a group and let α: $Q \to Q$ be a second-order automorphism of $Q(+)$, i.e., $\alpha^2 = \varepsilon$. If $f(x, y) = x + \alpha y$ for every $x, y \in Q$ and $\Sigma = \{+, f\}$ then the invertible algebra $(Q; \Sigma)$ satisfies the $\forall\exists(\forall)$-identity of associativity (20.1.6).

Example 20.3.6. Let $Q(+)$ be a group and let $\alpha : Q \to Q$ be a second-order automorphism of $Q(+)$. If $f_1(x, y) = x + \alpha y$ and $f_2(x, y) = \alpha x + y$ for every $x, y \in Q$ and $\Sigma = \{+, f_1, f_2\}$ then the invertible algebra $(Q; \Sigma)$ satisfies the $\forall\exists(\forall)$-identity of associativity (20.1.6).

Example 20.3.7. Let $Q(+)$ be a group and let $\alpha : Q \to Q$ be a second-order automorphism of $Q(+)$. If $f_1(x, y) = x + \alpha y$, $f_2(x, y) = \alpha x + y$ and $f_3(x, y) = \alpha x + \alpha y$ for every $x, y \in Q$ and $\Sigma = \{+, f_1, f_2, f_3\}$ then the invertible algebra $(Q; \Sigma)$ satisfies the $\forall\exists(\forall)$-identity of associativity (20.1.6).

Theorem 20.3.8. *Invertible algebra $(Q; \Sigma)$ with the $\forall\exists(\forall)$-identity of associativity (20.1.7) is linear on a group.*

Proof. Again according to Lemma 20.3.1 the algebra $(Q; \Sigma)$ is isotopic to a group $Q(\cdot)$. For every operations $A, B \in \Sigma$ there exist $A', B' \in \Sigma$ with the following identity:

$$A(B'(x, y), z) = B(x, A'(y, z)). \tag{20.3.2}$$

Hence,

$$A(x, y) = \alpha^{-1}(\beta x \cdot \gamma y), \quad B(x, y) = \lambda^{-1}(\mu x \cdot \nu y),$$
$$A'(x, y) = \alpha_1^{-1}(\beta_1 x \cdot \gamma_1 y), \quad B'(x, y) = \lambda_1^{-1}(\mu_1 x \cdot \nu_1 y).$$

Substituting these values in the equality (20.3.2), we obtain

$$\alpha^{-1}(\beta(\lambda_1^{-1}(\mu_1 x \cdot \nu_1 y)) \cdot \gamma z) = \lambda^{-1}(\mu x \cdot \nu(\alpha_1^{-1}(\beta_1 y \cdot \gamma_1 z))),$$
$$\lambda\alpha^{-1}(\beta\lambda_1^{-1}(x \cdot y) \cdot z) = \mu(\mu_1^{-1} x) \cdot \nu\alpha_1^{-1}(\beta_1 \nu_1^{-1} y \cdot \gamma_1 \gamma^{-1} z).$$

So we have the following holomorphisms of the group $Q(\cdot)$: $\beta\lambda_1^{-1} = \theta_1$, $\lambda\alpha^{-1} = \theta$ and $\lambda_1\alpha^{-1} = \theta_2$ by fixed value $B = B' \in \Sigma$. Hence, $\beta = \theta_1\lambda_1$, $\alpha^{-1} = \lambda_1^{-1}\theta_2$ and

$$\begin{aligned}A(x,y) &= \alpha^{-1}(\beta x \cdot \gamma y) = \lambda_1^{-1}\theta_2(\theta_1\lambda_1 x \cdot \gamma y)\\ &= \lambda_1^{-1}(\theta_2\theta_1\lambda_1 x \cdot (\theta_2 e)^{-1} \cdot \theta_2\gamma y) = \lambda_1^{-1}(\psi(\lambda_1 x) \cdot \delta x),\end{aligned}$$

where ψ is an automorphism of $Q(\cdot)$ and $\delta : Q \to Q$ is a bijection. Let us define the new operation $(\circ)$ on the set Q:

$$x \circ y = \lambda_1^{-1}(\lambda_1 x \cdot \lambda_1 y),$$

i.e., $Q(\circ)$ is a group, and

$$\begin{aligned}A(x,y) &= \lambda_1^{-1}(\lambda_1(\lambda_1^{-1}\psi\lambda_1 x) \cdot \lambda_1(\lambda_1^{-1}\delta y))\\ &= \lambda_1^{-1}\psi\lambda_1 x \circ \lambda_1^{-1}\delta y = \varphi x \circ \sigma y,\end{aligned}$$

where $\sigma y = \lambda_1^{-1}\delta y$ and $\varphi x = \lambda_1^{-1}\psi\lambda_1 x$ and φ is an automorphism of $Q(\circ)$:

$$\begin{aligned}\varphi(x \circ y) &= \lambda_1^{-1}\psi\lambda_1\lambda_1^{-1}(\lambda_1 x \cdot \lambda_1 y)\\ &= \lambda_1^{-1}(\psi\lambda_1 x \cdot \psi\lambda_1 y) == \lambda_1^{-1}(\lambda_1\varphi x \cdot \lambda_1\varphi y) = \varphi x \circ \varphi y.\end{aligned}$$

Now, from (20.3.2) we have

$$\begin{aligned}\varphi B'(x,y) \circ \sigma z &= \tau x \circ qA'(y,z),\\ \varphi(\tau' x \circ q' y) \circ \sigma z &= \tau x \circ (\varphi' y \circ \sigma' z),\end{aligned}$$

where φ, φ', τ and τ' are automorphisms of $Q(\circ)$. If $x = f$ (f is the unit element of $Q(\circ)$) we have

$$\begin{aligned}\varphi q' y \circ \sigma z &= q(\varphi' y \circ \sigma' z),\\ \varphi q'(\varphi')^{-1} y \circ \sigma(\sigma')^{-1} z &= q(y \circ z),\end{aligned}$$

i.e., q is a holomorphism of the group $Q(\circ)$. Hence, $qx = t \circ q_0 x$, where q_0 is an automorphism of $Q(\circ)$ and $t \in Q$. □

Theorem 20.3.9. *Invertible algebra $(Q;\Sigma)$ with the $\forall\exists(\forall)$-identity of associativity (20.1.8) is linear on a group.*

Proof. In this case for every $A, B \in \Sigma$ there exist $A', B' \in \Sigma$ with the following identity:

$$B'(A(x,y),z) = A'(x, B(y,z)). \tag{20.3.3}$$

Substituting in (20.3.3) the values of $A(x,y)$, $A'(x,y)$, $B(x,y)$ and $B'(x,y)$ from the proofs of previous theorems we obtain

$$\lambda_1^{-1}(\mu_1\alpha^{-1}(\beta x \cdot \gamma y) \cdot \nu_1 z) = \alpha_1^{-1}(\beta_1 x \cdot \gamma_1\lambda^{-1}(\mu y \cdot \nu z)),$$
$$\alpha_1\lambda_1^{-1}(\mu_1\alpha^{-1}(x \cdot y) \cdot z) = \beta_1\beta^{-1}x \cdot \gamma_1\lambda^{-1}(\mu\gamma^{-1}y \cdot \nu\nu_1^{-1}z),$$

i.e., we have the following holomorphisms of group $Q(\cdot)$: $\mu_1\alpha^{-1} = \theta_1$, $\mu\gamma^{-1} = \theta$ and $\mu_1\gamma^{-1} = \theta_2$ by fixing value of $B = B' \in \Sigma$. Hence, $\alpha^{-1} = \mu_1^{-1}\theta_1$, $\gamma = \theta_2^{-1}\mu_1$ and

$$\begin{aligned} A(x,y) &= \alpha^{-1}(\beta x \cdot \gamma y) = \mu_1^{-1}\theta_1(\beta x \cdot \theta_2^{-1}\mu_1 y) \\ &= \mu_1^{-1}(\theta_1\beta x \cdot (\theta_1 e)^{-1} \cdot \theta_1\theta_2^{-1}\mu_1 y) = \mu_1^{-1}(\delta x \cdot \psi(\mu_1 y)) \\ &= \mu_1^{-1}(\mu_1(\mu_1^{-1}\delta x) \cdot \mu_1(\mu_1^{-1}\psi\mu_1 y)) = \mu_1^{-1}\delta x \circ \mu_1^{-1}\psi\mu_1 y = \sigma x \circ \varphi y, \end{aligned}$$

where ψ is an automorphism of $Q(\cdot)$, $\varphi = \mu_1^{-1}\psi\mu_1$ is an automorphism of group $Q(\circ)$ and

$$x \circ y = \mu_1^{-1}(\mu_1 x \cdot \mu_1 y).$$

Then by equality (20.3.3) we have

$$\tau'(\sigma x \circ \varphi y) \circ q' z = \sigma' x \circ \varphi'(\tau y \circ q z)$$

where φ, q', φ' and q are automorphisms of $Q(\circ)$. If $z = f$ we have

$$\tau'(\sigma x \circ \varphi y) = \sigma' x \circ \varphi'\tau y,$$
$$\tau'(x \circ y) = \sigma'\sigma^{-1}x \circ \varphi'\tau\varphi^{-1}y,$$

where $\varphi'\tau\varphi^{-1} = \theta$ is an holomorphism of $Q\circ)$, hence $\tau = (\varphi')^{-1}\theta\varphi$ is an holomorphism of $Q(\circ)$ too. □

Theorem 20.3.10. *Invertible algebras $(Q;\Sigma)$ with the $\forall\exists(\forall)$-identity of associativity (20.1.9) is linear on a group.*

Proof. For every $A, B \in \Sigma$ there exist $A', B' \in \Sigma'$ with the following identity:

$$B'(A(x,y),z) = B(x, A'(y,z)).$$

Now

$$\begin{aligned}
&\lambda_1^{-1}(\mu_1\alpha^{-1}(\beta x \cdot \gamma y)\cdot \nu_1 z) = \lambda^{-1}(\mu x \cdot \nu\alpha_1^{-1}(\beta_1 y \cdot \gamma_1 z)),\\
&\lambda\lambda_1^{-1}(\mu_1\alpha^{-1}(x\cdot y)\cdot z) = \mu\beta^{-1}x \cdot \nu\alpha_1^{-1}(\beta_1\gamma^{-1}y \cdot \gamma_1\nu_1^{-1}z).
\end{aligned}$$

So we have the following holomorphisms: $\mu_1\alpha^{-1} = \theta_1$, $\mu\beta^{-1} = \theta$ and $\mu_1\beta^{-1} = \theta_2$ by fixing value of $B = B' \in \Sigma$. Hence, $\alpha^{-1} = \mu_1^{-1} = \theta_1$, $\beta = \theta_2^{-1}\mu_1$ and:

$$\begin{aligned}
A(x,y) &= \alpha^{-1}(\beta x \cdot \gamma y) = \mu_1^{-1}\theta_1(\theta_2^{-1}\mu_1 x \cdot \gamma y)\\
&= \mu_1^{-1}(\theta_1\theta_2^{-1}\mu_1 x \cdot (\theta_1 e)^{-1} \cdot \theta_1\gamma y) = \mu_1^{-1}(\psi\mu_1 x \cdot \delta y)\\
&= \mu_1^{-1}(\mu_1(\mu_1^{-1}\psi\mu_1 x)\cdot \mu_1(\mu^{-1}\delta y)) = \mu_1^{-1}\psi\mu_1 x \circ \mu^{-1}\delta y = \varphi x \cdot \sigma y,
\end{aligned}$$

where ψ is an automorphism of the group $Q(\cdot)$, $\varphi = \mu_1^{-1}\psi\mu_1$ is an automorphism of the group $Q(\circ)$, where

$$x \circ y = \mu_1^{-1}(\mu_1 x \cdot \mu_1 y).$$

The rest of the proof is similar to the proof of previous theorem. □

20.4. Corollaries

Corollary 20.4.1. *The $\forall\exists(\forall)$-identity of associativity (20.1.6) is equivalent to the following non-trivial hyperidentity of associativity in systems of groups*

$$X(Y(x,y),z) = X(x,Y(y,z)). \qquad (ass)_2$$

Proof. We need to prove that if the formula (20.1.6) is satisfied in a system of groups $(Q;\Sigma)$, then $(Q;\Sigma)$ satisfies the hyperidentity $(ass)_2$ too (the converse is evident!). Let $(\circ) \in \Sigma$ and $Q(\circ)$ be a group. According to Theorem 20.3.3 we have

$$x \circ y = \varphi x \cdot t \cdot \psi y,$$

where $\varphi, \psi \in Aut\, Q(\cdot)$, $t \in Q$. If e is the unit element of $Q(\circ)$ and $x = e$, then

$$y = \varphi e \cdot t \cdot \psi y,$$
$$\psi y = a \cdot y,$$

where $a = (\varphi e \cdot t)^{-1}$. Similarly $\varphi x = x \cdot b$, where $b = (t \cdot \psi e)^{-1}$. Hence,

$$x \circ y = x \cdot b \cdot a \cdot y = x \cdot s \cdot y,$$

where $s = b \cdot a \in Q$. So any operation $A \in \Sigma$ is defined by the rule:

$$A(x, y) = x \cdot s_A \cdot y, \quad s_A \in Q.$$

Using this result in equality (20.3.1) we obtain

$$x \cdot s'_B \cdot y \cdot s_A \cdot z = x \cdot s'_A \cdot y \cdot s_B \cdot z,$$
$$s'_B \cdot y \cdot s_A = s'_A \cdot y \cdot s_B,$$
$$(s'_A)^{-1} \cdot s'_B \cdot y = y \cdot s_B \cdot s_A^{-1},$$

and $s_B \cdot s_A^{-1} = (s'_A)^{-1} \cdot s'_B$ is a central element of the group $Q(\cdot)$. Now if we fix the operation $A \in \Sigma$ we have that $s_B \cdot s_0^{-1} = t_B$ is a central element of the group $Q(\cdot)$. Hence, $s_B = t_B \cdot s_0$ and

$$A(x, y) = x \cdot t_A \cdot s_0 \cdot y,$$

where $t_A \in \mathbb{Z}(Q(\cdot))$ and $s_0 \in Q$ is a fixed element for any operation $A \in \Sigma$ and algebra $(Q; \Sigma)$ satisfies the hyperidentity $(ass)_2$. □

Corollary 20.4.2. *The* $\forall\exists(\forall)$*-identity of associativity (20.1.7) is equivalent to the following non-trivial hyperidentity of associativity in systems of groups:*

$$X(Y(x, y), z) = Y(x, X(y, z)). \qquad (ass)_1$$

Corollary 20.4.3. *The* $\forall\exists(\forall)$*-identity of associativity (20.1.8) is equivalent to the non-trivial hyperidentity of associativity* $(ass)_1$ *in systems of groups.*

Corollary 20.4.4. *The* $\forall\exists(\forall)$*-identity of associativity (20.1.9) is equivalent to the non-trivial hyperidentity of associativity* $(ass)_2$ *in systems of groups.*

20.5. Schauffler–Belousov style theorems

In this section, we prove the Schauffler–Belousov style theorems for $\forall\exists(\forall)$-identities of associativity (20.1.6)–(20.1.9).

Let Q be a non-empty set and Ω_Q be the set of all quasigroup operations defined on the set Q, and G_Q be the set of all (binary) operations on the set Q.

Theorem 20.5.1. *Invertible algebra $(Q; \Omega_Q)$ satisfies the $\forall\exists(\forall)$-identity of associativity (20.1.6) if and only if the cardinality of Q is $|Q| \leqslant 3$.*

Proof. If the cardinality $|Q| \geqslant 5$ then there exists a non-associative loop $Q(\circ)$ which is not isotopic to a group by the second Albert's theorem [12,13]. If $|Q| = 4$, then there exist two non-isomorphic groups on the set Q, which are not isotopic to the same group. If $|Q| = 3$ we have 12 quasigroup operations on the set Q and invertible algebra $(Q; \Omega_Q)$ is linear on a group [254].

Namely, if $Q = \{0, a, b\}$, we have the following 12 quasigroup operations on the set Q:

$$\begin{array}{llll} x + y, & \alpha x + y, & x + \alpha y, & \alpha x + \alpha y, \\ x + a + y, & \alpha x + a + y, & x + a + \alpha y, & \alpha x + a + \alpha y, \\ x + b + y, & \alpha x + b + y, & x + b + \alpha y, & \alpha x + b + \alpha y, \end{array}$$

where $Q(+)$ is a cyclic group of order 3 and α is the unique second-order automorphism of $Q(+)$,

$$\alpha = \begin{pmatrix} 0 & a & b \\ 0 & b & a \end{pmatrix}.$$

We check straightforwardly that in the invertible algebra $(Q; \Omega_Q)$ the $\forall\exists(\forall)$-identity of associativity (20.1.6) is satisfied. For example, if $A(x, y) = \alpha x + a + y$ and $B(x, y) = \alpha x + b + \alpha y$ then $A'(x, y) = x + \alpha y$ and $B'(x, y) = \alpha x + \alpha y$.

If $|Q| = 2$, then in the invertible algebra $(Q; \Omega_Q)$ the non-trivial hyperidentity of associativity $(ass)_2$ is satisfied. □

Theorem 20.5.2. *Invertible algebra $(Q; \Omega_Q)$ satisfies the $\forall\exists(\forall)$-identity of associativity (20.1.7) if and only if the cardinality of Q is $|Q| \leqslant 3$.*

Theorem 20.5.3. *Invertible algebra $(Q; \Omega_Q)$ satisfies the $\forall\exists(\forall)$-identity of associativity (20.1.8) if and only if the cardinality of Q is $|Q| \leqslant 3$.*

Theorem 20.5.4. *Invertible algebra $(Q; \Omega_Q)$ satisfies the $\forall\exists(\forall)$-identity of associativity (20.1.9) if and only if the cardinality of Q is $|Q| \leqslant 3$.*

We need the following semantics of $\forall\exists(\forall)$-identities, too [254]. We say that a set Q $(*, *)$-satisfies the $\forall\exists(\forall)$-identity of associativity (20.1.6) if for every quasigroup operations $A, B \in \Omega_Q$ there exist (binary) operations $A', B' \in G_Q$ with identity:

$$A(B'(x, y), z) = A'(x, B(y, z)).$$

Analogously, we understand the $(*, *)$-satisfiability of a set Q of one of the $\forall\exists(\forall)$-identities of associativity (20.1.7)–(20.1.9).

Theorem 20.5.5. *An arbitrary non-empty set Q $(*, *)$-satisfies the $\forall\exists(\forall)$-identity of associativity (20.1.6).*

Theorem 20.5.6. *A non-empty set Q $(*, *)$-satisfies the $\forall\exists(\forall)$-identity of associativity (20.1.7) if and only if the cardinality of Q is $|Q| \leqslant 3$.*

Theorem 20.5.7. *A non-empty set Q $(*, *)$-satisfies the $\forall\exists(\forall)$-identity of associativity (20.1.8) if and only if the cardinality of Q is $|Q| \leqslant 3$.*

Theorem 20.5.8. *A non-empty set Q $(*, *)$-satisfies the $\forall\exists(\forall)$-identity of associativity (20.1.9) if and only if the cardinality of Q is $|Q| \leqslant 3$.*

20.6. Remarks

The functional *rank* of the $\forall\exists(\forall)$-identity of associativity is the number of functional variables which are under quantifier $\forall$ by the definition. The $\forall\exists(\forall)$-identities of associativity (20.1.6)–(20.1.9) in the current chapter and the $\forall\exists(\forall)$-identities of associativity in the Belousov paper [24] have the functional rank 2.

Therefore, the structure of invertible algebra with any associative $\forall\exists(\forall)$-identity of the functional rank 2 reduce to the structure of invertible algebra with one of these $\forall\exists(\forall)$-identity of associativity. For instance, an algebra $(Q; \Sigma)$ with the $\forall\exists(\forall)$-identity of associativity of functional rank 2

$$\forall X, Y \exists Z \forall x, y, z (X(X(x, y), z) = Z(x, Y(y, z)))$$

is an algebra with the $\forall\exists(\forall)$-identity of associativity (20.1.6). So we obtain the following result.

Theorem 20.6.1. *Invertible algebra* $(Q;\Sigma)$ *with the* $\forall\exists(\forall)$*-identity of associativity of functional* rank 2 *is linear on a group.*

From this result, it follows that invertible algebra $(Q;\Sigma)$ with the $\forall\exists(\forall)$-identity of associativity of functional rank 3 is linear on a group, too.

However, this result for the $\forall\exists(\forall)$-identities of associativity of functional rank 1 is false.

Problem 20.6.2. *Characterize pre-invertible algebras satisfying any of the following* $\forall\exists(\forall)$*-identities of associativity: (20.1.2)–(20.1.3), (20.1.6)–(20.1.9).*

Chapter 21

Distributive Formulae with Functional Variables

21.1. Introduction

In this chapter, the structure of invertible algebras with distributive second-order formulae with specialized quantifiers is given. As a consequence, the applications for solutions of the some functional equations of distributivity on quasigroups are provided.

For functional equations in algebra, logics, real analysis and topology, see [4,5,8,25,182–186,222–224].

The problems of solution of the following two general functional equations of distributivity:

$$A(x, B(y, z)) = H(G(x, y), K(x, y)),$$
$$A(B(y, z), x) = H(G(y, x), K(z, x))$$

on quasigroups are the oldest unsolved problems of quasigroup theory [4,5,8,25,182]. Moreover, the following functional equations still remain unsolved on quasigroups, as well:

$$A(x, A(y, z)) = H(G(x, y), K(x, y)),$$
$$A(A(y, z), x) = H(G(y, x), K(z, x)),$$
$$A(x, A(y, z)) = H(K(x, y), K(x, y)),$$
$$A(A(y, z), x) = H(K(y, x), K(z, x)),$$
$$A(x, B(y, z)) = B(A(x, y), A(x, z)),$$
$$A(B(y, z), x) = B(A(y, x), A(z, x)).$$

Here, the solution is unknown even in the case $A = B$.

Note that the solution of the following systems of functional equations are also unknown on quasigroups:

$$\begin{cases} A(x, B(y, z)) = B(A(x, y), A(x, z)), \\ A(B(y, z), x) = B(A(y, x), A(z, x)), \end{cases}$$

and

$$\begin{cases} A(x, B(y, z)) = B(A(x, y), A(x, z)), \\ B(A(y, z), x) = A(B(y, x), B(z, x)). \end{cases}$$

The following results concerning to the solution of the related functional equations are consequences from the main result of current chapter.

(1) If quasigroups $Q(A)$, $Q(K)$ and groupoid $Q(H)$ satisfy the following identities:

$$A(x, A(y, z)) = A(A(x, y), A(x, y)),$$
$$A(A(y, z), x) = H(K(y, x), K(z, x)),$$

then $Q(A)$, $Q(H)$ and $Q(K)$ are isotopic to a commutative Moufang loop.

(2) If quasigroups $Q(A)$, $Q(K)$ and groupoid $Q(H)$ satisfy the following identities:

$$A(A(y, z), x) = A(A(y, x), A(z, x)),$$
$$A(x, A(y, z)) = H(K(x, y), K(x, z)),$$

then $Q(A)$, $Q(H)$ and $Q(K)$ are isotopic to a commutative Moufang loop.

(3) If quasigroups $Q(A)$, $Q(K)$, $Q(K')$ and groupoids $Q(H)$ and $Q(H')$ satisfy the following identities:

$$A(x, x) = x,$$
$$A(x, A(y, z)) = H(K(x, y), K(x, z)),$$
$$A(A(y, z), x) = H'(K'(y, x), K'(z, x)),$$

then $Q(A)$, $Q(H)$, $Q(H')$, $Q(K)$ and $Q(K')$ are isotopic to a commutative Moufang loop.

21.2. Auxiliary results and concepts

Lemma 21.2.1. *If a binary algebra $Q(A, B, H, K)$ satisfies the following identity:*

$$A(x, B(y, z)) = H(K(x, y), K(x, z)), \tag{21.2.1}$$

where $Q(A)$ and $Q(K)$ are quasigroups, $Q(B)$ and $Q(H)$ are groupoids, then A and K are isotopic to a quasigroup operation A_0, and the operations B and H are isotopic to an idempotent operation B_0 such that:

$$A_0(x, B_0(y, z)) = B_0(A_0(x, y), A_0(x, z)). \tag{21.2.2}$$

Besides:

$$A(x, B_0(y, z)) = B_0^{L_A^{-1}}(A(x, y), A(x, z)), \tag{21.2.3}$$

where $L_A(x) = A(0, x)$, while the element 0 is an arbitrary fixed element of Q, and

$$B_0^{\varphi}(x, y) = \varphi^{-1} B_0(\varphi x, \varphi y).$$

In particular, if the operation B is idempotent, then

$$A(x, B(y, z)) = B^{L_A^{-1}}(A(x, y), A(x, z)). \tag{21.2.4}$$

In addition, if $Q(B)$ is a quasigroup, then $Q(H)$ also is a quasigroup.

Proof. If making the substitution $y = z$ in the equality (21.2.1), we obtain

$$A(x, \Theta_B(y)) = \Theta_H K(x, y), \tag{21.2.5}$$

where $\Theta_B(y) = B(y, y)$. Let $R_A(x) = A(x, 0)$.

If in equality (21.2.5) $y = 0$, then

$$A(x, \Theta_B(0)) = \Theta_H K(x, 0),$$

i.e.,

$$S(x) = \Theta_H R_K(x),$$

where $S(x) = A(x, \Theta_B(0))$ is a bijection. Hence, $\Theta_H = SR_K^{-1}$ also is a bijection. If in equality (21.2.5) $x = 0$, then

$$L_A \Theta_B(y) = \Theta_H L_K(y),$$
$$\Theta_B = L_A^{-1} \Theta_H L_K,$$

i.e., Θ_B is a bijection. According to (21.2.5), the operations A and K are isotopic and

$$K(x,y) = \Theta_H^{-1} A(x, \Theta_B y).$$

If in the equality (21.2.1) $x = 0$, then

$$L_A B(y,z) = H(L_k y, L_K z),$$
$$H(y,z) = L_A B\left(L_K^{-1} y, L_K^{-1} z\right).$$

Let us consider the following new operations:

$$B_0(x,y) = B\left(\Theta_B^{-1} x, \Theta_B^{-1} y\right),$$

and

$$A_0(x,y) = L_A^{-1} A(x,y).$$

Since $x = B\left(\Theta_B^{-1} x, \Theta_B^{-1} x\right)$, the operation B_0 is idempotent (and isotopic to B).

Substituting the values of operations K and H in identity (21.2.1), we obtain the equalities (21.2.2) and (21.2.3). If, in addition, the operation B is idempotent, then Θ_B is the identical mapping and $B_0 = B$. Hence, from (21.2.3) follows (21.2.4). □

Lemma 21.2.2. *If a binary algebra $Q(A,B,H,K)$ satisfies the following identity:*

$$A(B(y,z),x) = H(K(y,x),K(z,x)), \tag{21.2.6}$$

where $Q(A)$ and $Q(K)$ are quasigroups, $Q(B)$ and $Q(H)$ are groupoids, then A and K are isotopic to a quasigroup operation A_0, and operations B and H are isotopic to an idempotent operation B_0 such that

$$A_0(B_0(y,z),x) = B_0(A_0(y,x),A_0(z,x)). \tag{21.2.7}$$

Besides:

$$A(B_0(y,z),x) = B_0^{R_A^{-1}}(A(y,x),A(z,x)), \tag{21.2.8}$$

where $R_A(x) = A(x,0)$, while the element 0 is an arbitrary fixed element of Q. In particular, if the operation B is idempotent, then

$$A(B(y,z),x) = B^{R_A^{-1}}(A(y,x),A(z,x)). \tag{21.2.9}$$

In addition, if $Q(B)$ is a quasigroup, then $Q(H)$ also is a quasigroup.

21.3. Main results

An invertible algebra $(Q;\Sigma)$ is called D_l-algebra (D_r-algebra) if it satisfies the following two conditions:

(a) In the algebra $(Q;\Sigma)$ the following hyperidentity of the left (right) distributivity is satisfied:

$$X(x, X(y,z)) = X(X(x,y), X(x,z)) \tag{21.3.1}$$
$$(X(X(y,z),x) = X(X(y,x), X(z,x))). \tag{21.3.2}$$

(b) In the algebra $(Q;\Sigma)$ the following $\forall\exists^*\exists^{**}(\forall)$-identity of the left (right) distributivity is satisfied:

$$\forall X, Y \exists^* X' \exists^{**} Y' \forall x, y, z (X(Y(y,z),x) = X'(Y'(y,x), Y'(z,x))) \tag{21.3.3}$$

$$(\forall X, Y \exists^* X' \exists^{**} Y' \forall x, y, z (X(x, Y(y,z)) = X'(Y'(x,y), Y'(x,z)))), \tag{21.3.4}$$

where $\forall X, Y$ means "for every values of $X, Y \in \Sigma$", $\exists^* X'$ means "there exists an operation on Q" and $\exists^{**} Y'$ means "there exists a quasigroup operation on Q". Thus, (21.3.3) and (21.3.4) are the second-order formulae with specialized quantifiers (see [213]).

Examples.

(1) Let $Q(+,\cdot)$ be a field and for every $a \in Q$:

$$A_a(x,y) = (1-a)x + ay.$$

If $\Sigma = \{A_a | a \in Q\}$ then the algebra $(Q;\Sigma)$ is an D_l- and D_r-algebra.

(2) Let $Q(A)$ be a distributive quasigroup.
If $\Sigma = \{A, A^{-1}, {}^{-1}A, {}^{-1}(A^{-1}), ({}^{-1}A)^{-1}, A^*\}$, where $A^*(x,y) = A(y,x)$ for all $x, y \in Q$, then the algebra $(Q;\Sigma)$ is an D_l- and D_r-algebra (also see [296]).

(3) Let $Q(+,\cdot)$ be a field and

$$A_i(x,y) = a_i x + b_i y + c_i,$$

where $a_i, b_i, c_i \in Q$ and $a_i \neq 0$, $b_i \neq 0$, $a_i + b_i \neq 0$. If $\Sigma_I = \{A_i | i \in I\}$, then the algebra $(Q;\Sigma_I)$ satisfies the formulae (21.3.3) and (21.3.4).

Special cases of D_l-algebras and D_r-algebras were considered in Sec. 17.3. Namely, in Sec. 17.3, the invertible algebras with the following hyperidentities:

$$X(x, X(y, z)) = X(X(x, y), X(x, z)),$$
$$X(Y(y, z), x) = Y(X(y, x), X(z, x)),$$

and the invertible algebras with the following hyperidentities:

$$X(X(y, z), x) = X(X(y, x), X(z, x)),$$
$$X(x, Y(y, z)) = Y(X(x, y), X(x, z))$$

are characterized.

Now we prove the following more general results.

Theorem 21.3.1. (1) *If* $(Q; \Sigma)$ *is a* D_l*-algebra, then the quasigroups* $Q(A)$, $A \in \Sigma$, *are distributive and hence are isotopic to commutative Moufang loops. Every* D_l*-algebra satisfies the following non-trivial hyperidentity:*

$$X(Y(X(y, z), z), x) = X(Y(X(y, x), X(z, x)), X(z, x)). \tag{21.3.5}$$

(2) *If* $(Q; \Sigma)$ *is a* D_r*-algebra, then the quasigroups* $Q(A)$, $A \in \Sigma$ *are distributive and hence are isotopic to commutative Moufang loops. Every* D_r*-algebra satisfies the following non-trivial hyperidentity:*

$$X(x, Y(y, X(y, z))) = X(X(x, y), Y(X(x, y)X(x, z))). \tag{21.3.6}$$

Proof. (1) From the hyperidentity (21.3.1) of the left distributivity it follows that every quasigroup $Q(A)$, $A \in \Sigma$, is idempotent. If in the formula (21.3.3) $X = A \in \Sigma$, $Y = B \in \Sigma$, $X' = H$, $Y' = K$, we obtain

$$A(B(y, z), x)) = H(K(y, x), K(z, x)).$$

According to Lemma 21.2.2 (Eq. (21.2.9)) we have

$$A(B(y, z), x) = B^{R_A^{-1}}(A(y, x), A(z, x)). \tag{21.3.7}$$

If we substitute $z = x$ in (21.3.7), we obtain

$$A(B(y, x), x) = B^{R_A^{-1}}(A(y, x), x). \tag{21.3.8}$$

In particular, by putting $B = A$, we have

$$A(A(y, x), x) = A^{R_A^{-1}}(A(y, x), x),$$

or

$$A(u,x) = A^{R_A^{-1}}(u,x),$$

i.e., $A^{R_A^{-1}} = A$. Then substituting $B = A$ in (21.3.7) we obtain a right distributive identity for any quasigroup operation $A \in \Sigma$:

$$A(A(y,z),x) = A(A(y,x),A(z,x)).$$

Hence, the quasigroup $Q(A)$ is distributive for every operation $A \in \Sigma$.

According to [23] every distributive quasigroup $Q(A)$ is isotopic to certain commutative Moufang loop $Q(\underset{A}{+})$:

$$x \underset{A}{+} y = A\left(R_A^{-1}x, L_A^{-1}y\right),$$

where R_A and L_A are automorphisms of $Q(A)$ and automorphisms of the loop $Q(\underset{A}{+})$, $R_A L_A = L_A R_A$ and $R_A(x) \underset{A}{+} L_A(x) = x$ for any $x \in Q$.

Now we prove that hyperidentity (21.3.5) is satisfied in the D_ℓ-algebra $(Q;\Sigma)$.

From equality (21.3.8) we have

$$A \circ B = B^{R_A^{-1}} \circ A,$$

and

$$B^{R_A^{-1}} = A \circ B \circ {}^{-1}\!A.$$

Hence, according to equality (21.3.7), we obtain

$$\begin{aligned}
A(B(y,z),x) &= (A \circ B \circ {}^{-1}\!A)(A(y,x),A(z,x)),\\
A(B(y,z),x) &= A(B \circ {}^{-1}\!A(A(y,x),A(z,x)),A(z,x)),\\
A(B(y,z),x) &= A(B({}^{-1}\!A(A(y,x),A(z,x)),A(z,x)),A(z,x)),\\
A(B(y,z),x) &= A(B(A({}^{-1}\!A(A(y,z),x),A(z,x)),A(z,x)),\\
A(B(A(y,z),z),x) &= A(B(A({}^{-1}\!A(A(y,z),z),x),A(z,x)),A(z,x)),\\
A(B(A(y,z),z),x) &= A(B(A(y,x),A(z,x)),A(z,x)).
\end{aligned}$$

Thus, in the algebra $(Q;\Sigma)$, the hyperidentity (21.3.5) is satisfied.

(2) Analogously, by using the equality $B^{L_A^{-1}} = A \cdot B \cdot A^{-1}$, we prove property (2). □

Corollary 21.3.2. *If quasigroups $Q(A)$, $Q(K)$ and groupoid $Q(H)$ satisfy the following identities*:

$$A(x, A(y, z)) = A(A(x, y), A(x, z)),$$
$$A(A(y, z), x) = H(K(y, x), K(z, x)),$$

then $Q(A)$, $Q(H)$ and $Q(K)$ are isotopic to a commutative Moufang loop.

Proof. We applied Theorem 21.3.1 for $\Sigma = \{A\}$, and used Lemma 21.2.2. □

Corollary 21.3.3. *If quasigroups $Q(A)$, $Q(K)$ and groupoid $Q(H)$ satisfy the following identities*:

$$A(A(y, z), x) = A(A(y, x), A(z, x)),$$
$$A(x, A(y, z)) = H(K(x, y), K(x, z)),$$

then $Q(A)$, $Q(H)$ and $Q(K)$ are isotopic to a commutative Moufang loop.

Proof. We applied Theorem 21.3.1 when $\Sigma = \{A\}$, and used Lemma 21.2.1. □

Corollary 21.3.4. *If quasigroups $Q(A)$, $Q(K)$, $Q(K')$ and groupoids $Q(H)$ and $Q(H')$ satisfy the following identities*:

$$A(x, x) = x,$$
$$A(x, A(y, z)) = H(K(x, y), K(x, z)),$$
$$A(A(y, z), x) = H'(K'(y, x), K'(z, x)),$$

then $Q(A)$, $Q(H)$, $Q(H')$, $Q(K)$ and $Q(K')$ are isotopic to a commutative Moufang loop.

Proof. We applied the proof of Theorem 21.3.1 when $\Sigma = \{A\}$, and use Lemmas 21.2.1 and 21.2.2. Note that in the proof of Theorem 21.3.1 we used the idempotency of the operations of Σ. □

Let Σ_l be the set of commutative Moufang loop operations corresponding to the quasigroup operations from D_l-algebra $(Q; \Sigma)$ according to the previous theorem. We obtain a new algebra $(Q; \Sigma_l)$.

Let Σ_r be the set of commutative Moufang loop operations corresponding to the quasigroup operations from D_r-algebra $(Q; \Sigma)$ according to the previous theorem, too. We obtain an algebra $(Q; \Sigma_r)$.

Our final result shows the connection (through the hyperidentity) between the operations from Σ_l and Σ_r.

Theorem 21.3.5. *If $(Q;\Sigma)$ is a D_l-algebra (D_r-algebra), then the algebra $(Q;\Sigma_l)$ (algebra $(Q;\Sigma_r)$) satisfies the following non-trivial hyperidentity:*

$$X(x, Y(x, X(y, z))) = X(Y(x, y), Y(x, z)). \tag{21.3.9}$$

Proof. (1) Let $(Q;\Sigma)$ be a D_l-algebra. First, we prove the following equality:

$$B^{R_A^{-1}R_A(a)} = B$$

for every $A, B \in \Sigma$, where $R_A(a)$ and R_A are defined for any element $a \in Q$ and the arbitrary fixed element $0 \in Q$ by the rule:

$$R_A(a)(x) = A(x, a),$$
$$R_A(x) = A(x, 0).$$

By putting $x = a$ in identity (21.3.7), we have

$$A(B(y, z), a) = B^{R_A^{-1}}(A(y, a), A(z, a)),$$
$$R_A(a)B(y, z) = B^{R_A^{-1}}(R_A(a)y, R_A(a)z),$$
$$R_A(a)B\left(R_A^{-1}(a)y, R_A^{-1}(a)z\right) = B^{R_A^{-1}}(y, z),$$
$$B^{R_A^{-1}(a)} = B^{R_A^{-1}},$$
$$B^{R_A^{-1}R_A(a)} = B.$$

Thus, the mapping $\varphi = R_A^{-1}R_A(a)$ is an automorphism of quasigroup $Q(B)$ for any operation $B \in \Sigma$. We have

$$R_A\varphi x = R_A(a)(x) = A(x, a) = R_A(x) \underset{A}{+} L_A(a),$$
$$\varphi(x) = x \underset{A}{+} R_A^{-1}L_A(a) = x \underset{A}{+} b,$$

where $b = R_A^{-1}L_A(a)$. Then

$$\varphi B(x, y) = B(\varphi x, \varphi y),$$
$$B(x, y) \underset{A}{+} b = B(x \underset{A}{+} b, y \underset{A}{+} b),$$
$$B(x, y) \underset{A}{+} z = B(x \underset{A}{+} z, y \underset{A}{+} z),$$

where $z = b$ is an arbitrary element of Q. Thus,

$$(R_Bx \underset{B}{+} L_By) \underset{A}{+} z = R_B(x \underset{A}{+} z) \underset{B}{+} L_B(y \underset{A}{+} z). \tag{21.3.10}$$

The identity element of the loop $Q(\underset{A}{+})$ is the element $A(0,0) = 0$, and the identity element of the loop $Q(\underset{B}{+})$ is the element $B(0,0) = 0$. Substituting $x = 0$ or $y = 0$ in equality (21.3.10), we obtain

$$L_B y \underset{A}{+} z = R_B z \underset{B}{+} L_B(y \underset{A}{+} z),$$

$$R_B x \underset{A}{+} z = R_B(x \underset{A}{+} z) \underset{B}{+} L_B z,$$

$$L_B(y \underset{A}{+} z) = (-R_B z) \underset{B}{+} (L_B y \underset{A}{+} z),$$

$$R_B(x \underset{A}{+} z) = (-L_B z) \underset{B}{+} (R_B x \underset{A}{+} z),$$

where $(-x) \underset{B}{+} (x \underset{B}{+} y) = y$. From equality (21.3.10) we obtain

$$(R_B x \underset{B}{+} L_B y) \underset{A}{+} z = ((-L_B z) \underset{B}{+} (R_B x \underset{A}{+} z)) \underset{B}{+} ((-R_B z) \underset{B}{+} (L_B y \underset{A}{+} z)),$$

$$(x \underset{B}{+} y) \underset{A}{+} z = (L_B(-z) \underset{B}{+} (x \underset{A}{+} z)) \underset{B}{+} (R_B(-z) \underset{B}{+} (y \underset{A}{+} z)). \tag{21.3.11}$$

In the commutative Moufang loop $Q(\underset{B}{+})$ the following equality is satisfied:

$$(L_B u \underset{B}{+} v) \underset{B}{+} (R_B u \underset{B}{+} w) = u \underset{B}{+} (v \underset{B}{+} w)$$

for any $u, v, w \in Q$ (see Sec. 17.3).

Applying this identity in equality (21.3.11) we obtain

$$(x \underset{B}{+} y) \underset{A}{+} z = (-z) \underset{B}{+} ((x \underset{A}{+} z) \underset{B}{+} (y \underset{A}{+} z)),$$

$$z \underset{B}{+} ((x \underset{B}{+} y) \underset{A}{+} z) = (x \underset{A}{+} z) \underset{B}{+} (y \underset{A}{+} z),$$

$$z \underset{B}{+} (z \underset{A}{+} (x \underset{B}{+} y)) = (z \underset{A}{+} x) \underset{B}{+} (z \underset{A}{+} y).$$

Hence, in the algebra $(Q; \Sigma_l)$ the hyperidentity (21.3.9) is satisfied.

(2) Analogously, by using the equality:

$$B^{L_A^{-1} L_A(a)} = B,$$

it is proved that if $(Q; \Sigma)$ is a D_r-algebra, then the algebra $(Q; \Sigma_r)$ satisfies the hyperidentity (21.3.9). □

Problem 21.3.6. *Characterize pre-invertible algebras satisfying the following hyperidentities of distributivity:*

$$X(x, X(y, z)) = X(X(x, y), X(x, z)),$$
$$X(Y(x, y), z) = Y(X(x, z), X(y, z)).$$

Problem 21.3.7. *Characterize pre-invertible algebras satisfying the following hyperidentities of distributivity:*

$$X(x, Y(y, z)) = Y(X(x, y), X(x, z)),$$
$$X(X(x, y), z) = X(X(x, z), X(y, z)).$$

Problem 21.3.8. *Characterize pre-invertible algebras satisfying the hyperidentity* (21.3.9).

Chapter 22

Other Open Problems

Along with the open problems in the book, naturally there arises a set of other problems about characterization of hyperidentities and termal hyperidentities of varieties related to classical varieties of groups, rings, lattices, Boolean algebras, etc. Solutions of these problems could serve for development of the next steps of results and concepts involved in the current book.

1. Characterize $\{1,2\}$-algebras with hyperidentities of commutativity (5.7.2), associativity (5.7.3) and Robbins's hyperidentity (5.7.8). Can every $\{1,2\}$-algebra with these three hyperidentities and one unary operation be extended to an algebra with hyperidentities of Boolean algebras? Characterize subdirectly irreducible $T = \{1,2\}$-algebras with these three hyperidentities.
2. Let Gr be the class of all Grätzer algebras, and let qGr be the class of all Grätzer q-algebras [268] (see Chapter 15).

   Characterize hyperidentities and termal hyperidentities of Gr.
   Characterize hyperidentities and termal hyperidentities of qGr.
   Characterize algebras with hyperidentities (termal hyperidentities) of Gr.
   Characterize algebras with hyperidentities (termal hyperidentities) of qGr.

3. Characterize hyperidentities and termal hyperidentities of the variety of De Morgan bisemigroups (see Chapter 15).

   Characterize algebras with hyperidentities (termal hyperidentities) of the variety of De Morgan bisemigroups.

4. Characterize hyperidentities and termal hyperidentities of the variety of Boole–De Morgan algebras (see Chapter 3).

   Characterize algebras with hyperidentities (termal hyperidentities) of the variety of Boole–De Morgan algebras.

5. A lattice ordered group (briefly an l-group) is an algebra $\mathfrak{A} = Q(+, \cdot, \circ)$ with tree binary operations such that $Q(+, \cdot)$ is a lattice, $A(\circ)$ is a group, and the group multiplication is isotone in each of its argument. Actually, the lattice reduct $Q(+, \cdot)$ of an l-group in always distributive, and the group operation $\circ$ distributes over lattice joins and meets. Indeed, the maps $x \to a \circ x$ and $x \to x \circ a$ are ordinary automorphisms of $Q(+, \cdot)$. Let L_p be the variety of all l-groups.

   Characterize hyperidentities and termal hyperidentities of the variety L_g. Is every hyperidentity of the variety L_g the consequence of hyperidentities with functional rank $\leqslant 2$ of this variety? (posed by V. Garbunov).

   Characterize algebras with hyperidentities (termal hyperidentities) of the variety L_g.

6. An algebra $Q(+, \cdot, \to)$ with three binary operations is called Heyting algebra (see H. Rasiowa [339], who calls them pseudo-Boolean algebras), if it satisfies the following conditions:

   $Q(+, \cdot)$ is a distributive lattice,
   $(x \to y) \cdot y = y$,
   $x \cdot (x \to y) = x \cdot y$,
   $x \to (y \cdot z) = (x \to y) \cdot (x \to z)$,
   $(x + y) \to z = (x \to z) \cdot (y \to z)$.

   These algebras were introduced by G. Birkhoff under a different name, Brouwerian algebras, and with a different notation ($v : u$ for $u \to v$). For example, if $Q(+, \cdot, ')$ is a Boolean algebra and $x \to y = x' + y$, then $Q(+, \cdot, \to)$ is a Heyting algebra. Let He be the variety of all Heyting algebras.

   Characterize hyperidentities and termal hyperidentities of the variety He. Is every hyperidentity of the variety He the consequence of hyperidentities with functional rank $\leqslant 2$ of this variety?

   Characterize algebras with hyperidentities (termal hyperidentities) of the variety He.

Characterize hyperidentities (termal hyperidentities) of polyadic Heyting algebras, and describe the algebras with these hyperidentities (termal hyperidentities).

Characterize hyperidentities (termal hyperidentities) of cylindric Heyting algebras, and describe the algebras with these hyperidentities (termal hyperidentities).

7. An $\mathcal{O}$ckham algebra (see [33,163]) is an algebra $\mathfrak{A} = Q(+, \cdot, f)$ such that $Q(+, \cdot)$ is a distributive lattice and f is an antiendomorphism of $Q(+, \cdot)$, i.e. for every $x, y \in Q$:

$$f(x + y) = f(x) \cdot f(y),$$
$$f(x \cdot y) = f(x) + f(y).$$

Let $\mathcal{O}$ be the variety of all $\mathcal{O}$ckham algebras.

Characterize hyperidentities and termal hyperidentities of the variety $\mathcal{O}$.

Characterize algebras with hyperidentities (termal hyperidentities) of the variety $\mathcal{O}$.

The equation

$$f^2(x) = x$$

defines the subvariety of De Morgan algebras.

Characterize termal hyperidentities of the variety of De Morgan algebras.

Characterize algebras with termal hyperidentities of the variety of De Morgan algebras.

8. A modal algebra (see [42]) is an algebra $\mathfrak{A} = Q(+, \cdot, ', f, 0, 1)$ such that $Q(+, \cdot, ', 0, 1)$ is a Boolean algebra and f is a unary operation (the modal operator) that satisfies the identity $f(1) = 1$ and

$$f(x \cdot y) = f(x) \cdot f(y).$$

Modal algebras have been extensively investigated because of their connection with the modal logic. Let $\mathcal{M}$ be the variety of all modal algebras.

Characterize hyperidentities and termal hyperidentities of the variety $\mathcal{M}$.

Characterize algebras with hyperidentities (termal hyperidentities) of the variety $\mathcal{M}$.

Several varieties of modal algebras have received a great deal of attention in the investigations. We mention the varieties of interior algebras, monadic algebras, diagonalizable algebras, etc.

9. A relation algebra (see [161,164,203–210,231,232,397]) is an algebra $\mathfrak{A} = Q(+, \cdot, ', \circ, ^{-}, 0, 1)$ such that $Q(+, \cdot, ', 0, 1)$ is a Boolean algebra and $Q(\circ)$ is a monoid, and that the identities

$$(x + y) \circ z = (x \circ z) + (y \circ z),$$
$$\overline{x + y} = \overline{x} + \overline{y},$$
$$\overline{x \circ y} = \overline{y} \circ \overline{x},$$
$$\overline{\overline{x}} = x,$$
$$\overline{x} \circ (x \circ y)' \leqslant y'$$

hold. Let *Re* be the variety of all relation algebras.

Characterize hyperidentities and termal hyperidentities of the variety *Re*. Is every hyperidentity of the variety *Re* the consequence of hyperidentities with functional rank $\leqslant 2$ of this variety?

Characterize algebras with hyperidentities (termal hyperidentities) of the variety *Re*.

Characterize hyperidentities and termal hyperidentities of all symmetric relation algebras ($\overline{x} = x$).

Characterize hyperidentities and termal hyperidentities of all commutative relation algebras ($x \circ y = y \circ x$).

Characterize hyperidentities and termal hyperidentities of all (full) relation algebras of binary relations.

Characterize hyperidentities and termal hyperidentities of all representable relation algebras.

Clearly, every symmetric relation algebra is commutative and every Boolean algebra is a relation algebra with

$$x \circ y = x \cdot y \text{ and } \overline{x} = x.$$

10. Characterize hyperidentities of the variety of all rings. Are all hyperidentities, satisfied by the variety of all rings, the consequences of one object variable hyperidentities of this variety? (posed by B. I. Plotkin).

Characterize binary algebras with hyperidentities of the variety of all rings.

Characterize hyperidentities of all fields.

Characterize hyperidentities of all fields with a fix characteristic.

Characterize hyperidentities of all finite fields.

11. The loop $Q(\cdot)$ is called a Moufang loop (see [27,52,190,243,249,385]) if one of the following equivalent identities:

$$(zx \cdot y)x = z(x \cdot yx),$$
$$x(y \cdot xz) = (xy \cdot x)z,$$
$$xy \cdot zx = x(yz \cdot x)$$

is valid in it. Commutative Moufang loops are characterized by one identity:

$$x^2 \cdot yz = xy \cdot xz.$$

Moufang's theorem (see [27,52,243]). *If the relation $a \cdot bc = ab \cdot c$ is valid in a Moufang loop for its some three elements a, b, c, then the subloop, generated by these three elements, is a group.*

Characterize termal hyperidentities of the variety of all Moufang loops (commutative Moufang loops) (posed by V. D. Belousov).

Characterize algebras with termal hyperidentities of the variety of all Moufang loops (commutative Moufang loops).

Characterize termal hyperidentities of G-loops (see G. M. Bergman [32]).

12. Let P be a linearly ordered field. The interval $(0,\ 1) \subseteq P$ is a semigroup under multiplication of field P. The binary algebra $\mathfrak{A} = (Q,\ \Sigma)$ is called a stochastic algebra over P (in other terminology is a convexor (see [376]) or a barycentric algebra (see [345]), if $\mathfrak{A}$ is a Ω-algebra, under $\Omega = (0,\ 1)$, and the following equations are true:

S1. $\alpha(x,\ x) = x$,
S2. $\alpha(x,\ y) = (1-\alpha)(y,\ x)$,
S3. $\alpha(x,\ \beta(y,\ z)) = \alpha\beta(\frac{\alpha(1-\beta)}{1-\alpha\beta}(x,\ y),\ z)$

for any $x,\ y,\ z \in Q$ and $\alpha,\ \beta \in (0,\ 1)$. The first axiom is a hyperidentity of idempotency, while the second and third equations are $\forall\exists(\forall)$-identities of commutativity and associativity. Non-trivial hyperidentities of left and right distributivity are satisfied in every stochastic algebra:

$$\begin{aligned}\beta\left(\alpha(x,\,y),\,\alpha(x,\,z)\right) &= \beta\alpha\left(\frac{\beta(1-\alpha)}{1-\beta\alpha}\left(\alpha(x,\,y),\,x\right),\,z\right)\\ &= \beta\alpha\left(\frac{1-\beta\alpha-\beta+\beta\alpha}{1-\beta\alpha}\left(x,\,\alpha(x,\,y)\right),\,z\right)\\ &= \beta\alpha\left(\frac{(1-\beta)\alpha}{1-\beta\alpha}(x,\,y),\,z\right) = \alpha\left(x,\,\beta(y,\,z)\right),\end{aligned}$$

while from this hyperidentity and axiom S.2 implies the hyperidentity (5.7.5).

In addition, the solution of L. A. Skornyakov's problem about the characterization of ideals' lattices of stochastic algebras is obtained taking into account these facts.

Theorem 22.0.9. *A lattice of ideals of every stochastic algebra is isomorphic to a lattice of ideals of some semilattice.*

Theorem 22.0.10. *Every ideal of a stochastic algebra is an intersection of its simple ideals* (*cf.* [14]).

Let Stoch be the variety of all stochastic algebras.

Characterize hyperidentities (termal hyperidentities) of the variety Stoch (posed by L. A. Skornyakov).

Characterize algebras with hyperidentities (termal hyperidentities) of the variety Stoch.

13. Characterize termal hyperidentities of the class of BCK-algebras (BCH-algebras, BCC-algebras) (see [155,156]) (posed by K. Iseki). Characterize algebras with termal hyperidentities of BCK-algebras (BCH-algebras, BCC-algebras).
14. Obtain categorical characterization of
    (a) hypervarieties of Ω-algebras;
    (b) hypervarieties of T-algebras;
    (c) solid varieties of Ω-algebras;
    (d) solid hypervarieties of T-algebras;
    (posed by J.D.H. Smith).
    See [194,329,412].
15. (a) Is elementary theory of any De–Morgan algebra decidable?
    (b) Is elementary theory of the variety of De–Morgan algebras decidable?

(c) Is elementary theory of any super-De Morgan algebra with one unary and one binary operations is decidable?
(d) Is elementary theory of the variety of super-De Morgan algebras with one unary and one binary operations is decidable?
(e) Is elementary theory of any super-De Morgan algebra with one unary and two binary operations is decidable?
(f) Is elementary theory of the variety of super-De Morgan algebras with one unary and two binary operations is decidable?
See [101,354,397].

16. In the paper [433] free algebras of the variety of algebras with two binary operations satisfying the hyperidentity of associativity $(ass)_1$ were characterized.
    (a) Characterize free algebras of the variety of algebras with two binary operations satisfying the hyperidentity of associativity $(ass)_2$.
    (b) Characterize free algebras of the variety of algebras with two binary operations satisfying the hyperidentity of associativity $(ass)_3$.
17. Characterize the Burnside varieties of distributive (idempotent) quasigroups.
18. Characterize the Burnside varieties of abelian semigroups.
19. Characterize the Burnside varieties of abelian and hyperassociative algebras.
19. Prove a Cayley-style theorem for hyperassociative algebras.
20. Prove a Cayley-style theorem for hyperalternative algebras.

Bibliography

[1] L. R. Abrahamyan, Characterization of hyperidentities defined by the equalities $((x, y), u, v) = (x, (y, u), v)$ and $((x, y), u, v) = (x, y, (u, v))$, *Emil Artin International Conference dedicated to the 120th Anniversary of Emil Artin*, Yerevan, Armenia, May 27–June 2, 2018, Abstracts, pp. 9–10.

[2] L. R. Abrahamyan, Hyperidentities of ternary associativity, *Proceedings of the Yerevan State University, Physical and Mathematical Sciences* **3**(2003), pp. 36–44.

[3] J. Aczél, Remarques algebriques sur la solution donner par M. Frechet a l'equation de Kolmogoroff, *Pupl. Math. Debrecen*, **4**(1955), pp. 33–42.

[4] J. Aczél, *Forlesungen über Functionalgleichungen und ihre Anwendungem*, Berlin, VEB Deutsch. Verl. Wiss., 1961.

[5] J. Aczél, *Quelque problèmes non-résolus dans la théorie des équations fonctionelles*, in "Neki Nerešeni Problemi u Matematici", Beograd, 1963, pp. 149–152.

[6] J. Aczél, Proof of a theorem of distributive type hyperidentities, *Algebra Universalis* **1**(1971), pp. 1–6.

[7] J. Aczél, V. D. Belousov, M. Hosszu, Generalized associativity and bisymmetry on quasigroups, *Acta Math. Acad. Sci. Hungar.* **11**(1960), pp. 127–136.

[8] J. Aczél, J. Dhombres, *Functional Equations in Several Variables with Applications to Mathematics, Information Theory and to the Natural and Social Sciences*, Cambridge University Press, Cambridge, 1991.

[9] S. I. Adian, *The Burnside Problem and Identities in Groups*, Springer, Berlin/ New York, 1979.

[10] S. I. Adian, The Burnside Problem and related topics, *Russian Math. Surveys* **65(5)**(2010), pp. 805–855.

[11] M. Aguiar, M. Livernet, The associative operad and the weak order on the symmetric groups, *J. Homotopy Relat. Struct.* **2**(2007), pp. 57–84.

[12] A. A. Albert, Quasigroup I, *Trans. Amer. Math. Soc.* **54**(1943), pp. 507–519.

[13] A. A. Albert, Quasigroup II, *Trans. Amer. Math. Soc.* **55**(1944), pp. 401–419.

[14] V. A. Andrunakievich, Yu. M. Ryabukhin, On decompositions of elements of partially ordered algebraic systems into intersections of prime elements, *Dokl. Akad. Nauk SSSR* **198**(1) (1971), pp. 13–15.

[15] A. D. Anosov, On homomorphisms of many-sorted algebraic systems in connection with cryptographic applications, *Discrete Math. Appl.* **17(4)**(2007), pp. 331–347.

[16] O. Arieli, A. Avron, The value of four values, *Artificial Intelligence* **102**(1998), 97–141.

[17] B. H. Arnold, Distributive lattices with a third operation defined, *Pacific J. Math.* **1**(1951), pp. 33–41.
[18] A. Avron, The structure of interlaced bilattices, *Math. Struct. Comp. Sci.* **6**(1996), pp. 287–299.
[19] R. Balbes, Ph. Dwinger, *Distributive Lattices*, University of Missouri Press, 1974.
[20] W. E. Barnes, On Γ-rings of Nobusawa, *Pacific J. Math.* **3**(1966), pp. 411–422.
[21] N. D. Belnap, A useful for valued logic, in: G. Epstein, J.M. Dunn (Eds)., *Modern Uses of Multiple-Valued Logic*, Reidel Publishing Company, Boston, (1977), pp. 7–77.
[22] N. D. Belnap, *How Computer Should Think*, in: G. Ryle (Ed.), Contemporary Aspects of Philosophy, Oriel Press, (1977), pp. 30–56.
[23] V. D. Belousov, The structure of distributive quasigroups, *Mat. Sb. (N. S.)* **50(92)(3)**(1960), pp. 267–298.
[24] V. D. Belousov, Globally associative systems of quasigroups, *Mat. Sb. (N. S.)*, **55(97)(2)** (1961), pp. 221–236.
[25] V. D. Belousov, Systems of quasigroups with generalized identities, *Russian Math. Surveys* **20**(1965), pp. 73–143.
[26] V. D. Belousov, Gratzer algebra and S-systems of quasigroups, *Mat. Issled. Akad. Nauk. Mold.* **1**(1966), pp. 55–81.
[27] V. D. Belousov, *Foundations of the Theory of Quasigroups and Loops*, Nauka, Moscow, 1967 (in Russian).
[28] V. D. Belousov, Yu. M. Movsisyan, On the rank of generalized identities, *Izv. Akad. Nauk. Arm., Ser. Mat.* **9**(1974), pp. 135–142.
[29] G. B. Belyavskaya, A. M. Cheban, S-systems of arbitrary index I, *Mat. Issled. Akad. Nauk. Mold.* **7(1(23))**(1972), pp. 23–43.
[30] G. B. Belyavskaya, A. M. Cheban, S-systems of arbitrary index II, *Mat. Issled. Akad. Nauk. Mold.* **7(2(24))**(1972), pp. 3–13.
[31] G. M. Bergman, Hyperidentities of groups and semigroups, *Aequationes Math.* **23**(1981), pp. 50–65.
[32] G. M. Bergman, *An Invitation on General Algebra and Universal Constructions*, 2nd edn., Springer, 2015.
[33] J. Berman, Distributive lattices with an additional operation, *Aequationes Math.* **16**(1977), pp. 165–171.
[34] J. Berman, A proof of Lyndon's finite basis theorem, *Discrete Math.* **52**(1980), pp. 229–233.
[35] A. Bialynicki-Birula, H. Rasiowa, On the representation of quasi-Boolean algebras, *Bull. Acad. Polon. Sci., Ser. Math. Astronom. Phys.* **5**(1957), pp. 259–261.
[36] G. Birkhoff, On the structure of abstract algebras, *Proc. Cambridge Phil. Soc.* **31**(1935), pp. 433–454.
[37] G. Birkhoff, Subdirect unions in universal algebra, *Bull. Amer. Math. Soc.* **50**(1944), pp. 764–768.
[38] G. Birkhoff, *Lattice Theory*, 3rd edn., American Mathematical Society, Providence, Rhode Island, 1979.
[39] G. Birkhoff, Rings of sets, *Duke Math. J.* **3**(3)(1937), pp. 443–454.
[40] G. Birkhoff, Jr. O. Frink, Representations of lattices by sets, *Trans. Amer. Math. Soc.* **64**(2)(1948), pp. 299–316.
[41] G. Birkhoff, J. D. Lipson, Heterogeneous algebras, *J. Combin. Theory* **8**(1970), pp. 115–133.

[42] W. J. Blok, The lattice of modal logics, an algebraic investigation, *J. Symbolic Logic* **45**(1980), pp. 221–236.
[43] S. L. Bloom, Z. Ésik, E. G. Manes, A Cayley theorem for Boolean algebras, *Amer. Math. Monthly* **97**(1990), pp. 831–833.
[44] G. Boole, The Laws of Thought, Dover, 1958. Original edition 1853.
[45] F. Bou, U. Rivieccio, The logic of distributive bilattices, *Log. J. IGPL*, 19(2011), pp. 183–216.
[46] N. Bourbaki, *Elements of Mathematics, Algebra (Algebraic Structures)*, Addison-Wesley, 1974.
[47] S. J. Boyd, M. Gould, Interassociativity and isomorphism, *Pure Math. Appl.* **10(1)** (1999), pp. 23–30.
[48] S. J. Boyd, M. Gould, A. W. Nelson, Interassociativity of semigroups, *Proceedings of the Tennessee Topology Conference*, Nashville, TN, USA, 1996. Singapore: World Scientific, (1997), pp. 33–51.
[49] T. C. Brown, On the finiteness of semigroups in which $x^r = x$. *Proc. Camb. Philos. Soc.* **60**(1964), pp. 1028–1029.
[50] T. Brown, E. Lazerson, On finitely generated idempotent semigroups, *Semigroup Forum*, **78**(2009), pp. 183–186.
[51] R. H. Bruck, Contributions to the theory of loops, *Trans. Amer. Math. Soc.* **60**(1946), pp. 245–354.
[52] R. H. Bruck, *A Survey of Binary Systems.* Berlin, Heidelberg, Göttingen, Springer, 1971.
[53] R. H. Bruck and E. Kleinfeld, The structure of alternative division rings, *Proc. Amer. Math. Soc.* **2**(1951), pp. 878–890.
[54] J. A. Brzozowski, De Morgan bisemilattices, *The 30th IEEE International Sympozium on Multiple-Valued Logic*, Portland, Oregon, May 2000, IEEE Press, Los Alamitos, California, (2000), pp. 23–25.
[55] J. A. Brzozowski, A characterization of De Morgan algebras, *Int. J. Algebra Comput.* **11(5)**(2001), pp. 525–537.
[56] J. A. Brzozowski, Partially ordered structures for hazard detection, *Special Session: The Many Lives of Lattice Theory, Joint Mathematics Meetings*, San Diego, CA, January 6–9, 2002.
[57] J. A. Brzozowski, C.-J. H. Seger, *Asynchronous Circuits*, Springer-Verlag, 1995.
[58] L. Budaghyan, *Construction and Analysis of Cryptographic Functions*, Springer, 2014.
[59] L. Budaghyan, *The Equivalence of APN and AB Functions and Their Generalizations*, Scholars' Press, 2018.
[60] W. Burnside. On an unsettled question in the theory of discontinuous groups, *Quart. J. Pure and Appl. Math.* **33**(1902), pp. 230–238.
[61] S. Burris, H. P. Sankappanavar, *A Course in Universal Algebra*, Springer-Verlag, 1981.
[62] L. M. Cabrer, A. Craig, H. A. Priestley, Product representation for default bilattices: An application of natural duality theory. *J. Pure Appl. Algebra* **219**(7)(2015), pp. 2962–2988.
[63] L. M. Cabrer, H. A. Priestley, Distributive bilattices from the perspective of natural duality theory, *Algebra Universalies* **73**(2015), pp. 103–141.
[64] C. C. Chang, H. J. Keisler, *Model Theory*, North-Holland, Amsterdam, 1973; Russian transl., "Mir", Moscow, 1977.

[65] O. Chein, H. O. Pflugfelder, J. D. H. Smith, *Quasigroups and Loops: Theory and Applications*, Sigma Series in Pure Mathematics, Vol. 9, Heldermann Verlag, 1990.
[66] A. Church, *Introduction to Mathematical Logic*, Vol. 1. Princeton University Press, Princeton, NJ, 1956.
[67] A. H. Clifford, G. B. Preston, The Algebraic Theory of Semigroups, Vol. II, American Mathematical Society, 1967.
[68] P. Clote, E. Kranakis, *Boolean Functions and Computation Models*, Springer, 2002.
[69] P. M. Cohn, *Universal Algebra*, Harper & Row Publishers, New York, Evanston and London, 1965.
[70] D. Cowan, S. L. Wismath, *Unari hyperidentities for varieties of inverse semigroups*, Preprint, 1996.
[71] A. Craig, L. M. Cabrer, H. A. Priestley, Beyond FOUR: representations of non-interlaced bilattices using natural duality, *Research Workshop on Duality Theory in Algebra, Logic and Computer Science*, University of Oxford, June 13–14, 2012, pp. 16–16.
[72] Y. Crama, P. L. Hammer, *Boolean Functions: Theory, Algorithms, and Applications*, Cambridge University Press, New York, 2011.
[73] G. Čupona, "On the relation of distributivity between binary operations", *Fac. Philos. Univ. Skopje Sect. Sci. Nat. Annuaire* **9**(1956), pp. 23–29.
[74] G. Čupona, Za finitarnite operazii. Godishen zbornik univerzitetot Skopje, *Prirodno-Matematichki Fakultet* **1(12)**(1959–1961), pp. 7–49.
[75] B. A. Davey, The product representation theorem for interlaced pre-bilattices: some historical remarks, *Algebra Universalis* **70**(2013), pp. 403–409.
[76] S. S. Davidov, On invertible T-algebras, *J. Math. Sci. (N.Y.)* **193(3)**(2013), pp. 394–400.
[77] S. S. Davidov, A characterization of binary invertible algebras linear over a group, *Quasigroups Related Syst.* **19**(2011), pp. 207–222.
[78] S. S. Davidov, A characterization of binary invertible algebras of various type of linearity, *Quasigroups Related Syst.* **20**(2012), pp. 169–176.
[79] S. S. Davidov, On regular medial division algebras, *Quasigroups Related Syst.* **21**(2013), pp. 157–166.
[80] S. S. Davidov, A characterization of invertible algebras linear over a group by second-order formulas, *J. Algebra Appl.* (2017); DOI:10.1142/S0219498817502383.
[81] S. S. Davidov, A. Krapež, Yu. M. Movsisyan, Functional Equations with Division and regular Operations, *Asian-Eur. J. Math.* **11**(3)(2018) 1850033; doi.org/10.1142/S179355711850033X.
[82] D. S. Davidova, Yu. M. Movsisyan, Weakly idempotent lattices and bilattices, non-idempotent Plonka functions, *Demonstratio Math.* **48**(4)(2015), pp. 509–535.
[83] A. DeMorgan, *Formal Logic*, Taylor and Walton, London, 1847.
[84] R. Dedekind, Über Zerlegungen von Zahlen durch ihre größten gemeinsamen Teiler, *Festschrift der Techn. Hochsch. Braunschwig bei Gelegenheit der* **69(1897)**. Versammlung deutscher Naturforscher und Ärzte, pp. 1–40.
[85] K. Denecke, Hyperequational theory, *Math. Japonica* **47**(1998), pp. 333–356.
[86] K. Denecke, The partial clone of linear terms, *Siberian Math. J.* **57(4)**(2016), pp. 589–598.
[87] K. Denecke, J. Koppitz, Presolid varieties of commutative semigroups, *Tatra Mt. Math. Publ.* **5**(1995), pp. 41–48.

[88] K. Denecke, J. Koppitz, Hyperassociative Varieties of Semigroups, *Semigroup Forum* **49**(1994), pp. 41–48.

[89] K. Denecke, J. Koppitz, *M-solid varieties of Algebras* Advances in Mathematic, Vol. 10, Springer-Science+Business Media, New York, 2006.

[90] K. Denecke, D. Lau, R. Pöschel, D. Schweigert, Hyperidentities, hyperequational classes, and clone congruences. Contribution to Gen. Algebra **7**, Verlag Hölder-Pichler-Tempsky, Wien, (1991), pp. 97–118.

[91] K. Denecke, R. Pöschel, A characterization of Sheffer functions by hyperidentities, *Semigroup Forum* **37**(1988), pp. 351–362.

[92] K. Denecke, R. Pöschel, The characterization of primal algebras by hyperidentities, *Contributions Gen. Algebra* **6**(1988), pp. 67–87.

[93] K. Denecke, S. L. Wismath, *Hyperidentities and Clones*. Gordon and Breach Science Publishers, 2000.

[94] K. Denecke, S. L. Wismath, *Universal Algebra and Applications in Theoretical Computer Science*. Chapman and Hall/CRC, 2002.

[95] J. Denes, A. D. Keedwell, *Lattin Squares and Their Applications*, Akademiai Kiado, Budapest, 1974 (also Academic Press, New York,1974, and English Universities Press, London, 1974).

[96] R. M. Dicker, A set of independent axioms for a field and a condition for a group to be the multiplicative group of a field, *Proc. London Math. Soc.* **18**(1968), pp. 114–124.

[97] L. E. Dickson, Definition of a group and a field by independent postulates, *Trans. Amer. Math. Soc.* **6**(1905), pp. 198–204.

[98] M. Drouzy, La structuration des ensembles de semigroupes dordre 2, 3 et 4 par la relation dinterassociativite. Manuscript (1986).

[99] A. Ehsani, A. Krapez, Yu. M. Movsisyan, Algebras with parastropically uncancellable quasigroup equations, *Bull. Acad. Stiinte Repub. Mold. Mat.* **1(80)**(2016), pp. 41–63.

[100] A. Ehsani, Yu. Movsisyan, A representation of paramedial n-ary groupoids, *Asian-Eur. J. Math.* **7**(1)(2014), 1450020 (17 pages).

[101] Yu. L. Ershov, *Decidability Problems and Constractive Models*, Nauka, Moscow, 1980 (in Russian).

[102] Yu. L. Ershov et al., Elementary theories, *Uspekhi Matem. Nauk* **20(4)**(124)(1965), pp. 37–108; English transl. in *Russian Math. Surveys* **20**(1965), pp. 35–105.

[103] Z. Ésik, A Cayley theorem for ternary algebras, *Int. J. Algebra Comput.* **8(3)**(1998), pp. 311–316.

[104] Z. Ésik, Free De Morgan bisemigroups and bisemilattices, *Algebra Colloquium* **10**(2003), pp. 23–32.

[105] T. Evans, Universal algebra and Euler's officer problem, *Amer. Math. Monthly* **86**(1979), pp. 466–473.

[106] T. Evans, Properties of algebras almost equivalent to identities, *J. London Math. Soc.* **35**(1962), pp. 53–59.

[107] M. C. Fitting, *Logic programming on a topological bilattice*, *Fund. Inform.* **11**(1988), pp. 209–218.

[108] M. C. Fitting, Bilattices in logic programming. in: G. Epstein ed., *Proc. 20th Internat. Symp. on Multi-Valued Logic*, IEEE, New York, 1990, pp. 63–70.

[109] M. C. Fitting, Bilattices and the semantics of logic programming, *J. Logic Programming* **11**(1991), pp. 91–116.

[110] M. C. Fitting, Bilattices and the theory of truth, *J. Philos. Logic*, **18**(1989), pp. 225–256.

[111] J. M. Font, M. Moussavi, Notes on a six-valued extension of tree-valued logic, *J. Appl. Non-Classical Logic* **3**(1993), pp. 173–187.

[112] E. Freid, Weakly Associative Lattices with congruence extension property, *Algebra Universalis* **4**(1974), pp. 151–162.

[113] E. Fried, G. Gratzer, A nonassociative extension of the class of distributive *Lattices, Pacific J. Math.* **49(1)**(1973), pp. 59–78.

[114] E. Friedgut, Boolean functions with low average sensitivity depend on few coordinates, *Combinatorica* **18**(1)(1998), pp. 27–35.

[115] L. Fuchs, *Infinite Abelian Groups*, Vol. II, Academic Press, New York and London, 1973.

[116] B. Ganter, H. Werner, Equational classes of Steiner systems, *Algebra Universalis* **5**(1975), pp. 125–140.

[117] G. Gargov, Knowledge, Uncertainty and ignorance in logic: Bilattices and beyond, *J. Appl. Non-Classical Logics* **9(2–3)**(1999), pp. 195–283.

[118] M. Gehrke, C. Walker, E. Walker, A mathematical setting for fuzzy logics, *Int. J. Uncertainty, Fuzziness Knowledge-based Syst.*, **5**(3) (1997), pp. 223–238.

[119] M. Gehrke, C. Walker, E. Walker, Some comments on fuzzy normal forms, *Proc. Ninth IEEE Int. Conf. Fuzzy Systems, FUZZ-IEEE*, Vol. 7 (2000), pp. 593–598.

[120] I. M. Gelfand, A. N. Kolmogoroff, On the rings of continuous functions on topological spaces, *Doklady Akad. Nauk SSSR* **22**(1939), pp. 11–15.

[121] A. Gevorgyan, On medial structures, *Emil Artin International Conference dedicated to the 120th Anniversary of Emil Artin*, Yerevan, Armenia, May 27–June 2, 2018, Abstracts, pp. 50–51.

[122] P. S. Gevorgyan, On binary G-spaces, *Math. Notes* **96(4)**(2014), pp. 600–602.

[123] H. Ghumashyan, On {2,3}-hyperidentities in the invertible {2,3}-algebras with a ternary group-operation, *J. Contemp. Math. Anal.* **45(3)**(2010), pp. 47–56.

[124] L. Gillman, M. Jerison, *Rings of continuous functions*, Van Nostrand, Preston, 1960 (Second edition, 1976).

[125] M. L. Ginsberg, Bilattices. Preprint, Department of Computer Science, Stanford University, 1986.

[126] M. L. Ginsberg, Multi-valued logics. Technical report 84-17, KSL, Stanford University, 1986.

[127] M. L. Ginsberg, Multi-valued inference. Preprint, Department of Computer Science, Stanford University, 1986.

[128] M. L. Ginsberg, Multi-valued logics: a uniform approach to reasoning in artificial intelligence, *Computat. Intelligence* **4**(1988), pp. 265–316.

[129] B. N. Givens, K. Linton, A. Rosin, L. Dishman, Interassociates of the free commutative semigroup on n generators, *Semigroup Forum* **74**(2007), pp. 370–378.

[130] B. N. Givens, A. Rosin, K. Linton, Interassociates of the bicyclic semigroup, *Semigroup Forum* **94**(2017), pp. 104–122.

[131] A. B. Gorbatkov, Interassociates of a free semigroup on two generators, *Mat. Stud.* **41**(2014), pp. 139–145.

[132] A. B. Gorbatkov, Interassociates of the free commutative semigroup, *Sib. Math. J.* **54(3)**, pp. 441–445.

[133] M. Gould, K. A. Linton, A. W. Nelson, Interassociates of monogenic semigroups, *Semigroup Forum* **68**(2004), pp. 186–201.

[134] M. M. Glukhov, Some applications of quasigroups in cryptography, *Prikl. Diskr. Math.* **2(2)**(2008), pp. 28–32.

[135] E. Graczynska, On normal and regular identities, *Algebra Universalis* **27**(1990), pp. 387–397.

[136] E. Graczynska, *Algebras of M-solid Quasivarieties*, Siatras International Bookshop, Athens, 2014.
[137] E. Graczynska, On the problem on M-hyperquasivarieties, arXiv:math/06096000v3 [math.GM], 2007.
[138] E. Graczynska, D. Schweigert, Hypervarieties of a given type, *Algebra Universalis* **27**(1990), pp. 305–318.
[139] E. Graczynska, R. Pöschel, M. V. Volkov, Solid pseudovarieties, in *General Algebra and Applications in Discrete Mathematics*, Shaker-Verlag, Aachen, (1997), pp. 93–100.
[140] G. Grätzer, A theorem on doubly transitive permutation groups with application to universal algebra, *Fund. Math.* **53**(1963), pp. 25–41.
[141] G. Grätzer, Two Mal'tsev-type theorems in universal algebra, *J. Combin. Theory* **8**(1970), pp. 334–342.
[142] G. Grätzer, *Universal Algebra*, 2nd edn, Springer-Verlag, 1979.
[143] G. Grätzer, *Lattice Theory: Foundation*, Springer, Basel AG, 2011.
[144] J. A. Green, D. Rees, On semi-groups in which $x^r = x$, *Proc. Camb. Philos. Soc.* **48**(1952), pp. 35–40.
[145] H. P. Gumm, A cancellation theorem for finite algebras, *Colloquia Math. Soc., J. Bolyai* **29**(1977), pp. 341–344.
[146] A. E. Guterman, D. K. Kudryavtsev, Characteristic sequences of non-associative algebras, *Communi. Algebra* **48(4)** (2020), pp. 1713–1725.
[147] M. Hall (Jr.), *The theory of groups*, New York, 1959.
[148] M. Hall, On a theorem of Jordan, *Pacific J. Math.* **4(2)**(1954), pp. 291–226.
[149] H. Hedayati, K. P. Shum, An introduction to Γ-semirings, *Int. J. Algebra* **5**(2011), pp. 709–726.
[150] E. Hewitt, Rings of real-valued continuous functions. I, *Trans. Amer. Math. Soc.* **64**(1948), pp. 45–99.
[151] P. J. Higgins, Algebras with a scheme of operators, *Math. Nachr.* **27**(1963), pp. 115–132.
[152] N. J. S. Hughes, A theorem on isotopic groupoids, *J. London Math. Soc.* **32**(1957), pp. 510–511.
[153] L. Humberstone, Note on extending congruential modal logics, *Notre Dame J. Formal Logic* **57(1)**(2016), pp. 95–100.
[154] J. Hyndman, R. McKenzie, W. Taylor, k-ary monoids of term operations, *Semigroup Forum* **44**(1992), pp. 31–52.
[155] K. Iseki, A way to BCK and related systems, *Math. Japonica* **5**(2000), pp. 163–170.
[156] K. Iseki, Sh. Tanaka, An introduction to the theory of BCK-algebras, *Math. Japonica* **23**(1978), pp. 1–26.
[157] H. Israyelyan, About medial pairs of continuous and strictly monotonic binary functions, *Emil Artin International Conference Dedicated to the 120th Anniversary of Emil Artin*, Yerevan, Armenia, May 27–June 2, 2018, Abstracts, pp. 78–78.
[158] J. Jakubik, M. Kolibiar, On some properties of pair lattices, *Czechosl. Math. J.* **4(79)(1)**(1954), pp. 1–27.
[159] J. Jakubik, M. Kolibiar, Lattices with a third distributive operation, *Math. Slovaca* **27**(1977), pp. 287–292.
[160] J. Ježek, T. Kepka, P. Němec, *Distributive Groupoids*, Praha, 1981.
[161] B. Jonsson, Varieties of relation algebras, *Algebra Universalis* **15**(1982), pp. 273–298.
[162] B. Jonsson, *Distributive Bilattices*, Preprint, Vanderbilt University, Nashville, 1994.
[163] B. Jonsson, Congruence distributive varieties, *Math. Japonica* **42**(1995), pp. 353–401.

[164] B. Jonsson, A. Tarski, Representation problems for relation algebras. Abstract 89, *Bull. Amer. Math. Soc.* **54**(1948), pp. 80–80.

[165] C. Jordan, Recherches sur les substitutions, *J. Math. Pures Appl.* **2**(17)(1872), pp. 351–363.

[166] A. Jung, U. Rivieccio, Priestley duality for bilattices, *Studia Logica* **100**(2012), pp. 223–252.

[167] J. A. Kalman, Lattices with involution, *Trans. Amer. Math. Soc.* **87**(1958), pp. 485–491.

[168] J. Kahn, G. Kalai, and N. Linial, The influence of variables on boolean functions, in *Proc. 29th IEEE Symp. on Foundations of Comp. Sci.* (1988), pp. 68–80.

[169] I. Kaplansky, Topological rings, *Amer. Math. J.* **69**(1947), pp. 153–183.

[170] M. I. Kargapolov, Yu. I. Merzlyakov, *Foundation of Group Theory*, Nauka, Moscow, 1977 (in Russian).

[171] H. Karzel, Inzidenzgruppen. Lecture Notes by Pieper, I. and Sorensen, K., University of Hamburg (1965), pp. 123–135.

[172] N. Kehayopulu, On regular duo po-Γ-semigroups, *Math. Slovaca* **6(61)**(2011), pp. 871–884.

[173] T. Kepka, Distributive Steiner quasigroups of order 3^5, *Comment. Math. Univ. Carolinae* **195**(1978), pp. 389–401.

[174] W. Kerby, Uber eine scharf 3-fach transitiven Gruppen zugeordnete algebraische Structur, *Abh. Math. Sem. Univ. Hamburg* **37**(1972), pp. 225–235.

[175] S. A. Kiss, *Transformations on Lattices and Structure of Logic*, Stevens & Co., New York, 1947.

[176] A. Knoebel, A. Romanowska, Distributive Multisemilattices. Dissertationes Mathematicae, *Rozprawy Marhematyczne*, 1991, Warszawa.

[177] S. R. Kogalovskii, On multiplicative semigroups of rings, *Dokl. Akad. Nauk SSSR* **140**(1961), pp. 1005–1007 (in Russian).

[178] M. Kondo, On the structures of weak interlaced bilattice, in *32nd IEEE International Symposium on Multiple-Valued Logic* (*ISMVL 2002*), May 15–18, 2002, Boston, MA, USA, pages 23, IEEE Computer Society, 2002.

[179] J. Koppitz, K. Denecke, *M-Solid Varieties of Algebras*, Springer, 2006.

[180] A. D. Korshunov, Monotone Boolean functions, *Russian Math. Surv.* **58(5)**(2003), pp. 928–1001.

[181] H. V. Krainichuk, Classification of group isotopes according to their symmetry groups, *Folia Mathematica* **19(1)**(2017), pp. 84–98.

[182] A. Krapež, Functional equations of generalized associativity, bisymmetry, transitivity and distributivity, *Publ. Inst. Math.* (*Beograd*) **30**(1981), pp. 81–87.

[183] A. Krapež, Generalized associativity on groupoids, *Publ. Inst. Math.* (*Beograd*) **28**(1980), pp. 105–112.

[184] A. Krapež, Cryptographically Suitable Quasigroups via Functional Equations, in S. Markovski and M. Gusevarphi (Eds.), *ICT Innovations 2012*, AISC 207, Springer-Verlag, Berlin, Heidelberg, (2013), pp. 264–274.

[185] A. Krapež, Quadratic level quasigroup equations with four variables, *I. Publ. Inst. Math.* (*Beograd*) (*N.S.*) **81(95)**(2007), pp. 53–67.

[186] A. Krapež, Quadratic level quasigroup equations with four variables, II, *Publ. Inst. Math.* (*Beograd*) (*N.S.*) **93(107)**(2013), pp. 29–47.

[187] A. Komorowski, A. B. Romanowska and J. D. H. Smith, Barycentric algebras and beyond, *Algebra Universalis* **80(20)**(2019), https://doi.org/10.1007/s00012-019-0595-3.

[188] N. A. Koreshkov, Associative n-tuple algebras, *Math. Notes* **96(1)**(2014), pp. 38–49.
[189] A. G. Kurosh, *Lectures on General Algebra*, Chelsea, New York, 1963.
[190] A. G. Kurosh, *General Algebra*, Nauka, Moscow, 1974 (in Russian).
[191] A. V. Kuznetsov, On the problems of identities and functional completeness for algebraic systems, *Proceedings of All Union Math. Congress*, Vol. 2, Academy of Sciences of USSR Publisher, 1956, pp. 145–146.
[192] G. Lallement, *Semigroups and Combinatorial Applications*, Wiley, New York, 1979.
[193] D. Lau, *Functional Algebras on Finite Sets, A Basic Course on Many-Valued Logic and Clone Theory*, Springer-Verlag, Berlin, Heidelberg, 2006.
[194] F. W. Lawvere, Functorial semantics of algebraic theories, *Proc. Natl. Acad. Sci.* **50**(1963), pp. 869–872.
[195] E. Lazerson, Idempotent semigroups, Working Paper No. 38, Communications Research Division, Institute for Defense Analyses, von Neumann Hall, Princeton, NJ, 1961.
[196] W. Leissner, Ein Stufenaufbau der Fasthereiche, Fastkorper and Korper aus ihrer multiplikativen Gruppe, *Abh. Math. Sem. Univ.* **46**(1977), pp. 55–89.
[197] J.-L. Loday, M. O. Ronco, Order structure on the algebra of permutations and of planar binary trees, *J. Algebraic Combin.* **15**(2002), pp. 253–270.
[198] J. Łoś, Quelques remarques, théoremès et problemès sur les classes définissables d'algèbres. in *Mathematical Interpretation of Formula Systems*, pp. 98–113. North-Holland Publishing Co., Amsterdam, 1955.
[199] M. Lothaire, *Combinatorics On Words*. Cambridge University Press, Cambridge, 1997.
[200] J. Luh, On the theory of simple Γ-rings, *Michigan Math. J.* **16**(1969), pp. 65–75.
[201] I. G. Lysenok, The infinitude of Burnside groups of period $2k$ for $k \geq 13$. *Russian Math. Surv.*, **47(2)**(1992), pp. 229–230.
[202] S. Maclane, *Homology*, Die Grundlehren der Mathematischen Wissenschaften, Band 114, Springer-Verlag, Berlin, 1963.
[203] R. D. Maddux, Some sufficient conditions for the representability of relation algebras, *Algebra Universalis*, **8**(1978), pp. 162–172.
[204] R. D. Maddux, Nonfinite axiomatizability results for cylindric and relation algebras, *J. Symbolic Logic* **54**(1989), pp. 951–974
[205] R. D. Maddux, Introductory course on relation algebras, finite-dimensional cylindric algebras, and their interconnections, *Colloq. Math. Soc. J. Bolyai* **54**(1991), pp. 361–392. ("Algebraic logic (Budapest,1988)") North-Holland, Amsterdam.
[206] R. D. Maddux, Finitary algebraic logic, *Math. Logic Quart.* **39**(1993), pp. 566–569.
[207] R. D. Maddux, A perspective on the theory of relation algebras, *Algebra Universalis* **31**(1994), pp. 456–465.
[208] R. D. Maddux, Relation-algebraic semantics, *Theoret. Comput. Sci.* **160**(1996), pp. 1–85.
[209] R. D. Maddux, relation algebras, in *Relational Methods in Computer Science*, Springer-Verlag, Wien, (1997), pp. 22–38.
[210] R. D. Maddux, Relation algebras of formulas, in *Logic at Work*, Physica, Heidelberg, (1999), pp. 613–636.
[211] A. I. Mal'tsev, On the immersion of associative systems in groups, *Mat. Sb.* (*N. S.*) **6**(1939), pp. 331–336.
[212] A. I. Mal'tsev, On the general theory of algebraic systems, *Amer. Math. Soc. Trans.* **27(2)**(1963), pp. 125–142.
[213] A. I. Mal'tsev, Model correspondences, *AMS Selected Trans.* **23**(1959), pp. 313–336.

[214] A. I. Mal'tsev, Some questions of the theory of classes of models, *Proceedings of the IVth All-Union Mathematical Congress*, **1**(1963), pp. 169–198 (in Russian).

[215] A. I. Malcev, *Algorithms and Recursive Functions*, Nauka, Moscow, 1965 (in Russian).

[216] A. I. Mal'tsev, *Algebraic Systems*, Springer-Verlag, Berlin, 1973.

[217] A. I. Mal'tsev, I. A. Mal'tsev, *Iterative Post Algebras*, Nauka, Moscow, 2012 (in Russian).

[218] I. A. Mal'tsev, D. Schveigert, Hyperidentitäten von AZ-algebren. Preprint, Fachbereich Mathematik, Universität Kaiserslautern, 1989.

[219] M. Mando, Regular $\forall\exists(\forall)$-identities of length four in algebras. Ph.D. Thesis, Yerevan State University, Yerevan, Armenia, 1990.

[220] Yu. I. Manin, *Cubic Forms*, 2nd edn., North-Holland Mathematical Library, Vol. 4, North-Holland, Amsterdam, 1986.

[221] H. B. Mann, The construction of orthogonal Latin squares, *Ann. Math. Statist.* **13**(1942), pp. 418–423.

[222] H. B. Marandjian, On computable solutions to functional equations, *CSIT* (2011), pp. 20–22.

[223] H. B. Marandjian, *Selected Topics in Recursive Function Theory in Computer Science*, DTH, Lyngby, 1990.

[224] H. B. Marandjian, General form recursive equations I., in *Computer Science Logic. Selected Papers*, Lecture Notes in Computer Science, Vol. 933, Springer, Berlin, 1995.

[225] E. Marczewski, Independence and homomorphisms in abstract algebras, *Fund. Math.* **50**(1961), pp. 45–61.

[226] A. A. Markov, Constructive logic, *Uspekhi Mat. Nauk* **5**(1950), p. –187. (Russian).

[227] T. Martinsen, Correlation Immune and Resilient Generalized Boolean Functions, *The 3rd International Workshop on Boolean Functions and Their Applications* (*BFA*), ABSTRACTS, June 17–22, 2018, Loen, Norway.

[228] S. V. Matveev, Distributive groupoids in knot theory, *Mat. Sb.* **119(1)**(1982), pp. 78–88.

[229] V. D. Mazurov, On faithful doubly transitive groups, *Problems Algebra Logic, Proc. Siberian Branch RAS* **30**(1996), pp. 114–118.

[230] V. D. Mazurov, On doubly transitive permutation groups, *Siberian Math. J.* **31(4)**(1990), pp. 778–780.

[231] J. C. C. McKinsey, Postulates for the calculus of binary relations, *J. Symbolic Logic* **5**(1940), pp. 85–97.

[232] R. N. McKenzie, Representation of relation algebras, *Michigan J. Math.* **17**(1970), pp. 279–287.

[233] R. N. McKenzie,Equational bases for lattice theories, *Math. Scand.* **27**(1970) pp. 24–38.

[234] R. N. McKenzie, G. F. McNulty, W. T. Taylor, *Algebras, Lattices, Varieties, Volume I*, Wadsworth, 1987.

[235] D. McLean, Idempotent semigroups, *Amer. Math. Monthly* **61**(1954), pp. 110–113.

[236] V. G. Melkonian, Quasi-identities and conditional hyperidentities in distributive lattices, *Dokl. Akad. Nauk Armenii* **94**(1993), pp. 195–199 (Russian).

[237] V. G. Melkonian, Circuit integrating through lattice hyperterms, *Discrete Math. Algorithms Appl.* **3**(2011), pp. 101–119.

[238] I.I. Melnik, Nilpotent shift of manifolds, *Math. Notes*, **14**(1973), pp. 387–397.

[239] B. Mobasher, D. Pigozzi, G. Slutzki, Multi-valued logic programming semantics: An algebraic approach, *Theoret. Comput. Sci.* **171**(1997), pp. 77–109.

[240] B. Mobasher, D. Pigozzi, G. Slutzki, G. Voutsadakis, A duality theory for bilattices, *Algebra Universalis* **43**(2000), pp. 109–125.
[241] G. C. Moisil, Recherches sur l'algebre de la logique, *Ann. Sci. Univ. Jassy* **22**(1935), pp. 1–117.
[242] C. Moor, Quasi-linear cellular automata, *Physica D* **103**(1997), pp. 100–132.
[243] R. Moufang, Zur structure von Alternative Körpern, *Math. Ann.* **110**(1935), pp. 416–430.
[244] Yu. M. Movsisyan, Biprimitive classes of algebras of second degree, *Mat. Issledovaniya* **9**(1974), pp. 70–84 (in Russian).
[245] Yu. M. Movsisyan, Hyperidentities of associativity in systems of semigroups, *Dokl. Akad. Nauk Arm. SSR* **62**(1976), pp. 282–286.
[246] Yu. M. Movsisyan, Balanced hyperidentities of length three, algebras. Dokl *Dokl. Akad. Nauk Arm. SSR* **76**(1983), pp. 55–60.
[247] Yu. M. Movsisyan, Coidentities in algebras, *Dokl. Akad. Nauk Arm. SSR* **77**(1983), pp. 51–54.
[248] Yu. M. Movsisyan, A theorem of Albert in the category of binary algebras, *Dokl. Akad. Nauk Arm. SSR* **78**(1984), pp. 3–7.
[249] Yu. M. Movsisyan, *Introduction to the Theory of Algebras with Hyperidentities*, Yerevan State University Press, Yerevan, 1986 (in Russian).
[250] Yu. M. Movsisyan, Hyperidentities and groups, *Dokl. Akad. Nauk Arm. SSR* **87**(1988), pp. 104–105.
[251] Yu. M. Movsisyan, The multiplicative group of a field and hyperidentities, *Math. USSR Izv.* **35(2)**(1990), pp. 377–391.
[252] Yu. M. Movsisyan, *Hyperidentities and Hypervarieties in Algebras*, Yerevan State University Press, Yerevan, 1990 (in Russian).
[253] Yu. M. Movsisyan, Hyperidentities of Boolean algebras, *Russ. Acad. Sci. Izv. Math.* **40(3)**(1993), pp. 607–622.
[254] Yu. M. Movsisyan, On a theorem of Schaufflerl, *Math. Notes* **53**(1993), pp. 172–179.
[255] Yu. M. Movsisyan, Hyperidentities in the variety of lattices, *Math. Notes* **59(6)**(1996), pp. 686–688.
[256] Yu. M. Movsisyan, Algebras with hyperidentities of the variety of Boolean algebras, *Russ. Acad. Sci. Izv. Math.* **60(6)**(1996), pp. 1219–1260.
[257] Yu. M. Movsisyan, Hyperidentities in algebras and varieties, *Russian Math. Surv.* **53(1)**(1998), pp. 57–108.
[258] Yu. M. Movsisyan, Generalization of Toyoda theorem, Loop's 99, Praha July 27–August 1, (1999), pp. 26–27.
[259] Yu. M. Movsisyan, Hyperidentities and hypervarieties, *Sci. Math. Jpn.* **54**(2001), pp. 595–640.
[260] Yu. M. Movsisyan, Boolean Bisemigroups, Bigroups and Local Bigroups. Computer Science and Information Technologies, *Proceedings of the Conference*, Yerevan, (2005), pp. 97–104.
[261] Yu. M. Movsisyan, *On the Representations of De Morgan Algebras*, Trends in logic III, Studialogica, Warsaw, 2005, http://www.ifispan.waw.pl/studialogica/Movsisyan.pdf
[262] Yu. M. Movsisyan, Interlaced, modular, distributive and Boolean bilattices, *Armen. J. Math.* **1(3)**(2008), pp. 7–13.
[263] Yu. M. Movsisyan, Binary representation of algebras with at most two binary operations. A Cayley theorem for distributive lattices, *Internat. J. Algebra Comput.* **19(1)**(2009), pp. 97–106.

[264] Yu. M. Movsisyan, Bilattices and hyperidentities, *Proc. Steklov Inst. Math.* **274(1)**(2011), pp. 174–192.
[265] Yu. M. Movsisyan, Idempotent algebras with semigroup operations. AMS Spring Western Sectional Meeting, University of Utah, Salt Lake City, UT, April 9–10, 2016 (Saturday-Sunday) Meeting # 1119.
[266] Yu. M. Movsisyan, On Burnside varieties. VII International Joint Conference of the Georgian Mathematical Union & Georgian Mechanical Union (Dedicated to 125-th birthday anniversary of academician N. Muskhelishvili), Batumi, 2016, September 5–9; Book of Abstracts, p. 172.
[267] Yu. M. Movsisyan, Dicson's theorem for Bol loops, *Algebra Discrete Math.* **24(2)**(2017), pp. 297–301.
[268] Yu. M. Movsisyan, Hyperidentities and related concepts I, *Armen. J. Math.* **2**(2017), pp. 146–222.
[269] Yu. M. Movsisyan, Hyperidentities and related concepts II, *Armen. J. Math.* **4**(2018), pp. 1–85.
[270] Yu. M. Movsisyan, On functional equations and distributive second order formulae with specialized quantifiers, *Algebra Discrete Math.* **25(2)**(2018), pp. 269–285.
[271] Yu. M. Movsisyan, Boole–De Morgan bilattices, *J. Multiple-Valued Log. Soft Compu.* **38**(1–2) (2022), pp. 137–152. Special Issue Dedicated to Ivo G. Rosenberg (Part 2) (Edited by B. Larose, H. Machide, R. Pöschel and Á. Szendrei).
[272] Yu. M. Movsisyan, V. A. Aslanyan, A functional representation of free De Morgan algebras. Proceedings of the Yerevan State University, *Phys. Math. Sci.* **3**(2012), pp. 14–16.
[273] Yu. M. Movsisyan, V. A. Aslanyan, Hyperidentities of De Morgan algebras, *Log. J. IGPL* **20**(2012), pp. 1153–1174.
[274] Yu. M. Movsisyan, V. A. Aslanyan, Algebras with hyperidentities of the variety of De Morgan algebras, *J. Contemp. Math. Anal.* **5**(2013), pp. 233–240.
[275] Yu. M. Movsisyan, V. A. Aslanyan, Subdirectly irreducible algebras with hyperidentities of the variety of De Morgan algebras, *J. Contemp. Math. Anal.* **6**(2013), pp. 241–246.
[276] Yu. M. Movsisyan, V. A. Aslanyan, A functional completeness theorem for De Morgan functions, *Discrete Appl. Math.* **162**(2014), pp. 1–16.
[277] Yu. M. Movsisyan, V. A. Aslanyan, De Morgan functions and free De Morgan algebras, *Demonstratio Math.* **2**(2014), pp. 271–283.
[278] Yu. M. Movsisyan, V. A. Aslanyan, Boole–De Morgan algebras and quasi-De Morgan functions, *Comm. Algebra* **42**(2014), pp. 4757–4777.
[279] Yu. M. Movsisyan, V. A. Aslanyan, Super-Boolean functions and free Boolean quasilattices, *Discrete Math. Algorithms Appl.* **6**(2014), pp. 1450024 (13 pages).
[280] Yu. M. Movsisyan, V. A. Aslanyan, Super-De Morgan functions and free De Morgan quasilattices, *Cent. Eur. J. Math.* **12**(2014), pp. 1749–1761.
[281] Yu. M. Movsisyan, A. G. Barkhudaryan, On hypervarieties of QB-algebras. Proceedings of the YSU, *Phys. Math.* **2**(1996), pp. 16–24.
[282] Yu. M. Movsisyan, L. M. Budaghyan, The elementary characterization of algebras with hyperidentities of Boolean algebras, *International Conference Mathematical Logic, Algebra And Set Theory" dedicated to the 100th anniversary of P.S. Novikov*, August 27–31, 2001.
[283] Yu. M. Movsisyan, L. M. Budaghyan, On elementary decidability of quasi-Boolean algebras, *International Conference Mathematics in Armenia, Advances and Perspectives*", September 30–October 7, Yerevan (2003), pp. 65–67.

[284] Yu. M. Movsisyan, S. S. Davidov, *Algebras that are Nearly Quasigroups*, Moscow, LENAND, 2018 (in Russian).

[285] Yu. M. Movsisyan, S. S. Davidov, M. Safaryan, Constraction of free g-dimonoids. *Algebra Discrete Math.* **18**(2014), pp. 140–150.

[286] Yu. M. Movsisyan, D. S. Davidova, Representation theorem for interlaced q-bilattices. Logic, Algebra and Truth Degrees, 10–14 September 2012, Kanazawa, Ishikawa, Japan, JAIST, (2012), pp. 110–113.

[287] Yu. M. Movsisyan, D. S. Davidova, Hyperidentities of weakly idempotent lattices, *J. Contemp. Math. Anal.* **6**(2015), pp. 259–264.

[288] Yu. M. Movsisyan, D. S. Davidova, A set-theoretical representation for weakly idempotent lattices and interlaced weakly idempotent bilattices, *European J. Math.* **2**(2016), pp. 863–873.

[289] Yu. Movsisyan, A. Gevorgyan, Invertible algebras satisfying associative identities with functional variables, *Asian-Eur. J. Mathematics* **14(1)**(2021), 2050155 (16 pages) https://doi.org/10.1142/S1793557120501557.

[290] Yu. M. Movsisyan, T. A. Hakobyan, Associative nontrivial hyperidentities in semigroups, *J. Contemp. Math. Anal.* **46(3)**(2011), pp. 121–130.

[291] Yu. M. Movsisyan, T. A. Hakobyan, Distributive hyperidentities in semigroups, *J. Contemp. Math. Anal.* **46**(6)(2011), pp. 293–298.

[292] Yu. M. Movsisyan, T. A. Hakobyan, Artin theorem for semigroups, *J. Algebra Appl.* **16(2)**(2017), pp. 1750034 (16 pages).

[293] Yu. M. Movsisyan, G. Kirakosyan, Interassociativity via hyperidentities, *J. Contemp. Math. Anal.* **54**(2019), pp. 191–194.

[294] Yu. M. Movsisyan, V. G. Melkonian, On hyperidentities of distributive lattices, *Dokl. Akad. Nauk Armenii* **95**(1995), pp. 75–78.

[295] Yu. M. Movsisyan, E. Nazari, A Cayley theorem for the multiplicative semigroup of a field, *J. Algebra Appl.* **11(2)**(2012), pp. 1–12.

[296] Yu. M. Movsisyan, E. Nazari, Transitive modes, *Demonstratio Math.* **44(3)**(2011), pp. 511–522.

[297] Yu. M. Movsisyan, V. Pambuccian, The geometry of point reflections and quasigroups, *Results Math.* **75(4)**(2020), Paper No. 132, 19 pp.

[298] Yu. M. Movsisyan, J. Pashazadeh, Characterization of binary polynomials of idempotent algebras, *J. Math. Sci. (NY)* **186(5)**(2012), pp. 802–804.

[299] Yu. M. Movsisyan, A. B. Romanowska, J. D. H. Smith, Hyperidentities and Ginsberg-Fitting type theorems, *Eighteenth Midwest Conference on Combinatorics, Cryptography, and Computing*, October 28–30, 2004, Rochester, New York.

[300] Yu. M. Movsisyan, A. B. Romanowska, J. D. H. Smith, Superproducts, hyperidentities, and algebraic structures of logic programming, *J. Combin. Math. Combin. Comput.* **58**(2006), pp. 101–111.

[301] Yu. M. Movsisyan, I. R. Simonyan, Hyperidentities of distributivity in varieties of semigroups, *Computer Science and Information Technologies*, Yerevan, (1999), pp. 24–26.

[302] Yu. M. Movsisyan, V. Toghanyan, On t-stochastic algebras, *Aequationes Math.* **76(3)**(2008), pp. 241–248.

[303] Yu. M. Movsisyan, V. Toghanyan, *On De Morgan Bimonoids. Computer Science and Information Technologies, Proceedings of the Conference*, (2007), pp. 57–57.

[304] Yu. M. Movsisyan, M. Yolchyan, On idempotent and hyperassociative structures, *Lobachevskii J. Math.*, **40**(2019), pp. 1113–1121.

[305] W. D. Neumann, Malcev conditions, spectra and Kronecker product, *J. Austral. Math. Soc.* **25**(1978), pp. 103–117.

[306] N. Nobusawa, On a generalization of the ring theory, *Osaka J. Math.* **1**(1964), pp. 81–89.

[307] P. S. Novikov, S. I. Adian. Infinite periodic groups, *Izv. Akad. Nauk SSSR Ser. Mat.* (**32**)(1968), pp. 212–244, 251–524, 709–731.

[308] R. Padmanabhan, Regular identities in lattices, *Trans. Amer. Math. Soc.* **158**(1971), pp. 179–188.

[309] R. Padmanabhan, P. Penner, Bases of hyperidentities of lattices and semilattices, *C.R. Math. Rep. Acad. Sci. Canada* **4**(1982), pp. 9–14.

[310] R. Padmanabhan, P. Penner, Binary hyperidentities of lattices, *Aequationes Math.* **44**(1992), pp. 154–167.

[311] R. Padmanabhann, P. Penner, Lattice ordered polynomial algebras, *Order* **15**(1998), pp. 75–86.

[312] V. Pambuccian, Euclidean geometry problems rephrased in terms of midpoints and point-reflections, *Elem. Math.* **60**(2005), pp. 19–24.

[313] G. Paseman, A small basis for hyperassociativity. Preprint, Berkley, 1993.

[314] G. Paseman, On two problems from "Hyperidentities and Clones". 2014, arxiv.org/abs/1408.XXXX.

[315] J. Pashazadeh, A characterization of De Morgan bisemigroup of binary functions, *Internat. J. Algebra Comput.* **18(5)**(2008), pp. 951–956.

[316] F. J. Pelletier, N. M. Martin, Post's functional completeness theorem, *Notre Dame J. Formal Logic* **31**(1990), pp. 462–475.

[317] P. Penner, Hyperidentities of semilattices, *Houston J. Math.* **10**(1984), pp. 81–108.

[318] T. Pirashvili, Sets with two associative operations, *Cent. Eur. J. Math.* **2**(2003), pp. 169–183.

[319] H. Pflugfelder, *Quasigroups and Loops, Introduction*, Heldermann Verlag, Berlin, 1990.

[320] A. G. Pinus, On lattices of quasiorders on universal algebras, *Algebra Logic* **34**(1995), pp. 180–181.

[321] A. G. Pinus, *Foundations of Universal Algebra*, Novosibirsk State Technical University, Novosibirsk, 2005 (in Russian).

[322] J. Plonka, On distributive quasi-lattices, *Fund. Math.* **60**(1967), pp. 191–200.

[323] J. Plonka, On a method of construction of abstract algebras, *Fund. Math.* **61**(1967), pp. 183–189.

[324] J. Plonka, On distributive n-lattices and n-quasilattices, *Fund. Math.* **62**(1968), pp. 293–300.

[325] J. Plonka, On varieties of algebras defined by identities of some special forms, *Houston J. Math.* **14**(1988), pp. 253–263.

[326] J. Plonka, On hyperidentities of some varieties, *Exposition Math.* **21**(1995). pp. 199–213.

[327] J. Plonka, A. B. Romanowska, Semilattice sums, in *Universal Algebra and Quasigroup Theory* Eds. A. Romanowska, J. D. H. Smith), (1992), pp. 123–158. Helderman Verlag, Berlin.

[328] B. I. Plotkin, *Automorphisms Groups of Algebraic Systems*, Nauka, Moscow, 1966. (Russian)

[329] B. I. Plotkin, *Universal Algebra, Algebraic Logic, and Databases*, Kluwer Academic Publisher, 1994.

[330] L. Polák, On hyperassociativity, *Algebra Universalis* **36**(1996), pp. 293–300.

[331] L. Polák, All solid varieties of semigroups, *J. Algebra* **219**(1999), pp. 421–436.

[332] L. S. Pontryagin, *Continuous Groups*, Nauka, Moscow, 1973 (in Russian).

[333] E. Post, Introduction to a general theory of elementary propositions, *Amer. J. Math.* **43**(1921), pp. 163–185.

[334] E. Post, The two-valued iterative systems of mathematical logic, *Ann. Math. Studies* **5**(1941).

[335] H. A. Priestley, Distributive bilattices and their cousins: representation via natural dualities, *Research Workshop on Duality Theory in Algebra, Logic and Computer Science*, University of Oxford, June 13–14 (2012), pp. 18–18.

[336] J. H. Przytycki, Knot and distributive homology: from arc colorings to Yang-Baxter homology, in *New Ideas in Low Dimensional Topology* (2015), pp. 413–488.

[337] A. Pynko, Regular bilattices, *J. Appl. Non-Classical Logics* **10**(2000), pp. 93–111.

[338] R. W. Quackenbush, Varieties of Steiner loops and Steiner quasigroups, *Can. J. Math.* **28**(1976), pp. 1187–1198.

[339] H. Rasiowa, *An Algebraic Approach to Non-classical Logics*, North-Holland, Amsterdam, 1974.

[340] M. Reichel, *Bi-homomorphismen und Hyperidentitaten.* Dissertation, University Potsdam, 1994.

[341] K. Reidemeister, Gewebe und Gruppen, *Math. Z.* **29**(1928), pp. 427–435.

[342] B. Richter, Dialgebren, Doppelalgebren und ihre Homologie, Diplomarbeit, Universitat Bonn (1997). http:/www.math.uni-hamburg.de/home/richter/publications.html.

[343] U. Rivieccio, An algebraic study of bilattice-based logics, *Ph.D. Dissertation*, University of Barcelona, 2010.

[344] H. Rogers, *Theory of Recursive Functions and Effective Computability*, MIT Press, 1987.

[345] A. B. Romanowska, J. D. H. Smith, *Modal Theory*, Helderman-Verlag, Berlin, 1985.

[346] A. B. Romanowska, J. D. H. Smith, *Modes*, World Scientific, Singapore, 2002.

[347] A. B. Romanowska, A. Trakul, On the structure of some bilattices, *Universal and Applied Algebra* Eds. K. Hatkowska, B. Stawski, World Scientific, 1989, pp. 235–253.

[348] I. G. Rosenberg, La structure des fonctions de plusieurs variables sur un ensemble fini, *C. R. Acad. Sci. Paris, Ser. A. B.* **260**(1965), pp. 3817–3819.

[349] I. G. Rosenberg, The set of maximal closed classes of operations on an infinite set A has cardinality $2^{2^{|A|}}$, *Arch. Math. (Basel)*, **27**(1976), pp. 561–568.

[350] G. Sabbagh, How not to characterize the multiplicative groups of fields, *J. London Math. Soc.* **1**(1969), pp. 369–370.

[351] A. Sade, Theorie des systemes demosiens de groupoides, *Pacific J. Math.* **10**(1960), pp. 625–660.

[352] A. Sade, Groupoides en relation associative et semigroupes mutuellements associatifs, *Ann. Soc. Sci. Bruxelies* **1(75)**(1961), pp. 52–57.

[353] H. P. Sankappanavar, A characterization of principal congruences of De Morgan algebras and its applications. Math Logic in Latin America, *Proc. IV Latin Amer. Symp. Math. Logic, Santiago*, (1978), pp. 341–349. North-Holland Pub. Co., Amsterdam, 1980.

[354] H. P. Sankappanavar, Decision Problems: History and Methods. In *Mathematical logic: Proceedings of the First Brazilian Conference*, Eds. A. I. Arruda, N. C. A. da Costa, and R. Chuaqui, Marcel Dekker, New York, 1978, pp. 241–290.

[355] H. P. Sankappanavar, Implication zroupoids: an abstraction from De Morgan algebras, *Emil Artin International Conference dedicated to the 120th Anniversary of Emil Artin*, Yerevan, Armenia, May 27–June 2, 2018, Abstracts, pp. 119–120.

[356] I. N. Sanov. Solution of Burnside's problem for exponent 4, *Leningrad State Univ. Annals* [*Uchenye Zapiski*] *Math. Ser.* **10**(1940), pp. 166–170.

[357] M. Sapir, *Combinatorial Algebra: Syntax and Semantics*. Springer Monographs in Mathematics. Springer, Charm, 2014.

[358] S. K. Sardar, S. Gupta, K. P. Shum, Γ-semigroups with unites and Morita equivalence for monoids, *European J. Pure Appl. Math.* **6**(2013), pp. 1–10.

[359] S. K. Sardar, B. C. Saha, K. P. Shum, h-Prime and h-Semiprime Ideals in Semirings and Γ-semirings, *Int. J. Algebra* **4**(2010), pp. 209–220.

[360] R. Schauffler, Eine Anwendung zyklischer Permutationen and ihre Theorie, Ph.D. Thesis, Marburg University, 1948.

[361] R. Schauffler, Uber die Bildung von Codewörtern, *Arch. elekt. Uebertragung* **10**(1956), pp. 303–314.

[362] R. Schauffler, Die Associativität im Ganzen, besonders bei Quasigruppen, *Math. Z.* **67**(1957), pp. 428–435.

[363] D. Schweigert, *Hyperidentities. Algebras and Orders*, Kluwer Academic Publishers, Dordrecht Boston, London, (1993), pp. 405–506.

[364] M. K. Sen, S. Chattopadhyay, Γ-Semigroups. A Survey, In *Book: Algebra and Its Applications*, 2016.

[365] M. K. Sen, On Γ-semigroup. In: *Algebra and Its Applications* (New Delhi, 1981). Lecture Notes in Pure and Applied Mathematics, Vol. 91, Dekker, New York, 1984, pp. 301–308.

[366] M. K. Sen, N. K. Saha, On Γ-semigroup-I, *Bull. Calcutta Math. Soc.* **78**(1986), pp. 180–186.

[367] J. R. Senft, On weak automorphisms of universal algebras, *Dissertationes Math., Rozprawy Mat.*, **74**(1970).

[368] J.–P. Serre, *Lie Algebras and Lie Groups*, Lectures given at Harvard University, 1964 (W. A. Benjamin, 1965).

[369] A. Seth, Γ-group congruences on regular Γ-semigroups, *Int. J. Math. Math. Sci.* **15(1)**(1992), pp. 103–106.

[370] V. Shcherbacov, *Elements of Quasigroup Theory and Applications*, Taylor & Francis Group, 2017.

[371] V. A. Shcherbakov, Elements of quasigroup theory and some its applications in code theory and cryptology. 2003. (https://www2.karlin.mff.cuni.cz/~drapal/speccurs.pdf)

[372] I. P. Shestakov, A. I. Shirshov, A. M. Slińko, K. A. Zhevlakov, Rings That Are Nearly Associative. Academic Press, New York, 1982.

[373] K. P. Shum, A. Yang, Interior operators on complete lattices, *Pu. M. A. Ser. A.* **3**(1992), pp. 73–80.

[374] J. Sichler, Weak automorphisms of universal algebras, *Algebra Universalis* **3**(1973), pp. 1–7.

[375] I. R. Simonyan, Lattices of distributive and hyperdistributive varieties of semigroups, *Proc. YSU, Phys. Math.* **2**(2004), pp. 7–19.

[376] L. A. Skornyakov, Stochastic algebra, *Soviet Math.* (*Iz. Vuz. Matematika*), **29(7)**(1985), pp. 1–12.

[377] L. A. Skornyakov (Editor), *General Algebra*, Vol.2, Nauka, Moscow, 1991 (in Russian).

[378] J. Slominski, On the satisfiabilities and varieties for abstract algebras induced by the cones and functor dinamics, *Demonstratio Math.* **26**(1993), pp. 11–22.

[379] D. M. Smirnov, *Varieties of Algebras*, Nauka, Novosibirsk, 1992 (in Russian).

[380] J. D. H. Smith, *Malcev Varieties*, Lecture notes in Mathematics, Vol. 554, Springer-Verlag, Berlin (1976).

[381] J. D. H. Smith, Finite codes and groupoid words, *European J. Combin.* **12**(1991), pp. 331–339.
[382] J. D. H. Smith, *An Introduction to Quasigroups and Their Representations*, Chapman & Hall/CRC, 2006.
[383] J. D. H. Smith, On groups of hypersubstitutions, *Algebra Universalis* **64**(2010), pp. 39–48.
[384] J. D. H. Smith, Belousov's Theorem and the quantum Yang-Baxter equation, *Bul. Acad. Stiinte a Republicii Moldova*, **1(80)**(2016), pp. 7–23.
[385] J. D. H. Smith, A. B. Romanowska, *Post-Modern Algebra*, Wiley, New York, 1999.
[386] E. Sperner, Ein Satz uber Untermengen einer endlichen Menge, *Math. Z.* **27**(1928), pp. 544–548.
[387] P. Stanica, Journey into differential and graph theoretical properties of generalized Boolean functions, *The 3rd International Workshop on Boolean Functions and Their Applications* (*BFA*), Abstracts, June 17–22 (2018), Loen, Norway.
[388] S. K. Stein, On the foundation of quasigroups, *Trans. Amer. Math. Soc.* **85**(1957), pp. 228-256.
[389] J. Steiner, Combinatorische aufgabe, *J. Reine Angew. Math.* **45**(1853), pp. 181–182.
[390] M. Stone, The theory of representations for Boolean algebras, *Trans. Amer. Math. Soc.* **40**(1)(1936), pp. 37–111.
[391] M. H. Stone, Applications of the theory of Boolean rings to general topology, *Trans. Amer. Math. Soc.* **41**(1937), pp. 375–481.
[392] M. M. Stronkowski, Cancelation in entropic algebras, *Algebra Universalis*, **60**(2009), pp. 439–468.
[393] A. Suschkewitsch, *Generalized Group Theory*, Kiev, 1937 (in Russian).
[394] A. Szendrei, *Clones in Universal Algebra*, Les presses de L'universite de Montreal, 1986.
[395] Z. Szylicka, Hyperidentities of some generalizations of lattices, *Algebra Universalis* **39**(1998), pp. 1–29.
[396] M. Talagrand, On russo's approximate zero-one law, *Ann. Probab.*, **22**(3)(1994), pp. 1576–1587.
[397] A. Tarski, On the calculus of binary relations, *J. Symbolic Logic* **6**(1941), pp. 73–89.
[398] A. Tarski, *Decision method for elementary algebra and geometry*, Berkeley and Los Angeles, 1951.
[399] A. Tarski, A. Mostowski, R. Robinson, *Undecidable Theories*, North-Holland, Amsterdam, 1953.
[400] W. Taylor, Characterizing Mal'tsev conditions, *Algebra Universalis* **3**(1973), pp. 351–397.
[401] W. Taylor, Mal'tsev conditions and spectra, *J. Austral. Math. Soc.* **29**(1980), pp. 143–152.
[402] W. Taylor, Hyperidentities and hypervarieties, *Aequationes Math.* **23**(1981), pp. 30–49.
[403] W. Taylor, *Abstract Clone Theory, Algebras and Orders*, Kluwer Academic Publishers, Dordrecht, (1993), pp. 507–530.
[404] J. Tits, Sur les groupen doublement transitif continus, *Comment. Math. Helv.* **26**(1952), pp. 203–224.
[405] K. Tent, Sharply 3-transitive groups, *Adv. Math.* **286**(2016), pp. 722–728.
[406] K. Tent, M. Ziegler, Sharply 2-transitive groups, *Adv. Geom.* **16(1)**(2016), pp. 131–134.

[407] J. V. Thatcher, J. B. Wright, Generalized finite automata theory with an application to a decision problem of second order logic, *MST* **2**(1968), pp. 57–81.

[408] V. H. Toghanyan, Subdirectly irreducible algebras with various equations, *Aequationes Math.* **68**(2004), pp. 98–107.

[409] N. Tokareva, Algebraic normal form of a bent function: what is it? *The 3rd International Workshop on Boolean Functions and their Applications* (*BFA*), ABSTRACTS, June 17–22, 2018, Loen, Norway.

[410] K. Toyoda, On axioms of linear functions, *Proc. Imp. Acad. Tokyo*, **17**(1941), pp. 221–227.

[411] T. Traczyk, Weak isomorphisms of Boolean and post algebras, *Colloq. Math.* **13(2)**(1965), pp. 159–164.

[412] V. Trnkova, Amazingly extensive use of Cook continuum, *Math. Japonica* **51**(2000), pp. 499–549.

[413] K. Urbanik, On algebraic operations in idempotent algebras, *Colloquium Math.* **XIII**(1965), pp. 129–157.

[414] J. Ušan, Globally associative systems of n-ary quasigroups, *Publ. Inst. Math.* (*Beograd*) **19**(1975), pp. 155–165.

[415] J. Ušan, Description of super associative algebras with n-quasigroup operations, *Math. Morav.* **5**(2001), pp. 129–157.

[416] J. Ušan, *n-Groups in the light of the neutral operations*, Electronic version (2006).

[417] J. Ušan, M. Žižović, n-ary analogy of Schauffler theorem, *Publ. Inst. Math.* (*Beograd*) **19**(1975), pp. 167–172.

[418] F. W. Wilke, Pseudo-fields and doubly transitive groups, *Bull. Austral. Math. Soc.* **7**(1972), pp. 163–168.

[419] S. L. Wismath, Hyperidentity bases for rectangular bands and other semigroup varieties, *J. Austrral. Math. Soc.* (*Ser. A*) **55**(1993), pp. 270–286.

[420] S. L. Wismath, A basis for the type $\langle n \rangle$ hyperidentities of the medial variety of semigroups, *Aequationes Math.* **49**(1995), pp. 205–213.

[421] Yu. I. Yanov, A. A. Muchnik, On the existence of k-valued closed classes withouth a finite basis, *Dokl. Acad. Nauk SSSR* **127**(1959), pp. 44–46 (in Russian).

[422] K. A. Zaretskii, An abstract characterization of the bisemigroup of binary operations, *Math. Notes* **1(5)**(1967), pp. 347–350.

[423] D. Žarkov, A remark on globally associative systems of n-quasigroups over $Q = 1, 2, 3, 4$, *Publ. Inst. Math.* (*Beograd*) **21**(1977), pp. 207–211.

[424] H. Zassenhaus, Uber endliche Fastkorper, *Abh. Math. Sem. Univ. Hamburg* **11**(1935/36), pp. 187–220.

[425] H. Zassenhaus, Kennzeichnung endlicher linearer Gruppen als Permutationsgruppen, *Abh. Math. Sem. Univ. Hamburg* **11**(1936), pp. 17–40.

[426] E. I. Zel'manov, The solution of the restricted Burnside problem for groups of odd exponent, *Izv. Akad. Nauk. SSSR. Ser. Mat.* **54(1)**(1990), pp. 42–59; Trans. in *Math. USSR-Izv.* **36(1)**(1991), pp. 41–60.

[427] E. I. Zel'manov, The solution of the restricted Burnside problem for 2-groups, *Mat. Sb.* (*N S*), **182(4)**(1991), pp. 568–592.

[428] E. I. Zel'manov, On additional laws in the Burnside problem for periodic groups, *Int. J. Algebra Comput.* **3(4)**(1993), pp. 583–600.

[429] E. I. Zel'manov, Some open problems in the theory of infinite dimensional algebras, *J. Korean Math. Soc.* **44(5)**(2007), pp. 1185–1195.

[430] J. L. Zemmer, Near-fields, planar and non-planar, *Math. Stud.* **31**(1964), pp. 145–150.

[431] M. R. Žižović, Associative systems of topological n-quasigroups, *Algebraic Conference, Novi Sad* (1981), pp. 55–60.

[432] I. I. Zhegalkin, On the technique of calculating propositions in symbolic logic, *Mat. Sb.* **43**(1927), pp. 9–28.

[433] A. V. Zhuchok, Free products of doppelsemigroups, *Algebra Universalis* **76**(2016), pp. 355–366.

[434] A. V. Zuchok, Free rectangular n-tuple semigroups, *Chebyshevskii Sb.* **20(3)**(2020), pp. 261–271.

[435] A. V. Zuchok, Free n-tuple semigroups, *Math. Notes* **103(5)**(2018), pp. 737–744.

[436] A. V. Zhuchok, Relatively Free Doppelsemigroups, Potsdam University Press, 2018.

[437] A. V. Zhuchok, Structure of free strong doppelsemigroups, *Commun. Algebra* **46**(2018), pp. 3262–3279.

[438] A. V. Zhuchok, M. Demko, Free n-dinilpotent doppelsemigroups, *Algebra Discrete Math.* **22**(2016), pp. 304–316.

[439] A. V. Zhuchok, K. Knauer, Abelian doppelsemigroups, *Algebra Discrete Math.*, **26(2)**(2018), pp. 290–304.

[440] A. V. Zuchok, J. Koppitz, Free products of N-tuple semigroups, *Ukr. Math. J.* **70(11)**(2019), pp. 1710–1726.

[441] A. V. Zuchok, Y. V. Zhuchok, J. Koppitz, Free rectangular doppelsemigroups, *J. Algebra Appl.* (September 2019); DOI:10.1142/SO219498820502059.

[442] D. Zupnik, On interassociativity and related questions, *Aequationes Math.* **6(2)**(2016), pp. 141–148.

[443] I. E. Zverovich, Characterization of closed classes of Boolean functions in terms of forbidden subfunctions and Post classes, *Discrete Appl. Math.* **149**(2005), pp. 200–218.

Index

K

L

M

N

O

P

Q

R

W

Z

www.ingramcontent.com/pod-product-compliance
Lightning Source LLC
LaVergne TN
LVHW081935200926
848484LV00009B/2438